Medicinal Plants in the Asia Pacific for Zoonotic Pandemics

Medicinal Plants in the Asia Pacific for Zoonotic Pandemics

Series Editor:
Christophe Wiart

This series provides an unprecedented comprehensive overview of the pharmacological activity of more than 100 Asian plants. It discusses their actions against viruses and bacteria representing a threat of epidemic and pandemic diseases, with an emphasis on the molecular basis and cellular pathways. Scientific names, botanical classifications and descriptions, medicinal uses, and chemical constituents are presented, as well as chemical structures, and a vast selection of bibliographical references. This series is a critical reference for anyone involved in the discovery of lead compounds for the prevention or treatment of pandemic viral, bacterial, or fungal infections.

Medicinal Plants in the Asia Pacific for Zoonotic Pandemics,
Volume 1: Family Amborellaceae to Vitaceae, Christophe Wiart

Medicinal Plants in the Asia Pacific for Zoonotic Pandemics,
Volume 2: Family Zygophyllaceae to Salvadoraceae, Christophe Wiart

Medicinal Plants in the Asia Pacific for Zoonotic Pandemics

Family Zygophyllaceae to Salvadoraceae

Volume 2

Christophe Wiart

CRC Press
Taylor & Francis Group
Boca Raton London New York

CRC Press is an imprint of the
Taylor & Francis Group, an **informa** business

First edition published 2022
by CRC Press
6000 Broken Sound Parkway NW, Suite 300, Boca Raton, FL 33487-2742

and by CRC Press
2 Park Square, Milton Park, Abingdon, Oxon, OX14 4RN

CRC Press is an imprint of Taylor & Francis Group, LLC

ISBN: 978-1-032-00928-5 (hbk)
ISBN: 978-1-032-00536-2 (pbk)
ISBN: 978-1-003-17639-8 (ebk)

DOI: 10.1201/9781003176398

Typeset in Times
by codeMantra

Contents

Foreword ... xi
Preface ... xiii
Author ... xv

Chapter 7 The Clade Fabids ... 1

7.1 Order Zygophyllales Link (1829) ... 1
7.1.1 Family Zygophyllaceae R. Brown (1814) ... 1
7.1.1.1 Tribulus terrestris L. ... 1
7.2 Order Celastrales Link (1829) ... 2
7.2.1 Family Celastraceae R. Brown (1814) ... 2
7.2.1.1 Celastrus paniculatus Willd. ... 2
7.2.1.2 Euonymus alatus (Thunb.) Siebold ... 4
7.2.1.3 Kokoona zeylanica Thwaites ... 6
7.2.1.4 Microtropis japonica (Franch. & Sav.) Hallier f. ... 7
7.2.1.5 Tripterygium wilfordii Hook. f. ... 8
7.3 Order Oxalidales Bercht. & J.Presl (1820) ... 12
7.3.1 Family Cunoniaceae R. Brown (1814) ... 12
7.3.1.1 Weinmannia blumei Planch. ... 12
7.3.2 Family Elaeocarpaceae Juss. ex DC. (1816) ... 12
7.3.2.1 Elaeocarpus grandiflorus Sm ... 12
7.3.2.2 Elaeocarpus petiolatus (Jack) Wall. ... 14
7.3.3 Family Oxalidaceae R. Brown (1818) ... 15
7.3.3.1 Averrhoa carambola L. ... 15
7.3.3.2 Oxalis corniculata L. ... 16
7.4 Order Malpighiales Juss. ex Bercht. & J.Presl (1820) ... 18
7.4.1 Family Clusiaceae Lindley (1836) ... 18
7.4.1.1 Calophyllum inophyllum L. ... 18
7.4.1.2 Garcinia cowa Roxb. ... 20
7.4.1.3 Garcinia dulcis (Roxb.) Kurz ... 23
7.4.1.4 Garcinia hanburyi Hook. f. ... 24
7.4.1.5 Garcinia mangostana L. ... 25
7.4.1.6 Garcinia multiflora Champ. ex Benth. ... 27
7.4.1.7 Garcinia nigrolineata Planch. ex T. Anderson ... 29
7.4.1.8 Garcinia oblongifolia Champ. ex Benth. ... 31
7.4.1.9 Garcinia paucinervis Chun ex F.C. How ... 31
7.4.1.10 Garcinia speciosa Wall. ... 32
7.4.1.11 Mesua ferrea L. ... 33
7.4.2 Family Euphorbiaceae A.L. de Jussieu (1789) ... 36
7.4.2.1 Acalypha australis L. ... 36
7.4.2.2 Acalypha hispida Burm.f. ... 36
7.4.2.3 Acalypha indica L. ... 37
7.4.2.4 Aleurites moluccanus (L.) Willd. ... 38
7.4.2.5 Croton tiglium L. ... 40
7.4.2.6 Euphorbia antiquorum L. ... 42
7.4.2.7 Euphorbia hirta L. ... 43

7.4.2.8 Euphorbia nivulia Buch.-Ham. 47
7.4.2.9 Euphorbia thymifolia L. 48
7.4.2.10 Excoecaria agallocha L. 50
7.4.2.11 Jatropha curcas L. 51
7.4.2.12 Macaranga peltata (Roxb.) Müll. Arg. 54
7.4.2.13 Mallotus philippensis (Lam.) Müll. Arg. 57
7.4.2.14 Pedilanthus tithymaloides (L.) Poit. 60
7.4.2.15 Ricinus communis L. 61
7.4.2.16 Tragia involucrata L. 63
7.4.2.17 Trigonostemon reidioides (Kurz) Craib 64
7.4.3 Family Hypericaceae A. L de Jussieu (1789) 65
7.4.3.1 Cratoxylum arborescens Bl. 65
7.4.3.2 Cratoxylum formosum (Jack) Dyer 67
7.4.3.3 Cratoxylum cochinchinense (Lour.) Blume 68
7.4.3.4 Hypericum acmosepalum N. Robson 70
7.4.3.5 Hypericum japonicum Thunb. 71
7.4.3.6 Hypericum patulum Thunb. 75
7.4.4 Family Malpighiaceae A.L. de Jussieu (1789) 75
7.4.4.1 Galphimia glauca Cav. 75
7.4.4.2 Hiptage benghalensis (L.) Kurz 76
7.4.5 Family Passifloraceae Juss. ex Roussel (1806) 77
7.4.5.1 Adenia cordifolia Gagnep. 77
7.4.5.2 Passiflora foetida L. 78
7.4.6 Family Phyllanthaceae Martynov (1820) 80
7.4.6.1 Antidesma acidum Retz. 80
7.4.6.2 Bridelia retusa (L.) A. Juss. 82
7.4.6.3 Flueggea virosa (Roxb. ex Willd.) Royle 82
7.4.6.4 Glochidion littorale Bl. 85
7.4.6.5 Phyllanthus emblica L. 87
7.4.6.6 Phyllanthus niruri L. 89
7.4.7 Family Rafflesiaceae Dumortier (1829) 91
7.4.7.1 Rafflesia hasseltii Suringar 92
7.4.8 Family Rhizophoraceae C. H. Persoon (1806) 92
7.4.8.1 Bruguiera sexangula (Lour.) Poir. 92
7.4.8.2 Carallia suffruticosa Ridl. 94
7.4.8.3 Ceriops tagal (Perr.) C.B. Rob. 94
7.4.8.4 Rhizophora apiculata Bl. 96
7.4.9 Family Salicaceae Mirbel (1815) 96
7.4.9.1 Flacourtia indica (Burm.f.) Merr. 96
7.4.9.2 Flacourtia rukam Zoll. & Moritzi 97
7.4.9.3 Gynocardia odorata R.Br. 98
7.4.9.4 Flacourtia jangomas (Lour.) Raeusch 98
7.4.9.5 Hydnocarpus kurzii (King) Warb. 99
7.4.9.6 Xylosma longifolia Clos 100
7.4.10 Family Violaceae Batsch (1802) 101
7.4.10.1 Rinorea horneri (Korth.) O. Kuntze 101
7.4.10.2 Viola diffusa Ging. 102
7.5 Cucurbitales Juss. ex Bercht. & J.Presl (1820) 103
7.5.1 Family Begoniaceae Berchtold et J. Presl (1820) 103
7.5.1.1 Begonia roxburghii (Miq.) A. DC 103
7.5.1.2 Begonia fimbristipulata Hance 104

7.5.2 Family Cucurbitaceae A.L. de Jussieu (1789) ... 105
7.5.2.1 Benincasa hispida (Thunb.) Cogn. ... 105
7.5.2.2 Bryonopsis laciniosa (L.) Naudin ... 106
7.5.2.3 Cucumis sativus L. ... 107
7.5.2.4 Citrullus colocynthis (L.) Schrad. ... 108
7.5.2.5 Ecballium elaterium (L.) A. Rich. ... 112
7.5.2.6 Herpetospermum pedunculosum (Ser.) C.B. Clarke ... 113
7.5.2.7 Momordica charantia L. ... 115
7.5.2.8 Siraitia grosvenorii (Swingle) C. Jeffrey ex A.M. Lu & Z.Y. Zhang ... 116
7.5.2.9 Trichosanthes anguina L. ... 117
7.5.2.10 Trichosanthes kirilowii Maxim. ... 117
7.6 Order Fabales Bromhead (1838) ... 118
7.6.1 Family Fabaceae Lindley (1836) ... 118
7.6.1.1 Abrus precatorius L. ... 119
7.6.1.2 Acacia arabica (Lam.) Willd. ... 122
7.6.1.3 Acacia catechu (L. f.) Willd. ... 124
7.6.1.4 Acacia concinna (Willd.) DC. ... 125
7.6.1.5 Acacia farnesiana (L.) Willd. ... 126
7.6.1.6 Acacia pennata (L.) Willd. ... 128
7.6.1.7 Adenanthera pavonina L. ... 129
7.6.1.8 Albizia lebbeck (L.) Benth. ... 131
7.6.1.9 Albizia odoratissima (L.f.) Benth. ... 135
7.6.1.10 Albizia procera (Roxb.) Benth. ... 135
7.6.1.11 Astragalus mongholicus Bunge ... 137
7.6.1.12 Butea monosperma (Lam.) Taub. ... 139
7.6.1.13 Caesalpinia bonduc (L) Roxb. ... 142
7.6.1.14 Cajanus cajan (L.) Huth ... 146
7.6.1.15 Cassia fistula L. ... 150
7.6.1.16 Cassia alata L. ... 153
7.6.1.17 Cassia siamea Lam. ... 156
7.6.1.18 Cassia tora L. ... 158
7.6.1.19 Crotalaria pallida Aiton ... 162
7.6.1.20 Dalbergia pinnata (Lour.) Prain ... 164
7.6.1.21 Derris trifoliata Lour. ... 166
7.6.1.22 Desmodium gangeticum (L.) DC. ... 170
7.6.1.23 Erythrina orientalis (L.) Murr. ... 175
7.6.1.24 Glycyrrhiza glabra L. ... 177
7.6.1.25 Indigofera tinctoria L. ... 182
7.6.1.26 Lablab purpureus (L.) Sweet ... 183
7.6.1.27 Lupinus albus L. ... 185
7.6.1.28 Medicago sativa L. ... 186
7.6.1.29 Moghania strobilifera (L.) J. St.-Hil. ex O. Kuntze ... 187
7.6.1.30 Mucuna pruriens (L.) DC. ... 189
7.6.1.31 Parkia speciosa Hassk. ... 191
7.6.1.32 Pithecellobium dulce (Roxb.) Benth. ... 191
7.6.1.33 Psoralea corylifolia L. ... 194
7.6.1.34 Pterocarpus indicus Willd. ... 197
7.6.1.35 Pueraria lobata (Willd.) Ohwi ... 199
7.6.1.36 Saraca asoca (Roxb.) De Wilde ... 200
7.6.1.37 Sesbania sesban (L.) Merr. ... 202

7.6.1.38 Tamarindus indica L. 204
7.6.1.39 Tephrosia purpurea (L.) Pers. 204
7.7 Order Fagales Engl. (1892) 206
7.7.1 Family Fagaceae Dumortier (1829) 206
7.7.1.1 Castanea crenata Siebold & Zucc. 207
7.7.1.2 Castanea mollissima Bl. 208
7.7.1.3 Castanea sativa Mill. 208
7.7.1.4 Castanea tungurrut Bl. 209
7.7.1.5 Castanopsis motleyana King 210
7.7.1.6 Castanopsis sclerophylla (Lindl.) Schottky 210
7.7.1.7 Lithocarpus celebicus (Miq.) Rehder 211
7.7.1.8 Lithocarpus litseifolius (Hance) Chun 212
7.7.1.9 Quercus acutissima Carruth. 213
7.7.1.10 Quercus dentata Thunb. 214
7.7.1.11 Quercus dilata Lindl. ex A. DC. 214
7.7.1.12 Quercus ilex L. 215
7.7.1.13 Quercus infectoria Olivier 216
7.7.1.14 Quercus myrsinifolia Bl. 218
7.7.1.15 Quercus robur L. 219
7.7.2 Family Juglandaceae A.P. de Candolle ex Perleb (1818) 219
7.7.2.1 Engelhardia roxburghiana Wall. 219
7.7.2.2 Juglans mandshurica Maxim. 222
7.7.2.3 Juglans regia L. 224
7.7.2.4 Pterocarya stenoptera C. DC. 226
7.8 Order Rosales Bercht. & J.Presl (1820) 227
7.8.1 Family Cannabaceae Martinov (1820) 227
7.8.1.1 Cannabis sativa L. 227
7.8.1.2 Humulus scandens (Lour.) Merr. 229
7.8.2 Family Moraceae Link (1831) 230
7.8.2.1 Artocarpus heterophyllus Lamk. 230
7.8.2.2 Broussonetia papyrifera (L.) L'Hér. ex Vent. 233
7.8.2.3 Ficus benghalensis L. 234
7.8.2.4 Maclura cochinchinensis (Lour.) Corner 238
7.8.2.5 Morus alba L. 240
7.8.2.6 Streblus asper Lour. 243
7.8.3 Family Rosaceae A. L. de Jussieu (1789) 245
7.8.3.1 Agrimonia pilosa Ledeb. 245
7.8.3.2 Cydonia oblonga Mill. 248
7.8.3.3 Eriobotrya japonica (Thunb.) Lindl. 249
7.8.3.4 Prunus persica (L.) Batsch 252
7.8.3.5 Rosa damascena Mill. 253
7.8.3.6 Rubus glomeratus Bl. 256
7.8.4 Family Rhamnaceae A.L de Jussieu (1789) 258
7.8.4.1 Scutia myrtina (Burm. f.) Kurz 258
7.8.4.2 Ziziphus mauritiana Lam. 259
7.8.5 Family Ulmaceae Mirbel (1815) 261
7.8.5.1 Trema orientalis (L.) Bl. 261
7.8.6 Family Urticaceae A.L, de Jussieu (1789) 262
7.8.6.1 Boehmeria nivea (L.) Gaudich. 262

Chapter 8 The Clade Malvids 267

8.1 Order Geraniales Juss. ex Bercht. & J.Presl (1820) 267
8.1.1 Family Geraniaceae A.L de Jussieu (1789) 267
8.1.1.1 Geranium wallichianum D. Don ex Sweet 267
8.2 Order Myrtales Juss. ex Bercht. & J. Presl (1820) 271
8.2.1 Family Combretaceae R. Brown (1810) 271
8.2.1.1 Anogeissus acuminata (Roxb. ex DC.) Guill., Perr. & A. Rich. 271
8.2.1.2 Anogeissus latifolia (Roxb. ex DC.) Wall. ex Bedd. 272
8.2.1.3 Combretum trifoliatum Vent. 272
8.2.1.4 Terminalia arjuna (Roxb. ex DC.) Wight & Arn. 274
8.2.1.5 Terminalia bellirica (Gaertn.) Roxb. 275
8.2.1.6 Terminalia chebula Retz. 276
8.2.2 Family Lythraceae Jaume Saint-Hilaire (1805) 279
8.2.2.1 Lagerstroemia speciosa (L.) Pers. 279
8.2.2.2 Lawsonia inermis L. 281
8.2.2.3 Punica granatum L. 282
8.2.2.4 Sonneratia griffithii Kurtz. 284
8.2.2.5 Woodfordia fruticosa (L.) Kurz 286
8.2.3 Family Melastomataceae A. L. de Jussieu (1789) 287
8.2.3.1 Melastoma candidum D. Don 287
8.2.3.2 Osbeckia chinensis L. 289
8.2.4 Family Myrtaceae A.L de Jussieu (1789) 289
8.2.4.1 Decaspermum fruticosum Forst. 290
8.2.4.2 Eugenia aquea Burm.f. 290
8.2.4.3 Eugenia operculata Roxb. 296
8.2.4.4 Melaleuca cajuputi Roxb. 297
8.2.4.5 Psidium guajava L. 298
8.2.4.6 Syzygium cumini (L.) Skeels 303
8.2.5 Family Onagraceae A. L de Jussieu (1789) 306
8.2.5.1 Jussiaea repens L. 306
8.2.5.2 Ludwigia octovalvis (Jacq.) Raven 308
8.3 Order Brassicales Bromhead (1838) 310
8.3.1 Family Brassicaceae Burnett (1835) 310
8.3.1.1 Brassica alba L. 310
8.3.1.2 Brassica oleracea L. 311
8.3.1.3 Capsella bursa-pastoris (L.) Medik 312
8.3.1.4 Cardamine hirsuta L. 313
8.3.1.5 Cardaria draba (L.) Desv. 313
8.3.1.6 Descurainia sophia (L.) Webb ex Prantl 314
8.3.1.7 Draba nemorosa L. 315
8.3.1.8 Isatis tinctoria L. 315
8.3.1.9 Rorippa indica (L.) Hiern 316
8.3.1.10 Raphanus sativus L. 317
8.3.1.11 Wasabia japonica (Miq.) Matsum. 318
8.3.2 Family Capparaceae A. L. de Jussieu (1789) 319
8.3.2.1 Capparis micracantha DC 319
8.3.2.2 Cleome gynandra L. 320
8.3.2.3 Crateva religiosa Forts. 321
8.3.2.4 Stixis scortechinii (King) Jacobs 322

8.3.3 Family Caricaceae Dumortier (1829) 322
8.3.3.1 Carica papaya L. 322
8.3.4 Family Moringaceae Martinov (1820) 324
8.3.4.1 Moringa oleifera Lam. 324
8.3.5 Family Salvadoraceae Lindley (1836) 325
8.3.5.1 Salvadora persica L. 325

Bibliography 327

Index 335

Foreword

Those interested in pharmacy, traditional medicines, ethnopharmacology, pharmacology, medicinal plants, natural products, botany, and biodiversity in the Asia-Pacific region are impatiently awaiting the release of new titles by Professor Christophe Wiart. It is always a great pleasure to discover his works, whose unique phylogenetic presentation provides a real added value not only for chemotaxonomists but also for college students, post-doctoral fellows, academic and industrial scientists.

This new series – "Medicinal Plants in the Asia Pacific for Zoonotic Pandemics" – is particularly relevant in terms of subject, timing, and area. Indeed, with the eradication of smallpox in 1977 and the development of numerous vaccines, we were all convinced that infectious diseases would soon be under control and of less concern than cancers and cardiovascular diseases. However, the emergence of recent viral pandemics (AIDS, Ebola, SARS, Chikungunya, Zika, H5N1, H1N1) and now COVID-19 forces us to revise our optimism, as zoonoses represent an increasingly frequent and worrisome public health problem. Subject as we are to over a thousand parasites and pathogens, humans today constitute the most parasitized species.

The Neolithic revolution, which resulted in the sedentarization of hunter-gatherers, animal domestication and plant cultivation, lead to the first environmental changes and to consequences on human health. The extension of the agrarian way of life, the domestication of animals, and the increase in livestock were the source of new infections. For example, the emergence of measles, a serious pathology for humans, is due to the domestication of cattle which favored the transmission of the rinderpest virus to humans in the first Mesopotamian cities. In humans, this bovine virus has evolved into measles. According to Prof Kate E. Jones (2008) from University College London, 60% of emerging diseases between 1940 and 2004 are zoonoses, mostly of wild origin (72%). High up on the list come Ebola fevers, Marburg, Lassa and Coronavirus diseases (MERS, SARS, SARS-CoV-2 …). As early as 2015, the World Health Organization was already sounding the alarm about the risk of the appearance of new coronavirus pandemics.

For the past few decades, our behavior toward the environment has become increasingly inconsequential, particularly from a public health point of view. As mentioned by Professor Wiart in the preface to volume one, the increasing contact between humans and wild animals, due to deforestation, poaching, wet-markets for "bushmeat" consumption of living bats, primates and birds, is promoting the emergence of viral zoonotic pandemics in over-populated cities. Intensive industrial animal husbandry that cages, crams and confines pigs, poultry and cattle to the extreme is also the ideal breeding ground for the emergence of virulent strains. In practical terms, a mildly dangerous viral strain that affects overcrowded industrial animal production spreads very quickly. As it moves through the farm, replicating wildly, the virus mutates and gives rise to new, more dangerous strains. The increasingly low genetic diversity of industrial farms also favors epidemics by facilitating transmission from one animal to another.

Moreover, the unprecedented rise in the number of travelers, the speed and intensification of air travel and globalization in general, make the rapid spread of pandemics inevitable.

Thus, the SARS-CoV-2 (COVID-19) epidemic that started in November 2019 in Wuhan, imposed the confinement of more than three billion people, disrupted the economies of countries worldwide, counting almost four million victims to date, brings us face to face with reality. Such an acute health crisis should be interpreted as an alarm bell, inciting us to accelerate awareness and treatment of root causes. This microscopic bat virus should induce our omnipotent and omnipresent species on the planet to more modesty and propel us to act for truly sustainable development. *Homo sapiens* by its number, density and behaviors is now an invasive species, dangerous to its vegetal and animal co-species, its environment and, finally, to itself. The primordial option should be a holistic approach to sustainable development (One Health) that integrates the interrelated global issues of environmental, human, and animal health, while respecting climate and ecological objectives. Only by reasoning and acting on a global scale, can we reduce the risk of zoonotic and parasitic disease emergence.

It is by preserving biodiversity, studying traditional resources, that we should aim to continue using medicinal plants and develop this invaluable knowledge, particularly as more than 60% of the world's population relies on traditional medicine for its health care.

I am convinced that these four volumes will be an essential reference and a useful tool toward achieving this goal. It gives me great pleasure to write this foreword to volume two and I wish Professor Wiart's new work every success it deserves.

Dr. Bruno David, DPharm, PhD, HDR
Ex-Director of Phytochemistry and Biodiversity
Pierre Fabre Research Institute
Toulouse (France), June 6, 2021

Preface

The recent emergence of life-threatening waves of viral zoonosis of increasing severity, culminating today with the COVID-19 pandemic, marks the beginning of an era that might be the last before the ultimate disappearance of human beings. Throughout history, plagues threatened human survival. The Roman Empire was struck from about 166 to 189 A.D. by a smallpox pandemic known as "the Antonine plague" claiming the lives of half of the population of the Empire including the emperor Marcus Aurelius and contributing to the socioeconomic decline of the Western Roman Empire until its fall in 476 A.D. The zoonotic bacterium *Yersinia pestis* accounted for two major plagues in the Middle Ages: "the Justinian plague" (6th century) and the "Black Death" (14th century) claiming the lives of millions. The Spanish flu that followed World War I due to a zoonotic Influenza A (H1N1) virus took more than 20 million lifes. For the last about 50 years, novel life-threatening zoonotic viruses have emerged and spread globally including the Human immunodeficiency virus, the Ebola virus, the Zika virus, the Middle East respiratory syndrome coronavirus, and today the ignominious Severe acute respiratory syndrome coronavirus 2 (SARS-CoV-2) which according to the WHO has infected so far 124 million people, caused 2.72 million deaths, and paralyzes the whole world as never seen before. Among the reasons for the emergence of zoonotic pandemics is the increasing contact of humans with animal reservoirs due to deforestation and industrial animal farming. According to Wolfe et al. (2005), the richness of microbes in a region, environmental changes, and increased human and animal contact with wildlife result in the emergence of new zoonotic pathogenic microorganisms. Thus, one could envisage the increasing waves of zoonosis to be biological "negative feedback loops" meant to compel humans to put an end to primary rainforest burning, carbon dioxide emission, global warming, as well as intensive poultry, swine, and other animal farming. The pharmaceutical industry has managed to produce vaccines offering some levels of protection against SARS-CoV-2. However, if pressures on wildlife and nature continue, other new zoonotic pandemics will come with the possibility of the emergence of a completely untreatable fast killing virus, bacteria, or even fungi. Furthermore, between the time of appearance of a new virus and the time of production of a vaccine, people are left vulnerable. The intention behind the writing of this book is to offer researchers, academics, and students a comprehensive and interrelated set of botanical, ethnopharmacological, and pharmacological evidence to facilitate the discovery of natural products or even herbal remedies for the treatment or prevention of viral, bacterial, or fungal zoonotic pandemics. It is also to offer an intellectual rationale or tool to understand and thus cogently approach the fascinating realm of drug discovery from medicinal plants. There is a law for the rational selection and use of plants for the treatment of diseases. This second volume covers the medicinal plants traditionally used for the treatment of microbial infection in Asia and the Pacific classified in the Clades Fabids and Malvids. Within each clade, plants are presented according to their respective orders and families. Scientific and common names are givens as well as synonyms, habitat, distribution, botanical observation, pharmacology, commentaries, personally handmade botanical plates, as well as carefully selected references. We are now in lack of drugs for the treatment of COVID-19 and of lead molecules to face the upcoming pandemics. The American Kabbalist Rav Philip Berg (1927-2013) said "Remember, there is no such thing, such condition unless it is so severe, very rarely will we ever be placed in a position where there is no way out" and it is now time for us to join efforts and discover drugs from the medicinal plants in the Asia Pacific for the prevention and treatment of zoonotic pandemics.

Christophe Wiart
Kuala Lumpur
June 21, 2021

REFERENCES

Claas, E.C., 2000. Pandemic influenza is a zoonosis, as it requires introduction of avian-like gene segments in the human population. *Veterinary Microbiology*, *74*(1–2), pp. 133–139.

Littman, R.J. and Littman, M.L., 1973. Galen and the Antonine plague. *The American Journal of Philology*, *94*(3), pp. 243–255.

Morse, S.S., 2001. Factors in the emergence of infectious diseases. In *Plagues and Politics* (pp. 8–26). Palgrave Macmillan, London.

Sabbatani, S. and Fiorino, S., 2009. The antonine plague and the decline of the Roman Empire. *Le infezioni in medicina: rivista periodica di eziologia, epidemiologia, diagnostica, clinica e terapia delle patologie infettive*, *17*(4), pp. 261–275.

Slingenbergh, J., Gilbert, M., Balogh, K.D. and Wint, W., 2004. Ecological sources of zoonotic diseases. *Revue scientifique et technique-Office international des épizooties*, *23*(2), pp. 467–484.

Taylor, L.H., Latham, S.M. and Woolhouse, M.E., 2001. Risk factors for human disease emergence. *Philosophical Transactions of the Royal Society of London. Series B: Biological Sciences*, *356*(1411), pp. 983–989.

Wagner, D.M., Klunk, J., Harbeck, M., Devault, A., Waglechner, N., Sahl, J.W., Enk, J., Birdsell, D.N., Kuch, M., Lumibao, C. and Poinar, D., 2014. Yersinia pestis and the Plague of Justinian 541–543 AD: A genomic analysis. *The Lancet Infectious Diseases*, *14*(4), pp. 319–326.

Wolfe, N.D., Daszak, P., Kilpatrick, A.M. and Burke, D.S., 2005. Bushmeat hunting, deforestation, and prediction of zoonotic disease. *Emerging Infectious Diseases*, *11*(12), p. 1822.

Author

Christophe Wiart, PharmD, PhD, is an associate professor in the School of Pharmacy at the University of Nottingham, Malaysia Campus. His fields of expertise are Asian ethnopharmacology, chemotaxonomy, and ethnobotany. He has collected, identified, and classified several hundred species of medicinal plants from India, Southeast Asia, and China.

Dr. Wiart appeared on HBO's *Vice* (TV series) in season 3, episode 6 (episode 28 of the series) titled "The Post-Antibiotic World and Indonesia's Palm Bomb," April 17, 2015. It highlighted the need to find new treatments for infections that were previously treatable with antibiotics but are now resistant to multiple drugs. "The last hope for the human race's survival, I believe, is in the rainforests of tropical Asia," said ethnopharmacologist Christophe Wiart. "The pharmaceutical wealth of this land is immense."

7 The Clade Fabids

7.1 ORDER ZYGOPHYLLALES LINK (1829)

7.1.1 FAMILY ZYGOPHYLLACEAE R. BROWN (1814)

The family Zygophyllaceae consists of about 26 genera and 284 species of herbs or shrubs. The leaves are simple or compound, alternate or opposite, and stipulate. The flowers are solitary or arranged in racemes or cymes. The calyx comprises 4–5 sepals. The corolla comprises up to 5 petals. The androecium includes 4–5 stamens. The gynoecium presents 4-5 carpels. The fruits are capsules or drupes.

7.1.1.1 *Tribulus terrestris* L.

Synonyms: *Tribulus bimucronatus* Viv.; *Tribulus lanuginosus* L.; *Tribulus saharae* A. Chev.

Common names: Caltrop, puncture vine; ji li (China); goksura (India); kharkhasak (Iran); gokhru (Pakistan); khokkrasun (Thailand); çarık dikeni (Turkey); gai ma vương (Vietnam)

Habitat: Sandy and sun-exposed soils

Distribution: Turkey, Iran, Afghanistan, Pakistan, India, Bangladesh, and Thailand

Botanical observation: This prostrate herb with hairy and somewhat purplish stems. The leaves are paripinnate, alternate, and stipulate. The stipules are lanceolate and up to about 5 mm long. The blade is 2.5–5 cm long and comprises 5–6 pairs folioles, which are sessile, opposite, ovate to elliptic-oblong, 5–10 ×3–8 mm, and asymmetrical. The inflorescences are made of showy flowers. The calyx includes 5 sepals, which are lanceolate, hairy below, and up to 6 mm long. The corolla presents 5 yellow petals, which are obovate, up to 8 mm long, and round to somewhat notched at apex. A 10-lobed disc is present. The androecium includes 10 stamens. The schizocarps are about 1 cm in diameter and 8 mm long, tuberculate, hairy, and spiny.

Medicinal uses: Abscesses (Cambodia, Laos, Vietnam); cough, gonorrhea, leprosy (India)

Moderate broad-spectrum antibacterial polar extract: Ethanol extract of schizocarps inhibited the growth of *Staphylococcus aureus*, *Bacillus subtilis*, *Bacillus cereus*, *Corynebacterium diphtherae*, *Proteus vulgaris*, *Escherichia coli,* and *Klebsiella pneumoniae* (Al-Bayati & Al-Mola, 2008).

Moderate anticandidal polar extract: Ethanol extract of schizocarps inhibited the growth of *Candida albicans* (NCCLS, 2001) with an MIC value of 600 µg/mL (Al-Bayati & Al-Mola, 2008).

Strong antifungal (yeasts) steroidal saponins: TTS-12 inhibited *Candida albicans*, *Candida glabrata*, *Candida parapsilosis*, *Candida tropicalis*, *Cryptococcus neoformans*, and *Candida krusei* with MIC_{80} values of 4.4, 10.7, 24.0, 32.0, 10.7, and 8.8 µg/mL, respectively (Zhang et al., 2005). TTS-15 inhibited *Candida albicans*, *Candida glabrata*, *Candida parapsilosis*, *Candida tropicalis*, *Cryptococcus neoformans*, and *Candida krusei* with MIC_{80} values of 9.4, 18.7, 64.0, 64, 18.7, and 18.4 µg/mL, respectively (Zhang et al., 2005). The steroid saponin TTS-12 and TTS-15 inhibited fluconazole-resistant *Candida albicans* (Y0305433) with MIC_{80} values of 4.4 and 9.4 µg/mL, respectively (Zhang et al., 2005). TTS-12 altered cell membrane and inhibited hyphal formation in *Candida albicans* (SC5314) (Zhang et al., 2006).

Commentary: The plant generates a series of spirostanol and furostanol saponins (Bedir & Khan, 2000).

DOI: 10.1201/9781003176398-7

REFERENCES

Al-Bayati, F.A. and Al-Mola, H.F., 2008. Antibacterial and antifungal activities of different parts of Tribulus terrestris L. growing in Iraq. *Journal of Zhejiang University Science B*, *9*(2), pp. 154–159.

Bedir, E. and Khan, I.A., 2000. New steroidal glycosides from the fruits of Tribulus terrestris. *Journal of Natural Products*, *63*(12), pp. 1699–1701.

Zhang, J.D., Cao, Y.B., Xu, Z., Sun, H.H., An, M.M., Yan, L., Chen, H.S., Gao, P.H., Wang, Y., Jia, X.M. and Jiang, Y.Y., 2005. In vitro and in vivo antifungal activities of the eight steroid saponins from Tribulus terrestris L. with potent activity against fluconazole-resistant fungal. *Biological and Pharmaceutical Bulletin*, *28*(12), pp. 2211–2215.

Zhang, J.D., Xu, Z., Cao, Y.B., Chen, H.S., Yan, L., An, M.M., Gao, P.H., Wang, Y., Jia, X.M. and Jiang, Y.Y., 2006. Antifungal activities and action mechanisms of compounds from Tribulus terrestris L. *Journal of Ethnopharmacology*, *103*(1), pp. 76–84.

7.2 ORDER CELASTRALES LINK (1829)

7.2.1 FAMILY CELASTRACEAE R. BROWN (1814)

The family Celastraceae consists of about 90 genera and 1200 species of trees, climbers and shrubs. The leaves are simple, alternate or opposite. The inflorescences are axillary or terminal cymes, racemes, or fascicle or solitary. The calyx includes 4–5 sepals, and the corolla is made of 4–5 petals. A disc is present. The androecium includes 3–5 stamens. The gynoecium includes 3–5 carpels fused into a 2–5-locular ovary, each locule sheltering a pair of ovules growing on axile placentas. The fruit is a dehiscent capsule, a schizocarp, a drupe, a berry, or a samara.

7.2.1.1 *Celastrus paniculatus* Willd.

Synonyms: *Celastrus dependens* Wall.; *Celastrus multiflorus* Roxb.; *Diosma serrata* Blanco; *Euonymus euphlebiphyllus* Hayata

Common names: Black oil tree, climbing staff plant, intellect tree; may thee (Cambodia); deng you teng (China); sankhiran (Himalayas); myinkoungnayoung (Myanmar); amruta, adibaricham, munjui, palleru, thivva (India); day sang mau (Vietnam).

Habitat: Forests

Distribution: India, Sri Lanka, Nepal, Bhutan, Bangladesh, Cambodia, Laos, Vietnam, China, Thailand, Malaysia, Indonesia, Papua New Guinea, Australia, and Pacific Islands

Botanical observation: This magnificent climber grows up to about 18 m long. The stems are reddish brown and lenticelled. The bark is pale brown and the inner bark is pink and turns blue on exposure to light. The leaves are simple, spiral, and exstipulate. The petiole is up to 1.5 cm long. The blade is 6.3–10 × 3.8–7.5 cm, coriaceous, elliptic, ovate or obovate, cuneate at base, serrate, acuminate at apex, glossy, and marked with 5–7 pairs of secondary nerves. The inflorescences are terminal cymes, which are up to 20 cm long. The calyx include 5 sepals. The corolla includes 5 petals, which are triangular, minute, and light green to white. A slightly 5-lobed disc is present. The androecium comprises 5 stamens with white anthers. The gynoecium includes a globose ovary. The capsules are 3-lobed, bright yellow, sheltering 3–6 seeds, 1–1.3 cm in diameter, glossy, and globose. The seeds are enclosed in a red aril.

Medicinal uses: Dysentery (Indonesia); bronchitis (Himalaya); herpes (India)

Broad-spectrum antibacterial halo developed by polar extract: Ethanol extract of leaves (1 mg/mL per well) inhibited the growth of clinical strains of *Staphylococcus aureus*, *Klebsiella pneumoniae*, and *Pseudomonas aeruginosa* with inhibition zone diameters of 18.3, 11.4, and 11.9 mm, respectively (Harish et al., 2007).

Weak broad-spectrum antibacterial polar extract: Methanol extract of leaves inhibited the growth of *Staphylococcus aureus* (MTCC 1144), *Bacillus cereus* (clinical isolate), *Shigella flexneri* (clinical isolate), and *Vibrio cholerae* (MTCC 3904) with MIC values of 1000, 500, 1000, and 500 μg/mL, respectively (Panda et al., 2016).

Antifungal (filamentous) polar extract: Aqueous extract of seeds inhibited the growth of *Trichophyton rubrum* (Vonshak et al., 2003).

Very weak antileptospiral triterpene: Lupenone inhibited the growth of leptospiral strains belonging to 10 serovars in the range of 100–200 µg/mL. The range of minimum bactericidal concentrations was 400–800 µg/mL (Punnam Chander et al., 2015).

Commentaries: (i) The seeds yields a series of dihydro-*β*-agarofuran sesquiterpenes (Sasikumar et al., 2018), which could be examined for possible antiviral activities since 1α,2α,9β,15- tetracetoxy-8β-benzoyloxy-β-dihydroagarofuran inhibited the replication of the Human immunodeficiency virus (Gutierrez-Nicolas et al., 2014). Dihydroagarofurans of antimicrobial value are found in *Celastrus orbiculatus* Thunb. such as celaspene D, celaspene G, and celaspene F which inhibited the mycelial growth of *Gibberella zeae*, *Cladosporium cucumerinum*, and *Cercospora arachidicola*, respectively (Wang et al., 2014). Celaspene A, celaspene C, celaspene F, celaspene H, and celaspene I inhibited the growth of *Physalospora piricola* (Wang et al., 2014). Methanol extract of roots at a concentration of 100 µg/mL inhibited Human immunodeficiency virus-1 protease by 49.1% (Park, 2003).

(ii) *Glyptopetalum calocarpum* (Kurz) Prain used in India to break fever produces the sesquiterpene sclerocarpic acid (Log P=4.2; molecular weight=234.3 g/mol), which inhibited the replication of the Herpes simplex virus types 1 and 2 with the IC_{50} values of 5 and 12.5 µg/mL, respectively, and inhibited the growth of *Bacillus cereus*, *Bacillus subtilis*, *Sarcinia lutea*, *Staphylococcus aureus*, and *Microsporum gypseum* with MIC values of 50, 100, 200, 200, and 50 µg/mL, respectively (Sotanaphun et al., 1999).

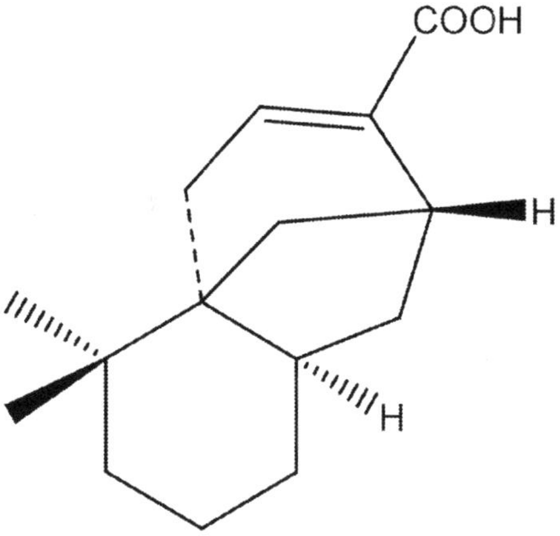

Sclerocarpic acid

REFERENCES

Gutierrez-Nicolas, F., Oberti, J.C., Ravelo, A.G. and Estevez-Braun, A., 2014. β-Agarofurans and sesquiterpene pyridine alkaloids from maytenus spinosa. *Journal of Natural Products*, *77*(8), pp. 1853–1863.

Harish, B.G., Krishna, V., Sharath, R., Kumara, S., Raja, N. and Mahadevan, K.M., 2007. Antibacterial activity of celapanin, a sesquiterpene isolated from the leaves of Celastrus paniculatus Willd. *International Journal of Biomedical and Pharmaceutical Sciences*, *1*, pp. 65–68.

Panda, S.K., Mohanta, Y.K., Padhi, L., Park, Y.H., Mohanta, T.K. and Bae, H., 2016. Large scale screening of ethnomedicinal plants for identification of potential antibacterial compounds. *Molecules*, *21*(3), p. 293.

Park, J.C., 2003. Inhibitory effects of Korean plant resources on human immunodeficiency virus type 1 protease activity. *Oriental Pharmacy and Experimental Medicine*, *3*(1), pp. 1–7.

Sasikumar, P., Sharathna, P., Prabha, B., Varughese, S., Kumar, A., Sivan, V.V., Sherin, D.R., Suresh, E., Manojkumar, T.K. and Radhakrishnan, K.V., 2018. Dihydro-β-agarofuran sesquiterpenoids from the seeds of Celastrus paniculatus Willd. and their α-glucosidase inhibitory activity. *Phytochemistry Letters*, *26*, pp. 1–8.
Vonshak, A., Barazani, O., Sathiyamoorthy, P., Shalev, R., Vardy, D. and Golan-Goldhirsh, A., 2003. Screening South Indian medicinal plants for antifungal activity against cutaneous pathogens. *Phytotherapy Research: An International Journal Devoted to Pharmacological and Toxicological Evaluation of Natural Product Derivatives*, *17*(9), pp. 1123–1125.
Wang, M., Zhang, Q., Ren, Q., Kong, X., Wang, L., Wang, H., Xu, J. and Guo, Y., 2014. Isolation and characterization of sesquiterpenes from Celastrus orbiculatus and their antifungal activities against phytopathogenic fungi. *Journal of Agricultural and Food Chemistry*, *62*(45), pp. 10945–10953.

7.2.1.2 *Euonymus alatus* (Thunb.) Siebold

Synonyms: *Celastrus alatus* Thunb.; *Celastrus striatus* Thunb.; *Euonymus alatus* Rupr.; *Euonymus arakianus* Koidz.; *Euonymus ellipticus* (C.H. Wang) C.Y. Cheng; *Euonymus kawachianus* Nakai; *Euonymus loeseneri* (Thunb.) Makino; *Euonymus sacrosanctus* Koidz.; *Euonymus striatus* (Thunb.) Loes.; *Euonymus subtriflorus* Blume; *Euonymus thunbergianus* Blume; *Microrhamnus taquetii* H. Lév.

Common names: Winged burning bush, winged Euonymus, cork bush, burning bush, winged spindle tree; wei mao (China); nishikigi (Japan); gui-jun woo (Korea)

Habitat: Forests, ornamental

Distribution: China, Korea, and Japan

Botanical observation: This beautiful shrub is often cultivated as an ornamental plant and is sacred in Japan. The bark is greyish, and the stems are characteristically winged. The leaves are simple, opposite, subsessile, and without stipules. The blade is 3.5×1.3 cm, very thin, orbiculate, often reddish, the apex is shortly acuminate, serrulate at margin, the midrib is raised on both surfaces, and the secondary nerves are hardly visible. The flowers are minute, solitary or cymose, and axillary. The calyx includes 4 sepals. The corolla comprises 4 irregular, ovate, yellow, small, and nerved petals. The androecium consists of 4 stamens, which alternate with the petals. A light green disc is present. The ovary comprises 4 locules. The capsule is ovoid, reddish, somewhat glossy, up to 1.3 cm long and contains a few seeds in red arils.

Medicinal uses: Dysentery and post-partum (China)

Antiviral (enveloped monopartite single-stranded (+) RNA virus) polar extract: Aqueous extract at a concentration of 250 μg/mL inhibited the replication of the Transmissible gastroenteritis virus by 100% (Kwon et al., 2003).

Antiviral (enveloped monopartite linear single-stranded (+) RNA) polar extract: Aqueous extract at a concentration of 250 μg/mL inhibited the replication of the Porcine Epidemic Diarrhea Virus by 64% (Kwon et al., 2003).

Viral enzyme inhibition by polar extract: Aqueous extract of leaves at 100 μg/mL inhibited Human immunodeficiency virus protease by 50% (Park et al., 2002).

Commentaries: (i) The plant makes lignans such as (–)-de-*O*-methylmagnolin, (+)-de-O-methylepimagnolin A, (+)-syringaresinol, (+)-pinoresinol, (+)-medioresinol, and (+)-lariciresinol 4′-O-β-d-glucopyranoside (Lee et al., 2016), as well as lupane and oleanane triterpenes (Kang et al., 2015; Yamashita et al., 2019). Note that triterpene quinones elaborated by members of the family Celastraceae, including celastrol, pristimerin, tingenone, and iguesterin (Log D=7.6 at pH 7.4; molecular mass=404.2 g/mol) isolated from *Tripterygium regelii* Sprague & Takeda inhibited the Severe acute respiratory syndrome-associated coronavirus 3-chymotrypsine-like protease with IC_{50} values of 10.3, 5.5, 9.9, and 2.6μM, respectively (Ryu et al., 2010). Thus, one could have interest to examine the anti-COVID-19 (and other pandemic coronaviruses) properties of triterpene quinones in this family. These triterpenes of Celastraceae can be added to the list of first line natural products to be tested for the coming viral pandemics. (ii) The Porcine Epidemic Diarrhea Virus is a member of the Coronaviridae, thus, phytomedications from this plant (should it be safe for consumption) could be of value in the face of viral coronaviridae pandemics. (iii) It is interesting but equally sad to observe that humans are looking for drugs from Nature to save them from the upcoming microbial tsunamis and at the same time continue to cut forests, pollute the seas and skies, and violate the rights of animals.

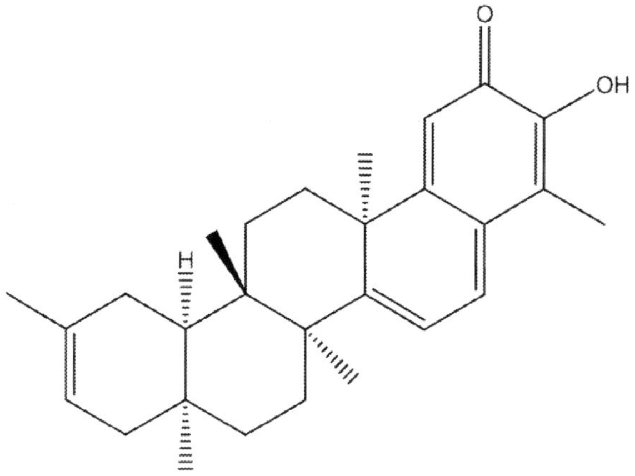

Iguesterin

(iv) Methanol extract of seeds of *Euonymus europaeus* L. inhibited the growth of *Bacillus. subtilis* (NCTC 9680), *Enterococcus faecalis* (NCIB 775), *Lactobacillus plantarum* (NCTC 6376), *Micrococcus luteus* (NCTC 9278), *Proteus mirabilis* (NCIB 60), *Serratia marcescens* (NCTC 1377), and *Staphylococcus aureus* (ATCC 10788) (Kumarasamy et al., 2003).

REFERENCES

Das, B., Tandon, V. and Saha, N., 2007. Genistein from Flemingia vestita (Fabaceae) enhances NO and its mediator (cGMP) production in a cestode parasite, Raillietina echinobothrida. *Parasitology*, *134*(10), pp. 1457–1463.

Figueiredo, J.N., Räz, B. and Séquin, U., 1998. Novel quinone methides from Salacia kraussii with in vitro antimalarial activity. *Journal of Natural Products*, *61*(6), pp. 718–723.

Ju, J.E., Hanee, S.E.O., Hyun, S.S. and Choong, K.Y., 2010. Triterpenoids from euonymus alatus leaves and twigs inhibited lipopolysaccharide-induced nitric oxide production in RAW264. 7 macrophage cells. *Bulletin of the Korean Chemical Society*, *35*(10), pp. 2945–2949.

Kang, H.R., Eom, H.J., Lee, S.R., Choi, S.U., Kang, K.S., Lee, K.R. and Kim, K.H., 2015. Bioassay-guided isolation of antiproliferative triterpenoids from Euonymus alatus twigs. *Natural product communications*, *10*(11), pp. 1929–1932.

Kumarasamy, Y., Cox, P.J., Jaspars, M., Nahar, L., Wilcock, C.C. and Sarker, S.D., 2003. Biological activity of Euonymus europaeus. *Fitoterapia*, *74*(3), pp. 305–307.

Kwon, D.H., Kim, M.B., Yoon, D.Y., Lee, Y.H., Kim, J.W., Lee, H.G., Choi, I.S., Lim, J.S. and Choe, Y.K., 2003. Screening of plant resources of anti-viral activity. *Korean Journal of Medicinal Crop Science*, *11*(1), pp. 24–30.

Lee, S., Moon, E., Choi, S.U. and Kim, K.H., 2016. Lignans from the Twigs of Euonymus alatus (Thunb.) Siebold and their biological evaluation. *Chemistry & Biodiversity*, *13*(10), pp. 1391–1396.

Li, X., Kang, D.G., Lee, J.K., Kim, S.J., Choi, D.H., Lee, K.B., Cui, H.Z., Yeom, K.B. and Lee, H.S., 2008. Study on the Vasorelaxant Mechanism of the Butanol Extract of Euonymus alatus. *Journal of Physiology & Pathology in Korean Medicine*, *22*(1), pp. 148–154.

Park, J.C., Hur, J.M., Park, J.G., Hatano, T., Yoshida, T., Miyashiro, H., Min, B.S. and Hattori, M., 2002. Inhibitory effects of Korean medicinal plants and camelliatannin H from Camellia japonica on human immunodeficiency virus type 1 protease. *Phytotherapy Research*, *16*(5), pp. 422–426.

Punnam Chander, M., Vinod Kumar, K., Shriram, A.N. and Vijayachari, P., 2015. Anti-leptospiral activities of an endemic plant Glyptopetalum calocarpum (Kurz.) Prain used as a medicinal plant by Nicobarese of Andaman and Nicobar Islands. *Natural Product Research*, *29*(16), pp. 1575–1577.

Ryu, Y.B., Park, S.J., Kim, Y.M., Lee, J.Y., Seo, W.D., Chang, J.S., Park, K.H., Rho, M.C. and Lee, W.S., 2010. Severe acute respiratory syndrome-acquired coronavirus 3CLpro inhibitory effects of quinone-methide triterpenes from Tripterygium regelii. *Bioorganic & Medicinal Chemistry Letters*, *20*(6), pp. 1873–1876.
Sotanaphun, U., Lipipun, V., Suttisri, R. and Bavovada, R., 1999. A new antiviral and antimicrobial sesquiterpene from Glyptopetalum sclerocarpum. *Planta medica*, *65*(3), pp. 257–258.
Yamashita, H., Matsuzaki, M., Kurokawa, Y., Nakane, T., Goto, M., Lee, K.H., Shibata, T., Bando, H. and Wada, K., 2019. Four new triterpenoids from the bark of Euonymus alatus forma ciliato-dentatus. *Phytochemistry Letters*, *31*, pp. 140–146.

7.2.1.3 *Kokoona zeylanica* Thwaites

Common name: Kokoon (Sri Lanka)

Habitat: Forests

Distribution: Sri Lanka

Botanical observation: This magnificent tree grows up to 35 m tall. The bark is rough and the inner bark is bright orangish. The leaves are simple, opposite, and stipulate. The petiole is up to about 1 cm long. The blade is obovate, coriaceous, cuneate at base, rounded at apex, 3.5–9×2.5–5 cm, and somewhat serrate. The inflorescences are axillary or cauliflorous panicles. The calyx is 5-lobed. The corolla presents 5 petals, which are rounded. A disc is present. The androecium consists of 5 stamens. The stigma is capitate. The capsules are oblong, up to 10cm long, 3-lobed, black and shelter numerous winged seeds, which are up to 8cm long.

Medicinal uses: Apparently none for infectious diseases

Strong antibacterial (Gram-positive) amphiphilic D:A-friedo-24-noroleanane triterpene: Zeylasterone (Gunaherath & Gunatilaka, 1983) (Log D=4.1 at pH 7.4; molecular mass=510.6 g/mol) inhibited the growth of *Bacillus alvei*, *Bacillus cereus*, *Bacillus megaterium*, *Bacillus pumilus*, *Bacillus subtilis*, *Micrococcus luteus*, and *Staphylococcus aureus* (De León & Moujir, 2008). Zeylasterone evoked permeability changes in the cytoplasmic membrane of *Bacillus cereus* as well as protein synthesis inhibition and had a bacteriostatic effect dependent on the growth phase and inoculum size (De León & Moujir, 2008).

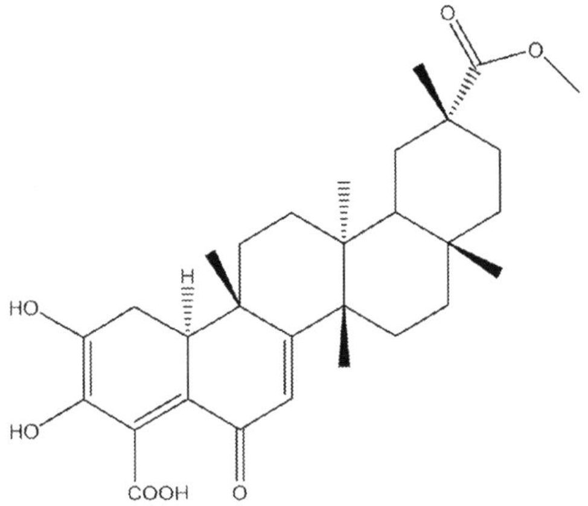

Zeylasterone

Strong anticandidal amphiphilic D:A-friedo-24-noroleanane triterpene: Zeylasterone inhibited the growth of *Candida albicans* with an MIC/MBC value of 10/>40 µg/mL (De León & Moujir, 2008).

Commentary: Natural products with strong *in vitro* antibacterial activity against Gram-positive bacteria from the plants belonging to the Clades covered in this volume include a broad-spectrum of medium to high molecular weight and hydrophilic or amphiphilic phenolic compounds, principally xanthones, coumarins, and flavones.

REFERENCES

De León, L. and Moujir, L., 2008. Activity and mechanism of the action of zeylasterone against Bacillus subtilis. *Journal of Applied Microbiology*, *104*(5), pp. 1266–1274.

Gunaherath, G.K.B. and Gunatilaka, A.L., 1983. Studies on terpenoids and steroids. Part 3. Structure and synthesis of a new phenolic D: A-friedo-24-noroleanane triterpenoid, zeylasterone, from kokoona zeylanica. *Journal of the Chemical Society, Perkin Transactions 1*, pp. 2845–2850.

7.2.1.4 *Microtropis japonica* (Franch. & Sav.) Hallier f.

Synonyms: *Cassine japonica* (Franch. & Sav.) Kuntze; *Cassine kotoensis* Hayata; *Elaeodendron japonicum* Franch. & Sav.; *Microtropis kotoensis* (Hayata) Koidz.; *Otherodendron japonicum* (Franch. & Sav.) Makino

Common names: Ri ben jia wei mao (China)

Habitat: Forests

Distribution: China, Taiwan, and Japan

Botanical observation: This handsome tree grows to about 5 m tall. The leaves are simple, opposite, and exstipulate. The petiole is about 1 cm long. The blade is broadly elliptic, 4.5–8×2–4 cm, coriaceous, cuneate or tapering at base, obtuse to acute at apex, and marked with 4–6 pairs of secondary nerves. The inflorescences are axillary or terminal cymes. The calyx develops 5 sepals persistent in fruit. The 5 petals are whitish, somewhat rutaceous at first glance, and minute. An annular and somewhat 5-lobed disc is present. The androecium presents 5 stamens. The capsule is oblong to obovate, 1.5–2 cm long, with a peculiar red, somewhat glossy, and dehiscent.

Medicinal use: Apparently none for microbial infections

Strong antimycobacterial dihydroagarofuran sesquiterpenes: Microjaponin isolated from the stems inhibited *Mycobacterium tuberculosis* (H37Rv) with an MIC value of 12.5 µg/mL (ethambutol=6.2 µg/mL) (Chen et al., 2014). The amphiphilic dihydroagarofuran sesquiterpenes celahin C (Log D=4.4 at pH 7.4, molecular mass=532.5 g/mol) and salasol A (Log D=4.4 at pH 7.4, molecular mass=532.5 g/mol) isolated from the roots inhibited the growth of *Mycobacterium tuberculosis* (H37Rv) with an MIC value of 15 µg/mL (ethambutol: 6.2 µg/mL) (Chou et al., 2008). 15-Acetoxyorbiculin G isolated from the stems inhibited the growth of *Mycobacterium tuberculosis* (H37Rv) with an MIC value of 39.6 µg/mL (Chen et al., 2008).

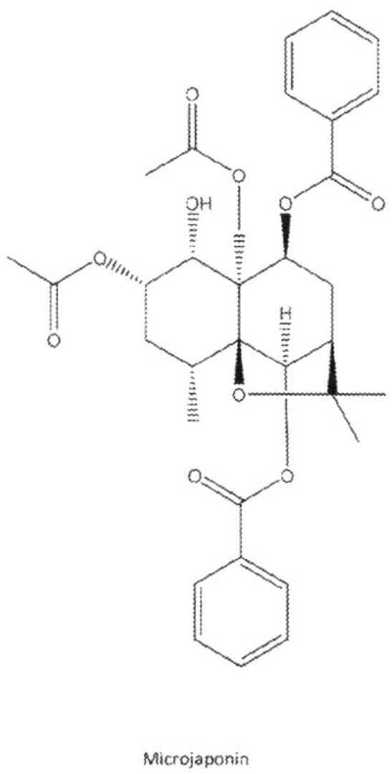

Microjaponin

Commentaries: (i) The dihydroagarofuran sesquiterpenes 8-acetoxymutangin and mutangin isolated from the stems of *Microtropis fokienensis* Dunn (local name: *Fu jian jia wei mao* in China) inhibited the growth of *Mycobacterium tuberculosis* (90-221387) with MIC values of 10 and 35 μg/mL, respectively (Ethambutol 6.2 μg/mL) (Chou et al., 2007). From this plant, 1α-acetoxy-2 α -hydroxy-6 β,9 β,15-tribenzoyloxy- β -dihydroagarofuran, 2 α -acetoxy-1 α -hydroxy-6 β,9 β,15-tribenzoyloxy β -dihydroagarofuran, orbiculin G, and triptogelin G-2, inhibited the growth of *Mycobacterium tuberculosis* (90-221387) with MIC values of 19.5, 15.8, 14.6, and 26 μM, respectively (Ethambutol: 30.6 μM) (Chen et al., 2007).

(ii) Consider that dihydroagarofurans are cytotoxic (Zhou et al., 2017) and cytotoxic principles are often antimycobacterial. (iii) The epoxide groups in dihydroagarofurans could be antimycobacterial pharmacophores.

REFERENCES

Chen, J.J., Chou, T.H., Peng, C.F., Chen, I.S. and Yang, S.Z., 2007. Antitubercular dihydroagarofuranoid sesquiterpenes from the roots of Microtropis fokienensis. *Journal of Natural Products*, *70*(2), pp. 202–205.

Chen, J.J., Kuo, W.L., Chen, I.S., Peng, C.F., Sung, P.J., Cheng, M.J. and Lim, Y.P., 2014. Microjaponin, a new dihydroagarofuranoid sesquiterpene from the stem of Microtropis japonica with antituberculosis activity. *Chemistry & Biodiversity*, *11*(8), pp. 1241–1246.

Chen, J.J., Yang, C.S., Peng, C.F., Chen, I.S. and Miaw, C.L., 2008. Dihydroagarofuranoid sesquiterpenes, a lignan derivative, a benzenoid, and antitubercular constituents from the stem of Microtropis japonica. *Journal of Natural Products*, *71*(6), pp. 1016–1021.

Chou, T.H., Chen, I.S., Peng, C.F., Sung, P.J. and Chen, J.J., 2008. A new dihydroagarofuranoid sesquiterpene and antituberculosis constituents from the root of Microtropis japonica. *Chemistry & Biodiversity*, *5*(7), pp. 1412–1418.

Chou, T.H., Chen, I.S., Sung, P.J., Peng, C.F., Shieh, P.C. and Chen, J.J., 2007. A new dihydroagarofuranoid sesquiterpene from Microtropis fokienensis with antituberculosis activity. *Chemistry & Biodiversity*, *4*(7), pp. 1594–1600.

Zhou, J., Han, N., Lv, G., Jia, L., Liu, Z. and Yin, J., 2017. Two new β-dihydroagarofuran sesquiterpenes from Celastrus orbiculatus thunb and their anti-proliferative activity. *Molecules*, *22*(6), p. 948.

7.2.1.5 *Tripterygium wilfordii* Hook. f.

Synonyms: *Aspidopterys hypoglauca* H. Lév.; *Tripterygium hypoglaucum* (H. Lév.) Hutch.; *Tripterygium regelii* Sprague & Takeda

Common name: Lei gong teng (China)

Habitat: Forest edges, scrubs

Distribution: Myanmar, China, Taiwan, Korea, Japan

Botanical observation: This graceful shrub grows up to a height of 3 m. The leaves are simple, opposite, and stipulate. The stipules are linear and ephemeral. The petiole is up to 2 cm long. The blade is elliptic, 8.5–12.5 × 5–10 cm, hairy below, acute at base, crenulate or not, and acuminate or acute at apex. The inflorescence is a showy thyrse. The calyx is 5-lobed. The corolla presents 5 petals, which are white and minute. The androecium presents 5 stamens. A disc is present. The capsule is 3-winged, reddish-green, and shelters a trigonal seed.

Medicinal use: Apparently none for the treatment of microbial infections

Strong broad-spectrum antibacterial prenylated flavanones: Tripteryol B isolated from the stems and roots inhibited the growth of *Pseudomonas aeruginosa* (ATCC 27853), low-level vancomycin-resistant *Enterococcus faecalis* (ATCC 51299), and methicillin-resistant *Staphylococcus aureus* (ATCC 43300 (methicillin-resistant *Staphylococcus aureus*) with IC_{50} values of 8.5, 4.3, and 4.4 μg/mL, respectively (Chen et al., 2017). (2*S*)-5,7,4′-trihydroxy-2′-methoxy-8,5′-di(3-methyl-2-butenyl)-6-methylflavanone inhibited the growth of *Staphylococcus aureus* (ATCC 43300 (methicillin-resistant *Staphylococcus aureus*) and *Staphylococcus aureus* (ATCC 29213) with IC_{50} values of 2 and 2.1 μg/mL, respectively (Chen et al., 2017). (±)-5,4′-Dihydroxy-2′-methoxy-6′,6″-dimethypyraro-(2″,3″:7,8)-6-methyflavanone from this plant inhibited methicillin-resistant *Staphylococcus* *aureus* (ATCC 43300 (methicillin-resistant *Staphylococcus aureus*) and *Staphylococcus aureus* (ATCC 29213) with IC_{50} values of 2 and 2.6 µg/mL, respectively (CLSI 2000) (Chen et al., 2017). (2*S*)-5,7,4′-trihydroxy-2′-methoxy-8,5′-di(3-methyl-2-butenyl)-6-methylflavanone inhibited the growth of *Staphylococcus aureus* (ATCC 43300 (methicillin-resistant *Staphylococcus aureus*) and *Staphylococcus aureus* (ATCC 29213) with IC_{50} values of 2 and 2.1 µg/mL, respectively (Chen et al., 2017). (±)-5,4′-Dihydroxy-2′-methoxy-6′,6″-dimethypyraro-(2″,3″:7,8)-6-methyflavanone from this plant inhibited methicillin-resistant *Staphylococcus aureus* (ATCC 43300 (methicillin-resistant *Staphylococcus aureus*) and *Staphylococcus aureus* (ATCC 29213) with IC_{50} values of 2 and 2.6 µg/mL, respectively (CLSI 2000) (Chen et al., 2017).

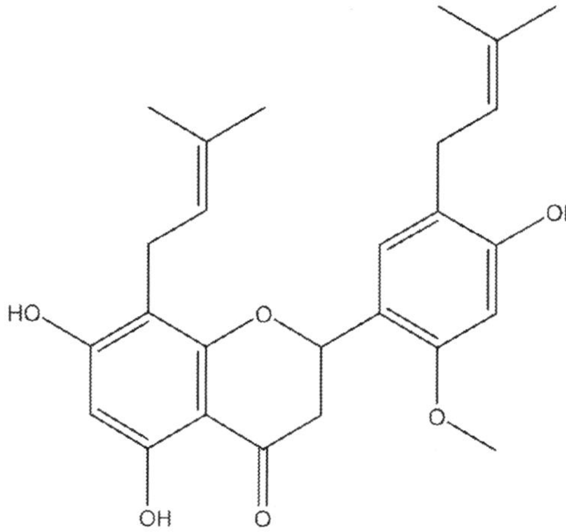

Tripteryol B

Strong antifungal (yeast) antifungal prenylated flavanones: Tripteryol B inhibited the growth of *Cryptococcus neoformans* (ATCC 90113) with an IC_{50} value of 2.9 µg/mL (Chen et al., 2017). (2*S*)-5,7,4′-trihydroxy-2′-methoxy-8,5′-di(3-methyl-2-butenyl)-6-methylflavanone inhibited the growth of *Cryptococcus neoformans* (ATCC 90113) with an IC_{50} value of 1 µg/mL (Chen et al., 2017).

Strong antiviral (enveloped monopartite linear single-stranded (+) RNA) amphiphilic D:A-friedo-24-noroleanane triterpene: Celastrol (Log D=4.0 at pH 7.4; molecular mass=450.6 g/mol) inhibited RNA replication of Dengue virus type 1, 2, 3, and 4 in Huh-cells with an IC_{50} value of 0.1 µM (Yu et al., 2017).

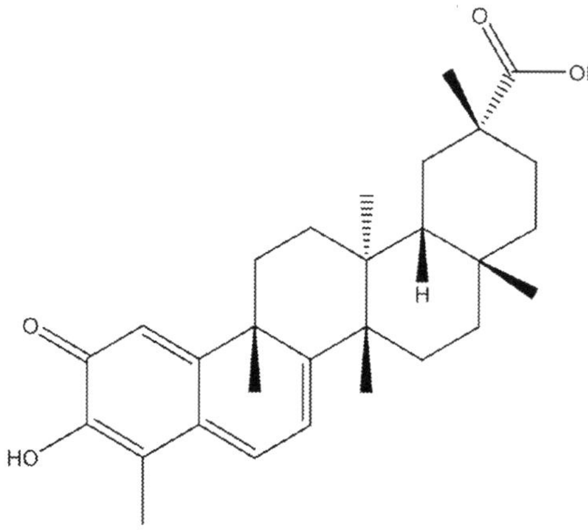

Celastrol

Strong antiviral (enveloped monopartite linear double-stranded DNA) alkaloid fractions: Alkaloid fraction of roots inhibited the replication of the Herpes simplex type-1 with an IC_{50} value of 6.5 µg/mL, which was notably lower than that of Acyclovir (15.4 µg/mL) (Ren et al., 2010).

Strong antiviral (enveloped monopartite linear dimeric single-stranded (+)RNA) hydrophilic diterpene: Triptolide (Log D=0.6 at pH 7.4; molecular mass=360.1 g/mol) inhibited the replication of Human immunodeficiency virus type-1 with an EC_{50} value of 0.3 nM (Wan et al., 2014). This diterpene inhibited Human immunodeficiency virus type-1 gene transcription and replication by prompting proteasomal degradation of Tat protein (Wan et al., 2014).

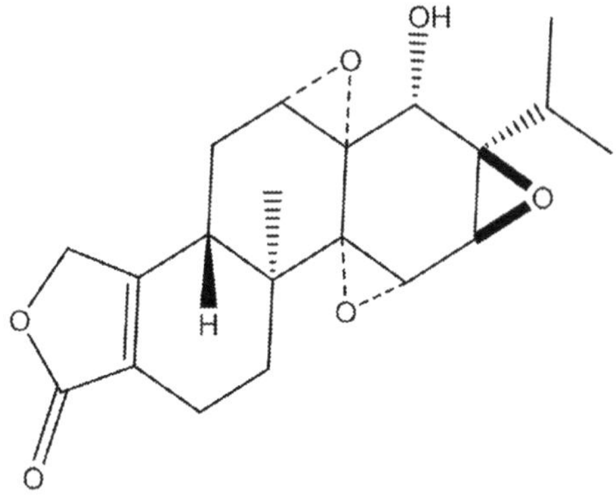

Triptolide

Strong antiviral (enveloped monopartite linear dimeric single-stranded (+)RNA) amphiphilic sesquiterpene alkaloids: Wilfortrine (LogD=3.6 at pH 7.4; molecular mass=873.8 g/mol), wilfordin (LogD=4.4 at pH 7.4; molecular mass=883.8 g/mol), triptonine B (LogD=2.0 at pH 7.4; molecular mass=961.9 g/mol) isolated from the root bark inhibited the replication of the Human immunodeficiency virus with an EC_{50} value of <0.1 µg/mL and a selectivity index above 1000 (Horiuch et al., 2006). From the same plant, hypoglaunine A (LogD=3.6 at pH 7.4; molecular weight=873.8 g/mol), hypoglaunine B (LogD=3.6 at pH 7.4; molecular weight=873.2 g/mol), and hyponin A inhibited the replication of the Human immunodeficiency virus with an EC_{50} value of 0.1 µg/mL (Horiuch et al., 2006). From this plant forrestine, cangoronine E-1, euonymine, neoeuonymine, and hyponine B (LogD=4.6 at pH 7.4; molecular weight=857.2 g/mol) inhibited the replication of the Human immunodeficiency virus with IC_{50} values of 0.4, 0.9, 0.2, 0.8, and 0.1 µg/mL, respectively (Horiuch et al., 2006).

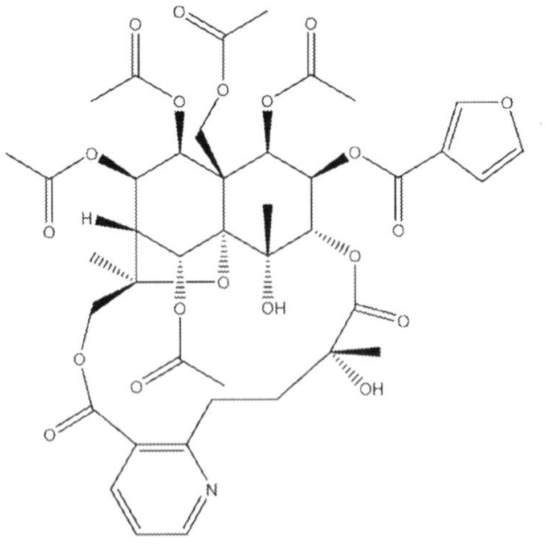

Wilfortrine

Strong antiviral (enveloped linear monopartite double-stranded DNA) lipophilic oleanane triterpene: Pristimerin (LogD=7.1 at pH 7.4; molecular weight=464.2 g/mol) inhibited Human cytomegalovirus replication in the human embryonic fibroblast cells. In a viral plaque-reduction assay, pristimerin showed dose-dependent inhibitory properties with a 50% inhibitory concentration of 0.5 μg/mL (selectivity index=27.9) (Murayama et al., 2007).

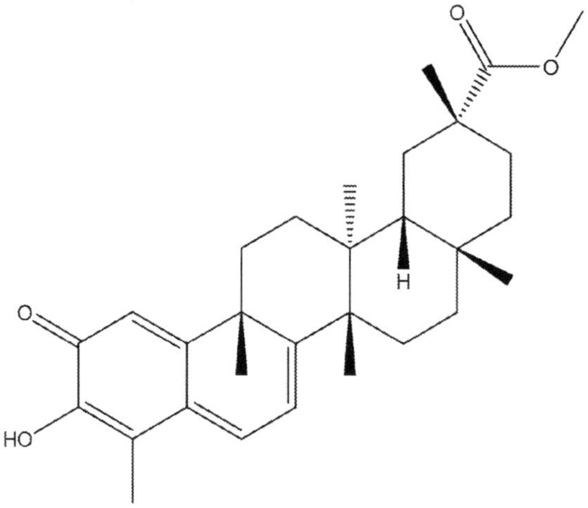

Pristimerin

In vivo (enveloped monopartite linear single-stranded (+) RNA) antiviral triterpene: Celastrol administered at 1, 3, and 5 days post-infection intracerebrally at a dose of 0.1 mg/Kg protected ICR mice infected with Dengue virus type 2 strain (Yu et al., 2017).

Moderate antiviral (enveloped monopartite linear double-stranded DNA) dihydroagarofuran sesquiterpene: Regelidine showed 31.7% inhibitory activity at 250 μg/mL against the Herpes simplex virus type-2 while acyclovir displayed a 60.6% inhibition at 250 μg/mL (Luo et al., 2011).

Commentary: Triterpenes in the family Celastraceae could be of value against coronaviruses.

REFERENCES

Chen, Y., Zhao, J., Qiu, Y., Yuan, H., Khan, S.I., Hussain, N., Choudhary, M.I., Zeng, F., Guo, D.A., Khan, I.A. and Wang, W., 2017. Prenylated flavonoids from the stems and roots of Tripterygium wilfordii. *Fitoterapia, 119*, pp. 64–68.

Horiuch, M., Murakami, C., Fukamiya, N., Yu, D., Chen, T.H., Bastow, K.F., Zhang, D.C., Takaishi, Y., Imakura, Y. and Lee, K.H., 2006. Tripfordines A–C, sesquiterpene pyridine alkaloids from Tripterygium wilfordii, and structure anti-HIV activity relationships of Tripterygium alkaloids. *Journal of Natural Products, 69*(9), pp. 1271–1274.

Luo, Y., Zhou, M., Ye, Q., Pu, Q. and Zhang, G., 2011. Dihydroagarofuran derivatives from the dried roots of Tripterygium wilfordii. *Journal of Natural Products, 75*(1), pp. 98–102.

Murayama, T., Eizuru, Y., Yamada, R., Sadanari, H., Matsubara, K., Rukung, G., Tolo, F.M., Mungai, G.M. and Kofi-Tsekpo, M., 2007. Anticytomegalovirus activity of pristimerin, a triterpenoid quinone methide isolated from Maytenus heterophylla (Eckl. & Zeyh.). *Antiviral Chemistry and Chemotherapy, 18*(3), pp. 133–139.

Ren, Z., Zhang, C.H., Wang, L.J., Cui, Y.X., Qi, R.B., Yang, C.R., Zhang, Y.J., Wei, X.Y., Lu, D.X. and Wang, Y.F., 2010. In vitro anti-viral activity of the total alkaloids from Tripterygium hypoglaucum against herpes simplex virus type 1. *Virologica Sinica, 25*(2), pp. 107–114.

Wan, Z. and Chen, X., 2014. Triptolide inhibits human immunodeficiency virus type 1 replication by promoting proteasomal degradation of Tat protein. *Retrovirology, 11*(1), p. 88.

Yu, J.S., Tseng, C.K., Lin, C.K., Hsu, Y.C., Wu, Y.H., Hsieh, C.L. and Lee, J.C., 2017. Celastrol inhibits dengue virus replication via up-regulating type I interferon and downstream interferon-stimulated responses. *Antiviral Research, 137*, pp. 49–57.

7.3 ORDER OXALIDALES BERCHT. & J.PRESL (1820)

7.3.1 Family Cunoniaceae R. Brown (1814)

The family Cunoniaceae consists of 27 genera and 350 species of shrubs and trees. The leaves are compound, opposite, and stipulate. The inflorescences are racemes or panicles. The calyx includes 4–5 sepals. The corolla presents 4–5 petals. The androecium is made of 4–10 stamens. The gynoecium consists of 2–5 carpels united to form a plurilocular ovary, each locule sheltering 2 to numerous ovules growing on axile placentas. The fruits are capsules containing numerous winged seeds. Plants in this family produce hydrolysable tannins.

7.3.1.1 *Weinmannia blumei* Planch.

Synonyms: *Wenmannia ledermannii* Schelcht.; *Weinmannia papuana* Schelcht.; *Weinmannia sundara* Miq.

Common names: Malayan mountain ash; antahasi, ki papatong (Indonesia)

Habitat: Forests

Distribution: Thailand, Malaysia, Indonesia, Papua New Guinea, and Solomon Islands

Botanical observation: This tree grows up to a height of 25 m. The wood is reddish brown. The stems are smooth, lenticelled, somewhat articulate, and dark brown. The leaves are compound, decussate, and stipulate. The stipules are interpetiolar, coriaceous, up to 2.3 cm long, and reniform. The folioles are opposite, somewhat asymmetrical, serrate, lanceolate, coriaceous, 1.1–3.5 × 4–12 cm, and shorter near the base of the rachis. The inflorescences are terminal and showy racemes. The calyx comprises 4 minute sepals. The corolla includes 4 petals which are pinkish and up to 1 cm long. The androecium includes 8 slender and showy stamens.

Medicinal uses: Fever, wounds (Indonesia)

Commentary: This plant and its antimicrobial principles, like so many plants from the Southeast Asian primary rainforest, is bound to vanish under the caterpillars of the palm oil industry and other logging activities looking for "quick" financial benefits. One must grasp that the more primary forest is cut the more lethal will be the zoonotic pandemic viruses. Antimicrobial frameworks of possible utility for the development of drugs for future pandemics are produced by these rare and precious trees. Organic chemists and in silico drug design do not match the power of creativity of nature.

7.3.2 Family Elaeocarpaceae Juss. ex DC. (1816)

The family Elaoecarpaceae consists of 12 genera and 625 species of elegant trees or shrubs. The leaves are simple, alternate, and stipulate or not. The petiole is often somewhat straight and swollen at both ends. The blade is not uncommonly serrate and glossy. The inflorescences are axillary or terminal, racemes, panicles, or corymbs. The calyx comprises 4 or 5 sepals, which are free or fused at base. The corolla includes 4 or 5 petals. The androecium comprises 8 to many stamens. A disc is present. The gynoecium comprises a 2 to many locular ovary, each locule sheltering 2 to many ovules. The fruits are olive-shaped.

7.3.2.1 *Elaeocarpus grandiflorus* Sm.

Common names: Lily of the valley tree; anyang (Indonesia); andoi (Malaysia); ye saga (Myanmar); mala (the Philippines); phi nai (Thailand)

Habitat: Forests, around villages, cultivated

Distribution: Southeast Asia

Botanical observation: This tree grows to about 20 m tall. The leaves are simple, spiral, and stipulate. The petiole is straight, up to 2.5 cm long and swollen at apices. The blade is lanceolate to narrowly elliptic, 1.5–5 × 5–15 cm, coriaceous, somewhat glossy, with about 5–8 pairs of secondary nerves, tapering at base, laxly serrate, and acute at apex. The flowers are pendulous, showy, bell-like, and arranged in axillary racemes which are up to 10 cm long. The 5 sepals are lanceolate, up to 2 cm long, and somewhat light pink. The 5 petals are as long as the sepals, oblong to obovate, pure white, and laciniate at apex. The androecium comprises up to about 60 stamens. A lobed disc is present. The gynoecium includes an ovoid hairy ovary and a single style. The drupe is olive-like, green, and up to about 4 cm long.

Medicinal uses: Ulcers (Indonesia); post-partum (Malaysia)

Antifungal (filamentous) halo developed by mid-polar extract: Methylene chloride extract inhibited the growth of *Sclerotium rolfsii* and *Aspergillus fumigatus* (Goun et al., 2003).

Antiviral (enveloped monopartite linear double-stranded DNA) polar extracts: Ethanol extract of leaves at a concentration of 100 µg/mL decreased plaque formation by Herpes simplex virus type-1 by about 25%, respectively (Lipipun et al., 2003).

Antiviral (non-enveloped monopartite linear dimeric single stranded (+)RNA) polar extract: Ethanol extract of leaves at a concentration of 100 µg/mL decreased plaque formation by Poliovirus by about 85%, respectively (Lipipun et al., 2003).

Antiviral (enveloped monopartite linear single-stranded (-)RNA) polar extracts: Ethanol extract of leaves at a concentration of 100 µg/mL decreased plaque formation by Measles virus by about 70% (Lipipun et al., 2003).

Viral enzyme inhibition by polar extract: Aqueous extract inhibited Human immunodeficiency reverse transcriptase with an IC50 value of 450 µg/mL (Kusumoto et al., 1992). Aqueous extract of fruits at a concentration of 250 µg/mL inhibited Human immunodeficiency type-1 protease by 53.9% (Xu et al., 1996).

Commentary: (i) Hydrolysable tannins could account for the medicinal uses (mild astringent) and its antiviral activities *in vitro*. This is because gallic acid, the ellagitannins corilagin, and chebulagic acid isolated from *Elaeocarpus tonkinensis* Aug. DC. inhibited the replication of the Influenza A virus (PR8 strain) with IC_{50} values of 8.1, 31.2 and 21 µg/mL, respectively (Dao et al., 2019). Note that hydrolysable tannins have a very low oral bioavailability and can't have any systemic antiviral effects (but their intestinal bacterial catabolites may). Tannins at about 4% or diet are hepatotoxic. Polar extract of plants producing tannins have broad-spectrum antiviral effects and inhibit almost all enzymes *in vitro* because tannins precipitate proteins. (ii) Plants in the family Elaeocarpaceae synthetize a fascinating array of indolizidine and pyrrolidine alkaloids (Katavic 2005) which could be evaluated for their antimicrobial properties, although these alkaloids are often hepatotoxic. (iii) In India, *Elaeocarpus lanceifolius* Roxb. (local name: Kharuan) and *Elaeocarpus serratus* L. (local name: *Uttraccham*) are used to treat dysentery. Plants used to treat dysentery and/or diarrhea are often tanniferous or accumulating phenolics. Tannins and phenolics are broad-spectrum antimicrobials *in vitro*. In Papua New Guinea, *Elaeocarpus sphaericus* (Gaertn.) K. Schum. is used to treat cough and pneumonia.

REFERENCES

Dao, N.T., Jang, Y., Kim, M., Nguyen, H.H., Pham, D.Q., Le Dang, Q., Van Nguyen, M., Yun, B.S., Pham, Q.M., Kim, J.C. and Hoang, V.D., 2019. Chemical constituents and anti-influenza viral activity of the leaves of vietnamese plant Elaeocarpus tonkinensis. *Records of Natural Products*, *13*(1), pp. 71–80.

Goun, E., Cunningham, G., Chu, D., Nguyen, C. and Miles, D., 2003. Antibacterial and antifungal activity of Indonesian ethnomedical plants. *Fitoterapia*, *74*(6), pp. 592–596.

Katavic, P.L., 2005. Chemical investigations of the alkaloids from the plants of the family Elaeocarpaceae. Natural Product Discovery (NPD), Faculty of Science, Griffith University, Australia.

Kusumoto, I.T., Shimada, I., Kakiuchi, N., Hattori, M., Namba, T. and Supriyatna, S., 1992. Inhibitory effects of Indonesian plant extracts on reverse transcriptase of an RNA tumour virus (I). *Phytotherapy Research*, *6*(5), pp. 241–244.

Lipipun, V., Kurokawa, M., Suttisri, R., Taweechotipatr, P., Pramyothin, P., Hattori, M. and Shiraki, K., 2003. Efficacy of Thai medicinal plant extracts against herpes simplex virus type 1 infection in vitro and in vivo. *Antiviral Research*, *60*(3), pp. 175–180.
Xu, H.X., Wan, M., Loh, B.N., Kon, O.L., Chow, P.W. and Sim, K.Y., 1996. Screening of Traditional Medicines for their Inhibitory Activity Against HIV-1 Protease. *Phytotherapy Research*, *10*(3), pp. 207–210.

7.3.2.2 *Elaeocarpus petiolatus* (Jack) Wall.

Synonym: *Monocera petiolata* Jack

Common name: Chang bing du ying (China)

Habitat: Forests

Distribution: India, Myanmar, Laos, Vietnam, China, Thailand, Malaysia, and Indonesia

Botanical observation: This tree grows to about 10 m tall. The leaves are simple, spiral, and stipulate. The petiole is straight, up to 6 cm long, and swollen at apices. The blade is oblong to broadly elliptic, 9–18 × 4–8 cm, coriaceous, glossy, with about 5–8 pairs of secondary nerves, rounded or obtuse at base, more or less serrate at margin, and acuminate at apex. The inflorescence is an axillary raceme, which is up to 10 cm long and many flowered, the flowers are somewhat pendulous. The 5 sepals are lanceolate, up to 7 mm long, and hairy below. The 5 petals are about as long as the sepals, oblong, membranous, whitish, and marked with 7–14 divisions at apex. A toothed disc is present. The gynoecium includes an ovoid ovary and a single style. The drupes are olive-like, green, and up to about 2 cm long.

Medicinal use: Fever (Malaysia)

Viral enzyme inhibition by megastigmane: Vomifoliol inhibited the Influenza A virus A/PR/8/34 (H1N1) neuraminidase activity with an IC_{50} values of 2.3 µg/mL (Xie et al., 2013).

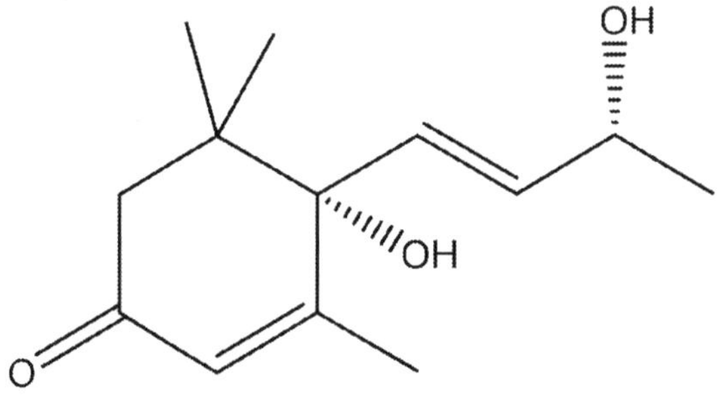

Vomifoliol

Commentaries: (i) The plant has apparently not been studied for its antimicrobial properties. It affords cucurbitane-type triterpenes (also called cucurbitacins) such as elaeocarpucin F as well as the megastigmane (terpene derived from the catabolism of abscisic acid) vomifoliol (Cho, 2019). Vomifoliol (also known as blumenol A) comprises a cyclohexenone (Mayekar et al., 2010) moiety and alkyl chain, which suggests that it may have antibacterial or antifungal effects. Cucurbitacin B isolated from *Ecballium elaterium* L. (family Cucurbitaceae) inhibited the growth of *Staphylococcus aureus* (ATCC 29213) and methicillin-resistant *Staphylococcus aureus* (ATCC 43300) (Hassan et al., 2017).

(ii) Ethanol extract of *Elaeocarpus sylvestris* (Loureiro) Poir. at a concentration of 10 µg/mL protected primary human foreskin fibroblasts from Human cytomegalovirus (To et al., 2014). In a subsequent study, ethyl acetate fraction extract at a concentration of 25 µg/mL decreased the expression of Varicella-Zoster virus DNA from infected primary human foreskin fibroblasts (Bae et al., 2017). From this extract, 1,2,3,4,6-penta-*O*-galloyl-β-glucose at a concentration of 10 µg/mL inhibited Varicella-Zoster virus by about 50% (Bae et al., 2017). Varicella-Zoster virus from the Herpesviridae is a double-stranded DNA-enveloped virus responsible for chicken pox in children and shingles in adults via penetration through the respiratory mucosa and body dissemination and preventable via vaccination (Cary & Lopez, 2011). Testing natural products or extracts against virus for which vaccination exist is not a waste of time because other pathogenic viruses from the same viral family and for which no vaccine exist could be targeted, and, secondly, understanding their mode of action could assist in finding antiviral targets. Also, viral mutants keep on reappearing.

REFERENCES

Bae, S., Kim, S.Y., Do, M.H., Lee, C.H. and Song, Y.J., 2017. 1, 2, 3, 4, 6-Penta-O-galloyl-ss-D-glucose, a bioactive compound in Elaeocarpus Sylvestris extract, inhibits varicella-zoster virus replication. *Antiviral Research*, *144*, pp. 266–272.

Chisholm, C. and Lopez, L., October 2011. Cutaneous infections caused by herpesviridae: A review. *Archives of Pathology & Laboratory Medicine*, *135*(10), pp. 1357–1362.

Cho, E.S., 2019. *Isolation, structure elucidation and cytotoxic activity of cucurbitacins from Elaeocarpus petiolatus* (Doctoral dissertation, University of Nottingham).

Hassan, S.T.S., Berchová-Bímová, K., Petráš, J. and Hassan, K.T.S., 2017. Cucurbitacin B interacts synergistically with antibiotics against Staphylococcus aureus clinical isolates and exhibits antiviral activity against HSV-1. *South African Journal of Botany*, *108*, pp. 90–94.

Mayekar, A.N., Li, H., Yathirajan, H.S., Narayana, B. and Kumari, N.S., 2010. Synthesis, characterization and antimicrobial study of some new cyclohexenone derivatives. *International Journal of Chemistry*, *2*(2), p. 114.

To, K.P., Kang, S.C. and Song, Y.J., 2014. The extract of Elaeocarpus sylvestris inhibits human cytomegalovirus immediate early gene expression and replication in vitro. *Molecular Medicine Reports*, *9*(2), pp. 744–748.

7.3.3 Family Oxalidaceae R. Brown (1818)

The family Oxalidaceae consists of 8 genera and about 800 species of herbs, shrubs, and trees. The leaves are compound, alternate or whorled, and exstipulate or minutely stipulate. The inflorescences are umbels, cymes, or racemes, or solitary. The calyx comprises 5 sepals. The corolla consists of 5 petals. The androecium includes 10 stamens. The gynoecium is made of 5 carpels united into a plurilocular ovary with 1–2 ovules born on axil placentas per locule. The fruit is an explosively dehiscent capsule or a berry. Plants in this small family are tanniferous.

7.3.3.1 *Averrhoa carambola* L.

Synonyms: *Averrhoa acutangula* Stokes; *Averrhoa pentandra* Blanco; *Connaropsis philippica* Fern.-Vill.; *Sarcotheca philippica* (Fern.-Vill.) Hallier f.

Common names: Star-fruit; kordoi, kam ranga (Bangladesh); spo (Cambodia); wu lien tsze (China); belimbing (Indonesia, Malaysia); karmal (India); mak hpung (Myanmar); macosembi (Papua New Guinea); balimbing, balingbing (the Philippines); phuang (Laos); ma phueang (Thailand); khe (Vietnam)

Habitat: Cultivated

Distribution: Tropical Asia and Pacific

Botanical observation: This tree native to South America grows to 5 m tall. The leaves are imparipinnate and spiral. The petiole is 2–8 cm long. The blade is 7–25 cm and comprises 4–7 subopposite pairs of folioles, which are elliptic, glossy 3–8 × 1.5–4.5 cm, obliquely rounded at base, apex acute to acuminate, and increasing in size from petiole to the apex of the rachis. The inflorescences are axillary, cymose, reddish, and support numerous small flowers. The 5 sepals are about 5 mm long. The 5 petals are purplish to burgundy, somewhat glossy, oblong, recurved, and up to about 1 cm long. The androecium includes 10 stamens. The ovary is hairy and develops 5 styles. The berry is yellow, palatable, oblong, 7–13×5–8 cm, deeply 5-ribbed, glossy, fleshy, and contains numerous minute seeds.

Medicinal uses: Boils, influenza (Bangladesh); cough, cuts, sores (Papua New Guinea); dysentery, chicken pox (India); fever (India, Bangladesh, Myanmar, China, the Philippines); jaundice (India, Bangladesh).

Moderate broad-spectrum polar extract: Ethanol extract of leaves inhibited the growth of *Staphylococcus aureus* (ATCC 29213), methicillin-resistant *Staphylococcus aureus* (clinical strain), *Staphylococcus saprophyticus* (ATCC 25352), *Enterococcus faecalis* (ATCC 29212), *Klebsiella pneumoniae* (ESBL, clinical), and *Acinatobacter baumannii* (ATCC 17978) (Silva et al., 2020).

Strong antiviral (non-enveloped segmented linear double-stranded RNA) polar extract: Aqueous extract of leaves at a concentration of 20 µg/mL inhibited the replication of the Human rotavirus (HCR3) and Simian rotavirus (SA-11) in African Rhesus monkey kidney cells (MA-104) by 66.8 and 43.8%, respectively (Gonçalves et al., 2005).

Viral enzyme inhibition by polar extract: Ethanol extract of leaves inhibited Human immunodeficiency virus type-1 integrase with an IC_{50} value of 47.6 µg/mL (Suedee et al., 2013).

Commentaries: (i) The plant brings forth proanthocyanidins (condensed tannins), flavonol glycosides, chalcone glycosides, gallic acid, (+)-epicatechin, and tetrahydroisoquinoline alkaloids (Yang et al., 2014; Jia et al., 2018). These might work synergistically for the traditional medicinal uses. As for anti-hepatitis and anti-influenza uses, validation remains warranted. (ii) In India, *Averrhoa bilimbi* L. (local name: *Karmaranga*) is used for fever, diarrhea, and hepatitis, in the Philippines (local name: *Kalamyas*) it is used to treat cough, chicken pox, and as postpartum protective remedy, in Yap it is used for diarrhea, and in Palau (local name: *Emgurs*) it is used to heal wounds.

REFERENCES

Gonçalves, J.L.S., Lopes, R.C., Oliveira, D.B., Costa, S.S., Miranda, M.M.F.S., Romanos, M.T.V., Santos, N.S.O. and Wigg, M.D., 2005. In vitro anti-rotavirus activity of some medicinal plants used in Brazil against diarrhea. *Journal of Ethnopharmacology*, *99*(3), pp. 403–407.

Silva, K.B., Pinheiro, C.T.S., Soares, C.R.M., Souza, M.A., Matos-Rocha, T.J., Fonseca, S.A., Pavão, J.M.S.J., Costa, J.G., Pires, L.L.S. and Santos, A.F., 2020. Phytochemical characterization, antioxidant potential and antimicrobial activity of Averrhoa carambola L.(Oxalidaceae) against multiresistant pathogens. *Brazilian Journal of Biology*, *81*, pp. 509–515.

Suedee, A., Tewtrakul, S. and Panichayupakaranant, P., 2013. Anti-HIV-1 integrase compound from Pometia pinnata leaves. *Pharmaceutical Biology*, *51*(10), pp. 1256–1261.

Yang, D., Xie, H., Yang, B. and Wei, X., 2014. Two tetrahydroisoquinoline alkaloids from the fruit of Averrhoa carambola. *Phytochemistry Letters*, *7*, pp. 217–220.

7.3.3.2 *Oxalis corniculata* L.

Synonyms: *Acetosella corniculata* (L.) Kuntze; *Oxalis albicans* Kunth; *Oxalis bradei* R. Knuth; *Oxalis filiformis* Kunth; *Oxalis foliosa* Blatt.; *Oxalis herpestica* Schltdl.; *Oxalis lupulina* Kunth; *Oxalis meridensis* Pittier; *Oxalis minima* Steud.; *Oxalis pilosiuscula* Kunth; *Oxalis procumbens* Steud. ex A. Rich.; *Oxalis pubescens* Stokes; *Oxalis radicosa* A. Rich.; *Oxalis repens*Thunb.; *Oxalis simulans* Baker; *Oxalis steudeliana* Kunth; *Oxalis taiwanensis* (Masam.) Masam.; *Oxalis tubistipula* Steud. ex Phil.; *Oxalis villosa* M. Bieb.; *Xanthoxalis corniculata* (L.) Small; *Xanthoxalis filiformis* (Kunth) Holub; *Xanthoxalis langloisii* Small; *Xanthoxalis repens* (Thunb.) Moldenke

Common names: Creeping lady's-sorrel, creeping wood-sorrel; amruli, amboti, amilani, amrul (Bangladesh); cu jiang cao, xián suān zǎi (China); chariamilo, khattibuti (India); calicing (Indonesia); chari amilo (Nepal); khithi mithi boti, kokavu (Papua New Guinea); khat kurla, khut korla, tarweka (Pakistan); yensil (Turkey); chua đất (Vietnam); gougus (Yap)

Habitat: Grasslands, vacant lands, and roadsides

Distribution: Pakistan, India, Bangladesh, Bhutan, Myanmar, Malaysia, Nepal, Pakistan, Thailand, Japan, Korea, Papua New Guinea, Pacific Islands

Botanical observation: This pretty creeping herb grows to about 50 cm long. It has an aura of positive vital force. The leaves are compound, clover-like at first glance, alternate, and stipulate. The stipules are minute. The petiole is slender and up to 10 cm long. The blade comprises 3 folioles, which are obcordate, 0.3–1.8 × 0.4–2.3 cm, and dull green. The inflorescences are umbellate, 1–7-flowered on slender peduncles. The calyx comprises 5 sepals, which are up to about 5 mm long, and oblong-lanceolate. The 5 petals are golden yellow, oblong to obovate, and up to about 1 cm long. The 10 stamens are minute. The gynoecium includes 5 carpels fused in a 5-locular ovary. The capsules are bullet-shaped, vertical, up to 2 cm long, pentagonal, dehiscent, hairy, and contain numerous minute seeds.

Medicinal uses: Abscesses, enteritis, jaundice, tympanite, pimples, toothache (India); antiseptic, gonorrhea, jaundice (Pakistan); burn, syphilis (Papua New Guinea) diarrhea (China, Nepal, Pakistan); dysentery (Pakistan, India, Nepal, Bangladesh); eye infection (India, Nepal); fever (Turkey, Pakistan, India, Nepal, Papua New Guinea); sore throat (China); toothache (India, China); wounds (Yap)

Broad-spectrum antibacterial halo developed by polar extract: Methanol extract of whole plant (100 μL of a 2 mg/mL solution/8 mm well) inhibited the growth of *Bacillus subtilis* (ATCC7966), *Escherichia coli* (ATCC8677), *Shigella dysenteriae* (ATCC29027), and *Salmonella typhi* (ATCC0650) with inhibition zone diameters of 24, 21.3, 17, and 11 mm, respectively (Rehman et al., 2015).

Moderate broad-spectrum antibacterial polar extract: Methanol extract inhibited the growth of *Escherichia coli* (ATCC 25923) and *Shigella flexneri* (2457T) with MIC values of 80 and 130 μg/mL, respectively (Mukherjee et al., 2013).

Commentaries: (i) *Oxalis corniculata* L. is an example of plant used throughout the Asia Pacific to treat microbial infections. Such plants are bound to contain broad-spectrum antimicrobial principles. In Nepal the plant is used to prevent COVID-19 (Khadka et al., 2020) as well as in Siddha medicine (Aishwarya et al., 2020). It produces the arylnaphtalide corniculin (Zhang et al., 2019) as well as the *C*-glycoside flavonols isoorientin, isovitexin, and swertisin (Mizokami et al., 2008). Consider that isoorientin inhibits the replication of the Respiratory syncytial virus (enveloped monopartite linear single-stranded (–) RNA) and is found in plants not uncommonly used to treat COVID-19. Hence, one could have interest to examine the anti-Severe acute respiratory syndrome-acquired coronavirus 2 activities (and other zoonotic pathogenic viruses) of isoorientin and other flavonol *C*-glycosides. (ii) *Oxalis acetosella* L. is used in Pakistan (local name: *Chhushin*) for fever and *Oxalis debilis* Kunth is used for boils in India (local name: *Khat-mitthi*). In Pakistan, *Oxalis corymbosa* DC is used for the treatment of jaundice. (iii) *Biophytum reinwardtii* Klotzsch is used in India (local names: *Lajjabati*, *Mukkutti*) for the treatment of cough, gonorrhea, and wounds and *Biophytum candolleanum* Wight (local name: *Perumanivatti*) is used to treat venereal diseases.

REFERENCES

Aishwarya, A., Kantham, T.L. and Meenakumari, R., 2020. Siddha dietary and lifestyle pattern: A strong shield and weapon to win the Covid-19 pandemic battle. *International Journal of Herbal Medicine*, 8(6), pp. 19–27.

Khadka, D., Dhamala, M.K., Li, F., Aryal, P.C., Magar, P.R., Bhatta, S., Thakur, M.S., Basnet, A., Shi, S. and Cui, D., 2020. The use of medicinal plant to prevent COVID-19 in Nepal. *Journal of Ethnobiology and Ethnomedicine*, *17*, pp. 1–17.

Mizokami, H., Tomita-Yokotani, K. and Yoshitama, K., 2008. Flavonoids in the leaves of Oxalis corniculata and sequestration of the flavonoids in the wing scales of the pale grass blue butterfly, Pseudozizeeria maha. *Journal of Plant Research*, *121*(1), pp. 133–136.

Mukherjee, S., Koley, H., Barman, S., Mitra, S., Datta, S., Ghosh, S., Paul, D. and Dhar, P., 2013. Oxalis corniculata (Oxalidaceae) leaf extract exerts in vitro antimicrobial and in vivo anticolonizing activities against Shigella dysenteriae 1 (NT4907) and Shigella flexneri 2a (2457T) in induced diarrhea in suckling mice. *Journal of Medicinal Food*, *16*(9), pp. 801–809.
Rehman, A., Rehman, A. and Ahmad, I., 2015. Antibacterial, antifungal, and insecticidal potentials of Oxalis corniculata and its isolated compounds. *International Journal of Analytical Chemistry*, *2015*. Article ID 842468.
Zhang, B., Jiang, L., Ma, X., Wang, A.M., Liu, T., Zhou, M., Liao, S.G., Wang, Y.L., Huang, Y. and Li, Y.J., 2019. A new arylnaphthalide lignan from oxalis corniculata. *Natural Product Communications*, *14*(9). p.1934578X19875885.

7.4 ORDER MALPIGHIALES JUSS. EX BERCHT. & J.PRESL (1820)

7.4.1 Family Clusiaceae Lindley (1836)

The family Clusiaceae consists of 28 genera and 470 species of handsome trees or shrubs yielding a yellow to orangish latex upon excision. The leaves are simple, exstipulate, opposite, or verticillate. The petiole is not uncommonly clasping at base. The blade is often coriaceous with parallel secondary nerves. The inflorescence is a cyme or a raceme. The flowers are actinomorphic. The calyx comprises 4–5 sepals. The corolla consists of free 4–5 free petals. The androecium comprises numerous showy stamens. The gynoecium is made of 2–5 carpels, which are united to form a 1–12 celled ovary containing numerous ovules on parietal or basal placentas, and 1–12 stigma. The fruits are drupes, berries, or capsules. Members of the family are known so far to produce antimicrobial coumarins, xanthones, biphenyls, benzophenones, biflavonoids, and triterpenes.

7.4.1.1 *Calophyllum inophyllum* L.

Synonyms: *Balsamaria inophyllum* (L.) Lour.; *Calophyllum bintagor* Roxb.

Common names: Alexandrian-laurel; kathalhi-chapa, hundal (Bangladesh); khtung (Cambodia); hong hou ke (China); punnaga, pinnai maram, sultan champa (India); nyamplung (Indonesia); bintangor laut, penaga laut (Malaysia); ponenyet (Myanmar); butalau, bitoc, dancalan (the Philippines); kra thing (Thailand); mu' u (Vietnam)

Distribution: India, Sri Lanka, Bangladesh, Myanmar, Cambodia, Laos, Vietnam, Thailand, Malaysia, Indonesia, the Philippines, Taiwan, and Papua New Guinea

Habitat: Sandy shores, beaches, cultivated

Botanical observation: This majestic tree grows to about 60m tall. The bole is not straight and the bark is rough, dark grey, and flaky. The stems are terete, striated, and yield a copious yellow latex upon incision. The leaves are simple, decussate, and exstipulate. The petiole is stout and up to 2.5cm long. The blade is elliptic to obovate, coriaceous, glossy, 4–8×8–15cm, with numerous slender, straight, and parallel pairs of fine secondary nerves, rounded or cuneate at base, margin entire, not uncommonly wavy, and round or retuse at apex. The inflorescences are up to about 10cm axillary lax racemes of slightly fragrant and ephemeral flowers. The flower pedicels are up to 4cm long. The calyx includes 4 sepals, which are about 8mm long. The 4 petals are obovate, pure white, and about 1cm long. The androecium consists of numerous showy stamens. The ovary is globose and develops a peltate stigma. The drupe is smooth, dull light green, laticiferous globose, and up to about 2.5cm long. The seed is globose and oily.

Medicinal uses: Boils (Papua New Guinea, Thailand); dysentery, cuts (Papua New Guinea); fungal infection, venereal diseases (India); leprosy (Bangladesh, India, Myanmar); sores, gonorrhea (Myanmar); sore eyes (India, Indonesia, Myanmar); ulcers, ringworm (Thailand); wounds (India, Myanmar)

Antibacterial (Gram-positive) halo elicited by mid-polar extract: Methanol-dichloromethane extract of seeds and bark (20 µg/disc) inhibited the growth of *Staphylococcus aureus* (ATCC 6538) with inhibition zone diameter of about 14mm (Yimdjo et al., 2004).

Antibacterial (Gram-positive) halo elicited by coumarins: The pyranocoumarins caloxanthone A, the coumarins calophynic acid, brasiliensic acid, inophylloidic acid, calaustralin, the pyranocoumarins calophyllolide, inophyllum C, and inophyllum E (20 μg/disc) inhibited the growth of *Staphylococcus aureus* (ATCC 6538) with inhibition zone diameters of 9, 10, 11, 9, 11, 16, 10, and 13 mm, respectively (Yimdjo et al., 2004).

Strong antibacterial (Gram-positive) polar extract: Methanol extract of latex inhibited the growth of *Staphylococcus aureus* (ATCC6538) with an IC_{50} of 1.1 μg/mL (Cuesta-Rubio et al., 2015).

Strong antifungal (filamentous) polar extract: Methanol extract of latex inhibited the growth of *Trichophyton rubrum* (B68183) with an IC_{50} of 3.3 μg/mL (Cuesta-Rubio et al., 2015).

Strong antiviral (enveloped monopartite linear dimeric single-stranded (+)RNA) non-polar extract: Hexane extract of seeds (containing mainly calophyllolide as well as including inophyllum B, C, P, and E) inhibited the replication of Human immunodeficiency virus type-1 (strain IIIB) at a concentration of 10 μg/mL in C8166 cells (Spino et al., 1998).

Strong antiviral (enveloped monopartite linear dimeric single-stranded (+)RNA) amphiphilic pyranocoumarins: Inophyllum B (LogD=5.2 at pH 7.4; molecular mass=404.4 g/mol) and P inhibited Human immunodeficiency virus type-1 replication with IC_{50} values of 1.4 and 1.6 μM, respectively, and the selectivity indices of 39 and 16, respectively (Patil et al., 1993).

Viral enzyme inhibition by pyranocoumarins: Inophyllum B, inophyllum B acetate, and inophyllum P from the leaves inhibited reverse transcriptase with IC_{50} values of 0.03, 0.7, and 0.1 μM, respectively (Patil et al., 1993).

Commentaries: (i) When a plant is used by several Asian systems of medicine to treat microbial infection it is an indication that it yields antimicrobial principles. (ii) Are pyranocoumarin targeting bacterial DNA? (iii) Consider that the coumarin calanolide A (LogD = 4.7 at pH 7.4; molecular mass=370.4 g/mol) isolated from the primary rainforest Sarawakian tree *Calophyllum teysmannii* var. *inophyloides* (King) P.F. Stevens exhibited strong activity against the Human immunodeficiency virus and reached preclinical trials, which were eventually interrupted (Creagh et al., 2001) this compound could be of value in the fight against pandemic coronaviruses including COVID-19.

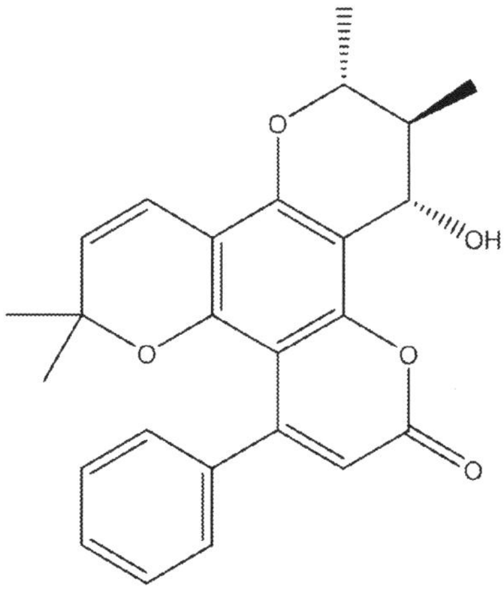

Inophyllum B

(iv) The xanthones blancoxanthone and pyranojacareubin isolated from the roots of *Calophyllum blancoi* Planch. & Triana inhibited the replication of Human coronavirus 229E with IC_{50} values of 3 and 15 µg/mL, respectively (Shen et al., 2005) suggesting that coumarins of *Calophyllum* L. could be of value as source of drugs for COVID-19 and other coronaviruses.

(v) Natural products with strong *in vitro* anti-coronavirus from the plants belonging to the Clades covered in this volume include a broad-spectrum of natural products.

REFERENCES

Creagh, T., Ruckle, J.L., Tolbert, D.T., Giltner, J., Eiznhamer, D.A., Dutta, B., Flavin, M.T. and Xu, Z.Q., 2001. Safety and pharmacokinetics of single doses of (+)-calanolide a, a novel, naturally occurring non-nucleoside reverse transcriptase inhibitor, in healthy, human immunodeficiency virus-negative human subjects. Antimicrobial *Agents* and *Chemotherapy*, 45(5), pp. 1379–1386.

Cuesta-Rubio, O., Oubada, A., Bello, A., Maes, L., Cos, P. and Monzote, L., 2015. Antimicrobial assessment of resins from Calophyllum antillanum and Calophyllum inophyllum. *Phytotherapy Research*, *29*(12), pp. 1991–1994.

Patil, A.D., Freyer, A.J., Eggleston, D.S., Haltiwanger, R.C., Bean, M.F., Taylor, P.B., Caranfa, M.J., Breen, A.L. and Bartus, H.R., 1993. The inophyllums, novel inhibitors of HIV-1 reverse transcriptase isolated from the Malaysian tree, Calophyllum inophyllum Linn. *Journal of Medicinal Chemistry*, *36*(26), pp. 4131–4138.

Shen, Y.C., Wang, L.T., Khalil, A.T., Chiang, L.C. and Cheng, P.W., 2005. Bioactive pyranoxanthones from the roots of Calophyllum blancoi. *Chemical and Pharmaceutical Bulletin*, *53*(2), pp. 244–247.

Spino, C., Dodier, M. and Sotheeswaran, S., 1998. Anti-HIV coumarins from Calophyllum seed oil. *Bioorganic & Medicinal Chemistry Letters*, *8*(24), pp. 3475–3478.

Yimdjo, M.C., Azebaze, A.G., Nkengfack, A.E., Meyer, A.M., Bodo, B. and Fomum, Z.T., 2004. Antimicrobial and cytotoxic agents from Calophyllum inophyllum. *Phytochemistry*, *65*(20), pp. 2789–2795.

7.4.1.2 *Garcinia cowa* Roxb.]

Synonyms: *Garcinia roxburghii* Wight; *Garcinia wallichii* Choisy; *Oxycarpus gangetica* Buch.-Ham

Common names: Baradhan, kau, kau-phol (Bangladesh); hao, yun shu, guomu bang, ma na (China); chengkek, imili, kowa (Bangladesh, India); kemenjing (Indonesia); kandis (Malaysia); pala-kye, tout ha tai (Myanmar); bilukau, sadungan (the Philippines); cha muang (Thailand); tai chua (Vietnam)

Habitat: Forests

Distribution: India, Sri Lanka, Bangladesh, Andamans, Cambodia, Laos, Vietnam, Thailand, Malaysia, and Indonesia

Botanical observation: This handsome tree grows to about 10 m tall. The bark is dark brown. The stems are terete, striated, and yield a copious yellow latex upon incision. The leaves are simple, decussate, and exstipulate. The petiole is up to 1.5 cm long. The blade is elliptic to lanceolate, coriaceous, glossy, 4–14 × 2–5 cm, with about 12–18 pairs of secondary nerves merging in an intra-marginal nerve, red when young, cuneate at base, and acuminate at apex. The inflorescences are terminal fascicles. The flower pedicels are up to 8 mm long. The calyx includes 4 sepals, which are yellowish, broadly ovate and about 5 mm long. The 4 petals are yellowish pink to red, oblong, about 1cm long, and somewhat fleshy. The androecium consists of numerous showy stamens. The ovary is globose, 6–8 celled, and developing a flat and sessile, 6–8 lobed stigma. The berry is edible (Andamans) yet sour , yellow to red, 6–8 ribbed, up to about 5 cm in diameters, marked at apex, and contain 2–4 seeds, which are bullet shaped and about 2.5 cm long.

Medicinal uses: Dysentery (India); fever (India, Thailand)

Strong antibacterial (Gram-positive) benzophenone: Cowanone isolated from the inflorescence inhibited the growth of *Staphyloccus aureus* (TISTR 1466) and methicillin-resistant *Staphylococcus aureus* (SK1) with MIC values of 2 and 0.5 µg/mL (Trisuwan & Ritthiwigrom, 2012).

Strong broad-spectrum antibacterial xanthones: β-Mangostin, α-mangostin, cowanin, 9-hydroxycalabaxanthone, garcinianone A, and cowanol isolated from the inflorescences inhibited the growth of *Staphylococcus aureus* (TISTR 1466) with MIC values of 64, 4, 128, 4, 26, and

8 μg/mL, respectively (Trisuwan & Ritthiwigrom, 2012). α-Mangostin inhibited *Staphylococcus aureus* (TISTR 1466), *Escherichia coli* (TISTR 780), and methicillin-resistant *Staphylococcus aureus* (TISTR 292) (Siridechakorn et al., 2012). From the fruits α-mangostin (LogD = 4.2 at pH 7.4; molecular mass=410.4 g/mol) inhibited *Bacillus cereus* (TISTR 688), *Bacillus subtilis* (TISTR 008), *Micrococcus luteus* (TISTR 884), *Staphylococcus aureus* (TISTR 1466), *Escherichia coli* (TISTR 780), *Pseudomonas aeruginosa* (TISTR 781), *Salmonella typhimurium,* (TISTR 292), and *Staphylococcus epidermidis* (ATCC 12228) with MIC values ranging from 0.5 to 16 μg/mL (Auranwiwat et al., 2014).

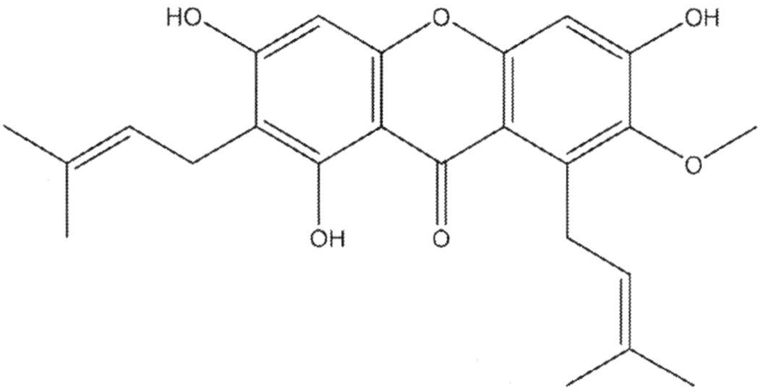

α-Mangostin

β-Mangostin, α-mangostin, cowanin, fuscaxanthone A, 9-hydroxycalabaxanthone, garcinianone A. and cowanol inhibited the growth of methicillin-resistant *Staphylococcus aureus* (SK1) with MIC values of 0.5, 8, 2, 2, 8, 4, 16, and 2 μg/mL, respectively (Trisuwan & Ritthiwigrom, 2012). β-mangostin inhibited *Bacillus cereus* (TISTR 688), *Bacillus subtilis* (TISTR 008), *Micrococcus luteus* (TISTR 884), *Staphylococcus aureus* (TISTR 1466), *Escherichia coli* (TISTR 780), *Pseudomonas aeruginosa* (TISTR 781), methicillin-resistant *Staphylococcus aureus* (TISTR 292), and *Staphylococcus epidermidis* (ATCC 12228) with the MIC values of 0.2, 0.2, 1, 64, 64, 128, 64, and 0.5 μg/mL, respectively (Auranwiwat et al., 2014).

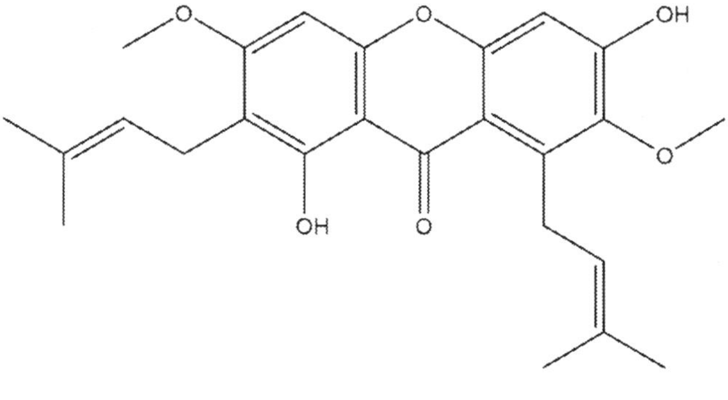

β-Mangostin

Cowanin inhibited *Staphylococcus aureus* (TISTR 1466), *Escherichia coli* (TISTR 780), and methicillin-resistant *Staphylococcus aureus* (TISTR 292) (Siridechakorn et al., 2012).

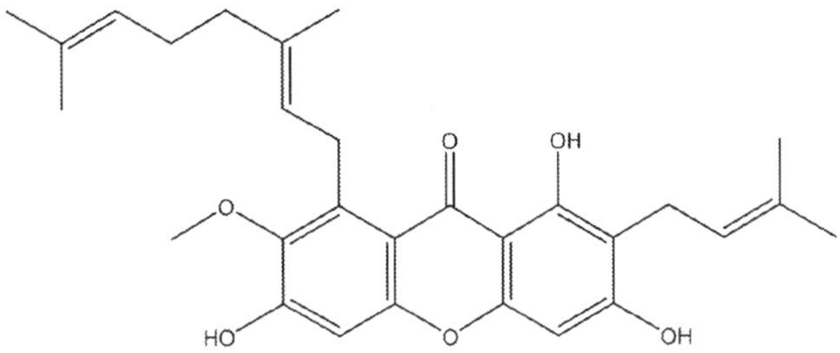

Cowanin

Cowanol inhibited *Staphylococcus aureus* (TISTR 1466), *Escherichia coli* (TISTR 780), and methicillin-resistant *Staphylococcus aureus* (TISTR 292) (Siridechakorn et al., 2012).

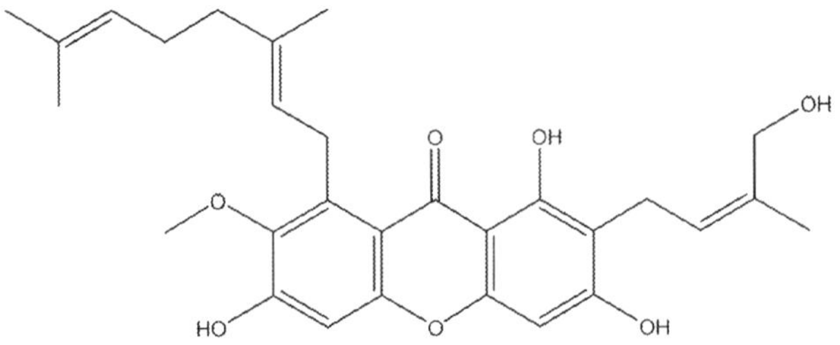

Cowanol

Strong antibacterial (Gram-positive) xanthone: Garciniacowone from the stem bark inhibited methicillin-resistant *Staphylococcus aureus*, *Staphylococcus aureus* (TISTR 1466), *Escherichia coli* (TISTR 780), and *Salmonella typhimurium* (TISTR 292) with MIC of 2, 2, 128, and 128 µg/mL, respectively (Siridechakorn et al., 2012).

Garcicowanone A isolated from fruits inhibited the growth of *Bacillus subtilis* (TISTR 688), *Bacillus subtilis* (TISTR 008), *Micrococcus luteus* (TISTR 884), *Staphylococcus aureus* (TISTR 1466), *Escherichia coli* (TISTR 780), *Pseudomonas aeruginosa* (TISTR 781), *Salmonella typhimurium* (TISTR 292), and *Staphylococcus epidermidis* (ATCC 12228) with MIC values of 0.2, 2, 4, 64, 64, 128, 64, and 4 µg/mL, respectively (Auranwiwat et al., 2014).

From the fruits rubraxanthone inhibited *Bacillus cereus* (TISTR 688), *Bacillus subtilis* (TISTR 008), *Micrococcus luteus* (TISTR 884), *Escherichia coli* (TISTR 780), *Pseudomonas aeruginosa* (TISTR 781), methicillin-resistant *Staphylococcus aureus* (TISTR 292), and *Staphylococcus epidermidis* (ATCC 12228) (Auranwiwat et al., 2014).

From the fruits chamuangone inhibited the growth of *Bacillus subtilis*, *Enterococcus* sp., *Staphylococcus aureus*, *Streptococcus pyogenes*, *Streptococcus viridans*, *Escherichia coli*, *Helicobacter pylori*, *Shigella sonnei*, and *Salmonella typhimurium* (Sakunpak et al., 2012).

Commentary: (i) Xanthones are cytotoxic for mammalian cells (Pattamadilok, 2016), indicating a possible mode of antibacterial effect on bacterial DNA machinery. Further, the presence of prenylated moieties could facilitate penetration of cytoplasmic membrane as well as increased affinity to enzymatic pockets via Van de Walls and other hydrophobic interactions. (ii) In India, *Garcinia indica* Choisy is used for dysentery, diarrhea, and fever and *Garcinia pedunculata* Roxb. is used for cough. In China, *Garcinia oblongifolia* Champ. ex Benth. (local name: *Jiusuan*) and *Garcinia paucinervis* Chun ex F.C. How are used to treat burns.

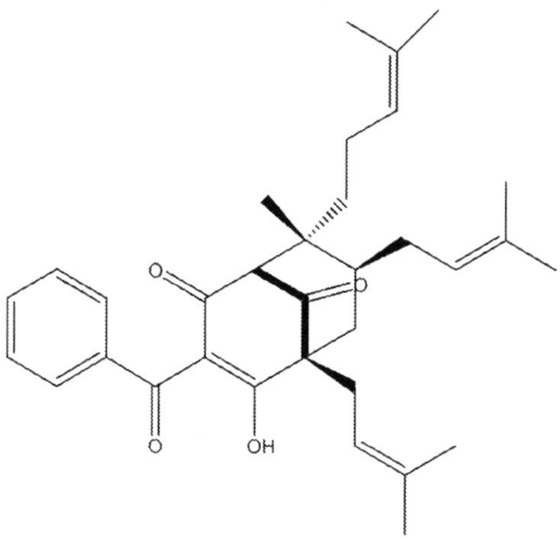

Chamuangone

REFERENCES

Auranwiwat, C., Trisuwan, K., Saiai, A., Pyne, S.G. and Ritthiwigrom, T., 2014. Antibacterial tetraoxygenated xanthones from the immature fruits of *cratoxylum*. *Fitoterapia*, *98*, pp. 179–183.

Pattamadilok, C., 2016. Xanthones from Garcinia cowa flowers and their cytotoxicity. *Thai Journal of Pharmaceutical Sciences (TJPS)*, *40*, pp. 84–87.

Sakunpak, A. and Panichayupakaranant, P., 2012. Antibacterial activity of Thai edible plants against gastro-intestinal pathogenic bacteria and isolation of a new broad-spectrum antibacterial polyisoprenylated benzophenone, chamuangone. *Food Chemistry*, *130*(4), pp. 826–831.

Siridechakorn, I., Phakhodee, W., Ritthiwigrom, T., Promgool, T., Deachathai, S., Cheenpracha, S., Prawat, U. and Laphookhieo, S., 2012. Antibacterial dihydrobenzopyran and xanthone derivatives from Garcinia cowa stem barks. *Fitoterapia*, *83*(8), pp. 1430–1434.

Trisuwan, K. and Ritthiwigrom, T., 2012. Benzophenone and xanthone derivatives from the inflorescences of Garcinia cowa. *Archives of Pharmacal Research*, *35*(10), pp. 1733–1738.

7.4.1.3 *Garcinia dulcis* (Roxb.) Kurz

Synonym: *Xanthochymus dulcis* Roxb.

Common names: Gourka, mundu (Indonesia, Malaysia); Daungyan, madaw (Myanmar); baniti, taklang anak (the Philippines); búa ngot (Vietnam); ma phut (Thailand)

Distribution: India, Bangladesh, Myanmar, Vietnam, Cambodia, Malaysia, Indonesia, and the Philippines

Habitat: Forests

Botanical observation: This tree grows to about 10 m tall. The bark is brown and rough. The stems are terete, smooth, striated, and yield a copious white latex upon incision. The leaves are simple, decussate, and exstipulate. The petiole is up to 3 cm long. The blade is lanceolate or elliptic, coriaceous, glossy, 7.5–30 × 3.2–18 cm, with 9–14 pairs of secondary nerves, round at base, and obtuse apex. The inflorescences are axillary racemes of flowers with sour smell. The 5 sepals are orbicular and about 5 mm long. The 5 petals are whitish, oblong, about 8 mm long, and somewhat fleshy. The androecium consists of numerous showy stamens. The ovary is globose, 1–5 celled, and developing a flat and sessile 5-lobed star-shaped stigma. The berries are edible yet sour, yellowish, smooth, up to about 8 cm in diameters, marked at apex, and containing 1–5 arillate seeds, which are about 2.5 cm long.

Medicinal uses: Mumps, tuberculosis (Indonesia); diarrhea, dysentery (Myanmar); antiseptic, cough, fever (Thailand)

Strong antibacterial (Gram-positive) amphiphilic xanthones: Cowanin and cowaxanthone isolated from the fruits inhibited the growth of *Staphylococcus aureus* (ATCC 25923) with MIC values of 32 and 16 μg/mL, respectively (Deachathai et al., 2005). Dulcisxanthone J, garciniaxanthone, 12b-hydroxy-des-D-garcigerrin A, globuxanthone inhibited *Staphylococcus aureus* (ATCC 25923) isolated from the stembark with MIC/MBC values of 16/>200, 32/>200, 4/128, and 32/>200 μg/mL, respectively (Thepthong et al., 2017). Dulcisxanthone J, garciniaxanthone, 12b-hydroxy-des-D-garcigerrin A, globuxanthone inhibited methicillin-resistant *Staphylococcus aureus* (SK1) with MIC/MBC values of 16/64, 32/>200, 4/>200, and 32/>200 μg/mL, respectively (Thepthong et al., 2017).

Strong antibacterial (gram-positive) amphiphilic isoflavone: Lupalbigenin (5,7,4′-Trihydroxy-6,3′-diprenylisoflavone; LogD = 4.5 at pH 7.4; molecular mass=406.4 g/mol)) isolated from the fruits inhibited the growth of *Staphylococcus aureus* (ATCC 25923) with an MIC value of 4 μg/mL (Deachathai et al., 2005).

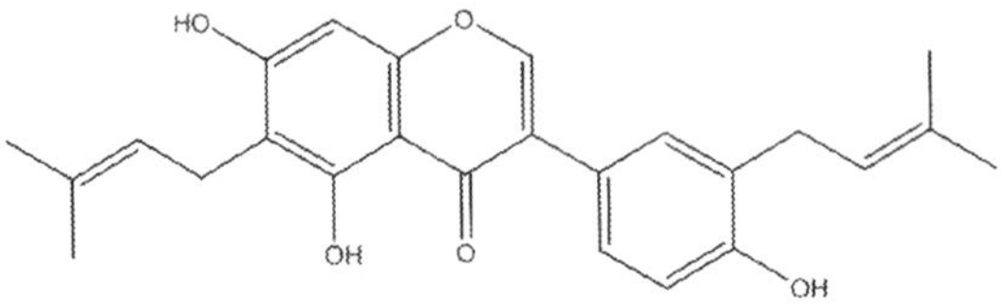

Lupalbigenin

REFERENCES

Abdullah, I., Phongpaichit, S., Voravuthikunchai, S.P. and Mahabusarakam, W., 2018. Prenylated biflavonoids from the green branches of Garcinia dulcis. *Phytochemistry Letters*, *23*, pp. 176–179.

Deachathai, S., Mahabusarakam, W., Phongpaichit, S. and Taylor, W.C., 2005. Phenolic compounds from the fruit of Garcinia dulcis. *Phytochemistry*, *66*(19), pp. 2368–2375.

Thepthong, P., Phongpaichit, S., Carroll, A.R., Voravuthikunchai, S.P. and Mahabusarakam, W., 2017. Prenylated xanthones from the stem bark of Garcinia dulcis. *Phytochemistry Letters*, *21*, pp. 32–37.

7.4.1.4 *Garcinia hanburyi* Hook. f.

Common names: Gamboge tree; rung (Cambodia); rong (Thailand); dang hoàng (Vietnam)

Distribution: Cambodia, Laos, Vietnam, Thailand, and Malaysia

Habitat: Forests

Botanical observation: This tree grows to about 15 m tall. The bark is grey and smooth. The stems are angled, smooth, striated, and yield a copious yellow latex upon incision. The leaves are simple, decussate, and exstipulate. The petiole is up to 2.5 cm long and stout. The blade is broadly elliptic, coriaceous, glossy, 10–25×3–10 cm, marked with up to about 25 pairs of secondary nerves, cuneate at base, and acuminate apex. The inflorescences are axillary clusters of fragrant flowers. The 4 sepals are orbicular and about 6 mm long. The 4 petals are fleshy, ovate, about 7 mm long, and light yellowish. The androecium consists of numerous short stamens. The ovary is globose, 1–4 celled, and developing a flat and sessile stigma. The berry is globose, yellowish, smooth, up to about 3 cm in diameter, marked at apex, and contain 1–4 seeds.

Medicinal use: Wounds (Thailand)

Strong antibacterial (Gram-positive) caged prenylated xanthones: Moreollic acid and the morellic acid isolated from the fruits inhibited methicillin-resistant *Staphylococcus aureus* with an MIC values of 25 μg/mL (Sukpondma et al., 2005).

Strong antiviral (enveloped monopartite linear dimeric single-stranded (+)RNA) lupane-type triterpenes: 2-Acetoxyalphitolic acid, 3-acetoxyalphitolic acid, betulinic acid, and betulin isolated from the latex inhibited the Human immunodeficiency virus type-1 syncitium formation with IC_{50} vlaues of 15.9, 19.8, 27.2, and 11.6 μg/mL, respectively, with selectivity indices of 2.8, 3.3, 2.6, and 1.9, respectively (Reutrakul et al., 2010).

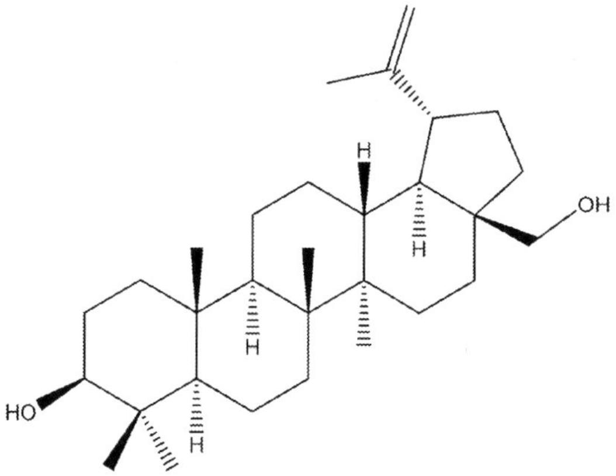

Betulin

Strong antiviral (enveloped monopartite linear dimeric single-stranded (+)RNA) caged prenylated xanthones: Hanburin and dihydroisomorellin isolated from the latex inhibited the Human immunodeficiency virus type-1 syncitium formation with EC_{50} values of 3 and 1.2 µg/mL and selectivity indices of 1.7 and 4.7 µg/mL, respectively (Reutrakul et al., 2007). Dihydroisomorellin inhibited Human immunodeficiency virus type-1 reverse transcriptase with an IC_{50} of 42.3 µg/mL (Reutrakul et al., 2007).

Viral enzyme inhibition by triterpenes: 3-Acetoxyalphitolic acid, betulinic acid, and betulin inhibited Human immunodeficiency virus type-1 reverse transcriptase activity with IC_{50} values of 16.3, 41.7, and 67.2 µg/mL, respectively (Reutrakul et al., 2010).

REFERENCES

Reutrakul, V., Anantachoke, N., Pohmakotr, M., Jaipetch, T., Sophasan, S., Yoosook, C., Kasisit, J., Napaswat, C., Santisuk, T. and Tuchinda, P., 2007. Cytotoxic and anti-HIV-1 caged xanthones from the resin and fruits of Garcinia hanburyi. *Planta Medica*, *73*(1), pp. 33–40.

Reutrakul, V., Anantachoke, N., Pohmakotr, M., Jaipetch, T., Yoosook, C., Kasisit, J., Napaswa, C., Panthong, A., Santisuk, T., Prabpai, S. and Kongsaeree, P., 2010. Anti-HIV-1 and Anti-Inflammatory Lupanes from the Leaves, Twigs, and Resin of Garcinia hanburyi. *Planta Medica*, *76*(4), pp. 368–371.

Sukpondma, Y., Rukachaisirikul, V. and Phongpaichit, S., 2005. Antibacterial caged-tetraprenylated xanthones from the fruits of Garcinia hanburyi. *Chemical and Pharmaceutical Bulletin*, *53*(7), pp. 850–852.

7.4.1.5 *Garcinia mangostana* L.

Synonym: *Mangostana garcinia* Gaertn.

Common names: Mangosteen; mang ji shi (China); sulum puli, mangusta (India); manggis (the Philippines, Malaysia); mingut (Myanmar); mang khut (Thailand)

Habitat: Cultivated

Distribution: India, Sri Lanka, Bangladesh, Myanmar, Laos, Cambodia, Vietnam, China, Thailand, Malaysia, Indonesia, the Philippines, Taiwan

Botanical observation: This handsome tree grows to about 20 m tall. The bark is grey and smooth. The stems are angled, smooth, and yield a copious yellow latex upon incision. The leaves are simple, decussate, and exstipulate. The petiole is up to 2 cm long and stout. The blade is elliptic, coriaceous, glossy, 14–25×5–10 cm, marked with up to about 50 pairs of secondary nerves, cuneate at base, wavy, and acuminate apex. The inflorescences are terminal clusters. The 4 sepals are orbicular and about 6 mm long. The 4 petals are elliptic, fleshy, about 7mm long, and light yellowish. The androecium consists of numerous short stamens. The ovary is globose, 5–8 celled, and developing a flat and multilobed stigma. This berry is subglobose dark purple, smooth, up to about 8 cm in diameters, marked at apex, and contains 4–5 seeds embedded in a white and delicious pulp enclosed in a spongy and juicy dark purple/brownish pericarp.

Medicinal uses: Wounds (Malaysia); dysentery (India, Myanmar); urinary tract infections, leucorrhea (India); diarrhea (India, Indonesia, Myanmar, the Philippines, Thailand)

Strong antibacterial (Gram-positive) polar extract: Ethanol extract inhibited the growth of 35 clinical isolates of methicillin-resistant *Staphylococcus aureus* with MIC/MBC values of 0.05–0.4/0.1–0.4 µg/mL(Voravuthikunchai & Kitpipit, 2005). Ethanol extract of pericarps inhibited methicillin-sensitive *Staphylococcus aureus* (ATCC 23235) and methicillin-resistant *Staphylococcus aureus* (DMST 20651) with MIC/MBC of 14/25 and 17 µg/mL/30 µg/mL, respectively (Tatiya-aphiradee et al., 2016).

Strong antibacterial (Gram-positive) xanthones: α-Mangostin and β-mangostin from the stem bark inhibited the growth of vancomycin-resistant *Enterococcus faecalis* (ATCC 51299) and vancomycin-sensitive *Enterococcus faecalis* (ATCC 8459) (Sakagami et al., 2005). α-Mangostin inhibited the growth of clinical strains of methicillin-resistant *Staphylococcus aureus* with an MIC value of 6.2 µg/mL (Sakagami et al., 2005). From the pericarp α-mangostin inhibited methicillin-sensitive *Staphylococcus aureus* (ATCC 23235) and methicillin-resistant *Staphylococcus aureus* (DMST 20651) (Tatiya-aphiradee et al., 2016). α-Mangostin inhibited *Bacillus subtilis* and *Staphylococcus aureus*, with MIC values of 3.9 and 7.8 µM, respectively (Al-Massarani et al., 2013). γ-Mangostin inhibited the growth of methicillin-resistant *Staphylococcus aureus*, methicillin-sensitive *Staphylococcus aureus*, vancomycin-resistant *Enterococcus* (VRE), and vancomycin-sensitive *Enterococcus*(VSE) strains with the MIC values of 3.1, 6.2, 6.2 and 6.2 µg/mL, respectively (Dharmaratne et al., 2013). Analogues of α-mangostin and β-mangostin were tested toward methicillin-resistant *Staphylococcus aureus* and vacomycin-resistant *Enterococcus* strains and a combination of C-6 and C-3 hydroxyl groups along with the prenyl side chain at C-2 in the 1,3,6,7-tetraoxygenated xanthones is essential for strong antibacterial activity (Dharmaratne et al., 2013). α-Mangostin inhibited clinical methicillin-resistant *Staphylococcus aureus* with an MIC of 1.9 µg/mL and MBC of 3.9 µg/mL (Chomnawang et al., 2009).

Weak antileptospiral(Gram-negative) xanthone: Garcinone C from the pericarp inhibited the growth of four species of *Leptospira* with an MIC ranging from 100 to 200 µM (Seesom et al., 2013).

Strong antimycobacterial xanthones: α-Mangostin and β-mangostin inhibited *Mycobacterium tuberculosis* (H37Ra) with an MIC of 6.2 µg/mL and garcinone B inhibited *Mycobacterium tuberculosis* (H37Ra) with an MIC of 12.7 µg/mL (Suksamrarn et al., 2003). α-Mangostin from the pericarp inhibited *Mycobacterium smegmatis*, *Mycobacterium cheleneoi*, *Mycobacterium xenopi*, and *Mycobacterium intracellulare* (Al-Massarani et al., 2013).

Broad-spectrum antibiotic potentiator xanthones: α-Mangostin was synergistic with gentamicin for vacomycin-resistant *Enterococcus* (Sakagami et al., 2005). α-Mangostin was synergistic with vancomycin toward methicillin-resistant *Staphylococcus aureus* (Sakagami et al., 2005). γ-Mangostin had a synergistic effect with penicillin G toward *Leptospira javanica* (Seesom et al., 2013).

In vivo antibacterial (Gram-positive) polar extract: In superficial skin infection model, the ethanol extract of pericarp (10% in propylene glycol) applied once a day for 9 days reduced the number of methicillin-resistant *Staphylococcus aureus* colonies after 24 hours of treatment and evoked complete healing after 9 days in ICR mice (Tatiya-aphiradee et al., 2016).

Weak broad-spectrum antifungal xanthones: γ-Mangostin inhibited the growth of *Fusarium oxysporum* var *infectum*, *Alternaria tenuis*, and *Dreschlera oryzae* (Gopalakrishnan et al., 1997). α-Mangostin inhibited the growth of *Candida albicans* with an MIC/MFC value of 1000/2000 µg/mL (Kaomongkolgit et al., 2009).

Viral enzyme inhibition by xanthones: Mangostin and γ-mangostin inhibited the Human immunodeficiency virus type-1 protease with IC_{50} values of 5.1 and 4.8 µM, respectively (Chen et al., 1996).

REFERENCES

Al-Massarani, S.M., El Gamal, A.A., Al-Musayeib, N.M., Mothana, R.A., Basudan, O.A., Al-Rehaily, A.J., Farag, M., Assaf, M.H., El Tahir, K.H. and Maes, L., 2013. Phytochemical, antimicrobial and antiprotozoal evaluation of Garcinia mangostana pericarp and α-mangostin, its major xanthone derivative. *Molecules*, *18*(9), pp. 10599–10608.

Chen, S.X., Wan, M. and Loh, B.N., 1996. Active constituents against HIV-1 protease from Garcinia mangostana. *Planta Medica*, *62*(04), pp. 381–382.

Chomnawang, M.T., Surassmo, S., Wongsariya, K. and Bunyapraphatsara, N., 2009. Antibacterial activity of Thai medicinal plants against methicillin-resistant Staphylococcus aureus. *Fitoterapia*, *80*(2), pp. 102–104.

Dharmaratne, H.R.W., Sakagami, Y., Piyasena, K.G.P. and Thevanesam, V., 2013. Antibacterial activity of xanthones from Garcinia mangostana (L.) and their structure–activity relationship studies. *Natural Product Research*, *27*(10), pp. 938–941.

Gopalakrishnan, G., Banumathi, B. and Suresh, G., 1997. Evaluation of the antifungal activity of natural xanthones from Garcinia mangostana and their synthetic derivatives. *Journal of Natural Products*, *60*(5), pp. 519–524.

Kaomongkolgit, R., Jamdee, K. and Chaisomboon, N., 2009. Antifungal activity of alpha-mangostin against Candida albicans. *Journal of Oral Science*, *51*(3), pp. 401–406.

Sakagami, Y., Iinuma, M., Piyasena, K.G.N.P. and Dharmaratne, H.R.W., 2005. Antibacterial activity of α-mangostin against vancomycin resistant Enterococci (VRE) and synergism with antibiotics. *Phytomedicine*, *12*(3), pp. 203–208.

Seesom, W., Jaratrungtawee, A., Suksamrarn, S., Mekseepralard, C., Ratananukul, P. and Sukhumsirichart, W., 2013. Antileptospiral activity of xanthones from Garcinia mangostana and synergy of gamma-mangostin with penicillin G. *BMC Complementary and Alternative Medicine*, *13*(1), p. 182.

Suksamrarn, S., Suwannapoch, N., Phakhodee, W., Thanuhiranlert, J., Ratananukul, P., Chimnoi, N. and Suksamrarn, A., 2003. Antimycobacterial activity of prenylated xanthones from the fruits of Garcinia mangostana. *Chemical and Pharmaceutical Bulletin*, *51*(7), pp. 857–859.

Tatiya-aphiradee, N., Chatuphonprasert, W. and Jarukamjorn, K., 2016. In vivo antibacterial activity of Garcinia mangostana pericarp extract against methicillin-resistant Staphylococcus aureus in a mouse superficial skin infection model. *Pharmaceutical Biology*, *54*(11), pp. 2606–2615.

Voravuthikunchai, S.P. and Kitpipit, L., 2005. Activity of medicinal plant extracts against hospital isolates of methicillin-resistant Staphylococcus aureus. *Clinical Microbiology and Infection*, *11*(6), pp. 510–512.

7.4.1.6 *Garcinia multiflora* Champ. ex Benth.

Synonym: *Garcinia hainanensis* Merr.

Common name: Cai cuo; mù zhú zi, shān pí pā, mikou, luo wang, bunang wa (China)

Habitat: Forests

Distribution: Vietnam, China, and Taiwan

Botanical observation: This tree grows to about 15 m tall. The bark is grey and rough. The stems are angled and yield a copious yellow latex upon incision. The leaves are simple, decussate, and exstipulate. The petiole is up to 1.2 cm long and stout. The blade is broadly elliptic to oblong or obovate, coriaceous, glossy, 7–16×3–6cm, marked with up to about 15 pairs of secondary nerves, cuneate at base and acuminate apex. The inflorescences are terminal panicles. The calyx includes 4 sepals, which are of irregular length and about 7 mm long. The corolla comprises 4 fleshy petals, which are obovate, about 1.5 cm long, and orangish-yellow. The androecium consists of 4 clusters of 50 stamens. The ovary is globose, 4 celled, somewhat quadrangular, and developing a 4-lobed and peltate stigma. The fruit is subglobose, laticiferous berry, which is bright yellow, smooth, up to about 3 cm in diameter, marked at apex, palatable, and containing 1–2 seeds.

Medicinal uses: Apparently none for the treatment of microbial infections

Strong antimycobacterial benzophenones: Garcimultiflorone A and garcimultiflorone C from the fruits inhibited *Mycobacterium tuberculosis* (H37Rv) with MIC values of 31.2 and 27.6 μM, respectively (Chen et al., 2008).

Strong antiviral (enveloped monopartite linear dimeric single-stranded (+)RNA) amphiphilic biflavonoid: Morelloflavone (LogD = 1.8 at pH 7.4; molecular mass = 556.4g/mol) isolated from leaves and stems inhibited the Human immunodeficiency virus type-1 (strain LAV-1) replication on peripheral blood mononuclear cells with an EC_{50} value of 6.9 µM and inhibited the Human immunodeficiency virus type-1 reverse transcriptase activity with an IC_{50} value of 116 µM (Lin et al., 1997).

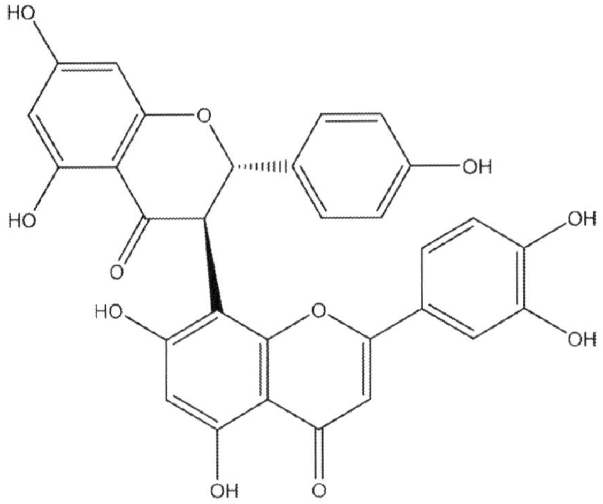

Morelloflavone

Strong antiviral (non-enveloped segmented linear double-stranded RNA) biphenyl: Multibiphenyl A, B, and C from stems and leaves inhibited the growth of Human rotavirus (Wa group) on MA104 cells with IC_{50} values of 11.5, 10.9, and 12.7 µg/mL, respectively, and the selectivity index values of 10.9. 12.3, and 20.1, respectively (Gao et al., 2016).

Commentary: Other antimycobacterial principles found in members of the genus *Garcinia* L. are bioflavonoids. For example, amentoflavone inhibited *Mycobacterium stegmatis* (ATCC 1441) with an MIC of 600 µg/mL (Kaikabo et al., 2011). Amentoflavone inhibited Severe acute respiratory syndrome-associated coronavirus 3-chymotrypsin-like protease ($3CL^{pro}$) with an IC_{50} value of 8.3 µM (Ryu et al., 2010). In this experiment luteolin inhibited Severe acute respiratory syndrome-associated coronavirus 3-chymotrypsin-like protease with an IC_{50} value of 20.2 µM (Ruy et al., 2010).

REFERENCES

Chen, J.J., Ting, C.W., Chen, I.S., Peng, C.F., Huang, W.T., Su, Y.C. and Lin, S.C., 2008. New polyisoprenyl benzophenone derivatives and antitubercular constituents from Garcinia multiflora. *Planta Medica*, *74*(09), p.PB48.

Gao, X.M., Ji, B.K., Li, Y.K., Ye, Y.Q., Jiang, Z.Y., Yang, H.Y., Du, G., Zhou, M., Pan, X.X., Liu, W.X. and Hu, Q.F., 2016. New biphenyls from Garcinia multiflora. *Journal of the Brazilian Chemical Society*, *27*(1), pp. 10–14.

Kaikabo, A.A. and Eloff, J.N., 2011. Antibacterial activity of two biflavonoids from Garcinia livingstonei leaves against Mycobacterium smegmatis. *Journal of Ethnopharmacology*, *138*(1), pp. 253–255.

Lin, Y.M., Anderson, H., Flavin, M.T., Pai, Y.H.S., Mata-Greenwood, E., Pengsuparp, T., Pezzuto, J.M., Schinazi, R.F., Hughes, S.H. and Chen, F.C., 1997. In vitro anti-HIV activity of biflavonoids isolated from Rhus succedanea and Garcinia multiflora. *Journal of Natural Products*, *60*(9), pp. 884–888.

Ryu, Y.B., Jeong, H.J., Kim, J.H., Kim, Y.M., Park, J.Y., Kim, D., Naguyen, T.T.H., Park, S.J., Chang, J.S., Park, K.H. and Rho, M.C., 2010. Biflavonoids from Torreya nucifera displaying Severe acute respiratory syndrome-acquired coronavirus 3CLpro inhibition. *Bioorganic & Medicinal Chemistry*, *18*(22), pp. 7940–7947.

7.4.1.7 *Garcinia nigrolineata* Planch. ex T. Anderson

Common names: Wild beaked kandis; kandis hutan (Malaysia); cha muang (Thailand)

Habitat: Forests

Distribution: Myanmar, Thailand, and Malaysia

Botanical observation: This tree grows to about 30 m tall. The bark is dark brown. The stems are somewhat articulated, smooth, squarrish, and yield a copious yellow latex upon incision. The leaves are simple, decussate, and exstipulate. The petiole is up to about 2.5 cm long and stout. The blade is oblanceolate to ovate-lanceolate, somewhat coriaceous, glossy, 12.5–20×3–7.5 cm, marked with up to about 15 pairs of secondary nerves, cuneate at base, and acuminate to caudate apex. The inflorescences are axillary clusters. The 4 sepals are about 3 mm long. The 4 petals are fleshy, obovate, about 1 cm long, and greenish white. The androecium consists of numerous stamens. The ovary is globose, 4–7 celled, and developing a 5–7-lobed and peltate stigma. The berries are subglobose, up to about 5 cm across, edible, laticiferous, glossy, orangish yellow, grooved, and contain 5–7 seeds.

Medicinal use: Syphilis (Malaysia)

Strong antibacterial (Gram-positive) xanthones: Nigrolineaxanthone N from the leaves inhibited methicillin-resistant *Staphylococcus aureus* with an MIC value of 4 μg/mL (vancomycin: MIC= 2 μg/mL) (Rukachaisirikul et al., 2003). From the same leaves, 8-desoxygartanin and ananixanthone inhibited methicillin-resistant *Staphylococcus aureus* an MIC of 16 and 32 μg/mL, respectively (Rukachaisirikul et al., 2003). Nigrolineaxanthone F, lastixanthone, and brasilixanthone from the stem bark inhibited methicillin-resistant *Staphylococcus aureus* with an MIC value of 2 μg/mL (Rukachaisirikul et al., 2005). Nigrolineaxanthones G and I from the stem bark and 6-deoxyjacareubin from stems and stem bark inhibited methicillin-resistant *Staphylococcus aureus* with an MIC value of 4 μg/mL (Rukachaisirikul et al., 2005). Nigrolineaxanthone Q inhibited the growth of *Micrococcus luteus*, *Streptococcus mutans*, *Staphylococcus epidermidis*, *Bacillus cereus*, and *Staphylococcus aureus* with MIC values of 8, 32, 128, 32 and 128 μg/mL, respectively (Raksat et al., 2019).

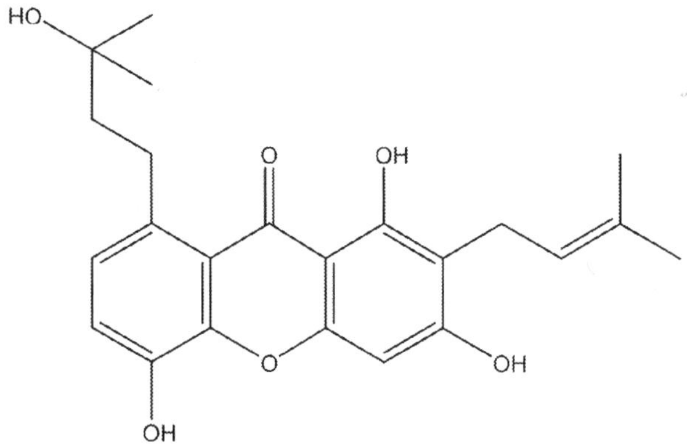

Nigrolineaxanthone N

Moderate antibacterial (Gram-negative) xanthone: Nigrolineaxanthone Q from the leaves inhibited the growth of *Salmonella typhimurium*, *Pseudomonas aeruginosa*, *Escherichia coli*, and *Shigella flexneri* (Raksat et al., 2019).

Broad-spectrum moderate antibacterial flavonol: 3′-Deoxyquercetin from the leaves inhibited the growth of *Micrococcus luteus*, *Streptococcus mutans*, *Staphylococcus epidermidis*, *Bacillus cereus*, *Staphylococcus aureus*, *Salmonella typhimurium*, *Pseudomonas aeruginosa*, *Escherichia coli*, and *Shigella flexneri* with MIC values of 32, 32, 128, 64, 32, 128, 128, 128, and 128 µg/mL, respectively (Raksat et al., 2019). 3′,3,4′,5,7-Pentahydroxyflavone from the leaves inhibited the growth of *Micrococcus luteus, Streptococcus mutans, Staphylococcus epidermidis, Bacillus cereus, Staphylococcus aureus, Salmonella typhimurium, Pseudomonas aeruginosa, Escherichia coli*, and *Shigella flexneri* with MIC values of 32, 32, 128, 64, 32, 128, 128, 128, and 128 µg/mL, respectively (Raksat et al., 2019).

Commentary: Prenylated xanthones produced by members of the genus *Garcinia* L. are more active against Gram-positive than Gram-negative bacteria and this is the case for most plant secondary metabolites. For instance, scortechinone B isolated from stem bark of *Garcinia scortechinii* King inhibited *Staphylococcus aureus* ATCC25923 and *Staphylococcus aureus* SK1 with the MIC values of 8 and 2 µg/mL, respectively (Rukachaisirikul et al., 2005). Scortechinone C isolated from stem bark of *Garcinia scortechinii* King inhibited *Staphylococcus aureus* (ATCC 25923) and *Staphylococcus aureus* (SK1) with the MIC values of 8 and 8 µg/mL, respectively (Rukachaisirikul et al., 2005). Scortechinone F isolated from stem bark of *Garcinia scortechinii* King inhibited *Staphylococcus aureus* (ATCC 25923) and *Staphylococcus aureus* (SK1) with the MIC values of 16 and 4 µg/mL (Rukachaisirikul et al., 2005).

Guttiferone A from a member of the *Garcinia* L. fresh fruits inhibited *Staphylococcus aureus* (ATCC6538) (Monzote et al., 2011). This compound had an MIC above 64 µM for *Escherichia coli* (ATCC6538) (Monzote et al., 2011).1,5,6-Trihydroxyxanthone and 1,6,7-trihydroxyxanthone from *Garcinia succifolia* Kurz inhibited the growth of *Staphylococcus aureus* (ATCC 25293) (Duangsrisai et al., 2014). 1,5,6-Trihydroxyxanthone and 1,6,7-trihydroxyxanthone inhibited *Bacillus subtilis* (ATCC 6633) with MIC/MBC of 64/>256 and 128 µg/mL />256 µg/mL, respectively (Duangsrisai et al., 2014). 1,7-Dihydroxyxanthone decreased the MIC of oxacillin and ampicillin toward *Staphylococcus aureus* B1 from 128 to 32 and 128 to 64 µg/mL, respectively (Duangsrisai et al., 2014). Morellic acid and gambogic acid inhibited the growth of methicillin-sensitive *Staphylococcus aureus* (ATCC 29213) with MIC/MBC of 12.5/12.5 and 12.5/25 µg/mL, respectively. Morellic acid and gambogic acid inhibited the growth of methicillin-resistant *Staphylococcus aureus* with MIC/MBC of 12.5/25 and 25/50 µg/mL, respectively (Chaiyakunvat et al., 2016). Xanthones are amphiphilic molecules that penetrate with difficulty through the outer membrane (via porins) of Gram-negative bacteria (Mahady, 2005). In addition, Gram-negative bacteria have resistance-nodulation-cell division(RND)-type pump and EmrAB-type pump, which extrude amphiphilic molecules across the outer membrane (Mahady, 2005). This can be extended to all low molecular weight secondary metabolites from medicinal plants.

REFERENCES

Duangsrisai, S., Choowongkomon, K., Bessa, L.J., Costa, P.M., Amat, N. and Kijjoa, A., 2014. Antibacterial and EGFR-tyrosine kinase inhibitory activities of polyhydroxylated xanthones from *Garcinia succifolia. Molecules, 19*(12), pp. 19923–19934.

Chaiyakunvat, P., Anantachoke, N., Reutrakul, V. and Jiarpinitnun, C., 2016. Caged xanthones: Potent inhibitors of global predominant MRSA USA300. *Bioorganic & Medicinal Chemistry Letters, 26*(13), pp. 2980–2983.

Mahady, G.B., 2005. Medicinal plants for the prevention and treatment of bacterial infections. *Current Pharmaceutical Design, 11*(19), pp. 2405–2427.

Monzote, L., Cuesta-Rubio, O., Matheeussen, A., Van Assche, T., Maes, L. and Cos, P., 2011. Antimicrobial evaluation of the polyisoprenylated benzophenones nemorosone and guttiferone A. *Phytotherapy Research, 25*(3), pp. 458–462.

Raksat, A., Maneerat, W., Andersen, R.J., Pyne, S.G. and Laphookhieo, S., 2019. A tocotrienol quinone dimer and xanthones from the leaf extract of Garcinia nigrolineata. *Fitoterapia, 136*, p. 104175.

Rukachaisirikul, V., Kamkaew, M., Sukavisit, D., Phongpaichit, S., Sawangchote, P. and Taylor, W.C., 2003. Antibacterial xanthones from the leaves of *Garcinia nigrolineata. Journal of Natural Products, 66*(12), pp. 1531–1535.

Rukachaisirikul, V., Phainuphong, P., Sukpondma, Y., Phongpaichit, S. and Taylor, W.C., 2005. Antibacterial caged-tetraprenylated xanthones from the stem bark of Garcinia scortechinii. *Planta medica*, *71*(02), pp. 165–170.
Rukachaisirikul, V., Tadpetch, K., Watthanaphanit, A., Saengsanae, N. and Phongpaichit, S., 2005. Benzopyran, biphenyl, and tetraoxygenated xanthone derivatives from the twigs of Garcinia nigrolineata. *Journal of Natural Products*, *68*(8), pp. 1218–1221.

7.4.1.8 *Garcinia oblongifolia* Champ. ex Benth.

Common names: China Mangosteen; lǐngnán shān zhú zi; huáng yá jú; chiguo, ling nan shan zhu zi (China)

Habitat: Forests

Distribution: China

Botanical observation: This tree grows to about 15 m tall. The bark is grey. The stems are somewhat articulated, smooth, and yield a copious yellow latex upon incision. The leaves are simple, decussate, and exstipulate. The petiole is up to 1 cm long and stout. The blade is oblong to spathulate, coriaceous, glossy, 5–10×2–3.5cm, marked with up to about 18 pairs of secondary nerves, tapering at base and acute, round to obtuse apex. The inflorescences are axillary clusters. The calyx includes 4 sepals, which are about 5 mm long, somewhat yellowish, and broadly lanceolate. The 4 petals are fleshy, oblong, fleshy, about 1 cm long, and orangish. The androecium consists of numerous stamens. The ovary is ovoid, 8–10 celled, lobed, and develops 8–10-lobed sessile and peltate stigma. The berries are ovoid, up to about 4 cm in diameter, laticiferous, edible, orangish yellow, grooved and containing a few seeds.

Medicinal use: Burns (China)

Moderate antiviral (non-enveloped monopartite linear single-stranded (+)RNA) phloroglucinols: Oblongifolin J and M from the leaves inhibited the replication of Enterovirus 71 (EV71) (1 strain BrCr) with IC_{50} values of 31.1 and 16.1 μM, respectively, and selectivity indices of 1.5 and 2.4 in Vero cells (ribavirin: IC_{50}: 253.1 μM and selectivity index>40) (Zhang et al., 2014). Oblongifolin M inhibited the growth of Enterovirus 71 (EV71) (SHZH98 strain) with an IC_{50} value of 2.3 μM and selectivity index value of 35.2 with inhibition of suppressed ERp57 expression in host cell (Wang et al., 2016).

Moderate antiviral (non-enveloped monopartite single-stranded (+)RNA) xanthone: Euxanthone isolated from the leaves inhibited the replication of Enterovirus 71 (EV71) (1 strain BrCr) with an IC_{50} value of 12.2 μM and a selectivity index of 3.0 (ribavirin: IC_{50}: 253.1 μM and selectivity index: >40) (Zhang et al., 2014).

REFERENCES

Wang, M., Dong, Q., Wang, H., He, Y., Chen, Y., Zhang, H., Wu, R., Chen, X., Zhou, B., He, J. and Kung, H.F., 2016. Oblongifolin M, an active compound isolated from a Chinese medical herb Garcinia oblongifolia, potently inhibits Enterovirus 71 reproduction through downregulation of ERp57. *Oncotarget*, *7*(8), p. 8797.
Zhang, H., Tao, L., Fu, W.W., Liang, S., Yang, Y.F., Yuan, Q.H., Yang, D.J., Lu, A.P. and Xu, H.X., 2014. Prenylated benzoylphloroglucinols and xanthones from the leaves of Garcinia oblongifolia with antienteroviral activity. *Journal of Natural Products*, *77*(4), pp. 1037–1046.

7.4.1.9 *Garcinia paucinervis* Chun ex F.C. How

Common names: Jīn sī lǐ, mai gui, miyou bo (China)

Habitat: Forests

Distribution: Vietnam, China

Botanical observation: This endangered tree grows to about 15 m tall. The bark is dark grey. The stems are squarrish, purplish, and yield a copious yellow latex upon incision. The leaves are simple, decussate, and exstipulate. The petiole is up to 1.5 cm long and stout. The blade is elliptic to oblong glossy, 8–14x2.5–6.5cm, reddish when young, marked with up to about 8 pairs of secondary nerves, cuneate at base, wavy, and acute or acuminate at apex. The inflorescences are axillary or terminal cymes. The calyx includes 4 sepals, which are broadly elliptic, and about 3 mm long. The 4 petals are fleshy, ovate, about 5 mm long, and yellowish. The androecium consists of numerous stamens. The ovary is ovoid, 1-celled, lobed, and develops a sessile, convex, and peltate stigma. The berries are ovoid, up to about 3.5 cm long, laticiferous, orangish yellow, and contain 2 seeds.

Medicinal uses: Burns, scalds (China)

Antiviral (non-enveloped linear monopartite single-stranded (+)RNA) xanthone: Paucinervin E from the leaves inhibited the replication of Tobacco Mosaic Virus with an IC_{50} value of 21.4 μM (Wu et al., 2013).

REFERENCE

Wu, Y.P., Zhao, W., Xia, Z.Y., Kong, G.H., Lu, X.P., Hu, Q.F. and Gao, X.M., 2013. Three novel xanthones from Garcinia paucinervis and their anti-TMV activity. *Molecules*, *18*(8), pp. 9663–9669.

7.4.1.10 *Garcinia speciosa* Wall.

Common name: Pha waa (Thailand)

Habitat: Forests

Distribution: India, Bangladesh, Myanmar, and Thailand

Botanical observation: This tree grows to about 15 m tall. The bark is dark greyish black and thin. The stems are quadrangular and yield a copious yellow latex upon incision. The leaves are simple, decussate, and exstipulate. The petiole is up to 5 cm long and slender. The blade is elliptic to oblong, glossy, 12.7–30.4 × 4.5–7.6 cm, marked with up to about 20 pairs of secondary nerves, cuneate at base, and acute or acuminate at apex. The inflorescences are axillary or terminal cymes of fragrant flowers. The 4 sepals are broadly ovate and about 1 cm long. The corolla comprises 4 fleshy petals, which are orbicular, fleshy, about 2 cm long, and yellowish. The androecium consists of 4 bundles of numerous stamens. The ovary is subglobose, 4-celled, lobed, and developing sessile, and peltate stigma. The berries are globose,up to about 5 cm long, laticiferous, red, and containing up to 4 seeds.

Medicinal use: Diarrhea (Thailand)

Strong antiviral (enveloped monopartite linear dimeric single-stranded (+)RNA) triterpenes: Garciosaterpene A and garciosaterpene C isolated from the bark inhibited syncytium formation by Human immunodeficiency virus in 1A2 cells with EC_{50} values of 5.8 and 37 μg/mL and selectivity indices of 3.4 and 1.9, respectively (Rukachaisirikul et al., 2003).

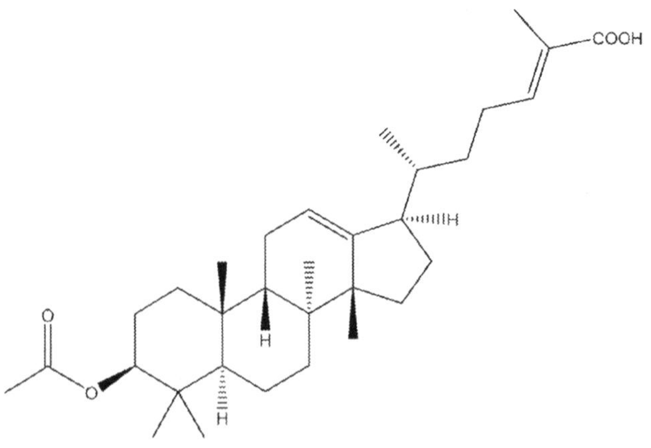

Garciosaterpene A

Moderate antiviral (enveloped monopartite linear dimeric single-stranded (+)RNA) biphenyl: Garciosine B inhibited the Human immunodeficiency virus type-1 induced syncytium formation in A12 cells with EC_{50} < 14.2 µM and selectivity index >4.7 (Pailee et al., 2018). At 200 µg/mL, garciosine B, 1,5-dihydroxyxanthone and 1,3,5-trihydroxyxanthone inhibited the Human immunodeficiency virus type-1 reverse transcriptase by 53.8, 21.9, and 72%, respectively (Pailee et al., 2018).

Moderate antiviral (enveloped monopartite linear dimeric single-stranded (+)RNA) xanthones: 1,5-Dihydroxyxanthone and 1,3,5-trihydroxyxanthone inhibited the replication of the Human immunodeficiency virus type-1 (Pailee et al., 2018).

Viral enzyme inhibition by triterpenes: Garciosaterpene A and garciosaterpene C inhibited the Human immunodeficiency virus type-1 reverse transcriptase with IC_{50} values of 15.5 and 12.2 µg/mL, respectively (Rukachaisirikul et al., 2003). (24E)-23-acetoxy-3-oxolanosta-9,24-dien-26-oic acid and garciosaterpene D from the fruits inhibited the Human immunodeficiency virus type-1 reverse transcriptase with IC_{50} values of 24.8 and 102.25 µg/mL, respectively (Pailee et al., 2017).

REFERENCES

Pailee, P., Kruahong, T., Hongthong, S., Kuhakarn, C., Jaipetch, T., Pohmakotr, M., Jariyawat, S., Suksen, K., Akkarawongsapat, R., Limthongkul, J. and Panthong, A., 2017. Cytotoxic, anti-HIV-1 and anti-inflammatory activities of lanostanes from fruits of Garcinia speciosa. *Phytochemistry Letters*, *20*, pp. 111–118.

Pailee, P., Kuhakarn, C., Sangsuwan, C., Hongthong, S., Piyachaturawat, P., Suksen, K., Jariyawat, S., Akkarawongsapat, R., Limthongkul, J., Napaswad, C. and Kongsaeree, P., 2018. Anti-HIV and cytotoxic biphenyls, benzophenones and xanthones from stems, leaves and twigs of Garcinia speciosa. *Phytochemistry*, *147*, pp. 68–79.

Rukachaisirikul, V., Pailee, P., Hiranrat, A., Tuchinda, P., Yoosook, C., Kasisit, J., Taylor, W.C. and Reutrakul, V., 2003. Anti-HIV-1 protostane triterpenes and digeranylbenzophenone from trunk bark and stems of Garcinia speciosa. *Planta Medica*, *69*(12), pp. 1141–1146.

7.4.1.11 *Mesua ferrea* L.

Synonyms: *Calophyllum nagassarium* Burm. f., *Mesua coromandelina* Wight, *Mesua nagassarium* (Burm. F.) Kosterm., *Mesua pedunculata* Wight, *Mesua roxburghii* Wight, *Mesua sclerophylla* Thw., *Mesua speciosa* Choisy

Common names: Ceylon Ironwood, Indian rose chestnut; nageshwar (Bangladesh); tie li mu (China); bos neak (Cambodia); nageswar, nagkashore, pinjara (India); nagasari (Indonesia); ka thang (Laos); penaga (Malaysia); ngaw (Myanmar)

Habitat: Ornamental

Distribution: India, Bangladesh, Myanmar, Laos, Cambodia, Thailand, Malaysia, Indonesia, Sri Lanka

Botanical observation: This handsome tree grows to about 30 m tall. The bole is straight and the wood is extremely hard. The bark is dark greyish black and fissured. The stems are terete, smooth, and yield a copious yellow latex upon incision. The leaves are simple, opposite, pendulous, and exstipulate. The petiole is up to 8 mm long. The blade is light red at birth, elliptic, glossy, 6–10×2–4cm, with numerous slender pairs of secondary nerves, glaucous below, somewhat coriaceous, base cuneate, and acute or acuminate at apex. The inflorescences are axillary and solitary. The 4 sepals are broadly ovate and about 1.5cm long. The 4 petals are membranous, obovate, about 3.5 cm long, and pure white. The androecium is showy, consists of numerous stamens, which are up to about 2 cm long. The ovary is somewhat pear-shaped and develops a 1.5 cm long stigma. The capsules are ovoid to globose, up to about 3 cm long, striated, pointed at apex, containing 1–4 seeds, and seated on persistent calyx.

Medicinal uses: Fungal infection (Bangladesh); leucorrhea (India)

Broad-spectrum antifungal halo elicited by polar extract: Methanol extract of seeds (125 µg/disc, 7 mm paper disc) inhibited the growth of *Candida albicans* (ATCC 2991), *Trichophyton beigelii* (NCIM 3404), *Aspergillus flavus* (NCIM 538), and *Aspergillus niger* (ATCC 6275) with inhibition zone diameters of 10, 10, 12, and 13 mm, respectively (Parekh & Chanda, 2008).

Strong antibacterial (Gram-positive) polar extract: Methanol extract of leaves and fruits inhibited the growth of *Staphylococcus aureus* (ATCC 25923) with an MIC value of 48 µg/mL via leakage of cytoplasmic contents (Aruldass et al., 2013).

Strong antibacterial (Gram-negative) polar extract: Methanol extract of flowers inhibited the growth of *Salmonella typhimurium* (NCTC 74) (Mazumder et al., 2005).

Strong broad-spectrum antibacterial polar extract: Methanol extract of flowers inhibited the growth of clinical isolates of *Staphylococcus aureus*, *Bacillus* spp., *Streptococcus pneumoniae*, *Sarcinia lutea*, *Escherichia coli*, *Salmonella* sp., *Shigella* sp., *Klebsiella pneumoniae*, *Proteus mirabilis*, *Lactophillus arabinosus*, *Vibrio cholerae*, and *Pseudomonas* sp. with an MIC value of 50 µg/mL (Mazumder et al., 2004).

Strong antibacterial (Gram-positive) amphiphilic coumarins: Isomammeisin (also known as Mammea A/BA; LogD = 3.5 at pH 7.4; molecular mass=406.4 g/mol) isolated from the flowers inhibited *Enterococcus faecalis* (18292) (resistant to ciprofloxacin, clindamycin, and teicoplanin) with an MIC of 8 µg/mL (Verotta et al., 2004). Isomammeisin inhibited *Staphylococcus aureus* (18110) resistant to ampicillin, erythromycin, gentamicin, ciprofloxacin, and clindamycin with an MIC value of 2 µg/mL. From the flowers, mesuol inhibited *Enterococcus faecalis* (18292) with an MIC value of 16 µg/mL (Verotta et al., 2004).

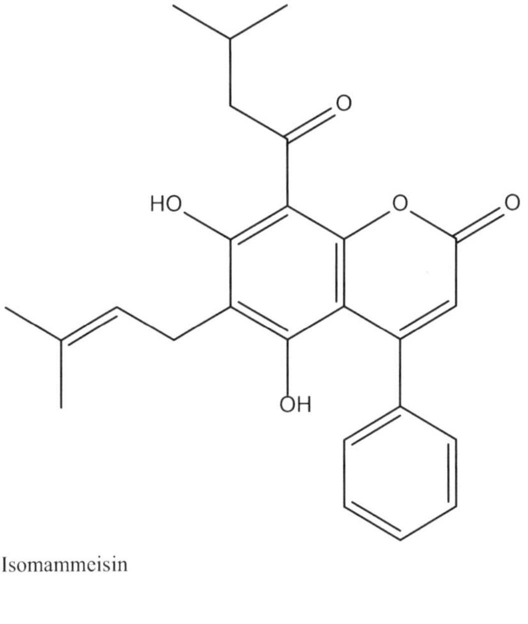

Isomammeisin

Mesuol inhibited *Staphylococcus aureus* (18110) resistant to ampicillin, erythromycin, gentamicin, ciprofloxacin, and clindamycin with an MIC value of 2 µg/mL (Verotta et al., 2004).

Strong antiviral (enveloped monopartite linear dimeric single-stranded (+)RNA) amphiphilic phenyl coumarin: Mesuol (LogD = 3.3 at pH 7.4; molecular mass=392.4 g/mol) completely inhibited the Human immunodeficiency virus type-1 replication in Jurkat cells at 15 µM (Márquez et al., 2005). This coumarin inhibited TNFα-induced Human immunodeficiency virus type-1 reverse transcriptase transcriptional activity by inhibiting the phosphorylation and the transcriptional activity of the NF-kB p65 subunit in TNFα-stimulated cells (Márquez et al., 2005).

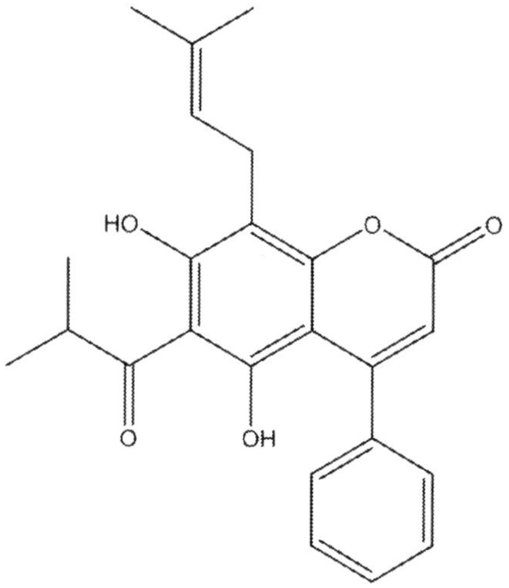

Mesuol

Antibiotic potentiator coumarin: 5,7-Dihydroxy-6-(2-methyl-butanoyl)-8-(3-methylbut-2-enyl)-4-phenyl-2*H*-chromen-2-one isolated from the flowering buds inhibited efflux at a subinhibitory concentration of 6.2 μg/mL in fluoroquinolone-resistant *Staphylococcus aureus* (1199B) and methicillin-resistant *Staphylococcus aureus* (1199B) (Roy et al., 2013).

Commentaries: (i) The antifungal property of the plant is, at least partially, owed to coumarins. These coumarins are cytotoxic for human cells *in vitro*. In general, natural products with cytotoxicity for mammalian or human cells are also, in most cases, but not always, toxic for fungi. It is for this reason that it can be advised to test antifungal agents isolated from plants in a panel of normal human cells to determine therapeutic index. The higher the index the most interesting will be the antifungal agent isolated. (ii) Mesuol being active against Human immunodeficiency virus type-1 might have some effects on coronaviruses. (iii) Natural products with strong *in vitro* anti-Human immunodeficiency virus activity from the plants belonging to the Clades covered in this volume are mainly diterpenes.

REFERENCES

Aruldass, C.A., Marimuthu, M.M., Ramanathan, S., Mansor, S.M. and Murugaiyah, V., 2013. Effects of Mesua ferrea leaf and fruit extracts on growth and morphology of Staphylococcus aureus. *Microscopy and Microanalysis*, *19*(1), pp. 254–260.

Márquez, N., Sancho, R., Bedoya, L.M., Alcamí, J., López-Pérez, J.L., San Feliciano, A., Fiebich, B.L. and Muñoz, E., 2005. Mesuol, a natural occurring 4-phenylcoumarin, inhibits HIV-1 replication by targeting the NF-κB pathway. *Antiviral Research*, *66*(2–3), pp. 137–145.

Mazumder, R., Dastidar, S.G., Basu, S.P. and Mazumder, A., 2005. Effect of Mesua ferrea Linn. flower extract on Salmonella. *IJEB*, *43*, pp. 566–586.

Mazumder, R., Dastidar, S.G., Basu, S.P., Mazumder, A. and Singh, S.K., 2004. Antibacterial potentiality of Mesua ferrea Linn. flowers. *Phytotherapy Research: An International Journal Devoted to Pharmacological and Toxicological Evaluation of Natural Product Derivatives*, *18*(10), pp. 824–826.

Parekh, J. and Chanda, S., 2008. In vitro antifungal activity of methanol extracts of some Indian medicinal plants against pathogenic yeast and moulds. *African Journal of Biotechnology*, *7*(23).

Roy, S.K., Kumari, N., Pahwa, S., Agrahari, U.C., Bhutani, K.K., Jachak, S.M. and Nandanwar, H., 2013. NorA efflux pump inhibitory activity of coumarins from Mesua ferrea. *Fitoterapia*, *90*, pp. 140–150.

Verotta, L., Lovaglio, E., Vidari, G., Finzi, P.V., Neri, M.G., Raimondi, A., Parapini, S., Taramelli, D., Riva, A. and Bombardelli, E., 2004. 4-Alkyl-and 4-phenylcoumarins from Mesua ferrea as promising multidrug resistant antibacterials. *Phytochemistry*, *65*(21), pp. 2867–2879.

7.4.2 Family Euphorbiaceae A.L. de Jussieu (1789)

The family Euphorbiaceae is a vast taxon of about 300 genera and about 7,500 species of trees, shrubs, herbs, climbers, and even cactus-shaped plants, often exuding a white latex upon incision. The leaves are simple or compound, alternate and stipulate or not. Several sorts of inflorescences occur in the family. The perianth comprises 5 tepals. The androecium consists of 5 or more stamens. A nectary disc is present. The gynoecium consists of 3 carpels forming a compound and 3-locular ovary with 3 distinct styles, each locules containing 1–2 ovules. The fruits are very characteristic dehiscent trilobed capsules.

7.4.2.1 *Acalypha australis* L.

Synonyms: *Acalypha chinensis* Roxb.; *Acalypha minima* H. Keng; *Acalypha pauciflora* Hornem.; *Urtica gemina* Lour.

Common names: Australian Acalypha; mukta borshi (Bangladesh); niǎo tà má (China); thiết hiện thái (Vietnam)

Habitat: Wastelands, roadsides

Distribution: Cambodia, Laos, Vietnam, Thailand, China, Taiwan, the Philippines, and Japan

Botanical observation: This erect herb grows to about 50 cm tall. The stems are hairy. The leaves are simple, spiral, and stipulate. The stipules are lanceolate and minute. The petiole is slender, 2–6 cm long, and straight. The blade is rhombic-ovate, 3–9 × 1–5 cm, membranous, cuneate at base, serrate, apex shortly acuminate, and with about 4 pairs of secondary nerves. The inflorescences are axillary or terminal, 1.5–5 cm, fascicles of minute bracteated flowers. The calyx produces 3–4 minute lobes. The androecium includes 8 stamens. The ovary is minute and pilose with 3 styles, which are laciniate. The fruits are capsules, which are 3-locular, about 5 mm in diameter, hairy and containing several minute seeds.

Medicinal use: Diarrhea (Bangladesh)

Moderate broad-spectrum antibacterial polar extract: Aqueous extract inhibited the growth of *Aeromonas hydrophila* (0388), *Aeromonas sobria* (0398), *Vibrio parahaemolyticus* (0394), *Vibrio anguillarum* (0387) *Staphylococcus aureus* (ATCC 25923), and *Bacillus subtilis* (ATCC 6633) (Xiao et al., 2013).

Commentary: The plant produces phenolic compounds such as gallic acid, protocatechuic acid, caffeic acid, rutin, isoquercitrin, and geraniin (Park et al., 1993), which may work synergistically as astringents for the traditional treatment of diarrhea. Consider that *Acalypha fruticosa* Forssk. is used in India for the treatment of diarrhea and cholera.

REFERENCES

Park, W.Y., Lee, S.C., Ahn, B.T., Lee, S.H., Ro, J.S. and Lee, K.S., 1993. Phenolic Compounds from Acalypha australis L. *Korean Journal of Pharmacognosy*, *24*(1), pp. 20–25.

Xiao, S., Zhang, L.F., Zhang, X., Li, S.M. and Xue, F.Q., 2013. Tracing antibacterial compounds from Acalypha australis Linn. by spectrum-effect relationships and semi-preparative HPLC. *Journal of Separation Science*, *36*(9–10), pp. 1667–1676.

7.4.2.2 *Acalypha hispida* Burm.f.

Common names: Chenille plant, red-hot cat's tail; aam-nanga, bara hatisur (Bangladesh); hong sui tie xian cai (China)

Habitat: Cultivated

Distribution: Tropical Asia and Pacific

Botanical observatiom: This shrub grows to about 3 m tall. The stems are hairy at apex. The leaves are simple, spiral, and stipulate. The stipules are hairy and up to 1 cm long. The petiole somewhat straight and up to 8 cm long. The blade is broadly lanceolate, 8–20 × 5–14 cm, membranous, somewhat cordate base, serrate, acuminate at apex, and with about 5 pairs of secondary nerves. The inflorescences are axillary or terminal, pendulous, up to 30 cm long, red and somewhat furry, like some sort of tail. The calyx produces 3–4 lobes, which are minute. The gynoecium includes 3 styles, which are up to 7 mm long.

Medicinal uses: Leprosy, diarrhea, ulcers, gonorrhea, bronchitis (Bangladesh)

Strong antibacterial (Gram-positive) hydrophylic ellagitannins: Corilagin (LogD = 0.9 at pH 7.4; molecular mass = 634.4 g/mol) inhibited the growth of *Staphylococcus aureus* (NCTC 6571) and *Bacillus subtilis* (NCTC 8236) with MIC values of 50 and 400 μg/mL, respectively (Adesina et al., 2000). Geraniin (LogD = −0.06 at pH 7.4; molecular mass = 952.6 g/mol) inhibited the growth of *Staphylococcus aureus* (NCTC 6571) and *Bacillus subtilis* (NCTC 8236) with MIC values of 25 and 100 μg/mL, respectively (Adesina et al., 2000).

Moderate antibacterial simple phenol: Gallic acid inhibited the growth of *Staphylococcus aureus* (NCTC 6571) with an MIC value of 100 μg/mL (Adesina et al., 2000).

Commentaries: The plant generates gallic acid (Reiersen et al., 2003) which is broad-spectrum antimicrobial agent (see earlier).

REFERENCE

Reiersen, B., Kiremire, B.T., Byamukama, R. and Andersen, Ø.M., 2003. Anthocyanins acylated with gallic acid from chenille plant, Acalypha hispida. *Phytochemistry*, *64*(4), pp. 867–871.

7.4.2.3 *Acalypha indica* L.

Synonyms: *Cupamenis indica* (L.) Raf.; *Ricinocarpus indicus* (L.) Kuntze

Common names: Common acalypha, Indian Acalypha; muka jhuri; muktajhuri, phool-jhuri (Bangladesh); re dai tie xian cai (China); rumput kokosonga (Indonesia); cika mas, cika emas (Malaysia); khokali, kuppi, kuppaimeni, (India); bugos (the Philippines); haan maeo (Thailand); tai tượng xanh (Vietnam)

Habitats: Roadsides, wastelands, villages

Distribution: Tropical Asia and Pacific

Botanical observation: This erect herb grows to about 50 cm tall. The plant looks like some sort of a strange dwarf tree. The leaves are simple, spiral, and stipulate. The stipules are minute. The petiole is slender, straight, and up to 3.5 cm long. The blade is rhombic-ovate, dull green, 2–3.5×1.5–2.5cm, membranous, cuneate at base, serrate, acute at apex, and presents about 5 pairs of secondary nerves. The inflorescences are axillary spikes, which are 2–7 cm long. The flowers are minute. The calyx comprises 4 sepals. The androecium includes 8 stamens. The ovary is pilose and develops 3 styles, which are laciniate. The capsules are 3-locular, minute and containing a few tiny seeds.

Medicinal uses: Fever, respiratory difficulties, cough (Bangladesh); bronchitis, pneumonia (India); sores, colds, infections, bronchitis (Myanmar); tuberculosis (Bangladesh; India); ringworm (India; Myanmar)

Moderate antibacterial (Gram-positive) polar extract: Methanol extract of leaves inhibited the growth of *Staphylococcus aureus, Staphylococcus epidermidis, Bacillus cereus, Streptococcus faecalis,* and *Pseudomonas aeruginosa* with MIC values of 156, 625, 156, 156, and 2500 μg/mL, respectively (Govindarajan et al., 2008).

Antimycobacterial polar extract: Aqueous extract of leaves inhibited the growth of *Mycobacterium tuberculosis* H37R (Gupta et al., 2010).

Strong antiviral (enveloped monopartite linear (-)RNA) polar extract: Ethanol extract inhibited the replication of the Vesicular stomatitis virus with an MIC value of 10 μg/mL (Hamidi et al., 1996).

Commentary: The plant produces flavonol glycosides (Nahrstedt et al., 2006) and is used in India for the treatment of Chikungunya virus infection (Viswanathan et al., 2008). One could have interest in isolating antimycobacterial or antiviral (RNA virus) principles from this common herb.

REFERENCES

Govindarajan, M., Jebanesan, A., Reetha, D., Amsath, R., Pushpanathan, T. and Samidurai, K., 2008. Antibacterial activity of Acalypha indica L. *European Review for Medical and Pharmacological Sciences*, *12*(5), pp. 299–302.

Gupta, R., Thakur, B., Singh, P., Singh, H.B., Sharma, V.D., Katoch, V.M. and Chauhan, S.V.S., 2010. Anti-tuberculosis activity of selected medicinal plants against multi-drug resistant Mycobacterium tuberculosis isolates. *Indian Journal of Medical Research*, *131*(6), p. 809.

Hamidi, J.A., Ismaili, N.H., Ahmadi, F.B. and Lajisi, N.H., 1996. Antiviral and cytotoxic activities of some plants used in Malaysian indigenous medicine. *Pertanika Journal of Tropical Agricultural Science*, *19*(2/3), pp. 129–136.

Nahrstedt, A., Hungeling, M. and Petereit, F., 2006. Flavonoids from Acalypha indica. *Fitoterapia*, *77*(6), pp. 484–486.

Viswanathan, M.V., Raja, D.K. and Khanna, S.D., 2008. Siddha way to cure Chikungunya. *Indian Journal of Traditional Knowledge*, *7*(2), pp. 345–346.

7.4.2.4 *Aleurites moluccanus* (L.) Willd.

Synonyms: *Aleurites moluccana* (L.) Willd.; *Aleurites triloba* J.R. Forst. & G. Forst.; *Camirium moluccanum* (L.) Kuntze; *Jatropha moluccana* L.; *Manihot moluccana* (L.) Crantz; *Rottlera moluccana* (L.) Scheff.

Common names: Belgaum walnut, candleberry tree, candlenut tree; bangla akrot (Bangladesh); shih leih (China); kuikui (Hawai); akhota, nattu-akhrotu (India); kemiri (Indonesia); wai-wai (Fiji); buah keras, kamiri (Malaysia); tosikyasi (Myanmar); lumbang (the Philippines); lai (Thailand)

Habitat: Villages, forests, cultivated

Distribution: From India to the Pacific Islands

Botanical observation: This handsome tree grows up to about 20 m tall. The bark exudes after cutting a clear watery sap. The stems, petioles, and blades are covered with a whitish starry pubescence. The leaves are simple, spiral, and stipulate. The stipules are minute. The petiole is 11.5–20 cm long, somewhat bent, and slender. The blade is coriaceous, 3–5-lobed, 7.5–12.5×4–16.5 cm, marked at base with 2, minute disc-shaped glands, of a characteristic dull green, and marked with 6–7 pairs of secondary nerves. The inflorescences are terminal panicles of whitish flowers. The calyx presents 3 lobes, which are ovate, about 3 mm long, and covered with white hairs. The 5 petals are spathulate, up to 8mm long, and whitish. The androecium includes up to 20 stamens fused at base into a 4 mm long column. The gynoecium includes a pair of styles. The capsules are woody, ovoid, 4–6.5x4–5cm, dull, light greyish-green, and contain 1 or 2 oily seeds, which are about 2 cm long. The oil expressed from the seeds has been used for the making of candles in India.

Medicinal uses: Diarrhea (Indonesia); cough (India); sore throat, thrush (Hawaii)

Moderate (Gram-positive) polar extract: Ethanol extract of bark inhibited the growth of *Staphylococcus aureus* and *Enterococcus faecalis* with MIC values of 256 and 512 µg/mL, respectively (Romulo et al., 2018).

Moderate anticandidal polar extract: Ethanol extract of bark inhibited the growth of *Candida albicans* with an MIC value of 256 µg/mL (Romulo et al., 2018).

Strong antiviral (enveloped monopartite linear dimeric single-stranded (+)RNA) polar extract: Acetonitrile: Water extract of leaves inhibited the replication of the Human immunodeficiency virus type-1 in MT-4 cells with an EC_{50} value of 7.8 and a selectivity index of 9 (Locher et al., 1996).

Commentary: The anti-Human immunodeficiency virus principle in *Aleurites moluccanus* (L.) Willd. is apparently unknown. Members of the genus *Aleurites* J.R. Forst. & G. Forst. bring to being a fascinating array of tigliane diterpenes, for instance, 13-*O*-myristyl-20-*O*-acetyl-12-deoxyphorbol from *Aleurites moluccanus* (L.) Willd. (Okuda et al., 1975; Satyanarayana et al., 2001). Consider that tigliane diterpenes are strongly antiviral *in vitro* for enveloped monopartite linear single-stranded (+) RNA viruses. The tigliane diterpene 12-*O*-tetradecanoylphorbol 13-acetate inhibited the replication of the Chikungunya virus and the Sindbis virus (strain HRsp) with EC_{50} values of 2.9 nM and 2.2 μM and selectivity indices of 1965 and above 162 (Bourjot et al., 2012). The tigliane diterpene prostatin inhibited the replication of the Chikungunya virus (strain 899) with an EC_{50} of 2.6 μM and a selectivity index of 30.3 (Bourjot et al., 2012). 4α-12-*O*-Tetradecanoylphorbol-13-acetate inhibited the replication of the Chikungunya virus (strain 899) with an EC_{50} of 2.8 μM and a selectivity index of 30.3 (Bourjot et al., 2012).

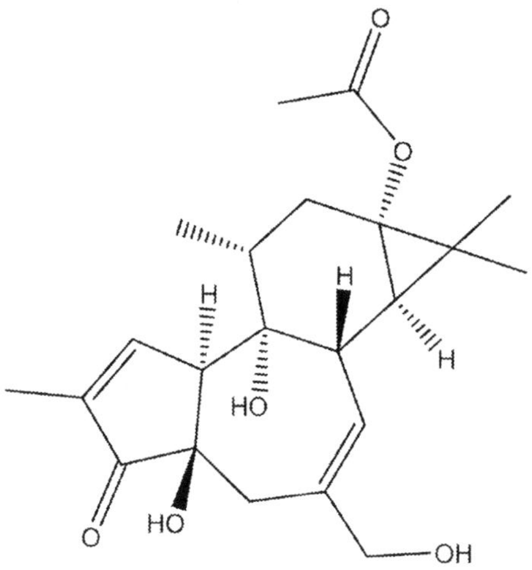

Prostatin

The tigliane diterpenes excoecafolins B and C isolated from the stems of *Excoecaria acerifolia* Didr inhibited the replication of the Human immunodeficiency virus type-1 with EC_{50} values of 0.03 and 0.04 μM and selectivity indices of 431.1 and 298.7, respectively (Huang et al., 2014). Although tigliane diterpenes have the tendency to promote carcinogenesis (Yasukawa et al., 2002), the examination of the anti-Severe acute respiratory syndrome-acquired coronavirus 2 activities of 12-*O*-tetradecanoylphorbol-13-acetate, 4α-12-*O*-tetradecanoylphorbol-13-acetate, prostatin, and excoecafolins B and C is warranted. Such compounds could be added to the list of first line natural products to be tested for the coming waves of pandemic coronaviruses.

REFERENCES

Bourjot, M., Delang, L., Nguyen, V.H., Neyts, J., Guéritte, F., Leyssen, P. and Litaudon, M., 2012. Prostratin and 12-O-tetradecanoylphorbol 13-acetate are potent and selective inhibitors of chikungunya virus replication. *Journal of natural products*, *75*(12), pp. 2183–2187.

Huang, S.Z., Zhang, X., Ma, Q.Y., Peng, H., Zheng, Y.T., Hu, J.M., Dai, H.F., Zhou, J. and Zhao, Y.X., 2014. Anti-HIV-1 tigliane diterpenoids from Excoecaria acertiflia Didr. *Fitoterapia*, *95*, pp. 34–41.

Locher, C.P., Witvrouw, M., De Béthune, M.P., Burch, M.T., Mower, H.F., Davis, H., Lasure, A., Pauwels, R., De Clercq, E. and Vlietinck, A.J., 1996. Antiviral activity of Hawaiian medicinal plants against human immunodeficiency virus type-1 (HIV-1). *Phytomedicine*, *2*(3), pp. 259–264.

Okuda, T., Yoshida, T., Koike, S. and Toh, N., 1975. New diterpene esters from Aleurites fordii fruits. *Phytochemistry*, *14*(2), pp. 509–515.

Quintão, N.L., Antonialli, C.S., da Silva, G.F., Rocha, L.W., de Souza, M.M., Malheiros, A., Meyre-Silva, C., Lucinda-Silva, R.M., Bresolin, T.M. and Cechinel Filho, V., 2012. Aleurites moluccana and its main active ingredient, the flavonoid 2″-O-rhamnosylswertisin, have promising antinociceptive effects in experimental models of hypersensitivity in mice. *Pharmacology Biochemistry and Behavior*, *102*(2), pp. 302–311.

Romulo, A., Zuhud, E.A., Rondevaldova, J. and Kokoska, L., 2018. Screening of in vitro antimicrobial activity of plants used in traditional Indonesian medicine. *Pharmaceutical Biology*, *56*(1), pp. 287–293.

Satyanarayana, P., Kumar, K.A., Singh, S.K. and Rao, G.N., 2001. A new phorbol diester from Aleurites moluccana. *Fitoterapia*, *72*(3), pp. 304–306.
Yasukawa, K., Kitanaka, S. and Seo, S., 2002. Inhibitory effect of stevioside on tumor promotion by 12-O-tetradecanoylphorbol-13-acetate in two-stage carcinogenesis in mouse skin. *Biological and Pharmaceutical Bulletin*, *25*(11), pp. 1488–1490.

7.4.2.5 *Croton tiglium* L.

Synonyms: *Alchornea vaniotii* H. Lév.; *Croton birmanicus* Müll. Arg.; *Croton himalaicus* D.G. Long; *Croton xiaopadou* (Y.T. Chang & S.Z. Huang) H.S. Kiu; *Tiglium officinale* Klotzsch

Common names: Purging croton, sultan's seeds, croton seeds; jypal (Bangladesh); pa run (China); jamalgota, jayapaala, nervalum (India); simalakian (Indonesia); mark tot; bidend jireh khatai, habb el salatin (Iran); (Laos); bua patu (Malaysia); kanakho, mai-hkang (Myanmar); camaisa, tuba (the Philippines); ma khaang (Thailand); ba dậu (Vietnam)

Habitat: Forests

Distribution: India, Sri Lanka, Bhutan, Nepal, Bangladesh, Myanmar, Cambodia, Thailand, Vietnam, China, Indonesia, the Philippines, Japan

Botanical observation: This tree grows up to about 8 m tall. The stems are hairy. The stipules are caducous and about 4 mm long. The leaves are simple, spiral, and stipulate. The petiole is up to 6 cm long. The blade is ovate to lanceolate, 5–15×2–7cm, cuneate to round at base, marked with a pair of discoid glands at base and 3–4 pairs of secondary nerves, somewhat serrate, and acute, acuminate, or caudate at apex. The inflorescences are terminal racemes, which are up to about 20 cm long. The 5 sepals are oblong-lanceolate and minute. The 5 petals are shorter than the sepals and whitish. The 15 stamens are showy. The gynoecium is hairy, 3-lobed, and develops 3 bifid stigmas. The capsules are trilobed, green turning red, smooth, 1–2 × 1–2 cm, somewhat hairy and containing a few seeds, which are oblong, about 1 cm long, greyish brown, and dreadfully cathartic.

Medicinal uses: Boils, fever, bronchitis (India); infection (Myanmar)

Moderate antimycobacterial tigliane diterpenes: Croton4-deoxy-20-oxophorbol 12-tiglyl 13-acetate and 7-oxo-5-ene-phorbol-13-(2-methylbutyrate) inhibited the growth of *Mycobacterium tuberculosis* (H37Ra) with IC_{50} values of 19.5 and 20.9 μM, respectively (isoniazid: 2.0 μM)(Zhao et al., 2016).

Commentary: (i) Consider that 12-*O*-decanoylphorbol13-acetate and 12-*O*-decanoyl-7-hydroperoxyphorbol-5-ene-13- acetate from the leaves of *Croton mauritianus* Lam inhibited the replication of the Chikungunya virus with EC_{50} values of 2.4 and 4 μM, respectively (Corlay et al., 2014). *ent*-18-Acetoxy-7α-hydroxykaur-16-en-15-one isolated from the leaves of *Croton tonkinensis* Gagnep. (used in Vietnam for the treatment of impetigo) inhibited the growth of *Mycobacterium tuberculosis* (H37Ra), *Mycobacterium tuberculosis* (H37Rv), multidrug-resistant *Mycobacterium tuberculosis*, isoniazid-resistant *Mycobacterium tuberculosis*, rifampicine-resistant *Mycobacterium tuberculosis*, Pyr-resistant *Mycobacterium tuberculosis*, and streptomycin-resistant *Mycobacterium tuberculosis* (Jang et al., 2016). *ent*-1β,14β-diacetoxy-7α-hydroxykaur-16-en-15-one isolated from the leaves of *Croton tonkinensis* Gagnep. inhibited the growth of *Mycobacterium tuberculosis* (H37Ra), *Mycobacterium tuberculosis* (H37Rv), multidrug-resistant *Mycobacterium tuberculosis*, isoniazid-resistant *Mycobacterium tuberculosis*, rifampicine-resistant *Mycobacterium tuberculosis*, Pyr-resistant *Mycobacterium tuberculosis*, and streptomycin-resistant *Mycobacterium tuberculosis* (Jang et al., 2016). *ent*-1β,7α,14β-triacetoxykaur-16-en-15-one isolated from the leaves of *Croton tonkinensis* Gagnep. inhibited the growth of *Mycobacterium tuberculosis* (H37Ra), *Mycobacterium tuberculosis* (H37Rv), multidrug-resistant *Mycobacterium tuberculosis*, isoniazid-resistant *Mycobacterium tuberculosis*, rifampicine-resistant *Mycobacterium tuberculosis*, Pyr-resistant *Mycobacterium tuberculosis*, and streptomycin-resistant *Mycobacterium tuberculosis* (Jang et al., 2016). *ent*-7α,18-dihydroxykaur-16-en-15-one isolated from the leaves of *Croton tonkinensis* Gagnep. inhibited the growth of *Mycobacterium tuberculosis* (H37Ra, H37Rv, MDR, isoniazid resistant, rifampicine resistant, Pyr resistant, and streptomycin resistant) (Jang et al., 2016). *ent*-16(S)-18-acetoxy-7α-hydroxykaur-15-one one isolated from the leaves of *Croton tonkinensis* Gagnep. inhibited the growth of *Mycobacterium tuberculosis* (H37Ra, H37Rv, MDR, isoniazid resistant, rifampicine resistant, Pyr resistant, and streptomycin resistant (Jang et al., 2016). *ent*-11β-acetoxy-7α-hydroxykaur-16-en-15-one isolated from the leaves of *Croton tonkinensis* Gagnep. inhibited the growth of sensitive *Mycobacterium tuberculosis* (H37Ra; ATCC 27294), *Mycobacterium tuberculosis* (H37Rv; ATCC 35835), *Mycobacterium tuberculosis* (MDR; KMRC-00116-00250), *Mycobacterium tuberculosis* (XRD KMRC-00203-00197), *Mycobacterium tuberculosis* (isoniazid-resistant KMR 00120-00137), *Mycobacterium tuberculosis* (rifampicine-resistant, KMRC 00121-00341), *Mycobacterium tuberculosis* (Pyrazinamide-resistant KMRC 00130-00064), and *Mycobacterium tuberculosis* (streptomycin-resistant, KMRC 00122-00123) (Jang et al., 2016).

(ii) The clerodane diterpene crotonolide G isolated from the stems of *Croton laui* Merr. & F.P. Metcalf was antibacterial against *Staphylococcus aureus* (ATCC 25923), *Staphylococcus epidermidis* (ATCC 12228), *Micrococcus luteus* (ATCC 9341), and *Bacillus subtilis* (CMCC 63501) (Liu et al., 2014).

(iii) The seeds have been known to Persian physicians as poison from a remote time and described by our peer the Portuguese physician Christoval Acosta (1525–1594) around 1578 as *pinones de Maluco* in his *Tractado de las drogas y medicinas de las Indias Orientales.* Consider that it is not the intention of the present work to make claims that the plants presented here are safe for human or animal consumption and usable as alternative medicine. Since the teaching of Pharmacognosy is being progressively demoted from pharmacy school curricula for the last 4 decades, the vast majority of pharmacy undergraduates are today not competent enough in terms of advising the public on the health benefits and dangers of medicinal plants and mushrooms (Sarker, 2012). Plants can cure diseases but can also be extremely poisonous, cause discomfort, and even bring death, and what defines plant morphology is their botanical features, and botany is not taught anymore in most Schools of Pharmacy (except in China, India, and a few other Asian countries). Today, almost everyone has a say or opinion about the virtues of plants but they did not undergo full pharmacognosy training, and thus are placing populations at risk. The Big Pharmas need monopoly of drugs sold in retailed pharmacy and hospitals via lobbies (Miller, 2013), which influence accreditation boards to ridicule the teaching of medicinal plants and label it as an obscure or very descriptive subject. The COVID-19 tragedy teaches us that not fully training pharmacy students in Pharmacognosy equals to weakening the resistance of humans in the face of global viral pandemics. Providing complete Pharmacognosy courses in schools of pharmacy globally would allow a safe use of medicinal plants to contribute toward synergy with modern medicines in preventing or even curing microbial pandemics. This course is also the guarantee that pharmacy graduates can embark into postgraduate studies in drug discovery from medicinal plants, leading to the isolation of antimicrobial drugs or the ethnopharmacological discovery of powerful antimicrobial herbs. It is time for Schools of Pharmacy to reimplement serious teaching in medicinal plants in the face of global microbial pandemics that are taking place and those that will follow.

REFERENCES

Corlay, N., Delang, L., Girard-Valenciennes, E., Neyts, J., Clerc, P., Smadja, J., Guéritte, F., Leyssen, P. and Litaudon, M., 2014. Tigliane diterpenes from Croton mauritianus as inhibitors of chikungunya virus replication. *Fitoterapia*, *97*, pp. 87–91.

Jang, W.S., Jyoti, M.A., Kim, S., Nam, K.W., Ha, T.K.Q., Oh, W.K. and Song, H.Y., 2016. In vitro antituberculosis activity of diterpenoids from the Vietnamese medicinal plant Croton tonkinensis. *Journal of Natural Medicines*, *70*(1), pp. 127–132.
Liu, C.P., Xu, J.B., Zhao, J.X., Xu, C.H., Dong, L., Ding, J. and Yue, J.M., 2014. Diterpenoids from Croton laui and their cytotoxic and antimicrobial activities. *Journal of Natural Products*, *77*(4), pp. 1013–1020.
Miller, J.E., 2013. From bad pharma to good pharma: aligning market forces with good and trustworthy practices through accreditation, certification, and rating. *The Journal of Law, Medicine & Ethics*, *41*(3), pp. 601–610.
Sarker, S.D., 2012. Pharmacognosy in modern pharmacy curricula. *Pharmacognosy Magazine*, *8*(30), p. 91.
Zhao, B.Q., Peng, S., He, W.J., Liu, Y.H., Wang, J.F. and Zhou, X.J., 2016. Antitubercular and cytotoxic tigliane-type diterpenoids from Croton tiglium. *Bioorganic & Medicinal Chemistry Letters*, *26*(20), pp. 4996–4999.

7.4.2.6 *Euphorbia antiquorum* L.

Synonym: *Tithymalus antiquorum* (L.) Moench

Common names: Milkhedge, fleshy spurge, triangular spurge; narsij (Bangladesh), pyathal (Myanmar); chanlat (Cambodia); huo yang le (China); chathurakkali, shadurak, simhunda (India)

Habitat: Wastelands; cultivated

Distribution: Tropical Asia and Pacific

Botanical observation: This cactus-shaped shrub grows to a height of 9 m tall. The plant produces a copious and dreadfully irritant milky latex. The stems are thorny and strongly 3-angled. The leaves are simple, ephemeral, 2–5 × 1–2 cm, sessile, and obovate. The stipular thorns are purple, in pairs, straight, and 2.5–3 mm long. The inflorescence is a terminal cyathium on slender pedicels. The gynoecium is 3-lobed and 3 mm diameter. The capsule is 3-lobed, 3.5–4 × 0.4–1.5 cm, and contains a few smooth and tiny seeds.

Medicinal use: Cholera (China); dysentery (Cambodia, India); bronchitis, tuberculosis, skin infection, syphilis (India)

Broad-antibacterial halo developed by mid-polar extract: Chloroform extract of latex (100 mg/mL in 10 mm wells) inhibited the growth of *Escherichia coli* (MTCC-2622), *Bacillus subtilis* (MTCC-2387) and *Proteus mirabilis* (MTCC-1429) with inhibition zone diameters of 14, 13.3, and 11 mm, respectively (Chandrasekaran et al., 2017).

Strong antiviral (enveloped monopartite linear dimeric single-stranded (+)RNA) diterpene: 8-*O*-Methylingol-3,12-diacetate-7-benzoate inhibited the replication of the Human immunodeficiency virus type-1 (strain NL4.3) in MT4 cells with an EC_{50} value of 1.3 μM (Dong et al., 2018).

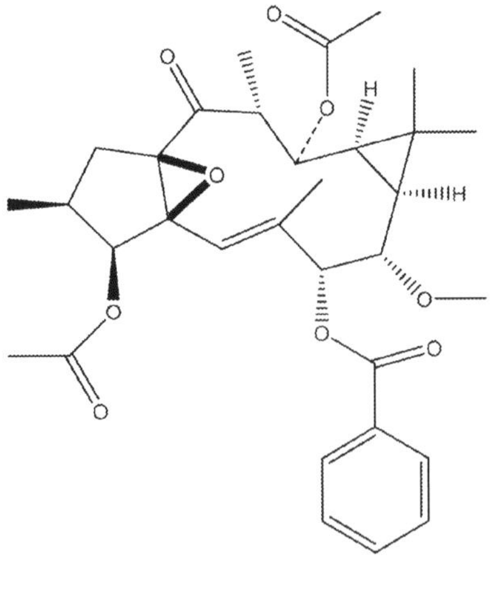

8-*O*-Methylingol-3,12-diacetate-7-benzoate

Commentary: The plant produces triterpenes such as taraxerol, epi-friedelanol, 3β-friedelanol, taraxerol, taraxerone, euphol, euphorbol, β-amyrin, and cycloartenol (Anjaneyulu & Ravi, K., 1989). Consider that *Euphorbia neriifolia* L. is used in India (local name: *Snuhi*) for the treatment of cough and ear infections. From this plant, 3β-friedelanol, 3β-acetoxy friedelane, friedelin, and epitaraxerol at a concentration of 5 μg/mL increased the viability of MRC-5 cells infected by Human coronavirus 229E by 132.4, 80.9, 109, and 111%, respectively (Chang et al., 2012). From this plant, euphorneroid D-1 and *ent*-3-oxoatisan-16α,17-acetonide inhibited the replication of the Human immunodeficiency virus (NL4.3 strain) with EC_{50} values of 34 and 24 μM and selectivity indices of 2.3 and 1.9, respectively (Yan et al., 2018).

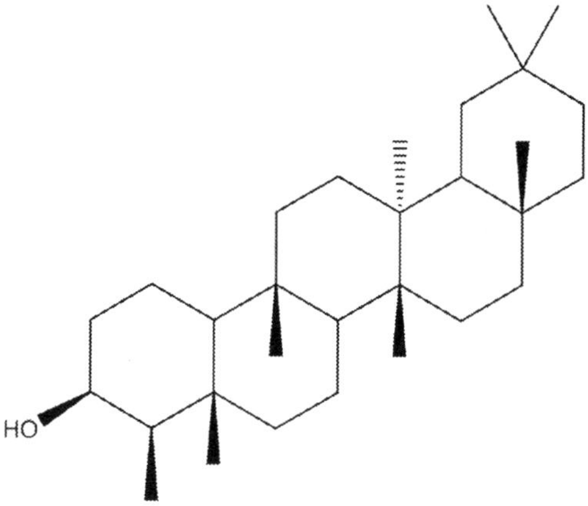

3β-Friedelanol

REFERENCES

Anjaneyulu, V. and Ravi, K., 1989. Terpenoids from Euphorbia antiquorum. *Phytochemistry*, *28*(6), pp. 1695–1697.

Chandrasekaran, R., Gnanasekar, S., Seetharaman, P., Krishnan, M. and Sivaperumal, S., 2017. Intrinsic studies of Euphorbia antiquorum L. latex extracts against human bacterial pathogens and mosquito vector Aedes aegypti, Culex quinquefasciatus (Diptera: Culicidae). *Biocatalysis and Agricultural Biotechnology*, *10*, pp. 75–82.

Chang, F.R., Yen, C.T., Ei-Shazly, M., Lin, W.H., Yen, M.H., Lin, K.H. and Wu, Y.C., 2012. Anti-human coronavirus (anti-HCoV) triterpenoids from the leaves of Euphorbia neriifolia. *Natural Product Communications*, *7*(11), p. 1934578X1200701103.

Dong, M., Chen, X.Q., Chen, C.H. and Li, R.T., 2018. Terpenes from Euphorbia antiquorum and Their in Vitro Anti-HIV Activity. *Chemistry & Biodiversity*, *15*(5), p. e1700560.

Yan, S.L., Li, Y.H., Chen, X.Q., Liu, D., Chen, C.H. and Li, R.T., 2018. Diterpenes from the stem bark of Euphorbia neriifolia and their in vitro anti-HIV activity. *Phytochemistry*, *145*, pp. 40–47.

7.4.2.7 *Euphorbia hirta* L.

Synonyms: *Anisophyllum piluliferum* (L.) Haw.; *Chamaesyce gemella* (Lag.) Small; *Chamaesyce hirta* (L.) Millsp.; *Chamaesyce hirta* (L.) Small; *Chamaesyce karwinskyi* (Boiss.) Millsp.; *Chamaesyce microcephala* (Boiss.) Croizat; *Chamaesyce pilulifera* (L.) Small; *Chamaesyce rosei* Millsp.; *Euphorbia bancana* Miq.; *Euphorbia capitata* Lam.; *Euphorbia chrysochaeta* W. Fitzg.; *Euphorbia gemella* Lag.; *Euphorbia globulifera* Kunth; *Euphorbia karwinskyi* Boiss.; *Euphorbia microcephala* Boiss.; *Euphorbia nodiflora* Steud.; *Euphorbia obliterata* Jacq.; *Euphorbia pilulifera* A. Chev.; *Euphorbia pilulifera* Jacq.; *Euphorbia pilulifera* L.; *Euphorbia verticillata* Vell.; *Tithymalus pilulifer* (L.) Moench; *Tithymalus piluliferus* (L.) Moench

Common names: Garden spurge, hairy spurge, euphorbia herb, asthma plant; lalkeru, dudhia, bara dudhe, dugdhika (Bangladesh); amumpatchai (India); gelang susu (Malaysia); kiki kana kuku (Papua New Guinea); gatas gatas (the Philippines); yaa nam muek (Thailand); cỏ sữa lá lớ (Vietnam)

Habitat: Parks, roadsides, grassy spots, gardens, vacant lands

Distribution: Tropical Asia

Botanical observation: This characteristic herb grows up to a height of 40 cm and produces an irritating milky latex. The stems are hirsute, reddish, subglabrous, scorpioid, somewhat bending, and pilose. The leaves are simple, opposite, and stipulate. The petiole is 1 mm long. The blade is asymmetrical, serrate, green with purplish margins, and 2–4 cm × 8 mm–1.5 cm. The inflorescence is an axillary cyme. The flowers and capsules are minute.

Medicinal uses: Dysentery (Bangladesh, India, Vietnam); diarrhea, cold, cough, ringworm, gonorrhea, sores (India)

Moderate broad-spectrum antibacterial polar extract: Ethanol extract inhibited the growth of *Escherichia coli*, *Proteus vulgaris*, *Pseudomonas aeruginosa*, *Bacillus subtilis*, *Bacillus pumilus*, *Staphylococcus aureus*, and *Streptococcus faecalis* with MIC values of 189, 200, 166, 296, 269, 216, and 241 µg/mL, respectively (Sudhakar et al., 2006).

Moderate antibacterial (Gram-negative) polar extract: Methanol extract of flowers inhibited the growth of *Shigella flexneri* and *Shigella dysenteriae* and with MIC values of 100 and 200 µg/mL, respectively (Vijaya et al., 1995).

Strong antibacterial (Gram-negative) amphiphilic flavanol: (–)-Epicatechin 3-*O*-gallate (also known as (–)-3-*O*-galloylepicatechin; LogD=2.1 at pH 7.4; molecular mass=442.3 g/mol) inhibited the growth *Pseudomonas aeruginosa* with an MIC value of 15.6 µg/mL via bactericidal effect (Perumal et al., 2015).

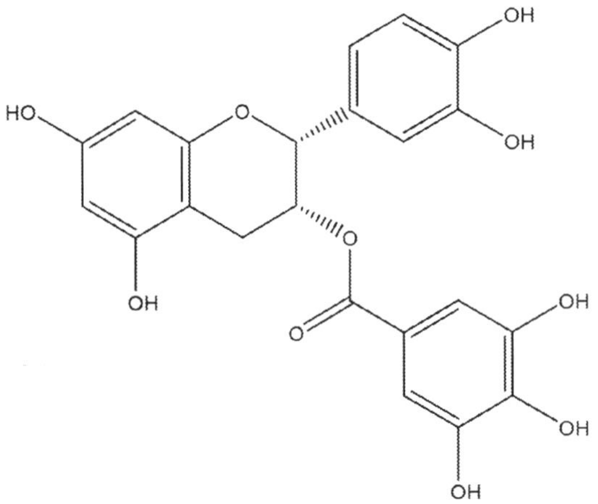

(–)-Epicatechin 3-*O*-gallate

Strong (Gram-negative) antibacterial hydrophilic hydroxycinnamic acid: Caffeic acid (also known as 3,4-dihydroxycinnamic acid; LogD=–1.7 at pH 7.4; molecular mass=180.1 g/mol) inhibited the growth of *Pseudomonas aeruginosa* with an MIC value of 31.3 µg/mL (Perumal et al., 2015).

Moderate broad-spectrum antifungal polar extract: Ethanol extract inhibited the growth of *Candida albicans*, *Aspergillus niger*, and *Rhizopus oligosporus* with MIC values of 275, 304, and 245 µg/mL, respectively (Sudhakar et al., 2006).

Strong antiviral (enveloped monopartite linear dimeric single-stranded (+)RNA) polar extracts: Ethanol extract inhibited the replication of the Human immunodeficiency virus type-1 with an MIC value of 100 µg/mL (Hamidi et al., 1996). Aqueous extract inhibited the replication of the Human immunodeficiency virus type-1 (strain IIIB) with an IC_{50} value of 9 μg/mL in MT4 cells and a selectivity index of 19 (Gyuris et al., 2009).

Strong antiviral (enveloped segmented single-stranded (+)RNA) hydrophilic hydroxycinnamic acid: Caffeic acid inhibited the replication of Human coronavirus NL63 with an IC_{50} value of 3.5 μM, plaque formation in LLC-MK2 cells with an IC_{50} value of 5.4 μM, and virus attachment to LLC-MK2 cells with an IC_{50} value of 8.1 μM (Weng et al., 2019).

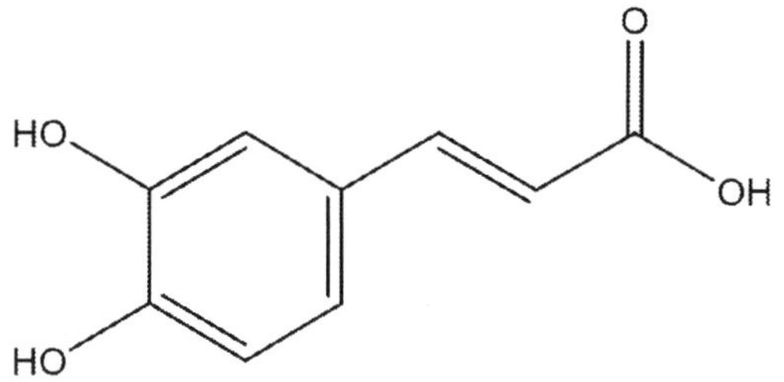

Caffeic acid

Strong (antiviral enveloped circular double-stranded DNA) hydrophilic hydroxycinnamic acid: Caffeic acid inhibited the replication of the Hepatitis B virus with an IC_{50} value of 3.9 μM and inhibited the expression of Hepatitis B surface antigen by in HepG2.2.15 cells with an IC_{50} value of 12.7 μM (Wang et al., 2009).

Strong (antiviral enveloped circular double stranded DNA) hydrophilic hydroxycinnamic acid: Chlorogenic acid (LogD = 1 at pH 7.4; molecular mass = 354.3 g/mol) inhibited the replication of the Hepatitis B virus with the IC_{50} value of 1.2 μM (Wang et al., 2009).

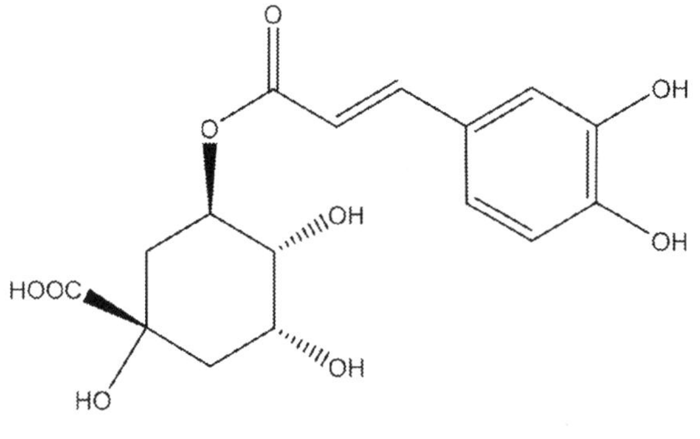

Chlorogenic acid

Strong antiviral (enveloped segmented linear single-stranded (−)RNA) gallotannin: 1,3,4,6-Tetra-*O*-galloyl-β-D-glucopyranoside inhibited the replication of the Influenza A H1N1 (A/Cal/079) and Influenza A H3N2 (A/Perth/16/09) with EC_{50} values of 0.3 and 1.1 μM and selectivity indices above 128.2 and 45, respectively (Chang et al., 2016). 1,3,4,6-Tetra-*O*-galloyl-β-D-glucopyranoside inhibited the replication of the Influenza virus B/FI/4/06 with an EC_{50} of 2.7 μM a selectivity index (Chang et al., 2016).

Strong antiviral (enveloped segmented linear single-stranded (−)RNA) amphiphilic simple phenolic: Ellagic acid (LogD = 1 at pH 7.4; molecular mass = 302.1 g/mol) inhibited the replication of the Influenza A H1N1 (A/Cal/079) and H3N2 (A/Perth/16/09) virus with EC_{50} values of 1.9 and 2.9 μM and selectivity indices above 25.5 and 17.1 (Chang et al., 2016). Ellagic acid inhibited Influenza B virus (B/FI/4/06) with an EC_{50} of 11.8 μM and SI >50 (Chang et al., 2016).

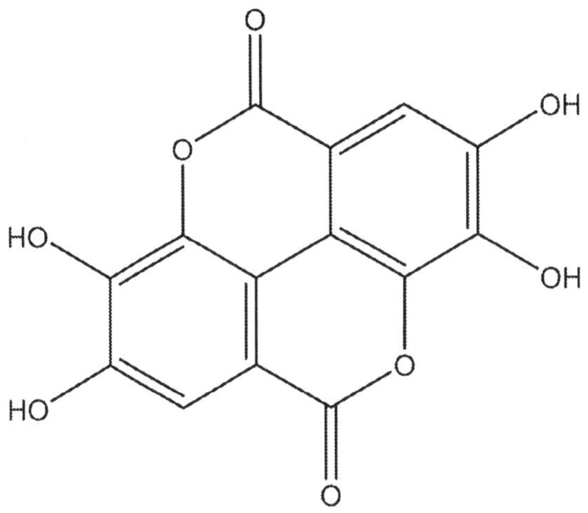

Ellagic acid

Strong antiviral (enveloped monopartite linear single-stranded (+)RNA) gallotannin: Tetra-*O*-galloyl-D-glucose inhibited the Severe acute respiratory syndrome-associated coronavirus (Wild-type virus BJ01 strain) infection in Vero cells with an EC_{50} value of 4.5 μM (and a selectivity index of 240 (Yi et al., 2004).

Weak antiviral (enveloped segmented single-stranded (+)RNA) polar simple phenol: Gallic acid inhibited the replication of the Human coronavirusNL63 with an IC_{50} value of 71.4 μM (Weng et al., 2019).

Weak antiviral (enveloped segmented single-stranded (+)RNA) polar hydroxycinnamic acid derivative: Chlorogenic acid inhibited the replication of the Human coronavirusNL63 with an IC_{50} value of 43.4 μM (Weng et al., 2019).

In vivo antiviral (enveloped circular double-stranded DNA) polar hydroxycinnamic acid: Caffeic acid given orally at the dose of 100 mg/Kg twice a day for 10 days protected ducklings against duck against duck Hepatitis B virus (Wang et al., 2009).

Commentaries: (i) The plant generates *ent*-kaurane-type diterpenes, triterpenes, and lignans (Li et al., 2015; Yan et al., 2011). One could have the curiosity to isolate anti-Human immunodeficiency virus principles from this common herb. Furthermore, aqueous extract of the plant could be of value against coronaviruses. (ii) Members of the genus *Euphorbia* L. produce antibacterial diterpenes and antiviral phenolics. For instance, the jatrophane-type diterpene euphoheliosnoid E inhibited the growth of *Streptococcus mutans* (ATCC 25175) and *Actinomycetes viscosus* (ATCC 27044) (Di et al., 2015).

(+)-(7′S,8′R)-erythro-7′-methylcarolignan E isolated from the aerial parts of *Euphorbia sikkimensis* Boiss. inhibited the replication of the Human immunodeficiency virus type-1 in MT4 cells with an EC_{50} value of 5.3 μM (Jiang et al., 2016).

(+)-Epicatechin 3-*O*-gallate isolated from *Euphorbia humifusa* Willd inhibited the replication of the Influenza virus B/FI/4/06 with an EC_{50} value of 19.8 μM and a selectivity index above 2.5 (Chang et al., 2016).

(iii) Natural products with strong *in vitro* anti-Influenza virus from the plants belonging to the Clades covered in this volume are mainly flavonoids and flavonoid glycosides.

REFERENCES

Chang, S.Y., Park, J.H., Kim, Y.H., Kang, J.S. and Min, J.Y., 2016. A natural component from Euphorbia humifusa Willd displays novel, broad-spectrum anti-influenza activity by blocking nuclear export of viral ribonucleoprotein. *Biochemical and Biophysical Research Communications, 471*(2), pp. 282–289.

Di, G.E.N.G., Li-Tao, Y.I., Yao, S.H.I. and Zhi-Da, M.I.N., 2015. Structure and antibacterial property of a new diterpenoid from Euphorbia helioscopia. *Chinese Journal of Natural Medicines, 13*(9), pp. 704–706.

Gupta, D.R. and Garg, S.K., 1966. A chemical examination of Euphorbia hirta Linn. *Bulletin of the Chemical Society of Japan, 39*(11), pp. 2532–2534.

Gyuris, A., Szlavik, L., Minarovits, J., Vasas, A., Molnar, J. and Hohmann, J., 2009. Antiviral activities of extracts of Euphorbia hirta L. against HIV-1, HIV-2 and SIVmac251. *In vivo, 23*(3), pp. 429–432.

Hamidi, J.A., Ismaili, N.H., Ahmadi, F.B. and Lajisi, N.H., 1996. Antiviral and cytotoxic activities of some plants used in Malaysian indigenous medicine. *Pertanika Journal of Tropical Agricultural Science, 19*(2/3), pp. 129–136.

Jiang, C., Luo, P., Zhao, Y., Hong, J., Morris-Natschke, S.L., Xu, J., Chen, C.H., Lee, K.H. and Gu, Q., 2016. Carolignans from the aerial parts of Euphorbia sikkimensis and their anti-HIV activity. *Journal of Natural Products, 79*(3), pp. 578–583.

Li, E.T., Liu, K.H., Zang, M.H., Zhang, X.L., Jiang, H.Q., Zhou, H.L., Wang, D.Y., Liu, J.G., Hu, Y.L. and Wu, Y., 2015. Chemical constituents from Euphorbia hirta. *Biochemical Systematics and Ecology, 62*, pp. 204–207.

Perumal, S., Mahmud, R. and Ramanathan, S., 2015. Anti-infective potential of caffeic acid and epicatechin 3-gallate isolated from methanol extract of Euphorbia hirta (L.) against Pseudomonas aeruginosa. *Natural Product Research, 29*(18), pp. 1766–1769.

Sudhakar, M., Rao, C.V., Rao, P.M., Raju, D.B. and Venkateswarlu, Y., 2006. Antimicrobial activity of Caesalpinia pulcherrima, Euphorbia hirta and Asystasia gangeticum. *Fitoterapia, 77*(5), pp. 378–380.

Vijaya, K., Ananthan, S. and Nalini, R., 1995. Antibacterial effect of theaflavin, polyphenon 60 (Camellia sinensis) and Euphorbia hirta on Shigella spp.—a cell culture study. *Journal of Ethnopharmacology, 49*(2), pp. 115–118.

Wang, G.F., Shi, L.P., Ren, Y.D., Liu, Q.F., Liu, H.F., Zhang, R.J., Li, Z., Zhu, F.H., He, P.L., Tang, W. and Tao, P.Z., 2009. Anti-hepatitis B virus activity of chlorogenic acid, quinic acid and caffeic acid in vivo and in vitro. *Antiviral Research, 83*(2), pp. 186–190.

Weng, J.R., Lin, C.S., Lai, H.C., Lin, Y.P., Wang, C.Y., Tsai, Y.C., Wu, K.C., Huang, S.H. and Lin, C.W., 2019. Antiviral activity of Sambucus FormosanaNakai ethanol extract and related phenolic acid constituents against human coronavirus NL63. *Virus Research, 273*, p. 197767.

Xie, Y.; Huang, B.; Yu, K.; Shi, F.; Liu, T.; Xu, W. Caffeic acid derivatives: A new type of influenza neuraminidase inhibitors. *Bioorg. Med. Chem. Lett.*, 2013, 23(12), pp. 3556–3560.

Yan, S., Ye, D., Wang, Y., Zhao, Y., Pu, J., Du, X., Luo, L. and Zhao, Y., 2011. Ent-Kaurane diterpenoids from Euphorbia hirta. *Records of Natural Products, 5*(4), p. 247.

Yi, L., Li, Z., Yuan, K., Qu, X., Chen, J., Wang, G., Zhang, H., Luo, H., Zhu, L., Jiang, P. and Chen, L., 2004. Small molecules blocking the entry of severe acute respiratory syndrome coronavirus into host cells. *Journal of Virology, 78*(20), pp. 11334–11339.

7.4.2.8 *Euphorbia nivulia* Buch.-Ham.

Synonym: *Euphorbia neriifolia* sensu Roxb.

Common names: Leafy milk hedge; shij (Bangladesh); akujamudu, patra-snuhi, ptoon (India)

Habitat: Ornamental

Distribution: Pakistan, India, Bangladesh, and Myanmar

Botanical observation: This fleshy, laticiferous, cactus-like, spiny tree grows to about 9m tall. The stems are terete, develop sharp stipular thorns, and exude a latex upon incision. The leaves are simple, sessile, spiral, and stipulate. The stipular thorns are about 5mm, darkish, long, and arranged in pairs. The blade is spathulate, 10–25×3–8cm, rounded at apex, tapered at base, with 6–8 pairs of secondary nerves, and somewhat fleshy. The inflorescences are axillary cymes of cyathia. The flowers are minute. The capsules are 6×13mm, smooth, and contain quadrangular, 4mm long, smooth seeds.

Medicinal uses: Rabies (Bangladesh); leprosy, bronchitis, syphilis (India)

Strong broad-spectrum antibacterial polar extract: Methanol extract of leaves inhibited the growth of *Arthrobacter citreus* (NCIM 2320), *Bacillus cereus* (MTCC 430), *Bacillus licheniformis* (MTCC 1520), *Bacillus polymixa* (NCIM 2188), *Clostridium* sp. (NCIM 2337), *Bacillus pumilus* (NCIM 2327), *Staphylococcus aureus*, *Bacillus subtilis*, *Streptococcus* sp. (NCIM 2611), *Escherichia coli* (NCIM 2345), *Klebsiella aerogenes* (NCIM 2386), *Pseudomonas aeruginosa*, *Salmonella typhymurium* (NCIM 1535), *Sarcina lutea* (NCIM 2103), with MIC values of 100, 12.5, 12.5, 25, 50, 12.5, 25, 12.5, 200, 25, 6.5, 6.2, 12.5, and 50 µg/mL, respectively (Annapurna et al., 2004).

Strong antifungal (yeasts) polar extract: Methanol extract of leaves inhibited the growth of *Candida albicans* (MTCC 227) and *Saccharomyces cerevisiae* (MTCC 170), with MIC values of 25 and 50 µg/mL, respectively (Annapurna et al., 2004).

Commentary: *Euphorbia nivulia* Buch.-Ham. synthetises ingol-type diterpenes (Ravikanth et al., 2002) that could be examined for their possible antiviral, leptospiral, or antimycobacterial activities.

REFERENCES

Annapurna, J., Chowdary, I.P., Lalitha, G., Ramakrishna, S.V. and Iyengar, D.S., 2004. Antimicrobial activity of Euphorbia nivulia leaf extract. *Pharmaceutical Biology*, *42*(2), pp. 91–93.

Ravikanth, V., Reddy, V.N., Rao, T.P., Diwan, P.V., Ramakrishna, S. and Venkateswarlu, Y., 2002. Macrocyclic diterpenes from Euphorbia nivulia. *Phytochemistry*, *59*(3), pp. 331–335.

7.4.2.9 *Euphorbia thymifolia* L.

Synonyms: *Anisophyllum thymifolium* (L.) Haw.; *Chamaesyce mauritiana* Comm. ex Denis; *Chamaesyce microphylla* (Lam.) Soják; *Chamaesyce rubrosperma* (Lotsy) Millsp.; *Chamaesyce thymifolia* (L.) Millsp.; *Euphorbia afzelii* N.E. Br.; Euphorbia botryoides Noronha; *Euphorbia foliata* Buch.-Ham. ex Dillwyn; *Euphorbia microphylla* Lam.; *Euphorbia philippina* J. Gay ex Boiss.; *Euphorbia rubicunda* Blume; *Euphorbia rubrosperma* Lotsy

Common names: Chicken weed, red caustic creeper, thyme-leafed spurge; gutedare, shwet kherua, dudhiya, swetkan (Bangladesh); qian gen cao (China); dudhi khurd, dughdika, sittra paladi, vajri (India); dudheejhar (Nepal); kheer wal (Pakistan)

Habitat: Sandy and sun-exposed vacant plots of land, open places, parking, roadsides

Distribution: Tropical Asia and Pacific

Botanical observation: This discrete creeping herb grows to a height of 20 cm from fibrous roots. The stem is terete, somewhat reddish brown, and hairy. The leaves are simple, opposite, and stipulate. The stipules are minute, ephemeral, and lanceolate. The petiole is minute. The blade is broadly elliptic, membranous, somewhat thyme-like, green with a reddish margin, cordate at base, acute at apex, and up to about 9 mm long. The inflorescence is an axillary cyathium of minute flowers. The perianth includes 5 lobes. The androecium comprises 1 stamen. The gynoecium includes a 3-lobed ovary and a bifid stigma. The capsules are 3-lobed, minute, hairy, and containing almond-shaped seeds.

Medicinal uses: Dysentery (Bangladesh); ringworm (Bangladesh, India, Nepal); fever (India)

Moderate antibacterial (Gram-negative) simple phenol: Methylgallate inhibited the growth of *Shigella dysenteriae* 1 (NT4907), *Shigella flexneri* 2a (B294), *Shigella boydii* 4 (BCH612), *Shigella sonnei* (1DH00968SS) with MIC/MBC values of 128/256, 128/512, 128/256, and 256/512 µg/mL, respectively (Acharyya et al., 2015). At 256 µg/mL, methyl gallate evoked fatal insults of internal and external membrane, inducing leaking of intracellular contents (Acharyya et al., 2015).

Strong antiviral (enveloped linear monopartite double-stranded DNA virus) hydrophilic simple phenols: Protocatechuic acid (LogD = −1.8 at pH 7.4; molecular mass = 154.1 g/mol) inhibited the replication of the Herpes simplex virus type-2 in Vero cells with an EC_{50} of 0.9 µg/mL and a selectivity index above 200) (Hassan et al., 2017). Methyl gallate (LogD=1.1 at pH 7.4; molecular mass=184.1 g/mol) inhibited the replication of the Herpes simplex virus type-2 (MS strain) and Herpes simplex virus type-1 (MacIntyre strain) with EC_{50} values of 0.2 and 0.6 µg/mL, respectively (Kane et al., 1988). Methyl gallate inhibited the replication of the Cytomegalovirus with an EC_{50} value of 6.9 µg/mL (Kane et al., 1988).

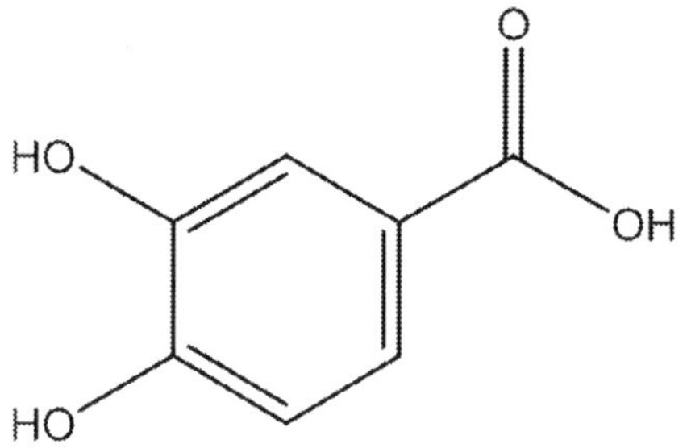

Protocatechuic acid

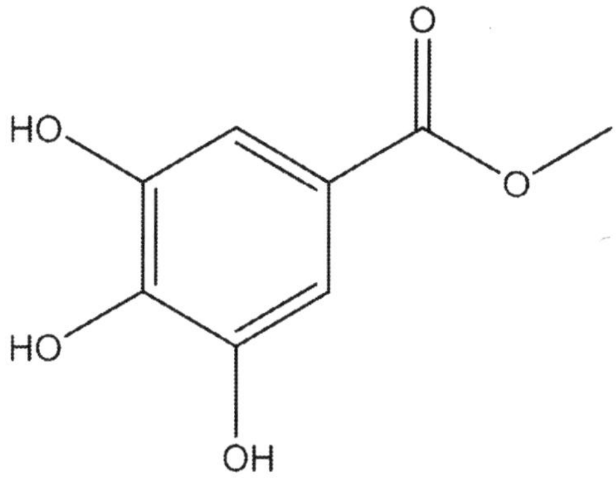

Methyl gallate

Very weak antiviral (enveloped segmented linear single-stranded (−)RNA) simple phenol: Methyl gallate inhibited the replication of the Influenza A virus with an EC_{50} value of 176 µg/mL (Kane et al., 1988).

Very weak (enveloped monopartite linear (−)RNA) antiviral simple phenol: Methyl gallate inhibited the replication of the Vesicular stomatitis virus with an EC_{50} value of 350 µg/mL (Kane et al., 1988).

In vivo antiviral (enveloped segmented linear single-stranded (−)RNA) simple phenolic: Protocatechuic acid given at a dose of 20 mg/Kg twice daily by oral gavages at 12-hour intervals for 7 days protected mice against the Influenza virus A/Chicken/Hebei/4/2008(H9N2) including decrease in lung inflammation (Ou et al., 2014).

Commentaries: (i) The plant brings to being phenolics of which protocatechuic acid, methylgallate, vanillic acid, and latifolicinin C, as well as ergostane-type triterpenes (Hu at al., 2018). One could have interest in testing protocatechuic acid, methylgallate, vanillic acid, and latifolicinin C against coronaviruses, including the Severe acute respiratory syndrome-acquired coronavirus 2. (ii) Consider that simple phenolics are often antiviral *in vitro*.

REFERENCES

Acharyya, S., Sarkar, P., Saha, D.R., Patra, A., Ramamurthy, T. and Bag, P.K., 2015. Intracellular and membrane-damaging activities of methyl gallate isolated from Terminalia chebula against multidrug-resistant Shigella spp. *Journal of Medical Microbiology*, *64*(8), pp. 901–909.
Hu, Y.K., Li, Y.Y., Li, M.J., Li, F., Xu, W. and Zhao, Y., 2018. Chemical constituents of Euphorbia thymifolia. *Chemistry of Natural Compounds*, *54*(6), pp. 1185–1186.
Kane, C.J., Menna, J.H., Sung, C.C. and Yeh, Y.C., 1988. Methyl gallate, methyl-3, 4, 5-trihydroxybenzoate, is a potent and highly specific inhibitor of herpes simplex virusin vitro. II. Antiviral activity of methyl gallate and its derivatives. *Bioscience Reports*, *8*(1), pp. 95–102.
Ou, C., Shi, N., Yang, Q., Zhang, Y., Wu, Z., Wang, B., Compans, R.W. and He, C., 2014. Protocatechuic acid, a novel active substance against avian Influenza virus H9N2 infection. *PLoS One*, *9*(10), p. e111004.

7.4.2.10 *Excoecaria agallocha* L.

Synonyms: *Commia cochinchinensis* Lour.; *Stillingia agallocha* (L.) Baill.

Common names: Blind-your-eye; geowa (Bangladesh; India); tatom (Cambodia); hai qi (China); sinu gaga (Fiji); kampeti (India); buta buta (Malaysia; the Philippines); bel-ay sisimet (Papua New Guinea); tatum thale (Thailand)

Habitat: Mangroves and tidal forests

Distribution: India, Sri Lanka, Bangladesh, Cambodia, Malaysia, Thailand, Vietnam, Indonesia, the Philippines, Japan, Papua New Guinea, Australia, Pacific islands

Botanical observation: This sinster tree (feared by Asian foresters) grows up to 15 m. The bark is greyish and upon cutting exudes a dreadfully vesicant milky sap (immediate anaphylactic shock possible), which raises blisters and can cause blindness (hence *buta*=blind). In India, the tree is sacred. The leaves are simple, spiral, and stipulate. The stipules are ephemeral, minute, and ovate. The petiole is up to about 3 cm long. The blade is dull green turning red before falling off, marked with a pair of glands at base as well as 9–13 pairs of secondary nerves, elliptic, 4.5–10×3–5 cm, obtuse at base, somewhat bending and fleshy, serrate, and acute at apex. The inflorescences are terminal and somewhat vertical, about 5 cm long, tail-shaped slender spikes. The male flowers includes 3 stamens and the female flowers a 3-lobed ovary and 2 slender stigmas. The capsule is green, somewhat glossy, dehiscent, 3-lobed, up to 1 cm across, smooth, and marked at apex with 3 remnant recurved styles. The seeds are globose and 4 mm across.

Medicinal uses: Leprosy (Fiji; India); sores (India)

Broad-spectrum antibacterial halo developed by polar extract: Ethanol extract of leaves (50 mg/mL/6 mm wells) inhibited the growth of *Staphylococcus aureus* (MTCC 1144), *Shigella flexneri*, *Bacillus licheniformis* (MTCC 7425), *Bacillus brevis* (MTCC 7404), *Pseudomonas aeruginosa* (MTCC 1034), *Streptococcus aureus*, *Staphylococcus epidermidis* (MTCC 3615), *Bacillus subtilis* (MTCC 7164), and *Escherichia coli* (MTCC 1089) with the inhibition zone diameters of 11, 18, 21, 7, 12, 17, 19, 16, and 16 mm, respectively (Patra et al., 2009).

Strong antiviral (non-enveloped monopartite linear single-stranded, (+)RNA) polar extract: Ethanol extract of leaves inhibited the replication of the Encephalomyocarditis virus in mouse fibroblast with an EC_{50} value of 16.7 μg/mL and a selectivity index of 8 (Premanathan et al., 1999).

Strong antiviral (enveloped monopartite linear dimeric single-stranded (+)RNA) polar extract: Ethanol extract of leaves inhibited the replication of the Human immunodeficiency virus in MT-4 cells with an EC_{50} value of 7.3 μg/mL and a selectivity index of 10.7 (Premanathan et al., 1999).

Strong antiviral (enveloped monopartite linear dimeric single-stranded (+)RNA) tigliane-type diterpene: 12-Deoxyphorbol 13-(3E,5E-decadienoate) inhibited the replication of the Human immunodeficiency virus type-1 with an IC_{50} value of 6 nM (Erickson et al., 1995).

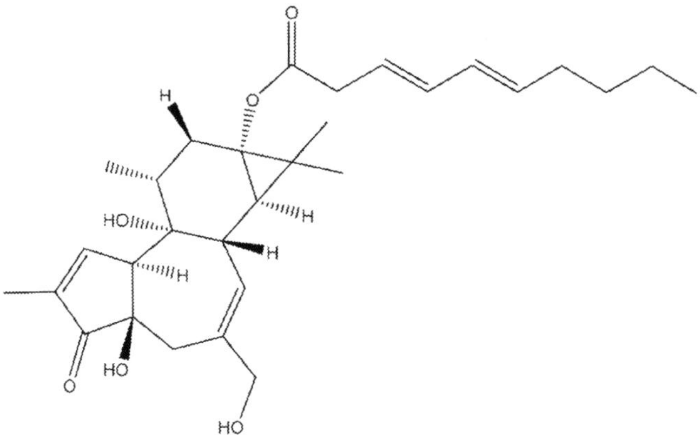

12-Deoxyphorbol 13-(3E,5E-decadienoate)

Commentary: *Excoecaria acerifolia* Didr. generates the tigliane-type diterpene excoecafolin B, which inhibited the replication of the Human immunodeficiency virus type-1 in C8166 cells with an EC_{50} value of 0.03 µM and a selectivity index of 431.1 (Huang et al., 2014). As discussed earlier, tigliane-type diterpenes have antimycobacterial properties. The traditional use of *Excoecaria agallocha* L. in Fiji could be mediated by tigliane-type diterpenes with activity against *Mycobacterium leprae*.

REFERENCES

Erickson, K.L., Beutler, J.A., Cardellina, J.H., McMahon, J.B., Newman, D.J. and Boyd, M.R., 1995. A novel phorbol ester from Excoecaria agallocha. *Journal of Natural Products*, *58*(5), pp. 769–772.

Huang, S.Z., Zhang, X., Ma, Q.Y., Peng, H., Zheng, Y.T., Hu, J.M., Dai, H.F., Zhou, J. and Zhao, Y.X., 2014. Anti-HIV-1 tigliane diterpenoids from Excoecaria acertiflia Didr. *Fitoterapia*, *95*, pp. 34–41.

Patra, J.K., Panigrahi, T.K., Rath, S.K., Dhal, N.K. and Thatoi, H., 2009. Phytochemical screening and antimicrobial assessment of leaf extracts of Excoecaria agallocha L.: A mangal species of Bhitarkanika, Orissa, India. *Advances in Natural and Applied Sciences*, *3*(2), pp. 241–246.

Premanathan, M., Kathiresan, K. and Nakashima, H., 1999. Mangrove halophytes: A source of antiviral substances. *South Pacific Study*, *19*(1–2), pp. 49–57.

7.4.2.11 *Jatropha curcas* L.

Synonyms: *Castiglionia lobata* Ruiz & Pav.; *Curcas adansonii* Endl.; *Curcas curcas* (L.) Britton & Millsp.; *Curcas drastica* Mart.; *Curcas indica* A. Rich.; *Curcas purgans* Medic.; *Curcas purgans* Medik.; *Jatropha acerifolia* Salisb.; *Jatropha afrocurcas* Pax; *Jatropha condor* Wall.; *Jatropha edulis* Cerv.; *Jatropha moluccana* Wall.; *Jatropha tuberosa* Elliot; *Jatropha yucatanensis* Briq.; *Manihot curcas* (L.) Crantz; *Ricinus americanus* Mill.; *Ricinus jarak* Thunb.

Common names: Physic nut, purging nut tree; bagh verenda (Bangladesh); lohong kvang sa (Cambodia);cha guo, ma feng shu (China); chitra, jamalgota, ratanjot, sada bharenda, (India); jarak belanda, jarak pagar (Malaysia); kesugi (Myanmar); rajani giri, saruwa (Nepal); katawa, tuba tuba (the Philippines); cây tra cúng (Vietnam)

Habitat: Cultivated

Distribution: Tropical Asia and Pacific

Botanical observation: This shrub grows to a height of 3 m. The stems are smooth, covered with very small whitish lenticels, and exude an abundant milky latex upon incision. The leaves are simple, spiral, and stipulate. The stipules are minute. The petiole grows up to about 20 cm long. The blade is 5-lobed, 7–18 × 6–16 cm, wavy, and marked with 2–3 pairs of secondary nerves. The inflorescences are axillary about 8 cm long. The 5 sepals are about 4mm long and the 5 petals are oblong, yellowish green, hairy, and about 6mm long. Five disc glands are present. The

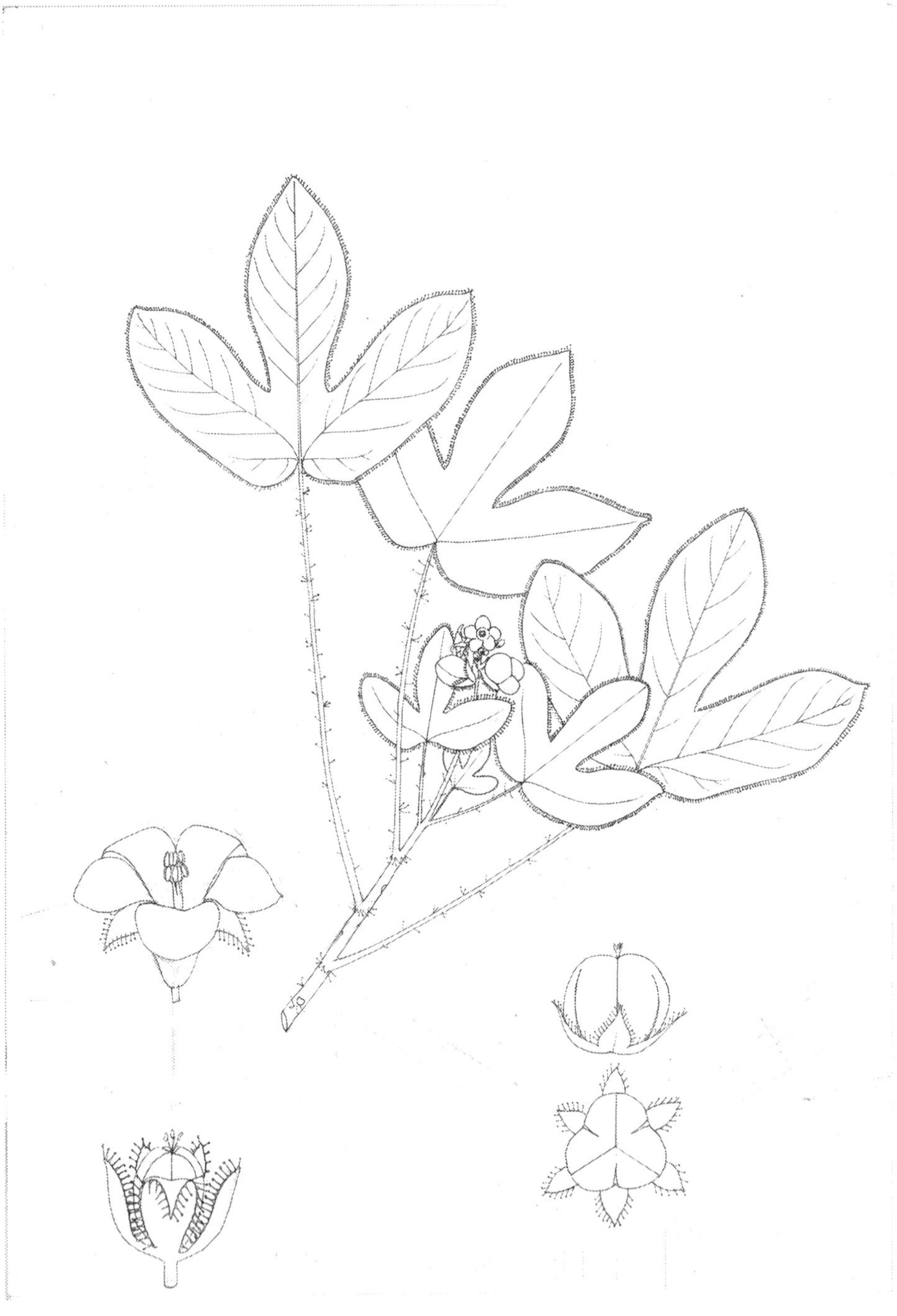

FIGURE 7.1 *Jatropha gossypifolia* L.

androecium comprises 10 stamens united at base. The ovary is 3-locular and the stigmas trifid. The capsules are 3-lobed, fleshy, about 2.5 cm long, of a characteristic dull green and contain 3 seeds which are oily and poisonous, brownish black, and up to 2 cm long.

Medicinal uses: Dysentery, cuts, diarrhea, fever, ophthalmia (India); ringworm, skin infection (Bangladesh); tooth infection (India, Nepal); thrush (China); syphilis (Bangladesh, Nepal); pneumonia (Nepal); wounds (Bangladesh; India; Nepal); herpes (Bangladesh, India)

Broad-spectrum antibacterial halo developed by polar extract: Methanol extract of seeds (50 µL of a 500 mg/mL solution in 8 mm well) inhibited the growth of *Escherichia coli*, *Pseudomonas aeruginosa*, and *Staphylococcus aureus* with inhibition zone diameters of 14, 25, and 25 mm, respectively (Rachana et al., 2012).

Moderate broad-spectrum antibacterial non-polar extract: Phorbol ester fraction of the seeds inhibited the growth of *Streptococcus pyogenes*, *Proteus mirabilis*, and *Pseudomonas putida* (Devappa et al., 2012).

Strong filamentous antifungal non-polar extract: Phorbol ester fraction of the seeds inhibited the growth of *Fusarium* sp. and *Aspergillus niger* (Devappa et al., 2012).

Strong antiviral (enveloped monopartite linear dimeric single-stranded (+)RNA) polar extract: Aqueous extract of stems inhibited the replication of the Human immunodeficiency virus type-1 strain IIIB in MT-4 cells with an IC_{50} value of 24 µg/mL and a selectivity index above 41.7 (Matsuse et al., 1988). Methanol extract of leaves inhibited the replication of the Human immunodeficiency virus type-1 (human T-cell lymphotropic virus (HTLV)-IIIB strain) in MT-4 cells with an IC_{50} value of 9 µg/mL and a selectivity index of 5.8 (Matsuse et al., 1988).

Viral enzyme inhibition by polar extract: Aqueous extract of leaves inhibited Human immunodeficiency virus reverse transcriptase with an IC_{50} value of 50 µg/mL (Matsuse et al., 1988).

Commentaries: (i) The seeds contain series of long-chain tigliane-type diterpenes, which activate protein kinase C in human cells by virtue of somewhat mimicking diacylglycerol hence triggering intracellular cascades resulting in tumor promotion (Roach et al., 2012). Consider that natural products modulating protein kinase C have the tendency to protect host cells against the Human immunodeficiency virus (Chowdhury et al., 1990; Warrilow et al., 2006). Protein kinase C inhibitors have protective effects against the Influenza virus (Root et al., 2000). (ii) The diterpene jatrophenone isolated from *Jatropha gossypifolia* L., a shrub traditionally used in Bangladesh (local name: *Beddha*; *Laljeol*) to treat dysentery, boils, carbuncles, and septiscaemia in India (local name: *Lal bharenda*), exhibited antibacterial activity against *Staphylococcus aureus* (Ravindranath et al., 2003).

REFERENCES

Chowdhury, M.I.H., Koyanagi, Y., Kobayashi, S., Hamamoto, Y., Yoshiyama, H., Yoshida, T. and Yamamoto, N., 1990. The phorbol ester TPA strongly inhibits HIV-1-induced syncytia formation but enhances virus production: possible involvement of protein kinase C pathway. *Virology*, *176*(1), pp. 126–132.

Devappa, R.K., Rajesh, S.K., Kumar, V., Makkar, H.P. and Becker, K., 2012. Activities of Jatropha curcas phorbol esters in various bioassays. *Ecotoxicology and Environmental Safety*, *78*, pp. 57–62.

Rachana, S., Tarun, A., Rinki, R., Neha, A. and Meghna, R., 2012. Comparative analysis of antibacterial activity of Jatropha curcas fruit parts. *Journal of Pharmaceutical and Biomedical Sciences (JPBMS)*, *15*(15), pp. 1–4.

Matsuse, I.T., Lim, Y.A., Hattori, M., Correa, M. and Gupta, M.P., 1998. A search for anti-viral properties in Panamanian medicinal plants.: The effects on HIV and its essential enzymes. *Journal of Ethnopharmacology*, *64*(1), pp. 15–22.

Ravindranath, N., Venkataiah, B., Ramesh, C., Jayaprakash, P. and Das, B., 2003. Jatrophenone, a novel macrocyclic bioactive diterpene from Jatropha gossypifolia. *Chemical and Pharmaceutical Bulletin*, *51*(7), pp. 870–871.

Roach, J.S., Devappa, R.K., Makkar, H.P. and Becker, K., 2012. Isolation, stability and bioactivity of Jatropha curcas phorbol esters. *Fitoterapia*, *83*(3), pp. 586–592.
Root, C.N., Wills, E.G., McNair, L.L. and Whittaker, G.R., 2000. Entry of influenza viruses into cells is inhibited by a highly specific protein kinase C inhibitor. *Journal of General Virology*, *81*(11), pp. 2697–2705.
Warrilow, D., Gardner, J., Darnell, G.A., Suhrbier, A. and Harrich, D., 2006. HIV type 1 inhibition by protein kinase C modulatory compounds. *AIDS Research & Human Retroviruses*, *22*(9), pp. 854–864.

7.4.2.12 *Macaranga peltata* (Roxb.) Müll. Arg.

Synonyms: *Mappa peltata* (Roxb.) Wight; *Osyris peltata* Roxb.;*Tanarius peltatus* (Roxb.) Kuntze

Common names: Parasol leaf tree; nainna bichi gaas (Bangladesh); manda, vattakanni (India); kenda (Sri Lanka)

Habitat: Roadside, vacant lands, villages

Distribution: India, Sri Lanka, Bangladesh, Andamans, and Thailand

Botanical observation: This fast-growing shrub can reach 10 m tall. The stems are green, articulated, smooth, hollowed (home to red ants) and with a reddish resin within. The leaves are simple, spiral, and stipulate. The stipules are ovate and up to 1 cm long. The petiole is slender, straight, and up to 19 cm long. The blade is soft, peltate, dull green, 13–25 × 12–21 cm, with about six pairs of secondary nerves, which are somewhat reddish, round at base, with prominent tertiary scalariform nervation below, marked with glands at margin, and acute to acuminate at apex. The inflorescences are cauliflorous, about 6 cm long racemes of minute flowers. The calyx comprises 3 sepals. The androecium includes 3 stamens. The ovary is minute and glandular. The fruits are about 5 mm in diameter, hairy, and trilobed.

Medicinal uses: Boils, piles (Bangladesh); skin infection (India)

Strong anticandidal hydrophilic simple phenolic: Bergenin (LogD=0.03 at pH 7.3; molecular weight=328.2 g/mol) inhibited the growth of *Candida albicans*, *Candida guillermondii*, and *Candida tropicalis* with MIC values below 10 μg/mL (Raj et al., 2012).

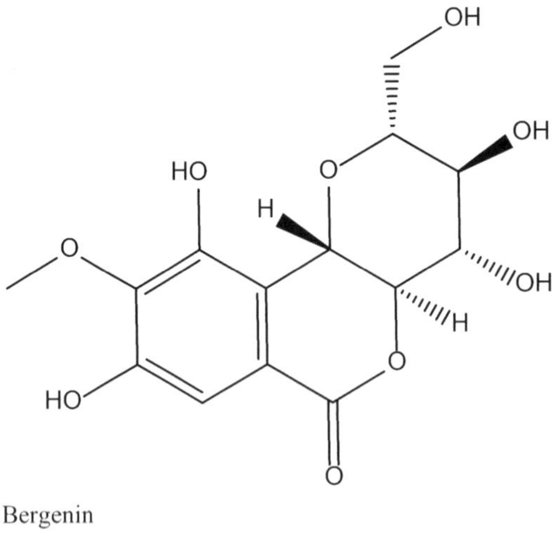

Bergenin

Strong broad-spectrum antifungal hydrophilic simple phenolic: Ellagic acid (LogD= −2 at pH 7.4; molecular mass = 302.1 g/mol) inhibited the growth of *Candida krusei* (ATCC 6258),

Candida parapsilosis (ATCC 20019), *Candida albicans* (ATCC 90028), and *Cryptococcus neoformans* (1640) with MIC values of 125, 7.8, 62.5 and 15.6 µg/mL, respectively (Rangkadilok et al., 2012).

Weak antifungal (filamentous) simple phenolic: Bergerin inhibited the growth of *Aspergillus flavus* and *Aspergillus nidulans* with MIC values of 625 and 321.5 µg/mL, respectively (Silva et al., 2009). In a subsequent study, bergerin inhibited the growth of *Trichophyton mentagrophytes*, *Epidermophyton floccosum*, *Trichophyton rubrum* (MTCC 296), *Aspergillus niger* (MTCC 1344), and *Botrydis cinerea* with MIC values of 125, 250, 500, 250, and 500 µg/mL, respectively (Raj et al., 2012).

Strong antiviral (enveloped monopartite linear dimeric single-stranded (+)RNA) hydrophlic simple phenolic: Ellagic acid inhibited the replication of the Human immunodeficiency virus type-1 (MN strain) with an EC_{50} value of 40 µg/mL and a selectivity index above 25 in C8166 cells (Piacente et al., 1996).

Strong antiviral (enveloped monopartite linear dimeric single-stranded (+)RNA) hydrophilc simple phenolic: Ellagic acid inhibited the replication of the Human immunodeficiency virus type-1 (strain NL4.3) with an EC_{50} value of 18.8 µg/mL and a selectivity index of 25 via inhibition of reverse transcriptase and protease (Nutan et al., 2013).

Strong antiviral (non-enveloped monopartite linear single-stranded (+)RNA) hydrophilic phenolic: Ellagic acid inhibited the replication of the Human rhinovirus type-2 (ATCC VR-1112AS/GP), Human rhinovirus type-3 (ATCC VR-1113), and Human rhinovirus type-4 (ATCC VR-1114AS/GP) in HeLa cells with the IC_{50} values of 41, 30, and 29 µg/mL, respectively, and selectivity indices of 2,4, 3.3, and 3.4, respectively (Park et al., 2014).

Strong antiviral (filamentous, monopartite, linear (−)RNA) hydrophilic phenolic: Ellagic acid inhibited the replication of the Ebola virus and the Marburg virus with IC_{50} values of 1.4 and 6.4 µM, respectively (Cui et al., 2018).

Strong antiviral (enveloped circular double stranded DNA) hydrophilic phenolic: Ellagic acid inhibited the secretion of the Hepatitis B virus e-antigen by HepG2 2.2.15 cells with an IC_{50} value of 0.07 µg/mL (Shin et al., 2005). Ellagic acid at a concentration of 200 µg/mL inhibited the secretion of the Hepatitis B virus surface antigen and e-antigen by HepG2 2.2.15 cells by 62.9% and 44.9%, respectively (Li et al., 2008).

In vivo antiviral (enveloped segmented linear single stranded (−)RNA) hydrophilic phenolic: Ellagic acid given intraperitoneally to BALB/c mice twice daily for 3 days evoked 10-35% against Influenza virus (rPR8-GFP strain) (Park et al., 2013).

Viral enzyme inhibition by phenolic: Ellagic acid inhibited the Hepatitis C virus protease with an IC_{50} value of 56.3 µM (Ajala et al., 2014).

Commentaries: (i) The red coloration of the sap within the stems is probably due to phenolics such as bergenin and its derivatives as well as ellagic acid (Ramaiah et al., 1979). (ii) Consider that natural products able to inhibit the replication of the Human immunodeficiency virus *in vitro* often to inhibit the replication of the Severe acute respiratory syndrome-acquired coronavirus 2 (Beg & Athar, 2020; Musarrat et al., 2020). Hence, bergerin could be examined for possible utility against COVID-19. It could also be part of a pool of molecules to be assessed against forthcoming pandemic coronaviruses. (iii) Is ellagic acid of value against COVID-19? (iv) *Macaranga tanarius* (L.) Muell.-Arg. is traditionally used to treat dysentery in Indonesia (local name: *Tutup ancur*) and to heal wounds in Malaysia (local name: *Mahang puteh*). In Malaysia, *Macaranga triloba* (Reinw.) Muell-Arg. is used to treat wounds. (v) Members of the genus *Macaranga* Thouars generate antibacterial prenylated flavanols such as macatrichocarpin A isolated from the leaves of *Macaranga trichocarpa* (Rchb. f. & Zoll.) Müll. Arg., which inhibited the growth of *Bacillus subtilis* with an MIC value of 26.5 µM (Fareza et al., 2014).

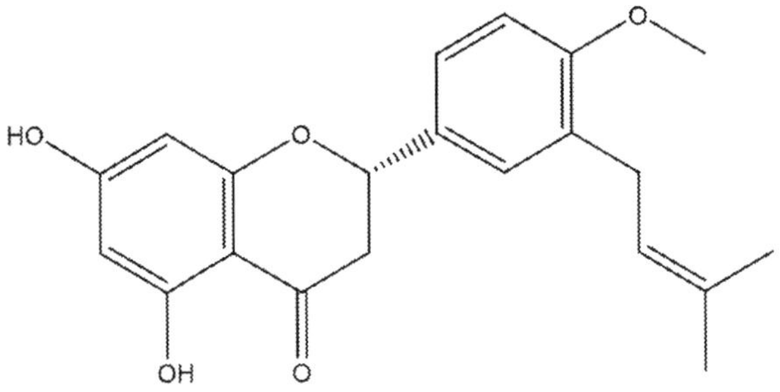

Macatrichocarpin A

REFERENCES

Ajala, O.S., Jukov, A. and Ma, C.M., 2014. Hepatitis C virus inhibitory hydrolysable tannins from the fruits of Terminalia chebula. *Fitoterapia*, *99*, pp. 117–123.

Beg, M.A. and Athar, F., 2020. Anti-HIV and Anti-HCV drugs are the putative inhibitors of RNA-dependent-RNA polymerase activity of NSP12 of the SARS CoV-2 (COVID-19). *Pharmacy & Pharmacology International Journal*, *8*(3), pp. 163–172.

Fareza, M.S., Syah, Y.M., Mujahidin, D., Juliawaty, L.D. and Kurniasih, I., 2014. Antibacterial flavanones and dihydrochalcones from Macaranga trichocarpa. *Zeitschrift für Naturforschung C*, *69*(9–10), pp. 375–380.

Li, J., Huang, H., Zhou, W., Feng, M. and Zhou, P., 2008. Anti-hepatitis B virus activities of Geranium carolinianum L. extracts and identification of the active components. *Biological and Pharmaceutical Bulletin*, *31*(4), pp. 743–747.

Musarrat, F., Chouljenko, V., Dahal, A., Nabi, R., Chouljenko, T., Jois, S.D. and Kousoulas, K.G., 2020. The anti-HIV Drug Nelfinavir Mesylate (Viracept) is a Potent Inhibitor of Cell Fusion Caused by the SARS-CoV-2 Spike (S) Glycoprotein Warranting further Evaluation as an Antiviral against COVID-19 infections. *Journal of Medical Virology*, *92*, pp. 2087–2095.

Nutan, M.M., Goel, T., Das, T., Malik, S., Suri, S., Rawat, A.K.S., Srivastava, S.K., Tuli, R., Malhotra, S. and Gupta, S.K., 2013. Ellagic acid & gallic acid from Lagerstroemia speciosa L. inhibit HIV-1 infection through inhibition of HIV-1 protease & reverse transcriptase activity. *The Indian Journal of Medical Research*, *137*(3), p. 540.

Park, S., Kim, J.I., Lee, I., Lee, S., Hwang, M.W., Bae, J.Y., Heo, J., Kim, D., Han, S.Z. and Park, M.S., 2013. Aronia melanocarpa and its components demonstrate antiviral activity against influenza viruses. *Biochemical and Biophysical Research Communications*, *440*(1), pp. 14–19.

Park, S.W., Kwon, M.J., Yoo, J.Y., Choi, H.J. and Ahn, Y.J., 2014. Antiviral activity and possible mode of action of ellagic acid identified in Lagerstroemia speciosa leaves toward human rhinoviruses. *BMC Complementary and Alternative Medicine*, *14*(1), p. 171.

Piacente, S., Pizza, C., De Tommasi, N. and Mahmood, N., 1996. Constituents of Ardisia japonica and their in vitro anti-HIV activity. *Journal of Natural Products*, *59*(6), pp. 565–569.

Raj, M.K., Duraipandiyan, V., Agustin, P. and Ignacimuthu, S., 2012. Antimicrobial activity of bergenin isolated from Peltophorum pterocarpum DC. flowers. *Asian Pacific Journal of Tropical Biomedicine*, *2*(2), pp. S901–S904.

Ramaiah, P.A., Row, L.R., Reddy, D.S., Anjaneyulu, A.S.R., Ward, R.S. and Pelter, A., 1979. Isolation and characterisation of bergenin derivatives from Macaranga peltata. *Journal of the Chemical Society, Perkin Transactions 1*, pp. 2313–2316.

Rangkadilok, N., Tongchusak, S., Boonhok, R., Chaiyaroj, S.C., Junyaprasert, V.B., Buajeeb, W., Akanimanee, J., Raksasuk, T., Suddhasthira, T. and Satayavivad, J., 2012. In vitro antifungal activities of longan (Dimocarpus longan Lour.) seed extract. *Fitoterapia*, *83*(3), pp. 545–553.

Shin, M.S., Kang, E.H. and Lee, Y.I., 2005. A flavonoid from medicinal plants blocks hepatitis B virus-e antigen secretion in HBV-infected hepatocytes. *Antiviral Research*, *67*(3), pp. 163–168.
Vinod, K.N., Ninge Gowda, K.N. and Sudhakar, R., 2010. Natural colorant from the bark of Macaranga peltata: Kinetic and adsorption studies on silk. *Coloration Technology*, *126*(1), pp. 48–53.

7.4.2.13 *Mallotus philippensis* (Lam.) Müll. Arg.

Synonyms: *Echinus philippinensis* Baill.; *Rottlera tinctoria* Roxb.

Common names: Kamala tree, monkey face tree; kuruar gaas, nainna bichi gaas ruda (Bangladesh); anada (Cambodia); bahupushpa, chenkolli, kamal, kampillaka, senduria (India); kapasan (Indonesia); kiz mun (Laos); rambai kuncing (Malaysia); hpawng-awn (Myanmar); royani (Nepal); tore (Papua New Guinea); tagusala (the Philippines); kham saet (Thailand)

Habitat: Open forests, waste lands, limestones

Distribution: West Himalayas, Sri Lanka, China, the Philippines, Taiwan, Papua New Guinea, Solomon Islands, and Australia.

Botanical observation: This bushy tree grows to a height of 9m. The stems, young leaves, and inflorescences are covered with scarlet glands. The leaves are simple, spiral, and stipulate. The petiole is 5 cm long and slender. The blade is ovate, trinerved, 5–20×1–5cm, glaucous below, covered with scarlet glands, acute at base, and acuminate at apex. The flowers are arranged in axillary racemes. The calyx is 5-lobed. The androecium comprises several stamens. The stigmas are characteristically plumose. The capsules are depressed, globose, 3 to 4-lobed, 6–7mm diameter, and covered with a dense and somewhat red powder.

Medicinal uses: Boils, herpes (Bangladesh); bronchitis, colds, cough, leprosy, post-partum (India); dysentery (Nepal); diarrhea, (Papua New Guinea, Nepal); wounds (Bangladesh, Papua New Guinea); ringworm (Bangladesh; India)

Broad-spectrum antibacterial mid-polar extract: Dichloromethane extract (1000 µg/mL in agar) abrogated the growth of *Bacillus cereus* var *mycoides* (ATCC 11778), *Bordetella bronchiseptica* (ATCC 4617), *Bacillus subtilis* (ATCC 6633), *Micrococcus luteus* (ATCC 9341), *Staphylococcus aureus* (ATCC 29737),), *Staphylococcus epidermidis* (ATCC 12228), and *Klebsiella pneumoniae* (ATCC 10031), and *Streptococcus faecalis* (MTCC 8043) (Kumar et al., 2006).

Antimycobacterial extract: Ethanol extract (dilution 1:40) inhibited the growth of *Mycobacterium tuberculosis* (H37Rv strain) (Grange & Davey, 1990).

Strong antimycobacterial lipophilic phloroglucinol: Rottlerin (LogD=5.8 at pH 7.4; molecular mass=516.5 g/mol) inhibited the growth of *Micobacterium tuberculosis* (H37Ra), *Mycobacterium tuberculosis* (BCG), and *Mycobacterium tuberculosis* (MS), with IC_{50} values of about 12.5, 12.5, and 50 µM, respectively, via inhibition of shikimate kinase (Pandey et al., 2016).

Strong filamentous antifungal amphiphilic prenylated chalcone: 1-(5,7-dihydroxy-2,2,6-trimethyl-2*H*-1-benzopyran-8-yl)-3-phenyl-2-propen-1-one isolated from the fresh whole uncrushed fruits inhibited the growth of *Cryptococcus neoformans* (PRL518), *Cryptococcus neoformans* (ATCC32045), and *Aspergillus fumigatus* with the IC_{50} values of 8, 4 and 16 µg/mL (Kulkarni et al., 2014).

Strong antiviral (enveloped monopartite linear dimeric single-stranded (+)RNA) lipophilic phloroglucinol: Rottlerin (LogD=5.8 at pH 7.4; molecular mass=516.5 g/mol) inhibited the replication of the Human immunodeficiency virus (NL4.3 strain) with an IC_{50} value of 2.2 µM in Jurkat cells via a mechanism involving, at least partially, protein kinase C_0 inhibition (López-Huertas, et al., 2011).

Commentaries: (i) Bergenin (see earlier) and 11-*O*-galloylbergenin occur in the wood. (ii) The red hairs covering the fruits haves been used to expel intestinal worms from a very remote period of time and used to dye silk in India. Greek and Middle Eastern physicians used these red hairs

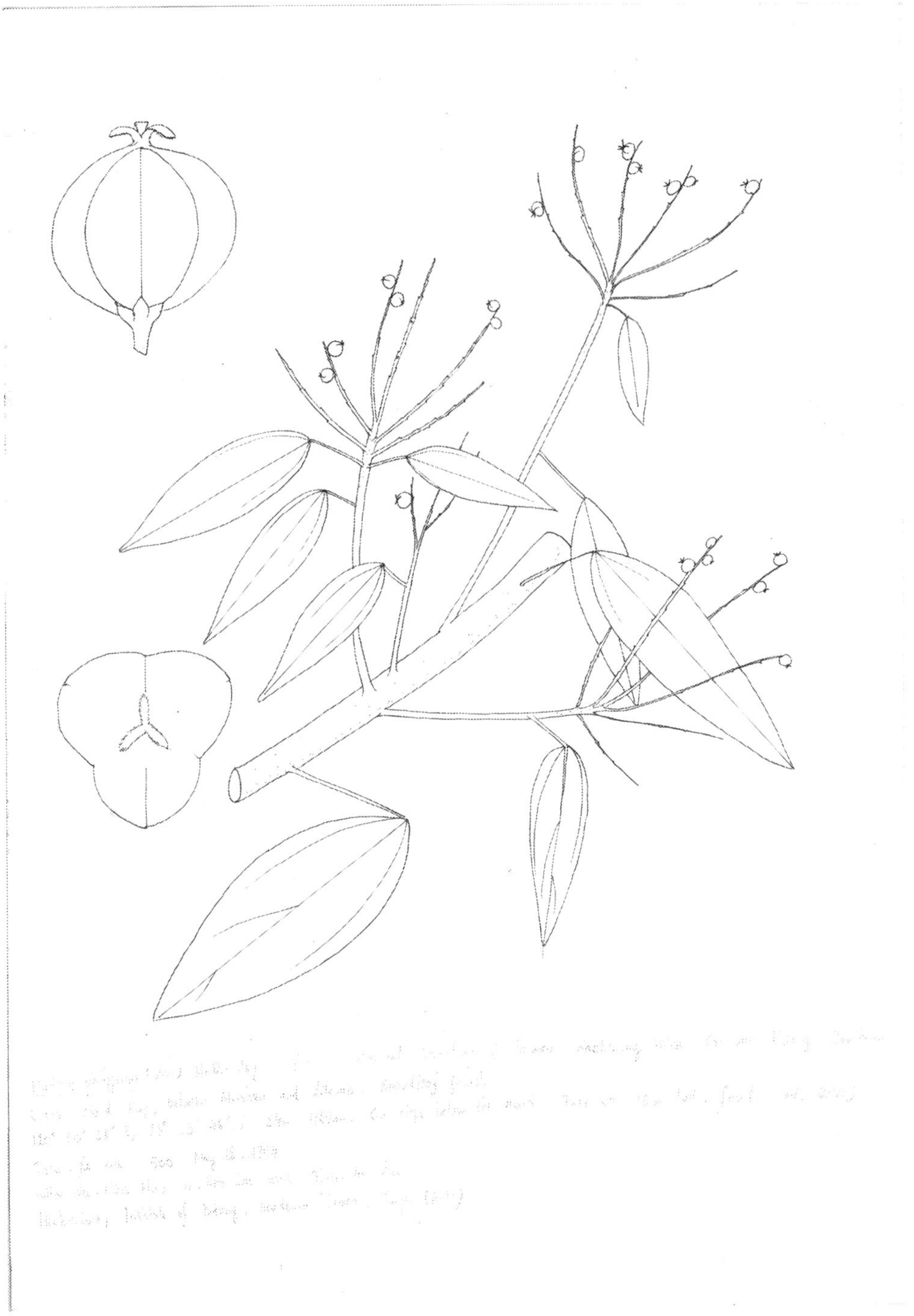

FIGURE 7.2 *Mallotus philippensis* (Lam.) Müll. Arg.

covering the capsules as an anthelminthic. (iii) Members of the genus *Mallotus* are not uncommonly used for the treatment or prevention of infectious diseases in Southeast Asia. In Malaysia, *Mallotus floribundus* (Bl.) Müll. Arg. (local name: *Maya maya)* is used to treat cholera, yaws, fever, and given at childbirth, *Mallotus macrostachyus* (Miq.) Müll. Arg. (local name: *balek angin*) is used to

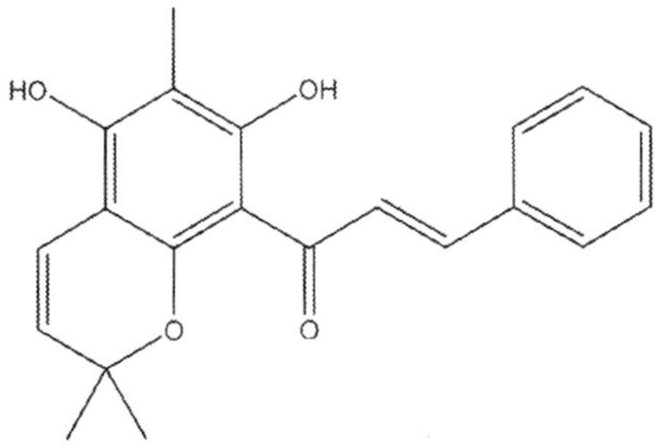

1-(5,7-dihydroxy-2,2,6-trimethyl-2*H*-1-benzopyran-8-yl)-3-phenyl-2-propen-1-one

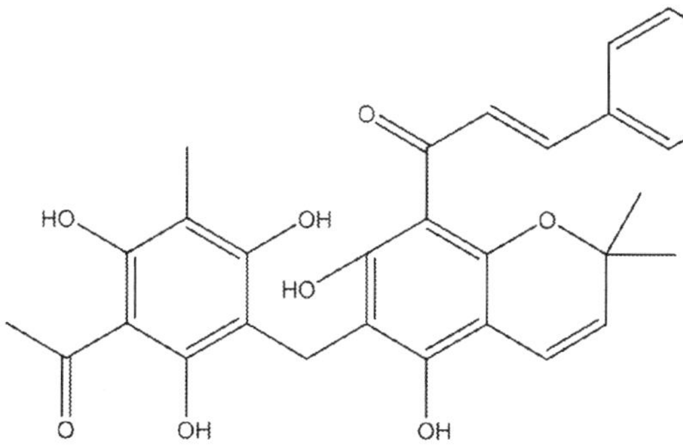

Rottlerin

treat boils, wounds, and fever, *Mallotus paniculatus* (Lam.) Müll. Arg. is used to heal wounds, to break fever, and given at childbirth, *Mallotus repandus* (Willd.) Müll. Arg. (local name: *Akar carek putih*) is used to treat colds. In Indonesia, *Mallotus blumeanus* Müll. Arg. (local name: *Calik angin*) is used at childbirth, whereas *Mallotus mollissimus* (Geiseler) Airy Shaw is used in Papua New Guinea (local names=*Di, Lossu*) to treat dysentery.

(iv) Consider that methanol extract of stems of *Mallotus japonicus* (Spreng.) Müll. Arg. at a concentration of 100 µg/mL inhibited RNA-dependent DNA polymerase and ribonuclease H activities of Human immunodeficiency virus type-1 reverse transcriptase by 94.4 and 13.7%, respectively (Min et al., 2001).

(v) Consider that natural products active against protein kinases C have the tendency to be antiviral. As seen previously anti-Human immunodeficiency virus principles are often active against coronaviruses, and one could examine the antiviral activity of rottlerin against COVID-19 and other pandemic coronaviruses. Here again, one can speak of reserve natural products to be tested in the case of new emerging viruses. Rottlerin is not toxic in humans and thus is an interesting candidate (?).

(vi) Rottlerin has apparently not been examined for antifungal properties. Being neuroactive, rottlerin might be able to inhibit bacterial and or fungal efflux pumps.

REFERENCES

Grange, J.M. and Davey, R.W., 1990. Detection of antituberculous activity in plant extracts. *Journal of Applied Bacteriology*, *68*(6), pp. 587–591.

Kulkarni, R.R., Tupe, S.G., Gample, S.P., Chandgude, M.G., Sarkar, D., Deshpande, M.V. and Joshi, S.P., 2014. Antifungal dimeric chalcone derivative kamalachalcone E from Mallotus philippinensis. *Natural Product Research*, *28*(4), pp. 245–250.

Kumar, V.P., Chauhan, N.S., Padh, H. and Rajani, M., 2006. Search for antibacterial and antifungal agents from selected Indian medicinal plants. *Journal of Ethnopharmacology*, *107*(2), pp. 182–188.
López-Huertas, M.R., Mateos, E., Díaz-Gil, G., Gómez-Esquer, F., del Cojo, M.S., Alcamí, J. and Coiras, M., 2011. Protein kinase Cθ is a specific target for inhibition of the HIV type 1 replication in CD4+ T lymphocytes. *Journal of Biological Chemistry*, *286*(31), pp. 27363–27377.
Min, B.S., Kim, Y.H., Tomiyama, M., Nakamura, N., Miyashiro, H., Otake, T. and Hattori, M., 2001. Inhibitory effects of Korean plants on HIV-1 activities. *Phytotherapy Research*, *15*(6), pp. 481–486.
Pandey, S., Chatterjee, A., Jaiswal, S., Kumar, S., Ramachandran, R. and Srivastava, K.K., 2016. Protein kinase C-δ inhibitor, Rottlerin inhibits growth and survival of mycobacteria exclusively through Shikimate kinase. *Biochemical and Biophysical Research Communications*, *478*(2), pp. 721–726.

7.4.2.14 *Pedilanthus tithymaloides* (L.) Poit.

Synonym: *Euphorbia tithymaloides* L.

Common names: Bird-cactus, devil's backbone, slipper plant, rang chita; patabahar (Bangladesh); dāo bǔ cǎo, yù dài gēn, (China); naagaphani, sij (India); tukdu (the Philippines); sa yaek (Thailand); thuốc giấu (Vietnam)

Habitat: Cultivated

Distribution: Tropical Asia and Pacific

Botanical observation: This fleshy, laticiferous shrub grows to about 1.5 m tall. The stems are terete, green, fleshy, zig-zag shaped, and yield after incision an acrid and white latex. The leaves are simple, sessile, alternate, and stipulate. The stipule is minute. The blade is lanceolate, fleshy, dull green, somewhat wavy at margin, the base attenuate, the apex obtuse, and with about 9 pairs of secondary nerves, which are inconspicuous. The cynthium is pink and about 1.5 cm long. The flowers are minute and much reduced. The capsules are trilobed, about 5 mm across. The plant is visited by olive-backed sunbirds.

Medicinal uses: Abscesses, sores (China); syphilis (Bangladesh; India); venereal diseases (India); wounds (Vietnam)

Moderate broad-spectrum polar extract: Ethanol extract of leaves inhibited the growth of *Staphylococcus aureus* (ATCC 25923), *Bacillus subtilis* (ATCC 6623), *Pseudomonas aeruginosa* (ATCC 15442), and *Escherichia coli* (ATCC 25922) with MIC values of 250, 125, 250, and 250 μg/mL, respectively (Vidotti et al., 2006).

Strong antimycobacterial mid-polar extract: Dichloromethane extract of latex inhibited the growth of *Mycobacterium tuberculosis* (H37Ra) with an MIC of 50 μg/mL (Mongkolvisut & Sutthivaiyakit, 2007).

Strong antimycobacterial jatrophane-type diterpene: 1α,13β,14 α- trihydroxy-3 β,7 β -dibenzoyloxy-9 β,15 β -diacetoxyjatropha-5,11 *E*-diene isolated from the latex inhibited the growth of *Mycobacterium tuberculosis* (H37Ra) with an MIC of 12.5 μM (Mongkolvisut & Sutthivaiyakit, 2007).

Strong antifungal (filamentous) coumarins: Pedilanthocoumarin A and B inhibited the conidial germination of *Magnaporthe oryzae* with MIC values of 5 and 2.5 μg/mL, respectively (Sandjo et al., 2012).

Commentaries: (i) The plant produces a fascinating array of jatrophane-type diterpenes, which inhibit P-glycoprotein efflux pump in HepG2 cells such as pedithithin D (Zhu et al., 2016). Consider that inhibitors of P-glycoprotein efflux pumps are not uncommonly able to inhibit bacterial efflux pumps and this is the case with verapamil (Tsuruo et al., 1981; Gupta et al., 2014). Hence, one could examine the antibiotic-potentiating effects of pedithithin D and other jatrophane-type diterpenes generated by this plant. (ii) One could feel that pedilanthocoumarin A and B would be better at home in the family Clusiaceae rather than Euphorbiaceae.

REFERENCES

Gupta, S., Cohen, K.A., Winglee, K., Maiga, M., Diarra, B. and Bishai, W.R., 2014. Efflux inhibition with verapamil potentiates bedaquiline in Mycobacterium tuberculosis. *Antimicrobial Agents and Chemotherapy*, *58*(1), pp. 574–576.

Mongkolvisut, W. and Sutthivaiyakit, S., 2007. Antimalarial and antituberculous poly-O-acylated jatrophane diterpenoids from Pedilanthus tithymaloides. *The Journal of Natural Products*, *70*(9), pp. 1434–1438.

Sandjo, L.P., Foster, A.J., Rheinheimer, J., Anke, H., Opatz, T. and Thines, E., 2012. Coumarin derivatives from Pedilanthus tithymaloides as inhibitors of conidial germination in Magnaporthe oryzae. *Tetrahedron Letters*, *53*(17), pp. 2153–2156.

Tsuruo, T., Iida, H., Tsukagoshi, S., and Sakurai, Y. 1981. Overcoming of vincristine resistance in P388 leukemia in vivo and in vitro through enhanced cytotoxicity of vincristine and vinblastine by verapamil. *Cancer Res*, *41*(1967), p. 72.

Vidotti, G.J., Zimmermann, A., Sarragiotto, M.H., Nakamura, C.V. and Dias Filho, B.P., 2006. Antimicrobial and phytochemical studies on Pedilanthus tithymaloides. *Fitoterapia*, *77*(1), pp. 43–46.

Zhu, J., Wang, R., Lou, L., Li, W., Tang, G., Bu, X. and Yin, S., 2016. Jatrophane diterpenoids as modulators of P-glycoprotein-dependent multidrug resistance (MDR): advances of structure–activity relationships and discovery of promising MDR reversal agents. *Journal of Medicinal Chemistry*, *59*(13), pp. 6353–6369.

7.4.2.15 *Ricinus communis* L.

Synonyms: *Ricinus africanus* Mill.; *Ricinus angulatus* Thunb.; *Ricinus armat*us Andrews; *Ricinus atropurpureus* Pax & K. Hoffm.; *Ricinus badius* Rchb.; *Ricinus borboniensis* Pax & K. Hoffm.; *Ricinus cambodgensis* Benary; *Ricinus digitatus* Noronha; *Ricinus europaeus* T. Nees; *Ricinus giganteus* Pax & K. Hoffm.; *Ricinus glaucus* Hoffmanns.; *Ricinus hybridus* Besser; *Ricinus inermis* Mill.; *Ricinus japonicus* Thunb.; *Ricinus krappa* Steud.; *Ricinus laevis* DC.; *Ricinus leucocarpus* Bertol.; *Ricinus lividus* Jacq., *Ricinus macrocarpus* Popova; *Ricinus macrophyllus* Bertol.; *Ricinus medius* J.F. Gmel.; *Ricinus megalospermus* Delile; *Ricinus messeniacus* Heldr.; *Ricinus metallicus* Pax & K. Hoffm.; *Ricinus microcarpus* Popova; *Ricinus minor* Mill.; *Ricinus nanus* Bald.; *Ricinus obermannii* Groenl.;*Ricinus peltatus* Noronha;*Ricinus perennis* Steud.; *Ricinus persicus* Popova; *Ricinus purpurascens* Bertol.;*Ricinus ruber* Miq.;*Ricinus rugosus* Mill.; *Ricinus rutilans* Müll. Arg.; *Ricinus sanguineus* Groenl.; Ricinus scaber Bertol. ex Moris; *Ricinus speciosus* Burm. f.; *Ricinus spectabilis* Blume; *Ricinus tunisensis* Desf.; *Ricinus undulatus* Besser; *Ricinus urens* Mill.; *Ricinus viridis* Willd.; *Ricinus vulgaris* Mill.; *Ricinus zanzibarinus* Popova

Common names: Castor bean, palma Christi, castor oil plant; arenda, reri, bherenda, gab-bherenda (Bangladesh); bi ma (China); jarak (Indonesia; Malaysia); amanda,arandi, am anakkam chedi, eramda (India); garchak farangi (Iran); pimaja (Korea); hungx saa (Laos); ander, andi (Nepal); randa (Pakistan); kaswel (Papua New Guinea); katawa tawa; tangan tangan (the Philippines); lahung daeng (Thailand); slung deng, ma puong, thau dau (Vietnam)

Habitat: Roadsides, villages, vacant lots of land

Distribution: Tropical Asia and Pacific

Botanical observation: This sinister shrub grows up to 3 m tall. The stems are terete, somewhat glaucous, and articulate. The leaves are simple, spiral, peltate, and stipulate. The petiole is 20 cm long, straight, and slender. The blade is palmate with 7–11 lobes and 7–11 nerves, which are straight and radiate from petiole insertion, serrate, up to about 50 cm diameter, glossy green to dark purplish brown above, and glaucous below. The inflorescences are terminal and about 30 cm long clusters. The 3–5 tepals are up to about 5mm long and broadly lanceolate. The 5 branched stamens are up to 8 mm long. The gynoecium consists of a 3 loculate, prickly, oblong, and about 4mm long ovary developing 3 bifid stigmas. The capsules are 3-lobed, prickly, up to about 2.5 cm in diameter, somewhat glaucous, and containing smooth, grey, brown, or black, somewhat tick-like seeds, which are up to 7 mm long.

Medicinal uses: Dysentery, syphilis (Bangladesh); boils (India); wounds (Bangladesh; India); sores (Bangladesh; China); fever (India; Nepal); bronchitis, leprosy (Pakistan); ringworm (Pakistan).

Moderate broad-spectrum antifungal mid-polar extract: Methanol: dichloromethane extract inhibited the growth of *Aspergillus parasiticus* (PP)RI 9153, *Aspergillus ochraceus* (PPRI 6816), *Aspergillus flavus* (PPRI 14636), *Fusarium graminearum* (PPRI 10340), *Fusarium verticilloides* (PPRI 10148), and *Fusarium oxysporum* (PPRI 10185) (Mongalo et al., 2018).

Strong antifungal (filamentous) lipophilic diterpene: Casbene (LogD = 8.5 at pH 7.4; molecular mass=272.4 g/mol) isolated from the germinating seeds (exposed to *Aspergillus niger*) inhibited the growth of *Aspergillus niger* with an MIC value of 10 μg/mL (Sitton & West, 1975).

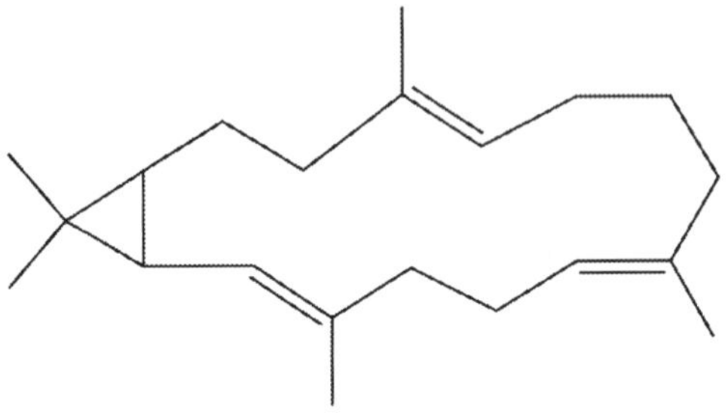

Casbene

Strong antiviral (enveloped monopartite linear dimeric single-stranded (+)RNA) polar extract: Ethanol extract inhibited the replication of the Human immunodeficiency virus type-1 with an MIC value of 10 μg/mL (Hamidi et al., 1996).

Strong antiviral (enveloped monopartite linear (−)RNA) polar extract: Ethanol extract inhibited the replication of the Vesicular stomatitis virus with an MIC value of 10 μg/mL (Hamidi et al., 1996).

Moderate antiviral (enveloped monopartite linear single-stranded (+) RNA) polar extract: Methanol extract of seeds inhibited the growth of the Sindbis virus with an IC_{50} value of 50 μg/mL (Mouhajir et al., 2001).

Moderate antiviral (enveloped monopartite linear dimeric single-stranded (+)RNA) polar extract: Methanol extract of seeds inhibited the growth of the Polio virus with an IC_{50} value of 50 μg/mL (Mouhajir et al., 2001).

Strong antiviral (enveloped linear monopartite double-stranded DNA virus) polar extract: Methanol extract of seeds inhibited the growth of the Herpes simplex virus with an IC_{50} value of 25 μg/mL (Mouhajir et al., 2001).

Commentaries: (i) This plant is one of the most commonly used plants in traditional Asian medical practice. (ii) The plant generates the pyridine alkaloids ricinine and *N*-demethylricinine (Kang et al., 1985), which could be examined for their antibacterial, antifungal, and antiviral effects. Consider that ricinine has central nervous effects (Ferraz et al., 1999) and as such could have antibiotic potentiating effects via bacterial efflux pump inhibition as discussed earlier. The plant generates the simple phenolic gentisic acid (Rana et al., 2012) and ellagic acid (Ribeiro et al., 2016). (iii) Casbene is an example of phytoalexin (Moesta & West, 1985).

REFERENCES

Ferraz, A.C., Angelucci, M.E.M., Da Costa, M.L., Batista, I.R., De Oliveira, B.H. and Da Cunha, C., 1999. Pharmacological evaluation of ricinine, a central nervous system stimulant isolated from Ricinus communis. *Pharmacology Biochemistry and Behavior, 63*(3), pp. 367–375.

Hamidi, J.A., Ismaili, N.H., Ahmadi, F.B. and Lajisi, N.H., 1996. Antiviral and cytotoxic activities of some plants used in Malaysian indigenous medicine. *Pertanika Journal of Tropical Agricultural Science, 19*(2/3), pp. 129–136.
Kang, S.S., Cordell, G.A., Soejarto, D.D. and Fong, H.H., 1985. Alkaloids and flavonoids from Ricinus communis. *Journal of Natural Products, 48*(1), pp. 155–156.
Moesta, P. and West, C.A., 1985. Casbene synthetase: Regulation of phytoalexin biosynthesis in Ricinus communis L. seedlings: purification of casbene synthetase and regulation of its biosynthesis during elicitation. *Archives of Biochemistry and Biophysics, 238*(1), pp. 325–333.
Mongalo, N.I., Dikhoba, P.M., Soyingbe, S.O. and Makhafola, T.J., 2018. Antifungal, anti-oxidant activity and cytotoxicity of South African medicinal plants against mycotoxigenic fungi. *Heliyon, 4*(11), p. e00973.
Mouhajir, F., Hudson, J.B., Rejdali, M. and Towers, G.H.N., 2001. Multiple antiviral activities of endemic medicinal plants used by Berber peoples of Morocco. *Pharmaceutical biology, 39*(5), pp. 364–374.
Rana, M., Dhamija, H., Prashar, B. and Sharma, S., 2012. Ricinus communis L. – A review. *International Journal of PharmTech Research, 4*(4), pp. 1706–1711.
Ribeiro, P.R., de Castro, R.D. and Fernandez, L.G., 2016. Chemical constituents of the oilseed crop Ricinus communis and their pharmacological activities: A review. *Industrial Crops and Products, 91*, pp. 358–376.
Sitton, D. and West, C.A., 1975. Casbene: An anti-fungal diterpene produced in cell-free extracts of Ricinus communis seedlings. *Phytochemistry, 14*(9), pp. 1921–1925.
Zarai, Z., Chobba, I.B., Mansour, R.B., Békir, A., Gharsallah, N. and Kadri, A., 2012. Essential oil of the leaves of Ricinus communis L.: In vitro cytotoxicity and antimicrobial properties. *Lipids in Health and Disease, 11*(1), p. 102.

7.4.2.16 *Tragia involucrata* L

Common names: Indian stinging nettle, canchorie root-plant; bichiti, sengal sing (Bangladesh); duralabha, kashagne, kodithuva, petari (India); welkahambiliya (Sri Lanka)

Habitat: Roadsides, abandoned lots of land, villages

Distribution: India, Sri Lanka, Bangladesh

Botanical observation: This erect herb grows to about 1 m tall. The stem is terete and hairy. The leaves are simple, spiral, and stipulate. The stipules are lanceolate, hairy, and up to about 5 mm long. The petiole grows up to 4 cm long. The blade is lanceolate 5–16 × 1.5–7.5 cm, hairy, serrate, glossy, with 3–7 pairs of conspicuous and sunken secondary nerves. The axillary racemes are 1.5–4 cm long. The 3 sepals are yellowish green. The androecium includes 3 stamens. The ovary is hairy and produces trifid styles. The capsules are 3-lobed, up to 1 cm long, and covered with white hairs.

Medicinal uses: Urinary tract infections (Bangladesh); bronchitis, fever, leprosy, wounds (India); skin infection (Bangladesh; India)

Antibacterial (Gram-negative) halo developed by polar extract: Acetone extract of leaves (20 mg/disc, 5 mm disc) inhibited the growth of *Escherichia coli* (NCIM 2065) with an inhibition zone diameter of 27.3 mm (Gobalakrishnan et al., 2013).

Commentaries: The plant has apparently not been examined for possible antiviral effects. Siddha healers in South India use the plant to treat COVID-19 (Sasikumar et al., 2020).

REFERENCES

Gobalakrishnan, R., Kulandaivelu, M., Bhuvaneswari, R., Kandavel, D. and Kannan, L., 2013. Screening of wild plant species for antibacterial activity and phytochemical analysis of Tragia involucrata L. *Journal of Pharmaceutical Analysis, 3*(6), pp. 460–465.
Sasikumar, R., Priya, S.D. and Jeganathan, C., 2020. A Case Study on Domestics Spread Of Severe acute respiratory syndrome-acquired coronavirus 2 Pandemic in India. *International Journal of Advanced Science and Technology, 29*(7), pp. 2570–2574.

7.4.2.17 *Trigonostemon reidioides* (Kurz) Craib

Synonyms: *Baliospermum reidioides* Kurz; *Trigonostemon hybridus* Gagnep.; *Trigonostemon rubescens* Gagnep.

Common names: Naat kham (Thailand); tin tranh (Vietnam)

Habitat: Forests

Distribution: Myanmar, Cambodia, Laos, Vietnam, and Thailand

Botanical observation: This graceful herb grows up to about 1.5 m tall. The stems are terete, hairy, and reddish at apex. The leaves are simple, alternate, and stipulate. The stipules are minute and linear. The petiole is up to about 2.5 cm long. The blade is oblong to oblanceolate, 3.5–15×0.5–6.3cm, round to acute and with a pair of glands at base, acute to acuminate at apex, hairy below, and marked with 2–8 pairs of secondary nerves. The panicles are terminal or axillary, lax, and grow up to 30cm long. The 5 sepals are up to 6 mm long, hairy, and oblong. The 5 petals are pure white, red, or dark purple, up to 6mm long, and obovate or oblong. An orangish disc is present. The androecium consists of 3 stamens. The ovary is trilobed. The capsules are 3-lobed, hairy, about 1 cm across, green, and containing a few seeds, which are about 5 mm across.

Medicinal uses: Abscesses, dysentery (Thailand)

Strong antimycobacterial daphnane-type diterpene: Rediocide A, rediocide B, rediocide C, rediocide F, and rediocide G isolated from the roots inhibited the growth of *Mycobacterium tuerculosis* (H37Ra) with MIC values of 15.7, 7.8, 3.8, 3.9, and 3.8 μM, respectively (Kaemchantuek et al., 2017).

Commentaries: (i) The plant generates a series of carboline alkaloids (Hu et al., 2009). (ii) *Trigonostemon longifolius* Baill. is used in the Andaman Islands to treat fever. (iii) Members of the genus *Trigonostemon* Bl. are not uncommonly able to inhibit the replication of viruses. Ethanol extract of stem of *Trigonostema xyphophylloides* (Croizat) L.K. Dai & T.L. Wu inhibited the replication of Human immunodeficiency virus type-1 at 10 μg/mL via inhibition the virus entry into host cells (Park et al., 2009). The daphnane-type diterpenes trigolin C, trigolin G, trigochinin A, trigonothyrin F, isolated from the stems of *Trigonostemon lii* Y.T Chang inhibited the replication of Human immunodeficiency virus type-1 with EC_{50} values of 2, 9.1, 11.4, and 9 μg/mL and selectivity indices of 26.4, >21.8, 9.3, and 9.5, respectively (Li et al., 2013). The daphnane diterpene trigonothyrins F isolated from *Trigonostemon thyrsoideus* Stapf inhibited the replication of the Human immunodeficiency virus type-1 with an IC_{50} value of 0.1 μg/mL and a selectivity index of 75.1 (Zhang et al., 2010).

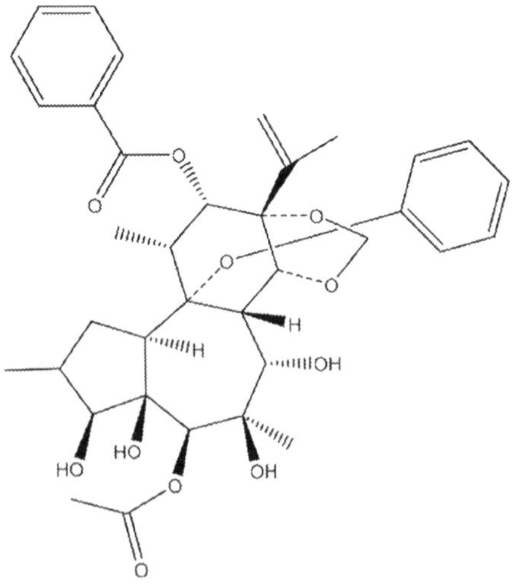

Trigonothyrin F

The diterpene 12-*O*-tetradecanoylphorbol 13-acetate inhibited the replication of the Chikungunya virus and the Sindbis virus as seen earlier (Bourjot et al., 2012).The daphnane diterpenes trigocherrierin A and trigocherriolide E isolated from *Trigonostemon cherrieri* J.M. Veillon inhibited Chikungunya virus -induced cell death with EC_{50} values of 0.6 and 0.7 μM, respectively, in Vero cells with the selectivity indices of 71.7 and 9.4 (Bourjot et al., 2014). The tigliane diterpene trigowiin A inhibited the replication of the Chikungunya virus (strain 899) with an EC_{50} of 43.2 μM and a selectivity index above 2.3 (Bourjot et al., 2014).

REFERENCES

Li, S.F., Zhang, Y., Huang, N., Zheng, Y.T., Di, Y.T., Li, S.L., Cheng, Y.Y., He, H.P. and Hao, X.J., 2013. Daphnane diterpenoids from the stems of Trigonostemon lii and their anti-HIV-1 activity. *Phytochemistry*, *93*, pp. 216–221.

Bourjot, M., Leyssen, P., Neyts, J., Dumontet, V. and Litaudon, M., 2014. Trigocherrierin A, a potent inhibitor of chikungunya virus replication. *Molecules*, *19*(3), pp. 3617–3627.

Bourjot M, Delang L, Nguyen VH, et al. 2012. Prostratin and 12-Otetradecanoylphorbol 13-acetate are potent and selective inhibitors of chikungunya virus replication. *Journal of Natural Products*, *75*, pp. 2183–2187.

Hu, X.J., Di, Y.T., Wang, Y.H., Kong, L.Y., Gao, S., Li, C.S., Liu, H.Y., He, H., Ding, J., Xie, H. and Hao, X., 2009. Carboline alkaloids from Trigonostemon lii. *Planta Medica*, *75*(10), pp. 1157–1161.

Kaemchantuek, P., Chokchaisiri, R., Prabpai, S., Kongsaeree, P., Chunglok, W., Utaipan, T., Chamulitrat, W. and Suksamrarn, A., 2017. Terpenoids with potent antimycobacterial activity against Mycobacterium tuberculosis from Trigonostemon reidioides roots. *Tetrahedron*, *73*(12), pp. 1594–1601.

Park, I.W., Han, C., Song, X., Green, L.A., Wang, T., Liu, Y., Cen, C., Song, X., Yang, B., Chen, G. and He, J.J., 2009. Inhibition of HIV-1 entry by extracts derived from traditional Chinese medicinal herbal plants. *BMC Complementary and Alternative Medicine*, *9*(1), p. 29.

Zhang, L., Luo, R.H., Wang, F., Dong, Z.J., Yang, L.M., Zheng, Y.T. and Liu, J.K., 2010. Daphnane diterpenoids isolated from Trigonostemon thyrsoideum as HIV-1 antivirals. *Phytochemistry*, *71*(16), pp. 1879–1883.

7.4.3 Family Hypericaceae A. L de Jussieu (1789)

The family Hypericaceae consists of 9 genera and 550 species of resinous trees, shrubs, and herbs. The leaves are simple, opposite, alternate, whorled or verticillate, and exstipulate. The blade is often dotted with tiny glands. The inflorescence is an axillary cyme or solitary. The flowers are actinomorphic. The calyx comprises 4–5 sepals. The corolla consists of 3–5 free petals. The androecium is made of numerous showy stamens.The gynoecium is made of 2–5 carpels united to form a multilocular ovary containing numerous ovules growing from on parietal or axile placentas. The fruits are septicidal capsules. Members of the family produce a bewildering array of antimicrobial xanthones, phloroglucinols, and benzophenones.

7.4.3.1 *Cratoxylum arborescens* Bl.

Synonyms: *Hypericum arborescens* Vahl; *Hypericum coccineum* Wall, *Cratoxylum cuneatum* Miq.

Common names: Geronggang, serungan, dat (Malaysia)

Habitat: Forests

Distribution: Myanmar, Thailand, Malaysia, Indonesia

Botanical observation: This tree grows to about 45 m tall. The bole is straight and the bark is grey, fissured, and yields a reddish resin upon incision. The stem is terete and smooth. The leaves are simple, opposite, and exstipulate. The petiole is up to 1 cm long. The blade is elliptic to obovate, 5–16×2–6cm, with numerous slender pairs of secondary nerves, somewhat glossy, coriaceous, cuneate at base, and acute or cuspidate at apex. The inflorescences are terminal panicles. The calyx includes 5 sepals, which are obovate to spathulate, and up to about 6 mm long. The 5 petals are burgundy red to whitish, obovate, inconspicuously nerved, and up to about 7mm long. The androecium comprises numerous anthers organized in fascicles, which are up to about 5 mm long. The ovary is minute. The capsules are bullet-shaped, seated on persistent calyx, up to about 1 cm long, dehiscent, and contain numerous tiny seeds.

Medicinal uses: Apparently none for the treatment of microbial infections

Broad-spectrum antibacterial halo developed by amphiphilic xanthone: α-Mangostin (LogD=4.2 at pH 7.4; molecular mass=410.4g/mol) isolated from the stem bark (10 μg/6 mm disc) inhibited the growth of *Bacillus subtilis*, *Bacillus cereus*, *Salmonella typhimurium*, and *Staphylococcus aureus* with inhibition zone diameters of 16, 20, 17, and 20mm, respectively (Yahayu et al., 2013). β-Mangostin (LogD=4.6 at pH 7.4; molecular mass=410.4g/mol) isolated from the stem bark (6mm diameter disc, 10 μg/disc) inhibited the growth of *Bacillus subtilis*, *Bacillus cereus*, *Salmonella typhimurium*, and *Staphylococcus aureus* with inhibition zone diameters of 7, 9, 11, and 11mm, respectively (Yahayu et al., 2013).

Moderate broad-spectrum amphiphilic lupane-type antibacterial triterpene: Betulinic acid inhibited the growth of *Enterobacter aerogenes* EA294 and *Klebsiella pneumonia* K2 with MIC values of 256 μg/mL (Tankeo et al., 2016).

Strong antiviral (enveloped monopartite linear dimeric single stranded (+)RNA) lupane-type triterpenes: Betulinic acid (LogD=5.1 at pH 7.4; molecular mass=456.7 g/mol) and lup-20(29)-ene-3β,30-diol isolated from leaves and twigs inhibited Human immunodeficiency virus type-1 syncytium formation with EC_{50} below 3.9 and 6.4 μg/mL, respectively (Reutrakul et al., 2006).

Viral enzyme inhibition by lupane-type triterpenes: Betulinic acid and lup-20(29)-ene-3β,30-diol inhibited Human immunodeficiency virus type-1 reverse transcriptase with an IC_{50} of 10.8 and 14 μg/mL, respectively (Reutrakul et al., 2006).

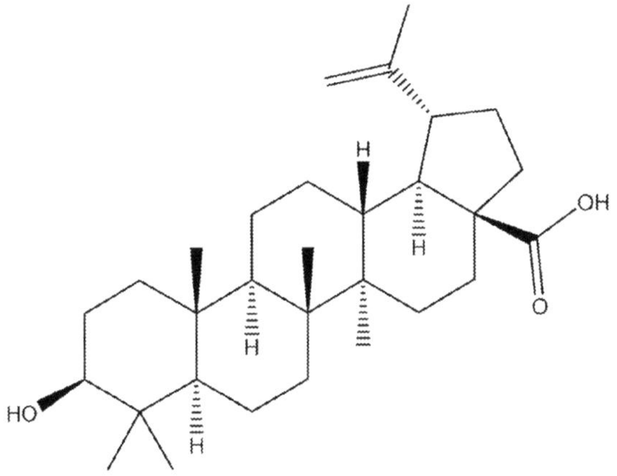

Betulinic acid

Commentary: Triterpenes like betulinic acid are common in medicinal plants. Although betulinic acid could have anti-Human immunodeficiency property by itself, if the compound isolated is not absolutely pure there is a possibility of having traces of very active principles and this is not uncommon especially with natural products isolated from marine organisms (Marc Butler, personal communication). In addition, natural products with chiral centers exist in racemic pairs and one of the conformers may be active or both compounds might act synergistically.

REFERENCES

Reutrakul, V., Chanakul, W., Pohmakotr, M., Jaipetch, T., Yoosook, C., Kasisit, J., Napaswat, C., Santisuk, T., Prabpai, S., Kongsaeree, P. and Tuchinda, P., 2006. Anti-HIV-1 constituents from leaves and twigs of Cratoxylum arborescens. *Planta Medica*, *72*(15), pp. 1433–1435.

Tankeo, S.B., Damen, F., Sandjo, L.P., Celik, I., Tane, P. and Kuete, V., 2016. Antibacterial activities of the methanol extracts, fractions and compounds from Harungana madagascariensis Lam. ex Poir. (Hypericaceae). *Journal of Ethnopharmacology*, *190*, pp. 100–105.
Yahayu, M.A., Rahmani, M., Hashim, N.M., Ee, G.C.L., Sukari, M.A. and Akim, A.M., 2013. Cytotoxic and Antimicrobial Xanthones from Cratoxylum arborescens (Guttiferae). *Malaysian Journal of Science*, *32*(1), pp. 53–60.

7.4.3.2 *Cratoxylum formosum* (Jack) Dyer

Synonym: *Elodes formosa* Jack

Common names: Yue nan huang niu mu (China); kemutul, mulun (Indonesia); tiou-tiou (Laos); derum, geronggang biabas (Malaysia); bamachet (Myanmar); salinggogon (the Philippines); tio khao, tiu khon tri (Thailand); thành ngạnh dẹp, thành ngạnh (Vietnam)

Habitat: Forests

Distribution: Cambodia, Laos, Myanmar, Thailand, Vietnam, Indonesia, Malaysia, the Philippines

Botanical observation: This tree grows to about 6 m tall. The bole is thorny at base and the bark is flaky. The stem is angular and smooth. The leaves are simple, opposite, and exstipulate. The petiole is up to 7 mm long. The blade is elliptic to oblong, 4–10×2–4cm, marked with 8–10 pairs of secondary nerves, dotted below, somewhat glossy, coriaceous, round at apex, and acute to obtuse at apex. The inflorescences are terminal cymes of showy flowers. The calyx includes 5 sepals which are elliptic, and up to about 6 mm long. The 5 petals are obovate, about 1.5 long and notched at apex. The androecium comprises numerous stamens organized in fascicles. The ovary is conical and about 4 mm long with a 3 mm long slender style. The capsules are bullet shaped, up to about 1.5 cm long, dehiscent, and contain numerous tiny seeds.

Medicinal use: Post-partum (Myanmar)

Moderate antibacterial (Gram-positive) polar extract: Methanol extract of leaves inhibited the growth of *Escherichia coli* (ATCC 25922), *Pseudomonas aeruginosa* (ATCC 9027), *Bacillus cereus* (ATCC 21768), *Bacilus subtilis* (ATCC 6633), and *Staphylococcus aureus* (ATCC 6538) with MIC/MBC values of 1000/2000, 2000/>2000, 125/125, 125/250, and 125/250 µg/mL, respectively (Vu et al., 2015).

Strong broad-spectrum antibacterial xanthones: Xanthone V1 from the roots inhibited *Pseudomonas aeruginosa*, *Bacillus subtilis*, *Staphylococcus aureus* (T1STR517), *Streptococcus faecalis*, and *Salmonella typhi* with the MIC values of 9.3, 1.1, 1.1, 1.1, and 1.1 µg/mL, respectively (Boonsri et al., 2006). Gerontoxanthone I (LogD=3.8 at pH 7.4; molecular mass=396.4 g/mol) from the roots inhibited *Bacillus subtilis*, *Staphylococcus aureus*, *Streptococcus faecalis*, and *Salmonella typhi* with the MIC values of 2.3, 1.1, 4.6, and 1.1 µg/mL (Boonsri et al., 2006): The prenylated xanthone 3-hydroxyblancoxanthone (also known as macluraxanthone; LogD=3.8 at pH 7.4; molecular mass=394.4 g/mol) isolated from the stem bark inhibited the growth of *Micrococcus luteus*, *Bacillus cereus*, *Bacillus subtilis*, *Staphylococcus aureus*, *Staphylococcus epidermidis*, *Escherichia coli*, and *Salmonella typhimurium* with MIC values of 4, 4, 8, 4, 32, 64, and 128 µg/mL, respectively (Raksat et al., 2014).

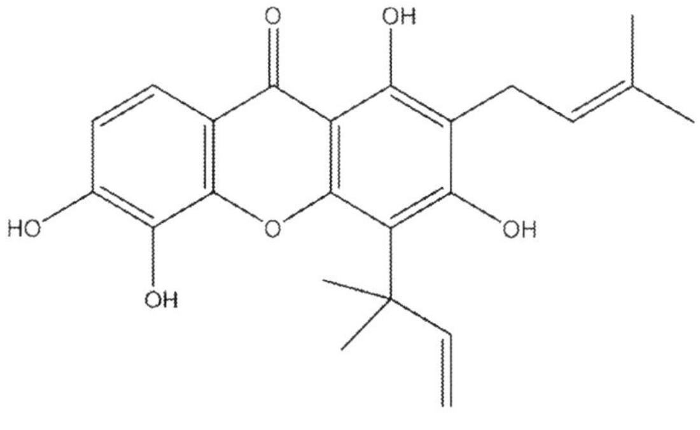

Gerontoxanthone I

REFERENCES

Boonsri, S., Karalai, C., Ponglimanont, C., Kanjana-Opas, A. and Chantrapromma, K., 2006. Antibacterial and cytotoxic xanthones from the roots of Cratoxylum formosum. *Phytochemistry*, *67*(7), pp. 723–727.

Raksat, A., Laphookhieo, S., Cheenpracha, S., Ritthiwigrom, T. and Maneerat, W., 2014. Antibacterial compounds from the roots of Cratoxylum formosum spp. pruniflorum. *Natural Product Communications*, *9*(10), p. 1934578X1400901020.

Vu, T.T., Kim, H., Tran, V.K., Le Dang, Q., Nguyen, H.T., Kim, H., Kim, I.S., Choi, G.J. and Kim, J.C., 2015. In vitro antibacterial activity of selected medicinal plants traditionally used in Vietnam against human pathogenic bacteria. *BMC Complementary and Alternative Medicine*, *16*(1), p. 32.

7.4.3.3 *Cratoxylum cochinchinense* (Lour.) Blume

Synonyms: *Ancistrolobus ligustrinus* Spach; *Cratoxylum biflorum* (Lam.) Turcz.; *Cratoxylum chinense* Merr.; *Cratoxylum ligustrinum* (Spach) Blume; *Cratoxylum petiolatum* Blume; *Cratoxylum polyanthum* Korth.; *Elodes chinensis* (Retz.) Hance; *Hypericum biflorum* Lam.; *Hypericum chinense* Retz.; *Hypericum cochinchinense* Lour.; *Oxycarpus cochinchinensis* Lour.; *Stalagmites erosipetalum* Miq.

Common names: Huáng níu jiăao, pun pa, huáng níu, zhu chi dou (China); geronggang bogoi, mampat (Malaysia)

Habitat: Disturbed forests

Distribution: Myanmar, Vietnam, Thailand, China, Malaysia, Indonesia, the Philippines

Botanical distribution: This tree grows to about 20 m tall. The bole is thorny at base and the bark is smooth, lenticelled, and greyish brown. The stem is angular and somewhat pinkish at apex. The leaves are simple, opposite, and exstipulate. The petiole is up to 3 mm long. The blade is elliptic to oblong, 3–10.5×1–4 cm, with dark glands below, marked with 8–12 pairs of secondary nerves, cuneate to obtuse at base and acute at apex. The inflorescences are terminal or axillary cymes of showy flowers. The calyx includes 5 sepals, which are oblong, with dark glands, and up to about 7 mm long. The 5 petals are up to about 1 cm long, obovate, deep crimson, and round at apex. The androecium comprises numerous stamens organized in fascicles. The ovary is conical and about 3 mm long with a 2 mm long style. The capsules are bullet-shaped, brown, up to about 1.2 cm long, dehiscent, and contain numerous tiny seeds.

Medicinal uses: External injuries, abscesses, fever, cold (China); skin diseases (Malaysia); fever, cough, diarrhea (Vietnam)

Strong antibacterial (Gram-positive) xanthones: Isocudraniaxanthone B and norathyriol from the roots inhibited *Staphylococcus aureus* (ATCC 25923) with an MIC of 16 and 64 µg/mL, respectively (Mahabusarakam et al., 2008). Isocudraniaxanthone B and norathyriol inhibited methicillin-resistant *Staphylococcus aureus* (SK1) with an MIC of 16 and 64 µg/mL, respectively (Mahabusarakam et al., 2008).

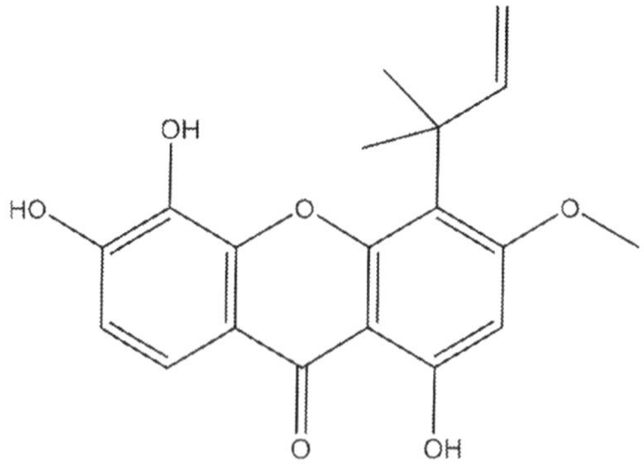

Isocudraniaxanthone B

Strong antibacterial (Gram-negative) lipophilic xanthone: The long-chain prenylated xanthone cochinchinone A (LogD=6.9 at pH 7.4; molecular mass=448.5 g/mol) isolated from the resin inhibited the growth of *Bacillus subtilis*, *Staphylococcus aureus*, methicillin-resistant *Staphylococcus aureus*, vancomycin-resistant *Enterococcus*, and *Pseudomonas aeruginosa* with MIC values of 150, 150, 150, 9.3, 150, and 4.7 μg/mL, respectively (Boonnak et al., 2009). Cochinchinone A evoked bacterial damage in *Pseudomonas aeruginosa* (Boonnak et al., 2009).

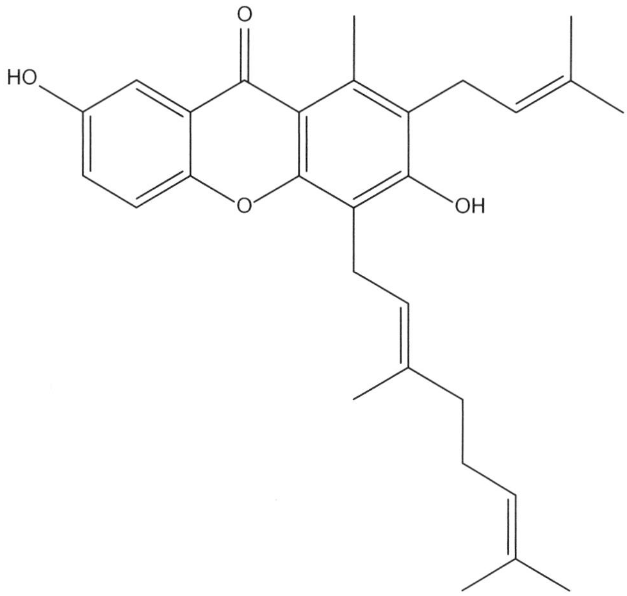

Cochinchinone A

Moderate anticandidal xanthone: Cochinchinone A inhibited the growth of *Candida albicans* with an MIC value of 75 μg/mL (Boonnak et al., 2009).

Commentary: Consider that the presence of long prenylated groups on natural products increases their lipophilicity and thus the ability to penetrate the phospholipid bilayer membrane of bacteria and fungi (Araya-Cloutier et al., 2018). Further, the presence of long prenylated groups increases the affinity of natural products to hydrophobic surfaces of enzymantic pockets with Van der Waals interactions. Cochinchinone A has the interesting ability to specifically inhibit the growth of *Pseudomonas aeruginosa* in the experiment described by Boonnak et al. (2009). Substances targeting the outer membrane of Gram-negative bacteria do not see their antibacterial activities decreased by efflux pumps. Thus, the development of the outer-membrane-targeting principle from medicinal plants as agents to treat multidrug-resistant bacteria could be a mean to curve bacterial resistance. The challenge is to find targets specific to bacterial outer membrane.

REFERENCES

Araya-Cloutier, C., Vincken, J.P., van de Schans, M.G., Hageman, J., Schaftenaar, G., den Besten, H.M. and Gruppen, H., 2018. QSAR-based molecular signatures of prenylated (iso) flavonoids underlying antimicrobial potency against and membrane-disruption in Gram positive and Gram negative bacteria. *Scientific Reports*, *8*(1), pp. 1–14.

Boonnak, N., Karalai, C., Chantrapromma, S., Ponglimanont, C., Fun, H.K., Kanjana-Opas, A., Chantrapromma, K. and Kato, S., 2009. Anti-Pseudomonas aeruginosa xanthones from the resin and green fruits of Cratoxylum cochinchinense. *Tetrahedron*, *65*(15), pp. 3003–3013.

Mahabusarakam, W., Rattanaburi, S., Phongpaichit, S. and Kanjana-Opas, A., 2008. Antibacterial and cytotoxic xanthones from Cratoxylum cochinchinense. *Phytochemistry Letters*, *1*(4), pp. 211–214.

7.4.3.4 *Hypericum acmosepalum* N. Robson

Common names: Jian e jin si tao, xiang ge da (China)

Habitat: Disturbed forests, roadsides, and riverbanks

Distribution: China

Botanical observation: This shrub grows about 2 m tall. The stem is quadrangular. The leaves are simple, opposite, sessile, and exstipulate. The blade is elliptic to oblong, 1.8– 6×0.5–1.5 cm, glaucous and gland dotted below, marked with 8–12 pairs of secondary nerves, cuneate at base, and acute to obtuse at apex. The inflorescences are terminal cymes of showy flowers. The calyx includes 5 sepals, which are obovate, dotted with glands, and up to about 1 cm long. The 5 petals are obovate, up to about 1 cm long, deep yellow, round at apex, and up to 2.5 cm long. The androecium comprises numerous up to 1.8 cm long stamens organized in fascicles. The ovary is conical and about 4.5 mm long. The capsules are ovoid, red, up to about 1.5cm long, dehiscent, and contain numerous tiny dark-orange seeds.

Medicinal use: Cough (China)

Strong antibacterial (Gram-positive) amphiphilic phloroglucinol: Hypercalin B (LogD=3.5 at pH 7.4; molecular mass=518.6 g/mol) inhibited the growth of multidrug-resistant strains of *Staphylococcus aureus* with MIC values ranging from 0.5 to 128 µg/mL (Osman et al., 2010). In a subsequent study, the same first author reports that hypercalin B inhibited the growth of *Staphylococcus aureus* (11998, expresses NorA multidrug-resistance efflux pump), methicillin-resistant *Staphylococcus aureus* type 15, methicillin-resistant *Staphylococcus aureus* type 16, *Staphylococcus aureus* (RN4220, methicillin-resistant *Staphylococcus aureus* macrolide efflux pump), *Staphylococcus aureus* (XU212, possess Tet(K) efflux pump), and *Staphylococcus aureus* (ATCC 25923) (Osman et al., 2012).

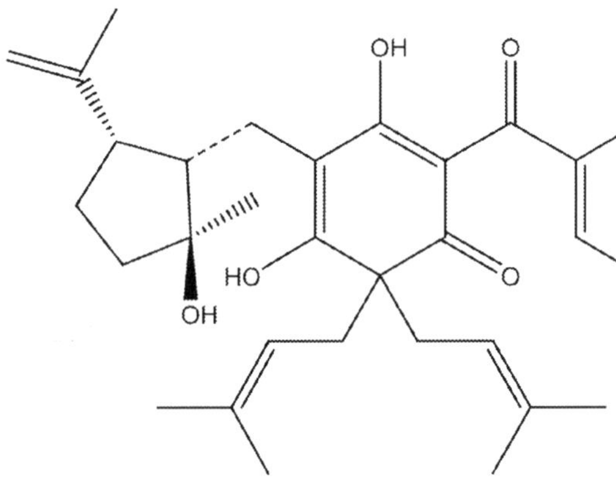

Hypercalin B

Strong antibacterial (Gram-positive) benzopyrone: Hyperenone A (molecular mass=271.1 g/mol) inhibited the growth of multidrug-resistant strains of *Staphylococcus aureus* with MIC values ranging from 2 to 128 µg/mL (Osman et al., 2010). In a subsequent study, the same group reports that hyperenone A inhibited the growth of *Staphylococcus aureus* (11998, expresses NorA multidrug-resistance efflux pump), methicillin-resistant *Staphylococcus aureus* type 15, methicillin-resistant *Staphylococcus aureus* type 16, *Staphylococcus aureus* (RN4220, methicillin-resistant *Staphylococcus aureus* macrolide efflux pump), *Staphylococcus aureus* (XU212, possess Tet(K) efflux pump), and *Staphylococcus aureus* (ATCC 25923) (Osman et al., 2012).

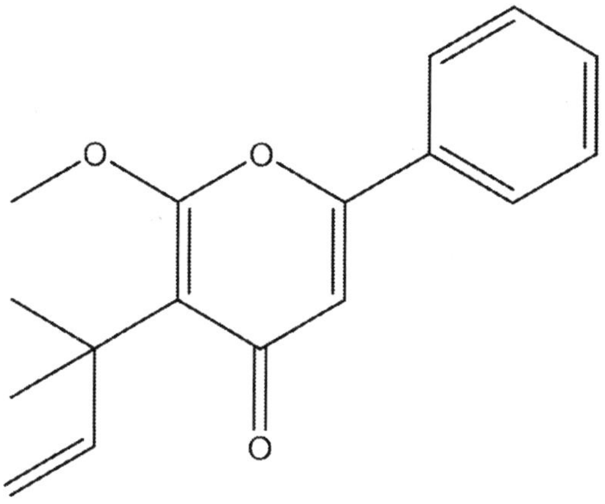

Hyperenone A

REFERENCES

Jayasuriya, H., McChesney, J.D., Swanson, S.M. and Pezzuto, J.M., 1989. Antimicrobial and cytotoxic activity of rottlerin-type compounds from Hypericum drummondii. *Journal of Natural Products*, *52*(2), pp. 325–331.

Osman, K., Basavannacharya, C., Envangelopoulos, D., Gupta, A., Bhakta, S. and Gibbons, S., 2010. Antibacterial from Hypericum acmosepalum showing inhibition of ATP dependent MurE ligase from Mycobacterium tuberculosis. *Planta Medica*, *76*(12), p. P411.

Osman, K., Evangelopoulos, D., Basavannacharya, C., Gupta, A., McHugh, T.D., Bhakta, S. and Gibbons, S., 2012. An antibacterial from Hypericum acmosepalum inhibits ATP-dependent MurE ligase from Mycobacterium tuberculosis. *International Journal of Antimicrobial Agents*, *39*(2), pp. 124–129.

7.4.3.5 *Hypericum japonicum* Thunb.

Synonyms: *Brathys japonica* (Thunb.) Wight; *Brathys laxa* Blume; *Hypericum cavaleriei* H. Lév.; *Hypericum chinense* Osbeck; *Hypericum laxum* (Blume) Koidz.; *Hypericum nervatum* Hance; *Hypericum thunbergii* Franch. & Sav.; *Sarothra japonica* (Thunb.) Y. Kimura; *Sarothra laxa* (Blume) Y. Kimura

Common names: Swamp hypericum, matted St. John's wort, Japanese St John wort; di er cao (China); asoybon, hatihuria, pikari-char (India); byouyanagi, hime-otô-giri (Japan); kugute jhar (Nepal); ngotokong (Papua New Guinea)

Habitat: Marshes, ditches, and open grasslands

Distribution: Tropical Asia, Korea, China, and Japan

Botanical observation: This herb grows to about 45 cm tall. The stem is angular and somewhat dotted with glands. The leaves are simple, sessile, amplexicaul, opposite, and exstipulate. The petiole is up to 3 mm long. The blade is ovate to oblong, 0.2–1.8 × 0.1–1 cm, with about 4 pairs of secondary nerves, with glands below, cordate at base, and acute to obtuse at apex. The inflorescences are terminal or axillary cymes of numerous flowers. The calyx includes 5 sepals, which are lanceolate, up to 5 mm long, veined and with dark glands. The 5 petals are obovate, up to about 1 cm long, and yellowish to orange. The androecium comprises up to about 30 stamens organized in fascicles. The ovary is ovoid, minute, and develops 3 tiny styles. The capsules are cylindrical, brown, up to about 6 mm long, dehiscent, containing numerous tiny seeds.

Medicinal use: Hepatitis (China)

Strong antibacterial (Gram-positive) flavanone glycoside: Taxifolin-7-*O*-α-L-rhamnopyranoside was tested toward 10 clinical strains of methicillin-resistant *Staphylococcus aureus* strains SCCmec III type and MIC ranged from 32 to 64 μg/mL and MBC ranged from

64 to 128 μg/mL and demonstrated MIC/MBC of 8/16 μg/mL for *Staphylococcus aureus* (ATCC 25923) (An et al., 2011).

Strong antibacterial (Gram-positive) amphiphilic phloroglucinol: Chinesin I (LogD = 1.6 at pH 7.4; molecular mass = 444.6 g/mol) isolated from the flowers inhibited the growth of *Staphylococcus aureus*, *Micrococcus luteus*, and *Bacillus subtilis* with MIC values of 3.1, 6.2, and 3.1 μg/mL, respectively (Nagai & Tada, 1987).

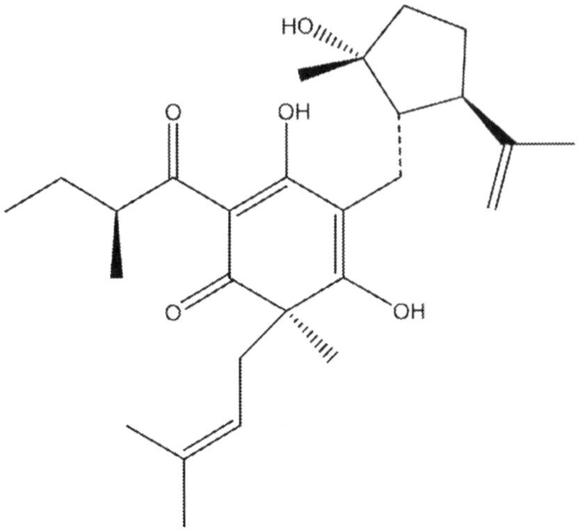

Chinesin I

Strong broad-spectrum antibacterial phloroglucinol: Hyperjaponicol C inhibited the growth of *Escherichia coli* (ATCC 11775), *Salmonella typhimurium* (ATCC 6539), and *Staphylococcus aureus* (ATCC 25922) with MIC values of 0.8, 0.8, and 3.3 μM, respectively (Li et al., 2018).

Strong antibacterial (Gram-positive) amphiphilic xanthone: Isojacareubin (LogD = 2.8 at pH 7.4; molecular mass = 328.3 g/mol) inhibited the growth of 10 clinical methicillin-resistant *Staphylococcus aureus* (strains of SCCmec III type) with MIC/MBC values ranging from 4/16 to 16/64 (Zuo et al., 2012). This xanthone increased the sensibility of clinical methicillin-resistant *Staphylococcus aureus* strains of SCCmec III type to ampicillin, ceftazidime, and levofloxacin (Zuo et al., 2012).

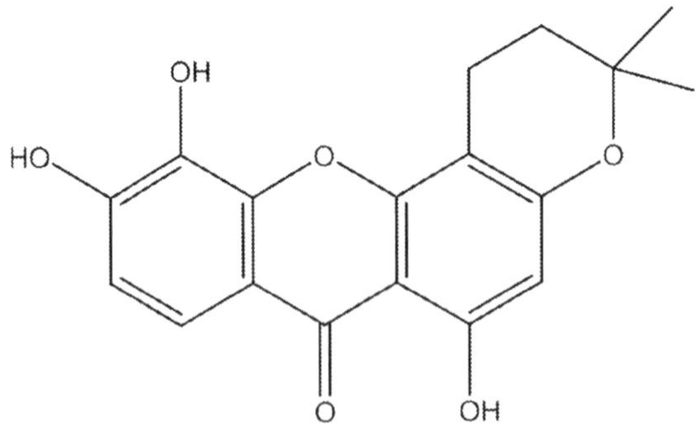

Isojacareubin

Antibiotic potentiator flavanone glycoside: Taxifolin-7-*O*-α-L-rhamnopyranoside had a synergistic effect with ceftazidime toward clinical methicillin-resistant *Staphylococcus aureus* (strains of SCCmec III type) (An et al., 2011).

Strong antiviral (enveloped monopartite linear dimeric single-stranded (+)RNA) lipophilic spirolactone: Biyouyanagin A (Log=7.4 at pH 7.4; molecular mass=474.6 g/mol) isolated from the leaves inhibited the replication of the Human immunodeficiency virus in H9 lymphocytes with an EC_{50} of 0.7 μg/mL and a selectivity index above 31.1 (Tanaka et al., 2005).

Weak antiviral spirolactones: (+)-Japonone A and (−)-japonone B isolated from aerial parts inhibited Kaposi's sarcoma associated herpesvirus with EC_{50} values of 166 and 251.2 μM, respectively (Hu et al., 2016).

Commentaries: (i) Prenylated phloroglucinols produced by members of the genus *Hypericum* L. are active against Gram-positive bacteria and myconacteria: Olympicin A (LogD=5.4 at pH 7.4; molecular mass=346.4 g/mol) isolated from the aerial parts of *Hypericum olympicum* L. cf *uniflorum* inhibited the growth of *Staphylococcus aureus* (1199B), *Staphylococcus aureus* (XU212), *Staphylococcus aureus* (RN4220), *Staphylococcus aureus* (ATCC 25923), *Staphylococcus aureus* (EMRSA-15), *Staphylococcus aureus* (EMRS-16) with the MIC values of 2.9, 2.9, 2.9, 2.9, 1.4, and 2.9 μM, respectively (Shiu et al., 2011). Olympicin A inhibited the growth of *Mycobacterium smegmatis* (ATCC 14468), *Mycobacterium fortuitum* (ATCC 6841), *Mycobacterium smegmatis* (MC_2 2700), and *Mycobacterium phlei* (ATCC 11758) with the MIC values of 11.6, 23.2, 11.6, and 11.6 μM, respectively (Shiu et al., 2011).

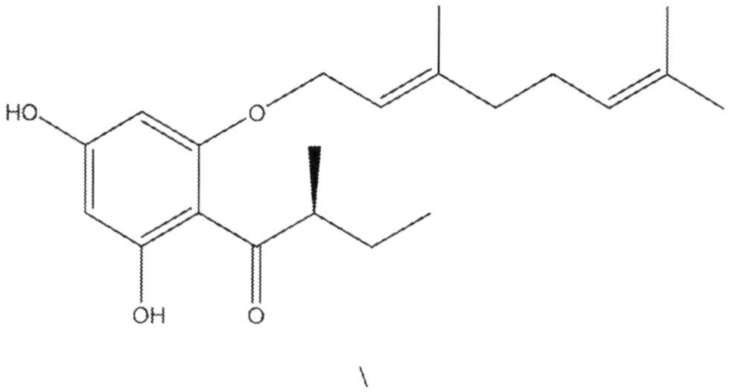

Olympicin A

(ii) Prenylated phloroglucinols built by members of the genus *Hypericum* L. have strong activities against methicillin-resistant *Staphylococcus aureus* such as geranyloxy-1-(2-methylpropanoyl) phloroglucinol and 2- geranyloxy-1-(2-methylbutanoyl)phloroglucinol (Sarkisian et al., 2012).

(iii) The antihepatitis property of the plant has not been substantiated but most probable as ethanol extract of *Hypericum perforatum* L. inhibited Hepatitis B virus replication in HepG22.2.15 cells at 40 μg/mL (Pang et al., 2010). This extract could decrease the secretion of Hepatitis B virus surface antigen by about 50% at 40 μg/mL in HepG22.2.15 cells (Pang et al., 2010).

(vi) The genus *Hypericum* L. is an interesting source of efflux pump inhibitors, for example, acylphloroglucinol isolated from *Hypericum olympicum* L. inhibited at 50 μM NorA multidrug efflux-pump in *Staphylococcus aureus*-1199B (overexpressing the NorA major facilitator superfamily (MFS) efflux pump (Shiu et al., 2013).

(v) Note that members of the genus *Hypericum* L. tend to produce essential oils with antibacterial activity. For example, essential oil of *Hypericum scabrum* L. (40 μL/5 mm disc) inhibited the growth of *Escherichia coli* (K12), *Bacillus brevis* (ATCC), *Pseudomonas aeruginosa* (DMC 66), *Staphylococcus aureus* (DMC70) with inhibition zone diameters of 18, 10, 16, and 16 mm, respectively (Kızıl et al., 2004).

(vi) Hydrophilic molecules do not cross the outer membrane of Gram-negative bacteria because they cannot dissolve in the hydrophobic core (van den Berg, 2010). The outer membrane of Gram-negative bacteria presents an inner layer of phospholipids and an outer layer of lipopolysaccharide (LPS) (van den Berg, 2010). Lipopolysaccharides consist of a hydrophobic moiety=the glycolipid lipid A, and a hydrophilic moiety made of oligosaccharides, which make it difficult for hydrophobic molecules to penetrate bacteria. Thus, the bacterial outer lipopolysaccharide layer is polar (and negatively charged according to pH in environment) and forms a barrier for hydrophobic and to less extent amphiphilic molecules (van den Berg, 2010). Hydrophilic and amphiphilic molecules that have a molecular mass below 600 g/mol cross the outer membrane of Gram-negative bacteria by passive diffusion via porins (van den Berg, 2010). Porins (or aquaporins) present charged and polar groups at the central, water-filled constriction of the channel, making it difficult for a hydrophobic molecule to traverse the channel (van den Berg, 2010). The entry of hydrophobic molecule in Gram-negative bacteria involves specialized porins such as the FadL channels as well as "lateral diffusion transport" (van den Berg, 2010). Thus, lipophilic natural products as well as hydrophilic natural products (with a molecular mass above 600 g/mol) inhibit the growth of Gram-negative bacteria by targeting the outer membrane.

(vii) Biyouyanagin A could be of value against coronaviruses.

REFERENCES

An, J., Zuo, G.Y., Hao, X.Y., Wang, G.C. and Li, Z.S., 2011. Antibacterial and synergy of a flavanol rhamnoside with antibiotics against clinical isolates of methicillin-resistant Staphylococcus aureus (MRSA). *Phytomedicine*, *18*(11), pp. 990–993.

Kızıl, G., Toker, Z., Özen, H.Ç. and Aytekin, Ç., 2004. The antimicrobial activity of essential oils of Hypericum scabrum, Hypericum scabroides and Hypericum triquetrifolium. *Phytotherapy Research*, *18*(4), pp. 339–341.

Hu, L., Zhu, H., Li, L., Huang, J., Sun, W., Liu, J., Li, H., Luo, Z., Wang, J., Xue, Y. and Zhang, Y., 2016. (±)-Japonones A and B, two pairs of new enantiomers with anti-KSHV activities from Hypericum japonicum. *Scientific Reports*, *6*, p. 27588.

Li, Y.P., Hu, K., Yang, X.W. and Xu, G., 2018. Antibacterial Dimeric Acylphloroglucinols from Hypericum japonicum. *Journal of Natural Products*, *81*(4), pp. 1098–1102.

Nagai, M. and Tada, M., 1987. Antimicrobial compounds, chinesin I and II from flowers of Hypericum chinense L. *Chemistry Letters*, *16*(7), pp. 1337–1340.

Pang, R., Tao, J., Zhang, S., Zhu, J., Yue, X., Zhao, L., Ye, P. and Zhu, Y., 2010. In vitro anti-hepatitis B virus effect of Hypericum perforatum L. *Journal of Huazhong University of Science and Technology [Medical Sciences]*, *30*(1), pp. 98–102.

Sarkisian, S.A., Janssen, M.J., Matta, H., Henry, G.E., LaPlante, K.L. and Rowley, D.C., 2012. Inhibition of bacterial growth and biofilm production by constituents from Hypericum spp. *Phytotherapy Research*, *26*(7), pp. 1012–1016.

Shiu, W.K., Rahman, M.M., Curry, J., Stapleton, P., Zloh, M., Malkinson, J.P. and Gibbons, S., 2011. Antibacterial acylphloroglucinols from Hypericum olympicum. *Journal of Natural Products*, *75*(3), pp. 336–343.

Shiu, W.K., Malkinson, J.P., Rahman, M.M., Curry, J., Stapleton, P., Gunaratnam, M., Neidle, S., Mushtaq, S., Warner, M., Livermore, D.M. and Evangelopoulos, D., 2013. A new plant-derived antibacterial is an inhibitor of efflux pumps in Staphylococcus aureus. *International Journal of Antimicrobial Agents*, *42*(6), pp. 513–518.

Tanaka, N., Okasaka, M., Ishimaru, Y., Takaishi, Y., Sato, M., Okamoto, M., Oshikawa, T., Ahmed, S.U., Consentino, L.M. and Lee, K.H., 2005. Biyouyanagin A, an Anti-HIV Agent from Hypericum c hinense L. var. s alicifolium. *Organic Letters*, *7*(14), pp. 2997–2999.

van den Berg, B., 2010. Going forward laterally: transmembrane passage of hydrophobic molecules through protein channel walls. *Chembiochem*, *11*(10), pp. 1339–1343.

Zuo, G.Y., An, J., Han, J., Zhang, Y.L., Wang, G.C., Hao, X.Y. and Bian, Z.Q., 2012. Isojacareubin from the Chinese Herb Hypericum japonicum: potent antibacterial and synergistic effects on clinical methicillin-resistant Staphylococcus aureus (MRSA). *International Journal of Molecular Sciences*, *13*(7), pp. 8210–8218.

7.4.3.6 *Hypericum patulum* Thunb.

Synonyms: *Hypericum argyi* H. Lév. & Vaniot; *Komana patula* (Thunb.) Y. Kimura ex Honda; *Norysca patula* (Thunb.) Voigt

Common names: Paharia (Bangladesh); jin si mei (China); la-syn-rit, tumbhul (India), urilo (Nepal)

Habitat: Open forests, roadsides

Distribution: China

Botanical observation: This shrub grows to about 1.5 m tall. The stem is quadrangular and somewhat dotted with glands. The leaves are simple, opposite, and exstipulate. The petiole is up to 3 mm long. The petiole is minute. The blade is oblanceolate, 1.5–6×0.5–3 cm, glaucous and dotted below, marked with 4–5 pairs of secondary nerves, cuneate at base, and obtuse to apiculate at apex. The inflorescences are terminal or axillary cymes of numerous flowers. The 5 sepals are reddish, broadly ovate, denticulate at margin, up to 1 cm long, veined and with dotted with dark glands. The 5 petals are oblong ovate, up to about 1.8 cm long, incised at margin, and golden yellow. The androecium comprises numerous stamens, which are up to about 1.2 cm long, and organized in 5 golden yellow fascicles. The ovary is ovoid, golden yellow, up to about 5 mm long, and develops 5 tiny styles. The capsules are red, glossy, broadly ovoid, brown, up to about 1.1cm long, dehiscent, and contain numerous tiny and dark brown seeds.

Medicinal uses: Wounds, fungal infection (India)

Broad-spectrum antibacterial halo elicited by polar extract: Methanol fraction of leaves inhibited the growth of *Bacillus subtilis* (NOM 2039), *Pseudomonas capacia* (NOM 2106), *Bacillus megaterium* (NCIM 208), *Bacillus coagulans* (NCIM 2323), *Escherichia coli* (NCIM 2345), and *Staphylococcus aureus* (NCIM 2492) with inhibition zone diameters (0.3 µL of 800 µg/L solution/ well) of 12, 14, 16, 20, 24, and 19 mm, respectively (Mukherjee et al., 2002)

Strong antibacterial (Gram-positive) lipophilic prenylated benzophenone: Hypatulin A (LogD=7.5 at pH 7.4; molecular mass=488.6 g/mol) isolated from the leaves inhibited the growth of *Bacillus subtilis* with an MIC value of 16 µg/mL (Tanaka et al., 2016).

REFERENCES

Mukherjee, P.K., Saritha, G.S. and Suresh, B., 2002. Antimicrobial potential of two different Hypericum species available in India. *Phytotherapy Research*, *16*(7), pp. 692–695.

Tanaka, N., Yano, Y., Tatano, Y. and Kashiwada, Y., 2016. Hypatulins A and B, meroterpenes from Hypericum patulum. *Organic Letters*, *18*(20), pp. 5360–5363.

7.4.4 Family Malpighiaceae A.L. de Jussieu (1789)

The family Malpighiaceae consists of 65 genera and about 1280 species of handsome trees, shrubs, or woody climbers. The leaves are simple, opposite, or whorled, and stipulate. The inflorescences are terminal or axillary racemes, corymbs, or umbels. The calyx comprises 5 sepals. The corolla includes 5 petals, which are often very characteristically clawed, dentate, and membranous. A disc is present. The androecium is made of 10 stamens. The gynoecium includes a 3-locular ovary, each locule containing 1 ovule on axile placenta. The fruits are capsules or follicles, with 1 seed per follicle.

7.4.4.1 *Galphimia glauca* Cav.

Synonyms: *Galphimia gracilis* Bartl.; *Galphimia humboldtiana* Bartl.; *Galphimia multicaulis* A. Juss.; *Malpighia glauca* (Cav.) Pers.; *Thryallis glauca* (Cav.) Kuntze; *Thryallis gracilis* (Bartl.) Kuntze

Common names: Galphimia; maiden's jealousy

Habitat: Villages, cultivated

Distribution: Tropical Asia and Pacific

Botanical observation: This shrub native to Central America grows to a height of about 2 m. The stem is somewhat reddish. The petiole is up to about 1.5 cm long. The leaves are simple, opposite, and stipulate. The stipules are about 3 mm long and linear. The blade is 2–5.5 × 1.3–3.5cm, lanceolate or elliptical, marked with a pair of glands at the margin near the base, and acute at base and apex. The inflorescence is a terminal raceme. The 5 sepals are about 4 mm long and lanceolate. The 5 petals are yellow, clawed, about 1 cm long, and with oblong to lanceolate limb. The androecium comprises 5 stamens of unequal length and about 5mm long. The 3 styles are about 7mm long.

Medicinal use: Wounds (Bangladesh).

Strong antiviral phenolic: Tetragalloyl quinic at a concentration of 6.2 μM acid inhibited the proliferation of the Human immunodeficiency virus (strain IIIB) by 52% (Nishizawa et al., 1989).

Viral enzyme inhibition by phenolic: Tetragalloyl quinic at a concentration of 30 μM acid inhibited the Human immunodeficiency virus reverse transcriptase by 84% (Nishizawa et al., 1989).

Commentary: The plant produces nor-3,4-seco-friedelanes triterpenes (known as galphimines) as well as gallic acid, methyl gallate, methyl 4-methoxygallate, ellagic acid, tetragalloylquinic acid, and flavonoid glycosides (Ortega et al., 2020). Consider that galphimines are sedatives (Tortoriello & Ortega, 1993) and as such could have antibiotic potentiating effects as discussed earlier. Consider that tetragalloylquinic acid is probably a broad-spectrum antiviral principle (Nakashima et al., 1992), which could be examined for its possible activity against Severe acute respiratory syndrome-acquired coronavirus. Furthermore, this phenolic substance is most probably antibacterial. Gallic acid, methyl gallate, methyl 4-methoxygallate, ellagic acid, tetragalloylquinic acid may work synergistically to heal wounds.

REFERENCES

Garcıa, V.N., Gonzalez, A., Fuentes, M., Aviles, M., Rios, M.Y., Zepeda, G. and Rojas, M.G., 2003. Antifungal activities of nine traditional Mexican medicinal plants. *Journal of Ethnopharmacology*, *87*(1), pp. 85–88.

Nakashima, H., Murakami, T., Yamamoto, N., Sakagami, H., Tanuma, S.I., Hatano, T., Yoshida, T. and Okuda, T., 1992. Inhibition of human immunodeficiency viral replication by tannins and related compounds. *Antiviral Research*, *18*(1), pp. 91–103.

Nishizawa, M., Yamagishi, T., Dutschman, G.E., Parker, W.B., Bodner, A.J., Kilkuskie, R.E., Cheng, Y.C. and Lee, K.H., 1989. Anti-AIDS agents, 1. Isolation and characterization of four new tetragalloylquinic acids as a new class of HIV reverse transcriptase inhibitors from tannic acid. *Journal of Natural Products*, *52*(4), pp. 762–768.

Ortega, A., Pastor-Palacios, G., Ortiz-Pastrana, N., Ávila-Cabezas, E., Toscano, R.A., Joseph-Nathan, P., Morales-Jiménez, J. and Bautista, E., 2020. Further galphimines from a new population of Galphimia glauca. *Phytochemistry*, *169*, p. 112180.

Tortoriello, J. and Ortega, A., 1993. Sedative effect of galphimine B, a nor-seco-triterpenoid from Galphimia glauca. *Planta Medica*, *59*(05), pp. 398–400.

7.4.4.2 *Hiptage benghalensis* (L.) Kurz

Synonyms: *Banisteria benghalensis* L.; *Gaertnera racemosa* (Cav.) Roxb.; *Hiptage madablota* Gaertn.

Common names: Hiptage; madhobi lota (Bangladesh); feng zheng guo (China); madhavilata, raisentur (India)

Habitat: Forests

Distribution: India, Nepal, Bhutan, Bangladesh, Cambodia, Laos, Vietnam, China, Thailand, Malaysia, Indonesia, the Philippines

Botanical observation: This stout climber grows to about 10m. The stems are hairy at apex. The leaves are simple, opposite, and exstipulate. The petiole is 5–10mm long and channeled above. The blade is coriaceous, elliptic-oblong, 9–18 × 3–7 cm, broadly cuneate or rounded at base, acuminate at apex and marked with 6 or 7 pairs of secondary nerves. The racemes are axillary or terminal and 5–10 cm long. The calyx comprises 5 sepals, which are ovate, and 5–6 mm long and hairy. The 5 petals are white, yellow at base, orbicular, up to 1.5 cm long, incised, and membranous. The 10 stamens are about 1.3cm long. The style is about 1cm long and curved at apex. Anthers and style are bent altogether is a characteristic manner giving to the plant (including opposite leaves) an aspect of a Melastonataceae. The samaras are dull brownish to greenish red, and develop 3 wings, the central up to about 4 cm long.

Medicinal uses: Skin infection, leprosy (Bangladesh); diarrhea (India)

Moderate broad-spectrum antibacterial polar extract: Methanol extract of root bark inhibited the growth of *Klebsiella pneumoniae* (MTCC 39), *Escherichia coli* (MTCC 40), *Micrococcus luteus* (MTCC 106), and *Pseudomonas aeruginosa* (MTCC 424) with MIC values of 625, 312.5, 625, and 625 μg/mL, respectively (Lalnundanga & Thanzami, 2015).

Commentary: Consider that the plant generates mangiferin, which is antiviral (see earlier) (Finnegan et al., 1968) as well as the toxin hiptagenic acid, a glycoside that includes 4 3-nitropropionic acid moieties that are released by bacterial enzymes in the gut of ruminant animals, resulting in neurologic disorders (Francis et al., 2013). One could have some interest in isolating antimicrobial principles from this climber.

REFERENCES

Finnegan, R.A., Stephani, R.A., Ganguli, G., Ganguly, S.N. and Bhattacharya, A.K., 1968. Occurrence of mangiferin in Hiptage madablota geartn. *Journal of Pharmaceutical Sciences*, *57*(6), pp. 1039–1040.

Francis, K., Smitherman, C., Nishino, S.F., Spain, J.C. and Gadda, G., 2013. The biochemistry of the metabolic poison propionate 3-nitronate and its conjugate acid, 3-nitropropionate. *IUBMB Life*, *65*(9), pp. 759–768.

Lalnundanga, L.N. and Thanzami, K., 2015. Antimicrobial Activity of Methanol Extract of Root Bark of Hiptage benghalensis (L) Kurz. *Journal of Pharmacognosy and Phytochemistry*, *3*(6), pp. 119–121.

7.4.5 Family Passifloraceae Juss. ex Roussel (1806)

The family Passifloraceae consists of about 16 genera and 660 species of climbers with coiled tendrils. The leaves are simple, spiral or alternate, and stipulate or not. The inflorescences are axillary cymes. The calyx includes 5 sepals. The corolla comprises 5 petals. A characteristic corona is present. The androecium comprises 5 stamens. The gynoecium consists of 3 carpels fused in a unilocular ovary containing many ovules on parietal placentas. The fruit is a berry or dehiscent capsule. Plants in this family produce flavonoids and cyanogenetic glycosides.

7.4.5.1 *Adenia cordifolia* Gagnep.

Common name: Akar kail (Malaysia)

Habitat: Forests

Distribution: Malaysia, Indonesia, and the Philippines

Botanical observation: This climber grows up to a length of 30 m. The stem is terete and develops axillary tendrils. The leaves are simple, alternate, and stipulate. The stipules are triangular and minute. The petiole presents a pair of glands at apex. The blade is cordate, glossy, dark green, glaucous below, somewhat membranous, 3–10.5 × 2–6 cm, marked with up to about 10 pairs of secondary nerves, and acute or acuminate at apex. The cymes are axillary, up to about 7 cm long, and present many yellow flowers. A hypanthium is present. The calyx is tubular and develops 5 lobes. The corolla includes 5 petals originating from the apex of the hypanthium. A corona is present. The androecium comprises 5 stamens. The ovary develops 3 styles. The capsules are pendulous, bright red, dehiscent, 3-valved, bullet shaped, up to about 12cm long, and containing up to 30 reniform seeds, which are up to about 1 cm long, black, and embedded in an aril.

Medicinal uses: Eye infection (Indonesia); fever (the Philippines)

Antibacterial (Gram-positive) polar extract: Methanol extract inhibited the growth of *Staphylococcus aureus* (Grosvenor et al., 1995).

Commentaries: In Malaysia, *Adenia populifolia* Engl. is used to treat ringworm. Note that polyacetylenes are known to occur in members of the genus *Adenia* Forssk. (Fullas et al., 1995).

REFERENCES

Fullas, F., Brown, D.M., Wani, M.C., Wall, M.R., Chagwedera, T.E., Farnsworth, N.R., Pezzuto, J.M. and Kinghorn, A.D., 1995. Gummiferol, a cytotoxic polyacetylene from the leaves of Adenia gummifera. *Journal of Natural Products*, *58*(10), pp. 1625–1628.

Grosvenor, P.W., Supriono, A. and Gray, D.O., 1995. Medicinal plants from Riau Province, Sumatra, Indonesia. Part 2: Antibacterial and antifungal activity. *Journal of Ethnopharmacology*, *45*(2), pp. 97–111.

7.4.5.2 *Passiflora foetida* L.

Synonyms: *Dysosmia foetida* (L.) M. Roem.; *Granadilla foetida* (L.) Gaertn.; *Passiflora hispida* DC. ex Triana & Planch.; *Tripsilina foetida* (L.) Raf.

Common names: Fetid passion flower, Stinking passion flower; Jhumkolata; Hurhuna; Powmachi (Bangladesh); Jumkalata (India); Rajutan (Indonesia); Timun hutan (Malaysia); Pasionariangmabaho, Taungon (the Philippines); k" by "Fetid passion flower, stinking passion flower; jhumkolata, hurhuna, powmachi (Bangladesh); jumkalata (India); rajutan (Indonesia); timun hutan (Malaysia); pasionariangmabaho, taungon (the Philippines); kra prong thong (Thailand); and lac tien (Vietnam)

Habitat: Wastelands, roadsides

Distribution: Native to South America, Tropical Asia and Pacific

Botanical observation: This slender, dreadfully poisonous, flexuous, herbaceous, and slightly malodorant climber grows up to 4 m long. The stems are terete and hairy. The leaves are simple, alternate, and stipulate. The stipules are falcate and tripinnatisect. The petiole is slender, hairy, and 2–4cm long. The blade is 3-lobed, membranous, cordate at base, 4–7×3–5cm, and ciliate at the margin. The inflorescence is axillary, solitary on up to 4cm long peduncles. The calyx comprises 5 sepals which are oblong, 2–2.5cm long, and white. The 5 petals are oblong, as long as the sepals, obtuse, mucronate, and white. A multiseriate corona, which is purple to bluish purple, is present. The androecium comprises 5 stamens. The gynoecium includes a globose ovary, which is hairy and develops clavate styles. The berries are ovoid, orange yellow, soft, enclosed by enlarged dissected bracts and about 2cm in diameter.

Medicinal uses: Boils (India); wounds (the Philippines)

Broad-spectrum antibacterial halo elicited by polar extract: Ethanol extract (200 μL of a 200 μg/mL solution/5mm well) inhibited the growth of *Pseudomonas putida*, *Vibrio cholerae*, *Shigella flexneri*, and *Streptococcus pyogenes* with the inhibition zone diameters of 12, 15, 19, and 20mm, respectively (Mohanasundari et al., 2007).

Broad-spectrum antifungal polar extract: Ethanol extract of leaves at a concentration of 200 μg/mL inhibited the growth of *Trichophyton mentagrophytes*, *Trichophyton rubrum*, and *Candida albicans* (Natarajan et al., 2011).

Antimycobacterial amphiphilic flavonol: Ermanin (kaempferol 3,4′-dimethyl ether) at a concentration of 100μg/mL evoked some levels of growth inhibition of *Mycobacterium tuberculosis* (Murillo et al., 2003).

Strong antiviral (non-enveloped linear monopartite single-stranded (+)RNA) amphiphilic flavonol: Ermanin (also known as 5,7-dihydroxy-3,4′-dimethoxyflavone; LogD = 1.5 at pH 7.4; molecular mass = 314.2 g/mol) inhibited the growth of Coxsackie virus B3 with an IC_{50} value of 2μg/mL (Elsohly et al., 1997)

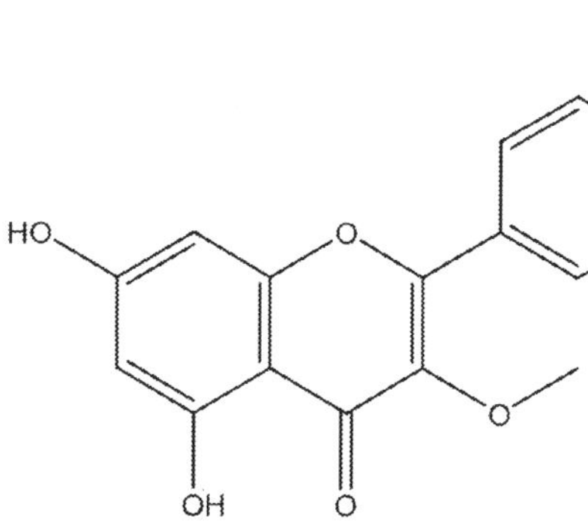

Ermanin

Antiviral (non-enveloped monopartite linear single-stranded (+)RNA) flavonol: Ermanin inhibited the growth of Poliovirus (Robin et al., 1998).

Commentaries: (i) The plant accumulates cyanogenic glycosides (Andersen et al., 1998). (ii) Note that 4-hydroxy-2-cyclopentenone from the leaves of *Passiflora tetrandra* Banks ex DC, inhibited *Pseudomonas aeruginosa, Escherichia coli,* and *Bacillus subtilis* with the minimum inhibiting dose of 10 μg/disc (Perry et al., 1991). (iii) Consider the presence of the flavone chrysin in members of the genus *Passiflora* L. (Mani & Natesan, 2018), which has affinity for benzodiazepine receptors in humans (Wolfman et al., 1994). Chrysin inhibited the growth of *Bacillus subtilis, Bacillus sphaericus, Staphylococcus aureus, Klebsiella aerogenes,* and *Chromobacterium violaceum* with MIC values of 50, 25, 50, 50, and 25 μg/mL, respectively (Babu et al., 2006). Chrysin at a concentration of about 10 μM inhibited the growth of *Fusarium solani* and *Rhizoctonia repens* by about 20% (Shimura et al., 2007). Chrysin (LogD = 1.8 at pH 7.4; molecular mass = 254.2 g/mol) inhibited the replication of the non-enveloped monopartite single-stranded (+)-RNA viruses: Enterovirus 71 (EV71) by suppressing viral 3C protease activity (Wang et al., 2017) as well as the Coxsackievirus B3 (ATCC VR-30) in Vero cells with an EC_{50} value of about 2 μM (Song et al., 2015). Thus, chrysin could be examined for its activity against the Severe acute respiratory syndrome-acquired coronavirus and other coronaviruses.

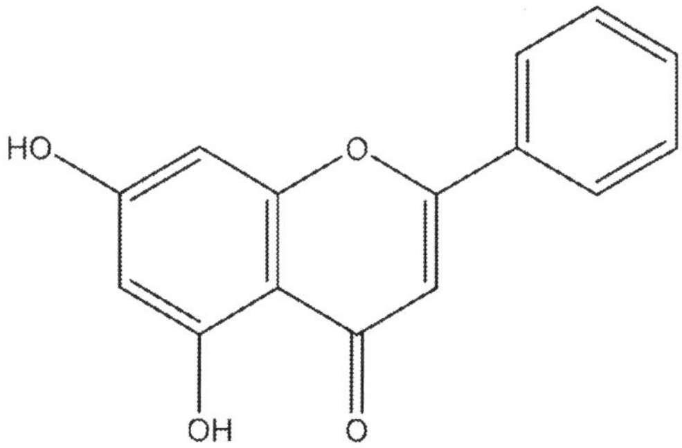

Chrysin

REFERENCES

Andersen, L., Adsersen, A. and Jaroszewski, J.W., 1998. Cyanogenesis of Passiflora foetida. *Phytochemistry, 47*(6), pp. 1049–1050.

Babu, K.S., Babu, T.H., Srinivas, P.V., Kishore, K.H., Murthy, U.S.N. and Rao, J.M., 2006. Synthesis and biological evaluation of novel C (7) modified chrysin analogues as antibacterial agents. *Bioorganic & Medicinal Chemistry Letters, 16*(1), pp. 221–224.

Elsohly, H.N., El-Feraly, F.S., Joshi, A.S. and Walker, L.A., 1997. Antiviral flavonoids from Alkanna orientalis. *Planta Medica*, *63*(04), pp. 384–384.
Mani, R. and Natesan, V., 2018. Chrysin: Sources, beneficial pharmacological activities, and molecular mechanism of action. *Phytochemistry*, *145*, pp. 187–196.
Mohanasundari, C., Natarajan, D., Srinivasan, K., Umamaheswari, S. and Ramachandran, A., 2007. Antibacterial properties of Passiflora foetida L.–a common exotic medicinal plant. *African Journal of Biotechnology*, *6*(23).
Murillo, J.I., Encarnación-Dimayuga, R., Malmstrøm, J., Christophersen, C. and Franzblau, S.G., 2003. Antimycobacterial flavones from Haplopappus sonorensis. *Fitoterapia*, *74*(3), pp. 226–230.
Natarajan, D., Mohanasundari, C. and Srinivasan, K., 2011. Anti-dermatophytic activity of Passiflora foetida L: An exotic plant. *International Journal of Phytopharmacy Research*, *2*, pp. 72–74.
Perry, N.B., Albertson, G.D., Blunt, J.W., Cole, A.L.J., Munro, M.H.G. and Walker, J.R.L., 1991. 4-Hydroxy-2-cyclopentenone: An anti-pseudomonas and cytotoxic component from Passiflora tetrandral. *Planta Médica*, *57*(02), pp. 129–131.
Shimura, H., Matsuura, M., Takada, N. and Koda, Y., 2007. An antifungal compound involved in symbiotic germination of Cypripedium macranthos var. rebunense (Orchidaceae). *Phytochemistry*, *68*(10), pp. 1442–1447.
Song, J.H., Kwon, B.E., Jang, H., Kang, H., Cho, S., Park, K., Ko, H.J. and Kim, H., 2015. Antiviral activity of chrysin derivatives against coxsackievirus B3 in vitro and in vivo. *Biomolecules & Therapeutics*, *23*(5), p. 465.
Wang, J., Zhang, T., Du, J., Cui, S., Yang, F. and Jin, Q., 2014. Anti-Enterovirus 71 effects of chrysin and its phosphate ester. *PLoS One*, *9*(3), p.e89668.
Wolfman, C., Viola, H., Paladini, A., Dajas, F. and Medina, J.H., 1994. Possible anxiolytic effects of chrysin, a central benzodiazepine receptor ligand isolated from Passiflora coerulea. *Pharmacology Biochemistry and Behavior*, *47*(1), pp. 1–4.

7.4.6 Family Phyllanthaceae Martynov (1820)

The family Phyllanthaceae consists of about 60 genera and 200 species of herbs, shrubs, or trees. The leaves are simple, characteristically alternate (at first glance not uncommonly looking like pinnate), and stipulate. The inflorescences are cymes, panicles, racemes, or solitary and axillary. The calyx includes 4–6 sepals. The corolla comprises 4–6 petals or none. The androecium comprises mostly 2–5 stamens. The gynoecium includes 3–10 carpels fused into a multilocular ovary, each locule enclosing 2 ovules on axile placentas. The fruit is mostly a dehiscent capsule.

7.4.6.1 *Antidesma acidum* Retz.

Synonyms: *Antidesma diandrum* (Roxb.) Roth; *Antidesma diandrum* (Roxb.) Spreng.; *Antidesma wallichianum* C. Presl; *Stilago diandra* Roxb.; *Stilago lanceolaria* Roxb.

Common names: Elena, muta (Bangladesh); xi nan wu yue cha (China); pulleru, rohitaka (India); mao soi (Thailand)

Habitat: Forests

Distribution: India, Nepal, Bangladesh, Bhutan, Myanmar, Cambodia, Laos, Vietnam, Chona, Malaysia, Indonesia

Botanical observation: This shrub grows to about 3 m tall. The stems are hairy at apex. The leaves are simple, alternate, and stipulate. The stipules are linear and up to 1 cm long. The petiole is up to 5 mm long. The blade is obovate to elliptic-oblong, 4–10 × 2.5–5 cm, glossy, cuneate at base, somewhat dark green, rounded at apex, and with 4–9 pairs of secondary nerves. The spikes are terminal or axillary,and up to 15cm long (somewhat *Piper*-like). The flowers are minute. The calyx is 4-lobed. A disc is present. The androecium includes 2 stamens. The ovary is ellipsoid and produces 3 stigmas. The drupes are ellipsoid, about 5 mm long, reddish-purple, sour (edible Thailand), and glossy.

Medicinal uses: Dysentery, pneumonia, rabies, jaundice (Bangladesh)

Weak antibacterial (Gram-positive) flavanone: Taxifolin inhibited the growth of *Streptococcus sobrinus* (Kuspradini et al., 2009).

Strong antiviral (enveloped monopartite linear dimeric single-stranded (+)RNA) amphiphilic flavanone: Taxifolin (LogD = 1.1 at pH 7.4; molecular mass = 304.2 g/mol) isolated from the stem-bark inhibited the cytopathic effect of the Human immunodeficiency virus type-1 in MT-4 cells with an IC_{50} value of 25 μg/mL (Min et al., 2002).

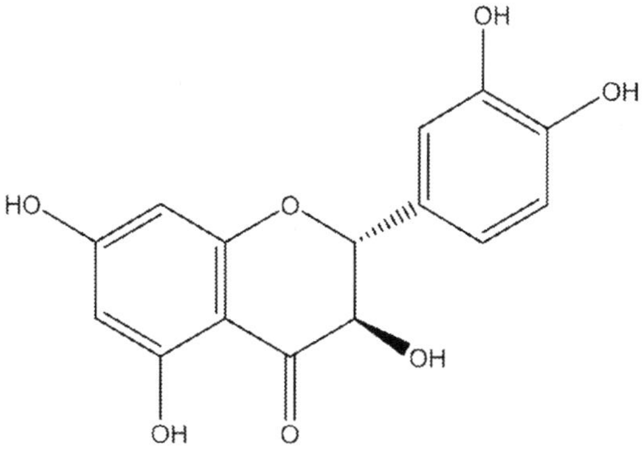

Taxifolin

Moderate antiviral (enveloped linear monorpatite double-stranded DNA) flavanone: Taxifolin inhibited the replication of the Herpes simplex virus type-1 (F strain) with an IC_{50} value of 48.1 μM (Wu et al., 2011).

Commentaries: (i) The plant synthetises: acidumonate, 3-(1,1-dimethylallyl)-scopoletin, (–)-5,7-dihydroxy-2-eicosyl-chromone, the coumarin barbatumol A, *N*-trans-feruloyltyramine, syringic aldehyde, p-hydroxybenzoic acid, taxifolin, (+)-catechin, gallocatechin, and 2,5-dimethoxy-1,4-benzoquinone (Kaennakam et al., 2013), which may work synergistically as mild astringents to check dysentery. The presence of phenolics in plants is often a sign of antiviral properties. (ii) 1–4 Benzoquinone is a broad-spectrum antibacterial pharmacophore in natural products. By itself, it inhibited the growth of *Staphylococcus aureus*, *Streptococcus pyogenes*, *Salmonella typhimurium*, and *Escherichia coli* with MIC values of 64, 64, 32, and 32 μg/mL, respectively (Lana et al., 2006). 2,5-Dihydroxy 1–4 benzoquinone reacts with secondary amines and thiol moieties of amino acids (Hettegger et al., 2020). The plant being used to treat rabies and jaundice might generate antiviral principles, which are apparently yet to be isolated. (+)-Catechin inhibited the Human immunodeficiency virus type-1 integrase with an IC_{50} value of 46.3 μM (Panthong et al., 2015). (iii) Consider that members of the genus *Antidesma* L. are often used for the treatment of infectious diseases in traditional Southeast Asian medicines. *Antidesma montanum* Bl. is used in Malaysia (local name: *Berunai*) to treat measles and chicken pox, *Antidesma ghaesembilla* Gaertn. is used to treat fever in China (local name: *Aó yè*) and *Antidesma bunius* (L.) Spreng. is used to treat syphilis in India (local name: *Nolai-tali*).

REFERENCES

Hettegger, H., Steinkellner, K., Zwirchmayr, N.S., Potthast, A., Edgar, K.J. and Rosenau, T., 2020. Reaction of 2, 5-dihydroxy-[1, 4]-benzoquinone with nucleophiles–ipso-substitution vs. addition/elimination. *Chemical Communications*, *56*(84), pp. 12845–12848.

Kaennakam, S., Sichaem, J., Siripong, P. and Tip-pyang, S., 2013. A new cytotoxic phenolic derivative from the roots of Antidesma acidum. *Natural Product Communications*, *8*(8), p. 1934578X1300800820.

Kuspradini, H., Mitsunaga, T. and Ohashi, H., 2009. Antimicrobial activity against Streptococcus sobrinus and glucosyltransferase inhibitory activity of taxifolin and some flavanol rhamnosides from kempas (Koompassia malaccensis) extracts. *Journal of Wood Science*, *55*(4), pp. 308–313.

Lana, E.J., Carazza, F. and Takahashi, J.A., 2006. Antibacterial evaluation of 1, 4-benzoquinone derivatives. *Journal of Agricultural and Food Chemistry*, *54*(6), pp. 2053–2056.

Panthong, P., Bunluepuech, K., Boonnak, N., Chaniad, P., Pianwanit, S., Wattanapiromsakul, C. and Tewtrakul, S., 2015. Anti-HIV-1 integrase activity and molecular docking of compounds from Albizia procera bark. *Pharmaceutical Biology*, *53*(12), pp. 1861–1866.

Wu, N., Kong, Y., Zu, Y., Fu, Y., Liu, Z., Meng, R., Liu, X. and Efferth, T., 2011. Activity investigation of pinostrobin towards herpes simplex virus-1 as determined by atomic force microscopy. *Phytomedicine*, *18*(2–3), pp. 110–118.

7.4.6.2 *Bridelia retusa* (L.) A. Juss.

Synonyms: *Clutia squamosa* Lam; *Bridelia spinosa* (Roxb.) Willd.; *Bridelia cambodiana* Gagnep.; *Bridelia pierrei* Gagnep; *Bridelia montana* auct. non Willd.

Common names: Spinous kino tree; geio, shukkuja gaas (Bangladesh); asana, ekavira (India); hle-kanan, kasi, (Myanmar); gayo (Nepal); teng nam (Thailand)

Habitats: Forests and open lands

Distribution: From India to Indonesia

Botanical observation: This tree grows up to a height of 20 m tall. The bole presents massive, somewhat sinister, and straight woody thorns, which are up to about 15 cm long. The leaves are simple, alternate, and stipulate. The stipules are caducous. The petiole is up to 1.2 cm long. The blade is elliptic-oblong to oblanceolate, 10–20 × 4–10 cm, rounded or cuneate at base, acute at apex, somewhat coriaceous, marked with 15–20 pairs of secondary nerves, and glaucous below. The inflorescences are axillary fascicles. The flowers have something like a *Passiflora* look. The calyx includes 5 sepals, which are triangular, green, fleshy, and minute. The 5 petals are shorter than petals, whitish, and flabelliform. A conspicuous purplish disc is present. The androecium consists of 5 stamens fused at base into a column. The ovary is globose and presents a bifid style. The berries are globose, 7–9 mm in diameter, fleshy, edible (Nepal), on persistent calyx, and containing a few seeds.

Medicinal uses: Skin infection, wounds (Bangladesh); diarrhea, dysentery (Nepal)

Antifungal (filamentous) bisabolene-type sesquiterpene: 4-(1,5-dimethyl-3-oxo-4-hexenyl)benzoic acid isolated from the stem bark inhibited the growth of *Cladosporium cladosporioides* at 5 µg (on thin layer chromatography) (Jayasinghe et al., 2003).

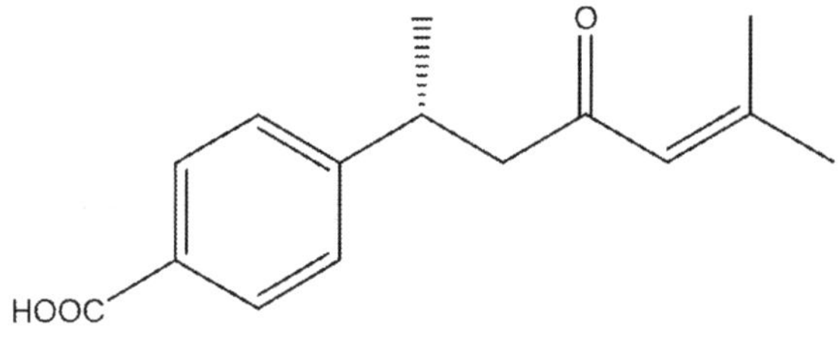

4-(1,5-dimethyl-3-oxo-4-hexenyl)benzoic acid

Commentary: *Bridelia stipularis* (L.) Bl. is used in Bangladesh (local name: *Bangari gach*) to treat skin infections.

REFERENCE

Jayasinghe, L., Kumarihamy, B.M., Jayarathna, K.N., Udishani, N.G., Bandara, B.R., Hara, N. and Fujimoto, Y., 2003. Antifungal constituents of the stem bark of Bridelia retusa. *Phytochemistry*, *62*(4), pp. 637–641.

7.4.6.3 *Flueggea virosa* (Roxb. ex Willd.) Royle

Synonyms: *Acidoton virosus* (Roxb. ex Willd.) Kuntze; *Securinega virosa* (Roxb. ex Willd.) Baill.

Common names: Snow berry tree; bai fan shu (China); pani koita (India); sigar jalak (Indonesia); botolan, tulita-ngalong (the Philippines); ma taek, kang pla khao, ma taek, (Thailand); bong nô'(Vietnam)

Habitats: Roadsides, vacant lots of lands, villages, slopy shrublands

Distribution: From Pakistan to the Pacific Islands

Botanical observation: This shrub grows up to a height of 6 m. The stems are somewhat angled, reddish, and smooth when young. The leaves are simple, alternate, and stipulate. The stipules are lanceolate, up to 3 mm long, and laciniate at margin. The petiole is up to about 1 cm long. The blade is elliptic-oblong to obovate or spathulate, 2–5 × 1–3 cm, coriaceous, glossy, without conspicuous secondary nerves, obtuse or cuneate at base, acute or round, to emarginate at apex, marked with 5–8 pairs of secondary nerves, and glaucous below. The inflorescences are axillary fascicles. The calyx includes 5 sepals, which are minute, greenish, and serrate. The corolla is absent. A pentagonal disc is present. The androecium consists of 5 stamens, which are minute. The ovary is 3 locular and develops 3 bifid stigma. The berries are trilobate, globose to slightly oblate, pure white, seated on a persistent calyx, up to 5mm in diameter, and marked at apex vestigial stigmas.

Medicinal use: Boils (the Philippines)

Moderate broad-spectrum antibacterial mid-polar extract: Chloroform extract of root bark inhibited the growth of *Staphylococcus aureus*, multidrug-resistant *Staphylococcus aureus*, *Bacillus subtilis*, *Micrococcus flavus*, *Streptococcus faecalis*, and *Klebsiella aerogenes* with MIC values of 125, 64, 250, 250, 500, and 500 µg/mL, respectively (Mensah et al., 1990).

Strong broad-spectrum securinega alkaloid: Viroallosecurinine (molecular mass = 217.2 g/mol) inhibited the growth of *Escherichia coli*, *Enterococcus faecalis*, *Pseudomonas aeruginosa*, and *Staphylococcus aureus* with MIC/MBC values of 780/1560, 700/1560, 48/192, and 48/96 µg/mL, respectively (Mensah et al., 1990).

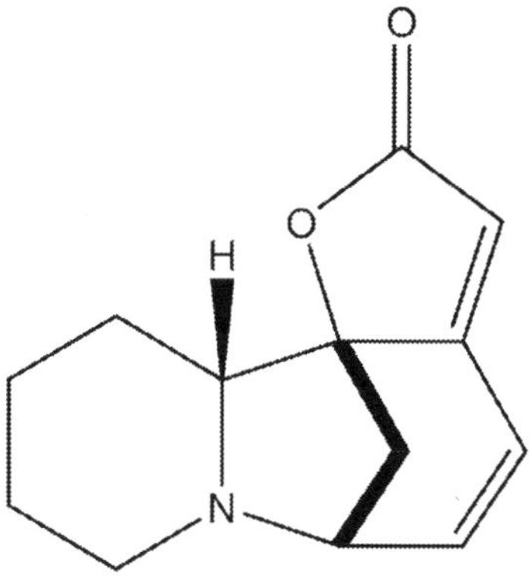

(+)-Viroallosecurinine

Moderate antimycobacterial securinega alkaloid: Viroallosecurinine inhibited the growth of *Mycobacterium smegmatis* with MIC/MBC values of 96/192 µg/mL, respectively (Mensah et al., 1990).

Moderate antifungal (filamentous) mid-polar extract: Chloroform extract of root bark inhibited the growth of *Trichophyton interdigitale* and *Microsporum floccosum* with MIC values of 125 and 250 µg/mL, respectively (Dickson et al., 2006).

Strong antiviral (enveloped monopartite linear dimeric single-stranded (+)RNA) securinega alkaloids: Fluevirosinine B inhibited the replication of the Human immunodeficiency virus type-1 (strain NL 43) in MT-4 cells with an EC_{50} value of 14.1 µM and a CC_{50} (cytotoxic concentration) value above 100 (Zhang et al., 2015). Flueggenine D and virosecurinine inhibited the replication of the Human immunodeficiency virus type-1 with IC_{50} values of 7.8 and 19.3µM and CC_{50} values of 97.9 and >100, respectively (Zhang et al., 2015).

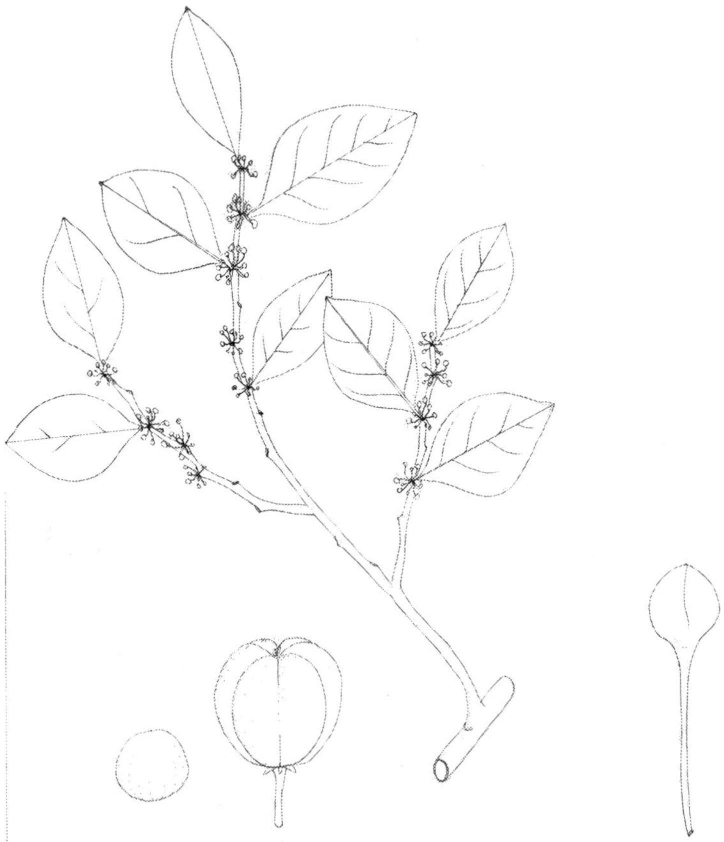

FIGURE 7.3 *Flueggea virosa* (Roxb. ex Willd.) Royle

Viral enzyme inhibition by polar extract: Ethanol extract at the concentration of 200 μg/mL inhibited the Human immunodeficiency virus type-1 reverse transcriptase by 88.2% (Wiwat & Kwantrairat, 2014).

Commentaries: (i) The plant brings to being bergenin (see earlier) (Agber et al., 2020). (ii) Consider that indolizidine alkaloids inhibit the replication of coronaviruses (Yang et al., 2010).

Extract of *Securinega durissima* J.F. Gmel. inhibited the replication of Chikungunya virus with an IC_{50} value of 15.9 μg/mL (Ledoux et al., 2018). Thus fluevirosinine B and congeners could be examined for their anti-Severe acute respiratory syndrome-acquired coronavirus 2 and other enveloped single-stranded (+)-RNA viruses.

REFERENCES

Agber, C.T., Tor-Anyii, T.A., Igoli, J.O. and Anyam, J.V., 2020. Isolation and characterisation of bergenin from ethyl acetate extract of flueggea virosa leaves. *Journal of Chemical Society of Nigeria*, *45*(6).

Dickson, R.A., Houghton, P.J., Hylands, P.J. and Gibbons, S., 2006. Antimicrobial, resistance-modifying effects, antioxidant and free radical scavenging activities of Mezoneuron benthamianum Baill., Securinega virosa Roxb. &Wlld. and Microglossa pyrifolia Lam. *Phytotherapy Research: An International Journal Devoted to Pharmacological and Toxicological Evaluation of Natural Product Derivatives*, *20*(1), pp. 41–45.

Ledoux, A., Cao, M., Jansen, O., Mamede, L., Campos, P.E., Payet, B., Clerc, P., Grondin, I., Girard-Valenciennes, E., Hermann, T. and Litaudon, M., 2018. Antiplasmodial, anti-chikungunya virus and antioxidant activities of 64 endemic plants from the Mascarene Islands. *International journal of antimicrobial agents*, *52*(5), pp. 622–628.

Mensah, J.L., Lagarde, I., Ceschin, C., Michelb, G., Gleye, J. and Fouraste, I., 1990. Antibacterial activity of the leaves of Phyllanthus discoideus. *Journal of ethnopharmacology*, *28*(1), pp. 129–133.

Tatematsu, H., Mori, M., Yang, T.H., Chang, J.J., Lee, T.T.Y. and Lee, K.H., 1991. Cytotoxic principles of Securinega virosa: Virosecurinine and viroallosecurinine and related derivatives. *Journal of Pharmaceutical Sciences*, *80*(4), pp. 325–327.

Yang, C.W., Lee, Y.Z., Kang, I.J., Barnard, D.L., Jan, J.T., Lin, D., Huang, C.W., Yeh, T.K., Chao, Y.S. and Lee, S.J., 2010. Identification of phenanthroindolizines and phenanthroquinolizidines as novel potent anti-coronaviral agents for porcine enteropathogenic coronavirus transmissible gastroenteritis virus and human severe acute respiratory syndrome coronavirus. *Antiviral Research*, *88*(2), pp. 160–168.

Zhang, H., Han, Y.S., Wainberg, M.A. and Yue, J.M., 2015. Anti-HIV Securinega alkaloid oligomers from Flueggea virosa. *Tetrahedron*, *71*(22), pp. 3671–3679.

Zhang, H., Zhang, C.R., Han, Y.S., Wainberg, M.A. and Yue, J.M., 2015. New Securinega alkaloids with anti-HIV activity from Flueggea virosa. *RSC Advances*, *5*(129), pp. 107045–107053.

Wiwat, C. and Kwantrairat, S., 2014. HIV-1 Reverse Transcriptase Inhibitors fromThai Medicinal Plants and Elephantopus scaber Linn. *Mahidol U-niversity Journal of Pharmaceutical Sciences*, *40*(3), pp. 35–44.

7.4.6.4 *Glochidion littorale* Bl.

Synonyms: *Bradleia littorea* (Bl.) Steud.; *Diasperus litoralis* (Bl.) Kuntze; *Phyllanthus littoralis* (Bl.) Müll. Arg.

Common names: Monkey apple; jambu kera, selensur, ubah (Malaysia) kapo-kapo, ketumbang (Indonesia), bagnang-lalake, kayong, nigad, padi-padi, sagasa, tabañgo (the Philippines), mun poo (Thailand); mui (Vietnam)

Habitats: Seashores

Distribution: India, Sri Lanka, Myanmar, Malaysia, Cambodia, Laos, Vietnam, Thailand, Vietnam, the Philippines, and Indonesia.

Botanical observation: This handsome tree grows up to 6m tall. The stems are terete and somewhat slightly zig-zag shaped. The leaves are simple, alternate, and stipulate. The stipules are triangular, minute, and caducous. The petiole is fissured, up to 5mm long, and glabrous. The blade is dark green and glossy above, coriaceous, elliptic to obovate, 2.2–9.2 × 1.9–5 cm, rounded to shortly attenuate at base, slightly asymmetric, emarginate to rounded at apex, and marked with 4–6 pairs of secondary nerves. The inflorescences are axillary fascicles. The calyx comprises 6 sepals basally united, and the margin of which is usually hyaline. The calyx develops 5 minute and triangular sepals. The androecium consists of 5–6 stamens. The gynoecium includes a 10–14-locular ovary longitudinally ribbed and stigmas, which form a cone 11–14 dentate. The capsules are oblate, edible, somewhat like minute pumpkins, up to 1.4cm in diameter, lobed, dull pinkish whitish to dark pink, smooth, and containing red seeds which are about 6mm long.

Medicinal uses: Dysentery, post-partum (Indonesia; Malaysia); tonsillitis (Indonesia)

Commentaries: (i) The plant has apparently not been examined for antimicrobial effects. The triterpene lup-20 (29)-ene-3α, 23-diol isolated from *Glochidion daltonii* (Müll. Arg.) Kurz evoked weak antibacterial activities against *Escherichia coli* (ATCC 25922), *Pseudomonas aeruginosa* (ATCC 27853), and *Serratia marcescens*, and was synergistic with tetracycline against *Pseudomonas aeruginosa* (ATCC 27853) (Kongcharoensuntorn et al., 2019). Other group of natural

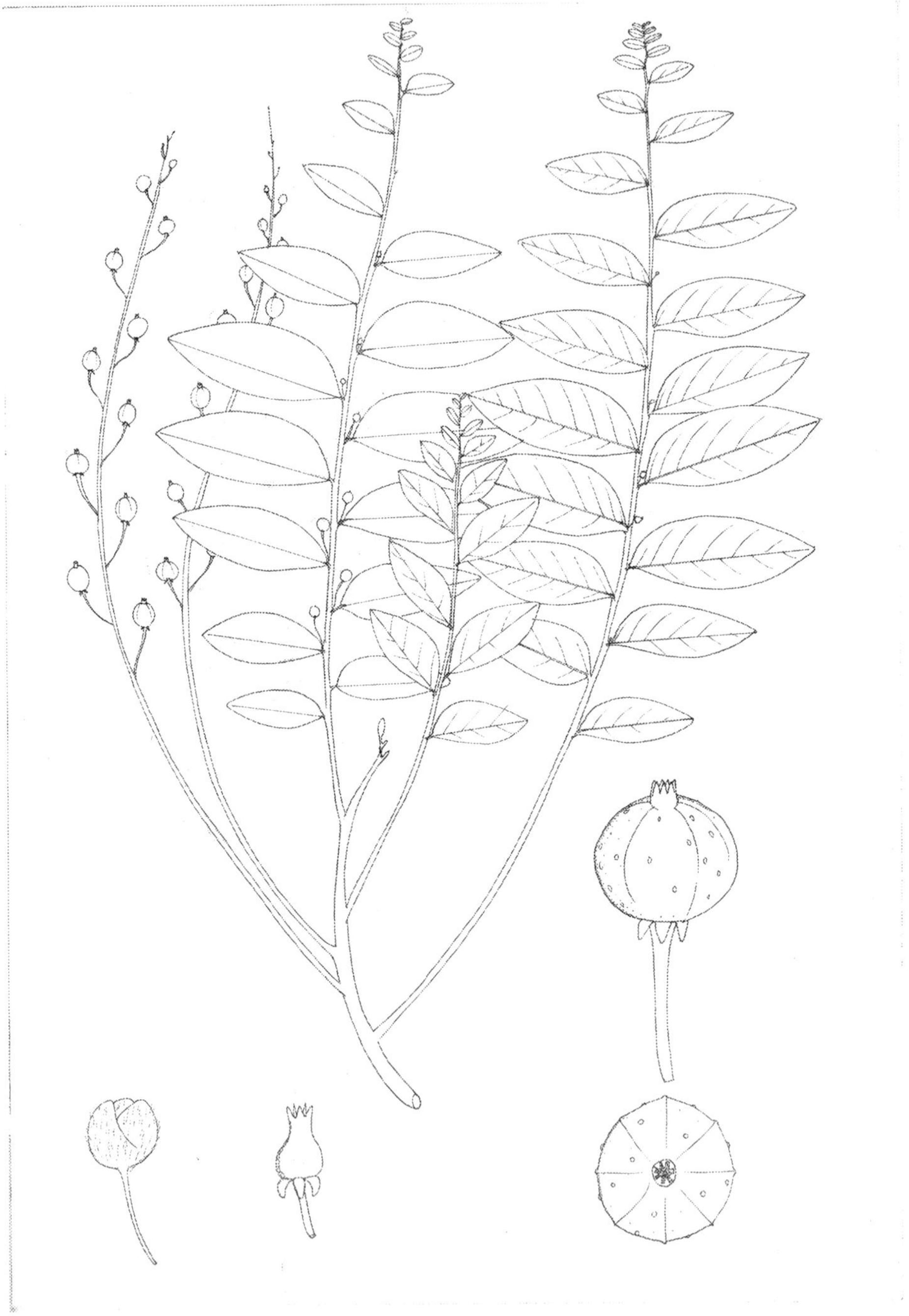

FIGURE 7.4 *Glochidion obscurum* (Roxb. ex Willd.) Bl.

products of interest generated by plants in the genus *Glochidion* J.R. Forst. & G. Forst. are flavanol gallates, such as dihydromyricetin-4′-*O*-(3″-*O*-methyl)-gallate, as well as dihydromyricetin (Yin et al., 2010) and glochidioboside (Takeda et al., 1991). Glochidioboside inhibited the growth of *Candida albicans* (ATCC 90028), *Candida parapsilosis* (ATCC 22019), *Trichophyton beigelii* (KTC 7707), and *Malassezia furfur* (KTC 77444) with MIC values of 6.5, 3.3, 6.5, and 6.5 μg/mL (Lee et al., 2014). Glochidioboside inhibited the growth of *Enterococcus faecalis* (ATCC 29212), *Staphylococcus aureus* (ATCC 25923), methicillin-resistant *Staphylococcus aureus*, *Escherichia coli* O157 (ATCC 43895), and *Pseudomonas aeruginosa* (ATCC 27853) with MIC values of 10, 5, 10, 5, and 5 μg/mL, respectively, via a mechanism involving the formation of cytoplasmic membrane pores and the subsequent leakage of cytoplasmic constituents (Lee et al., 2015).

(ii) Members of the genus *Glochidion* J.R. Forst. & G. Forst. are often used to treat microbial infections in Asia and Pacific. *Glochidion obscurum* (Roxb. ex Willd.) Bl. is traditionally used in Malaysia to treat diarrhea (local name: *Dulang dulang*) and in Indonesia for dysentery and after childbirth (local name: *Ki pare lalaki*). In Indonesia, *Glochidium rubrum* Bl. is used to treat cough (local name: *Dempul*). In Papua New Guinea, *Glochidion submolle* (K. Schum. & Lauterb.) Airy Shaw (local name: *Hin*) is used to treat tooth ache. *Glochidion multiloculare* (Rottler ex Willd.) Voigt is used in Bangladesh (local name: *Kudurpala*) to treat diarrhea.

REFERENCES

Kongcharoensuntorn, W., Naengchomnong, W. and Samae, A., 2019. Synergistic Antibacterial Effect of Lup-20 (29)-ene-3α, 23-diol from Glochidion daltonii (MÜll. Arg.) Kurz. and Antibiotics on Opportunistic Bacteria. *Chonburi Hospital Journal-วารสาร โรง พยาบาล ชลบุรี*, *44*(3), pp. 207–207.

Lee, H., Choi, H., Ko, H.J., Woo, E.R. and Lee, D.G., 2014. Antifungal effect and mode of action of glochidioboside against Candida albicans membranes. *Biochemical and Biophysical Research Communications*, *444*(1), pp. 30–35.

Takeda, Y., Mima, C., Masuda, T., Hirata, E., Takushi, A. and Otsuka, H., 1998. Glochidioboside, a glucoside of (7S, 8R)-dihydrodehydrodiconiferyl alcohol from leaves of Glochidion obovatum. *Phytochemistry*, *49*(7), pp. 2137–2139.

Yin, S., Sykes, M.L., Davis, R.A., Shelper, T., Avery, V.M., Camp, D. and Quinn, R.J., 2010. New galloylated flavanols from the Australian plant Glochidion sumatranum. *Planta Medica*, *76*(16), pp. 1877–1881.

7.4.6.5 *Phyllanthus emblica* L.

Synonyms: *Diasperus emblica* (L.) Kuntze; *Dichelactina nodicaulis* Hance; *Emblica officinalis* Gaertn.; *Phyllanthus mairei* H. Lév.

Common names: Emblic myrobalan; amla, amloki; amlokhi; kada mola; khulu; pyandhum (Bangladesh); kam lam (Cambodia); pokok melaka (Malaysia); alma khushk, amla (Indian); malaka (Indonesia); ta sah pen (Myanmar); neli (the Philippines); mak kam pom (Laos); bing gnot (Vietnam)

Habitat: Roadsides, villages, cultivated

Distribution: India to South China

Botanical observation: This graceful tree grows to a height of about 8 m. The bark is greyish red. The inner bark is sappy and pink, and the heart wood is purple brown. The stems are more or less rusty and scurfy. The leaves are simple, alternate, looking like pinnate, and up to about 1 cm long. The petiole is minute. The blade is linear, dull green, 0.5–1.6×0.1–0.3 cm, asymmetrical at base, and round to acute at apex. The flowers are minute and arranged in dense axillary clusters. The calyx comprises 6 sepals, which are white, oblong, and spathulate. Perianth none. A nectary disc is present. The androecium includes 3 stamens fused in a column. The gynoecium comprises a 3-locular ovary developing 3 styles, which are free and spreading. The drupes are globose, fleshy, up to about 2 cm in diameter, smooth, greenish yellow, and edible, sour at first then sweet, and sheltering a woody and somewhat winged stone containing a few seeds.

Medicinal uses: Leucorrhea, tympanites, urinary tract infections, jaundice, fever (Bangladesh); diarrhea, dysentery (India)

Broad-spectrum antibacterial polar extract: Methanol extract of leaves (30 µL of a 2000 µg/mL solution in 6 mm diameter wells) inhibited the growth of *Escherichia coli* (MTCC 739), *Bacillus subtilis* (MTCC 441), *Staphylococcus aureus* (MTCC 2940), *Proteus vulgaris* (MTCC 426), and *Enterococcus aerogenes* (MTCC 111) with the inhibition zone diameters of 12, 13, 12, 15, and 16 mm, respectively (Ghosh et al., 2008).

Moderate antibacterial (Gram-positive) simple phenolic acid: Gallic acid inhibited the growth of 17 strains of methicillin-resistant *Staphylococcus aureus* and *Staphylococcus aureus* (ATCC 25923) with MIC values of 63–125 and 125 µg/mL, respectively (Chusri & Voravuthikunchai, 2009).

Strong antiviral (enveloped monopartite linear dimeric single-stranded (+)RNA) mid-polar extract: Ethyl acetate fraction of fruits inhibited the replication of the Human immunodeficiency virus type-1 (NL4.3 strain) in CD4+ T cell line CEM-GFP by 43.8% with a maximum non-cytotoxic concentration of 15 µg/mL (Sabde et al., 2011).

Strong antiviral (enveloped linear monopartite double-stranded DNA) norbisabolane-type sesquiterpene glycoside: Phyllaemblicin H6 isolated from the roots inhibited the Herpes simplex virus type-1 with an IC_{50} of 10.5 µg/mL and a CC_{50} of 26.7 µg/mL in Vero cells (Lv et al., 2015).

Strong antiviral (non-enveloped linear monopartite single-stranded (+)RNA) norbisabolane-type sesquiterpene glycosides: Phyllaemblicin B, phyllaemblicin C, and phyllaemblic acid methyl ester isolated from the roots inhibited the replication of Coxsackie virus B3 (Nacy strain) in HeLa cell with IC_{50} values of 7.8, 11, and 21.8 µg/mL and the selectivity indices of 6.4, 6.2, and 9.8, respectively (Liu et al., 2009). Phyllaemblicin H12 isolated from the roots inhibited Coxsackievirus B3 with IC_{50} of 21.0 and CC_{50} of 55.1 µg/mL in Vero cells (Lv et al., 2015).

Strong antiviral (enveloped segmented linear single-stranded (−)RNA) norbisabolane-type sesquiterpene glycoside: Phyllaemblicin B isolated from the roots inhibited the Influenza H3N2 virus with an IC_{50} of 2.6 and a CC_{50} of 6.9 µg/mL in Madin-Darby canine kidney cells (Lv et al., 2015). Glochicoccinoside D isolated from the roots inhibited the Influenza H3N2 virus with an IC_{50} of 4.5 µg/mL and a CC_{50} of 14.3 µg/mL in Madin-Darby canine kidney cells (Lv et al., 2015).

Strong antiviral (non-enveloped monopartite single-stranded (+)RNA) norbisabolane-type sesquiterpene glycoside: Glochicoccinoside D isolated from the roots inhibited the Enterovirus 71 with an IC_{50} of 2.6 µg/mL and a CC_{50} of 9.9 µg/mL in Vero cells (Lv et al., 2015).

Strong antiviral (enveloped circular double-stranded DNA) norbisabolane-type sesquiterpene: Phyllaemblicin G6 inhibited the Hepatitis B virus surface antigen and e-antigen with IC_{50} values of 8.5 and 5.6 µM, respectively (Lv et al., 2014).

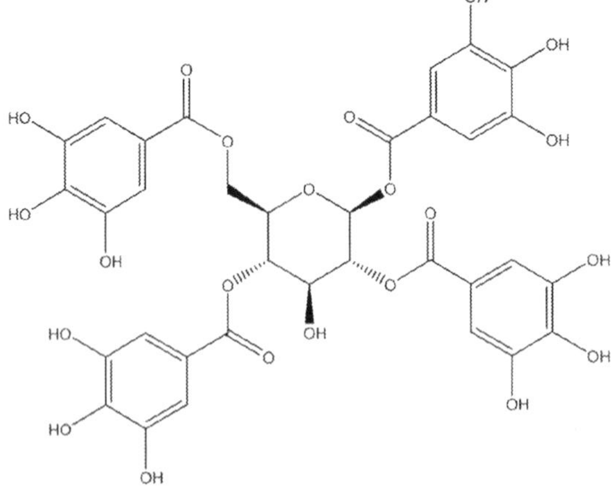

1,2,4,6-tetra-*O*-galloyl-β-D-glucose

Strong antiviral (enveloped linear monopartite double-stranded DNA) gallotannin: 1,2,4,6-tetra-*O*-galloyl-β-D-glucose isolated from the leaves and stems inhibited the replication of the Herpes simplex virus type-1 (ATCC VR-733) and Herpes simplex virus type-2 (strain 333) in Vero cells via virucidal effect and inhibiting virus synthesis on host cells (Xiang et al., 2011).

Viral enzyme inhibition by ellagitannin: Chebulagic acid inhibited the Hepatitis C virus protease with an IC_{50} value of 5.2 μM, respectively (Ajala et al., 2014).

Commentaries: (i) The plant generates ellagitannins and gallotannins as well as gallic acid (Yang & Liu, 2014). (ii) The potentials of extracts from this plant or fruits as herbal products to prevent or manage COVID-19 and other zoonotic pandemic viruses need to be examined.

REFERENCES

Ajala, O.S., Jukov, A. and Ma, C.M., 2014. Hepatitis C virus inhibitory hydrolysable tannins from the fruits of Terminalia chebula. *Fitoterapia*, *99*, pp. 117–123.

Ghosh, A., Das, B.K., Roy, A., Mandal, B. and Chandra, G., 2008. Antibacterial activity of some medicinal plant extracts. *Journal of Natural Medicines*, *62*(2), pp. 259–262.

Liu, Q., Wang, Y.F., Chen, R.J., Zhang, M.Y., Wang, Y.F., Yang, C.R. and Zhang, Y.J., 2009. Anti-coxsackie virus B3 norsesquiterpenoids from the roots of Phyllanthus emblica. *Journal of Natural Products*, *72*(5), pp. 969–972.

Lv, J.J., Wang, Y.F., Zhang, J.M., Yu, S., Wang, D., Zhu, H.T., Cheng, R.R., Yang, C.R., Xu, M. and Zhang, Y.J., 2014. Anti-hepatitis B virus activities and absolute configurations of sesquiterpenoid glycosides from Phyllanthus emblica. *Organic & Biomolecular Chemistry*, *12*(43), pp. 8764–8774.

Lv, J.J., Yu, S., Xin, Y., Cheng, R.R., Zhu, H.T., Wang, D., Yang, C.R., Xu, M. and Zhang, Y.J., 2015. Anti-viral and cytotoxic norbisabolane sesquiterpenoid glycosides from Phyllanthus emblica and their absolute configurations. *Phytochemistry*, *117*, pp. 123–134.

Sabde, S., Bodiwala, H.S., Karmase, A., Deshpande, P.J., Kaur, A., Ahmed, N., Chauthe, S.K., Brahmbhatt, K.G., Phadke, R.U., Mitra, D. and Bhutani, K.K., 2011. Anti-HIV activity of Indian medicinal plants. *Journal of Natural Medicines*, *65*(3–4), pp. 662–669.

Voravuthikunchai, S.P., Suwalak, S. and Mitranan, W., 2012. Ellagitannin from Quercus infectoria eradicates intestinal colonization and prevents renal injuries in mice infected with Escherichia coli O157: H7. *Journal of Medical Microbiology*, *61*(10), pp. 1366–1372.

Lv, J.J., Yu, S., Xin, Y., Cheng, R.R., Zhu, H.T., Wang, D., Yang, C.R., Xu, M. and Zhang, Y.J., 2015. Anti-viral and cytotoxic norbisabolane sesquiterpenoid glycosides from Phyllanthus emblica and their absolute configurations. *Phytochemistry*, *117*, pp. 123–134.

Lv, J.J., Yu, S., Xin, Y., Cheng, R.R., Zhu, H.T., Wang, D., Yang, C.R., Xu, M. and Zhang, Y.J., 2015. Anti-viral and cytotoxic norbisabolane sesquiterpenoid glycosides from Phyllanthus emblica and their absolute configurations. *Phytochemistry*, *117*, pp. 123–134.

7.4.6.6 *Phyllanthus niruri* L.

Synonyms: *Diasperus niruri* (L.) Kuntze; *Phyllanthus asperulatus* Hutch.; *Phyllanthus filiformis* Pavon ex Baillon; *Phyllanthus lathyroides* Kunth; *Phyllanthus microphyllus* Mart.; *Phyllanthus niruri* var. *amarus* (Schumach. & Thonn.) Leandri

Common names: Niruri, stonebreaker; ban amlaki, bhui amla (Bangladesh); preak phle (Cambodia); ku wei ye xia zhu (China); memeniran (Indonesia); dukong anak (Malaysia); bhumy ámali, Bhui amla, sada hazarmani (India); manjinimbi (Papua New Guinea); sampa sampalukan (the Philippines); yaa tai bai (Thailand); cay cho de (Vietnam)

Habitat: Roadsides, vacant lots, in town, parkings, in garden pots

Distribution: Tropical Asia and Pacific

Botanical observation: This gracile herb grows to a height of 50 cm from somewhat fibrous roots. The stems are glabrous, green, somewhat coriaceous, and reddish or green. The leaves are simple, alternate, subsessile, and stipulate. The stipules are minute and acute. The blade is 5–9 mm × 3–4 mm, oblong-elliptic, somewhat asymmetrical, round at base and apex, and membranous. The inflorescences are axillary fascicles or solitary. The calyx includes 5 ovate sepals.

The androecium comprises 3 stamens fused in a column. A 5-lobed and flat disc is present. The gynoecium includes a lobed ovary developing 3 stigmas. The capsules are minute, depressed globose, 3-lobed, on persistent calyx, smooth, and contain a few longitudinally ribbed seeds.

Medicinal uses: Dysentery (Bangladesh; India); bronchitis, fever, infections, gonorrhea, ringworm, jaundice (Bangladesh); jaundice (Bangladesh; Malaysia); AIDS (Thailand)

Antibacterial (Gram-positive) polar extract: Ethanol extract (200 μg/6 mm disc) inhibited the growth of methicillin-resistant *Staphylococcus aureus* (clinical strain) and vancomycin-resistant *Enterococcus* with inhibition zone diameters of 11 and 13 mm, respectively (Valle et al., 2015).

Strong antiviral (enveloped monopartite linear single-stranded (+) RNA) polar extract: Methanol extract inhibited the replication of the Dengue virus type-1, -2, -3, and -4 with IC_{50} of 3.7, 8.3, 20.7, and 11.1 μg/mL, respectively (Sood et al., 2015).

Strong antiviral (enveloped monopartite linear single-stranded (+)RNA) amphiphilic apocarotenoid lactone: Loliolide (LogD = 1.4 at the pH 7.4; molecular mass = 196.2 g/mol) inhibited the replication of the Hepatitis C virus with an EC_{50} of 2.4 μg/mL and a selectivity index of 62.6 via inactivation of free Hepatitis C virus, blockage of virus attachment to Huh-7.5 cells as well as fusion (Chung et al., 2016).

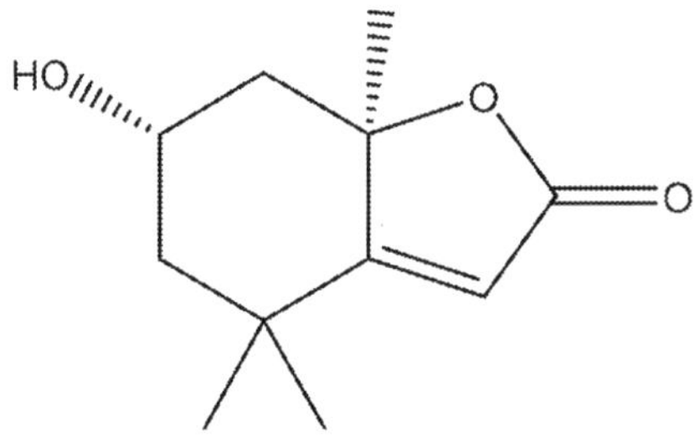

Loliolide

Strong antiviral (non-enveloped linear monopartite single-stranded (+)RNA) ellagitannin: Corilagin inhibited the replication of Coxsackievirus A16 and Enterovirus 71 in Vero cells with IC_{50} values of 32.3 and 5.6 μg/mL, respectively, and selectivity indices of 3.3 and 58.3, respectively (Yeo et al., 2015).

Strong antiviral (enveloped circular double-stranded DNA) aryltetralin-type lignan: Nirtetralin B inhibited the secretion of the Hepatitis B virus surface antigen and e-antigen secretion from HepG2.2.215 cells with IC_{50} values of 9.5 and 17.4 μM, respectively and inhibited Hepatitis B virus DNA level by about 50% at a concentration of 16.3 μM (Liu et al., 2014).

Strong antiviral (enveloped circular double-stranded DNA) dibenzyl butan lignan: Niranthin inhibited the secretion of the Hepatitis B virus surface antigen and e-antigen secretion from HepG2.2.215 cells with IC_{50} values of 15.6 and 25.1 μM, respectively and inhibited Hepatitis B virus DNA level by about 50% at a concentration of 16.3 μM (Liu et al., 2014a).

In vivo antiviral (enveloped circular double-stranded DNA) aryltetralin-type lignan: Nirtetralin B given intragastrically at a dose 100 mg/Kg/day for 14 days to ducks experimentally infected with the Hepatitis B virus reduced plasma Hepatitis B surface and e-antigen secretion by 64.9 and 54.5 %, respectively and evoked some levels of hepatic protection (Liu et al., 2014).

Niranthin given intragastrically at a dose 100 mg/Kg/day for 14 days to ducks experimentally infected with the Hepatitis B virus reduced plasma Hepatitis B surface and e-antigen secretion by 65.3% and 50% (Liu et al., 2014a).

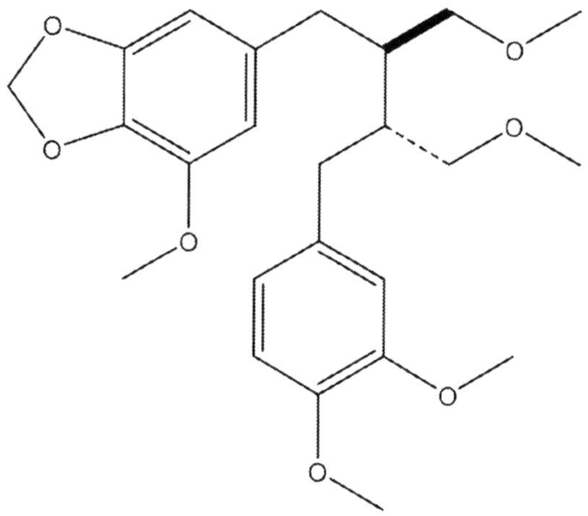

Niranthin

Viral enzyme inhibition by cinnamic acid glycosides: Niruriside inhibited the binding of the Human immunodeficiency virus regulation of virion expression protein to the virion expression protein responsive element RNA with an IC_{50} value of 3.3 µM (Qian-Cutrone et al., 1996).

Commentary: *Phyllanthus reticulatus* Poir. is traditionally used in Bangladesh (local name: *Simikdare*) to sanitize the teeth.

REFERENCES

Chung, C.Y., Liu, C.H., Burnouf, T., Wang, G.H., Chang, S.P., Jassey, A., Tai, C.J., Tai, C.J., Huang, C.J., Richardson, C.D. and Yen, M.H., 2016. Activity-based and fraction-guided analysis of Phyllanthus urinaria identifies loliolide as a potent inhibitor of hepatitis C virus entry. *Antiviral Research, 130*, pp. 58–68.

Liu, S., Wei, W., Li, Y., Lin, X., Shi, K., Cao, X. and Zhou, M., 2014. In vitro and in vivo anti-hepatitis B virus activities of the lignan nirtetralin B isolated from Phyllanthus niruri L. *Journal of Ethnopharmacology, 157*, pp. 62–68.

Liu, S., Wei, W., Shi, K., Cao, X., Zhou, M. and Liu, Z., 2014a. In vitro and in vivo anti-hepatitis B virus activities of the lignan niranthin isolated from Phyllanthus niruri L. *Journal of Ethnopharmacology, 155*(2), pp. 1061–1067.

Qian-Cutrone, J., Huang, S., Trimble, J., Li, H., Lin, P.F., Alam, M., Klohr, S.E. and Kadow, K.F., 1996. Niruriside, a new HIV REV/RRE binding inhibitor from Phyllanthus niruri. *Journal of Natural Products, 59*(2), pp. 196–199.

Sood, R., Raut, R., Tyagi, P., Pareek, P.K., Barman, T.K., Singhal, S., Shirumalla, R.K., Kanoje, V., Subbarayan, R., Rajerethinam, R. and Sharma, N., 2015. Cissampelos pareira Linn: Natural source of potent antiviral activity against all four dengue virus serotypes. *PLOS Neglected Tropical Diseases, 9*(12), p. e0004255.

Valle Jr, D.L., Andrade, J.I., Puzon, J.J.M., Cabrera, E.C. and Rivera, W.L., 2015. Antibacterial activities of ethanol extracts of Philippine medicinal plants against multidrug-resistant bacteria. *Asian Pacific Journal of Tropical Biomedicine, 5*(7), pp. 532–540.

Wei, W., Li, X., Wang, K., Zheng, Z. and Zhou, M., 2012. Lignans with anti-hepatitis B virus activities from Phyllanthus niruri L. *Phytotherapy Research, 26*(7), pp. 964–968.

Yeo, S.G., Song, J.H., Hong, E.H., Lee, B.R., Kwon, Y.S., Chang, S.Y., Kim, S.H., won Lee, S., Park, J.H. and Ko, H.J., 2015. Antiviral effects of Phyllanthus urinaria containing corilagin against human Enterovirus 71 and Coxsackievirus A16 in vitro. *Archives of Pharmacal Research, 38*(2), pp. 193–202.

7.4.7 Family Rafflesiaceae Dumortier (1829)

The family Rafflesiaceae comprise 3 genera and 20 species of fleshy parasites of roots of trees. These are only found in tropical Asia. The leaves are minute or vestigial. The flowers are often gigantic. The calyx presents 4 or more sepals. The androecium includes 5 or more stamens. The gynoecium includes 4 or more carpels and contains numerous ovules. The fruits are berry like. Members of this family are tanniferous.

7.4.7.1 *Rafflesia hasseltii* Suringar

Common name: Bunga patma (Malaysia)

Habitat: Primary lowland dipterocarp forests to lower montane forests

Distribution: Peninsular Malaysia and Sumatra

Botanical observation: This magnificent flower (growing on the roots of *Tetrastigma leucostaphylum* (Dennst.) Mab) has a diameter of about 45 cm. The perianth develops 5 lobes, which are 10–13 × 14–18 cm, thick, oblong to round, revolute somewhat chocolate reddish brown, with 7–9 whitish spots. A central disc is present and supports about 20 elongated finger-like bodies. The lower surface of the corona presents numerous bristles. The ramenta are a linear and swollen apices.

Medicinal use: Post-partum (Malaysia)

Broad-spectrum antibacterial polar extract: Methanol extract of the whole plant (1 mg/disc, 6 mm diameter) inhibited the growth of *Bacillus cereus*, *Bacillus subtilis*, *Pseudomonas aeruginosa*, and *Staphylococcus aureus* with inhibition zone diameters of 7, 12, 9, and 7 mm, respectively (Wiart et al., 2004).

Commentary: Members of the genus *Rafflesia* R. Br. ex Gray abound with hydrolysable tannins (Kanchanapoom et al., 2007), which may account for the antibacterial property and traditional use of *Rafflesia hasseltii* Suringar.

REFERENCES

Kanchanapoom, T., Kamel, M.S., Picheansoonthon, C., Luecha, P., Kasai, R. and Yamasaki, K., 2007. Hydrolyzable tannins and phenylpropanoid from Rafflesia kerrii Meijer (Rafflesiaceae). *Journal of Natural Medicines*, *61*(4), pp. 478–479.

Wiart, C., Mogana, S., Khalifah, S., Mahan, M., Ismail, S., Buckle, M., Narayana, A.K. and Sulaiman, M., 2004. Antimicrobial screening of plants used for traditional medicine in the state of Perak, Peninsular Malaysia. *Fitoterapia*, *75*(1), pp. 68–73.

7.4.8 Family Rhizophoraceae C. H. Persoon (1806)

The Family Rhizophoraceae consists of about 14 genera and about 100 species of shrubs and trees not uncommonly found in mangroves and developing aerial roots. The leaves are simple, entire, opposite, and stipulate. The flowers are solitary or in cymes, and actinomorphic. The perianth comprises 4–5 or more sepals and petals which are valvate and fleshy. The stamens are twice, thrice, or five times as numerous as the petals. A nectary disc is present. The gynoecium comprises 2–6 carpels united to form a compound ovary with as many locules as carpels. Each locule contains 2 ovules attached to apical-axillary placentas.

7.4.8.1 *Bruguiera sexangula* (Lour.) Poir.

Synonyms: *Bruguiera eriopetala* Wight & Arn.; *Bruguiera sexangula* var. *rhynchopetala* W.C. Ko; *Rhizophora sexangula* Lour.

Common names: Plaông prâsak (Cambodia); hai lian (China); bandari (India); bakau tampusing; mata buaya (India); saung (Myanmar); tagasa (the Philippines); prasak (Thailand); vet (Vietnam)

Habitat: Mangroves

Distribution: India, Sri Lanka, Bangladesh, Myanmar, Cambodia, Thailand, Vietnam, Malaysia, Indonesia, the Philippines, and Pacific islands

Botanical observation: This handsome tree grows to 8 m tall. The bark is greyish and smooth. The apical stems are rough and somewhat annulated. The leaves are simple, spiral, and stipulate.

The stipule is up to about 4cm long. The petiole is up to about 3.5cm long. The blade is elliptic to oblanceolate, coriaceous, 8–13×3–6cm, cuneate at base, and acute at apex. The flowers are solitary on about 1 cm long peduncles. The calyx is tubular, conical, somewhat fleshy, about 1.5cm long, and develops about 10–12 somewhat sharp and about 2cm long lobes. The 10–12 petals are about 1.5cm long, whitish yellow, and bifid at apex. The about 22 stamens are 1.5cm long. The style is slender and about 2cm long. The fruit is elongated, coriaceous, designed to fall and plug into the mud straight, up to about 10cm long and seated on a 1.8 cm long calyx.

Medicinal uses: Burns (India); shingles (Malaysia)

Antibacterial (Gram-negative) polar extract: Ethanol extract of leaves inhibited the growth of *Escherichia coli* (clinical strain) (Abeysinghe et al., 2010).

Commentaries: (i) A curious fact about mangrove Rhizophoraceae is that they are the home of symbiotic filamentous fungi including *Penicillium*, *Streptomyces*, and *Xylaria* species, which produce antibacterial and antifungal principles: meroterpenes, alkaloids, and benzophenones (Caijuan et al., 2014; Xu et al., 2016, 2019), which may shield the plant against pathogenic microorganisms. A point to grasp is that all life forms are interdependent, and this is particularly the case between plants and fungi or bacteria for nutrient exchange. In the soil, fungi and bacteria release minute amounts of antibiotic and antifungal substances that seem to be at concentrations below MICs, raising the question of what the second role of antifungal and antibiotic agents is. One may think of communication substance. Thus, when one is looking for a specific antibacterial or antifungal agent for crops, there is the possibility to isolate such agents from the symbiotic microorganisms living on that plant. Consider that the antimicrobial substances generated by symbiotic microorganisms may be fungal or bacterial metabolites of initial molecules generated by the plant itself. One could speak of a universal microbial homeostasis, which by the way is being unbalanced by the use of pesticides and antibiotics and antifungals in the environment, which will result in the emergence of untreatable human pathogenic or phytopathogenic bacteria or fungi. Consider that the use of pesticides may result in the selection of untreatable phytopathogenic microorganisms, hence catastrophic consequence in terms of crop loss and risks of global famine (Lucas, 2015).

(ii) *Bruguiera cylindrica* (L.) Bl. is used to treat cough in Thailand (local name: *Thua khao*) and *Bruguiera gymnorhiza* (L.) Savigny is used in India to treat dysentery and diarrhea and in Palau (local name: *Adege*) to treat otitis media. The cyanogenic glycosides menisdaurin B, menisdaurin, and coclauril isolated from *Bruguiera gymnorhiza* (L.) Savigny inhibited the replication of the Hepatitis B virus with IC_{50} values of 8.7, 5.1, and 7.6 µg/mL and selectivity indices of 8.3, 11.3, and 11, respectively (Yi et al., 2015). Are these cyanogenic glycosides of fungal origin?

REFERENCES

Abeysinghe, P.D., 2010. Antibacterial activity of some medicinal mangroves against antibiotic resistant pathogenic bacteria. *Indian Journal of Pharmaceutical Sciences*, *72*(2), p.167.

Caijuan, Z., Guolei, H., Xiongzhao, T., Deneng, W., Xiaolu, G., Qiang, Z., Xiaoping, S. and Guangying, C., 2014. Secondary Metabolites and Antibacterial Activities of a Bruguiera sexangula var. Rhynchopetala-Derived Fungus Penicillium sp (J41221). *Chinese Journal of Organic Chemistry*, *34*(6), pp. 1172–1176.

Lucas, J.A., Hawkins, N.J. and Fraaije, B.A., 2015. The evolution of fungicide resistance. Advances in *Applied Microbiology*, 90, pp. 29–92.

Xu, R., Li, X.M. and Wang, B.G., 2016. Penicisimpins A C, three new dihydroisocoumarins from Penicillium simplicissimum MA-332, a marine fungus derived from the rhizosphere of the mangrove plant Bruguiera sexangula var. rhynchopetala. *Phytochemistry Letters*, *17*, pp. 114–118.

Xu, Z., Xiong, B. and Xu, J., 2019. Chemical Investigation of Secondary Metabolites Produced By Mangrove Endophytic Fungus Phyllosticta Capitalensis. *Natural Product Research*, pp. 1–5.

Yi, X.X., Deng, J.G., Gao, C.H., Hou, X.T., Li, F., Wang, Z.P., Hao, E.W., Xie, Y., Du, Z.C., Huang, H.X. and Huang, R.M., 2015. Four new cyclohexylideneacetonitrile derivatives from the hypocotyl of mangrove (Bruguiera gymnorrhiza). *Molecules*, *20*(8), pp. 14565–14575.

7.4.8.2 *Carallia suffruticosa* Ridl.

Synonym: *Carallia fascicularis* Guill

Common names: Sisek puyu, redip pepuyuh, tulang daeng (Malaysia); sang ma (Vietnam)

Habitat: Forests, cultivated

Distribution: Vietnam, Thailand, and Malaysia

Botanical observation: This treelet grows to 4 m tall. The bark is greenish brown with prominent lenticels. The stems are slender. The leaves are decussate, simple, and stipulate. The stipule is narrowly lanceolate and about 1 cm long. The petiole is up to 1.5 cm long and channeled above. The blade is papery, dull green, smooth, lanceolate, up to about 3–6 × 10–18 cm, marked with black dots at base, sharply and finely toothed, and shows 8–13 pairs of secondary nerves, a few tertiary nerves, and a midrib which is sunken above. The inflorescences are axillary cymes of yellow flowers. The calyx comprises 5–7 sepals, which are about 4 mm long. The corolla includes 5–8 petals, which are about 4 mm long. The androecium includes 16 stamens. The berries are red and ovoid to oblate, marked at apex by vestiges or perianth, about 1.5 cm × 7 mm, and sheltering a few 1 cm long seeds.

Medicinal uses: Boils (Malaysia, Vietnam); post-partum, fever (Malaysia)

Commentaries: (i) The plant has apparently not been examined for antimicrobial effects and may disappear soon owed to constant palm oil deforestation and logging. *Carallia brachiata* (Lour.) Merr. is traditionally used in India (local name: *Vallabham*) to heal wounds and to treat stomatitis. In Myanmar, it is used for wounds, eye infection, and fever (local name: *Yat*). The pyrrolidine alkaloid (+)-hygroline has been isolated from this plant (Fitzgerald, 1965), (+)-hygroline is a cholinergic antagonist (Morales et al., 2003). As discussed earlier, neuroactive substances are not uncommonly able to inhibit bacterial efflux pumps. This is exemplified with the anticholinergic drug Dicyclomine used in the treatment of irritable bowel syndrome and efflux pump inhibitor (Marshall et al., 2020). One could examine the antibiotic-potentiating properties of (+)-hygroline and congeners in the family Rhizophoraceae.

REFERENCES

Fitzgerald, J.S., 1965. (+)-Hygroline, the major alkaloid of Carallia brachiata (Rhizophoraceae). *Australian Journal of Chemistry*, *18*(4), pp. 589–590.

Marshall, R.L., Lloyd, G.S., Lawler, A.J., Element, S.J., Kaur, J., Ciusa, M.L., Ricci, V., Tschumi, A., Kühne, H., Alderwick, L.J. and Piddock, L.J., 2020. New multidrug efflux inhibitors for gram-negative bacteria. *Mbio*, *11*(4).

Morales, M.A., Ahumada, F., Castillo, E., Burgos, R., Christen, P., Bustos, V. and Muñoz, O., 2013. Inhibition of cholinergic contractions of rat ileum by tropane-type alkaloids present in Schizanthus hookeri. *Zeitschrift für Naturforschung C*, *68*(5–6), pp. 203–209.

7.4.8.3 *Ceriops tagal* (Perr.) C.B. Rob.

Synonym: *Ceriops candolleana* Arn.

Common names: Goran (Bangladesh; India); same (Cambodia); tengar (Indonesia); tengar putih (Malaysia); tangal (the Philippines); prong (Thailand); tagal dza do (Vietnam)

Habitat: Mangroves, muddy swamps

Distribution: India, Sri Lanka, Bangladesh, Myanmar, Cambodia, Thailand, Vietnam, China, Taiwan, Malaysia, Indonesia, the Philippines, Papua New Guinea, and Northern Australia

Botanical observation: This handsome tree grows to a height of 8 m. The bole is straight. The bark is brownish and inner bark somewhat reddish. The leaves are simple, opposite, and stipulate. The stipules are lanceolate and up to 2 cm long. The petiole is 1–3 cm long. The blade is fleshy, coriaceous, smooth, obovate to spathulate, 4–9 × 2–5 cm, cuneate to tapering at base, and obtuse at apex. The inflorescences are cymes. The 5 sepals are oblong, about 1 cm long, greenish yellow, reddish at base, stout, coriaceous, sharply pointed and recurved at apex. The 5 petals are oblong, about 5 mm long, wavy, and develop at apex 3 appendages, which look like some sort of stamens, A disc is present. The androecium includes 10 stamens. The fruits are pendulous, elongated, and up to 30 cm long.

Medicinal uses: Cuts, dysentery skin diseases, ulcers (India); post-partum (Malaysia)

Weak broad-spectrum antibacterial pimarane-type diterpene: 16-Hydroxypimar-8(14)-en-15-one isolated from the roots inhibited the growth of *Bacillus cereus*, *Staphylococcus aureus*, *Micrococcus kristinae*, *Streptococcus pyogenes*, and *Salmonella pooni* with MIC values 100, 100, 100, 500, to 250 μg/mL, respectively (Chacha et al., 2008).

In vivo antiviral (enveloped spiral double-stranded DNA) polar extract: Aqueous extract given orally at a dose of 10% of the body weight twice a day protected shrimps against the White spot syndrome virus (Sudheer et al., 2011).

Commentary: (i) Members of the genus *Ceriops* Arn. bring to being abietane-, dolabrane-, kaurane-, and pimmarane-type diterpenes as well as dammarane-, lupane-, and oleanane-type triterpenes (Wang et al., 2012), which could be examined for their antimicrobial properties. (ii) *Ceriops decandra* (Griff.) W. Theob is used in India to treat jaundice and ulcers (local names: *Ghran, Jhamti garani*). Ethanol extract of bark of *Ceriops decandra* (Griff.) W. Theob. inhibited the growth of *Staphylococcus aureus*, *Streptococcus pyogenes*, *Shigella dysenteriae*, *Salmonella typhi*, *Escherichia coli*, *Vibrio cholerae*, *Enterobacter aerogenes*, and *Pseudomonas aeruginosa* with inhibition zone diameters of 10, 12, 10, 12, 12, 18, 16, and 12 mm, respectively (0.5 mg/disc) (Uddin et al., 2008). Ethanol extract inhibited the growth of *Pseudomonas aeruginosa*, *Enterobacter* sp., *Staphylococcus aureus* (all clinical isolates) with inhibition zone diameters of 8.2, 6, and 13 mm, respectively (6 mm diameter impregnated with crude extracts (5 mg/mL) (Ravikumar et al., 2010). A benzene extract of wood of *Ceriops decandra* inhibited the growth of *Bacillus subtilis*, *Bacillus coagulans*, *Bacillus cereus*, *Bacllus polymyxa*, *Bacillus pumilus*, *Micrococcus luteus*, *Shigella sonnei*, *Klebsiella pneumoniae*, and *Saccharomyces cerevisiae* with MIC values of 62.5, 62.5, 125, 15.6, 125, 125, 125, 500, and 78.1 μg/mL, respectively (Simlai et al., 2016). Aqueous ethanol extract of leaves inhibited the replication of the Vaccinia virus, Semliki forest virus, and the Human immunodeficiency virus with EC_{50} values of 48.1 (selectivity index: 3.9), 18 (selectivity index: 11), and 13.4 μg/mL, respectively (selectivity index: 16.2) (Premanathan et al., 1999). Thus, one could endeavor to look for anti-Severe acute respiratory syndrome-acquired coronavirus 2 and other coronavirus principles from members of the genus *Ceriops* Arn. since both Semliki forest virus and Severe acute respiratory syndrome-acquired coronavirus 2 are enveloped monopartite, linear single-stranded (+) RNA viruses.

REFERENCES

Chacha M, Maptise R, Afolayan AJ, Majinda RRT. 2008. Antibacterial diterpenes from the roots of ceriops tagal. *Natural Product Communications*, *3*, p. 17.

Premanathan, M., Kathiresan, K. and Nakashima, H., 1999. Mangrove halophytes: A source of antiviral substances. *South Pacific Study*, *19*(1–2), pp. 49–57.

Ravikumar, S., Gnanadesigan, M., Suganthi, P. and Ramalakshmi, A., 2010. Antibacterial potential of chosen mangrove plants against isolated urinary tract infectious bacterial pathogens. *International Journal of Medicine and Medical Sciences*, *2*(3), pp. 94–99.

Simlai, A., Mukherjee, K., Mandal, A., Bhattacharya, K., Samanta, A. and Roy, A., 2016. Partial purification and characterization of an antimicrobial activity from the wood extract of mangrove plant Ceriops decandra. *EXCLI Journal*, *15*, p. 103.

Sudheer, N.S., Philip, R. and Singh, I.B., 2011. In vivo screening of mangrove plants for anti WSSV activity in Penaeus monodon, and evaluation of Ceriops tagal as a potential source of antiviral molecules. *Aquaculture*, *311*(1–4), pp. 36–41.

Uddin, S.J., Rouf, R., Shilpi, J.A., Alamgir, M., Nahar, L. and Sarker, S.D., 2008. Screening of some Bangladeshi medicinal plants for in vitro antibacterial activity. *Oriental Pharmacy and Experimental Medicine*, *8*(3), pp. 316–321.

Wang, H., Li, M.Y. and Wu, J., 2012. Chemical constituents and some biological activities of plants from the genus Ceriops. *Chemistry & Biodiversity*, *9*(1), pp. 1–11.

7.4.8.4 *Rhizophora apiculata* Bl.

Synonyms: *Rhizophora candelaria* DC.; *Rhizophora conjugata* L.

Common names: Tall-stilted mangrove; hong shu (China); kantal, rai (India); bako (Indonesia); bakau minyak (Malaysia); pyu (Myanmar); pan a (Papua New Guinea); uakatan (the Philippines); kongkang bailek (Thailand)

Habitat: Mangroves and deep soft mud estuaries

Distribution: Asia-Pacific

Botanical observation: This sinister tree grows to a height of 30 m. The bole produces numerous stilt roots. The bark is grey with shallow horizontal cracks. The stems are somehow swollen, marked by conspicuous leaf scars, 1 cm in diameter, and annular. The stipules are 4–8 cm long. The petiole is 1.5–3 cm long and reddish. The blade is elliptic oblong, up to about 7×18 cm, reddish at margin, cuneate at base, and apiculate at apex. The flowers are green to yellowish, and arranged in pairs on 5 mm–1.5 cm long pedicels arising from the leafless parts of the stems. The calyx includes 4 lobes. The 4 petals are up to about 1 cm long. A disc is present. The androecium consists of 12 stamens. The fruits are club-shaped, cylindrical, with blunt tips, smooth, up to 40 × 1.5 cm, and brown with a few large lenticels.

Medicinal uses: Dysentery, post-partum (Malaysia); fever (Thailand); typhoid, hepatitis (India)

Broad-spectrum antibacterial polar extract: Hydrolyzable tannins extract of bark inhibited the growth of *Acinobacter calcoaceticus*, *Bacillus licheniformis*, *Proteus mirabilis*, and *Staphylococcus saprophyticus* with MIC/MBC values of 1500/3130, 1560/3130, 3130/6250, and 3130/6250 μg/mL, respectively (Lim et al., 2006).

Commentary: Tannins by virtue of their astringency (protein precipitation) might participate in the antidysenteric property of the plant. In Iran, *Rhizophora mucronata* Lam. (local name: *Chandal*) is used to heal wounds, and to treat diarrhea in China (local name: *Hong qie don*) and Japan. Alkaloid fraction of leaves of *Rhizophora mucronata* Lam. inhibited clinical isolates of *Bacillus cereus*, *Staphylococcus aureus*, *Streptococcus faecalis*, *Streptococcus pyogenes*, *Escherichia coli*, and *Pseudomonas aeruginosa* (Gurudeeban et al., 2015).

REFERENCES

Gurudeeban, S., Ramanathan, T. and Satyavani, K., 2015. Antimicrobial and radical scavenging effects of alkaloid extracts from Rhizophora mucronata. *Pharmaceutical Chemistry Journal*, *49*(1), pp. 34–37.

Lim, S.H., Darah, I. and Jain, K., 2006. Antimicrobial activities of tannins extracted from Rhizophora apiculata barks. *Journal of Tropical Forest Science*, pp. 59–65.

Joel, E.L. and Bhimba, V., 2010. Isolation and characterization of secondary metabolites from the mangrove plant Rhizophora mucronata. *Asian Pacific Journal of Tropical Medicine*, *3*(8), pp. 602–604.

7.4.9 Family Salicaceae Mirbel (1815)

The family Salicaceae consists of about 100 genera and about 1500 species of trees and shrubs. The leaves are alternate, simple, and stipulate. The inflorescences are spikes or racemes. The perianth is very much reduced. The androecium includes 2 stamens. The ovary is uni- or bi-locular, with several ovules of parietal placentas. These capsule enclosing seeds bearing long silky hairs or are berries, or capsules or drupes. The seeds are often embedded in an aril. Some taxonomists have recently incorporated the Flacourtiaceae in the Salicaceae although these are botanically and chemically distinct.

7.4.9.1 *Flacourtia indica* (Burm.f.) Merr.

Synonyms: *Flacourtia parvifolia* Merr.; *Flacourtia ramontchi* L'Hér.; *Gmelina indica* Burm. f.

Common names: Governor's Plum, Madagascar plum; bainchi (Bangladesh); krâk hôp nhii (Cambodia); ci li mu (China); aghori, baichi, bilangra (India); duri rukem (Indonesia); kerkup kecil (Malaysia); bitolgol (the Philippines); makwen pa (Thailand); an do (Vietnam)

Habitat: Open lands, villages

Distribution: Tropical Asia

Botanical observation: This treelet grows to 5 m tall. The stems are zigzag-shaped, present axillary spines, and are hairy. The leaves are simple, alternate, and stipulate. The petiole is red, short, 3–5 mm, and hairy. The blade is ovate, oblong-obovate, 2–4 × 1.5–3 cm, marked with 5–7 pairs of secondary nerves, somewhat coriaceous, serrulate, and acute at base and apex. The racemes are axillary. The 5–6 sepals are minute and hairy. The androecium includes numerous stamens. Petals are absent. A disc is present. The ovary is minute and produces 5–6 styles. The berries are globose, edible, somewhat glossy, blackish red, about 1cm across, and contain 5–6 seeds.

Medicinal uses: Cholera, jaundice, wounds (Bangladesh, India); post-partum (Cambodia, Laos, Vietnam)

Weak antibacterial (Gram-negative) polar extract: Aqueous decoction of leaves inhibited the growth of *Shigella* spp. and *Shigella typhi* (Chingwaru et al., 2020).

Commentary: The plant yields phenolics including mururin A and the flavone glycoside chrysoeriol-7-*O*-β-D-glucopyranoside (Sashidhara et al., 2013), which are probably, at least partially, involved in the medicinal uses.

REFERENCES

Chingwaru, C., Bagar, T. and Chingwaru, W., 2020. Aqueous extracts of Flacourtia indica, Swartzia madagascariensis and Ximenia caffra are strong antibacterial agents against Shigella spp., Salmonella typhi and Escherichia coli O157. *South African Journal of Botany*, *128*, pp. 119–127.

Sashidhara, K.V., Singh, S.P., Singh, S.V., Srivastava, R.K., Srivastava, K., Saxena, J.K. and Puri, S.K., 2013. Isolation and identification of β-hematin inhibitors from Flacourtia indica as promising antiplasmodial agents. *European journal of medicinal chemistry*, *60*, pp. 497–502.

7.4.9.2 *Flacourtia rukam* Zoll. & Moritzi

Common names: Indian plum; da ye ci li mu (China); ganda rukam (Indonesia); ken (Laos); rukam gajah (Malaysia); aganas, amait (the Philippines); khrop-dong (Thailand); mung guân ru'ng. (Vietnam)

Habitat: Forests, cultivated

Distribution: India to Pacific Islands

Botanical observation: This tree grows up to about 10 m tall. The bole presents some thorns. The leaves are simple, spiral, and exstipulate. The petiole is up to 8 mm long and curved. The blade is ovate-oblong, to broadly lanceolate, 6–16×4–7 cm, serrate, with 5–11 pairs of secondary nerves, obtuse to rounded at base, and acuminate at apex. The inflorescences are axillary fascicles. The calyx includes 4 sepals, which are ovate and minute. Petals none. The androecium includes numerous stamens. The ovary is bottle shaped and develops 4–6 styles, which are free. The berries are pinkish to red, glossy, globose, edible, up to 2.5 cm across, angled, and marked with remnant styles at apex.

Medicinal uses: Diarrhea, wounds (Malaysia); post-partum (the Philippines)

Moderate broad-spectrum antibacterial polar extract: Ethanol extract of leaves inhibited the growth of *Escherichia coli* and *Bacillus cereus* with MIC values of 125 μg/mL (Chusri et al., 2012).

REFERENCE

Chusri, S., Chaicoch, N., Thongza-ard, W., Limsuwan, S. and Voravuthikunchai, S.P., 2012. In vitro antibacterial activity of ethanol extracts of nine herbal formulas and its plant components used for skin infections in Southern Thailand. *Journal of Medicinal Plants Research*, *6*(44), pp. 5616–5623.

7.4.9.3 *Gynocardia odorata* R.Br.

Synonyms: *Chaulmoogra odorata* (R.Br.) Roxb.; *Chilmoria dodecandra* Buch.-Ham.

Common names: False chaulmoogra; man dan guo (China); chal-moghra (Bangladesh) India); cholmugra (India)

Habitat: Forests

Distribution: India, Bangladesh, and Myanmar

Botanical observation: This tree grows to about 40 m tall. The leaves are simple, alternate, and stipulate. The stipule is minute and caducous. The petiole is 1–3 cm long. The blade is oblong to elliptic, 13–20 × 5–10 cm, coriaceous, marked with 4–8 pairs of secondary nerves, wavy, rounded at base, and acuminate at apex. The inflorescences are cauliflorous corymbs of fragrant flowers. The flower peduncles are 2.5–5 cm long. The 5 sepals are oblong and 7 mm long. The 5 petals are yellowish green, oblong, and 1.5–2cm long. The stamens are numerous, filamentous, and about 1 cm long. The ovary is minute and develops 5 styles. The fruits are woody, globose, dull brown, 8–12 cm in diameter, and containing several seeds, which are ellipsoid and 2.5–3 cm long.

Medicinal uses: Fever, ulcers, bronchitis, syphilis, wounds (India); leprosy (Bangladesh; India)

Commentaries: (i) The plant produces friedelane-type triterpene lactones (Pradhan et al., 1984) and kaurane-type diterpene lactones (Pradhan et al., 1995). One could have some interest to examine this plant for antibacterial and antimycobacterial constituents. *Gynocardia odorata* R.Br. was believed to be the source of Chaulmoogra oil used traditionally in India, Bangladesh, and Myanmar for the treatment of leprosy (Banerjee, 1877). One of the sources of Chaulmoogra oil is *Hydnocarpus kurzii* (King) Warb. (ii) Consider the fact that treating patients infected with multi-drug-resistant *Mycobacterium tuberculosis* is becoming increasingly difficult if not in some cases impossible (Sanguinetti et al., 2001). As for *Mycobacterium leprae*, cases of resistance to dapsone, rifampicin, and ofloxacin have emerged around 1964, 1976, and 1996, respectively (Maeda et al., 2001).

REFERENCES

Maeda, S., Matsuoka, M., Nakata, N., Kai, M., Maeda, Y., Hashimoto, K., Kimura, H., Kobayashi, K. and Kashiwabara, Y., 2001. Multidrug resistant Mycobacterium leprae from patients with leprosy. *Antimicrobial agents and chemotherapy*, *45*(12), pp.3635-3639.

Pradhan, B.P., Hassan, A. and Shoolery, J.N., 1984. Three new friedelane lactones from the bark of gynocardia odorata (flacourtiaceae). *Tetrahedron letters*, *25*(8), pp. 865–868.

Pradhan, B.P., Chakraborty, S., Ghosh, R.K. and Roy, A., 1995. Diterpenoid lactones from the roots of Gynocardia odorata. *Phytochemistry*, *39*(6), pp. 1399–1402.

Banerjee, P.C., 1877. Chaulmugra (Gynocardia Odorata) in Leprosy. *The Indian Medical Gazette*, *12*(6), p. 166.

Sanguinetti, M., Ardito, F., Fiscarelli, E., La Sorda, M., D'Argenio, P., Ricciotti, G. and Fadda, G., 2001. Fatal pulmonary infection due to multidrug-resistant Mycobacterium abscessus in a patient with cystic fibrosis. Journal of *Clinical Microbiology*, 39(2), pp. 816–819.

7.4.9.4 *Flacourtia jangomas* (Lour.) Raeusch

Synonyms: *Flacourtia cataphracta* Roxb. ex Willd., *Roumea jangomas* Spreng., *Stigmarota jangomas* Lour., *Xylosma borneense* Ridley

Common names: Puneala plum; bainchi (Bangladesh); rukem (Indonesia); rerkup (Malaysia); khrop (Thailand); bo quan (Vietnam)

Habitat: Cultivated

Distribution: Tropical Asia

Botanical observation: This handsome tree grows to about 10 m tall. The stems develop sharp and straight woody thorns. The leaves are simple, alternate, and stipulate. The stipules are minute. The petiole is up to 8cm long. The blade is ovate-oblong, 7–14 × 2–5cm, marked with 3–6 pairs of secondary nerves, acute at base, serrate, and obtuse at apex. The inflorescences are axillary, racemose, and about up to 5 cm long. The calyx comprises 4 or 5 sepals, which are minute. The corolla is absent. The androecium is showy and includes numerous filamentous stamens. The ovary is bottle-shaped, minute, and develops 4–6 styles. The berries are brownish red or purple to blackish, globose, edible, 1.5–2.5 cm across, and contain 4 or 5 seeds.

Medicinal uses: Ringworm (Bangladesh)

Broad-spectrum antibacterial halo elicited by mid-polar extract: Ethyl acetate-methanol extract of root (250 µg/disc) inhibited the growth of *Shigella shiga*, *Bacillus cereus*, and *Bacillus megaterium* with the inhibition zone diameters of 8, 7, and 8 mm, respectively (Parvin et al., 2011).

Commentary: One could have some interest in looking for antibacterial and antifungal principle in this plants. Phenolics are most probably involved.

REFERENCE

Parvin, S., Kader, A., Sarkar, G.C. and Hosain, S.B., 2011. In-vitro studies of antibacterial and cytotoxic properties of Flacourtia jangomas. *International Journal of Pharmaceutical Sciences and Research*, 2(11), p. 2786.

7.4.9.5 *Hydnocarpus kurzii* (King) Warb.

Synonym: *Taraktogenos kurzii* King

Common names: Chaulmoogra tree; gupto mul (Bangladesh); chalmogra (India); kulau (Malaysia); kalaw (Myanmar); ma tuk (Thailand)

Habitat: Forests

Distribution: India, Bangladesh, Myanmar, Thailand, Malaysia

Botanical observation: This tree grows to 25 m tall. The stems are hairy. The leaves simple, alternate, and stipulate. The stipules are triangular and minute. The petiole is 3 cm long. The blade is oblong, 15–28×4–9 cm, cuneate at base, acuminate at apex, and with 5–7 pairs of secondary nerves. The inflorescences are axillary clusters, which are about 2 cm long and hairy. The 4 sepals are orbicular and about 7 mm long. The 5–9 petals are greenish white, and 5 mm long. The androecium includes 15–30 stamens. The ovary is hairy. The capsules are globose, brown, leprous, woody, 10 cm across, hairy, and contain numerous seeds, which are about 1.5 cm long.

Medicinal uses: Leprosy, skin diseases (India)

Broad-spectrum antibacterial halo elicited by mid-polar extract: Chloroform extract of leaves (400 µg/disc) inhibited the growth of *Bacillus cereus*, *Staphylococcus aureus*, *Escherichia coli*, *Pseudomonas aeruginosa*, *Salmonella paratyphi*, *Shigella dysenteriae*, and *Vibrio parahaemolyticus* with inhibition zone diameters of 14, 11.6, 12.6, 9, 11.6, 11.6, and 12.6 mm, respectively (Sidker et al., 2011).

Strong antimycobacterial amphiphilic cyclopentenyl fatty acid: Chaulmoogric acid (LogD = 3.6 at pH 7.4; molecular mass = 280.4 g/mol) inhibited the growth of *Mycobacterium bovis* and *Mycobacterium tuberculosis* with MIC values of 2.3 and 9.2 µM, respectively (Wang et al., 2010).

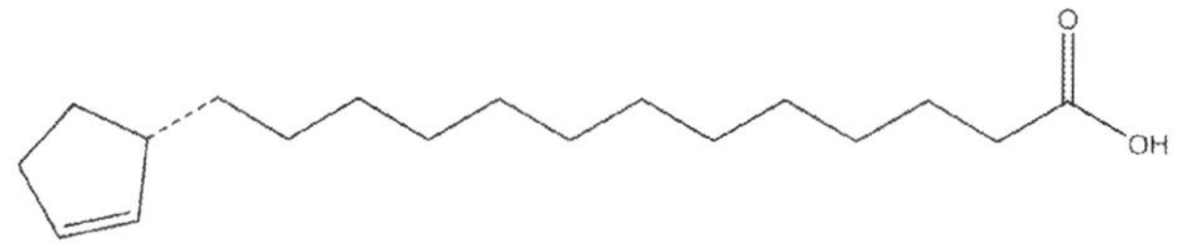

Chaulmoogric acid

Commentaries: (i) The oil expressed from the seeds has been used for the treatment of leprosy on account of chaulmoogric acid, which is active against *Mycobacterium leprae* (Levy, 1975). Other antimycobacterial cyclopentenyl fatty acids in chaulmoogra oil are hydnocarpic acid and gorlic acid (Sahoo et al., 2014). The mode of action of chaulmoogric acid and congeners involves their incorporation into the cytoplasmic membrane phospholipids and triacylglycerols of *Mycobacteria*, resulting in slower bacterial growth and membrane deficiency (Cabot et al., 1981). Furthermore, cyclopentenyl fatty acids act by blocking the co-enzyme activity of biotin (Sahoo et al., 2014). Consider that "totally drug-resistant" strains of *Mycobacterium tuberculosis* (TDR-TB) have recently emerged from Iran, India, and South Africa (Babu & Laxminarayan, 2012; Velayati et al., 2013; Dheda et al., 2014).

(ii) The seed oil of *Hydnocarpus wightiana* Bl. contains the cyclopentenyl fatty acids hydnocarpic acid, gorlic acid, and chaulmoogric acid (Sahoo et al., 2014).

(iii) The plant assembles also 5′-methoxyhydnocarpin D, which inhibited *Staphylococcus aureus* efflux multidrug-resistant pump (Stermitz et al., 2001).

REFERENCES

Babu, G.R. and Laxminarayan, R., 2012. The unsurprising story of MDR-TB resistance in India. *Tuberculosis*, *92*(4), pp. 301–306.

Cabot, M.C. and Goucher, C.R., 1981. Chaulmoogric acid: Assimilation into the complex lipids of mycobacteria. *Lipids*, *16*(2), pp. 146–148.

Dheda, K., Gumbo, T., Gandhi, N.R., Murray, M., Theron, G., Udwadia, Z., Migliori, G.B. and Warren, R., 2014. Global control of tuberculosis: From extensively drug-resistant to untreatable tuberculosis. *The Lancet Respiratory Medicine*, *2*(4), pp. 321–338.

Levy, L., 1975. The activity of chaulmoogra acids against Mycobacterium leprae. *American Review of Respiratory Disease*, *111*(5), pp. 703–705.

Sikder, M.A.A., Hossian, A.N., Siddique, A.B., Ahmed, M., Kaisar, M.A. and Rashid, M.A., 2011. In vitro antimicrobial screening of four reputed Bangladeshi medicinal plants. *Pharmacognosy Journal*, *3*(24), pp. 72–76.

Stermitz, F.R., Beeson, T.D., Mueller, P.J., Hsiang, J.F. and Lewis, K., 2001. Staphylococcus aureus MDR efflux pump inhibitors from a Berberis and a Mahonia (sensu strictu) species. *Biochemical Systematics and Habitat*, *29*(8), pp. 793–798.

Velayati, A.A., Farnia, P. and Masjedi, M.R., 2013. The totally drug resistant tuberculosis (TDR-TB). *International Journal of Clinical and Experimental Medicine*, *6*(4), p. 307.

Wang, J.F., Dai, H.Q., Wei, Y.L., Zhu, H.J., Yan, Y.M., Wang, Y.H., Long, C.L., Zhong, H.M., Zhang, L.X. and Cheng, Y.X., 2010. Antituberculosis agents and an inhibitor of the para-aminobenzoic acid biosynthetic pathway from Hydnocarpus anthelminthica seeds. *Chemistry & Biodiversity*, *7*(8), pp. 2046–2053.

7.4.9.6 *Xylosma longifolia* Clos

Synonym: *Xylosma longifolium* Clos

Common names: Katari (Bangadesh); chang ye zuo mu (China); dandal, hadda (India)

Habitat: Mountain forests

Distribution: India, Nepal, Bangladesh, Myanmar, Laos, Thailand, Vietnam

Botanical observation: This tree grows to about 7 m tall. The stems are spiny. The leaves are simple, alternate, and stipulate. The petiole is 5–8 mm long. The blade is narrowly elliptic, 4–15 × 2–7 cm, coriaceous, marked with 7–11 pairs of secondary nerves, cuneate at base, serrate, and acuminate at apex. The inflorescences are showy and axillary clusters of minute yellowish-white flowers. The 4 or 5 sepals are minute. The stamens are numerous and showy. A disc is present. The ovary is ovoid, minute, and develops 2–3 styles. The berries are red, globose, 4–6 mm across, and contain 4 or 5 seeds.

Medicinal use: Dysentery (India)

Antibacterial (Gram-positive) halo elicited by polar extract: Aqueous extract of leaves (100 μL/8 mm well) inhibited the growth of *Staphylococcus aureus* with an inhibition zone diameter of 11 mm (Parveen & Ghalib, 2012).

Strong antimycobacterial long chain isocoumarin: 8-Hydroxy-6-methoxy-3-pentylisocoumarin isolated from the stem bark inhibited the growth of *Mycobacterium tuberculosis* (H37Rv) with an MIC value of 40.5 μg/mL (Truong et al., 2011).

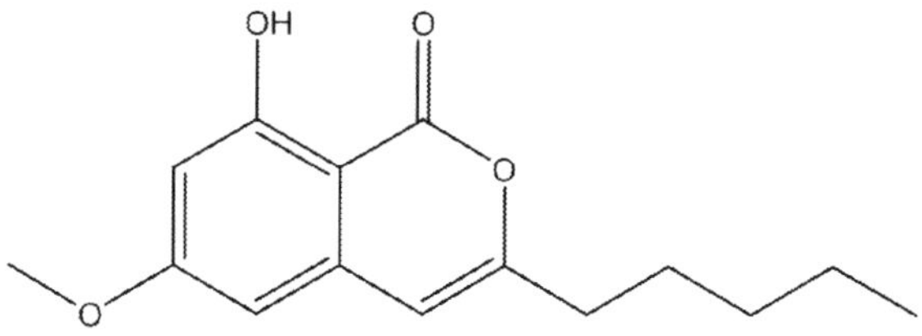

8-Hydroxy-6- methoxy-3-pentylisocoumarin

Antifungal (broad-spectrum) halo elicited by polar extract: Aqueous extract of leaves (100 µL/8mm well) inhibited the growth of *Candida albicans* and *Trichoderma viridae* with inhibition zone diameters of 11 and 7 mm, respectively (Parveen & Ghalib, 2012).

Commentary: The plant assembles phenolic glycosides (Swapana Truong et al., 2011) that could be tested for antiviral properties. Consider that phenolic glycosides from a member of the genus *Flacourtia* Comm. ex L'Her. inhibited Dengue virus RNA polymerase (Bourjot et al., 2012).

REFERENCES

Bourjot, M., Leyssen, P., Eydoux, C., Guillemot, J.C., Canard, B., Rasoanaivo, P., Guéritte, F. and Litaudon, M., 2012. Flacourtosides A–F, phenolic glycosides isolated from Flacourtia ramontchi. *Journal of Natural Products*, *75*(4), pp. 752–758.

Parveen, M. and Ghalib, R.M., 2012. Flavonoids and antimicrobial activity of leaves of Xylosma longifolium. *Journal of the Chilean Chemical Society*, *57*(1), pp. 989–991.

Swapana, N., Noji, M., Nishiuma, R., Izumi, M., Imagawa, H., Kasai, Y., Okamoto, Y., Iseki, K., Singh, C.B., Asakawa, Y. and Umeyama, A., 2017. A new diphenyl ether glycoside from xylosma longifolium collected from North-East India. *Natural Product Communications*, *12*(8), p. 1934578X1701200832.

Truong, B.N., Pham, V.C., Mai, H.D.T., Nguyen, M.C., Nguyen, T.H., Zhang, H.J., Fong, H.H., Franzblau, S.G., Soejarto, D.D. and Chau, V.M., 2011. Chemical constituents from Xylosma longifolia and their anti-tubercular activity. *Phytochemistry Letters*, *4*(3), pp. 250–253.

7.4.10 Family Violaceae Batsch (1802)

The family Violaceae consists of about 20 genera and 100 species of herbs. The leaves are simple, alternate, and stipulate. The calyx includes 5 sepals. The corolla comprises 5 petals of unequal size. The androecium comprises 5 stamens. The gynoecium consists of a 3-locular ovary with parietal placentas. The fruits are capsules.

7.4.10.1 *Rinorea horneri* (Korth.) O. Kuntze

Synonym: *Rinorea kunstleriana* (King) Taub.

Common names: Sigoh, gemotan pacat, meroyan minko (Malaysia); toei cha khru (Thailand)

Habitat: Limestones

Distribution: Thailand, Malaysia, Indonesia, the Philippines, Papua New Guinea

Botanical observation: This graceful treelet grows up to about 5 m tall. The stems are hairy. The leaves are simple, spiral, and stipulate. The petiole is up to 2.5 cm long and hairy. The blade is membranous, narrowly lanceolate, dark green above, glossy, 10–30×5–10 cm, tapering at base, marked with about 15–20 pairs of secondary nerves, and acuminate at apex. The inflorescences are axillary and up to 7 cm long racemes of minute yellowish, subsessile,and fragrant flowers. The calyx comprises 5 sepals, which are up to about 4 mm long and somewhat fleshy, and recurved. The 5 petals are up to 8 mm long. A 5-lobed disc is present. The androecium presents a tube that joins 5 stamens with brown anthers. The capsules are about 1 cm long, dehiscent, 3-lobed, dark green, coriaceous, and in persistent calyx. The seeds are few and about 8 mm long.

Medicinal use: Syphilis (Malaysia)

Commentary: The antimicrobial properties of this endangered plant have not been examined. The plant may produce antimicrobial ellagic acid derivatives because 3,3′,4,4′,5′-pentamethylcoruleoellagic acid was isolated from a member of the genus *Rinorea* Aubl. inhibited the growth of *Haemophilus influenza* with an MIC value of 9.3 µg/mL (Munvera et al., 2020). *Rinorea lanceolata* (Wallich) O. Kuntz is used in Malaysia (local name: *Babi kurus*) to treat fever. In Indonesia, *Rinorea anguifera* Kuntze is traditionally used to treat diarrhea.

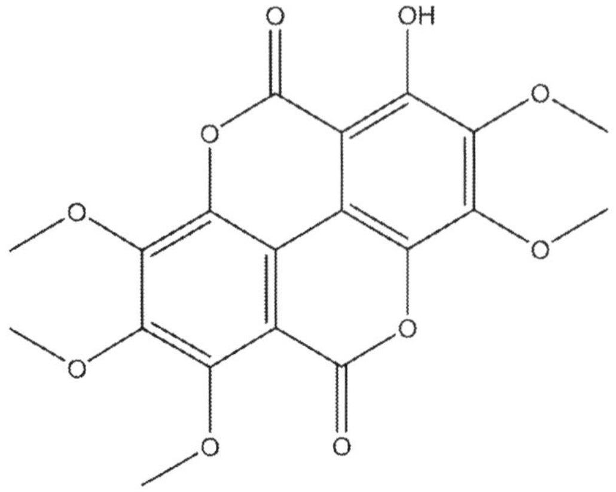

3,3′,4,4′,5′-Pentamethylcoruleoellagic acid

REFERENCE

Munvera, A.M., Ouahouo, B.M.W., Mkounga, P., Mbekou, M.I.K., Nuzhat, S., Choudhary, M.I. and Nkengfack, A.E., 2020. Chemical constituents from leaves and trunk bark of Rinorea oblongifolia (Violaceae). *Natural Product Research*, *34*, pp. 2014–2021.

7.4.10.2 *Viola diffusa* Ging.

Synonyms: *Viola diffusoides* C.J. Wang; *Viola kiusiana* Makino; *Viola tenuis* Benth.; *Viola wilsonii* W. Becker

Common name: Qi xing lian (China)

Habitat: Forests, grassy slopes, stream banks

Distribution: India, Nepal, Bhutan, Myanmar, Thailand, Vietnam, Malaysia, Indonesia, the Philippines, Japan, Papua New Guinea.

Botanical observation: This magnificent stoloniferous herb grows from a rhizome. The leaves are simple, basal, or alternate, and stipulate. The stipules are lanceolate, up to 1.2 cm long, and serrate. The petiole is up to 4.5 cm long and winged. The blade is ovate or ovate-oblong, 1.5–3.5 × 1–2 cm, hairy, cuneate at base, serrate, and acute at apex. The inflorescences are solitary and axillary on up to about 8 cm long peduncle. The calyx includes 5 sepals auriculate at base, which are lanceolate and up to 5 mm long. The corolla presents 5 petals, which are yellowish to purple, the lateral petals obovate, up to 8 mm long, the anterior one about 6 mm. The androecium includes 5 stamens. The capsules are oblong, up to about 7 mm long, and marked with persistent style.

Medicinal use: Cough (Vietnam)

Weak antiviral seco-friedelolactone-type triterpenes: Violaic A and B inhibited the secretion of Hepatitis B surface antigen by HepG2.2.15 cells with IC_{50} values of 26.2 (selectivity index: 1.2) and

33.7 μM (selectivity index: 2.5), respectively (Dai et al., 2015). Violaic acid A and B inhibited the secretion of Hepatitis B virus e-antigen by HepG2.2.15 cells with IC_{50} values of 8 (selectivity index: 3.8) and 15.2 μM (selectivity index: 5.6), respectively (Dai et al., 2015).

Commentaries: (i) In Asia plants of the genus *Viola* L. are often used to treat infections. Notably, in Korea, *Viola lactiflora* Nakai is used to treat syphilis. In China, *Viola patrinii* DC. and *Viola pinnata* L. are used to treat skin infections, whereas *Viola chinensis* G. Don is used to stop diarrhea. However, the main antibacterial principles are yet to be identified. (ii) The coumarin esculetin from *Viola prionantha* Bunge (local name in China: *Zao kai jin cai*) inhibited the growth of *Bacillus cereus* with an MIC of 50 μg/mL (Xie et al., 2004).

REFERENCES

Dai, J.J., Tao, H.M., Min, Q.X. and Zhu, Q.H., 2015. Anti-hepatitis B virus activities of friedelolactones from Viola diffusa Ging. *Phytomedicine*, *22*(7–8), pp. 724–729.

Xie, C., Kokubun, T., Houghton, P.J. and Simmonds, M.S., 2004. Antibacterial activity of the Chinese traditional medicine, Zi Hua Di Ding. *Phytotherapy Research*, *18*(6), pp. 497–500.

7.5 CUCURBITALES JUSS. EX BERCHT. & J.PRESL (1820)

7.5.1 Family Begoniaceae Berchtold et J. Presl (1820)

The family Begoniaceae comprises 4 genera and 950 species of graceful herbs. The stem is erect and not uncommonly fleshy. The leaves are simple, alternate, and stipulate. The blade is characteristically and strongly asymmetrical at base, serrate, and with secondary nerves originating from the base. The inflorescences are cymes or panicles. The perianth includes 2 or 5 tepals. The androecium includes numerous stamens. The ovary is often 3-loculed, each locule containing numerous ovules on axile or parietal placentas. The fruits are 3 winged capsules containing numerous seeds.

7.5.1.1 *Begonia roxburghii* (Miq.) A. DC.

Synonyms: *Begonia malabarica* auct. non Lam.; *Diploclinium roxburghii* Miq.

Common names: Malabar Begonia; kayalapuli, siltetoi (India)

Habitat: Forests

Distribution: India, Nepal, Sikkim, Bangladesh, Myanmar

Botanical observation: This shrub grows up to about 1.2 m. The stem is reddish, fleshy, terete, and glabrous. The leaves are simple, alternate, and stipulate. The stipules are lanceolate and about 1.5 cm long. The petiole is up to about 20 cm long. The blade is strongly asymmetrical at base, somewhat glossy, cordate, 15–30 × 5–20 cm, with secondary nerves originating from the base, and acuminate at apex. The inflorescences are axillary cymes. The 4 tepals are oblong-elliptic, up to about 1 cm long, and white or pink. The androecium comprises up to about 60 stamens. The ovary is 4-locular. The capsules are glossy, somewhat fleshy, up to 8 mm long, reddish, and 4-lobed.

Broad-spectrum antibacterial halo elicited by mid-polar extract: Chloroform extract of leaves (300 μL of a 1.5 mg/mL solution/10 mm well) inhibited the growth of *Aeromonas hydrophila, Chromobacterium violaceum, Escherichia coli, Klebsiella pneumoniae, Shigella typhi, Vibrio cholerae, Vibrio parahaemolyticus, Bacillus subtilis*, and *Staphylococcus aureus* (Ramesh et al., 2002).

Commentary: The antibacterial principle(s) in this plant is (are) apparently not yet known. Members of the genus *Begonia* L. produce quercetin and the quercetin glycosides: rutin and quercitrin, as well as luteolin, catechin (Vereskovski et al., 1987), and cucurbitane-type steroids (cucurbitacins) (Frei et al., 1998)

REFERENCES

Frei, B., Heinrich, M., Herrmann, D., Orjala, J.E., Schmitt, J. and Sticher, O., 1998. Phytochemical and biological investigation of Begonia heracleifolia. *Planta Medica*, *64*(04), pp. 385–386

Ramesh, N., Viswanathan, M.B., Saraswathy, A., Balakrishna, K., Brindha, P. and Lakshmanaperumalsamy, P., 2002. Phytochemical and antimicrobial studies of Begonia malabarica. *Journal of Ethnopharmacology*, *79*(1), pp. 129–132.

Vereskovskii, V.V., Gorlenko, S.V., Kuznetsova, Z.P. and Dovnar, T.V., 1987. Flavonoids of the leaves of Begonia erythrophylla. I. *Chemistry of Natural Compounds*, *23*(4), pp. 505–505.

7.5.1.2 *Begonia fimbristipulata* Hance

Synonym: *Begonia cyclophylla* Hook. f.

Common name: Zi bei tian kui (China)

Habitat: Forests, among stone, cultivated

Distribution: China

Botanical observation: This begonia grows up to 60 cm tall from a tuber. The leaves are simple and solitary. The stipules are up to about 7 mm long. The petiole is up to about 12 cm long and hairy. The blade is asymmetrical at base, ovate, serrate, cordate, 5–12×5–10 cm, with secondary nerves originating from the base, hairy and purple below, and acuminate or acute at apex. The inflorescences are terminal cymes on a pedicel which is up to about 15 cm tall. The 3–4 tepals are ovate, up to about 1.2 cm long, and purplish or pink. The androecium comprises up to about 60 stamens. The ovary is 3-locular and develops 3 styles. The capsules are ovoid, up to about 1 cm long, 3-winged, up to about 1.5 cm long, and contain numerous seeds.

Medicinal use: Tuberculosis (China)

Commentary: The antimycobacterial properties of this plant have apparently not been examined and the antimycobacterial properties of cucurbitacins (which may occur in this plant) have been discussed earlier. Note that the antimicrobial properties of plants in the genus *Begonia* L. have apparently not been investigated much despite interesting findings. One such finding is that 3β,22α-dihydroxyolean-12-en-29-oic acid, indole-3-carboxylic acid, 5,7-dihydroxychromone, and (-)-catechin isolated from the roots of *Begonia nantoensis* M.J. Lai & N.J. Chung (local name in China: *Nan tou qiu hai tang*) inhibited Human immunodeficiency virus replication in H9 lymphocytic cells with EC_{50} values of 5.6, 2.4, 18.6, and 14.3 µg/mL and selectivity indices of 4.4, 6.7, 1.3, and 1.7, respectively (Wu et al., 2004).

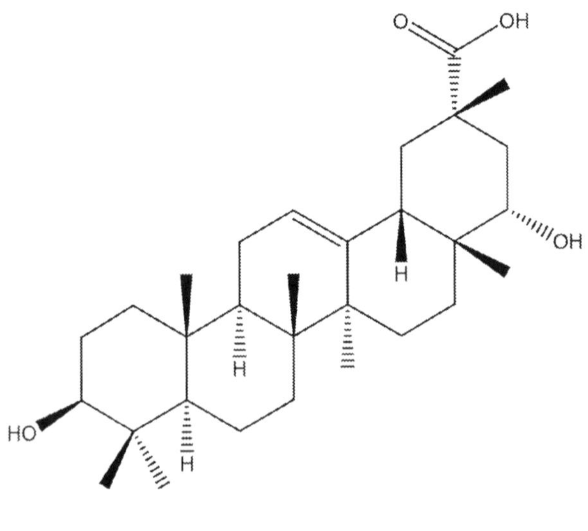

3β,22α-Dihydroxyolean-12-en-29-oic acid

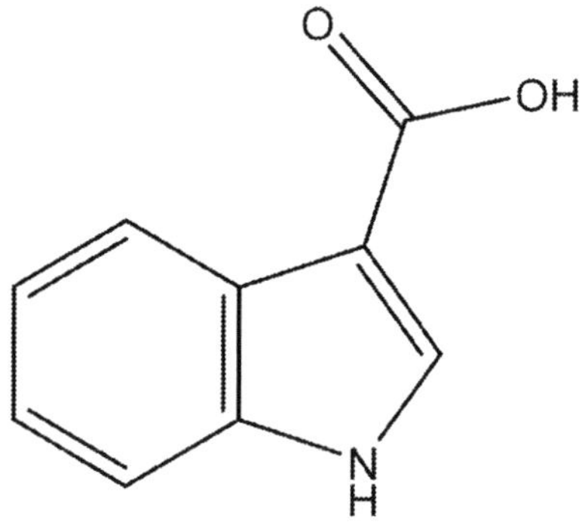

Indole-3-carboxylic acid

REFERENCE

Wu, P.L., Lin, F.W., Wu, T.S., Kuoh, C.S., Lee, K.H. and Lee, S.J., 2004. Cytotoxic and anti-HIV principles from the rhizomes of Begonia nantoensis. *Chemical and Pharmaceutical Bulletin*, *52*(3), pp. 345–349.

7.5.2 Family Cucurbitaceae A.L. de Jussieu (1789)

The family Cucurbitaceae consists of about 90 genera and 700 species of tendriliferous and often succulent climbers. The stems are often covered with rigid hairs, fleshy, and somewhat zig-zag shaped. The leaves are simple, alternate, and without stipules. The blade is often palmately lobed and rugose. The flowers are axillary, showy, unisexual, and actinomorphic. The calyx is tubular, and the corolla is tubular or consisting of 5 petals which are ephemeral and membranaceous. The androecium consists of several stamens, which are free or variously united. The gynoecium consists of 3 carpels united to form a compound, unilocular inferior ovary containing several ovules attached to parietal placentas. The fruits are characteristic berries, often large and palatable, or capsules containing several flattened seeds. Members of this family produce a fascinating array of antimicrobial cucurbitan-type triterpenes (cucurbitacins).

7.5.2.1 *Benincasa hispida* (Thunb.) Cogn.

Synonyms: *Benincasa cerifera* Savi; *Benincasa pruriens* (Sol. ex Seem.) W.J. de Wilde & Duyfjes; *Cucurbita hispida* Thunb.

Common names: Ash gourd, ash pumpkin, tallow gourd; chal kumra (Bangladesh); bi dao (Cambodia, Laos, Vietnam); kuushmaandanaadi, petha, pushini, suphala (India) ; tung kua (China); tougan (Japan); terak bileng, terak sayak, lepo ga (Malaysia); kyaukpayon (Myanmar); kubhindo (Nepal)

Habitat: Cultivated

Distribution: Tropical Asia and Pacific

Botanical observation: This smelly climber native to tropical Asia grows to about 6 m. The stems are angular, hispid and develop tendrils. The leaves are simple, alternate, and without stipules. The petiole is 6–8 cm long and hispid. The blade is palmately 5-lobed, very thin, and hispid beneath. The blade is serrate and marked with 5–6 pairs of secondary nerves. The flowers are simple, axillary, and showy. The flower pedicels are 5–10 cm long and hispid. The calyx consists of 5 linear, 8 mm long and hispid sepals. The 5 petals are yellow, orbicular, membranous, and showily nerved. The berries are massive, edible, up to 40 cm long, ovate, smooth, beautiful, and covered with a dense chalky white powder.

Medicinal uses: Cough, tuberculosis (Myanmar, Nepal); wounds (Nepal)

Commentary: (i) The plant abounds with triterpene saponins (Han et al., 2013) and displayed antipyretic and anti-inflammatory effects (Qadrie et al., 2009; Gill et al., 2010). Triterpenes saponins are often anti-inflammatory. (ii) The Hippocratic school of humorism of ancient Greeks has been used by Europeans physicians until the 19th century and is still used today in Asia in the traditional system of medicines of Iran, Afghanistan, Pakistan, India, and Bangladesh. This system, or the Unani system, of medicine classifies plants as being cold or hot and humid or dry of the first, second, third, or fourth degree, and some members of the family Cucurbitaceae are classified as "cold" and "humid" medicines whereas inflammation and fever are "hot" and "dry" symptoms (Saad & Said, 2011). Thus in this system a cold and humid plant will treat a hot and dry condition (although in some European schools medieval physicians used hot and dry plants to treat hot and dry pathologies (Girres, 1981). The knowledge we have now about the pharmacology of natural products tends to (not always) confirm this ancient concept. In Pakistan and India, some hakeems manage to treat (not always) pathologies that do not respond to Western medicine. Thus, although highly controversial, one could suggest the use the Hippocratic system of medicine as one of the means to facilitate the discovery of preventive, mitigating, or curative COVID-19 herbal remedies that could be taken orally alone or in combination with Western medicines. Hippocratic theory of humorism might be needed for the coming pandemic waves. *Benincasa hispida* (Thunb.) Cogn. is used for the treatment of COVID-19 in China (Wang et al., 2020).

REFERENCES

Gill, N.S., Dhiman, K., Bajwa, J., Sharma, P. and Sood, S., 2010. Evaluation of free radical scavenging, anti-inflammatory and analgesic potential of Benincasa hispida seed extract. *IJP-International Journal of Pharmacology*, *6*(5), pp. 652–657.

Girres L., 1981. *La médecine par les plantes à travers les ages*. Rennes: Ouest France.

Han, X.N., Liu, C.Y., Liu, Y.L., Xu, Q.M., Li, X.R. and Yang, S.L., 2013. New triterpenoids and other constituents from the fruits of Benincasa hispida (Thunb.) Cogn. *Journal of Agricultural and Food Chemistry*, *61*(51), pp. 12692–12699.

Qadrie, Z.L., Hawisa, N.T., Khan, M., Ali, W., Samuel, M. and Anandan, R., 2009. Antinociceptive and anti-pyretic activity of Benincasa hispida (thunb.) cogn. in Wistar albino rats. *Pakistan Journal of Pharmaceutical Sciences*, *22*(3), pp. 287–290.

Saad, B. and Said, O., 2011. *Greco-Arab and Islamic Herbal Medicine: Traditional System, Ethics, Safety, Efficacy, and Regulatory Issues*. Hoboken, NJ: John Wiley & Sons.

Wang, S.X., Wang, Y., Lu, Y.B., Li, J.Y., Song, Y.J., Nyamgerelt, M. and Wang, X.X., 2020. Diagnosis and treatment of novel coronavirus pneumonia based on the theory of traditional Chinese medicine. *Journal of Integrative Medicine*, *18*, pp. 275–283.

7.5.2.2 *Bryonopsis laciniosa* (L.) Naudin

Synonyms: *Bryonia palmata* L., *Cayaponia laciniosa* (L.) C. Jeffrey; *Diplocyclos palmatus* (L.) C. Jeffrey

Common names: Bryony; mala (Bangladesh); du gua (China); chitraphalaa; shivlingi (India)

Habitat: Forests, roadsides

Distribution: Tropical Asia and Pacific

Botanical observation: This poisonous climber grows up to about 6 m long from a tuber. The stems present bifid tendrils. The leaves are simple, alternate, and exstipulate. The petiole is up to 6 cm long. The blade is palmately 5-lobed, 8–10 × 2–3.5 cm, dull green, the lobes linear-lanceolate to elliptic, and acute or acuminate at apex. The inflorescences are axillary clusters of green-yellow flowers. The calyx tube is 5-lobed and up to 6 mm long. The corolla is tubular, about 1 cm across, yellowish-green, and develops 5 triangular lobes. The androecium comprises 3 stamens. The ovary is ovoid and 5 mm long. The berries are spherical, dark green to dull reddish purple with longitudinal white lines, with a sinester aura, 1.5–2.5 cm across, and contain few seeds, which are ovate, dark brown, and up to 5 mm long.

Medicinal use: Venereal diseases (India)

Moderate broad-spectrum antibacterial polar extract: Ethanol extract of leaves inhibited the growth of *Staphylococcus aureus, Micrococcus luteus, Pseudomonas aeruginosa*, and *Bacillus cereus* with MIC values of 625, 625, 1000, and 1250 μg/mL, respectively (Ehsan et al., 2009).

Commentary: The plant has apparently not been studies for its antimicrobial constituents. It produces cucurbitan-type triterpenes (Patel et al., 2020) such as the toxic glycoside bryonin (Lewis, 1998) as well as the unsaturated fatty acid punicic acid (Gowrikumar et al., 1981).

REFERENCES

Chavhan, S.A., Shinde, S.A. and Ambhore, J.P., 2020. Phytopharmacognostic review on Bryonia laciniosa (Shivlingi Beej). *Journal of Current Pharma Research, 10*, pp. 3724–3734.

Ehsan, B., Vital, A. and Bipinraj, N., 2009. Antimicrobial activity of the ethanolic extract of Bryonopsis laciniosa leaf, stem, fruit and seed. *African Journal of Biotechnology, 8*(15).

Gowrikumar, G., Mani, V.V.S., Rao, T.C., Kaimal, T.N.B. and Lakshminarayana, G., 1981. Diplocyclos palmatus L.: A new seed source of punicic acid. *Lipids, 16*(7), pp. 558–559.

Lewis, R.A., 1998. *Lewis' Dictionary of Toxicology*. Boca Raton, FL: CRC Press.

Patel, S.B., Attar, U.A., Sakate, D.M. and Ghane, S.G., 2020. Efficient extraction of cucurbitacins from Diplocyclos palmatus (L.) C. Jeffrey: Optimization using response surface methodology, extraction methods and study of some important bioactivities. *Scientific Reports, 10*(1), pp. 1–12.

7.5.2.3 *Cucumis sativus* L.

Common names: Cucumber; trasak (Cambodia); huang gua (China); traapusha, vellarikkai (India); timun (Indonesia, Malaysia); teng (Laos); takwa (Myanmar); khira (Pakistan); kukamba (Papua New Guinea); kalabaga (the Philippines); taegkwa (Thailand); dua chot (Vietnam)

Habitat: Cultivated

Distribution: Tropical and subtropical Asia and Pacific

Botanical observation: This climber grows to a length of about 6 m. The stems are fleshy, angular, hairy, and develop tendrils. The leaves are simple, alternate, and exstipulate. The petiole is up to 2 cm long. The blade is 3–8 lobed, serrate, up to about 18 cm across, cordate at base, dull green, and hairy. The inflorescences are solitary and axillary fascicles. The calyx is tubular and develops 5 lobes. The corolla is yellow, up to about 2.5 cm long, infundibuliform, 5-lobed, and membranous. The androecium comprises 3 stamens, which are up to about 4 mm long. The ovary is bullet-shaped and 3-locular. The berries are smooth, green, oblong, whitish within, edible, up to 40 cm long, and contain numerous smooth and compressed seeds.

Medicinal use: Dysentery (Cambodia, Laos, Vietnam)

Broad-spectrum antibacterial aliphatic aldehyde: (*E,Z*)-2,6-Nonadienal (produced by cucumber in response to phytopathogenic microorganisms) inhibited at the concentration of 500 ppm the growth of *Bacillus cereus* (ATCC 11778) and *Salmonella typhimurium* (ATCC 14028) (Cho et al., 2004). At this concentration (*E,Z*)-2,6-nonadienal evoked a 5.8-log reduction in *Escherichia coli* (O157:H7) cells, whereas *Listeria monocytogenes* (U.S. Food and Drug Administration serotype 1/2a) had a 2-log decrease (Cho et al., 2004). (*E*)-2-nonadienal at 1000 ppm was bactericidal for *Escherichia coli* (O157:H7), *Listeria monocytogenes* (U.S. Food and Drug Administration serotype 1/2a) and *Salmonella typhimurium* (ATCC 14028) (Cho et al., 2004).

Moderate antibacterial (Gram-positive) sphingolipids: (2*S*,3*S*,4R,10*E*)-2-[(2′*R*)-2-hydroxytetracosanoylamino]-1,3,4-octadecanetriol-10-ene, 1-*O*-β-D-glucopyranosyl-(2*S*,3*S*,4*R*,10*E*)-2-[(2′*R*)-2-hydroxytetracosanoylamino]-1,3,4-octadecanetriol-10-ene, and soya-cerebroside I isolated from the stems inhibited the growth of *Bacillus cereus* with MIC values of 50.2, 87.9, and 110.9 μg/mL, respectively (Tang et al., 2010).

Strong antibacterial (Gram-negative) sphingolipids: (2*S*,3*S*,4*R*,10*E*)-2-[(2′*R*)-2-hydroxytetracosanoylamino]-1,3,4-octadecanetriol-10-ene, 1-*O*-β-D-glucopyranosyl-(2S,3S,4R,10E)-2-[(2′R)-2-hydroxytetracosanoylamino]-1,3,4-octadecanetriol-10-ene, and soya-cerebroside I isolated from the stems inhibited the growth of *Xanthomonas vesicatoria* with MIC values of 25.6, 32.4, and 64.5 μg/mL, respectively (Tang et al., 2010). (2*S*,3*S*,4*R*,10*E*)-2-[(2′*R*)-2-hydroxytetracosanoylamino]-1,3,4-octadecanetriol-10-ene, 1-*O*-β-D-glucopyranosyl-(2*S*,3*S*,4*R*,10*E*)-2-[(2′*R*)-2-hydroxytetracosanoylamino]-1,3,4-octadecanetriol-10-ene, and soya-cerebroside I isolated from the stems inhibited the growth of *Pseudomonas lacrymans* with MIC values of 15.3, 17.4, and 37.3 μg/mL, respectively (Tang et al., 2010).

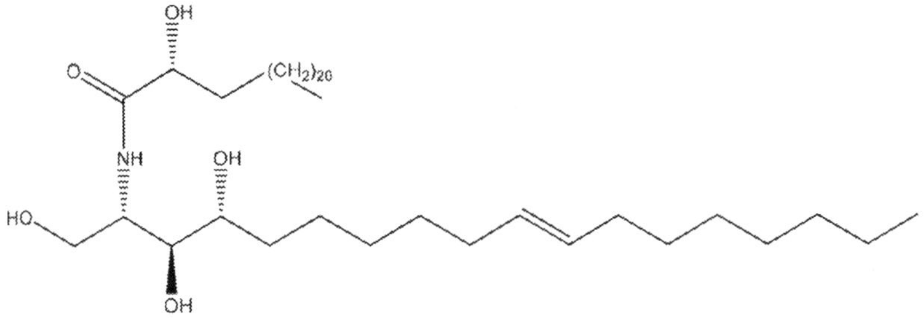

(2S,3S,4R,10E)-2-[(2'R)-2-hydroxytetracosanoylamino]-1,3,4-octadecanetriol-10-ene

Broad-spectrum antifungal halo elicited by polar extract: Ethanol extract of leaves and stems (80 μg/disc) inhibited the growth of *Aspergillus niger, Blastomyces dermatitidis, Candida albicans, Pytosporum ovale, Trichophyton* sp., and *Microsporum* sp. with inhibition zone diameters of about 10, 9, 8, 8, 8, and 7 mm, respectively (Biddyanagar & Chittagong, 2012).

Strong antifungal (filamentous) sphingolipid: (2*S*,3*S*,4*R*,10*E*)-2-[(2′*R*)-2-hydroxytetracosanoylamino]-1,3,4-octadecanetriol-10-ene at the concentration of 100 μg/mL inhibited the growth of *Pythium aphanidermatum, Botryosphaeria dothidea, Botrydis cinerea*, and *Fusarium oxysporum* f. *cucumerinum* by 100, 22.3, 48.4, and 10.7% respectively (Tan et al., 2012).

Commentary: (*E*,*Z*)-2,6-Nonadienal is an example of volatile phytoalexin. Other phytoalexins produced by this plant are *C*-glycosyl flavonoids vitexin, isovitèxin, orientin, isoorientin, and cucumerin A and B in response to *Podosphaera xanthii* infestation (McNally et al., 2003).

REFERENCES

Biddyanagar, C. and Chittagong, B., 2012. Cytotoxicity and antifungal activities of ethanolic and chloroform extracts of Cucumis sativus Linn (Cucurbitaceae) leaves and stems. *Research Journal of Phytochemistry*, *6*(1), pp. 25–30.

Cho, M.J., Buescher, R.W., Johnson, M. and Janes, M., 2004. Inactivation of pathogenic bacteria by cucumber volatiles (E, Z)-2, 6-nonadienal and (E)-2-nonenal. *Journal of Food Protection*, *67*(5), pp. 1014–1016.

McNally, D.J., Wurms, K.V., Labbé, C. and Bélanger, R.R., 2003. Synthesis of C-glycosyl flavonoid phytoalexins as a site-specific response to fungal penetration in cucumber. *Physiological and Molecular Plant Pathology*, *63*(6), pp. 293–303.

Tang, J., Meng, X., Liu, H., Zhao, J., Zhou, L., Qiu, M., Zhang, X., Yu, Z. and Yang, F., 2010. Antimicrobial activity of sphingolipids isolated from the stems of cucumber (Cucumis sativus L.). *Molecules*, *15*(12), pp. 9288–9297.

7.5.2.4 *Citrullus colocynthis* (L.) Schrad.

Synonyms: *Colocynthis vulgaris* Schrad.; *Cucumis colocynthis* L.

Common names: Bitter-apple, colocynth; makhal (Bangladesh); hanzal, indrazana (India); khiar kabiste talkh (Iran); trooh, tumba, tumma (Pakistan); ebucehil karpuzu (Turkey)

Habitat: Sun-exposed desertic land, sandy soils, plantations, roadsides, cultivated

Distribution: Turkey, Iran, Afghanistan, Pakistan, India, and Bangladesh

Botanical observation: This climber grows up to 3 m long. The stems are hairy and develop tendrils. The leaves are simple, alternate, and exstipulate. The petiole is up to 20 cm long and hairy. The blade is deeply 3–5-lobed, laxly dentate, 10–60 × 8–50 cm, cordate at base, dull green, and hairy below. The inflorescences are solitary and axillary fascicles. The calyx is tubular, about 5 mm long, and develops 5 lobes. The corolla is yellow, develops 5 oblong lobes, which are up to about 8 mm long and membranous. The androecium comprises 3 stamens. The ovary is ovate. The berries are smooth, globose, up to about 20 cm across, yellow or green with yellowish marks, on a stout pedicel, and contain numerous oblong seeds, which are up to about 6 mm long.

Medicinal uses: Fever (Iran, Pakistan); leprosy, toothache (India, Pakistan); jaundice (India, Pakistan); tuberculosis (India)

Moderate broad-spectrum polar extract: Aqueous extract of immature fruits inhibited the growth of *Escherichia coli* (ATCC 25922), *Pseudomonas aeruginosa* (ATCC 27853), *Staphylococcus aureus* (ATCC 25293), and *Enterococcis faecalis* (ATCC 29212) with the MIC/MBC values of 200/400, 200/400, 400/800, and 800/1600 µg/mL, respectively (Marzouk et al., 2009).

Antimycobacterial polar extract: Ethanol extract (1:80) inhibited the growth of *Mycobacterium tuberculosis* (H37Rv) (Grange & Davey, 1990).

Strong antimycobacterial amphiphilic cucurbitane-type steroid glycoside: Cucurbitacin E 2-*O*-β-D-glucopyranoside (molecular mass = 718.8 g/mol) inhibited the growth of *Mycobacterium tuberculosis* (H37Rv) with MIC/MBC values of 25/62.5 µg/mL, and inhibited the growth of 9 clinical strains of *Mycobacterium tuberculosis* with MIC ranging from 50 to 62.5 µg/mL (Mehta et al., 2013).

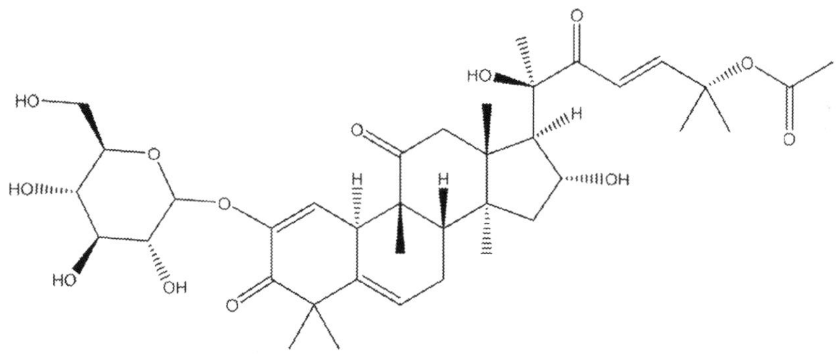

Cucurbitacin E 2-O-β-D-glucopyranoside

Strong antimycobacterial amphiphilic ursane-type triterpene: Ursolic acid (LogD = 5.8 at pH 7.4; molecular mass = 456.7 g/mol) inhibited the growth of *Mycobacterium tuberculosis* (strain H37Rv, ATCC 27294) with MIC/MBC values of 50/>125 µg/mL (Mehta et al., 2013). Ursolic acid inhibited the growth of 9 clinical strains of *Mycobacterium tuberculosis* with MIC ranging from 62.5 to 125 µg/mL, whereas all MBC were above 125 µg/mL (Mehta et al., 2013).

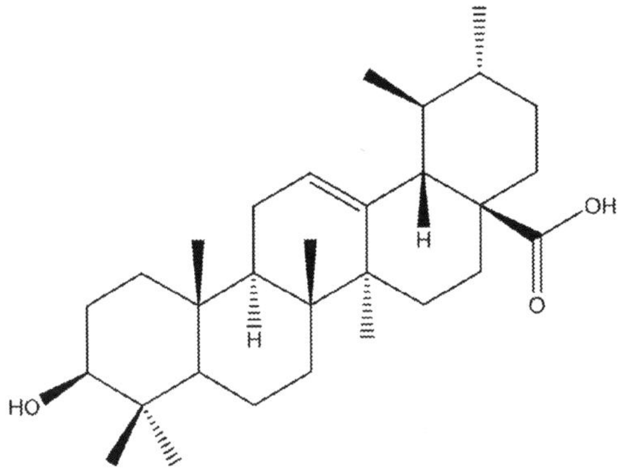

Ursolic acid

Moderate antibacterial (Gram-positive) ursane-type triterpene: Ursolic acid inhibited the growth of *Staphylococcus aureus* (ATCC 25923), *Staphylococcus aureus* (ATCC 33591), *Staphylococcus epidermidis* (ATCC12228), and *Enterococcus faecalis* (ATCC 29212) with MIC_{50} values of 62.5, 62.5, 15.6, and 250 µg/mL, respectively (Bonvicini et al., 2017).

Strong antibacterial (Gram-positive) amphiphilic quinoline alkaloid: 4-Methylquinoline (also known as lepidine; LogD=2.5 at pH 7.4; molecular mass=143.1 g/mol) isolated from the ethanol extract of fruits inhibited the growth of *Bacillus cereus* (ATCC 14579), *Listeria monocytogenes* (ATCC 15313), *Staphylococcus aureus* (KCCM 11335), *Salmonella typhimurium* (IFO 14193), and *Shigella sonnei* (ATCC 25931) with MIC/MBC values of 50/100, 25/75, 12.2/50, 75/<100, and 100/100 µg/mL, respectively (Kim et al., 2014).

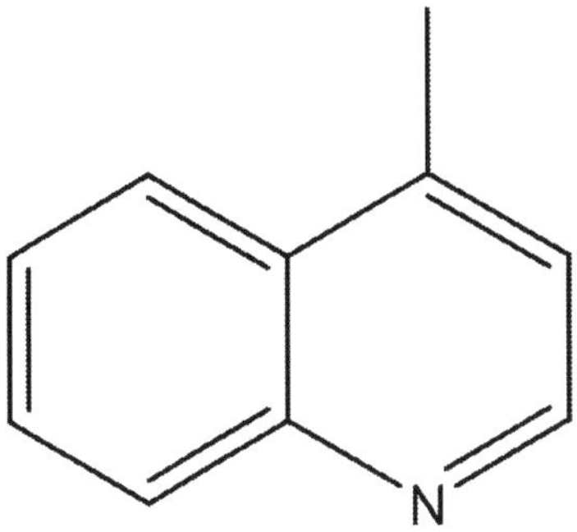

4-Methylquinoline

Commentary: Consider the presence of quinoline moiety in bedaquilin, a drug used to treat multidrug-resistant *Mycobacterial tuberculosis* and in synthetic molecules with antifungal activity (Machado et al., 2020). The quinoline derivative chloroquine (used for the treatment of malaria and derived from the quinoline alkaloid quinine from the bark of the South American *Cinchona ledgeriana* (Howard) Bern. Moens ex Trimen in the Rubiaceae family) has recently attracted attention and fueled endless debates as a possible treatment for the treatment of COVID-19. Chloroquine (LogD = 1.7 at pH 7.4; molecular mass=319.8 g/mol) inhibited the replication of Severe acute respiratory syndrome-associated coronavirus (Frankfurt 1 strain) and Severe acute respiratory syndrome-acquired coronavirus 2 in Vero E6 cells with an IC_{50} value of 8.8 µM and a selectivity index of 30 (Keyaerts et al., 2004) (Cortegiani et al., 2020). Chloroquine hampers the entry of the Severe acute respiratory syndrome-acquired coronavirus 2 in this host cell by, at least partially, interfering with the glycosylation of the angiotensin-converting enzyme 2 receptor and virus-endosome fusion by increasing endosomal pH. Is 4-methylquinoline antiviral?

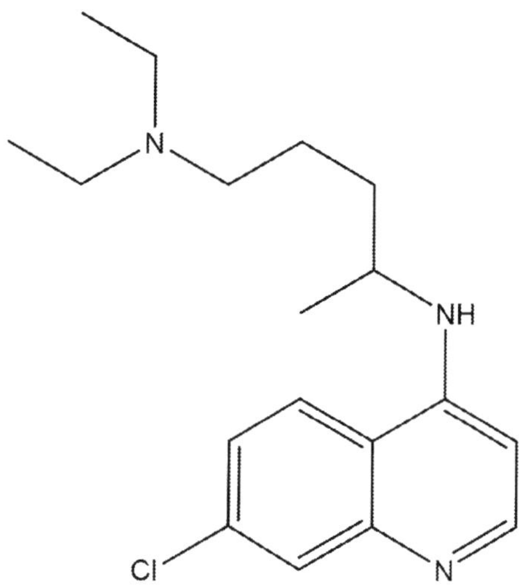

Chloroquine

REFERENCES

Al Bonvicini, F., Antognoni, F., Mandrone, M., Protti, M., Mercolini, L., Lianza, M., Gentilomi, G.A. and Poli, F., 2017. Phytochemical analysis and antibacterial activity towards methicillin-resistant Staphylococcus aureus of leaf extracts from Argania spinosa (L.) Skeels. *Plant Biosystems-An International Journal Dealing with all Aspects of Plant Biology*, *151*(4), pp. 649–656.

Cortegiani, A., Ingoglia, G., Ippolito, M., Giarratano, A. and Einav, S., 2020. A systematic review on the efficacy and safety of chloroquine for the treatment of COVID-19. *Journal of Critical Care*, *57*, pp. 279–283.

Grange, J.M. and Davey, R.W., 1990. Detection of antituberculous activity in plant extracts. *Journal of Applied Bacteriology*, *68*(6), pp. 587–591.

Hashem, A.M., Alghamdi, B.S., Algaissi, A.A., Alshehri, F.S., Bukhari, A., Alfaleh, M.A. and Memish, Z.A., 2020. Therapeutic use of chloroquine and hydroxychloroquine in COVID-19 and other viral infections: A narrative review. *Travel Medicine and Infectious Disease*, p. 101735.

Keyaerts, E., Vijgen, L., Maes, P., Neyts, J. and Van Ranst, M., 2004. In vitro inhibition of severe acute respiratory syndrome coronavirus by chloroquine. *Biochemical and Biophysical Research Communications*, *323*(1), pp. 264–268.

Kim, M.G., Lee, S.E., Yang, J.Y. and Lee, H.S., 2014. Antimicrobial potentials of active component isolated from Citrullus colocynthis fruits and structure–activity relationships of its analogues against foodborne bacteria. *Journal of the Science of Food and Agriculture*, *94*(12), pp. 2529–2533.

Machado, G.D.R.M., Diedrich, D., Ruaro, T.C., Zimmer, A.R., Teixeira, M.L., de Oliveira, L.F., Jean, M., Van de Weghe, P., de Andrade, S.F., Gnoatto, S.C.B. and Fuentefria, A.M., 2020. Quinolines derivatives as promising new antifungal candidates for the treatment of candidiasis and dermatophytosis. *Brazilian Journal of Microbiology*, *51*, pp. 1691–1701.

Marzouk, B., Marzouk, Z., Décor, R., Edziri, H., Haloui, E., Fenina, N. and Aouni, M., 2009. Antibacterial and anticandidal screening of Tunisian Citrullus colocynthis Schrad. from Medenine. *Journal of Ethnopharmacology*, *125*(2), pp. 344–349.

Mehta, A., Srivastva, G., Kachhwaha, S., Sharma, M. and Kothari, S.L., 2013. Antimycobacterial activity of Citrullus colocynthis (L.) Schrad. against drug sensitive and drug resistant Mycobacterium tuberculosis and MOTT clinical isolates. *Journal of Ethnopharmacology*, *149*(1), pp. 195–200.

7.5.2.5 *Ecballium elaterium* (L.) A. Rich.

Synonym: *Momordica elaterium* L.

Common names: Squirting cucumber; pen gua (China); kantaki (India); khiar vahshi (Iran); acı kelek, acıkavun, düvelek, seytankelegi (Turkey)

Habitat: Mountains, deserts

Distribution: Turkey, Iran, Afghanistan, India, China

Botanical observation: This poisonous climber grows to about 2 m long from a stout root. The stem is hairy and develops tendrils. The leaves are simple, alternate, and estipulate. The petiole is up to 15 cm long, stout, and hairy. The blade is cordate, somewhat triangular, 8–20×6–15 cm, hairy, incised to something serrate, wavy, and obtuse at apex. The flowers are axillary and solitary. The calyx presents 5 lanceolate lobes, which are about 5 mm long, and hairy. The corolla is yellow, membranous, 5-lobed, 3 cm in diameter, the lobes broadly lanceolate and nerved. The androecium comprises 3 stamens. The capsules are ellipsoid, green, hairy, up to 5 cm long, explosively dehiscent, suspended from a vertical peduncle, bending down (at very first glance the whole looking Papaveraceous), and containing numerous seeds, which are about 4 mm long.

Medicinal uses: Diarrhea, jaundice, sinusitis (Turkey)

Moderate broad-spectrum antibacterial polar extract: Ethanol extract of fruits inhibited the growth of *Staphylococcus aureus* (ATCC 25923) with an MIC value of 1950 μg/mL (Adwan et al., 2011). Methanol extract of leaves inhibited the growth of *Salmonella typhimurium, Listeria monocytogenes, Enterococcus faecalis* (ATCC 29212), and *Staphylococcus aureus* (ATCC 25923) with MIC values of 1250, 150, 78, and 2500 μg/mL, respectively (Abbassi et al., 2014).

Moderate antibacterial (Gram-positive) cucurbitane-type steroid: Cucurbitacin B inhibited the growth of *Staphylococcus aureus* (ATCC 29213) and methicillin-resistant *Staphylococcus aureus* (ATCC 43300) with the MIC values of 190 and 160 μg/mL, respectively, and 4 clinical strains of *Staphylococcus aureus* with MIC values ranging from 120 to 420 μg/mL (Hassan et al., 2017)

Antibiotic potentiator polar extract: Ethanol extract of fruits increased the sensitivity of *Staphylococcus aureus* (ATCC 25293) and all the strains of methicillin-resistant *Staphylococcus aureus* to penicillin (Adwan et al., 2011).

Antibiotic potentiator cucurbitane-type steroid: Cucurbitacin B increased the sensitivity of methicillin-resistant *Staphylococcus aureus* (ATCC 43300) to tetracycline and oxacillin (Hassan et al., 2017).

Strong anticandidal polar extract: Ethanol extract of fruits inhibited the growth of *Candida albicans* (ATCC 10231) with an MIC value of 97 μg/mL (Adwan et al., 2011).

Weak broad-spectrum antifungal polar extract: Ethanol extract of seeds inhibited the growth of *Aspergillus fumigatus* and *Candida albicans* with MIC values of 1000 μg/mL (Farahani et al., 2016).

Strong antifungal (filamentous) amphiphilic cucurbitane-type steroid: Cucurbitacin E (LogD = 3.4 at pH 7.4; molecular mass = 556.6 g/mol) at a concentration of 2×10^{-4} M inhibited gallic acid-induced extracellular laccase by 82.7% by *Botrytis cinerea* after 7 days (Viterbo et al., 1992).

Strong antiviral (enveloped monopartite linear double-stranded DNA) amphiphilic cucurbitane-type steroid: Cucurbitacin B (LogD = 3.0 at pH 7.4; molecular mass = 558.7 g/mol) inhibited the replication of Herpes simplex virus type-1 (KOS strain) with an IC_{50} value of 0.9 μM and a selectivity index of 127.7 in Vero cells (Hassan et al., 2017).

Commentaries: (i) The plant is dreadfully poisonous (Salhab, A.S., 2013). (ii) The plant is used to treat jaundice and one could look into the anti-Hepatitis B virus property of cucurbitacin B. Note that cucurbitacin IIA (LogD = 3.2 at pH 7.4; molecular mass = 562.7 g/mol) from a member of the genus *Hemsleya* Cogn. ex F.B. Forbes & Hemsl. inhibited Hepatitis B DNA replication in HepG 2.2.15 cells with an IC_{50} value of 11.4 μM and a selectivity index of 5.8 (Guo et al., 2012). What is the mode of action of cucurbitacin B against Herpes simplex virus type-1?

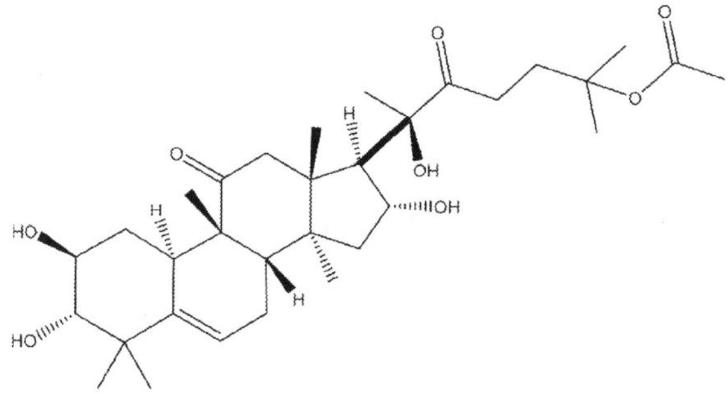

Cucurbitacin IIA

(iii) Cucurbitacins are phytoalexins (Bar-Nun & Mayer, 1990).

REFERENCES

Abbassi, F., Ayari, B., Mhamdi, B. and Toumi, L., 2014. Phenolic contents and antimicrobial activity of squirting cucumber (Ecballium elaterium) extracts against food-borne pathogens. *Pakistan Journal of Pharmaceutical Sciences*, *27*(3), pp. 475–479.

Adwan, G., Salameh, Y. and Adwan, K., 2011. Effect of ethanolic extract of Ecballium elaterium against Staphylococcus aureus and Candida albicans. *Asian Pacific Journal of Tropical Biomedicine*, *1*(6), pp. 456–460.

Bar-Nun, N. and Mayer, A.M., 1990. Cucurbitacins protect cucumber tissue against infection by Botrytis cinerea. Phytochemistry, 29(3), pp. 787–791.

Farahani, Y.F., Amin, G., Sardari, S. and Ostad, N., 2016. Effect of ecballium elaterium fruit on Candida albicans, Aspergillus fumigatus and Escherchia coli. *International Journal of Biological Research*, *4*, pp. 44–45.

Hassan, S.T.S., Berchová-Bímová, K., Petráš, J. and Hassan, K.T.S., 2017. Cucurbitacin B interacts synergistically with antibiotics against Staphylococcus aureus clinical isolates and exhibits antiviral activity against HSV-1. *South African Journal of Botany*, *108*, pp. 90–94.

Salhab, A.S., 2013. Human exposure to Ecballium elaterium fruit juice: fatal toxicity and possible remedy. *Pharmacology & Pharmacy*, *2013*. Article ID:36044.

Viterbo, A., Yagen, B. and Mayer, A.M., 1992. Cucurbitacins,'attack'enzymes and laccase in Botrytis cinerea. *Phytochemistry*, *32*(1), pp. 61–65.

7.5.2.6 *Herpetospermum pedunculosum* (Ser.) C.B. Clarke

Synonyms: *Bryonia pedunculosa* Ser.; *Herpetospermum caudigerum* Wall. ex Chakrav.; *Herpetospermum grandiflorum* Cogn.

Common name: Bo leng gua (China)

Habitat: Mountain forest margins

Distribution: India, Nepal, Bhutan, and China

Botanical observation: This climber grows up to 2 m long. The stems are slender, hairy, and develop bifid tendrils. The leaves are simple, alternate, and exstipulate. The petiole is up to 8 cm long. The blade is cordate, 6–12 × 4–9 cm, membranous, hairy, dentate, and acuminate at apex. The inflorescence is an axillary raceme, which is up to 40 cm long. The calyx is tubular, up to 2.5 cm long, hairy, and develops 5 linear lobes. The corolla is yellow, up to about 2 cm long, membranous, funnel-shaped, 5-lobed, the lobes broadly lanceolate. The androecium comprises 3 stamens. The ovary is oblong and 3-locular and develops a trifid stigma. The berries are ellipsoid, up to 8 cm long, green, and containing a few rough and woody seeds, which are up to about 1 cm long.

Medicinal use: Fever (Tibet)

Moderate antiviral (enveloped circular double stranded DNA) benzofuran lignans: 3-Benzofuranmethanol-2,3-dihyro-2-(4-hydroxyl-3-methoxyphenyl(-4-methoxy-6-[tetrahydro-2-

(3-hydroxy-4-methoxyphenyl)-3-methanol]-2-furanmethyl inhibited Hepatitis B virus e-antigen and surface antigen from HepG2.2.15 cells transfected with the Hepatitis B virus with IC_{50} values of 139.6 and 170.5 µg/mL and selectivity indices of 8.3 and 7.8, respectively (Yuan et al., 2006). Herpepropenal isolated from the seeds inhibited Hepatitis B virus e-antigen and Hepatitis B virus surface antigen from HepG2.2.15 cells with IC_{50} values of 139.6 and 156.5 µg/mL and selectivity indices of 8.3 and 7.8, respectively (Yang et al., 2010). The benzofuran lignan herpetin inhibited the Hepatitis B virus e-antigen and surface antigen with IC_{50} values of 114.4 and 176.9 µg/mL and selectivity indices 1.2 and 2.6, respectively (Gong et al., 2016).

Strong antiviral (enveloped circular double-stranded DNA) tetrahydrofuran lignans: (+)-(7′*S*, 7″*S*, 8′*R*, 8″*R*)-4, 4′, 4″-Trihydroxy-3, 5′, 3″-trimethoxy-7-oxo-8-ene [8-3′, 7′-*O*-9″, 8′-8″, 9′-*O*-7″] lignoid, (1*S*)-4-hydroxy-3-[2-(4-hydroxy-3-methoxy-phenyl)-1-hydroxymethyl-2-oxo-ethyl]-5-methoxy-benzaldehyde, and herpetetrone displayed inhibitory activities on Hepatitis virus B surface antigen secretion with IC_{50} values of 20.5, 0.3, and 4.8 µM, and exhibited inhibitory activities on Hepatitis B virus e-antigen secretion with IC_{50} values of 3.5, 4.8×10^{-4}, and 8 µM, respectively (Yu et al., 2014).

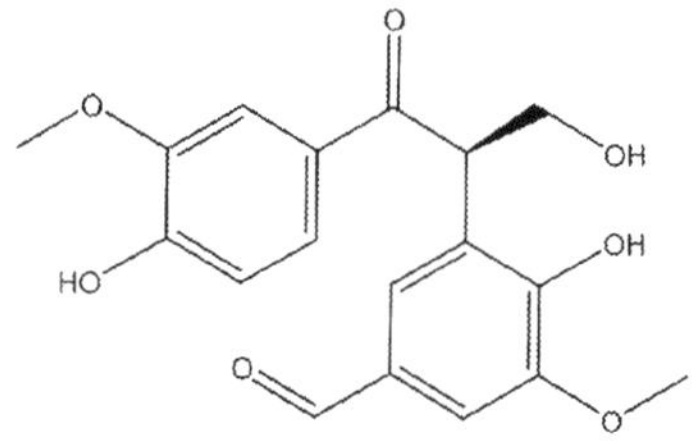

(1*S*)-4-Hydroxy-3-[2-(4-hydroxy-3-methoxy-phenyl)-1-hydroxymethyl-2-oxo-ethyl]-5-methoxy-benzaldehyde

Viral enzyme inhibition by benzofuran lignan: Herpetofluorenone at the concentration of 100 µg/mL inhibited Hepatitis B virus DNA replication by 53.8% (Gong et al., 2016).

Commentaries: (i) Note that the phenylpropanoids para-hydroxycinnamique acid, coniferylic acid, and synapilic acids are monomers used by the plant to synthetize the polymer of lignin that accumulates in the woody and lignose parts of a plant (probably like the seed coat of *Herpetospermum pedunculosum* (Ser.) C.B. Clarke) (Bruneton, 1989). These phenylpropanoids are also precursors of lignans, explaining why lignans are often (not always) accumulated in woody or lignose parts of plants. In general, lignans have interesting antimicrobial properties. (ii) Consider that anti-Hepatitis B virus natural products are not uncommonly able to inhibit the replication of coronaviruses, thus, one could examine the anti-Severe acute respiratory syndrome-acquired coronavirus 2 properties of (1*S*)-4-Hydroxy-3-[2-(4-hydroxy-3-methoxy-phenyl)-1-hydroxymethyl-2-oxo-ethyl]-5-methoxy-benzaldehyde and add this substance to the list of natural products to be tested against the upcoming waves of coronavirus and DNA viruses pandemics.

REFERENCES

Bruneton, J., 1995. *Pharmacognosy, Phytochemistry, Medicinal Plants*. Paris: Lavoisier Publishing.

Gong, P.Y., Yuan, Z.X., Gu, J., Tan, R., Li, J.C., Ren, Y. and Hu, S., 2016. Anti-HBV activities of three compounds extracted and purified from herpetospermum seeds. *Molecules*, *22*(1), p. 14.

Yang, F., Zhang, H.J., Zhang, Y.Y., Chen, W.S., Yuan, H.L. and Lin, H.W., 2010. A hepatitis B virus inhibitory neolignan from Herpetospermum caudigerum. *Chemical and Pharmaceutical Bulletin*, *58*(3), pp. 402–404.

Yuan, H.L., Yang, M., Li, X.Y., You, R.H., Liu, Y., Zhu, J., Xie, H. and Xiao, X.H., 2006. Hepatitis B virus inhibiting constituents from Herpetospermum caudigerum. *Chemical and Pharmaceutical Bulletin*, *54*(11), pp. 1592–1594.

Yu, J.Q., Hang, W., Duan, W.J., Wang, X., Wang, D.J. and Qin, X.M., 2014. Two new anti-HBV lignans from Herpetospermum caudigerum. *Phytochemistry Letters*, *10*, pp. 230–234.
Zhang, M., Deng, Y., Zhang, H.B., Su, X.L., Chen, H.L., Yu, T. and Guo, P., 2008. Two new coumarins from Herpetospermum caudigerum. *Chemical and Pharmaceutical Bulletin*, *56*(2), pp. 192–193.

7.5.2.7 *Momordica charantia* L.

Synonyms: *Cucumis argyi* H. Lév.; *Momordica chinensis* Spreng.; *Momordica indica* L., *Momordica muricata* Willd., *Sicyos fauriei* H. Lév.

Common names: Basalm apple, bitter gourd; karala (Bangladesh); moreas (Cambodia); ku gua (China); karela (India); peria (Indonesia, Malaysia); ampalia (the Philippines); muop dong (Vietnam)

Habitat: Cultivated

Distribution: Tropical Asia and Pacific

Botanical observation: This slender climber grows up to 5 long. The stems are wiry and hairy. The tendrils are up to 20 cm. The leaves are simple, spiral, and exstipulate. The petiole is slender and 4–6 cm long. The blade is 5-lobed, 4–12×4–12 cm, dull green, the lobes ovate-oblong, and the margin crenate. The flowers are axillary on 3–12 cm long pedicels. The calyx comprises 5 sepals. The corolla is yellow, membranous, 5-lobed, the lobes up to 2 cm. The androecium includes 3 stamens. The ovary is fusiform, verrucose, and expands in a 2-lobed stigma. The berries are pendulous, bitter yet edible, elliptical, 10–20 cm long, verrucose to leprous, dehiscent, green turning yellowish orange, and contain numerous seeds embedded in a bright red aril.

Medicinal uses: Ulcers, leprosy, chicken pox (Bangladesh); dengue, rabies, fever, cholera (Myanmar); AIDS (Thailand)

Moderate antibacterial (Gram-positive) essential oil: Essential oil of seeds inhibited the growth of *Staphylococcus aureus* (ATCC 6538) with an MIC value of 125 μg/mL and inhibited the growth of 12 *Staphylococcus aureus* (clinical isolates) with MIC values ranging from 125 to 500 μg/mL (Braca et al., 2008).

Moderate broad-spectrum polar extract: Ethanol extract of fresh leaves inhibited the growth of *Staphylococcus aureus* (ATCC 12692), *Bacillus cereus* (ATCC 33018), *Staphylococcus aureus* (clinical strain), and *Escherichia coli* (clinical strain) with MIC values of 64, 512, 256, and 512 μg/mL, respectively (Costa et al., 2010). Ethanol extract of dried leaves was less effective (Costa et al., 2010).

Antibiotic potentiator polar extract: Ethanol extract of leaves at 32 μg/mL decreased the resistance of *Staphylococcus aureus* (SA358, resistant to several aminoglycosides) to amikacin, gentamicin, kanamycin, neomycin, and tobramycin (Coutinho et al., 2010

Antiviral (enveloped monopartite linear single-stranded (+) RNA) polar extract: Methanol extract at 200 μg/mL inhibited the cytopathic effect of the Dengue virus type 1 on Vero E6 cells by about 50% (Tang et al., 2010).

Commentary: Since methanol extract was active *in vitro* against Dengue virus, it might have activity against coronaviruses.

REFERENCES

Braca, A., Siciliano, T., D'Arrigo, M. and Germanò, M.P., 2008. Chemical composition and antimicrobial activity of Momordica charantia seed essential oil. *Fitoterapia*, *79*(2), pp. 123–125.
Costa, J.G.M., Nascimento, E.M., Campos, A.R. and Rodrigues, F.F., 2010. Antibacterial activity of Momordica charantia (Curcubitaceae) extracts and fractions. *Journal of Basic and Clinical Pharmacy*, *2*(1), p. 45.
Coutinho, H.D., Costa, J.G., Falcão-Silva, V.S., Siqueira-Júnior, J.P. and Lima, E.O., 2010. Effect of Momordica charantia L. in the resistance to aminoglycosides in methicilin-resistant Staphylococcus aureus. *Comparative Immunology, Microbiology and Infectious Diseases*, *33*(6), pp. 467–471.
Tang, L.I., Ling, A.P., Koh, R.Y., Chye, S.M. and Voon, K.G., 2012. Screening of anti-dengue activity in Methanol extracts of medicinal plants. *BMC Complementary and Alternative Medicine*, *12*(1), p. 3.

7.5.2.8 *Siraitia grosvenorii* (Swingle) C. Jeffrey ex A.M. Lu & Z.Y. Zhang

Synonyms: *Momordica grosvenorii* Swingle; *Thladiantha grosvenorii* (Swingle) C. Jeffrey

Common names: Monksfruit; luo han guo (China)

Habitat: Forest margin, roadsides, cultivated

Distribution: Vietnam, Thailand, and China

Botanical observation: This climber grows to a length of 5 m. The stems are stout, hairy, and develop tendrils. The leaves are simple, alternate, and exstipulate. The petiole is up to 10 cm long. The blade is cordate, 12–25 × 5–20 cm, membranous, with about 5 pairs of secondary nerves, and acuminate at apex. The inflorescence is an axillary raceme. The calyx is tubular, hairy, up to 8 mm long, and develop 5 lobes. The corolla is yellow, presents 5 oblong lobes, which are up to 1.5 cm long and veined. The androecium comprises 3 stamens. The ovary is oblong, about 1 cm long, hairy, and develops a trifid stigma. The berries are globose to ellipsoid, dull green, smooth, up to about 12 cm across, and contain numerous compressed seeds, which are about 1.5 cm long.

Medicinal use: Sore throat (China)

Strong antibacterial (Gram-positive) lipophilic cholestane-type sterol: 5α,8α-epidioxy-24(*R*)-methylcholesta-6,22-dien-3β-ol (also known as ergosterol peroxide; LogD = 7.2 at pH 7.4; molecular mass = 430.6 g/mol) inhibited the growth of *Streptococcus mutans* (ATCC 25175) with an MIC value of 4.8 μg/mL (Zheng et al., 2011).

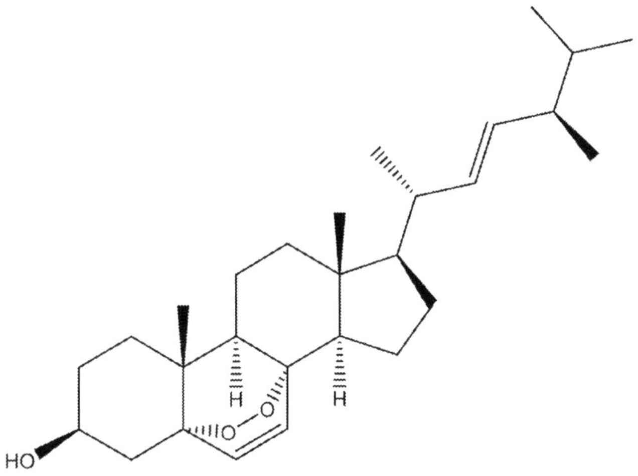

5α,8α-Epidioxy-24(R)-methylcholesta-6,22-dien-3β-ol

Strong antibacterial (Gram-positive) simple phenolic: para-Hydroxyl benzyl acid from the leaves inhibited the growth of *Streptococcus mutans* (ATCC 25175) with an MIC value of 12.2 μg/mL, (Zheng et al., 2011).

Strong antifungal (filamentous) lipophilic cholestane-type sterol: 5α,8α-epidioxy-24(*R*)-methylcholesta-6,22-dien-3β-ol at a concentration of 0.9 μg/mL inhibited the mycelial growth of *Pyricularia oryzae, Verticillium* sp., *Septoria* sp., *Rhizoctonia* sp., and *Fusarium* sp. (You et al., 2009).

Commentaries: (i) The plant abounds with cucurbitane-type saponins (Chaturvedula & Prakash, 2011), which could be involved in medicinal use. (ii) 5α,8α-Epidioxy-24(*R*)-methylcholesta-6,22-dien-3β-ol, bearing an endoperoxide moiety and being lipophilic, is probably antimycobacterial.

REFERENCES

Chaturvedula, V.S.P. and Prakash, I., 2011. Cucurbitane glycosides from Siraitia grosvenorii. *Journal of Carbohydrate Chemistry*, *30*(1), pp. 16–26.

You, F., Han, T., Wu, J.Z., Huang, B.K. and Qin, L.P., 2009. Antifungal secondary metabolites from endophytic Verticillium sp. *Biochemical Systematics and Ecology*, *37*(3), pp. 162–165.
Zheng, Y., Huang, W., Yoo, J.G., Ebersole, J.L. and Huang, C.B., 2011. Antibacterial compounds from Siraitia grosvenorii leaves. *Natural Product Research*, *25*(9), pp. 890–897.

7.5.2.9 *Trichosanthes anguina* L.

Synonyms: *Cucumis anguina* L.; *Trichosanthes cucumerina* L. var. *anguina* (L.) Haines

Common names: Snake gourd; chichinga (Bangladesh); che gua (China); petola ular (Indonesia, Malaysia); pudalankai (India); hebiuri (Japan); ngo ngews (Laos); melon-daga (the Philippines); ma noi (Thailand); yilan kabagi (Turkey); day na tay (Vietnam)

Habitat: Cultivated

Distribution: Tropical Asia

Botanical observation: This strange climber grows to a length of 5 m. The stems are slender and angular. The leaves are simple, spiral, and exstipulate. The petiole is 3–7 cm long. The blade is 5-lobed, 8–16 × 6–18 cm, the lobes obovate, and the margin finely denticulate. The racemes are axillary, 1–15-flowered on a peduncle that is 1–18 cm long. The calyx tube is about 5 mm long and develops 5 lobes. The corolla is infundibuliform, pure white, 5-lobed, about 3 cm across, and with an appearance of somewhat a lace. The androecium includes 3 stamens. The ovary is fusiform and minute. The berries are about 1.5 m long, smooth, somewhat like hanging eerie or hellish greenish fat snakes or gigantic leeches, and edible. The seeds are oblong and up to 1.7 cm long.

Medicinal uses: Fever, syphilis (Bangladesh); cough (Indonesia); diarrhea (China)

Broad-spectrum antibacterial halo elicited by non-polar extract: Petroleum ether extract of seeds (300 µg/disc) inhibited the growth of *Staphylococcus aureus*, *Streptococcus β-haemolyticus*, *Bacillus subtilis*, *Klebsiella* sp., *Shigella shiga*, and *Shigella boydii* with inhibition zone diameters of 8, 6, 11, 9, 8, and 7 mm, respectively (Ali et al., 2011).

Broad-spectrum antifungal halo elicited by non-polar extract: Petroleum ether extract of seeds (300 µg/disc) inhibited the growth of *Penicillum* sp., *Aspergillus niger*, *Trichoderma viride*, *Candida albicans*, and *Helminthosporium sativum* with inhibition zone diameters of 10, 14, 8, 8, and 11 mm, respectively (Ali et al., 2011).

Commentary: The plant has apparently not been examined for its antibacterial constituents, and especially antimycobacterial principles. In India, *Trichosanthes dioica* Roxb. is used to treat leprosy and fever (local name: *Patol*). Aqueous extract of leaves of *Trichosanthes dioica* Roxb. inhibited the growth of *Mycobacterium smegmatis* (6 mm disc impregnated with 75 mg/mL solution) with an inhibition zone diameter of 30 mm (Rai et al., 2010). Antimycobacterial agents await discovery in members of the genus *Trichosanthes* L. Although highly controversial, one could see the laciniated flowers of these plants as signs for the treatment of tuberculosis as per the theory of signatures.

REFERENCES

Ali, M.A., Sayeed, M.A., Islam, M.S., Yeasmin, M.S., Khan, G.R.M.A.M. and Muhamad, I.I., 2011. Physicochemical and antimicrobial properties of Trichosanthes anguina and Swietenia mahagoni seeds. *Bulletin of the Chemical Society of Ethiopia*, *25*(3).
Rai, P.K., Mehta, S., Gupta, R.K. and Watal, G., 2010. A novel antimicrobial agents Trichosanthes dioica. *International Journal of Pharma and Bio Sciences*, *1*(3), pp. 202–209.

7.5.2.10 *Trichosanthes kirilowii* Maxim.

Synonym: *Trichosanthes obtusiloba* C.Y. Wu ex C.Y. Wu & C.H. Yueh

Common names: Lota-mohakaal (Bangladesh); gua lou (China); timun dendang (Malaysia)

Habitat: Forest borders, villages, cultivated

Distribution: Tropical Asia, China, Korea, and Japan

Botanical observation: This climber with slender and angular stems grows to a height of 5 m. The leaves are simple, spiral, and exstipulate. The petiole is up to 6 cm long and hairy. The blade is

5-lobed, 5–20 cm across, the lobes obovate, the margin somewhat lobed, and obtuse at apex. The racemes are axillary, 5–8 flowered, on a peduncle, which is up to 20 cm long, or the flowers are solitary. The calyx tube is 5-lobed, up to about 1 cm long, the lobes lanceolate. The corolla is infundibuliform, pure white, 5-lobed, and up to about 7 cm across, and with an appearance of somewhat a lace. The androecium includes 3 stamens. The ovary is fusiform and minute. The berries are orange to red, smooth, oblong to globose, and containing numerous light brown and about 1 cm long seeds.

Medicinal uses: Jaundice, tuberculosis (Bangladesh)

Weak antibacterial (Gram-positive) polar extract: Aqueous extract of fruits inhibited the growth of *Streptococcus mutans* (MT 5091) and *Streptococcus mutans* (OMZ 176) with MIC values of 1250 and 1250 μg/mL, respectively (Chen et al., 1989).

Viral enzyme inhibition by polar extract: Methanol extract at a concentration of 100 μg/mL inhibited the Human immunodeficiency virus type-1 protease by 24.5% (Min et al., 2001).

Commentaries: (i) Consider that Asian traditional systems of medicine not uncommonly use medicinal plants with yellow wood, exudates, or fruits for the treatment of jaundice. (ii) The plant is used in China for the treatment of COVID-19 in combination with oseltamivir phosphate (Luo, 2020). The principle(s) responsible for the anti-Severe acute respiratory syndrome-acquired coronavirus 2 effects is/are unknown and might involve antiretroviral proteins such as trichosanthin (Lee-Huang et al., 1991). However, peptides and proteins, due to the action of proteolic enzymes in the guts, do not have favorable oral bioavailability (Werle & Bernkop-Schnürch, 2006) although some absorption is possible via M-cells of intestinal Peyer patches (Muranichi et al., 1997). Methanol precipitates proteins (Levine, 1999), hence the *in vitro* anti-Human immunodeficiency virus effects observed by Min et al. (2001) results from nonprotein principles. Such principles could be cucurbitan-type steroids such as, for instance, cucurbitacin D (Takahashi et al., 2009).

REFERENCES

Chen, C.P., Lin, C.C. and Tsuneo, N., 1989. Screening of Taiwanese crude drugs for antibacterial activity against Streptococcus mutans. *Journal of Ethnopharmacology*, *27*(3), pp. 285–295.

Lee-Huang, S., Huang, P.L., Kung, H.F., Li, B.Q., Huang, P., Huang, H.I. and Chen, H.C., 1991. TAP 29: an anti-human immunodeficiency virus protein from Trichosanthes kirilowii that is nontoxic to intact cells. *Proceedings of the National Academy of Sciences*, *88*(15), pp. 6570–6574.

Levine, B. ed., 1999. *Principles of Forensic Toxicology* (Vol. 2). Washington, DC: American Association for Clinical Chemistry.

Luo, A., 2020. Positive Severe acute respiratory syndrome-acquired coronavirus 2 test in a woman with COVID-19 at 22 days after hospital discharge: A case report. *Journal of Traditional Chinese Medical Sciences.*

Min, B.S., Kim, Y.H., Tomiyama, M., Nakamura, N., Miyashiro, H., Otake, T. and Hattori, M., 2001. Inhibitory effects of Korean plants on HIV-1 activities. *Phytotherapy Research*, *15*(6), pp. 481–486.

Muranishi, S., Fujita, T., Murakami, M. and Yamamoto, A., 1997. Potential for lymphatic targeting of peptides. *Journal of Controlled Release*, *46*(1–2), pp. 157–164.

Takahashi, N., Yoshida, Y., Sugiura, T., Matsuno, K., Fujino, A. and Yamashita, U., 2009. Cucurbitacin D isolated from Trichosanthes kirilowii induces apoptosis in human hepatocellular carcinoma cells in vitro. *International Immunopharmacology*, *9*(4), pp. 508–513.

Werle, M. and Bernkop-Schnürch, A., 2006. Strategies to improve plasma half life time of peptide and protein drugs. *Amino Acids*, *30*(4), pp. 351–367.

7.6 ORDER FABALES BROMHEAD (1838)

7.6.1 Family Fabaceae Lindley (1836)

The family Fabaceae is a vast taxon of about 650 genera and 18000 species of trees, shrubs, climbers, or herbs. This family used to be divided into the Caesalpiniaceae, Mimosaceae, and Papilionaceae, a division that was in fact reflecting perfectly the phytochemical differences among these 3 groups. Taxonomists keep and changing family, genera, and species name creating confusion for phytochemists and pharmacologists. The use of genetic instead of simple but logical observation of morphological plant features to classify plants will create even more confusion. The leaves are simple or compound, alternate or spiral, and stipulate. The flowers are actinomorphic or zygomorphic. The inflorescences are racemes, corymbs, spikes, heads, or panicles. The calyx is tubular or comprises 5 sepals. The corolla comprises 5 petals, which are either similar or comprising a standard, a pair of wings and pair of lower petals forming a keel. The androecium includes 10 stamens free or partially fused. The gynoecium includes a single carpel containing 1 to numerous ovules on adaxial placentas. The fruit is a pod. The seeds are often smooth and glossy. Members of this family produce a bewildering array of antimicrobial flavonoids.

7.6.1.1 *Abrus precatorius* L.

Synonyms: *Abrus abrus* (L.) W. Wight; *Abrus maculatus* Noronha; *Abrus minor* Desv.; *Abrus pauciflorus* Desv.; *Abrus squamulosus* E. Mey.; *Abrus tunguensis* P. Lima; *Abrus wittei* Baker f.; *Glycine abrus* L.

Common names: Wild licorice, Jamaica wild licorice, Indian licorice, jequirity bean, jumble beads, crab-eyes vine, coral pea, prayer beads, rosary pea; kunch, gunch, kawet (Bangladesh); xiang si zi (China); gunga, kunch, kawet, kunni, kuntrimani, lalgeri, ratti (India); saga manis (Indonesia); chashm-i-khurus (Iran); makam (Laos); pokok memanjat, akar saga betina, akar saga (Malaysia); chek-awn, ywe nge (Myanmar); karjani (Nepal);saga (the Philippines); klam khruea, maklam ta nu (Thailand); cam thao day, day chi chi (Vietnam).

Habitat: Seashores, beach forests

Distribution: Tropical Asia and Pacific

Botanical observation: This slender climbing shrub grows to a length of 4 m. The stems are slender and flexuous. The leaves are pinnate, spiral, and stipulate. The blade is and 5–8 cm long, and comprises 8–13 pairs of folioles, which are hairy below, ephemeral, membranous, oblong, acute at base, paddle shaped or oblong-elliptic, bifid at apex, and 2–3 cm long. The flowers are small and arranged in dense axillary racemes, which grow up to 8 cm long. The calyx is campanulate and 4-lobed. The corolla is reddish, purplish, or whitish, much exerted, the standard ovate, acute and adhering below the staminal tube, the wings are narrow and the keel arcuate. The androecium consists of 9 stamens united into a tube. The ovary is hairy, the style is short, and the stigma capitate. The pods are ellipsoid, light brown, up to about 4 cm long, somewhat coriaceous, dehiscent, and contain 2–6 round, red and black, glossy (looking like some sort of red beetles with black heads) hard, dreadfully toxic seeds, which are used for making necklaces and rosaries and used in the past by Indian goldsmiths as standard weights. Open pods displaying the seeds have altogether a sinister aura.

Medicinal uses: Cough (Bangladesh, India, and Myanmar); abscesses, bronchitis, hepatitis, pneumonia, tonsilitis (India); ringworm (Thailand); leucorrhea (Myanmar); fever (India, Thailand); sore throat (India)

Antibacterial (Gram-positive) halo developed by polar extract: Methanol extract of roots (5.4 mm disc impregnated with a solution at 2500 µg/mL) inhibited the growth of *Staphylococcus aureus* with an inhibition zone diameter of about 15 mm (Mistry et al., 2010).

Antibacterial (Gram-positive) halo developed by non-polar extract: Petroleum extract of roots (5.4 mm paper disc impregnated with a solution at 1100 µg/mL) inhibited the growth of *Staphylococcus aureus* with an inhibition zone diameter of about 12 mm (Mistry et al., 2010).

Moderate broad-spectrum antibacterial polar extract: Methanol extract of leaves inhibited the growth of *Staphylococcus aureus* (ATCC 25923), *Staphylococcus epidermidis*, *Bacillus subtilis*, *Corynebacterium* sp., and *Pseudomonas aeruginosa* with MIC values of 128, 256, 128. 512, and 256 µg/mL, respectively (Adelowotan et al., 2008).

Antimycobacterial amphiphilic isoflavan: Abruquinone B (LogD = 1.6 at pH 7.4; molecular mass = 390.3 g/mol) isolated from the aerial parts inhibited the growth of *Mycobacterium tuberculosis* (H32Ra) with an MIC value of 12.5 μg/mL (Limmatvapirat et al., 2004).

Strong antiviral (enveloped monopartite linear double-stranded DNA) amphiphilic oleanane-type saponin: Glycyrrhizin (LogD = 1.0 at pH 7.4; molecular mass = 822.9 g/mol) inhibited the replication of the Herpes simplex virus type-1 in HEF cells with an IC_{50} value of 3.6 μM (Hirabayashi et al., 1991).

Strong antiviral (enveloped monopartite linear double-stranded DNA) isoflavan: Abruquinone G inhibited the replication of the Herpes simplex virus type-1 in Vero cell with an IC_{50} value of 20–50 μg/mL (Limmatvapirat et al., 2004).

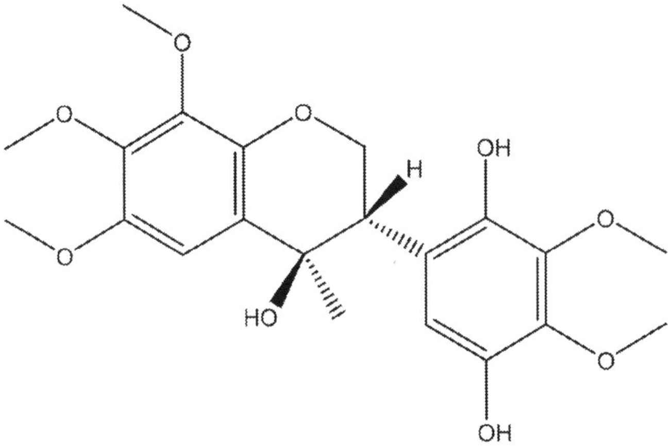

Abruquinone G

Very weak antiviral (enveloped monopartite linear single-stranded (+)RNA) amphiphilic oleanane-type saponin: Glycyrrhizin (LogD = 1.0 at pH 7.4; molecular mass = 822.9 g/mol) inhibited the replication of the Severe acute respiratory syndrome associated coronavirus (clinical isolate) in Vero cells with an IC_{50} values of 641 μg/mL and a selectivity index of 1.2 (Loizzo et al., 2008).

Very weak antiviral (enveloped monopartite linear single-stranded (+) RNA) amphiphilic oleanane-type saponin: Glycyrrhizin inhibited the replication of the Dengue type-2 virus in Vero cells with an IC_{50} value of 174.2 μg/mL and a selectivity index of 1.2 (Crance et al., 2003).

Very weak antiviral (enveloped monopartite linear dimeric single-stranded (+)RNA) amphiphilic oleanane-type saponin: Glycyrrhizin inhibited the replication of the West Nile virus in cells in Vero cells with an IC_{50} value of 228.9 μg/mL and a selectivity index of 11 (Crance et al., 2003).

Very weak antiviral (enveloped monopartite linear single-stranded (+) RNA) amphiphilic oleanane-type saponin: Glycyrrhizin inhibited the replication of the Zika virus in Vero cells with the IC_{50} value of 383.7 μg/mL and a selectivity index of 7 (Crance et al., 2003).

Very weak antiviral (enveloped monopartite linear dimeric single-stranded (+)RNA) amphiphilic oleanane-type saponin: Glycyrrhizin inhibited the replication of the Human immunodeficiency virus type-1 in MT-4 cells with an IC_{50} value of 200 μg/mL (Harada, 2005).

Very weak antiviral (enveloped segmented linear single-stranded (−)RNA) amphiphilic oleanane-type saponin: Glycyrrhizin inhibited the replication of the Influenza A virus (Influenza A/Aichi/2/68) with an IC_{50} value of 200 μg/mL (Harada, 2005).

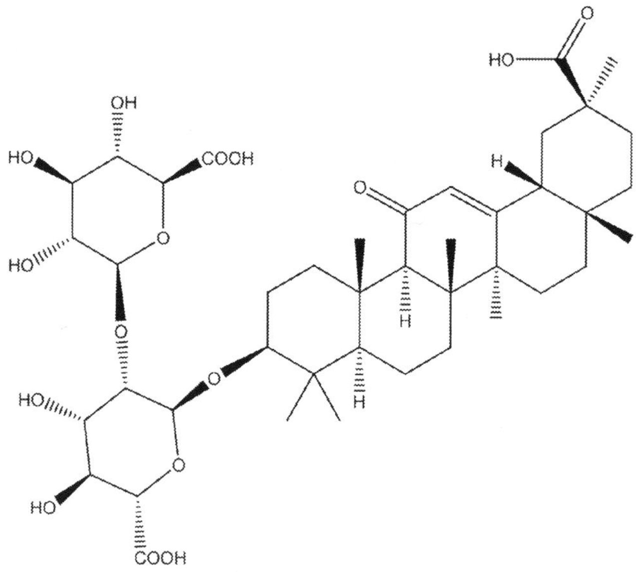

Glycyrrhizin

Commentaries: (i) The seeds contain abrin, which is, for humans, one of the most poisonous protein generated by flowering plants (Dickers et al., 2003). The plant generates oleanane-type triterpenes, as well as *C*-glycoside flavonols and indole alkaloids (Nakamura & Hattori, 1998) and an interesting series of isoflavan quinones, such as abruquinone B (Chun-Qing & Zhi-Bi, 1998). Consider that plants accumulating saponins are often traditionally used in Asia for the treatment of cough on probable account of their amphiphilic properties and thus ability to dissolve phlegm.

(ii) Glycyrrhizin (also known as glycyrrhizic acid) very weakly inhibits *in vitro* the replication of a broad-spectrum of enveloped RNA viruses, and one could examine its effect(s) on the envelope of (Wolkerstorfer et al., 2009) and/or the fusion of enveloped viruses with host cells. According to Bone and Mills (2013), natural products with Lethal Doses 50% (LD_{50}) above 15,000 mg are nontoxic, from 500 to 5,000 mg/Kg are slightly toxic, from 50 to 500 mg/Kg are highly toxic, and below 50 mg/Kg are extremely toxic. The oral LD_{50} value of glycyrrhizin is 12,700 mg/Kg and chronic administration evokes among other things hypertension and hypokalemia (Nazari et al., 2017). Thus, this saponin, if active against the Severe acute respiratory syndrome-acquired coronavirus 2 *in vitro*, might be of value for the prevention or mitigation of COVID-19. Glycyrrhizin is an example of natural products to be included in first-line compounds to be tested when new zoonotic viral pandemics emerge.

REFERENCES

Adelowotan, O., Aibinu, I. and Adenipekun, E., 2008. The In-Vitro Antimicrobial Activity of Abrus precatorius. *Nigerian Postgraduate Medical Journal*, *15*(1), p. 33.

Chun-Qing, S.O.N.G. and Zhi-Bi, H.U., 1998. Abruquinone A, B, D, E, F and G from the root of Abrus precatorius. *Journal of Integrative Plant Biology*, *40*(8), pp. 734–739.

Crance, J.M., Scaramozzino, N., Jouan, A. and Garin, D., 2003. Interferon, ribavirin, 6-azauridine and glycyrrhizin: antiviral compounds active against pathogenic flaviviruses. *Antiviral Research*, *58*(1), pp. 73–79.

Dickers, K.J., Bradberry, S.M., Rice, P., Griffiths, G.D. and Vale, J.A., 2003. Abrin poisoning. *Toxicological Reviews*, *22*(3), pp. 137–142.

Harada, S., 2005. The broad anti-viral agent glycyrrhizin directly modulates the fluidity of plasma membrane and HIV-1 envelope. *Biochemical Journal*, *392*(1), pp. 191–199.
Hirabayashi, K., Iwata, S., Matsumoto, H., Mori, T., Shibata, S., Baba, M., Ito, M., Shigeta, S., Nakashima, H. and Yamamoto, N., 1991. Antiviral activities of glycyrrhizin and its modified compounds against human immunodeficiency virus type 1 (HIV-1) and herpes simplex virus type 1 (HSV-1) in vitro. *Chemical and Pharmaceutical Bulletin*, *39*(1), pp. 112–115.
Limmatvapirat, C., Sirisopanaporn, S. and Kittakoop, P., 2004. Antitubercular and antiplasmodial constituents of Abrus precatorius. *Planta Medica*, *70*(3), pp. 276–278.
Loizzo, M.R., Saab, A.M., Tundis, R., Statti, G.A., Menichini, F., Lampronti, I., Gambari, R., Cinatl, J. and Doerr, H.W., 2008. Phytochemical analysis and in vitro antiviral activities of the essential oils of seven Lebanon species. *Chemistry & biodiversity*, *5*(3), pp. 461–470.
Mistry, K., Mehta, M., Mendpara, N., Gamit, S. and Shah, G., 2010. Determination of antibacterial activity and MIC of crude extract of Abrus precatorius L. *Adv Biotech*, *10*(2), pp. 25–27.
NAKAMURA, N. and HATTORI, M., 1998. Saponins and C-glycosyl flavones from the seeds of Abrus precatorius. *Chemical and Pharmaceutical Bulletin*, *46*(6), pp. 982–987.
Nazari, S., Rameshrad, M. and Hosseinzadeh, H., 2017. Toxicological effects of Glycyrrhiza glabra (licorice): a review. *Phytotherapy Research*, *31*(11), pp. 1635–1650.
Wolkerstorfer, A., Kurz, H., Bachhofner, N. and Szolar, O.H., 2009. Glycyrrhizin inhibits influenza A virus uptake into the cell. *Antiviral Research*, *83*(2), pp. 171–178.

7.6.1.2 *Acacia arabica* (Lam.) Willd.

Synonyms: *Acacia nilotica* Lam.; *Minosa arabica* Lam.; *Mimosa nilotica* L.; *Vachellia nilotica* (L.) P.J.H. Hurter & Mabb.

Common names: Indian Gum Arabic tree; blabla, babbuula, baul (Bangladesh); aqaaqia, babla, karu-velamaram, karuvelam, nalla thumma, (India); babur, khayar (Nepal); kikar (Pakistan)

Habitat: Deserts, dry sunny sandy spots, cultivated

Distribution: Pakistan, India, Sri Lanka, Nepal, Bangladesh

Botanical observation: This tree grows to about 10 m tall. The bark yield a light brown and translucent gum upon incision. The leaves are bipinnate, spiral, and stipulate. The stipules are spinescent and up to 8 cm long. The blade presents 2–11 pairs of pinnae with 7–25 pairs of 5 mm long folioles per pinnae. The inflorescences are axillary pedunculate 6–15 mm in diameter heads of minute flowers. The calyx is tubular, 5-lobed, and minute. The corolla comprises 5 petals. The pods are constricted between the seeds and 4–22 × 1.3–2.2 cm. The seeds are blackish brown and up to about 8 mm long.

Medicinal uses: Cough (Pakistan); bronchitis, mouth sores, mouth wash (India); venereal diseases (Nepal); tuberculosis (Pakistan); diarrhea, dysentery (India; Pakistan, Sri Lanka)

Broad-spectrum antibacterial halo developed by polar extract: Ethanol extract of stem bark (30 mg/mL/5 mm well) inhibited the growth of *Streptococcus viridans*, *Bacillus subtillis*, *Escherichia coli*, and *Shigella sonnei* (Banso, 2009).

Weak antibacterial (Gram-negative) polar extract: Ethanol extract of leaves inhibited the growth of *Escherichia coli, Salmonella enterica*, *Salmonella enteridis*, and *Shigella typhimurium* (all multidrug-resistant isolated from beef and poultry meat samples) with MIC/MBC values of 3120/6250, 1560/3120, 1560/3120, and 1560/3120 μg/mL, respectively (Sadiq et al., 2017).

Strong broad-spectrum antibacterial amphiphilic cassane-type diterpene: Niloticane (Log D = 3.8 at pH 7.4; molecular mass = 318.4 g/mol) isolated from the bark inhibited the growth of *Bacillus subtilis*, *Staphylococcus aureus*, *Klebsiella pneumonia*, and *Escherichia coli*, with MIC values of 4, 8, 16, and 33 μg/mL, respectively (Eldeen et al., 2010).

Strong antiviral (enveloped monopartite linear single-stranded (+)RNA) polar extract: Acetone extract of leaves at a concentration of 100 μg/mL inhibited the replication of the Hepatitis C virus in Huh-7 cells by about 82% (Rehman et al., 2011).

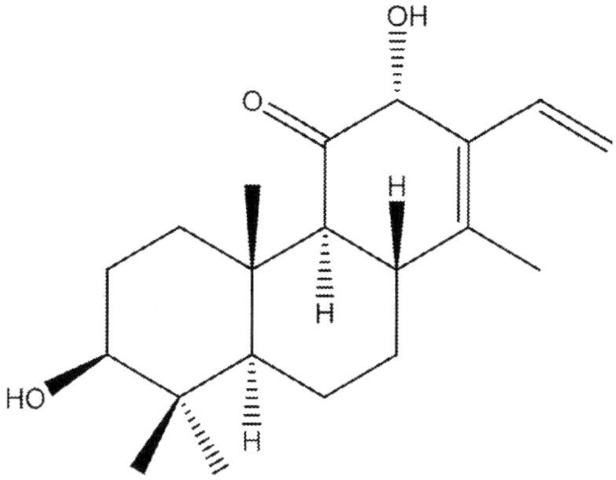

Niloticane

Strong antiviral (enveloped monopartite linear double-stranded (+)RNA) polar extract: Aqueous extract of leaves at a concentration of 156 µg/mL protected Madin-Darby canine kidney cells against the Bovine herpes virus type-1 by 61.1% (Gupta et al., 2010).

Strong antiviral (enveloped monopartite linear double-stranded DNA) polar extract: Methanol extract of bark inhibited the replication of the Herpes simplex virus type-2 and acyclovir-resistant Herpes simplex type-2 with EC_{50} values of 4.7 and 6.7 µg/mL, respectively (Donalisio et al., 2018).

Strong antiviral (non-enveloped circular double-stranded DNA) polar extract: Methanol extract of bark inhibited the replication of the Human papilloma virus type-16 with an EC_{50} value of 1.8 µg/mL (Donalisio et al., 2018).

Commentaries: The plant builds ellagitannins (Bhanu et al., 1964; Mueller-Harvey, 2001), (+)-catechin-5,7-digallate, as well as rutin (quercetin 3-*O*-rutinoside) (Seigler, 2003), which probably account for the antiviral effects and traditional uses. (ii) *Acacia modesta* Wall. is traditionally used for the treatment of dysentery in Pakistan (local name: *Phulai*). *Acacia torta* (Roxb.) Craib is used in India (local name: *Kallinja*) to treat bronchitis, cough, measles, and tuberculosis. *Acacia auriculiformis* A. Cunn. ex Benth. is used in Bangladesh (local name: *Akash-moni*) for abscesses.

REFERENCES

Bhanu, K.U., Rajadurai, S. and Nayudamma, Y., 1964. Studies on the tannins of babul, Acacia arabica, bark. *Australian Journal of Chemistry*, *17*(7), pp. 803–809.

Banso, A., 2009. Phytochemical and antibacterial investigation of bark extracts of Acacia nilotica. *Journal of Medicinal Plants Research*, *3*(2), pp. 082–085.

Donalisio, M., Cagno, V., Civra, A., Gibellini, D., Musumeci, G., Rittà, M., Ghosh, M. and Lembo, D., 2018. The traditional use of Vachellia nilotica for sexually transmitted diseases is substantiated by the antiviral activity of its bark extract against sexually transmitted viruses. *Journal of Ethnopharmacology*, *213*, pp. 403–408.

Eldeen, I.M.S., Van Heerden, F.R. and Van Staden, J., 2010. In vitro biological activities of niloticane, a new bioactive cassane diterpene from the bark of Acacia nilotica subsp. kraussiana. *Journal of Ethnopharmacology*, *128*(3), pp. 555–560.

Gupta, D., Goel, A. and Bhatia, A.K., 2010. Studies on antiviral property of Acacia nilotica. *Journal of Environmental Research And Development*, *5*, pp. 141–152.

Mueller-Harvey, I., 2001. Analysis of hydrolysable tannins. *Animal Feed Science and Technology*, *91*(1–2), pp. 3–20.

Ojha, D., Singh, G. and Upadhyaya, Y.N., 1969. Clinical evaluation of Acacia catechu, Willd.(Khadira) in the treatment of lepromatous leprosy. *International Journal of Leprosy and Other Mycobacterial Diseases: Official Organ of the International Leprosy Association*, *37*(3), pp. 302–307.

Rehman, S., Ashfaq, U.A., Riaz, S., Javed, T. and Riazuddin, S., 2011. Antiviral activity of Acacia nilotica against Hepatitis C Virus in liver infected cells. *Virology Journal*, *8*(1), p. 220.

Sadiq, M.B., Tarning, J., Aye Cho, T.Z. and Anal, A.K., 2017. Antibacterial activities and possible modes of action of Acacia nilotica (L.) Del. against multidrug-resistant Escherichia coli and Salmonella. *Molecules*, *22*(1), p.47.

Seigler, D.S., 2003. Phytochemistry of Acacia—sensu lato. *Biochemical Systematics and Ecology*, *31*(8), pp. 845–873.

7.6.1.3 *Acacia catechu* (L. f.) Willd.

Synonyms: *Acacia wallichiana* DC., *Mimosa catechu* L. f.

Common names Cutch tree, catechu; khayer (Bangladesh); er cha (China); khair (India); mung-ting (Myanmar); khayar (Nepal)

Habitat: Deserts, dry sunny sandy spots, or cultivated

Distribution: Pakistan, India, Bhutan, Sri Lanka, Nepal, Myanmar, Bangladesh, and Thailand

Botanical observation: This tree grows up to a height of 10 m. The wood is dark reddish brown and source of catechu. The stems are spiny. The leaves are bipinnate, spiral, and stipulate. The stipules are spinescent. The blade consists of 10–30 pairs of pinnae with 20–50 pairs of folioles per pinnae. The folioles are about 5mm long. The spikes are axillary and grow up to 10 cm long. The calyx is tubular, 5-lobed, and minute. The corolla comprises 5 petals, which are lanceolate and up to 3 mm long. The androecium comprises numerous stamens longer than the petals. The pods are flat, thin, somewhat glossy, 12–15 cm long and sheltering 3–10 seeds.

Medicinal uses: Dysentery, gonorrhea, bronchitis, pharyngitis (India); leprosy, fever, ringworm (Bangladesh); dysentery, leucorrhea (Nepal); diarrhea (Bangladesh, India, Nepal); ulcers (Myanmar)

Strong antibacterial (Gram-positive) polar extract: Ethanol extract inhibited the growth of 35 clinical isolates of methicillin-resistant *Staphylococcus aureus* and *Staphylococcus aureus* with MIC/MBC values of 1.6–3.2/25 and 1.6/12.5 µg/mL, respectively (Voravuthikunchai & Kitpipit, 2005). Ethanol extract of wood inhibited the growth of *Escherichia coli* (O157:H7) and *Escherichia coli* (ATCC ATCC 25922) with MIC/MBC values of 980/3910 and 1950/3910 µg/mL, respectively (Voravuthikunchai & Limsuwan, 2006).

Antimycobacterial extract: The antileprosy effect of the plant has been clinically investigated (Ojha et al., 1969).

Antifungal (filamentous) mid-polar extract: Ethyl acetate extract of wood (5% solution in 5 mm diameter wells) inhibited the growth of *Fusarium oxysporium*, *Fusarium moniliformis*, *Fusarium proliferatum*, and *Exserohilum turticum* with inhibition zone diameters of 17, 9, 9, and 14 mm, respectively (Joshi et al., 2011).

Strong antiviral (enveloped monopartite linear dimeric single-stranded (+)RNA) polar extract: Aqueous extract of stem bark inhibited the replication of the Human immunodeficiency virus type-1 (NL4.3 strain) with an IC_{50} value of 1.8 µg/mL (Modi et al., 2013).

Viral enzyme inhibition by polar extract: Aqueous extract of stem bark inhibited the Human immunodeficiency virus type-1 protease with an IC_{50} value of 12.9 µg/mL (Moodi et al., 2013).

Commentary: The plant yields (+)-catechin as well as a series of condensed tannins (Duval & Averous, 2016).

REFERENCES

Duval, A. and Averous, L., 2016. Characterization and physicochemical properties of condensed tannins from Acacia catechu. *Journal of Agricultural and Food Chemistry*, *64*(8), pp. 1751–1760.

Joshi, S., Subedi, Y.P. and Paudel, S.K., 2011. Antibacterial and antifungal activity of heartwood of Acacia catechu of Nepal. *Journal of Nepal Chemical Society*, *27*, pp. 94–99.

Modi, M., Dezzutti, C.S., Kulshreshtha, S., Rawat, A.K.S., Srivastava, S.K., Malhotra, S., Verma, A., Ranga, U. and Gupta, S.K., 2013. Extracts from Acacia catechu suppress HIV-1 replication by inhibiting the activities of the viral protease and Tat. *Virology Journal*, *10*(1), p.309.

Ojha, D., Singh, G. and Upadhyaya, Y.N., 1969. Clinical evaluation of *Acacia catechu*, Will d.(Khadira) in the treatment of lepromatous leprosy. *International Journal of Leprosy and Other Mycobacterial Diseases: Official Organ of the International Leprosy Association*, *37*(3), pp. 302–307.

Voravuthikunchai, S.P. and Kitpipit, L., 2005. Activity of medicinal plant extracts against hospital isolates of methicillin-resistant Staphylococcus aureus. *Clinical Microbiology and Infection*, *11*(6), pp. 510–512.

Voravuthikunchai, S.P. and Limsuwan, S., 2006. Medicinal plant extracts as anti–Escherichia coli O157: H7 agents and their effects on bacterial cell aggregation. *Journal of Food Protection*, *69*(10), pp. 2336–2341.

7.6.1.4 *Acacia concinna* (Willd.) DC.

Synonyms: *Acacia sinuata* (Lour.) Merr.; *Mimosa concinna* Willd.; *Mimosa sinuata* Lour.

Common names: Soap pod; rita; sikaki (Bangladesh); banla saot (Cambodia); saptala, shika kai (India); kate kate kecil (Indonesia); kin mum gyin (Myanmar); keo me (the Philippines); som khon (Thailand)

Habitats: Forests

Distribution: Tropical Asia

Botanical observation: This scrambling shrub grows up to about 20 m long. The stems are hairy and minutely prickly. The leaves are bipinnate, alternate, and stipulate. The stipules are ovate-cordate, and up to 8 mm long. The blade is 10–20 cm and presents 6–18 pairs of pinnae. The pinnae are 8–12 cm long and present 15–25 pairs of folioles. The folioles are glaucous below, membranous, edible (India) linear-oblong, asymmtrical, and 8–12 × 2–3 mm. The panicles are axillary and present globose and about 1 cm diameter heads. The calyx is tubular, reddish, 5-lobed, and minute. The corolla is tubular, 5-lobed, and whitish yellow. The ovary is stipitate. The pods are strap-shaped, 8–15 × 2–3 cm, smooth, stout, and dark brown.

Medicinal uses: Skin infection, diarrhea (Myanmar); fever (India, Thailand); shampoo (India, Myanmar, the Philippines); cough, leprosy (India)

Broad-spectrum antibacterial halo developed by polar extract: Ethanol extract of pods (100 mg/mL/6 mm well) inhibited the growth of *Bacillus subtilis* (MTCC 121), *Staphylococcus aureus* (MTCC 3160), *Micrococcus luteus* (MTCC 106), *Escherichia coli* (MTCC 443), and *Proteus vulgaris* (MTCC 426) with inhibition zone diameters of 12.5. 15, 15.5, 14, and 14 mm, respectively (Medisetti et al., 2017).

Strong broad-spectrum antibacterial mid-polar extract: Chloroform extract inhibited the growth of *Escherichia coli* (ATCC 25822), *Salmonella enterica* serovar *typhimurium* (strain U302, DT1046), *Salmonella enterica* serovar *enteritidis* (human), *Salmonella enterica* serotype 4,5,12: i-(human) US clone, *Bacillus cereus*, and *Listeria monocytogenes* (10403S) (Yasurin & Piya-Isarakul, 2015).

Broad-spectrum antifungal halo developed by polar extract: Ethanol extract of pods (140 μg/disc) inhibited the growth of *Penicillium marneffei*, *Candida albicans*, and *Cryptococcus neoformans* with inhibition zone diameters of 18.7, 10.5, and 12 mm, respectively (Wuthi-udomlert & Vallisuta, 2011).

Commentaries: (i) The pods are used to make shampoo because they abound with triterpene saponins (Gafur et al., 1997; Tezuka et al., 2000). The precise molecular mechanisms behind the antibacterial and antifungal properties of saponins is yet not fully delineated but evidence converge to the fact that (a) saponins are amphiphilic surfactant agents that disrupt bacterial, fungal, and viral membranes (Dong et al., 2020; Zhao et al., 2020); (b) saponins form complexes with sterols in the cell membrane of bacterial and fungi leading to the formation of pores and subsequent loss of cytoplasmic constituents and weakening antibacterial resistance (Mert-Türk, 2006); and (c) in Gram-negative bacteria, saponins bind to the lipid A part of lipopolysaccharides resulting in an increase in Gram-negative bacterial outer membrane permeabilities (Arabski et al., 2009). Consider that saponins by virtue of their surfactant properties are virucidal for enveloped viruses (Apers, 2001). (ii) Chloroform does not dissolve saponins, thus, non-saponin and antibacterial principles are generated by this plant. (iii) The plant produces the simple isoquinoline (+)-calycotomine (Menachery et al., 1986), which could be examined for its antimicrobial effects.

REFERENCES

Apers, S., Baronikova, S., Sindambiwe, J.B., Witvrouw, M., De Clercq, E., Berghe, D.V., Van Marck, E., Vlietinck, A. and Pieters, L., 2001. Antiviral, haemolytic and molluscicidal activities of triterpenoid saponins from Maesa lanceolata: establishment of structure-activity relationships. *Planta Medica*, *67*(06), pp. 528–532.

Arabski, M., Wąsik, S., Dworecki, K. and Kaca, W., 2009. Laser interferometric and cultivation methods for measurement of colistin/ampicilin and saponin interactions with smooth and rough of Proteus mirabilis lipopolysaccharides and cells. *Journal of Microbiological Methods*, *77*(2), pp. 178–183.

Dong, S., Yang, X., Zhao, L., Zhang, F., Hou, Z. and Xue, P., 2020. Antibacterial activity and mechanism of action saponins from Chenopodium quinoa willd. husks against foodborne pathogenic bacteria. *Industrial Crops and Products*, *149*, p. 112350.

Gafur, M.A., Obata, T., Kiuchi, F. and Tsuda, Y., 1997. Acacia concinna saponins. I. Structures of prosapogenols, concinnosides AF, isolated from the alkaline hydrolysate of the highly polar saponin fraction. *Chemical and Pharmaceutical Bulletin*, *45*(4), pp. 620–625.

Medisetti, V., Battu, G.R., Sheik, P.B., Samineni, S., Ruthala, P. and Pothula, R., 2017. Antibacterial and anthelmintic activities of aqueous and alcoholic extracts of Acacia concinna (Shikakai) pods. *Inventi Rapid: Ethnopharmacology*, *2017*(3).

Menachery, M.D., Lavanier, G.L., Wetherly, M.L., Guinaudeau, H. and Shamma, M., 1986. Simple isoquinoline alkaloids. *Journal of Natural Products*, *49*(5), pp. 745–778.

Mert-Türk, F., 2006. Saponins versus plant fungal pathogens. *Journal of Cellular and Molecular Biology*, *5*, pp. 13–17.

Tezuka, Y., Honda, K., Banskota, A.H., Thet, M.M. and Kadota, S., 2000. Kinmoonosides A– C, Three New Cytotoxic Saponins from the Fruits of Acacia c oncinna, a Medicinal Plant Collected in Myanmar. *Journal of Natural Products*, *63*(12), pp. 1658–1664.

Wuthi-udomlert, M. and Vallisuta, O., 2011. In vitro effectiveness of Acacia concinna extract against dermatomycotic pathogens. *Pharmacognosy Journal*, *3*(19), pp. 69–73.

Yasurin, P. and Piya-Isarakul, S., 2015. In-vitro antibacterial activity screening of herb extracts against foodborne pathogenic bacteria from Thailand. *Journal of Pure and Applied Microbiology*, *9*(3), pp. 2175–2184.

Zhao, Y., Su, R., Zhang, W., Yao, G.L. and Chen, J., 2020. Antibacterial activity of tea saponin from Camellia oleifera shell by novel extraction method. *Industrial Crops and Products*, *153*, p. 112604.

7.6.1.5 *Acacia farnesiana* (L.) Willd.

Synonyms: *Acacia acicularis* Humb. & Bonpl. ex Willd.; *Acacia acicularis* R. Br.; *Acacia densiflora* (Alexander ex Small) Cory; *Acacia edulis* Humb. & Bonpl. ex Willd.; *Acacia ferox* M. Martens & Galeotti; *Acacia lenticellata* F. Muell.; *Acacia leptophylla* DC.; *Acacia pedunculata* Willd.; *Acacia smallii* Isely; *Farnesia odora* Gasp.; *Mimosa arcuata* M. Martens & Galeotti; *Mimosa farnesiana* L.; *Mimosa scorpioides* Forssk.; *Pithecellobium minutum* M.E. Jones; *Poponax farnesiana* (L.) Raf.; *Vachellia farnesiana* (L.) Wight & Arn.; *Vachellia guanacastensis* (H.D. Clarke, Seigler & Ebinger) Seigler & Ebinger

Common names: Sponge tree, sweet acacia, West Indian blackthorn; guya babula; kanta naksha (Bangladesh); baranga daru, gabur, gandi babul, (India); bunga bandara (Indonesia); kan thin nam (Laos); poko laksana (Malaysia); sambue mye (Myanmar); kambang (the Philippines); krathin hom (Thailand); cay keo ta (Vietnam)

Habitat: Cultivated

Distribution: Tropical Asia and Pacific

Botanical observation: This handsome tree grows to 4 m tall. The stems are zig-zag shaped. The leaves are bipinnate, alternate, and stipulate. The stipules are spines, which are up to 2 cm long. The rachis is 1.2–2.5 cm long and bears 2–8 pairs of pinnae. The pinnae are 1.5–3.5 cm long andinclude 10–20 pairs of folioles. The folioles are 2.5–5.5 mm × 1–1.5 mm, linear-oblong and asymmetrical. The fascicles of fragrant and graceful yellow heads are axillary. The calyx is campanulate, minute, and 5-lobed, the lobes triangular. The corolla is tubular, about 3 mm long, and develops 5 elliptical lobes. The androecium is showy and consists of numerous stamens, which are about twice as long as the corolla. The ovary is oblong and hairy. The pods are characteristic, resembling fleshy and cylindrical curved caterpillars, 4.5–7.5 ×. 0.8–1.2 cm, and containing numerous seeds, which are brown, ovoid, and about 5 mm across.

Medicinal uses: Leucorrhea, infection of the gums, gonorrhea (Myanmar); dysentery (India)

Moderate (Gram-negative) polar extract: Methanol extract of bark inhibited the growth of *Vibrio cholerae* (O1 strain 569-B) and *Vibrio cholerae* (O139 strain AI-1837) with MBC values of 500 and 900 μg/mL, respectively (Sánchez et al., 2010).

Moderate antibacterial (Gram-positive) polar extract: Ethanol extract of leaves inhibited the growth of *Bacillus subtilis* (ATCC 6633) with an MIC value of 800 μg/mL (Ramli et al., 2011).

Strong antibacterial (Gram-negative) amphiphilic simple phenol: Methyl gallate (LogD = 1.1 at pH 7.4; molecular mass = 184.1 g/mol) isolated from the bark inhibited the growth of *Vibrio cholerae* (O1 strain 569-B) and *Vibrio cholerae* (O139 strain AI-1837) with MBC values of 30 and 50 μg/mL, respectively, with induction of cytoplasmic membrane hyperpolarization, increase in cytoplasmic pH, and reduction of ATP production (Sánchez et al., 2013). In a previous study methyl gallate inhibited the growth of *Escherichia coli* (NCTC 1048) and *Staphylococcus aureus* (NCTC 6571) with MIC values of 400 and 100 μg/mL, respectively (Adesina et al., 2000).

Moderate antimycobacterial polar extract: Aqueous extract of pods inhibited the growth of *Mycobacterium tuberculosis* (ATCC 27294 (H37Rv) and *Mycobacterium tuberculosis* (multidrug-resistant clinical isolate (MDR) G122 (resistant to isoniazid, rifampicin, and ethambutol) with MIC values of 200 and 100 μg/mL, respectively (Hernández-García et al., 2019).

Moderate antimycobacterial amphiphilic simple phenol: Methyl gallate inhibited the growth of *Mycobacterium tuberculosis* (ATCC 27294 (H37Rv) with an MIC value of 50 μg/mL (Hernández-García et al., 2019).

Weak broad-spectrum antifungal polar extract: Ethanol extract of leaves inhibited the growth of *Saccharomyces cerevisiae* (ATCC 9763) with an MIC value of 2500 μg/mL (Ramli et al., 2011). Hydroalcoholic extract inhibited the growth of *Histoplasma capsulatum* (1526) and *Histoplasma capsulatum* (1591) with MIC values of 500 and 500 μg/mL, respectively (Alanís-Garza et al., 2007).

Commentary: Methyl gallate is amphiphilic and as such probably penetrate (depending on its concentration) in the cytoplasmic membrane and destabilize it, causing malfunction of the respiratory chain resulting in decrease in both ATP production and cytoplasmic pH (since the respiratory chain expulses H^+ from the cytoplasm to the periplasm of Gram-negative bacteria). Consider that pH homeostasis in bacteria is necessary for physiological enzymatic activity, and a decrease or increase in pH inhibits enzymatic catalytic activity (Booth, 1985). Does methyl gallate act on the outer envelop of viruses?

REFERENCES

Adesina, S.K., Idowu, O., Ogundaini, A.O., Oladimeji, H., Olugbade, T.A., Onawunmi, G.O. and Pais, M., 2000. Antimicrobial constituents of the leaves of Acalypha wilkesiana and Acalypha hispida. *Phytotherapy Research*, *14*(5), pp. 371–374.

Alanís-Garza, B.A., González-González, G.M., Salazar-Aranda, R., de Torres, N.W. and Rivas-Galindo, V.M., 2007. Screening of antifungal activity of plants from the northeast of Mexico. *Journal of Ethnopharmacology*, *114*(3), pp. 468–471.
Booth, I.R., 1985. Regulation of cytoplasmic pH in bacteria. *Microbiological Reviews*, *49*(4), p.359.
Hernández-García, E., García, A., Garza-González, E., Avalos-Alanís, F.G., Rivas-Galindo, V.M., Rodríguez-Rodríguez, J., Alcantar-Rosales, V.M., Delgadillo-Puga, C. and del Rayo Camacho-Corona, M., 2019. Chemical composition of Acacia farnesiana (L) wild fruits and its activity against Mycobacterium tuberculosis and dysentery bacteria. *Journal of Ethnopharmacology*, *230*, pp. 74–80.
Ramli, S., Harada, K.I. and Ruangrungsi, N., 2011. Antioxidant, antimicrobial and cytotoxicity activities of Acacia farnesiana (L.) Willd. Leaves ethanolic extract. *Pharmacognosy Journal*, *3*(23), pp. 50–58.
Sánchez, E., Heredia, N. and Garcıa, S. (2010) Extracts of edible and medicinal plants damage membranes of Vibrio cholerae. *Applied and Environmental Microbiology*, 76, pp. 6888–6894.
Sánchez, E., Heredia, N., Camacho-Corona, M.D.R. and García, S., 2013. Isolation, characterization and mode of antimicrobial action against Vibrio cholerae of methyl gallate isolated from Acacia farnesiana. *Journal of Applied Microbiology*, *115*(6), pp. 1307–1316

7.6.1.6 *Acacia pennata* (L.) Willd.

Synonyms: *Acacia arrophula* D. Don; *Acacia hainanensis* Hayata; *Acacia pentagona* (Schumach. & Thonn.) Hook. f.; *Acacia philippinarum* Benth.; *Albizia tenerrima* de Vriese; *Mimosa ferruginea* Rottler; *Mimosa pennata* L.; *Mimosa torta* Roxb.

Common names: Rusty mimosa; thma roeb (Cambodia); han ba teng, yu ye jin he huan (China); rembete (Indonesia); sijrai, valli khadira (India); han (Laos); cha om (Thailand)

Habitat: Scrubs, thickets

Distribution: India, Bhutan, Myanmar, Cambodia, Laos, Vietnam, Thailand, and Indonesia

Botanical observation: This woody climber grows up to about 10 m long. The stems are hairy at apex. The bark is used in India for the making of soap. The leaves are bipinnate, edible (Thailand), alternate, and stipulate. The stipules are sharply and narrowly triangulate and up to about 3 cm long. The rachis of the blade is hairy. The blade is 10–20 cm and presents 8–24 pairs of pinnae, each of them bearing 30–50 pairs of folioles, which are papery, linear-oblong, asymmetrical, and 5–10 × 0.5–1.5 mm. The inflorescences are axillary or terminal globose flower heads, which are about 1 cm across. The calyx is tubular, 5-lobed, and minute. The corolla is tubular, 5-lobed, up to about 4 mm long, and whitish yellow. The androecium is showy and consists of 10 or more stamens, which are about twice as long as the corolla. The ovary is stipitate. The pods are reddish brown, flat, 9–20 × 1.5–3.5 cm, somewhat twisted, glossy, oblong, and shelter 6–12 seeds.

Medicinal uses: Burns (China); cough, fever (India)

Antibacterial (Gram-positive) polar extract: Methanol extract (15 μL of a 400 mg/mL solution on 6 mm disc) inhibited the growth of *Bacillus cereus* (DMST 5040) and *Lactobacillus plantarum* (TISTR 050) with inhibition zone diameters of 8.5 and 7.7 mm, respectively (Nanasombat & Teckchuen, 2009).

Viral enzyme inhibition by triterpene saponin: 21-β-*O*-[(2*E*)-6-hydroxyl-2,6-dimethyl-2,7- octadienoyl] pitheduloside G inhibited the Human immunodeficiency type-1 protease activity, with an IC_{50} value of 2 μM (Nguyen et al., 2018).

Commentary: (i) The plant brings to being quercetin, kaempferol, pinocembrin glycosides and proanthocyanidins (Kim et al., 2015) as well as the diterpene taepeenin D (Rifai et al., 2010), which could be examined for possible antibacterial, antifungal, or antiviral effects because cassane-type furanoditerpenes are not uncommonly antimicrobial. (ii) Saponins in this plant may account for the antitussive medicinal use. (iii) Consider that pitheduloside G inhibited the Human immunodeficiency type-1 protease activity, with an IC_{50} value of 18 μM (Nguyen et al., 2018) implying that the anti-Human immunodeficiency virus type-1 protease activity of triterpene saponins could be mainly owed to the glycosidic moiety.

REFERENCES

Kim, A., Choi, J., Htwe, K.M., Chin, Y.W., Kim, J. and Yoon, K.D., 2015. Flavonoid glycosides from the aerial parts of Acacia pennata in Myanmar. *Phytochemistry*, *118*, pp. 17–22.

Nanasombat, S. and Teckchuen, N., 2009. Antimicrobial, antioxidant and anticancer activities of Thai local vegetables. *Journal of Medicinal Plants Research*, *3*(5), pp. 443–449.

Nguyen, V.D., Nguyen, H.L.T., Do, L.C., Van Tuan, V., Thuong, P.T. and Phan, T.N., 2018. A New Saponin with Anti-HIV-1 Protease Activity from Acacia pennata. *Natural Product Communications*, *13*(4), p.1934578X1801300408.

Rifai, Y., Arai, M.A., Koyano, T., Kowithayakorn, T. and Ishibashi, M., 2010. Terpenoids and a flavonoid glycoside from Acacia pennata leaves as hedgehog/GLI-mediated transcriptional inhibitors. *Journal of Natural Products*, *73*(5), pp. 995–997.

7.6.1.7 *Adenanthera pavonina* L.

Common names: coral pea, Red wood tree; Circassian seeds tree; Raktochandan (Bangladesh); chan' trèi (Cambodia); hai hong dou (China); manchadai, rakta kanchana, ruyka chundun (India); saga telik (Indonesia); lam ta khouay, (Laos); pokok saga (Malaysia); mai-chek, ywe-ni (Myanmar); alalangat, baguiroro (the Philippines), sem (India); ma hok daeng (Thailand); muô`ng ràng rang (Vietnam)

Habitat: Forests, cultivated

Distribution: Tropical Asia

Botanical observation: This handsome tree grows up to 30 m tall and provides a peculiarly cool shade. The wood is hard and red. The leaves are bipinnate, alternate, and stipulate. The stipules are filiform, caducous, and up to about 5 mm long. The rachis is up to 40 cm long and presents 2–6 pairs of pinnae. The pinnae are about 12 cm long and present 7–15 pairs of folioles. The folioles are alternate, 1.5–4.0 cm × 5–25 mm, and somewhat characteristically elliptical, dull green, and membranous. The racemes are axillary or terminal and 7.5–20 cm long. The calyx is tubular, somewhat cup-shaped, hairy, 5-lobed, and minute. The corolla consists of 5 petals, which are elliptic, acute, pale yellow, and about 3 mm long. The androecium includes 10 stamens, which are about 5 mm long. The ovary is sessile and develops a capitate stigma. The pods are dehiscent, curved, strap-shaped, 10–22.5 cm long, coriaceous, and contain numerous seeds. The seeds are magnificent, bright red, glossy ellipsoid, and about 5 mm across.

Medicinal uses: Abscesses, conjunctivitis, leucorrhea (Bangladesh); diarrhea, dysentery (India); boils (Bangladesh; India)

Moderate antibacterial (Gram-positive) polar extract: Extract of leaves inhibited the growth of *Campylobacter jejuni* (ATCC 29428) with MIC/MBC values of 62.5/125 µg/mL (Dholvitayakhu et al., 2012).

Moderate broad-spectrum antibacterial amphiphilic flavonol: Robinetin inhibited the growth of *Staphylococcus aureus* and *Proteus vulgaris* with MIC values of 100 and 100 µg/mL (Mori et al., 1987).

Strong broad-spectrum antifungal amphiphilic flavonol: Robinetin (LogD = 1.1 at pH 7.4; molecular mass = 302.2 g/mol) inhibited the growth of *Rhizopus oryzae* and *Cryptococcus neoformans* with an MIC_{75} of 15 µg/mL (Tie et al., 2004).

Viral enzyme inhibition by flavonol: Robinetin inhibited the Hepatitis C virus RNA-dependent RNA polymerase activity by 55.9% at a concentration of 20 µM (Ahmed-Belkacem et al., 2014).

Commentaries: (i) The plant generates quercetin and kaempferol glycosides (Mohammed et al., 2014) as well as robinetin (Devi et al., 2007), a series of oleanane-type triterpene saponins (Ara et al., 2020). (ii) In India, a dye from the wood is used by Hindus for marking the forehead. (iii) Consider that polyhydroxylated flavones like baicalein, scutellarein, or myricetin are not uncommonly able to inhibit reverse transcriptase and cellular DNA and RNA polymerase (Ono et al., 1990). Baicalein

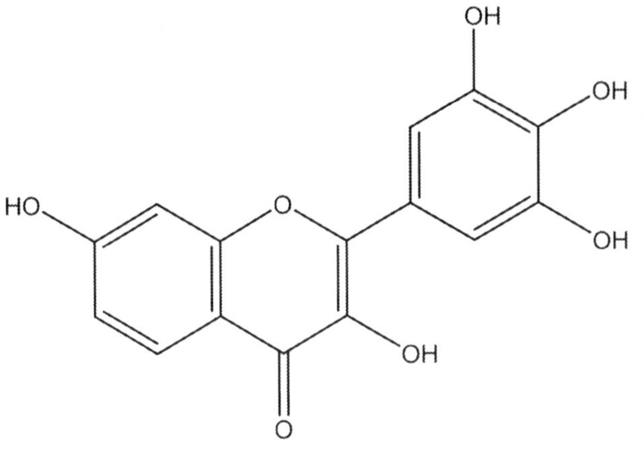

Robinetin

inhibited the Severe acute respiratory syndrome-acquired coronavirus 3-chymotrypsine-like protease activity with an IC_{50} of 0.3 µM (Liu et al., 2020). Scutellarein and myricetin inhibited Severe acute respiratory syndrome-acquired coronavirus helicase and nsP13 ATPase with IC_{50} values 0.8 and 2.7 µM (Yu et al., 2012). Thus, one could examine the anti-Severe acute respiratory syndrome-acquired coronavirus 2 properties of polyhydroxylated flavones, including baicalein, myricetin, scutellarein, and robinetin. These flavonoids could be added to the list of first line molecules to be tested for upcoming viral pandemics.

REFERENCES

Ahmed-Belkacem, A., Guichou, J.F., Brillet, R., Ahnou, N., Hernandez, E., Pallier, C. and Pawlotsky, J.M., 2014. Inhibition of RNA binding to hepatitis C virus RNA-dependent RNA polymerase: a new mechanism for antiviral intervention. *Nucleic acids research*, *42*(14), pp. 9399–9409.

Ara, A., Saleh-E-In, M.M., Ahmad, M., Hashem, M.A. and Hasan, C.M., 2020. Isolation and Characterization of Compounds from the Methanol Bark Extract of Adenanthera pavonina L. *Analytical Chemistry Letters*, *10*(1), pp. 49–59.

Devi, V.G., Biju, P.G., Devi, D.G., Cibin, T.R., Lija, Y. and Abraham, A., 2007. Isolation of robinetin from Adenanthera pavonina and its protective effect on perchlorate induced oxidative damage. *J Trop Med Plants*, *7*(1), p.160.

Dholvitayakhun, A., Cushnie, T.T. and Trachoo, N., 2012. Antibacterial activity of three medicinal Thai plants against Campylobacter jejuni and other foodborne pathogens. *Natural product research*, *26*(4), pp. 356–363.

Mohammed, R.S., Abou Zeid, A.H., El-Kashoury, E.A., Sleem, A.A. and Waly, D.A., 2014. A new flavonol glycoside and biological activities of Adenanthera pavonina L. leaves. *Natural Product Research*, *28*(5), pp. 282–289.

Mori, A., Nishino, C., Enoki, N. and Tawata, S., 1987. Antibacterial activity and mode of action of plant flavonoids against Proteus vulgaris and Staphylococcus aureus. *Phytochemistry*, *26*(8), pp. 2231–2234.

Ono, K., Nakane, H., Fukushima, M., Chermann, J.C. and Barré-Sinoussi, F., 1990. Differential inhibitory effects of various flavonoids on the activities of reverse transcriptase and cellular DNA and RNA polymerases. *European Journal of Biochemistry*, *190*(3), pp. 469–476.

Tie, S.H., Santhanam, J., Wahab, I.A., Marzuki, A. and Weber, J.F.F., 2004. Pharmaco-chemistry of Malaysian timber trees. X. evaluation of the antifungal activity of phenolic constituents of the wood of some tropical timber species. In *18. Seminar of the Malaysian Natural Products Society, 18, Kota Kinabalu, Sabah (Malaysia), 21–24 Oct 2002*. Universiti Malaysia Sabah.

Yu, M.S., Lee, J., Lee, J.M., Kim, Y., Chin, Y.W., Jee, J.G., Keum, Y.S. and Jeong, Y.J., 2012. Identification of myricetin and scutellarein as novel chemical inhibitors of the SARS coronavirus helicase, nsP13. *Bioorganic & Medicinal Chemistry Letters*, *22*(12), pp. 4049–4054.

7.6.1.8 *Albizia lebbeck* (L.) Benth.

Synonyms: *Acacia lebbeck* (L.) Willd.; *Albizzia lebbeck* Benth.; *Feuilleea lebbeck* (L.) Kuntze; *Mimosa lebbeck* L.; *Mimosa lebbek* Forssk.; *Mimosa sirissa* Roxb.; *Mimosa speciosa* Jacq.

Common names: Siris tree, woman's tongue; sherisha; siris (Bangladesh; India); kuo jia he huan (China); kala shareen (India); anya-kokk, kokko (Myanmar); shireen (Pakistan); gmogmol (Yap)

Habitat: Cultivated

Distribution: Pakistan, India, Sri Lanka, Nepal, Bhutan, Bangladesh, Myanmar, China, Pacific Islands

Botanical observation: This graceful tree grows to about 10 m tall. The stems are hairy. The leaves are bipinnate, alternate, and stipulate. The stipule is about 5 mm long, linear, caducous, and hairy. The rachis is 7.5–15 cm long and presents 1–4 pairs of pinnae. The pinnae are 5–20 cm long and comprise 3–9 pairs of folioles. The folioles are oblong and 2–4.5 × 1.3–2 cm. The inflorescences are pedunculate heads of fragrant flowers, solitary or fasciculated on 3.5–10 cm long peduncles. The calyx is campanulate, about 5 mm long, and 5-lobed. The corolla is 7–8 mm long, funnel shaped, with 5 deltoid-ovate lobes. The androecium is showy, spreading, and includes numerous filiform stamens, which are 2.5–3.5 cm long. The ovary is glabrous and sessile. The pods are 15–30 ×2.5–5 cm, dehiscent, flat, pendulous, characteristically light brown and flat, and contain 6–12 seeds.

Medicinal uses: Animal bites, piles (Myanmar); bronchitis, carbuncles, gonorrhea, sore throat, syphilis, mouth ulcers (India); boils (India, Myanmar); cough (India, Pakistan); diarrhea, dysentery (India, Cambodia, Laos, Vietnam)

Broad-spectrum antibacterial non-polar extract: Petroleum ether extract (5 mm/disc/ 300 µg/disc) of bark inhibited the growth of *Bacillus polymyxa*, *Bacillus subtilis*, *Staphylococcus aureus*, *Vibrio mimicus*, *Vibrio cholerae*, *Salmonella typhi*, *Shigella boydii*, *Shigella flexneri*, *Shigella dysenteriae*, and *Klebsiella pneumoniae* with inhibition zone diameters of 10, 11, 13, 11, 8, 13, 10, 8, 14, and 12 mm, respectively (Ali et al., 2018).

Broad-spectrum antibacterial mid-polar extract: Ethyl acetate extract of bark (5 mm/disc/ 300 µg/disc) inhibited the growth of *Bacillus subtilis*, *Bacillus megaterium*, *Staphylococcus aureus*, *Vibrio mimicus*, *Vibrio cholerae*, *Salmonella typhi*, *Shigella boydii*, *Shigella flexneri*, *Shigella dysenteriae* (Ali et al., 2018).

Broad-spectrum antifungal mid-polar extract: Ethyl acetate extract of bark (300 µg/ 5mm disc) inhibited the growth of *Candida arrizae*, *Aspergillus niger*, *Candida albicans*, and *Saccharomyces cerevisiae* with inhibition zone diameters of 10, 9, 8, and 7 mm, respectively (Ali et al., 2018).

Commentary: Antimicrobial principles have not apparently been isolated from this plant, which is known to generate a series of oleanane-type saponins (Desai et al., 2019) as well as catechin (Singh, 2019) and quercetin and kaempferol glycosides (El-Mousallamy, 1998). Members of the genus *Albizia* Durraz. are not uncommonly used to treat microbial infections in Asia. *Albizia amara* (Roxb.) Boivin is used in India (local name: *Shirish*) for the treatment of boils. *Albizia corniculata* (Lour.) Druce is used in China (local name: *Rì luó hè*) for the treatment of sore throat and stomatitis. In Nepal, *Albizia chinensis* (Osbeck) Merr. is used to heal wounds and cuts (local name: *Rato siris*), and in Myanmar it is used as shampoo and to treat burns (local name: *M'bang*). In India, *Albizia julibrissin* Durazz is used to treat bronchitis, asthma, leprosy, and tuberculous glands. From this plant, a flavonol glycoside with antimycobacterial activity was isolated (Yadava & Reddy, 2001). *Albizzia myriophylla* Benth. is used to treat fever in Malaysia

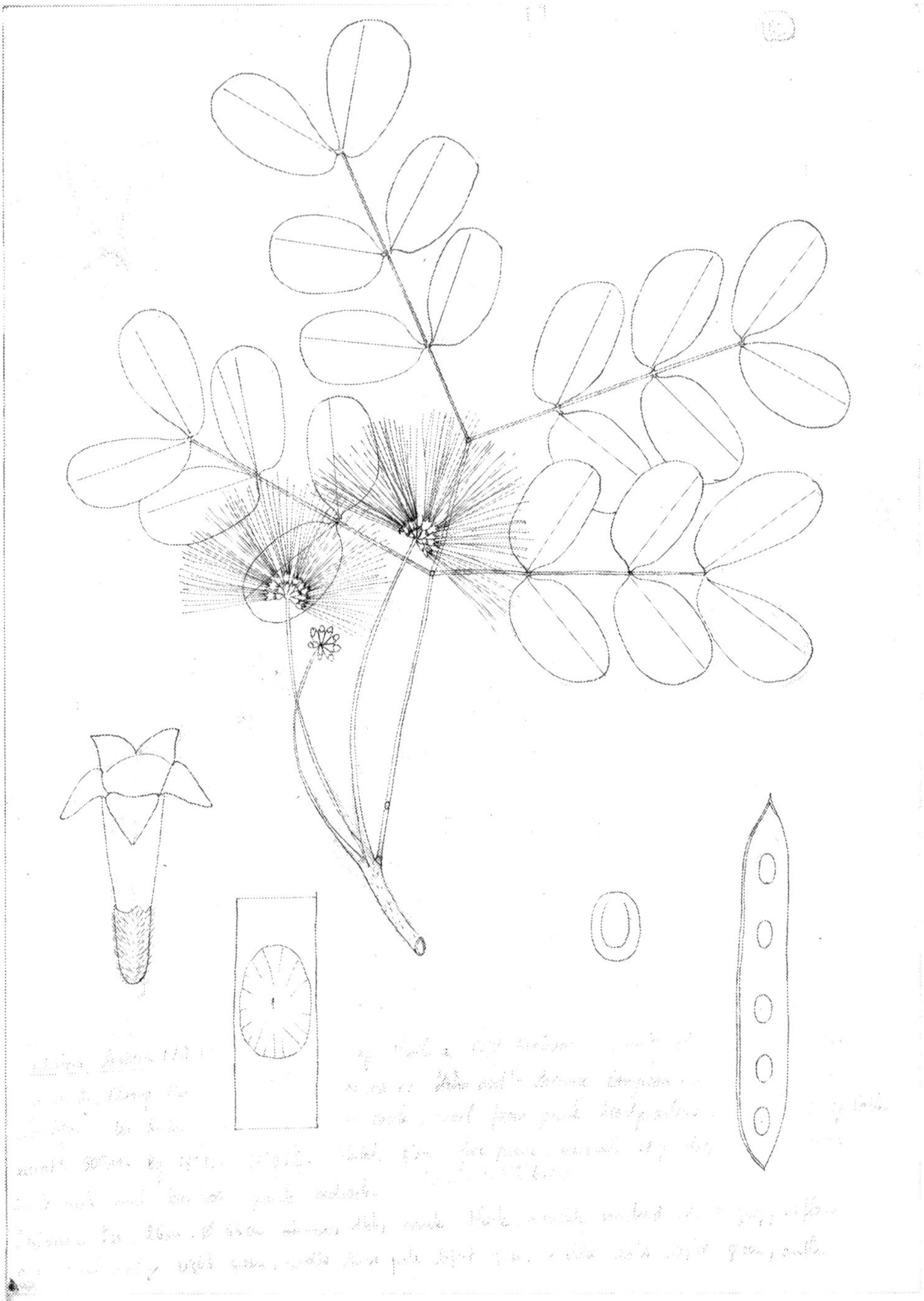

FIGURE 7.5 *Albizia lebbeck* (L.) Benth

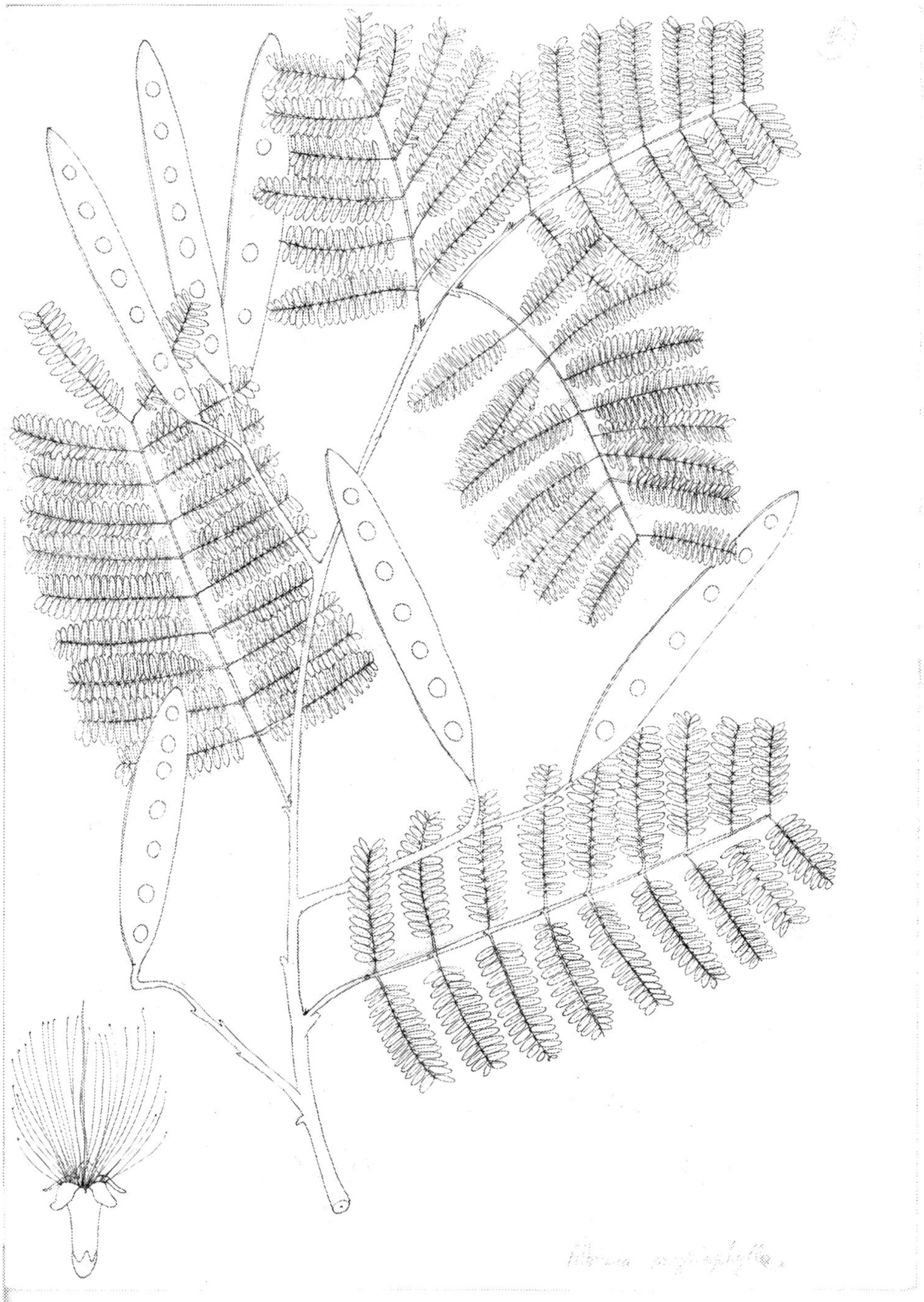

FIGURE 7.6 *Albizzia myriophylla* Benth

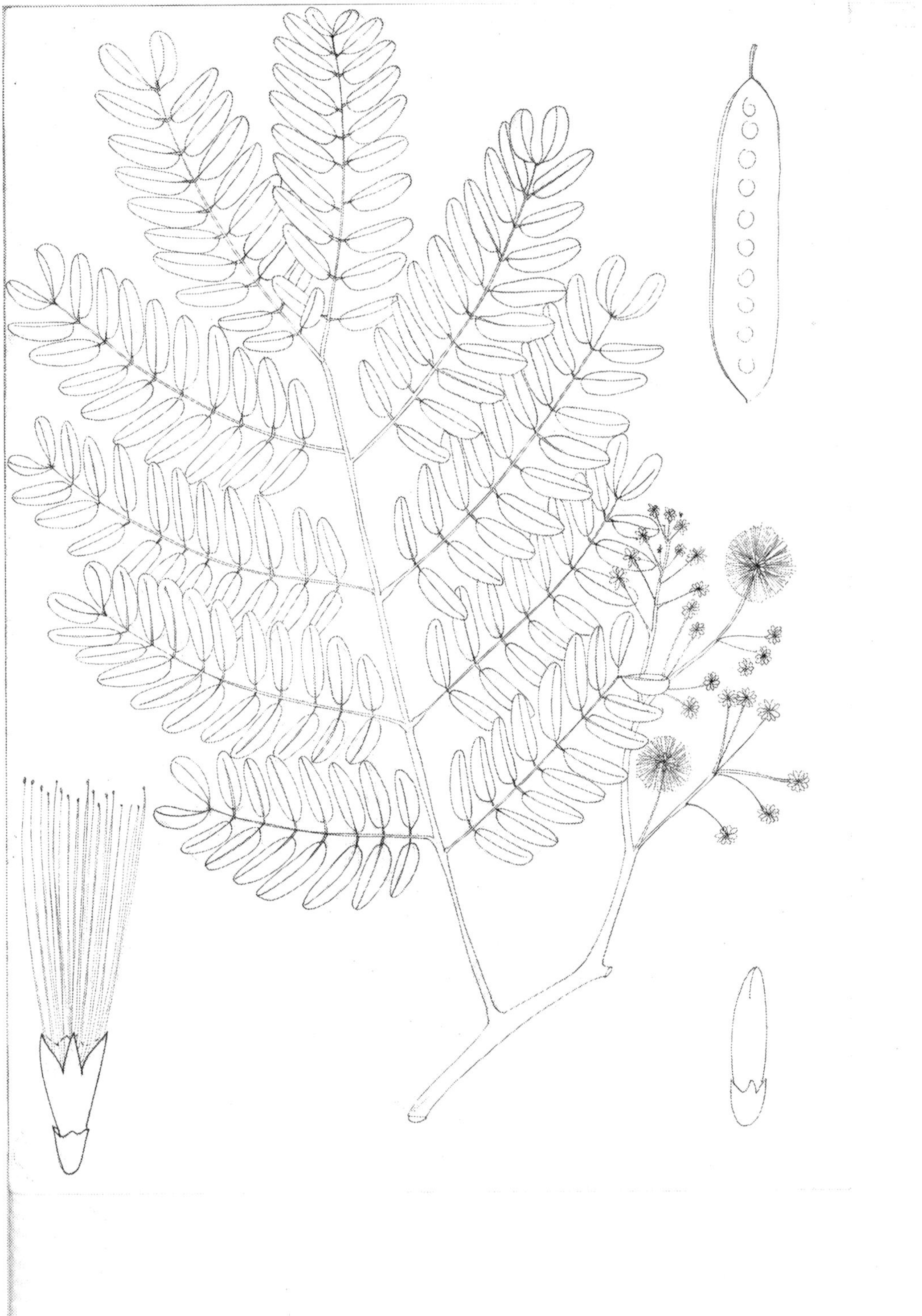

FIGURE 7.7 *Albizia odoratissima* (L.f.) Benth.

(local name: *Akar kulit manis)* and in Cambodia (local name: *Voe aem*), Laos (local name: *Khua khang hung*), and Vietnam to treat bronchitis, cough, and wounds. In Thailand (local name: *Som poi*) the plant is used to treat cough.

REFERENCES

Ali, M.T., Haque, S.T., Kabir, M.L., Rana, S. and Haque, M.E., 2018. A comparative study of in vitro antimicrobial, antioxidant and cytotoxic activity of Albizia lebbeck and Acacia nilotica stem bark. *Bulletin of faculty of Pharmacy, Cairo University*, *56*(1), pp. 34–38.

Desai, T.H. and Joshi, S.V., 2019. Anticancer activity of saponin isolated from Albizia lebbeck using various in vitro models. *Journal of Ethnopharmacology*, *231*, pp. 494–502.

El-Mousallamy, A.M., 1998. Leaf flavonoids of Albizia lebbeck. *Phytochemistry*, *48*(4), pp. 759–761.

Singh, P., 2019. Isolation of catechin from bark of Albizia lebbeck (L.) Benth. *International Journal of Scientific Research in Chemical Sciences*, *6*(2), pp. 8–12.

Yadava, R.N. and Reddy, V.M.S., 2001. A biologically active flavonol glycoside of seeds of Albizzia julibrissin Durazz. *Journal-Institution of Chemists India*, *73*(5), pp. 195–199.

7.6.1.9 *Albizia odoratissima* (L.f.) Benth.

Synonyms: *Acacia odoratissima* Willd.; *Feuilleea odoratissima* (L. f.) Kuntze; *Mimosa odoratissima* L.f.

Common names: Ceylon rosewood; jatkorai (Bangladesh); chichwa, koroi, usila maram (India); mai kying lwai (Myanmar); kali siri (Pakistan)

Habitat: Forests

Distribution: Pakistan, India, Sri Lanka, Bangladesh, Myanmar, Laos, Vietnam, Thailand, and China

Botanical observation: This magnificent tree grows to about 10 m tall. The stems are hairy. The leaves are bipinnate, alternate, and stipulate. The stipules are filiform and minute. The rachis is 10–20 cm long. The blade presents 3–8 pairs of pinnae, which are 4.5–9 cm long, and with 8–20 pairs of folioles. The folioles are 1.7–2.5 × 5–10 mm asymmetrical, oblong, and hairy below. The inflorescences are hairy panicles of fragrant flower heads. The calyx is cup-shaped, hairy, and minute. The corolla is pale yellow, tubular, about 6 mm long, and develops 5 triangular lobes. The androecium is showy, spreading, and includes 20 stamens, which are about 1.5 cm long and fused at base into a tube. The ovary is hairy and develop at filiform style, which grows up to 1.5 cm long. The pods are 10–30 × 1.5–4 cm, hairy, reddish brown, and containing 6–12 seeds.

Medicinal uses: Cough, leprosy, ulcers (Myanmar)

Moderate antibacterial broad-spectrum polar extract: Methanol extract of leaves inhibited the growth of *Staphylococcus aureus*, *Escherichia coli*, *Klebsiella pneumoniae*, *Proteus vulgaris*, and *Pseudomonas aeruginosa* with MIC/MBC values of 136/273, 273/546, 273/546, 546/1093, and 546/1093 µg/mL, respectively (Banothu et al., 2017).

Commentary: Antimicrobial principles have not yet been isolated from this plant. It is known to produce flavones (Rao et al., 2002).

REFERENCES

Banothu, V., Neelagiri, C., Adepally, U., Lingam, J. and Bommareddy, K., 2017. Phytochemical screening and evaluation of in vitro antioxidant and antimicrobial activities of the indigenous medicinal plant Albizia odoratissima. *Pharmaceutical Biology*, *55*(1), pp. 1155–1161.

Rao, Y.K., Reddy, M.V.B., Rao, C.V., Gunasekar, D., Blond, A., Caux, C. and Bodo, B., 2002. Two new 5-deoxyflavones from Albizia odoratissima. *Chemical and Pharmaceutical Bulletin*, *50*(9), pp. 1271–1272.

7.6.1.10 *Albizia procera* (Roxb.) Benth.

Synonyms: *Acacia procera* (Roxb.) Willd.; *Feuilleea procera* (Roxb.) Kuntze; *Mimosa procera* Roxb.,

Common names: Forest siris, white siris, tall albizia; kori gach, shil koroi, sada sirish, (Bangladesh); tramkâng (Cambodia); huang dou shu (China); karai, peela shareen, shveta shirisha, usila maram (India); wangkal (Indonesia); thon (Laos); oriang (Malaysia); sitpen (Myanmar); akleng parang (the Philippines); thing thon (Thailand); muong xanh (Vietnam)

Habitat: Forest margins

Distribution: India, Bangladesh, Myanmar, Cambodia, Laos, Vietnam, Thailand, China, Malaysia, Indonesia, the Philippines, Papua New Guinea, Northern Australia

Botanical observation: This tree grows to 15 m tall. The leaves are bipinnate, alternate, and stipulate. The rachis is 30–45 cm long and presents 2–6 pairs of pinnae, which are up to 25 cm long. The pinnae include 4–16 pairs of folioles, which are oblong-ovate, hairy below, obtuse, asymmetrical, with about seven pairs of secondary nerves, and round at apex. The inflorescences are axillary or terminal panicles of globose heads. The calyx is minute, tubular, and 5 lobed. The corolla is tubular, 5-lobed, yellow-white, and about 6 mm long. The androecium is showy, spreading, fused at base and includes numerous stamens. The ovary is glabrous and develops a slender style. The pods are flat, 10–15 × 1.5–2.5 cm, and contain 8–12 seeds.

Medicinal uses: Ulcers (Bangladesh; India); bronchial problems (India); cough, diarrhea (Thailand)

Broad-spectrum antibacterial mid-polar extract: Ethyl acetate extract of leaves (800 μg/6 mm disc) inhibited the growth of *Staphylococcus aureus*, *Bacillus cereus*, *Pseudomonas aeruginosa*, *Escherichia coli*, *Shigella sonnei*, and *Shigella boydii* with inhibition zone diameters of 10, 8, 9, 10, 10, and 8 mm, respectively (Khatoon et al., 2014).

Strong antiviral (enveloped monopartite linear dimeric single-stranded (+) RNA) hydrophilic flavanol: (+)-Catechin (LogD = 0.5 at pH 7.4; molecular mass = 290.2 g/mol) inhibited the replication of the Human immunodeficiency virus type-1 in C8166 lymphocytes with an EC_{50} value of 14.4 μg/mL and a selectivity index >13.8, (Lin et al., 2010).

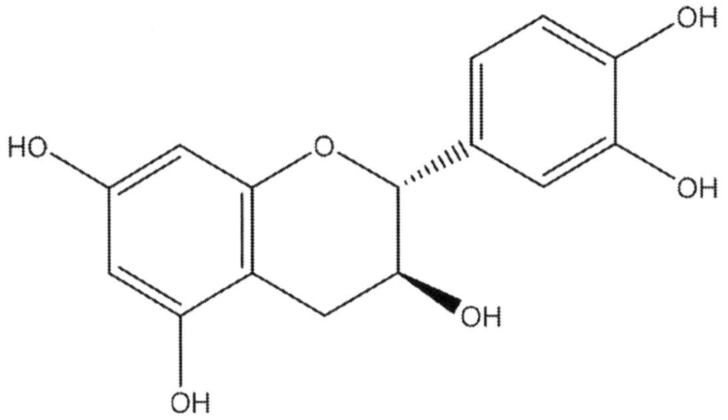

(+)-Catechin

Moderate antiviral (enveloped monopartite linear single-stranded (+)RNA) hydrophilic flavanol: (+)-Catechin at concentrations in the range of 20–40 μM inhibited the replication of the Transmissible gastroenteritis coronavirus in swine testicular cells by about 90% with concomitant decrease in RNA synthesis (Liang et al., 2015).

Strong antiviral (enveloped segmented linear single-stranded (−)RNA) hydrophilic flavanol: (+)-Catechin inhibited the replication of the Influenza A (H1N1) virus (A/California/07/2009) replication in Madin-Darby canine kidney cells with an EC_{50} value of 18.4 μg/mL (You et al., 2018).

Viral enzyme inhibition by flavanol: (+)-Catechin isolated from the bark inhibited the Human immunodeficiency virus type-1 integrase activity with an IC_{50} value of 46.3 µM (Panthong et al., 2015).

Viral enzyme inhibition by simple phenolic: Protocatechuic acid isolated from the bark inhibited the Human immunodeficiency virus type-1 integrase by 46% at a concentration of 100 µM (Panthong et al., 2015).

Commentaries: (i) The plant produces triterpene saponins (Melek et al., 2007). (ii) One could anticipate that (+)-catechin inhibit RNA viruses by interfering with the replication of their genetic material.

REFERENCES

Hassan, S.T., Švajdlenka, E. and Berchová-Bímová, K., 2017. Hibiscus sabdariffa L. and its bioactive constituents exhibit antiviral activity against HSV-2 and anti-enzymatic properties against urease by an ESI-MS based assay. *Molecules, 22*(5), p. 722.

Khatoon, M.M., Khatun, M.H., Islam, M.E. and Parvin, M.S., 2014. Analgesic, antibacterial and central nervous system depressant activities of Albizia procera leaves. *Asian Pacific Journal of Tropical Biomedicine, 4*(4), pp. 279–284.

Liang, W., He, L., Ning, P., Lin, J., Li, H., Lin, Z., Kang, K. and Zhang, Y., 2015. (+)-Catechin inhibition of transmissible gastroenteritis coronavirus in swine testicular cells is involved its antioxidation. *Research in Veterinary Science, 103*, pp. 28–33.

Lin, H.W., Sun, M.X., Wang, Y.H., Yang, L.M., Yang, Y.R., Huang, N., Xuan, L.J., Xu, Y.M., Bai, D.L., Zheng, Y.T. and Xiao, K., 2010. Anti-HIV activities of the compounds isolated from Polygonum cuspidatum and Polygonum multiflorum. *Planta Medica, 76*(09), pp. 889–892.

Melek, F.R., Miyase, T., Ghaly, N.S. and Nabil, M., 2007. Triterpenoid saponins with N-acetyl sugar from the bark of Albizia procera. *Phytochemistry, 68*(9), pp. 1261–1266.

Panthong, P., Bunluepuech, K., Boonnak, N., Chaniad, P., Pianwanit, S., Wattanapiromsakul, C. and Tewtrakul, S., 2015. Anti-HIV-1 integrase activity and molecular docking of compounds from Albizia procera bark. *Pharmaceutical Biology, 53*(12), pp. 1861–1866.

You, H.L., Huang, C.C., Chen, C.J., Chang, C.C., Liao, P.L. and Huang, S.T., 2018. Anti-pandemic influenza A (H1N1) virus potential of catechin and gallic acid. *Journal of the Chinese Medical Association, 81*(5), pp. 458–468.

7.6.1.11 *Astragalus mongholicus* Bunge

Synonyms: *Astragalus borealimongolicus* Y.Z. Zhao, *Astragalus membranaceus* Bunge *Astragalus membranaceus* var. *mongholicus* (Bunge) P.G. Xiao, *Astragalus mongholicus* var. *dahuricus* (DC.) Podl., *Astragalus propinquus* Schischk., *Astragalus purdomii* N.D. Simpson *Phaca abbreviata* Ledeb., *Phaca alpina* var. *dahurica* DC., *Phaca macrostachys* Turcz.

Common names: Huangqi, meng gu huang qi (China); oh gi (Japan)

Habitats: Steppes, coniferous forests

Distribution: China, Mongolia, Korea, Japan

Botanical observation: This graceful herb grows up to about 60 cm tall. The stems are hairy. The leaves are compound, spiral, and stipulate. The stipules narrowly triangular and up to about 1 cm long. The blade is up to about 15 cm and presents 8–12 pairs of folioled. The folioles are narrowly ovate to narrowly elliptic, 5–22 × 3–11 mm, and rounded or emarginated at apex. The racemes are lax and about 5 cm long. The calyx is tubular, up to 9 mm long, hairy, and develops 5 lobes. The corolla is yellow and comprises a standard broadly elliptic and up to 2 cm long, 2 wings about 1.5 cm long, and a keel about 1.5 cm long. The androecium includes 10 slender stamens. The gynoecium comprises a unilocular ovary. The pods are up to 3 cm long, hairy, and beaked.

Medicinal use: Dysentery (China)

Broad-spectrum antibacterial polar extract: Methanol extract (5 mg/6 mm disc) inhibited the growth of *Escherichia coli*, *Salmonella enteridis*, *Shigella* sp., and *Campylobacter* with inhibition zone diameters of 22, 17, 19, and 12 mm, respectively (Balachandar et al., 2012).

Strong antimycobacterial amphiphilic isoflavonol: Formononetin (LogD = 2.4 at pH 7.4; molecular mass = 268.2 g/mol) at a concentration of 10 µM inhibited the growth of *Mycobacterium tuberculosis* (H37Rv) by 88% (Mutai et al., 2015).

Strong antifungal (filamentous) flavone glycoside: Isorhamnetin-3-*O*-(4-*O*-[*E*]-coumaroyl-3,6-α-L-*O*-fuco-pyranosyl)-β-D-galactopyranoside at a concentration of 35 µM inhibited the mycelial growth of *Fusarium oxysporum* f.sp. *dianthi* pathotypes (Fod race 4) by 32% (Wang et al., 2015a).

Antifungal isoflavanol: Isomucronulatol inhibited the growth of *Cladosporium herbarum* (Ingham, 1977).

Moderate antiviral (enveloped monopartite linear double-stranded DNA) polar extract: Aqueous extract inhibited the replication of the Herpes simplex virus type-1 in 2BS cells with an IC_{50} value 980 µg/mL and a selectivity index of 128 (Sun & Yang, 2004).

Strong antiviral (non-enveloped linear monopartite, single-stranded (+)RNA) hydrophilic isoflavone glycoside: Calycosin-7-*O*-β-D-glucopyranoside (LogD=0.1 at pH 7.4; molecular mass=446.4 g/mol) inhibited the replication of the Coxsackie virus B3 (Nancy strain) in Vero cells with the IC_{50} value 25 µg/mL and the selectivity index of 125 (Zhu et al., 2009).

Strong antiviral (non-enveloped linear monopatite single stranded (+)RNA) amphiphilic isoflavone: Calycosin (LogD=2.0 at pH 7.4; molecular mass=284.2 g/mol) inhibited the replication of the Coxsackie virus B3 (Nancy strain) in Hep-2 cells with an IC_{50} value 7.9 µg/mL and a selectivity index of 5.6 (Zhu et al., 2009).

Strong antiviral (enveloped monopartite linear dimeric single-stranded (+)RNA) amphiphilic isoflavonol: Calycosin (LogD = 2.0 at pH 7.4; molecular mass=284.2 g/mol) inhibited the replication of the Human immunodeficiency type-1 (strain IIIB) in C8166 cells with an EC_{50} value of 2.2 µg/mL and a selectivity index of 39.6 (Chen et al., 2011).

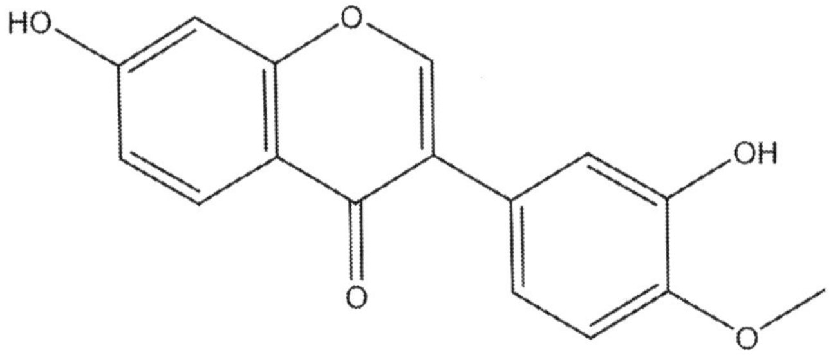

Calycosin

Strong antiviral (non-enveloped linear monopartite single-stranded (+)RNA) amphiphilic isoflavonol: Formononetin (LogD=2.4 at pH 7.4; molecular mass=268.2 g/mol) inhibited the replication of the Enterovirus 71 (SHZH98 strain) with an IC_{50} value of 3.4 µmol/L in Vero cells and 3.9 µmol/L in SK-NSH cells with selectivity indices of 38.1 in Vero cells and 43.0 in SK-N-SH cells (Wang et al., 2015).

In vivo antiviral (non-enveloped linear monopartite single-stranded (+)RNA) isoflavone glycoside: Calycosin-7-*O*-β-D-glucopyranoside given to BALB/c mice infected with the Coxsackie virus B3 (Nancy Strain) at the dose of 24 mg/Kg/day for 14 days increased the survival rate from 44.4 to 77.7% with protection of cardiac muscles, decrease in the virus titers in the heart, decreased heart indices, and improved left ventricular function (Zhu et al., 2009).

Viral enzyme inhibition by polar extract: Aqueous extract of roots at a concentration of 250 µg/mL inhibited Human immunodeficiency virus type-1 protease by 15.6% (Xu et al., 1996). Extract inhibited the Rauscher murine leukemia virus reverse transcriptase with an IC_{50} value of 500 µg/mL (Ono et al., 1989).

Commentaries: (i) The plant yields cycloartane-type triterpenoid saponins known as astragalosides (Kitagawa, et al., 1983) as well as isoflavone glycosides (Ma et al., 2003), and the cycloartane triterpene cycloastragenol (He & Findlay, 1991). (ii) In Turkey, *Astragalus aureus* Willd. is used to treat jaundice (local name: *Geven*) and *Astragalus fasciculifolius* subsp. *arbusculinus* (Bornm & Gauba)Tietz is used to treat jaundice in Iran (local name: *Anzerut*). In China, *Astragalus licentianus* (local name: *Gan su huang qi*) is used to treat diarrhea. (iii) The plant is used in China for the treatment of COVID-19 (Xu & Zhang, 2020), and one could examine the anti-Severe acute respiratory syndrome-acquired coronavirus 2 activity of calycosin. This molecule could be added to the list of natural products to test for the next viral pandemics.

REFERENCES

Balachandar, S., Jagadeeswari, M., Dhanabalan, R. and Meenachi, M., 2012. Antimicrobial activity of Astragalus membranaceus against diarrheal bacterial pathogens. *International Journal of Pharmacy*, *2*(2), pp. 416–418.

Chen, L., Li, Z., Tang, Y., Cui, X., Luo, R., Guo, S., Zheng, Y. and Huang, C., 2011. Isolation, identification and antiviral activities of metabolites of calycosin-7-O-β-d-glucopyranoside. *Journal of Pharmaceutical and Biomedical Analysis*, *56*(2), pp. 382–389.

He, Z.Q. and Findlay, J.A., 1991. Constituents of Astragalus membranaceus. *Journal of Natural Products*, *54*(3), pp. 810–815.

Ingham, J.L., 1977. Phytoalexins of hyacinth bean (Lablab niger). *Zeitschrift für Naturforschung C*, *32*(11–12), pp. 1018–1020.

Kitagawa, I., WANG, H., Saito, M., TAKAGI, A. and YOSHIKAWA, M., 1983. Saponin and sapogenol. XXXV. Chemical constituents of astragali radix, the root of Astragalus membranaceus Bunge. (2). Astragalosides I, II and IV, acetylastragaloside I and isoastragalosides I and II. *Chemical and Pharmaceutical Bulletin*, *31*(2), pp. 698–708.

Mutai, P., Pavadai, E., Wiid, I., Ngwane, A., Baker, B. and Chibale, K., 2015. Synthesis, antimycobacterial evaluation and pharmacophore modeling of analogues of the natural product formononetin. *Bioorganic & Medicinal Chemistry Letters*, *25*(12), pp. 2510–2513.

Ono, K., Nakane, H., MENG, Z.M., OSE, Y., SAKAI, Y. and MIZUNO, M., 1989. Differential inhibitory effects of various herb extracts on the activities of reverse transcriptase and various deoxyribonucleic acid (DNA) polymerases. *Chemical and Pharmaceutical Bulletin*, *37*(7), pp. 1810–1812.

Sun, Y. and Yang, J., 2004. Experimental study of the effect of Astragalus membranaceus against herpes simplex virus type 1. *Di 1 jun yi da xue xue bao= Academic Journal of the First Medical College of PLA*, *24*(1), pp. 57–58.

Wang, H., Zhang, D., Ge, M., Li, Z., Jiang, J. and Li, Y., 2015. Formononetin inhibits Enterovirus 71 replication by regulating COX-2/PGE 2 expression. *Virology Journal*, *12*(1), p.35.

Wang, Q., Wu, R., Wu, X., Tai, W., Dai, N., Wu, J. and Han, N., 2015a. Three flavonoids from the leaves of Astragalus membranaceus and their antifungal activity. *Monatshefte für Chemie-Chemical Monthly*, *146*(10), pp. 1771–1775.

Xu, H.X., Wan, M., Loh, B.N., Kon, O.L., Chow, P.W. and Sim, K.Y., 1996. Screening of Traditional Medicines for their Inhibitory Activity Against HIV-1 Protease. *Phytotherapy Research*, *10*(3), pp. 207–210.

Xu, J. and Zhang, Y., 2020. Traditional Chinese medicine treatment of COVID-19. *Complementary Therapies in Clinical Practice*, p.101165.

Zhu, H., Zhang, Y., Ye, G., Li, Z., Zhou, P. and Huang, C., 2009. In vivo and in vitro antiviral activities of calycosin-7-O-β-D-glucopyranoside against coxsackie virus B3. *Biological and Pharmaceutical Bulletin*, *32*(1), pp. 68–73.

7.6.1.12 *Butea monosperma* (Lam.) Taub.

Synonyms: *Butea frondosa* K.D. Koenig ex Roxb.; *Butea frondosa* Roxb. Ex Willd.; *Butea monosperma* Kuntze; *Erythrina monosperma* Lam.; *Plaso monosperma* (Lam.) Kuntze

Common names: Bastard teak, flame of the forest, Bengal kino tree; murut (Bangladesh; India), palas (Bangladesh); ma ru rtse (Bhutan); zhi kuang (China); moduga, palasa, palash, raktapushpaka (India); palasa (Indonesia); barg-i-hind, parakeh-i-hindi (Iran); chan (Laos); puk pen, paukpin (Myanmar); laharo (Nepal); chichra, dhak, palate (Pakistan); kela (Sri Lanka); tong kwoaw (Thailand); lam vo (Vietnam)

Habitat: Forests, cultivated

Distribution: Pakistan, India, Sri Lanka, Bhutan, Nepal, Bangladesh, Myanmar, Vietnam, China, Thailand, Malaysia, Indonesia, Papua New Guinea

Botanical observation: This characteristic tree, native to India, grows to 10 m tall. The bark exudes a red gum upon incision that dries to form beautiful glossy and dark purple chunks. The stems are hairy. The leaves are imparipinnate, alternate, and stipulate. The stipules are minute. The rachis is 10–20 cm long. The blade presents 3 folioles, which are coriaceous, somewhat glossy, about 12.5–20 × 11–17.5 cm, broadly ovate, and marked with about 5 pairs of secondary nerves. The inflorescences are axillary or terminal racemes, which are up to 17.5 cm long. The calyx is 1.2 cm long, campanulate, and hairy. The corolla is falcate, orangish to bright red, with a recurved standard that reaches 4.5 cm long. The androecium is diadelphous and includes 10 stamens. The ovary is hairy and the style slender. The pods are flat, 10–20 ×2.5–5 cm, and coriaceous. The seeds are reddish brown, reniform, flat, and about 3 cm long.

Medicinal uses: Boils, sores, ulcers (Cambodia, Laos, Vietnam); diarrhea (Bangladesh, Sri Lanka, Myanmar); leprosy (Bangladesh, India, Sri Lanka); leucorrhea, sores, urinary infection, (Bangladesh); sores (Bangladesh, Myanmar); dysentery (India, Sri Lanka); ulcers, sore throat (Sri Lanka); ringworm (Bangladesh, India); fever (India)

Very weak broad-spectrum antibacterial non-polar extract: Petroleum ether extract of leaves inhibited the growth of *Pseudomonas aeruginosa* (MTCC 1688) and *Enterococcus* sp. (MTCC 439) with MIC/MBC values of 520/1160 and 1160/2620 µg/mL, respectively (Sahu et al., 2013).

Moderate broad-spectrum polar extract: Methanol extract of flowers inhibited the growth of *Staphylococcus aureus*, *Bacillus subtilis*, *Escherichia coli*, and *Pseudomonas aeruginosa* with the MIC values of 750, 500, 750, and 1000 µg/mL, respectively (Suthar et al., 2016).

Moderate broad-spectrum antibacterial polar extract: Methanol extract of flowers inhibited the growth of *Staphylococcus aureus*, *Bacillus subtilis*, *Escherichia coli*, and *Pseudomonas aeruginosa* with MIC values of 750, 500, 750, and 1000 µg/mL, respectively (Suthar et al., 2016). Ethanol extract of gum inhibited the growth of *Staphylococcus aureus* (ATCC 25923), *Bacillus subtilis* (MTCC 441), *Bacillus cereus* (MTCC 430), and *Pseudomonas aeruginosa* (MTCC 424) with the MIC values of 300, 200, 200, and 500 µg/mL, respectively (Gurav ey al., 2008).

Strong antimycobacterial amphiphilic chalcone: Butein (LogD = 2.2 at pH 7.4; molecular mass = 272.2 g/mol) from the flowers inhibited the growth of *Mycobacterium tuberculosis* with an MIC value of 12.5 µg/mL (Chokchaisiri et al., 2009).

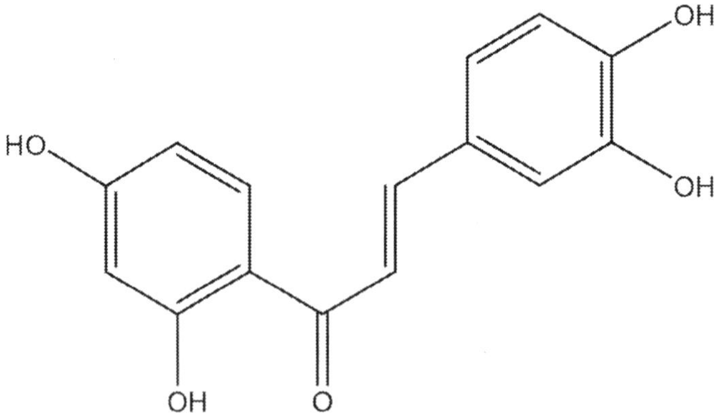

Butein

Moderate antifungal (yeast) polar extract: Ethanol extract of gum inhibited the growth of *Candida albicans* (MTCC 227) and *Saccharomyces cerevisiae* (MTCC 170) with MIC values of 200 and 200 μg/mL, respectively (Gurav ey al., 2008).

Antifungal (filamentous) pterocarpan flavonoid: Medicarpin (also known as (–)-3-Hydroxy-9-methoxypterocarpan) isolated from the stem bark inhibited the growth of *Cladosporium cladosporioides* (Bandara et al., 1989).

Antifungal halo developed by flavone glycoside: 5,7-Dihydroxy-3,6,4′ trimethoxyflavone-7-*O* α-L-xylopyranosyl-(1 → 3)-*O*-α-L-arabinopyranosyl-(1 → 4)-*O*-β-D-galactopyranoside (20% solution/5 mm disc) isolated from the flowers inhibited the growth of *Aspergillus niger, Fusarium oxysporum, Trichoderma viride*, and *Penicillium digitatum* (Yadava & Tiwari, 2007).

Moderate antiviral (enveloped monopartite linear double-stranded DNA) polar extract: Methanol extract of flowers inhibited the replication of the Herpes simplex virus type-1 (ATCC VR-733) and the Herpes simplex virus type-1 (clinical isolate, VU-09) in Vero cells with EC_{50} values of 107.5 and 111.3 μg/mL and selectivity indices 3 and 2.9, respectively (Goswami et al., 2016).

Strong antiviral (enveloped monopartite linear dimeric single-stranded (+)RNA) polar extract: Extract of stem at a concentration of 6 μg/mL inhibited the replication of the Human immunodeficiency virus type-1 by 34% (Sabde et al., 2011).

Commentaries: (i) Commentaries: (i) The flowers accumulate the flavanones butin, liquiritigenin, the flavanone glucosides butrin and isomonospermoside, the chalcone isoliquirigenin, flavones, and the isoflavonols formononetin (see earlier), afromosin, and formononetin-7-O-glucoside (Chokchaisiri et al., 2009). Among the isoflavonoid phytoalexins with the broadest antifungal activity are the pterocarpans (Jiménez-González et al., 2008). One such pterocarpan is medicarpin, which is notably produced by *Vicia faba* L. infected by *Botrytis* sp. (Dakora and Phillips, 1996). The precise antifungal and antibacterial mode of action of pterocarpans is not fully understood but evidence converge to suggest DNA strand scission, inhibition of DNA replication, and interaction with topoisomerase (Chaudhuri et al., 1995; Selvam et al., 2017; Tesauro et al., 2010). (ii) The plant is used in Nepal to prevent COVID-19 (Khadka et al., 2020). One could examine the *in vivo* activity of gum or flower extracts against COVID-19 and this material could be added to the list of possible phytomedication for other coronavirus zoonosis.

REFERENCES

Bandara, B.R., Kumar, N.S. and Samaranayake, K.S., 1989. An antifungal constituent from the stem bark of Butea monosperma. *Journal of Ethnopharmacology*, *25*(1), pp. 73–75.

Chaudhuri SK, Huang L, Fullas F, Brown DM, Wani MC, Wall ME, Tucker JC, Beecher CWW, Kinghorn AD (1995) Isolation and structure identification of an active DNA strand scission agent, (+)-3,4-dihydroxy8,9-methylenedioxypterocarpan. *Journal of Natural Products*, 58, pp. 1966–1969.

Chokchaisiri, R., Suaisom, C., Sriphota, S., Chindaduang, A., Chuprajob, T. and Suksamrarn, A., 2009. Bioactive flavonoids of the flowers of Butea monosperma. *Chemical and Pharmaceutical Bulletin*, *57*(4), pp. 428–432.

Dakora, F.D. and Phillips, D.A., 1996. Diverse functions of isoflavonoids in legumes transcend anti-microbial definitions of phytoalexins. *Physiological and Molecular Plant Pathology*, *49*(1), pp. 1–20.

Goswami, D., Mukherjee, P.K., Kar, A., Ojha, D., Roy, S. and Chattopadhyay, D., 2016. Screening of ethnomedicinal plants of diverse culture for antiviral potentials.

Gurav, S.S., Gulkari, V.D., Duragkar, N.J. and Patil, A.T., 2008. Antimicrobial activity of Butea monosperma Lam. gum. *Iranian Journal of pharmacology and Therapeutics*, *7*, pp. 21–24.

Jiménez-González, L., Álvarez-Corral, M., Muñoz-Dorado, M. and Rodríguez-García, I., 2008. Pterocarpans: interesting natural products with antifungal activity and other biological properties. *Phytochemistry Reviews*, *7*(1), pp. 125–154.

Khadka, D., Dhamala, M.K., Li, F., Aryal, P.C., Magar, P.R., Bhatta, S., Thakur, M.S., Basnet, A., Shi, S. and Cui, D., 2020. The use of medicinal plant to prevent COVID-19 in Nepal. *Journal of Ethnobiology and Ethnomedicine*, *17*, pp. 1–17.

Sabde, S., Bodiwala, H.S., Karmase, A., Deshpande, P.J., Kaur, A., Ahmed, N., Chauthe, S.K., Brahmbhatt, K.G., Phadke, R.U., Mitra, D. and Bhutani, K.K., 2011. Anti-HIV activity of Indian medicinal plants. *Journal of Natural Medicines*, *65*(3–4), pp. 662–669.
Sahu, M.C. and Padhy, R.N., 2013. In vitro antibacterial potency of Butea monosperma Lam. against 12 clinically isolated multidrug resistant bacteria. *Asian Pacific Journal of Tropical Disease*, *3*(3), pp. 217–226.
Selvam, C., Jordan, B.C., Prakash, S., Mutisya, D. and Thilagavathi, R., 2017. Pterocarpan scaffold: A natural lead molecule with diverse pharmacological properties. *European Journal of Medicinal Chemistry*, *128*, pp. 219–236.
Tesauro, C., Fiorani, P., d'Annessa, I., Chillemi, G., Turchi, G. and Desideri, A., 2010. Erybraedin C, a natural compound from the plant Bituminaria bituminosa, inhibits both the cleavage and religation activities of human topoisomerase I. *Biochemical Journal*, *425*(3), pp. 531–539.
Yadava, R.N. and Tiwari, L., 2007. New antifungal flavone glycoside from Butea monosperma O. Kuntze. *Journal of Enzyme Inhibition and Medicinal Chemistry*, *22*(4), pp. 497–500.

7.6.1.13 *Caesalpinia bonduc* (L) Roxb.

Synonyms: *Caesalpinia bonducella* (L.) Fleming; *Caesalpinia crista* L.; *Caesalpinia crista* Thunb.; *Guilandina bonduc* Aiton; *Guilandina bonduc* Griseb.; *Guilandina bonduc* L.; *Guilandina bonducella* L.; *Guilandina crista* (L.) Small

Common names: Bonduc nut, fever nut, Yellow nicker, nicker nut, Molucca bean; Natakaranj; nata; koranju; karanj, putikaranja (Bangladesh); soni (Fiji); gavin, kat karanj, kazhanji, latakaranja, nata, putikaranja (India); areuy mata hiyang (Indonesia); khaza-i-iblis, tukhm-i-iblis (Iran); gorek (Malaysia); kathgarer, kaande jhang (Nepal); kurere (Papua New Guinea); kalumbibit (the Philippines); waat (Thailand); vuốt hùm (Vietnam)

Habitat: Seashores

Distribution: Pakistan to Pacific Islands

Botanical observation: This woody climber grows to about 10 m long. The stems are prickly. The leaves are bipinnate, alternate, and stipulate. The stipules are deciduous, large, and about 2 cm long. The blade is 30–45 cm and made of 6–9 pairs of pinnae. The pinnae present 6–12 pairs of folioles. The folioles are oblong, 1.5–4 × 1.2–2 cm, membranous, oblique at base, rounded to acute at apex, and glossy. The racemes are axillary and up to 30cm long. The 5 sepals are 8mm long and hairy. The 5 petals are 1.5cm long, yellow, oblanceolate, clawed, the standard tinged with red spots. The androecium includes 10 stamens. The ovary is hairy. The pods are oblong, 5–7 × 4–5 cm, coriaceous, dehiscent, beaked at apex, spiny, with a sinister aura, the spines up to 1cm long and contain 2–3 seeds. The seeds are about 1 cm long, with a unique type of dull light grey, smooth, glossy and globose to broadly elliptic. The seeds are sacred in India and used for charm laces. Floating seeds collected in the shores of Scotland and England were valued as a protective charm for childbirth.

Medicinal uses: Boils, wounds (Bangladesh); fever (India, Nepal)

Broad-spectrum antibacterial mid-polar extract: Chloroform extract of leaves (2 mg/6 mm disc) inhibited the growth of *Staphylococcus aureus*, *Bacillus subtilis*, *Bacillus cereus*, *Bacillus megaterium*, *Salmonella typhi*, *Shigella dysenteriae*, *Pseudomonas aeruginosa*, *Escherichia coli*, and *Klebsiella* sp. with inhibition zone diameters of 12, 8, 9, 7, 10, 7.5, 7, 7, and 7 mm, respectively (Kaewpiboon et al., 2012).

Antibacterial (Gram-negative) halo developed by diterpene: Bondenolide isolated from the seeds (1 mg/mL/well) inhibited the growth of *Pseudomonas aeruginosa* with an inhibition zone diameter of 7 mm (Simin et al., 2001).

Weak broad-spectrum antibacterial polar extract: Methanol extract of seeds inhibited the growth of *Bacillus megaterium*, *Bacillus subtilis*, *Bacillus thurigiensis*, *Sarcina lutea*, *Staphylococcus albus*, *Staphylococcus aureus*, *Staphylococcus epidermidis*, *Klebsiella pneumoniae*, *Salmonella typhi*, *Shigella boydii*, *Shigella dysenteriae*, and *Shigella sonnei* (Saeed & Sabir, A.W., 2001).

Strong broad-spectrum antibacterial polar extract: Methanol extract of seed coat inhibited the growth of *Pseudomonas aeruginosa* (MTCCB 741), *Pseudomonas aeruginosa* (135) ampicillin resistant), *Pseudomonas aeruginosa* (1124) ampicillin resistant), *Pseudomonas aeruginosa* (325) ampicillin resistant), *Staphylococcus aureus* (MTCCB 737), *Staphylococcus aureus* (187 (methicillin-resistant), *Staphylococcus aureus* (349 (methicillin-resistant), *Staphylococcus aureus* (674 (methicillin-resistant), *Escherichia coli* (MTCCB 82), *Proteus mirabilis* (MTCCB 564), *Klebsiella pneumoniae* (MTCCB 109), *Salmonella typhi* (MTCCB 733), *Vibrio cholera* (MTCCB) 458), *Bacillus thuringiensis* (MTCCB 1824), and *Bacillus cereus* (MTCCB 1272) (Arif et al., 2009).

In vivo antibacterial polar extract: Methanol extract of seed coat given to Wistar rats subcutaneously at a dose of 25 mg/Kg body weight once a day for 10 days evoked reduction of lung abscesses induced by *Pseudomonas aeruginosa.* (Arif et al., 2009).

Strong broad-spectrum antibacterial amphiphilic cassane furano diterpene: Neocaesalpin P (LogD = 3.5 at pH 7.4; molecular mass = 360.4 g/mol) isolated from the seeds inhibited the growth of *Staphylococcus aureus*, *Streptococcus agalactiae*, and *Pseudomonas aeruginosa* with MIC values of 16, 16, and 32 μg/mL, respectively (Ata et al., 2009).

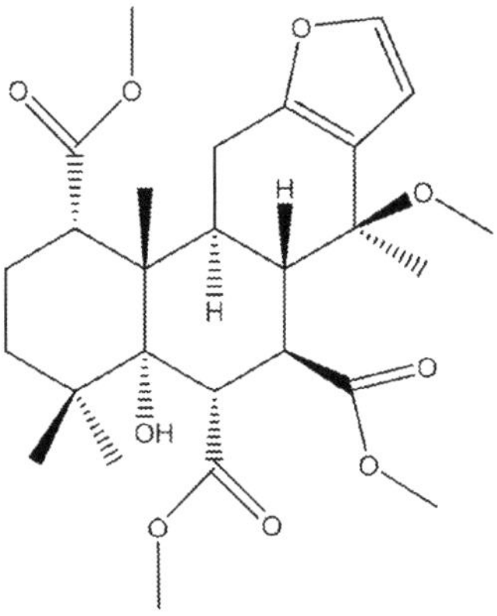

Caesalmin F

Broad-spectrum antifungal cassane-type diterpene: Bondenolide isolated from the seeds (200 μg/mL in potato dextrose agar) inhibited the growth of *Nigrospora oryzae, Stachybrotrys atra, Curvularia lunata, Dreschslera rostrata, Aspergillus niger, Pleuroaetus oustreatus, Allescheria boydii, Microsporum canis*, and *Epidermophyton flocossum* (Simin et al., 2001).

Commentary: (i) The plant yields homoisoflavonoids, including caesalpinianone (Ata et al., 2009), as well as an interesting series of cassane-type furanoditerpenes. Interesting because the furanone moiety opens and reacts with the thiol groups of microbial peptides and proteins (Jackson et al., 2017). The cassane furanoditerpene caesalmin C, D, E, and F isolated from the seeds of *Caesalpinia minax* Hance inhibited the proliferation of the Parainfluenza virus type 3 in Hep-2 cells with IC_{50} values of 8.2, 9.6, 10.3, and 7.8 μg/mL, respectively and selectivity indices of 23.9, 19.1, 16, and 17.5, respectively (Jiang et al., 2001). From the same plant, caesalmin B, bonducellpin D, 1α,5α,6α,7β,14β-pentahydroxyvouacapane, 1α,5α-dihydroxy-14β-methoxy-6α,7β-diacetoxyvouacapane as well as the triterpene friedelin inhibited the replication of the Parainfluenza type 3-induced cytopathogenicity (Jiang et al., 2002). Spirocaesalmin B, caesalpinin M_1, caesalpinin M_2, caesalmin E_1, caesalmin E_2, caesalmin E_3, and caesalpinin F_1 inhibited influenza neuraminidase with IC_{50} values of 56.8, 51.7,

87.4, 61.5, 64.3, 45.9, and 29.0 μM, respectively (Zanamivir at 6.3 μM) (Wu et al., 2014). Thus, one could endeavor to isolate antiviral cassanes from *Caesalpinia bonduc* (L) Roxb.

(ii) Administration of extracts parenterally to animals induces unnecessary suffering, and the oral route should be preferred because there is no way for a plant extract to be safely administered to humans parenterally (Schwarz et al., 2020). (iii) Members of the genus *Caesalpinia* L. are not uncommonly used for the treatment of microbial infectious diseases in Asia. This is the case for *Caesalpinia sappan* L., from the heartwood of which was isolated the chromane brazilin (LogD = 1.6 at pH 7.4; molecular mass = 286.2 g/mol), which inhibited the growth of vancomycin-resistant Enterococci, *Staphylococcus epidermidis* (ATCC 11490), *Staphylococcus epidermidis* (# 20), methicillin-resistant *Staphylococcus aureus* (9247922), *Streptococcus agalactiae* (Group B Strep) A 909), *Streptococcus pyogenes* (Group A Strep) M1), *Bulkholderia cepacia* (92.443), *Salmonella typhimurium* (ATCC 13311), and *Escherichia coli* (ATCC 25922) with MIC values of 32, 8, 64, 16, 8, 4, 64, 64, and 256 μg/mL, respectively (Yin et al., 2004). Brazilin was bactericidal for methicillin-resistant *Staphylococcus aureus* and inhibited the incorporation of [3H] thymidine and [3H] serine by methicillin-resistant *Staphylococcus aureus* into DNA (Yin et al., 2004).

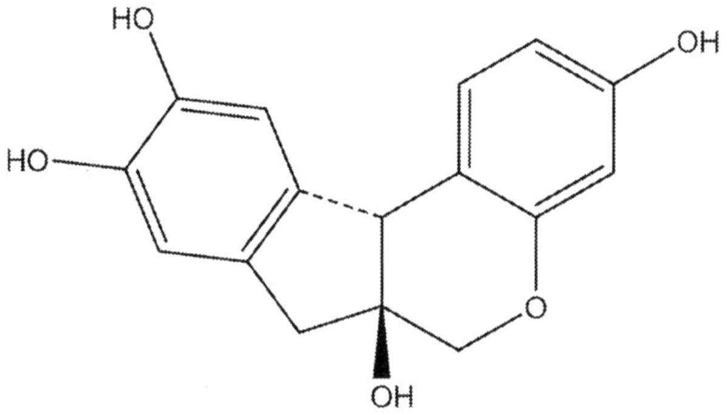

Brazilin

Brazilin inhibited the neuraminidase activity of the Influenza A/PR/8/34 (H1N1) (ATCC VR-1469), Influenza A virus A/Hong Kong/8/68 (H3N2) (ATCC VR-544), and Influenza virus A/Chichen/Korea/MS96 (H9N2) with the IC_{50} values of 0.7, 1.1, and 1 μM, respectively (Jeong et al., 2012). From the same plant, the homoisoflavonoid sappanone A inhibited the neuraminidase activity of the Influenza A/PR/8/34 (H1N1) (ATCC VR-1469), Influenza A virus A/Hong Kong/8/68 (H3N2) (ATCC VR-544), and Influenza virus A/Chicken/Korea/MS96 (H9N2) with IC_{50} values of 0.2, 0.3, and 0.4 μM, respectively (Jeong et al., 2012). Consider that neuraminidase catalyzes the hydrolysis of the α-(2,3)- or α-(2,6)-glycosidic linkage between a terminal sialic acid residue and its adjacent carbohydrate moiety on the host receptor allowing the release from the host cell of the newly formed virion, and Oseltamivir is an example of a neuraminidase inhibitor drug used for the treatment of Influenza (Moscona, 2005). The Middle East respiratory syndrome-acquired coronavirus spike proteins contain a N-terminal receptor-binding domain known as the "S1A" domain, which binds to host cell sialic acids (Qing et al., 2020). Thus, natural products inhibiting this binding could be of value as anti-coronavirus agents. This remains to be examined. Is brazilin useful against COVID-19?

(iii) The cassane-type furanoditerpenes 6*β*-cinnamoyl-7β-hydroxyvouacapen-5*α*-ol and 6*β*-benzoyl-7*β*-hydroxyvouacapen-5*α*-ol from the root of *Caesalpinia pulcherrima* (L.) Sw. inhibited *Mycobacterium* sp. with MIC values of 6.2 and 25 μg/mL, respectively (Promsawan et al., 2003).

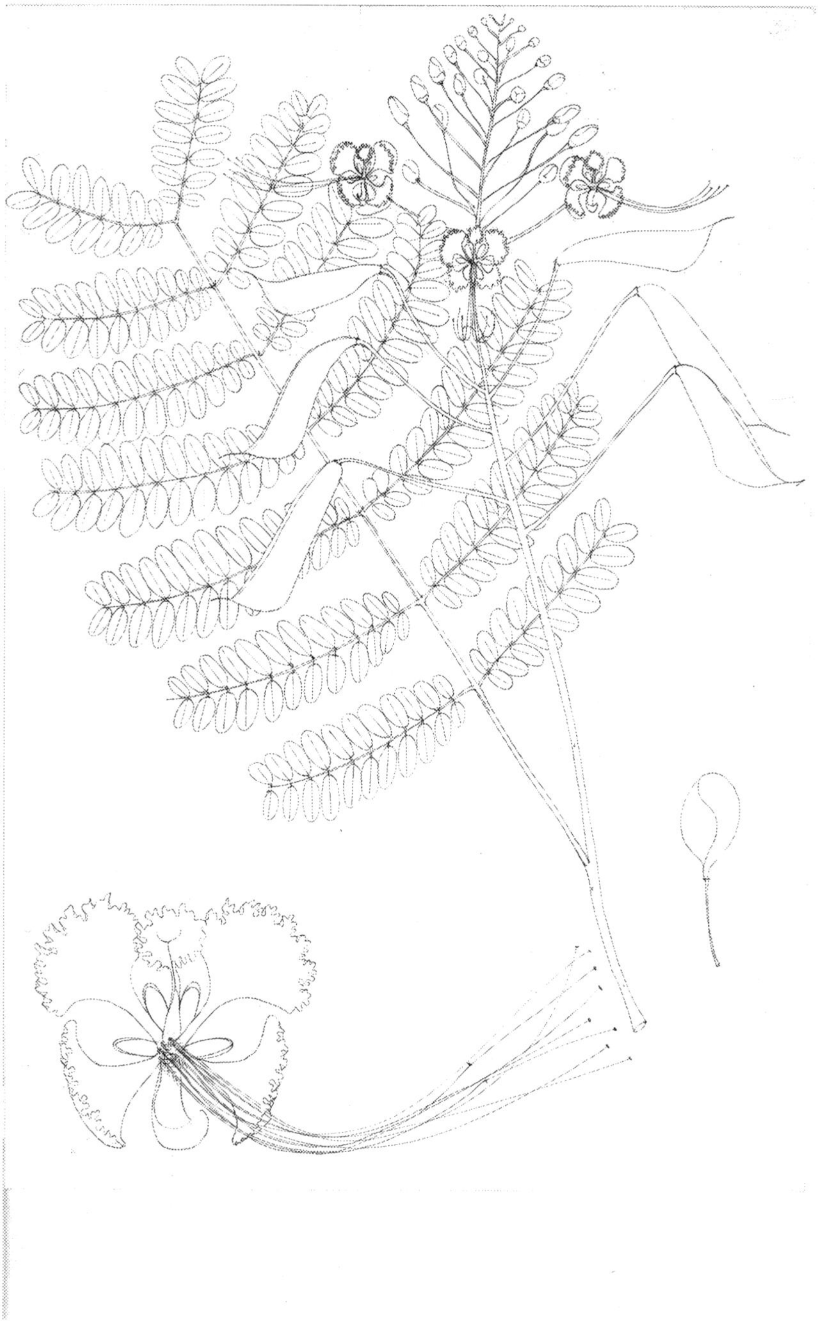

FIGURE 7.8 *Caesalpinia pulcherrima* (L.) Sw.

REFERENCES

Arif, T., Mandal, T.K., Kumar, N., Bhosale, J.D., Hole, A., Sharma, G.L., Padhi, M.M., Lavekar, G.S. and Dabur, R., 2009. In vitro and in vivo antimicrobial activities of seeds of Caesalpinia bonduc (Lin.) Roxb. *Journal of Ethnopharmacology*, *123*(1), pp. 177–180.

Ata, A., Gale, E.M. and Samarasekera, R., 2009. Bioactive chemical constituents of Caesalpinia bonduc (Fabaceae). *Phytochemistry Letters*, *2*(3), pp. 106–109.

Ata, A., Udenigwe, C.C., Gale, E.M. and Samarasekera, R., 2009. Minor chemical constituents of Caesalpinia bonduc. *Natural Product Communications*, *4*(3), p. 1934578X0900400302.

Jackson, P.A., Widen, J.C., Harki, D.A. and Brummond, K.M., 2017. Covalent modifiers: A chemical perspective on the reactivity of α, β-unsaturated carbonyls with thiols via hetero-Michael addition reactions. *Journal of Medicinal Chemistry*, *60*(3), pp. 839–885.

Jeong, H.J., Kim, Y.M., Kim, J.H., Kim, J.Y., Park, J.Y., Park, S.J., Ryu, Y.B. and Lee, W.S., 2012. Homoisoflavonoids from Caesalpinia sappan displaying viral neuraminidases inhibition. *Biological and Pharmaceutical Bulletin*, *35*(5), pp. 786–790.

Jiang, R.W., Ma, S.C., But, P.P.H. and Mak, T.C., 2001. New Antiviral Cassane Furanoditerpenes from Caesalpinia m inax. *Journal of Natural Products*, *64*(10), pp. 1266–1272.

Jiang, R.W., Ma, S.C., He, Z.D., Huang, X.S., But, P.P.H., Wang, H., Chan, S.P., Ooi, V.E.C., Xu, H.X. and Mak, T.C., 2002. Molecular structures and antiviral activities of naturally occurring and modified cassane furanoditerpenoids and friedelane triterpenoids from Caesalpinia minax. *Bioorganic & Medicinal Chemistry*, *10*(7), pp. 2161–2170.

Kaewpiboon, C., Lirdprapamongkol, K., Srisomsap, C., Winayanuwattikun, P., Yongvanich, T., Puwaprisirisan, P., Svasti, J. and Assavalapsakul, W., 2012. Studies of the in vitro cytotoxic, antioxidant, lipase inhibitory and antimicrobial activities of selected Thai medicinal plants. *BMC Complementary and Alternative Medicine*, *12*(1), p.217.

Moscona, A., 2005. Neuraminidase inhibitors for influenza. *New England Journal of Medicine*, *353*(13), pp. 1363–1373.

Promsawan, N., Kittakoop, P., Boonphong, S. and Nongkunsarn, P., 2003. Antitubercular cassane furanoditerpenoids from the roots of Caesalpinia pulcherrima. *Planta Medica*, *69*(08), pp. 776–777.

Qing, E., Hantak, M., Perlman, S. and Gallagher, T., 2020. Distinct roles for sialoside and protein receptors in coronavirus infection. *Mbio*, *11*(1).

Saeed, M.A. and Sabir, A.W., 2001. Antibacterial activity of Caesalpinia bonducella seeds. *Fitoterapia*, *72*(7), pp. 807–809.

Schwarz, S., Knauth, M., Schwab, S., Walter-Sack, I., Bonmann, E. and Storch-Hagenlocher, B., 2000. Acute disseminated encephalomyelitis after parenteral therapy with herbal extracts: A report of two cases. *Journal of Neurology, Neurosurgery & Psychiatry*, *69*(4), pp. 516–518.

Simin, K., Khaliq-uz-Zaman, S.M. and Ahmad, V.U., 2001. Antimicrobial activity of seed extracts and bondenolide from Caesalpinia bonduc (L.) Roxb. *Phytotherapy Research*, *15*(5), pp. 437–440.

Wu, J., Chen, G., Xu, X., Huo, X., Wu, S., Wu, Z. and Gao, H., 2014. Seven new cassane furanoditerpenes from the seeds of Caesalpinia minax. *Fitoterapia*, *92*, pp. 168–176.

7.6.1.14 *Cajanus cajan* (L.) Huth

Synonyms: *Cajan cajan* (L.) Huth; *Cajan inodorum* Medik.; *Cajan inodorum* Medik.; *Cajanum thora* Raf.; *Cajanus bicolor* DC.; *Cajanus cajan* (L.) Druce; *Cajanus cajan* (L.) Merr.*Cajanus cajan* (L.) Millsp.*Cajanus flavus* DC.; *Cajanus indicus* Spreng.; *Cajanus luteus* Bello; *Cajanus obcordifolia* Singh; *Cajanus pseudocajan* (Jacq.) Schinz & Guillaumin; *Cajanus striatus* Bojer; *Cytisus cajan* L.; *Cytisus guineensis* Schumach. & Thonn.; *Cytisus pseudocajan* Jacq.

Common names: Pigeon pea, arahor, baredare, orhor-kolai (Bangladesh); sandaek klon (Cambodia); mu dou (China); aadhaki, thuvarai (India); kacan bali (Indonesia); thwax he (Laos); kacang (Malaysia); tabios (the Philippines); ma hae (Thailand); dau thong (Vietnam)

Habitat: Roadsides, vacant plots of land, cultivated

Distribution: Tropical Asia and Pacific

Botanical observation: This handsome shrub native to India grows to 3 m tall. The stems are hairy. The leaves are imparipinnate, alternate, and stipulate. The stipules are minute, ovate-lanceolate, and about 3 mm long. The petiole is 1.5–5 cm long. The blade is 3-foliolate. The folioles are dull light green, lanceolate to elliptic, 2.8–10 × 0.5–3.5 cm, papery, hairy below, acute or acuminate at apex, and marked with about 8 pairs of secondary nerves. The racemes are terminal or axillary and 3–7 cm long. The calyx is campanulate, hairy, 5–7 mm long, and 5-lobed. The corolla is papilionaceous, yellow to reddish brown, and about 1.5 cm long, the standard suborbicular. The ovary is hairy, develops a long style and a capitate stigma. The pods are linear-oblong, 4–8.5 × 0.6–1.2 cm, pubescent, beaked at apex, and contain 3–6 seeds. The seeds are grey, subspherical, 5mm longs, and used in Pakistan, India, and Bangladesh as pulse for making of the delicious and nutritious dhal.

Medicinal uses: Jaundice (Bangladesh); bronchitis, cholera, cough, measles (India)

Strong antibacterial (Gram-positive) coumarin: Cajanuslactone isolated from the leaves inhibited the growth of *Staphylococcus epidermidis* (ATCC 12228), *Staphylococcus aureus* (ATCC 6538), and *Bacillus subtilis* (ATCC 6633) with MIC/MBC values of 125/250, 31/125, and 125/250 μg/mL (Kong et al., 2010).

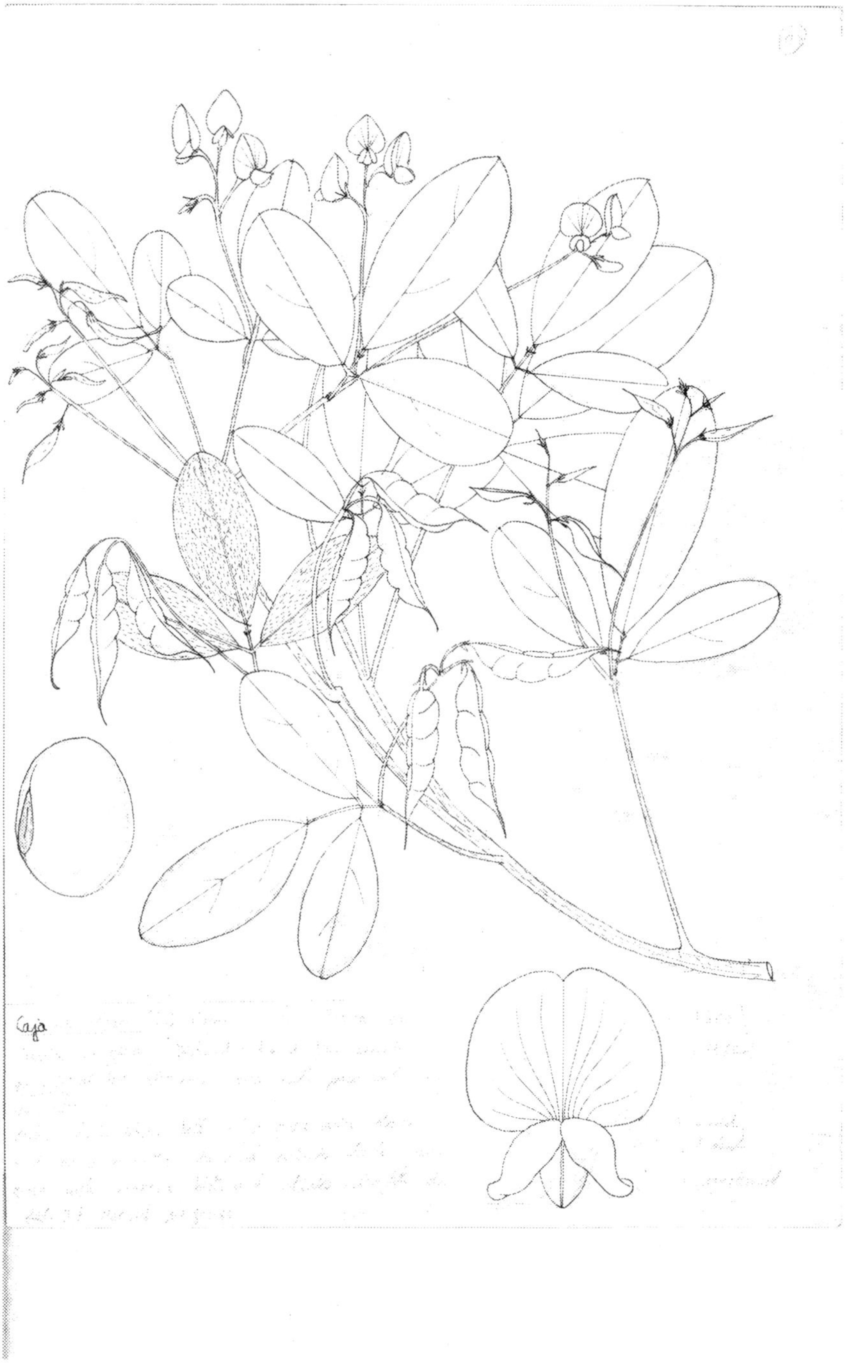

FIGURE 7.9 *Cajanus cajan* (L.) Huth

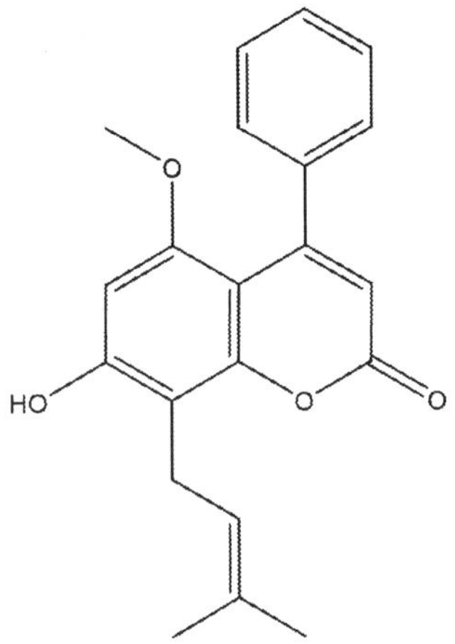

Cajanuslactone

Strong antibacterial (Gram-positive) stilbene: Cajaninstilbene acid isolated from the leaves inhibited the growth of *Staphylococcus epidermidis* (ATCC 12228), *Staphylococcus aureus* (ATCC 6538), and *Bacillus subtilis* (ATCC 6633) with MIC/MBC values of 13/100, 25/105, and 25/250 μg/mL (Kong et al., 2010).

Strong antiviral (enveloped monopartite linear double-stranded DNA) polar extract: Ethanol extract inhibited the replication of the Herpes simplex virus type-1 and type-2 viruses with IC_{50} values of 0.02 and 0.1 μg/mL, respectively (Zu et al., 2010).

Strong antiviral (enveloped monopartite linear double-stranded DNA) amphiphilic flavanol: Pinostrobin (LogD=3.5 at pH 7.4; molecular weight=270.2 g/mol) inhibited the replication of the Herpes simplex virus type-1 (F strain) in Vero cells with an EC_{50} value of 22.7 μg/mL and an selectivity index above 4.4 (Wu et al., 2011).

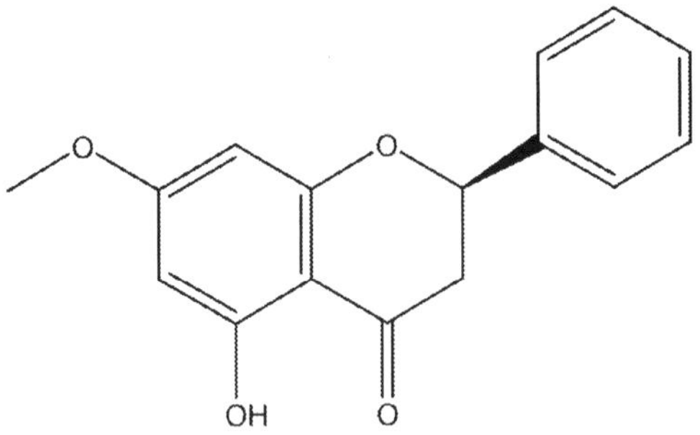

(S)-Pinostrobin

In vivo antiviral (enveloped monopartite linear double-stranded DNA) flavanol: Pinostrobin given orally to mice at a dose of 50 mg/Kg/day for 7 days abrogated lethality evoked by the Herpes simplex virus type-1 (Wu et al., 2011).

Commentaries: (i) Bacterial, fungal, or viral infection in plants in the family Fabaceae (especially members of the subfamily Faboideae) stimulate the transcription of stilbene synthetases, chalcone isomerases, and/or isoflavone synthetases, resulting in increased production and accumulation of low molecular weight antimicrobial phenolics: stilbenes, isoflavonoids,

flavonoid phytoalexins and/or precursors of phytoalexins (Paxton
hytoanticipins are constitutive antimicrobial compounds present in
s or saponins, or alkaloids" before microbial infection (Tiku, 2020).
cursors of phytoalexins are termed "pro-phytoalexins". Examples
tance flavonoid glycosides stored in healthy plant tissues that get
aglycones as reaction to microbial infections or plant tissues dam-
of plant material hydrolyzing enzymes are activated and convert
nto antimicrobial aglycones (phytoalexins). Pro-phytoanticipins in
flavonoid glycosides, which, when the plant is challenged by phy-
are hydrolyzed into antimicrobial isoflavones that can in parallel
antimicrobial pterocarpans phytoalexins. Pterocarpans are phy-
et al., 1990; van Etten et al., 1994; Siyabonga et al., 2016). The
roduced by *Pisum sativum* L. in the presence of *Fusarium solani*
Perrin & Bottomley, 1961) or the pterocarpan phaseolin and the
by *Phaseolus vulgaris* L. in presence of *Phytophtora infestans*,
bacco necrosis virus (Dakora & Phillips, 1996) are examples of
avone isolated from *Arachis hypogaea* L inhibits the growth of
1975). The isoflavonoids coumestrol, daidzein, lyceollin genis-
, and glycitein are examples of Fabaceous phytoalexins (Dakora
of natural products could be scrutinized for their antimicrobial
s as they regulate microbial genes that promote the development
ants (Dakora & Phillips, 1996). Isoflavonoids are not uncom-
wonder if natural products with hormonal properties may have
or even fungal efflux pumps. In fact, isoflavonoid phytoalexins,
ing property to inhibit the NorA efflux pump in *Staphylococcus*
thermore, the isoflavonols puerarin and dadzein inhibited the
e-acquired coronavirus 3-Chymotrypsin-like protease with an
en et al., 2005) implying that isoflavonoid phytoalexins could
nical frameworks of interest for the development of drugs for
ronaviruses. (ii) Pinostrobin is neuroactive (Nicholson et al.,
ntial as an efflux pump inhibitor, as discussed earlier.

REFERENCES

Chen, C.N., Lin, C.P., Huang, K.K., Chen, W.C., Hsieh, H.P., Liang, P.H. and Hsu, J.T.A., 2005. Inhibition of SARS-CoV 3C-like protease activity by theaflavin-3, 3′-digallate (TF3). *Evidence-Based Complementary and Alternative Medicine*, *2*(2), pp. 209–215.

Dakora, F.D. and Phillips, D.A., 1996. Diverse functions of isoflavonoids in legumes transcend anti-microbial definitions of phytoalexins. *Physiological and Molecular Plant Pathology*, *49*(1), pp. 1–20.

Nicholson, R.A., David, L.S., Le Pan, R. and Liu, X.M., 2010. Pinostrobin from Cajanus cajan (L.) Millsp. inhibits sodium channel-activated depolarization of mouse brain synaptoneurosomes. *Fitoterapia*, *81*(7), pp. 826–829.

Perrin, D.R. and Bottomley, W., 1961. Pisatin: an antifungal substance from Pisum sativum L. *Nature*, *191*(4783), p.76.

Sharma, A., Gupta, V.K. and Pathania, R., 2019. Efflux pump inhibitors for bacterial pathogens: From bench to bedside. *The Indian Journal of Medical Research*, *149*(2), p.129.

Siyabonga, S.J., Ayanda, M., Sydney, M.T. and Oluwole, S.F., 2016. Comparative evaluation of antibacterial activity of induced and non-induced Cajanus cajan seed extract against selected gastrointestinal tract bacteria. *African Journal of Microbiology Research*, *10*(10), pp. 319–323.

Sobolev, V.S., Neff, S.A., Gloer, J.B., Khan, S.I., Tabanca, N., De Lucca, A.J. and Wedge, D.E., 2010. Pterocarpenes elicited by Aspergillus caelatus in peanut (Arachis hypogaea) seeds. *Phytochemistry*, *71*(17–18), pp. 2099–2107.

Tiku, A.R., 2020. Antimicrobial Compounds (Phytoanticipins and Phytoalexins) and Their Role in Plant Defense. *Co-Evolution of Secondary Metabolites*, pp. 845–868.

Turner, R.B., Lindsey, D.L., Davis, D.D. and Bishop, R.D., 1975. Isolation and identification of 5, 7-dimethoxyisoflavone, an inhibitor of Aspergillus flavus from peanuts. *Mycopathologia*, *57*(1), pp. 39–40.
Wu, N., Kong, Y., Zu, Y., Fu, Y., Liu, Z., Meng, R., Liu, X. and Efferth, T., 2011. Activity investigation of pinostrobin towards herpes simplex virus-1 as determined by atomic force microscopy. *Phytomedicine*, *18*(2–3), pp. 110–118.
Zu, Y., Fu, Y., Wang, W., Wu, N., Liu, W., Kong, Y., Schiebel, H.M., Schwarz, G., Schnitzler, P. and Reichling, J., 2010. Comparative study on the antiherpetic activity of aqueous and ethanolic extracts derived from Cajanus cajan (L.) Millsp. *Complementary Medicine Research*, *17*(1), pp. 15–20.

7.6.1.15 *Cassia fistula* L.

Synonyms: *Bactyrilobium fistula* (L.) Willd., *Cassia bonplandiana* DC., *Cassia excelsa* Kunth., *Cassia fistuloides* Collad., *Cassia rhombifolia* Roxb., *Cathartocarpus excelsus* G.Don, *Cathartocarpus fistula* (L.) Pers., *Cathartocarpus fistuloides* (Collad.) G.Don, *Cathartocarpus rhombifolius* G Don

Common names: Golden shower tree; bandar lathi, daad, delong pada, sonamukhi (Bangladesh); reach chhpus (Cambodia), a po le (China), amaha, amaltasa, kanikonna, raj briksha, sampaka (India), klobop (Indonesia), fulus (Iran); khoun, lom leng (Laos), bereksa (Malaysia), ngu pin (Myanmar), rajbriksha (Nepal); amaltas, gardanali (Pakistan); ancherhan (the Philippines); kavani (Sri Lanka), khuun (Thailand); bo cap muoc, muô`ng hoàng yê´n (Vietnam)

Habitat: Cultivated

Distribution: Tropical Asia and Pacific

Botanical observation: This magnificent tree grows to 10 m tall. The bark is whitish to brownish and smooth. The stem is terete and hairy at apex. The leaves are pinnate, spiral, and stipulate. The stipule is deltoid, acute, and minute. The rachis is enlarged at base and 12–25 cm long with 3–8 pairs of folioles. The folioles are sub-opposite, ovate to lanceolate, somewhat dull green, soft, thin, round at base, acute at apex, and marked with 12–15 pairs of secondary nerves. The racemes are pendulous, axillary, graceful, and up to 50 cm long. The calyx is 5-lobed, green, hairy, and about 1 cm long. The corolla shows 5 petals, which are obovate, blunt, orbicular, membranous and conspicuously veined. The stamens and the style are curved and conspicuous. The pod is somewhat cylindrical, pendulous, glabrous, indehiscent, to 60 cm long, dull green to black and glossy, and contains numerous seeds in a pulp.

Medicinal uses: Bronchitis (India); dysentery (Nepal, Pakistan); leprosy, ringworm (Pakistan, Myanmar); wounds (India); jaundice (Iran); syphilis (Nepal, India); sore throat (Nepal); tuberculosis (Pakistan); fever (Iran, Pakistan); wounds (Laos, Vietnam)

Moderate broad-spectrum antibacterial mid-polar extract: Ethyl acetate extract of flowers inhibited the growth of *Bacillus subtilis* (MTCC 441), *Staphylococcus aureus* (ATCC 25923), *Staphylococcus epidermidis* (MTCC 3615), *Enterococcus faecalis* (ATCC 29212), and *Pseudomonas aeruginosa* (ATCC 27853) with MIC values of 78, 312, 78, 1250, and 625 µg/mL, respectively (Duraipandiyan & Ignacimuthu, 2007).

Moderate broad-spectrum antifungal mid-polar extract: Ethyl acetate extract of flowers inhibited the growth of *Trichophyton rubrum* (296), *Trichophyton rubrum* (57/01), *Trichophyton mentagrophytes*, *Trichophyton simii*, *Epidermophyton floccosum*, and *Scopulariopsis* sp. with MIC values of 1000, 500, 250, 1000, 500, and 500 µg/mL (Duraipandiyan & Ignacimuthu, 2007). Ethanol extract of seed inhibited the growth of *Candida albicans* with an MIC value of 500 µg/mL (Sony et al., 2018).

Very weak anticandidal polar extract: Methanol extract of pod pulp inhibited the growth of *Candida albicans* (ATCC 10261), *Candida tropicalis* (ATCC 750), and *Candida glabrata* (ATCC 90030) with MIC values of 150, 250, and 100 µg/mL, respectively, with simultaneous decrease in ergosterol contents (Irshad et al., 2011).

Strong broad-spectrum antibacterial (Gram-negative) amphiphilic isoflavonol: Biochanin A (also known as genistein 4′-methyl ether or olmelin or 5,7-dihydroxy-4′-methoxyisoflavone, LogD=1.9 at pH 7.4; molecular mass=284.2 g/mol) inhibited the growth of *Chlamydia pneumoniae* (clinical strain K7) with an IC_{50} value of 12 μM (Pohjala et al., 2012), and *Bacillus cereus*, *Listeria monocytogenes*, and *Streptococcus pyogenes* with MIC values of 64, 64, and 32 μg/mL, respectively (Hummelova et al., 2015).

FIGURE 7.10 *Cassia fistula* L.

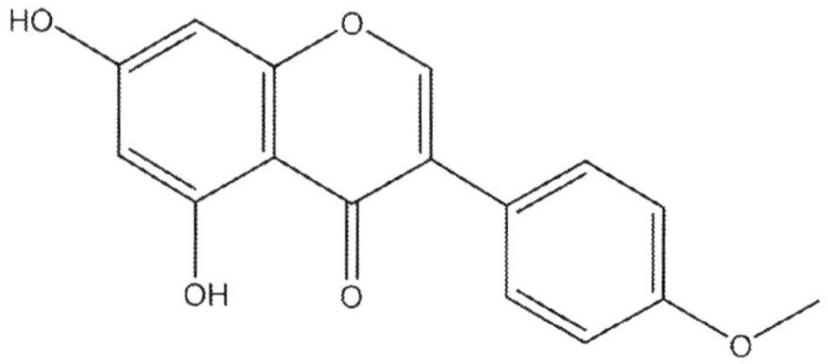

Biochanin A

Very weak antifungal (filamentous) simple phenolic acid: 4-Hydroxybenzoic acid isolated from the flowers inhibited the growth of *Trichophyton mentagrophytes* and *Epidermophyton floccosum* with MIC values of 500 and 500 µg/mL (Duraipandiyan & Ignacimuthu, 2007).

Antifungal amphiphilic isoflavonol: Biochanin A at the concentration of 200 µM inhibited the growth of *Rhozoctonia solani* and *Scerotium rolfsii* (Weidenbörner et al., 1990).

Bacterial efflux pump inhibitor isoflavonol: Biochanin A inhibited the NorA efflux pump in *Staphylococcus aureus* (Sharma et al., 2019) as well as ethidium bromide efflux pump *Mycobacerium tuberculosis* (Lechner et al., 2008).

Moderate antiviral (enveloped monopartite linear dimeric single-stranded (+)RNA) polar extract: Aqueous extract of bark at a concentration of 250 µg/mL inhibited the Human immunodeficiency virus type-1 protease by 64.8% (Xu et al., 1996).

Strong antiviral (enveloped monopartite linear dimeric single-stranded (+)RNA) amphiphilic isoflavonol: Biochanin A reduced the fusion of the Human immunodeficiency virus type-1 to host cell with an EC_{50} value of 5.1 µM, and selectivity index of 39, the expression of p24 with an EC_{50} value of 38 µmol/L and selectivity index of 5.2, and the early activation of antigen CD69 on human $CD4^+$ lymphocytes at a concentration of 50 µM (Lin et al., 2007).

Strong antiviral (enveloped segmented linear single-stranded (−)RNA) amphiphilic isoflavonol: Biochanin A inhibited the replication of the Influenza A (H5N1 strain A/Thailand/1(Kan-1)/04) in A549 cells with an IC_{50} value of 8.9 µM and a selectivity index of 5.6 (Sithisarn et al., 2013).

Commentaries: (i) In the "*Compendium de épidimia*" written by the medical faculty of Paris around 1348, *Cassia fistula* L. is recommended against the Black Death. As for COVID-19 and the coronaviruses that will follow, biochanin A binds to angiotensin-converting enzyme 2 (Xue et al., 2019) and as such might have potentials. Natural products able to inhibit the replication of the Human immunodeficiency virus type-1 have the tendency to inhibit the replication of coronaviruses. Thus, the anti-Severe acute respiratory syndrome-acquired coronavirus 2 properties of biochanin A need to be examined. Furthermore, this interesting isoflavonol could be added to an armamentarium of compounds to be ready to be tested the other waves of zoonotic virus on our way.

(ii) The plant belongs to the subfamily Caesalpinioideae in the Tribe Cassiaeae, the members of which have the ability to generate anthraquinones (Sakulpanich & Gritsanapan, 2009), which are not uncommonly antimicrobial. *Cassia sophora* is used to treat dysentery in Yap where it is called *Gigiol*. (Yap).

REFERENCES

Duraipandiyan, V. and Ignacimuthu, S., 2007. Antibacterial and antifungal activity of Cassia fistula L.: An ethnomedicinal plant. *Journal of Ethnopharmacology*, *112*(3), pp. 590–594.

Hummelova, J., Rondevaldova, J., Balastikova, A., Lapcik, O. and Kokoska, L., 2015. The relationship between structure and in vitro antibacterial activity of selected isoflavones and their metabolites with special focus on antistaphylococcal effect of demethyltexasin. *Letters in Applied Microbiology*, *60*(3), pp. 242–247.

Irshad, M., Shreaz, S., Manzoor, N., Khan, L.A. and Rizvi, M.M.A., 2011. Anticandidal activity of Cassia fistula and its effect on ergosterol biosynthesis. *Pharmaceutical Biology*, *49*(7), pp. 727–733.
Lechner, D., Gibbons, S. and Bucar, F., 2008. Plant phenolic compounds as ethidium bromide efflux inhibitors in Mycobacterium smegmatis. *Journal of Antimicrobial Chemotherapy*, *62*(2), pp. 345–348.
Lin, C.L., Zeng, Y.Y., Zeng, X.F., Zhao, L.Z., Wang, T., Li, H.X. and Li, L.P., 2007. Inhibitory effect of biochanin A on HIV-1 and early activation of human CD4^+ lymphocytes. *Chinese Pharmacological Bulletin*, *23*(2), p. 214.
Pohjala, L., Uvell, H., Hakala, E., Gylfe, Å., Elofsson, M. and Vuorela, P., 2012. The isoflavone biochanin a inhibits the growth of the intracellular bacteria Chlamydia trachomatis and Chlamydia pneumoniae. *Planta Medica*, *78*(11), p. PD132.
Sakulpanich, A. and Gritsanapan, W., 2009. Determination of anthraquinone contents in Cassia fistula leaves for alternative source of laxative drugs. *Planta Medica*, *75*(09), p. PG11.
Sharma, A., Gupta, V.K. and Pathania, R., 2019. Efflux pump inhibitors for bacterial pathogens: From bench to bedside. *The Indian Journal of Medical Research*, *149*(2), p. 129.
Sithisarn, P., Michaelis, M., Schubert-Zsilavecz, M. and Cinatl Jr, J., 2013. Differential antiviral and anti-inflammatory mechanisms of the flavonoids biochanin A and baicalein in H5N1 influenza A virus-infected cells. *Antiviral Research*, *97*(1), pp. 41–48.
Sony, P., Kalyani, M., Jeyakumari, D., Kannan, I. and Sukumar, R.G., 2018. In vitro antifungal activity of cassia fistula extracts against fluconazole resistant strains of Candida species from HIV patients. *Journal de mycologie medicale*, *28*(1), pp. 193–200.
Weidenbörner, M., Hindorf, H., Jha, H.C., Tsotsonos, P. and Egge, H., 1990. Antifungal activity of isoflavonoids in different reduced stages on Rhizoctonia solani and Sclerotium rolfsii. *Phytochemistry*, *29*(3), pp. 801–803.
Xue, H.X., Kong, H., Yu, Y.G., Zhou, J.W., Chen, H.Q. and Yin, Y.Y., 2019. Biochanin A protects against angiotensin II-induced damage of dopaminergic neurons in rats associated with the increased endophilin A2 expression. Behavioural Pharmacology, 30(8), pp. 699–710.

7.6.1.16 *Cassia alata* L.

Synonyms: *Cassia bracteata* L. f.; *Cassia herpetica* Jacq.; *Herpetica alata* (L.) Raf.; *Senna alata* (L.) Roxb.

Common names: Candle bush, ringworm cassia; dang het (Cambodia); dadrughna, damadar, seemai agathi (India); ketepeng (Indonesia); khi let ban (Laos) gelenggang besar (Malaysia); yult (Palau); rari (Papua new Guinea); akapulko, kapulkuru, lingwa, sunting (the Philippines); bakua (Solomon Islands); chem het tet (Thailand); muong (Vietnam)

Habitat: Wastelands, roadsides, cultivated

Distribution: Tropical Asia and Pacific

Botanical observation: This characteristic upright shrub native to South America grows up to about 3m tall and has an aura of life force. The leaves are paripinnate, spiral, and stipulate. The stipules are 6–9 mm long, persistent, and triangular. The petiole is 1.5–2.5 cm long and winged. The blade comprises a winged rachis, which is 30–60 cm long and supporting 8–24 pairs of folioles, which are, 3–15 × 2–7 cm, oblong-dull light green, parallel, and rounded at apex. The raceme is 15–70 cm long, somewhat oblong, erect, and showy. The calyx comprises 5 sepals, which are about 2 cm long and yellow. The 5 petals are about 2 cm long and orangish to yellow. The androecium includes about 10 stamens. The pod is coriaceous, straight, sharply squarish across, up to 20 cm long, and shelters up to 60 seeds.

Medicinal uses: Ringworm (India, Malaysia); athlete's foot (the Philippines); skin diseases (Bangladesh, India); cuts, leprosy, wounds (India)

Broad-spectrum antibacterial mid-polar extract: Dichloromethane extract of flowers (4 mg/disc) inhibited the growth of *Bacillus cereus*, *Bacillus subtilis*, *Micrococcus luteus*, *Staphylococcus aureus*, *Staphylococcus epidermidis*, *Streptococcus faecalis*, *Streptococcus pneumoniae*, *Citrobacter freundii*, *Enterobacter aerogenes*, *Klebsiella pneumoniae*, *Neisseria gonorrhoeae*, *Proteus mirabilis*, *Pseudomonas aeruginosa*, *Salmonella typhi*, *Salmonella typhimurium*, and *Serratia marscencens* with inhibition zone diameters of 18, 14, 18, 16, 20, 16, 18, 16, 18, 16, 20, 18, 16, 20, 18, and 16 mm, respectively (Khan et al., 2001).

Weak antibacterial polar extracts: Aqueous extract inhibited the growth of *Escherichia coli* with an MIC value of 1600 μg/mL (Crockett et al., 1992).

Antibacterial (Gram-positive) halo elicited by anthraquinone: Emodin (50 μg/disc) inhibited the growth of methicillin-sensitive *Staphylococcus aureus* and methicillin-resistant *Staphylococcus aureus* with the zones of inhibition of 19 and 30 mm, respectively (Joung et al., 2012).

Strong antibacterial (Gram-positive) amphiphilic anthraquinones: Emodin (LogD = 1.7 at pH 7.4; molecular mass = 270.2 g/mol) inhibited the growth of *Bacillus cereus* (TISTR 687), methicillin-resistant *Staphylococcus aureus*, and *Staphylococcus aureus* (TISTR 1466) with MIC values of 16, 4, and 16 μg/mL (Promgool et al., 2014). Aloe-emodin (LogD = 1.2 at pH 7.4; molecular mass = 270.2 g/mol) inhibited the growth of methicillin-resistant *Staphylococcus aureus* with an MIC_{50} value of 12 μg/mL (Hazni et al., 2008).

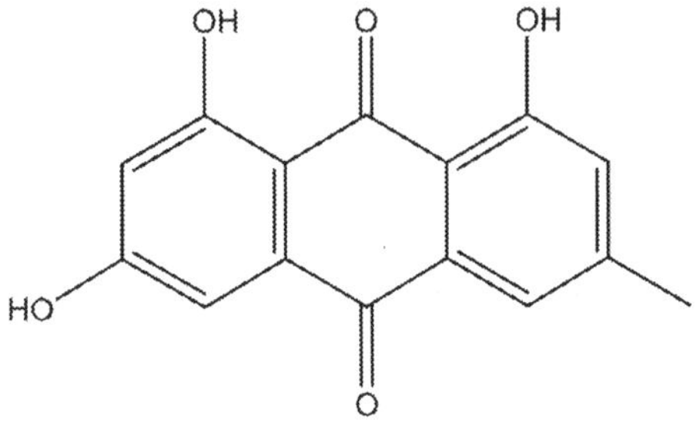

Emodin

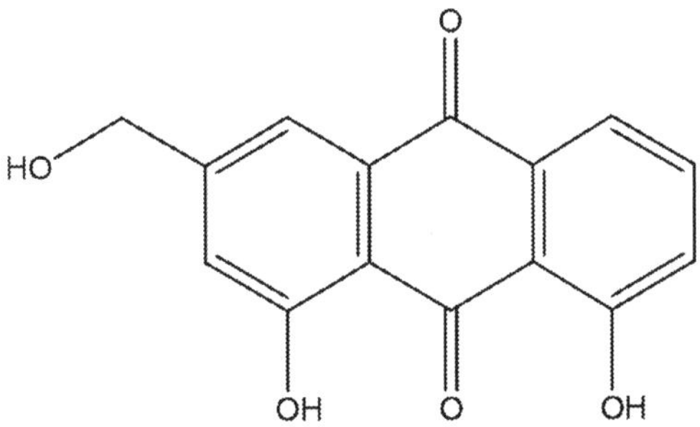

Aloe-emodin

Strong antibacterial (Gram-positive) hydrophilic flavonol: Kaempferol (LogD = 0.8 at pH 7.4; molecular mass = 286.2 g/mol) inhibited methicillin-resistant *Staphylococcus aureus* with an MIC_{50} value of 13 μg/mL (Hazni et al., 2008).

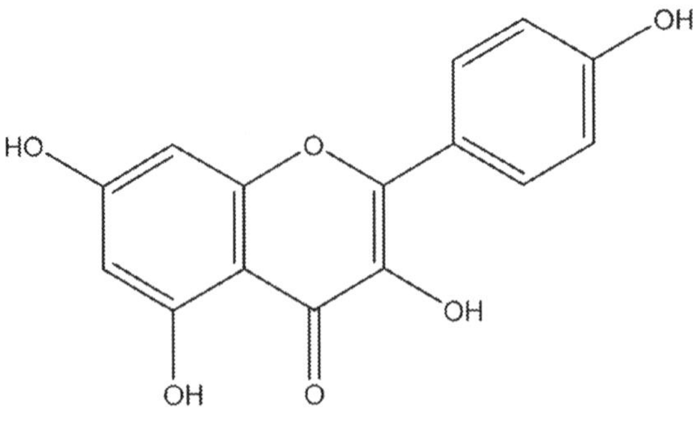

Kaempferol

Strong antimycobacterial amphiphilic anthraquinone: Emodin inhibited *Mycobacterium tuberculosis* (H37Ra) with MIC/MBC values of 4/8 µg/mL (Dey et al., 2014).

Broad-spectrum antifungal halo developed by mid-polar extract: Ethyl acetate extract of leaves (150 µL of a 50 µg/µL solution/ disc) inhibited the growth of *Candida albicans* and *Trichophyton mentagrophytes* with the inhibition zones of 15 and 16mm, respectively (Villaseñor et al., 2002).

Anticandidal polar extract: Aqueous extract of bark (6mm paper disc impregnated with 30 µg/µL) inhibited the growth of *Candida albicans* with an inhibition zone diameter of 16.8mm (Somchit et al., 2003).

Strong anticandidal amphiphilic anthraquinone: Emodin inhibited the growth of *Candida albicans*, *Candida krusei*, *Candida parapsilosis*, and *Candida tropicalis* with MIC/MFC values of 12.5/25, 25/25, 25/25, and 50/50 µg/mL, respectively (Janeczko et al., 2017).

Strong antiviral (non-enveloped monopartite linear single-stranded (+)RNA) amphiphilic anthraquinone: Emodin inhibited the cytopathic effects of the Poliovirus type 3 in Buffalo green monkey kidney cells with an IC_{50} value of 0.2 µg/mL and selectivity index of 12.4 (Semple et al., 2001).

Strong antiviral (enveloped monopartite linear single stranded (+)RNA) amphiphilic anthraquinone: Aloe-emodin inhibited Japanese encephalitis virus (T1P1) plaque reduction in BHK-21 with IC_{50} of 17.3 µg/mL (Chang et al., 2014). Aloe-emodin was virucidal for Japanese encephalitis virus (T1P1) with an IC_{50} of 0.4 µg/mL (Chang et al., 2014).

Strong antiviral (enveloped monopartite linear double-stranded DNA) amphiphilic anthraquinone: Aloe-emodin inhibited the growth of the Herpes simplex virus type-1 (KOS), the Herpes simplex virus type -2 (ATCC VR-734), pseudorabies virus (PSV), and varicella-zoster virus (VZV) (ATCC VR-916) with IC_{50} values of 1.6, 1.5, 5.0, and 6 µg/mL, respectively (Sydiscis et al., 1991). Emodin inhibited the replication of the Herpes simplex virus type-1 and Herpes simplex virus type -2 in cells at 50 µg/mL with antiviral indices of 2 and 3.5, respectively (Xiong et al., 2011).

Strong antiviral (enveloped segmented linear single-stranded (−)RNA) amphiphilic anthraquinone: Aloe-emodin inhibited the replication of the Influenza virus A (INF strain) with an IC_{50} value of 4.5 µg/Ml (Sydiscis et al., 1991).

Strong antiviral (enveloped monopartite linear single-stranded (+)RNA) amphiphilic anthraquinone: Emodin blocked the binding of S protein to angiotensin-converting enzyme type-2 with an IC_{50} value of 200µM and at a concentration of 50 µg/mL protected by 94.1% Vero cells against SARS-CoV infectivity (Ho et al., 2007).

Strong antiviral (enveloped segmented linear single-stranded (−)RNA) hydrophilic flavonol: Kaempferol (LogD = 0.8 at pH 7.4; molecular weight = 286.2 g/mol) inhibited Influenza virus A/chicken/Rostock/34 (H7N1) in CEF cells with an EC_{50} value of 12.2 µg/mL and a selectivity index of 4.9 (Pantev et al., 2006).

Viral enzyme inhibition by anthraquinone: Aloe-emodin inhibited the Human immunodeficiency virus type-1 Ribonuclease H with an IC_{50} value of 21 µM, respectively (Esposito et al., 2016).

Commentary: Emodin is planar and interferes with microbial DNA and RNA (Wink et al., 2020) and is an example of broad-spectrum antimicrobial principle. Emodin interacts with DNA in mammalian cells and is thus cytotoxic and mutagenic (Müller et al., 1996).

REFERENCES

Crockett, C.O., Guede-Guina, F., Pugh, D., Vangah-Manda, M., Robinson, T.J., Olubadewo, J.O. and Ochillo, R.F., 1992. Cassia alata and the preclinical search for therapeutic agents for the treatment of opportunistic infections in AIDS patients. *Cellular and Molecular Biology (Noisy-le-Grand, France)*, *38*(7), pp. 799–802.

Esposito, F., Carli, I., Del Vecchio, C., Xu, L., Corona, A., Grandi, N., Piano, D., Maccioni, E., Distinto, S., Parolin, C. and Tramontano, E., 2016. Sennoside A, derived from the traditional chinese medicine plant Rheum L., is a new dual HIV-1 inhibitor effective on HIV-1 replication. *Phytomedicine*, *23*(12), pp. 1383–1391.

Hazni, H., Ahmad, N., Hitotsuyanagi, Y., Takeya, K. and Choo, C.Y., 2008. Phytochemical constituents from Cassia alata with inhibition against methicillin-resistant Staphylococcus aureus (MRSA). *Planta Medica*, *74*(15), pp. 1802–1805.
Ho, T.Y., Wu, S.L., Chen, J.C., Li, C.C. and Hsiang, C.Y., 2007. Emodin blocks the SARS coronavirus spike protein and angiotensin-converting enzyme 2 interaction. *Antiviral Research*, *74*(2), pp. 92–101.
Janeczko, M., Masłyk, M., Kubiński, K. and Golczyk, H., 2017. Emodin, a natural inhibitor of protein kinase CK2, suppresses growth, hyphal development, and biofilm formation of Candida albicans. *Yeast*, *34*(6), pp. 253–265.
Joung, D.K., Joung, H., Yang, D.W., Kwon, D.Y., Choi, J.G., Woo, S., Shin, D.Y., Kweon, O.H., Kweon, K.T. and Shin, D.W., 2012. Synergistic effect of rhein in combination with ampicillin or oxacillin against methicillin-resistant Staphylococcus aureus. *Experimental and Therapeutic Medicine*, *3*(4), pp. 608–612.
Khan, M.R., Kihara, M. and Omoloso, A.D., 2001. Antimicrobial activity of Cassia alata. *Fitoterapia*, *72*(5), pp. 561–564.
Müller, S.O., Eckert, I., Lutz, W.K. and Stopper, H., 1996. Genotoxicity of the laxative drug components emodin, aloe-emodin and danthron in mammalian cells: topoisomerase II mediated? *Mutation Research/Genetic Toxicology*, *371*(3–4), pp. 165–173.
Pantev, A., Ivancheva, S., Staneva, L. and Serkedjieva, J., 2006. Biologically active constituents of a polyphenol extract from Geranium sanguineum L. with anti-influenza activity. *Zeitschrift für Naturforschung C*, *61*(7–8), pp. 508–516.
Promgool, T., Pancharoen, O. and Deachathai, S., 2014. Antibacterial and antioxidative compounds from Cassia alata Linn. *Songklanakarin Journal of Science and Technology*, *36*(4), pp. 459–463.
Semple, S.J., Pyke, S.M., Reynolds, G.D. and Flower, R.L., 2001. In vitro antiviral activity of the anthraquinone chrysophanic acid against poliovirus. *Antiviral Research*, *49*(3), pp. 169–178.
Somchit, M.N., Reezal, I., Nur, I.E. and Mutalib, A.R., 2003. In vitro antimicrobial activity of ethanol and water extracts of Cassia alata. *Journal of Ethnopharmacology*, *84*(1), pp. 1–4.
Sydiscis, R.J., Owen, D.G., Lohr, J.L., Rosler, K.H. and Blomster, R.N., 1991. Inactivation of enveloped viruses by anthraquinones extracted from plants. *Antimicrobial Agents and Chemotherapy*, *35*(12), pp. 2463–2466.
Villaseñor, I.M., Canlas, A.P., Pascua, M.P.I., Sabando, M.N. and Soliven, L.A.P., 2002. Bioactivity studies on Cassia alata Linn. leaf extracts. *Phytotherapy Research*, *16*(S1), pp. 93–96.
Xiong, H.R., Luo, J., Hou, W., Xiao, H. and Yang, Z.Q., 2011. The effect of emodin, an anthraquinone derivative extracted from the roots of Rheum tanguticum, against herpes simplex virus in vitro and in vivo. *Journal of Ethnopharmacology*, *133*(2), pp. 718–723.

7.6.1.17 *Cassia siamea* Lam.

Synonyms: *Cassia arborea* Macfad., *Cassia florida* Vahl., *Cassia gigantea* Bertero ex DC, *Cassia sumatrana* Roxb., *Chamaefistula gigantea* G.Don, *Sciacassia siamea* (Lam.) Britton, *Senna siamea* (Lam.) H.S. Irwin & Barn., *Senna sumatrana* Roxb.

Common names: Kassod tree, Siamese cassia, Thai copper pod; ângkanh (Cambodia); ki lek ba (Laos); kasod (India); johar (Indonesia); johor, petai belalang (Malaysia); mai-mye-sili, mejari, mezali, mai mye sili, taw-mezali (Myanmar); phak chili (Thailand); muong (Vietnam)

Habitat: Roadsides, villages, cultivated

Distribution: Tropical Asia and Pacific

Botanical observation: This tree grows to 20 m tall. The stem is terete, straight, lenticelled, and velvety by apex. The leaves are compound, spiral, and stipulated. The rachis is flattish at base, triangular, presents a series of glands and 7–16 pairs of folioles. The folioles are dark green, glossy, coriaceous, marked with 8–10 pairs of secondary nerves, oblong, and 5–10 × 1.5–2.5 cm. The panicles are terminal and hairy. The flowers are 2.5–3 cm wide and bright yellow. The calyx consists of 5 sepals, which are about 5 mm long. The 5 petals are plain yellow, about 5 cm long, and orbicular to spoon shaped. The androecium consists of 10 stamens with brownish anthers. The pods are flat, curved, glossy, with a few hairs, brown, up to 30cm long and contain up to 30 seeds.

Medicinal use: Rhinitis (Cambodia, Laos, Vietnam)

Moderate antibacterial (Gram-positive) polar extract: Ethanol extract of stem bark inhibited the growth of *Micrococcus luteus* (MTCC (106), *Bacillus subtilis* (MTCC (121), *Bacillus cereus*

(MTCC (430), *Staphylococcus aureus* (MTCC (96), and *Streptococcus pneumoniae* (MTCC (2672) (Singh et al., 2010).

Strong antiviral (enveloped monopartite linear dimeric single-stranded (+)RNA) chromone: Siamchromone D isolated from the stems inhibited the replication of the Human immunodeficiency

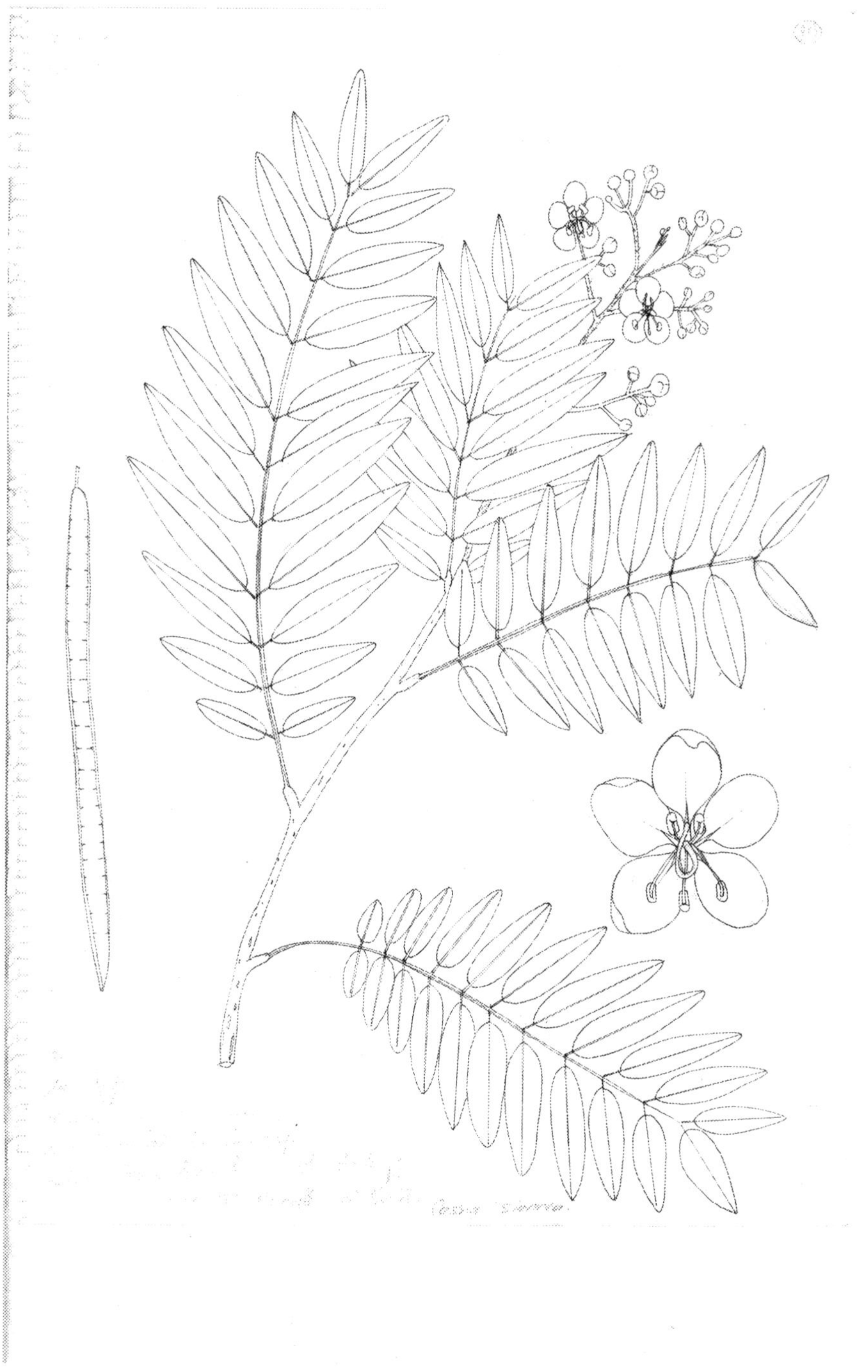

FIGURE 7.11 *Cassia siamea* Lam.

virus type-1 in C8166 cells with an IC_{50} value of 1.8 µg/mL and a selectivity index above 106.9 (Hu et al., 2012).

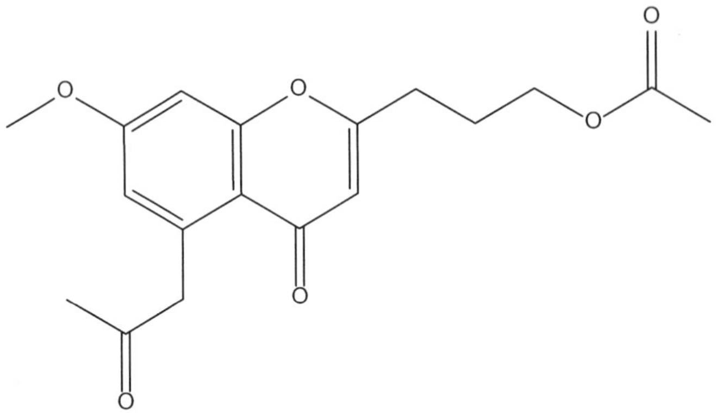

Siamchromone D

Strong antiviral (enveloped monopartite linear dimeric single-stranded (+)RNA) chromone glycosides: 11-hydroxy-sec-*O*-glucosylhamaudol, sec-*O*-glucosylhamaudol, and 2-methyl-5-(2′-hydroxypropyl)-7-hydroxychromone-2′-*O*-D-glucopyranoside isolated from the stems inhibited the replication of the Human immunodeficiency virus type-1 in C8166 cells with IC_{50} values of 1.8, 1.7, and 1.8 µg/mL and selectivity indices of 108.7, 49.6, and 63.3, respectively (Hu et al., 2012).

Commentary: Is siamchromone D active against coronaviruses?

REFERENCES

Hu, Q.F., Zhou, B., Gao, X.M., Yang, L.Y., Shu, L.D., Shen, Y., Li, G.P., Che, C.T. and Yang, G.Y., 2012. Antiviral chromones from the stem of Cassia siamea. *Journal of Natural Products*, *75*(11), pp. 1909–1914.

Singh, M., Khatoon, S., Singh, S., Kumar, V., Rawat, A.K.S. and Mehrotra, S., 2010. Antimicrobial screening of ethnobotanically important stem bark of medicinal plants. *Pharmacognosy Research*, *2*(4), p. 254.

7.6.1.18 *Cassia tora* L.

Synonyms: *Cassia candenatensis* Dennst.; *Cassia contorta* Vogel; *Cassia foetida* Salisb.; *Cassia gallinaria* Collad.; *Cassia humilis* Collad.; *Cassia obtusifolia* L.; *Cassia pentagonia* Mill.; *Cassia sunsub* Forssk.; *Cassia tagera* Lam.; *Cassia tala* Desv.; *Cassia toroides* Roxb.; *Chamaefistula contorta* G. Don; *Diallobus falcatus* Raf.; *Diallobus uniflorus* Raf.; *Emelista tora* (L.) Britton & Rose; *Senna tora* (L.) Roxb.; *Senna toroides* Roxb.

Common names: Sickle senna, sickle pod, coffee weed, foetid cassia; chueh ming, jue míng zi, tsao chueh (China); aayroun, ayudham, bonelach, chakonda, chakramarda, loki haedma, puadia, palakarpo tochay (India); gelenggang kecil (Malaysia); dangwe (Myanmar); chhinchhine, tapre (Nepal); khilek jued (Thailand); dau ma; dau giau, thao quyet minh, muong ngu (Vietnam).

Habitat: Wastelands, roadsides

Distribution: Tropical Asia and Pacific

Botanical observation: This herb grows to a height of 1 m. The leaves are compound, spiral, foetid, stipulate, and bear 6 pairs of folioles. The stipules are linear and caducous. The folioles are obovate, apiculate, and 2 cm–5 cm long (edible India and Nepal). A gland is present between the folioles of the first and second pairs. The flowers are arranged in axillary bracteated racemes. The calyx comprises 5 sepals, which are ovate, acute, and 8 mm long. The 5 petals are obovate and 1.2 cm–1.5cm long. The androecium comprises 7 stamens and 3 staminodes. The ovary is long, conspicuous, and characteristically sickle-shaped. The pods are linear, quadrangular, 15 cm × 3 mm, and contain 25–30 seeds, which are very small (used to make coffee in Nepal).

Medicinal uses: Abscesses, cough, cuts, dysentery, ringworm (India); fever (Nepal, Thailand); ringworm (Nepal)

Strong antibacterial (Gram-positive) hydrophilic naphthalene: Torachrysone (LogD=0.6 at pH 7.4; molecular mass=246.2 g/mol) inhibited the growth of *Escherichia coli* (K12), methicillin-sensitive *Staphylococcus aureus* (209P), and methicillin-resistant *Staphylococcus aureus* (OM481) with MIC values of 512, 32, and 32 μg/mL, respectively, and was inactive for *Pseudomonas aeruginosa* (PA01) (Hatano et al., 1999).

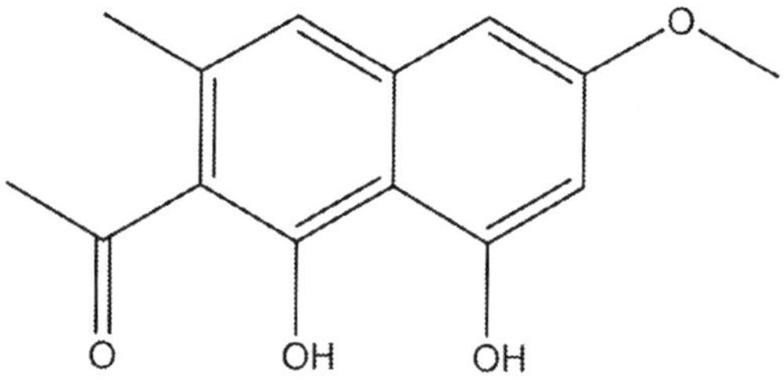

Torachrysone

Weak antibacterial (Gram-positive) polar extract: Aqueous extract of seeds inhibited the growth of *Streptococcus mutans* (MT 5091) and *Streptococcus mutans* (OMZ 176) with MIC values of 2500 and 2500 μg/mL, respectively (Chen et al., 1989).

Strong antibacterial (Gram-positive) *amphiphilic anthraquinone:* Aloe-emodin (LogD=1.2 at pH 7.4; molecular mass=270.2 g/mol) from the seeds inhibited the growth of methicillin-sensitive *Staphylococcus aureus* (209P) and methicillin-resistant *Staphylococcus aureus* (OM481) with an MIC value of 2 μg/mL (Hatano et al., 1999).

Strong antibacterial (Gram-positive) hydrophilic anthraquinone: Rhein (LogD=−0.5 at pH 7.4; molecular mass=284.2 g/mol) from the seeds inhibited the growth of *Escherichia coli* (K12), methicillin-sensitive *Staphylococcus aureus* (209P), and methicillin-resistant *Staphylococcus aureus* (OM481) with MIC values of 128, 32, and 32 μg/mL (Hatano et al., 1999). Rhein inhibited the growth of *Bacteroides fragilis* and *Clostridium perfringens* with the MIC of 1.5 and 50 μg/mL, respectively (Jong-Chol et al., 1987).

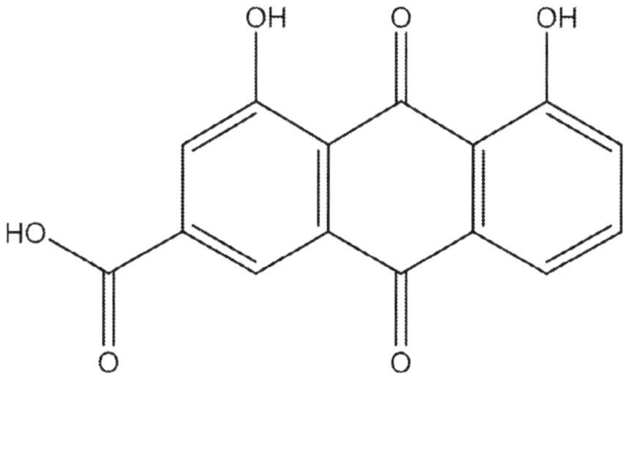

Rhein

Strong antiviral (enveloped monopartite linear single-stranded (+)RNA) non-polar extract: Hexane extract of seeds inhibited the replication of the Severe acute respiratory syndrome-acquired coronavirus in Vero cells with the IC_{50} value of 8.4 µg/mL and a selectivity index above 59.3 (Wen et al., 2011).

Moderate antiviral (enveloped monopartite linear single-stranded (+) RNA) anthraquinone: Rhein at a concentration of 100µM inhibited the interaction between Severe acute respiratory syndrome-acquired coronavirus S protein and angiotensin-converting enzyme 2 by about 20% (Ho et al., 2007).

Viral enzyme inhibition by anthraquinone: Rhein inhibited the Human immunodeficiency virus type-1 ribonuclease with an IC_{50} value of 60µM (Esposito et al., 2016).

Commentaries: (i) The plant brings to being the anthraquinones aloe-emodin, emodin, physcion, and rhein (Kim et al., 2004), which may contribute synergistically to medicinal uses. (ii) Severe acute respiratory syndrome-acquired coronavirus S protein is the membrane glycoprotein that binds the host cell surface angiotensin-converting enzyme 2 (Ho et al., 2007). (ii) Consider that rhein suppresses lung inflammatory injury induced by human respiratory syncytial virus (Shen et al., 2020) and as such could be examined for its possible anti-COVID-19 properties. (iii) Members of the genus *Cassia* L. are not uncommonly used to treat microbial infections in Asia. *Cassia auriculata* L. is used in India (local name: *Sona gash*) for the treatment of skin diseases. Methanol extract of leaves and flowers of *Cassia auriculata* L. (6mg/well) inhibited the growth of *Bacillus cereus*, *Listeria monocytogenes*, *Staphylococcus aureus*, *Escherichia coli*, and *Salmonella anatum* with inhibition zone diameters of 12.7, 13.2, 17.4, 7.1, and 10.6mm, respectively (Shan et al., 2007). The magnificent tree *Cassia javanica* L. is used in Thailand (local name: *Chaiaphruk*) to treat fever, and from its leaves was isolated the proanthocyanidin ent-epiafzelechin-(4α→8)-epiafzelechin, which inhibited the replication and plaque formation of the Herpes simplex virus type-2 (strain 196) in Vero cells via inhibition of penetration in cells and late stage replication (Cheng et al., 2006). *Cassia occidentalis* L. is traditionally used in India for the treatment of conjunctivitis (local name: *Anwal* or *kasari*), in Pakistan to treat cough and fever (local name: *Amla*), and in both countries to treat ringworm. Methanol extract of roots of this handsome shrub inhibited the replication of the Human immunodeficiency virus type-1 (NL4.3 strain) in Human $CD4^{+}$ T by 68.6 % with a maximum non-cytotoxic concentration of 5 µg/mL (Sabde et al., 2011). *Cassia occidentalis* var. *sophera* (L.) Kuntze (also known as *Cassia sophera* L.) is used in India to treat ringworm (local name: *Kasamarda*). Methanol extract of this shrub inhibited the growth of *Mycobacterium tuberculosis* (strains H37RV), multidrug-resistant *Mycobacterium tuberculosis*, *Mycobacterium tuberculosis* (strain CL-1), and *Mycobacterium tuberculosis* (strain CL-2) with MIC values of 500, 500, 250, and 125 µg/mL, respectively (Singh et al., 2013). *Cassia garrettiana* Craib builds the stilbene piceatannol (LogD = 1.9 at pH 7.4; molecular mass = 244.2 g/mol), which inhibited the angiotensine-converting enzyme with an IC_{50} values of 8.4µM (Madaka et al., 2018).

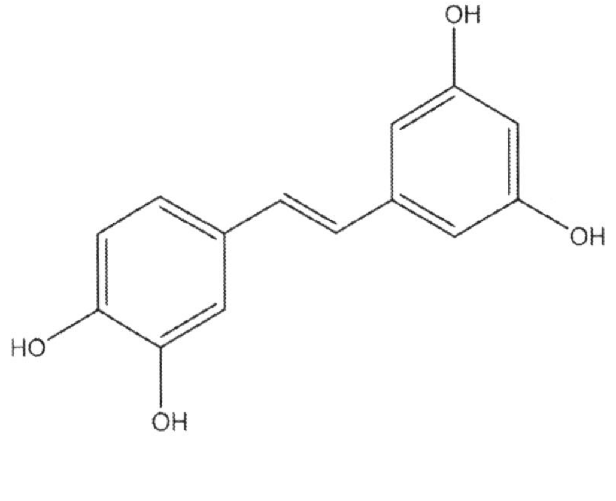

Piceatannol

Ethyl acetate extract of wood of this plant inhibited the growth of *Streptococcus mutans* (DMST 26095), *Streptococcus mitis* (ATCC 49456T), and *Streptococcus pyogenes* (DNAT 17020) with MIC/MBC values of 312.5/625.0, 78.1/312.5, and 156/312.5 µg/mL, respectively (Meerungrueang and Panichayupakaranant, 2014. The stilbene dimers scirpusin A and scirpusin B from the aerial parts of *Caragana rosea* Turcz. ex Maxim. (Local name in China: *Hong hua jin ji er*) inhibited the Human immunodeficiency virus type-1 (strain IIIB) with EC_{50} values of 10 and 7 µg/mL, respectively (Yang et al., 2005). The stilbene dimers scirpusin A (LogD = 3.7 at pH 7.4; molecular mass = 470.4 g/mol) and scirpusin B from the aerial parts of *Caragana rosea* (Tribe Hedysareae) inhibited Human immunodeficiency virus type-1 (strain IIIB) with EC_{50} values of 10 and 7 µg/mL, respectively (Yang et al., 2005)

REFERENCES

Chen, C.P., Lin, C.C. and Tsuneo, N., 1989. Screening of Taiwanese crude drugs for antibacterial activity against Streptococcus mutans. *Journal of Ethnopharmacology*, *27*(3), pp. 285–295.

Cheng, H.Y., Yang, C.M., Lin, T.C., Shieh, D.E. and Lin, C.C., 2006. ent-Epiafzelechin-(4α→ 8)-epiafzelechin extracted from Cassia javanica inhibits herpes simplex virus type 2 replication. *Journal of Medical Microbiology*, *55*(2), pp. 201–206.

Esposito, F., Carli, I., Del Vecchio, C., Xu, L., Corona, A., Grandi, N., Piano, D., Maccioni, E., Distinto, S., Parolin, C. and Tramontano, E., 2016. Sennoside A, derived from the traditional chinese medicine plant Rheum L., is a new dual HIV-1 inhibitor effective on HIV-1 replication. *Phytomedicine*, *23*(12), pp. 1383–1391.

Hatano, T., Uebayashi, H., Ito, H., Shiota, S., Tsuchiya, T. and Yoshida, T., 1999. Phenolic constituents of Cassia seeds and antibacterial effect of some naphthalenes and anthraquinones on methicillin-resistant Staphylococcus aureus. *Chemical and Pharmaceutical Bulletin*, *47*(8), pp. 1121–1127.

Jong-Chol, C., Tsukasa, M., Kazuo, A., Hiroaki, K., Haruki, Y. and Yasuo, O., 1987. Anti-Bacteroides fragilis substance from rhubarb. *Journal of Ethnopharmacology*, *19*(3), pp. 279–283.

Kim, Y.M., Lee, C.H., Kim, H.G. and Lee, H.S., 2004. Anthraquinones isolated from Cassia tora (Leguminosae) seed show an antifungal property against phytopathogenic fungi. *Journal of Agricultural and Food Chemistry*, *52*(20), pp. 6096–6100.

Madaka, F. and Charoonratana, T., 2018. Angiotensin-converting enzyme inhibitory activity of Senna garrettiana active compounds: Potential markers for standardized herbal medicines. *Pharmacognosy Magazine*, *14*(57), p.335.

Meerungrueang, W. and Panichayupakaranant, P., 2014. Antimicrobial activities of some Thai traditional medical longevity formulations from plants and antibacterial compounds from Ficus foveolata. *Pharmaceutical Biology*, *52*(9), pp. 1104–1109.

Sabde, S., Bodiwala, H.S., Karmase, A., Deshpande, P.J., Kaur, A., Ahmed, N., Chauthe, S.K., Brahmbhatt, K.G., Phadke, R.U., Mitra, D. and Bhutani, K.K., 2011. Anti-HIV activity of Indian medicinal plants. *Journal of Natural Medicines*, *65*(3–4), pp. 662–669.

Shan, B., Cai, Y.Z., Brooks, J.D. and Corke, H., 2007. The in vitro antibacterial activity of dietary spice and medicinal herb extracts. *International Journal of Food Microbiology*, *117*(1), pp. 112–119.

Shen, C., Zhang, Z., Xie, T., Ji, J., Xu, J., Lin, L., Yan, J., Kang, A., Dai, Q., Dong, Y. and Shan, J., 2020. Rhein suppresses lung inflammatory injury induced by human respiratory syncytial virus through inhibiting NLRP3 inflammasome activation via NF-κB pathway in mice. *Frontiers in Pharmacology*, *10*, p. 1600.

Singh, R., Hussain, S., Verma, R. and Sharma, P., 2013. Anti-mycobacterial screening of five Indian medicinal plants and partial purification of active extracts of Cassia sophera and Urtica dioica. *Asian Pacific Journal of Tropical Medicine*, *6*(5), pp. 366–371.

Wen, C.C., Shyur, L.F., Jan, J.T., Liang, P.H., Kuo, C.J., Arulselvan, P., Wu, J.B., Kuo, S.C. and Yang, N.S., 2011. Traditional Chinese medicine herbal extracts of Cibotium barometz, Gentiana scabra, Dioscorea batatas, Cassia tora, and Taxillus chinensis inhibit SARS-CoV replication. *Journal of Traditional and Complementary Medicine*, *1*(1), pp. 41–50.

Xu, H.X., Wan, M., Loh, B.N., Kon, O.L., Chow, P.W. and Sim, K.Y., 1996. Screening of Traditional Medicines for their Inhibitory Activity Against HIV-1 Protease. *Phytotherapy Research*, *10*(3), pp. 207–210.

Yang, G.X., Zhou, J.T., Li, Y.Z. and Hu, C.Q., 2005. Anti-HIV bioactive stilbene dimers of Caragana rosea. *Planta Medica*, *71*(06), pp. 569–571.

7.6.1.19 *Crotalaria pallida* Aiton

Synonyms: *Crotalaria obovata* G. Don; *Crotalaria saltiana* Prain ex King; *Crotalaria striata* DC.

Common names: Smooth rattlebox; junjuni, kudug jhunjuni, prosharini (Bangladesh); zhu shi dou, wong bong long duan (China); atasi (India); kekecrekan (Indonesia); hingx hay (Laos); giring-giring (Malaysia); jhunjhuna, runche (Nepal); gorung-gorung (the Philippines); cà phê rung (Vietnam)

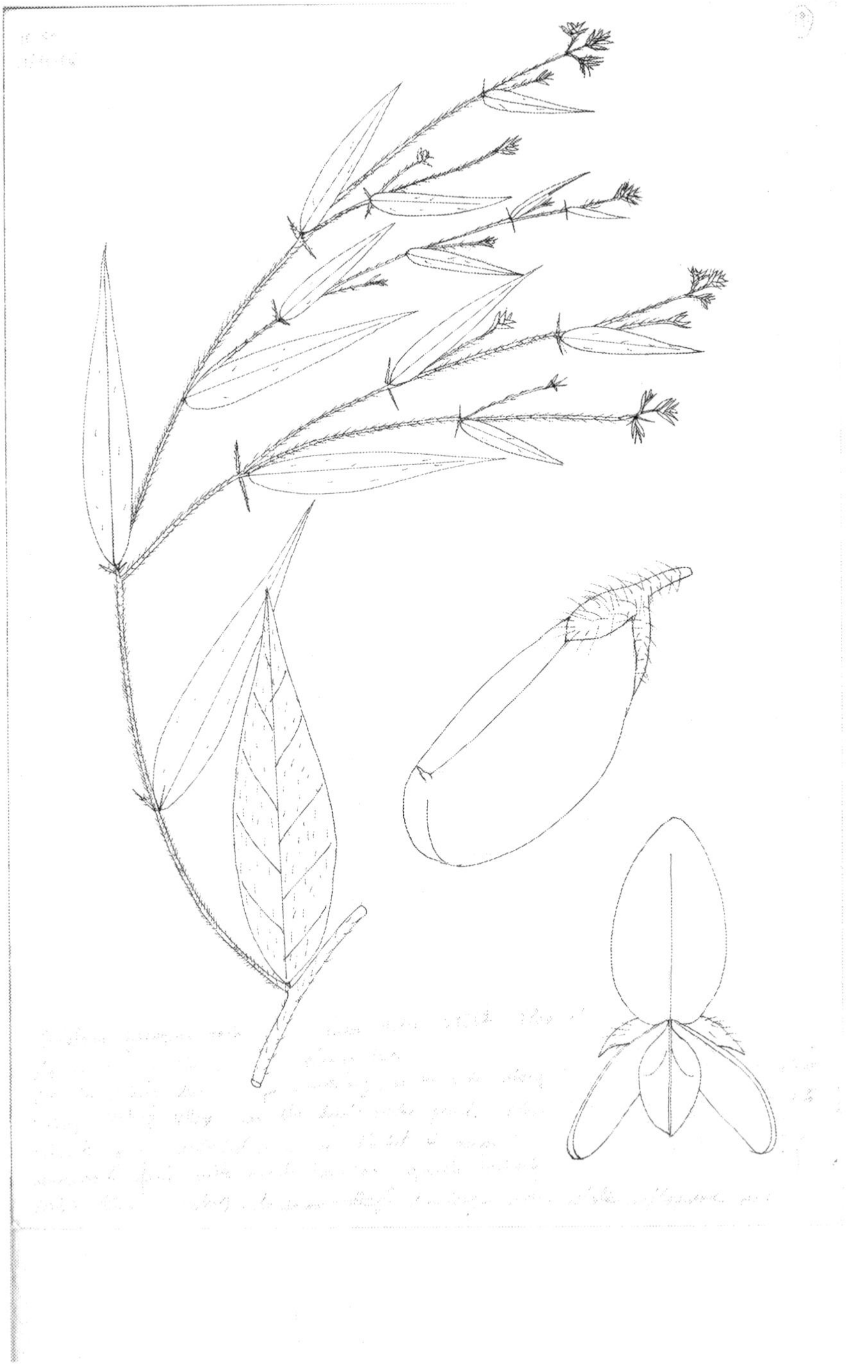

FIGURE 7.12 *Crotalaria ferruginea* Graham ex Benth.

Habitat: Roadsides, vacant plots, riverbanks, around lakes

Distribution: Tropical Asia

Botanical observation: This graceful herb probably native to Africa grows up to a height of 1.5 m. The stems are terete, ribbed, and hairy. The leaves are 3-lobed, spiral, and stipulate. The stipules are linear minute, and caducous. The petiole is 2–4 cm long. The folioles are oblong,

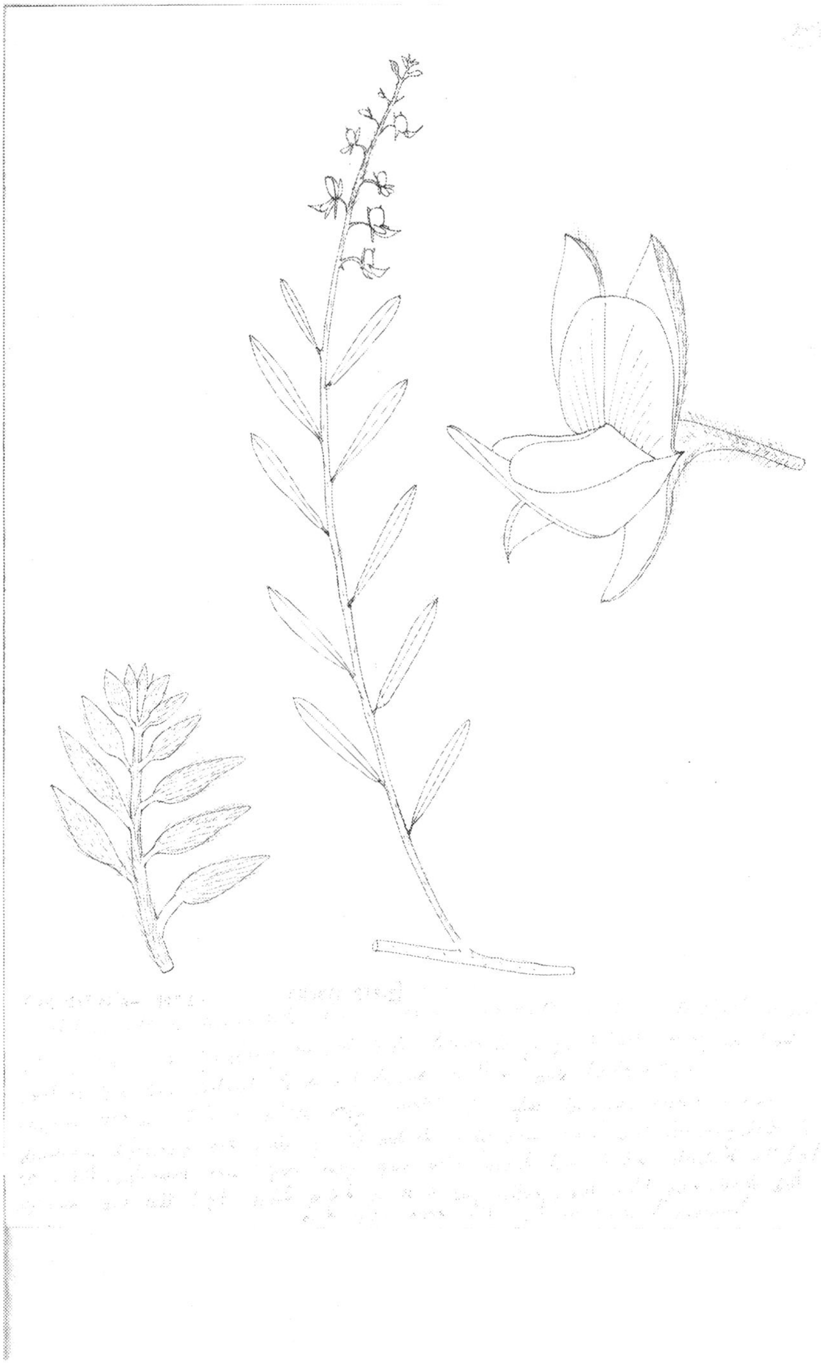

FIGURE 7.13 *Crotalaria albida* Heyne ex Roth

oblanceolate, to elliptic, 3–6 × 1.5–3 cm, dull light green, hairy below, cuneate at base, and obtuse to retuse at apex. The racemes are terminal, up to 25 cm long and many flowered. The calyx is tubular, about 5 mm long, hairy, and develops 5 triangular lobes. The 5 petals are yellow, including an orbicular to elliptic standard, a pair of oblong wings, and a keel, which is about 1.2 cm long. The androecium includes 10 stamens of irregular lengths and fused at base. The ovary is narrowly oblong and hairy. The pods are oblong, about 4 cm long, and sheltering numerous seeds.

Medicinal uses: Dysentery (Bangladesh); measles, hepatitis, urethritis (China)

Broad-spectrum antibacterial non-polar extract: Petroleum ether extract of stems (4 mm disc impregnated with a solution at 4 mg/mL) inhibited the growth of *Staphylococcus aureus* (ATCC 25923), *Escherichia coli* (ATCC 25922), and *Pseudomonas aeruginosa* (ATCC 27853) with inhibition zone diameters of 17, 20, and 14 mm, respectively (Islam et al., 2018).

Commentaries: (i) The plant generates quite an unusual series of homoisoflavonoids, flavonoids, and poisonous alkaloids (Hu et al., 2017) and has apparently not been examined for antimicrobial properties. The chalcones crotmadine and crotmarine from the leaves and stems of *Crotalaria madurensis* R. Wight inhibited the growth of *Trichophyton mentagrophytes* at a concentration of 62.5 μg/mL (Bhakuni & Chaturvedi, 1984). (ii) In India, *Crotalaria retusa* L. is used to treat leprosy and fever (local name: *Atasi*) and *Crotalaria juncea* L. (local name: *Shana*) is used to treat diarrhea and dysentery. In Bangladesh, *Crotalaria verrucosa* L. (local name: *Kuduk jhunjhuni*) is used to heal wounds and to treat tetanus, and in Thailand *Crotalaria albida* Heyne ex Roth (local name: *Hinghai*) is used for fever. *Crotalaria ferruginea* Graham ex Benth. (local name: *Jia di lan*) is used in China for the treatment of cough, tuberculosis, and fever.

REFERENCES

Bhakuni, D.S. and Chaturvedi, R., 1984. Chemical constituents of Crotalaria madurensis. *Journal of Natural Products*, *47*(4), pp. 585–591.

Hu, X.R., Chou, G.X. and Zhang, C.G., 2017. Flavonoids, alkaloids from the seeds of Crotalaria pallida and their cytotoxicity and anti-inflammatory activities. *Phytochemistry*, *143*, pp. 64–71.

Islam, M.Z., Hossain, M.T., Hossen, F., Mukharjee, S.K., Sultana, N. and Paul, S.C., 2018. Evaluation of antioxidant and antibacterial activities of Crotalaria pallida stem extract. *Clinical Phytoscience*, *4*(1), pp. 1–7.

7.6.1.20 *Dalbergia pinnata* (Lour.) Prain

Synonyms: *Dalbergia dubia* Elmer, *Dalbergia livida* Wall., *Dalbergia pinatubensis* Elmer, *Dalbergia rufa* Graham, *Dalbergia tamarindifolia* Roxb., *Derris pinnata* Lour., *Endespermum scandens* Bl.

Common names: Java polisander; xie ye huang tan (China); jampak luyak (Indonesia); cham bia ab trau (Cambodia, Laos, and Vietnam); jak (Iran); emelit jangkar, lorotan haji (Malaysia); damar (Nepal); tikos maiadon (the Philippines); trac la me (Vietnam)

Habitat: Forests

Distribution: India, Bangladesh, Myanmar Cambodia, Laos, Vietnam, Malaysia, Indonesia, the Philippines, Pacific Island

Botanical observation: This climbing shrub grows to a length of about 12 m. The stem is hairy, lenticelled, and zig-zag shaped. The leaves are compound, spiral, and stipulate. The stipules are lanceolate and 5 mm long. The rachis is velvety, 10 –15cm long, straight, and holds about 25–41 pairs of folioles. The folioles are sub-opposite to alternate, oblong, 1.5cm long, round with a minute tip at apex, and round and asymmetrical at base. The racemes are axillary, velvety, and 1.5 –2.5cm long. The calyx is campanulate, minute, 5–lobed, and with triangular lobes. The corolla is 1 cm long, made of 5 petals, which are white with a broad standard. The pods are flat, glabrous, dull, linear, covered with a pattern of nerves, –reddish-brown, and up to 7.5 cm long.

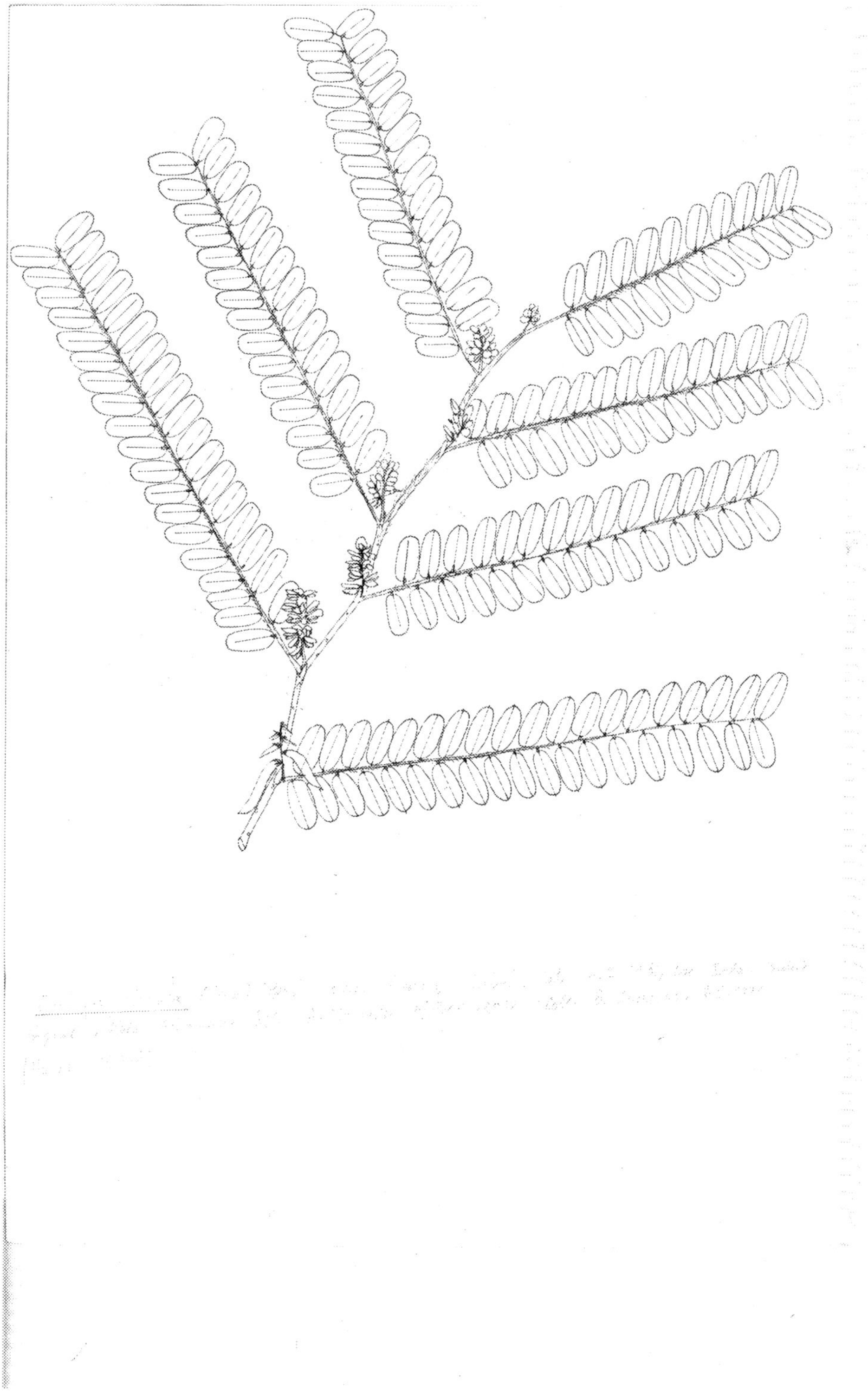

FIGURE 7.14 *Dalbergia pinnata* (Lour.) Prain

Medicinal uses: Fever (Indonesia); diarrhea (Iran)

Strong antibacterial essential oil: Essential oil inhibited the growth of *Staphylococcus aureus* (ATCC 25923) and *Streptococcus pyogenes* (ATCC 19615) with MIC/MBC values of 0.7/25, and 6.2/25 μL/mL, respectively (Zhou et al., 2020).

Strong anticandidal essential oil: Essential oil inhibited the growth of *Candida albicans* (ATCC 10231) with an MIC/MFC value of 12.5/50 μL/mL (Zhou et al., 2020).

Commentary: Members of the genus *Dalbergia* L.f are not uncommonly used to treat microbial infections in the Indian subcontinent. In Bangladesh, *Dalbergia sissoo* Roxb. ex DC. (local name: *Sissoo*) is used to treat gonorrhea, wounds, and abscesses and is used to treat diarrhea in Nepal (local name: *Sisau*). *Dalbergia latifolia* Roxb. is used to treat dysentery and leprosy in India (local name: *Eeti*) and Bangladesh. In Bangladesh, *Dalbergia lanceolaria* L.f. is used to treat jaundice and diarrhea.

REFERENCE

Zhou, W., He, Y., Lei, X., Liao, L., Fu, T., Yuan, Y., Huang, X., Zou, L., Liu, Y., Ruan, R. and Li, J., 2020. Chemical composition and evaluation of antioxidant activities, antimicrobial, and anti-melanogenesis effect of the essential oils extracted from Dalbergia pinnata (Lour.) Prain. *Journal of Ethnopharmacology*, p. 112731.

7.6.1.21 *Derris trifoliata* Lour.

Synonyms: *Deguelia trifoliata* (Lour.) Taub.; *Deguelia uliginosa* (Roxb. ex Willd.) Baill.; *Derris uliginosa* (Roxb. ex Willd.) Benth.; *Galedupa uliginosa* (Roxb. ex Willd.) Roxb.; *Pongamia madagascariensis* Bojer ex Oliver; *Pongamia uliginosa* (Roxb. ex Willd.) DC.; *Robinia uliginosa* Roxb. ex Willd.

Common names: Kalla lata (Bangladesh); yu teng (China); ketia nai, kirtana (India); gadel (Indonesia); akar ketuil, tuba bekut (Malaysia); kemokem (Palau); phak taep, thopthaep nam (Thailand); long ken (Vietnam); gamo (Papua New Guinea); sila sila (the Philippines)

Habitat: Muddy seashores, mangroves, and beaches

Distribution: From India to the Pacific Islands

Botanical observation: This large woody climber grows to a length of 15 m. The stems are smooth and angled at apex. The leaves are compound, spiral, and stipulate. The rachis is up to 15 long and bears 3 pairs of folioles and a terminal foliole, which are thinly coriaceous, broadly lanceolate, 5–10.5 × 2–6 cm, round at base and acuminate at apex. The racemes are axillary and up to 12 cm long. The calyx is campanulate and about 3 mm long. The corolla is pinkish white or pale lavender, up to about 1.3 cm long with a suborbicular standard. The androecium includes 10 stamens. The ovary is somewhat hairy. The pod is flat, ovoid or globose, or oblong, up to 3.5 cm, and containing 1 or 2 seeds, which are compressed and reniform.

Medicinal use: Dysentery (Palau)

Broad-spectrum antibacterial halo developed by mid-polar extracts: Ethyl acetate extract of roots (4 mg/disc) inhibited the growth of *Bacillus cereus*, *Bacillus megaterium*, *Lactobacillus casei*, *Micrococcus luteus*, *Micrococcus roseus*, *Staphylococcus aureus*, *Streptococcus pneumoniae*, *Escherichia coli*, *Klebsiella pneumoniae*, *Neisseria gonorrhea*, *Proteus murabilis*, *Pseudomonas aeruginosa*, and *Salmonella typhi* with inhibition zone diameters of 12, 14, 12, 14, 14, 16, 12, 14, 14, 14, 14, 14, and 12 mm, respectively (Khan et al., 2006). Chloroform extract of leaves (1500 μg/mL solution/6mm well) inhibited the growth of *Staphylococcus aureus* (MTCC 96), *Staphylococcus epidermidis* (MTCC 535), *Bacillus subtilis* (MTCC 441), *Escherichia coli* (MTCC 443), *Pseudomonas aeruginosa* (MTCC 741), and *Klebsiella pneumoniae* (MTCC 39) with inhibition zone diameters of 22, 22, 18, 20, 18, and 18 mm, respectively (Kumar et al., 2011).

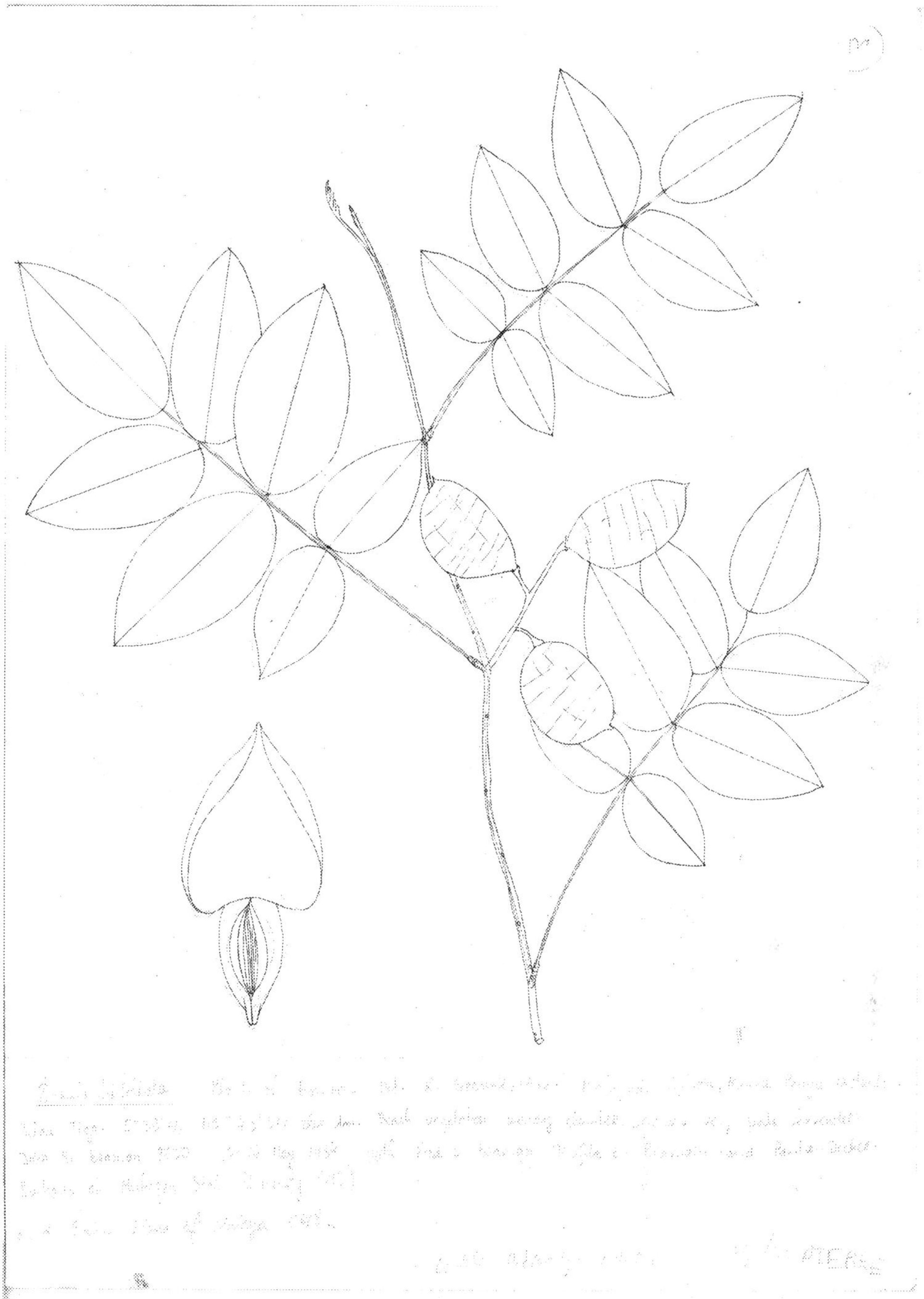

FIGURE 7.15 *Derris trifoliata* Lour.

FIGURE 7.16 *Derris scandens* Benth,

Broad-spectrum antifungal halo developed by mid-polar extract: Chloroform extract of leaves (1500 µg/mL solution/6mm well) inhibited the growth of *Aspergillus niger* (MTCC 134), *Rhizopus oryzae* (MTCC 262), *Candida albicans* (MTCC 227), and *Saccharomyces cerevisiae* (MTCC 170) with inhibition zone diameters of 20, 16, 16, and 20 mm, respectively (Kumar et al., 2011).

Strong broad-spectrum antibacterial lipophilic flavanone: Lupinifolin (LogD = 5.8 at pH 7.4; molecular weight = 406.4 g/mol) inhibited the growth of *Staphylococcus aureus*, *Bacillus subtilis*, *Escherichia coli*, and *Pseudomonas aeruginosa* with minimum mass of 0.5, 0.5, 10, and 0.5 µg (Mazimba et al., 2019). Lupinifolin inhibited the growth of *Staphylococcus aureus* with an MIC value of 8 µg/mL, with rupture to the bacterial cell membrane (Yusook et al., 2017)

Strong anticandidal lipophilic flavanone: Lupinifolin inhibited the growth of *Candida albicans* with minimum mass of 0.01 µg (Mazimba et al., 2019).

Strong antiviral (enveloped monopartite double-stranded DNA) amphiphilic rotenoid-type isoflavonoid: Deguelin (LogD = 4.7 at pH 7.4; molecular mass = 394.4 g/mol) inhibited the Human cytomegalovirus cell-associated viral DNA in NuFF-1 cells with an IC_{50} value of 55.8 nM (Phrutivorapongkul et al., 2002).

Moderate antiviral (enveloped monopartite linear double-stranded DNA) rotenoid-type isoflavonol: Rotenone inhibited the replication of the Herpes simplex virus type-1 (KOS) and Herpes simplex virus type-2 (186) by 15.4 and 24.4% at 50 µg/mL, respectively (Phrutivorapongkul et al., 2002).

Commentaries: (i) The plant yields flavonoids (Yenesew et al., 2009). Consider that prenylated isoflavones from members of the genus *Derris* L. are often strongly antibacterial against Gram-positive bacteria as illustrated with lupalbigenin (see earlier) isolated from the stems of *Derris scandens* Benth. (local name in India: *Noe lata*), which inhibited *Staphylococcus aureus* (ATCC 25293) and methicillin-resistant *Staphylococcus aureus* SK1 with MIC values of 2 and 4 µg/mL, respectively (Mahabusarakam et al., 2004). Derrisisoflavone A isolated from the stems of *Derris scandens* Benth. inhibited *Staphylococcus aureus* (ATCC 25293) and methicillin-resistant *Staphylococcus aureus* SK1 with MIC values of 16 and 4 µg/mL, respectively (Mahabusarakam et al., 2004).

One further example of antibacterial isoflavonoid in this genus is santal isolated from the stems of *Derris scandens* Benth. inhibited methicillin-resistant *Staphylococcus aureus* (SK1) with an MIC value of 2 µg/mL (Mahabusarakam et al., 2004).

(ii) Rotenone is a dreadful neurotoxin which induces Parkinson's disease (Alam & Schmid, 2002).

(iii) *Derris elliptica* (Roxb.) Benth. is used in Palau to treat ringworm and to heal cuts (local name: *Yubu*). *Derris robusta* Benth. is used to heal wounds in Bangladesh (local name: *Kuruar gash*).

(iv) From the Tribe Milletieae, *Millettia erythrocalyx* Gagnep. (local name in China: *Hong e ji xue teng*) generates in its leaves the flavones ovalifolin, pongol methyl ether, and milletocalyxin A, which inhibited the replication of the Herpes simplex virus type-1 with EC_{50} values of 97.7, 70.5, and 53.6 µM, respectively (Likhitwitayawuid et al., 2005). Ovalifolin, pongol methyl ether, and milletocalyxin A inhibited the replication of the Herpes simplex virus type-2 with EC_{50} values of 55.3, 108.7, and 132.8 µM (acyclovir 2.2 µM).

REFERENCES

Alam, M. and Schmidt, W.J., 2002. Rotenone destroys dopaminergic neurons and induces parkinsonian symptoms in rats. *Behavioural Brain Research*, *136*(1), pp. 317–324.

Khan, M.R., Omoloso, A.D. and Barewai, Y., 2006. Antimicrobial activity of the Derris elliptica, Derris indica and Derris trifoliata extractives. *Fitoterapia*, *77*(4), pp. 327–330.

Kumar, V.A., Ammani, K. and Siddhardha, B., 2011. In vitro antimicrobial activity of leaf extracts of certain mangrove plants collected from Godavari estuarine of Konaseema delta, India. *International Journal of Medicinal and Aromatic Plants*, *1*(2), pp. 132–136.

Mahabusarakam, W., Deachathai, S., Phongpaichit, S., Jansakul, C. and Taylor, W.C., 2004. A benzil and isoflavone derivatives from Derris scandens Benth. *Phytochemistry*, *65*(8), pp. 1185–1191.

Mazimba, O., Masesane, I.B. and Majinda, R.R., 2012. A flavanone and antimicrobial activities of the constituents of extracts from Mundulea sericea. *Natural Product Research*, *26*(19), pp. 1817–1823.

Phrutivorapongkul, A., Lipipun, V., Ruangrungsi, N., Watanabe, T. and Ishikawa, T., 2002. Studies on the constituents of seeds of Pachyrrhizus erosus and their anti herpes simplex virus (HSV) activities. *Chemical and Pharmaceutical Bulletin*, *50*(4), pp. 534–537.

Yenesew, A.B.I.Y., Twinomuhwezi, H., Kabaru, J.M., Akala, H.M., Kiremire, B.T., Heydenreich, M., Peter, M.G., Eyase, F.I., Waters, N.C. and Walsh, D.S., 2009. Antiplasmodial and larvicidal flavonoids from Derris trifoliata. *Bulletin of the Chemical Society of Ethiopia*, *23*(3).

Yusook, K., Weeranantanapan, O., Hua, Y., Kumkrai, P. and Chudapongse, N., 2017. Lupinifolin from Derris reticulata possesses bactericidal activity on Staphylococcus aureus by disrupting bacterial cell membrane. *Journal of Natural Medicines*, *71*(2), pp. 357–366.

7.6.1.22 *Desmodium gangeticum* (L.) DC.

Synonyms: *Aeschynomene gangetica* (L.) Poir., *Aeschynomene maculata* (L.) Poir., *Desmodium cavaleriei* H. Lev., *Desmodium lanceolatum* (Schum & Thonn.) Walp., *Desmodium polygonoides* Welw. ex Baker, *Hedysarum collinum* Roxb., *Hedysarum gangeticum* L., *Hedysarum lanceolatum* Schum. & Thonn., *Hedysarum maculatum* L., *Hedysarum ochroleucum* Moench, *Meibomia gangetica* (L.) O. Kuntze., *Meibomia polygonodes* (Welw. Ex Baker) Kuntze, *Pleurolobus gangeticus* (L.) J. St. Hil., *Pleurolobus maculatus* J St. Hil.

Common names: Salpani, Chalani (Bangladesh); da ye shan ma huang (China); devi, oterai, pulladi, salaparni (India); daun bulu ayam (Indonesia), tuk hma (Laos), dampate (Nepal); paiang-paiang (the Philippines), yaa tuet maeo, baw berchai (Thailand)

Habitat: Disturbed forests and wastelands

Distrbution: From India to the Pacific Islands

Botanical observation: This shrub grows to about 1 m tall. The stem is terete or angled, somewhat glossy, and subglabrous. The leaves are simple, spiral, and stipulate. The stipules are persistent, lanceolate–linear, and about 1×0.2 cm. The petiole is hairy, flat above and 0.8–1.7 cm long. The blade is dull dark green, glossy, membranous, translucent, acute at apex, 10.5–3×1.7–5cm, minutely hairy, and marked with 6–8 pairs of secondary nerves. The racemes are terminal and the flower pedicels are reddish. The calyx is 2mm long, light green to reddish, hairy, and with triangular lobes. The corolla is 4mm long. The standard is 3mm long and broadly cuneate at base. The wings are violet. The pod is 1.2–2cm long, hemiglobose, membranous, hairy, and about 2 cm long.

Medicinal uses: Bronchitis, burns, fever, typhoid (India)

Broad-spectrum antibacterial halos developed by polar extract: Ethanol extract of leaves inhibited the growth of *Streptococcus pyogenes*, *Shigella dysenteriae*, *Shigella boydii*, and *Escherichia coli* with inhibition zones of 8, 12, 9, and 8 mm, respectively (0.5 mg/disc) (Uddin et al., 2008).

Strong antiviral (enveloped monopartite linear single-stranded (+) RNA) polar extract: Methanol extract inhibited the replication of the Dengue virus serotype-1, -2, -3, and -4 with IC_{50} values of 12.5, 6.7, 14.6, and 12.5 µg/mL, respectively (Sood et al., 2015)

Commentaries: (i) Antimicrobial principles have apparently not been isolated from this plant. Consider that it has strong antiviral activity *in vitro* against the Dengue virus, which is an enveloped virus with monopartite, linear, single-stranded (+) RNA just like the Zika virus and the Yellow fever virus, which are zoonotic. The Severe acute respiratory syndrome-acquired coronavirus 2 is also an enveloped virus with monopartite, linear, single-stranded (+) RNA and thus, one could examine the effect of this plant against COVID-19. Members of the genus *Desmodium* Desv. produce prenylated flavonols inhibiting the growth of methicillin-resistant *Staphylococcus aureus* (Sasaki et al., 2012). Ueno et al. (1978) isolated the flavonol desmodol from *Desmodium caudatum* DC. Ethanol extract of *Desmodium canadense* (L.) DC inhibited he replication of the Avian infectious bronchitis virus

(Vero-adapted Beaudette strain) with an IC_{50} value of 0.01 μg/mL and a selectivity index of 17.1 in Vero cells (Lelešius et al., 2019).Thus, one could suggest *Desmodium gangeticum* (L.) DC. to generate antiviral prenylated flavonoid, and it needs to be examined.

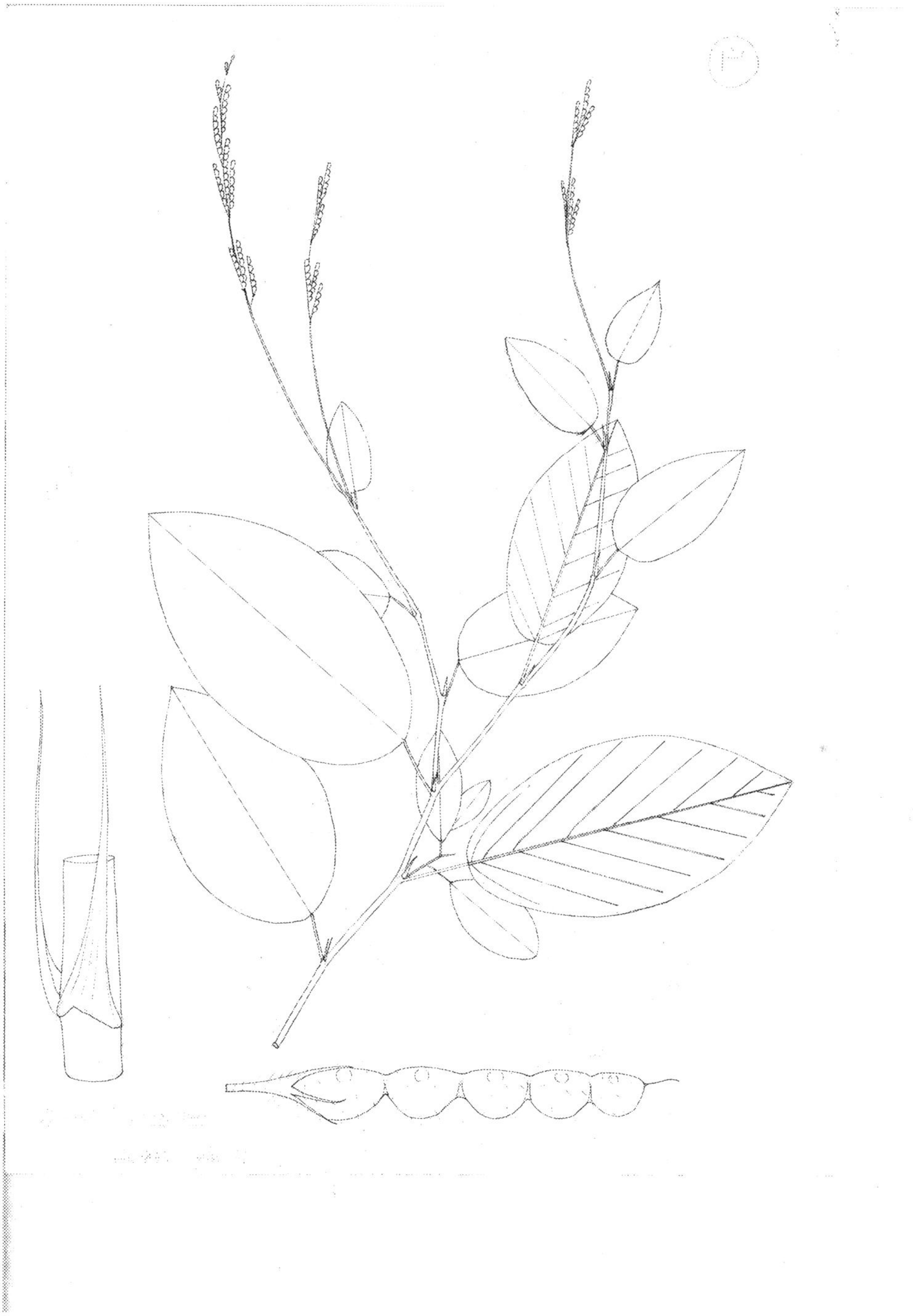

FIGURE 7.17 *Desmodium gangeticum* (L.) DC

(ii) Members of the genus *Desmodium* Desv. are often used for the treatment of microbial infections in India: *Desmodium triflorum* (L.) DC. (local name: *Cherupulladi*) is used to treat diarrhea, ulcers, and wounds, *Desmodium velutinum* (Willd) DC. (local name: *Latlati*) is used to treat stomatitis and diarrhea, and *Desmodium triflorum* (L.) DC. is used to treat cough, diarrhea, and wounds (local name: *Tripadi*). *Desmodium triflorum* (L.) DC. is used in Bangladesh to heal sores

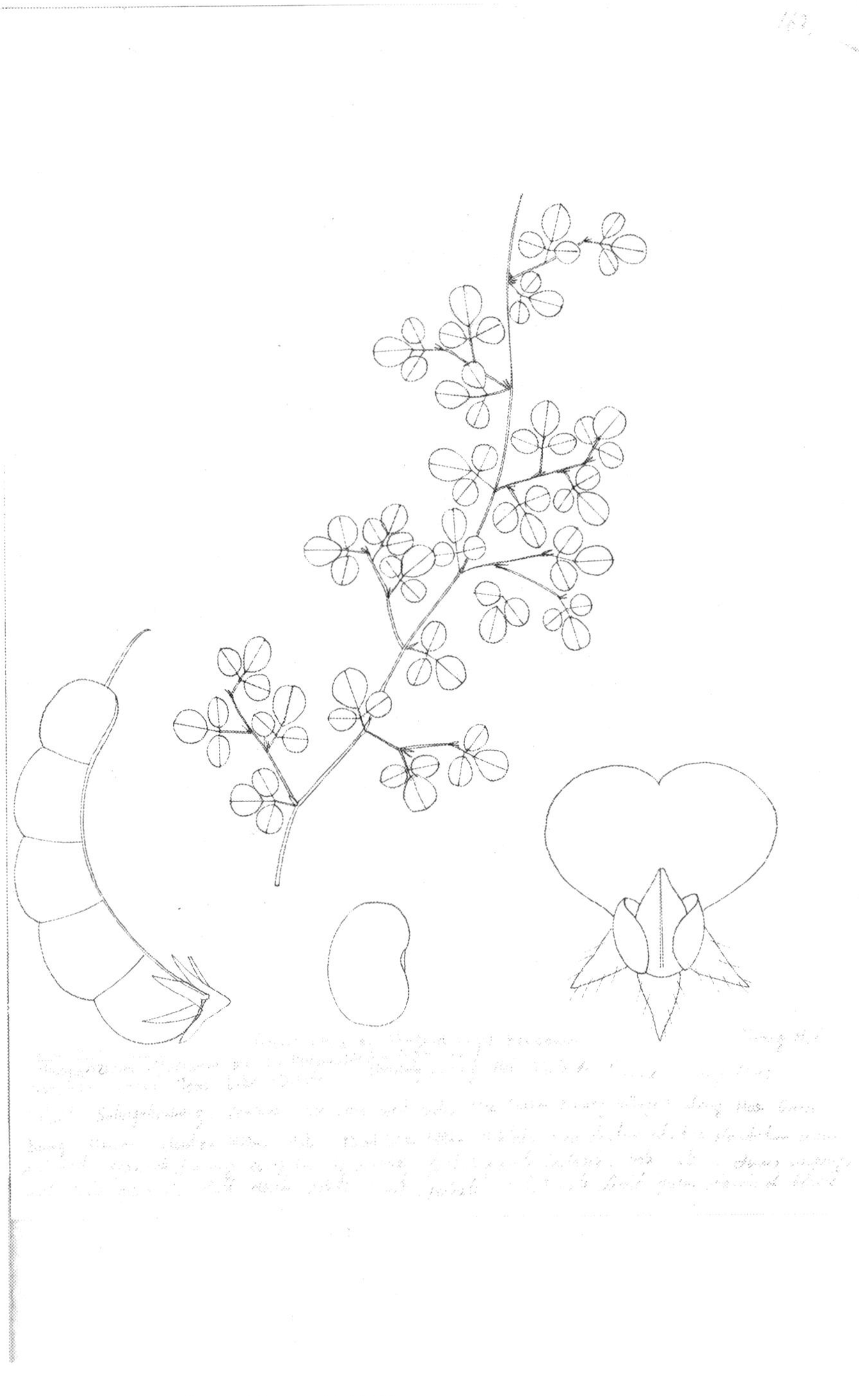

FIGURE 7.18 *Desmodium triflorum* (L.) DC.

(local name: *Kulaliya*), and in Thailand (local name: *Yak let hoi noi*) it is used to treat mouth ulcers and diarrhea. In Myanmar, *Desmodium triquetrum* (L.) DC. (local name: *Lauk-thay)* is used to treat dysentery, sores, fever, and otitis and in India it is used for sores and cold, and in China it is used to heal abscesses. *Desmodium heterocarpon* (L.) DC. (local name: *Bakhre ghans*) is used in Nepal to treat cough and diarrhea and *Desmodium multiflorum* DC. (local name: *Bhatte*) is used to heal

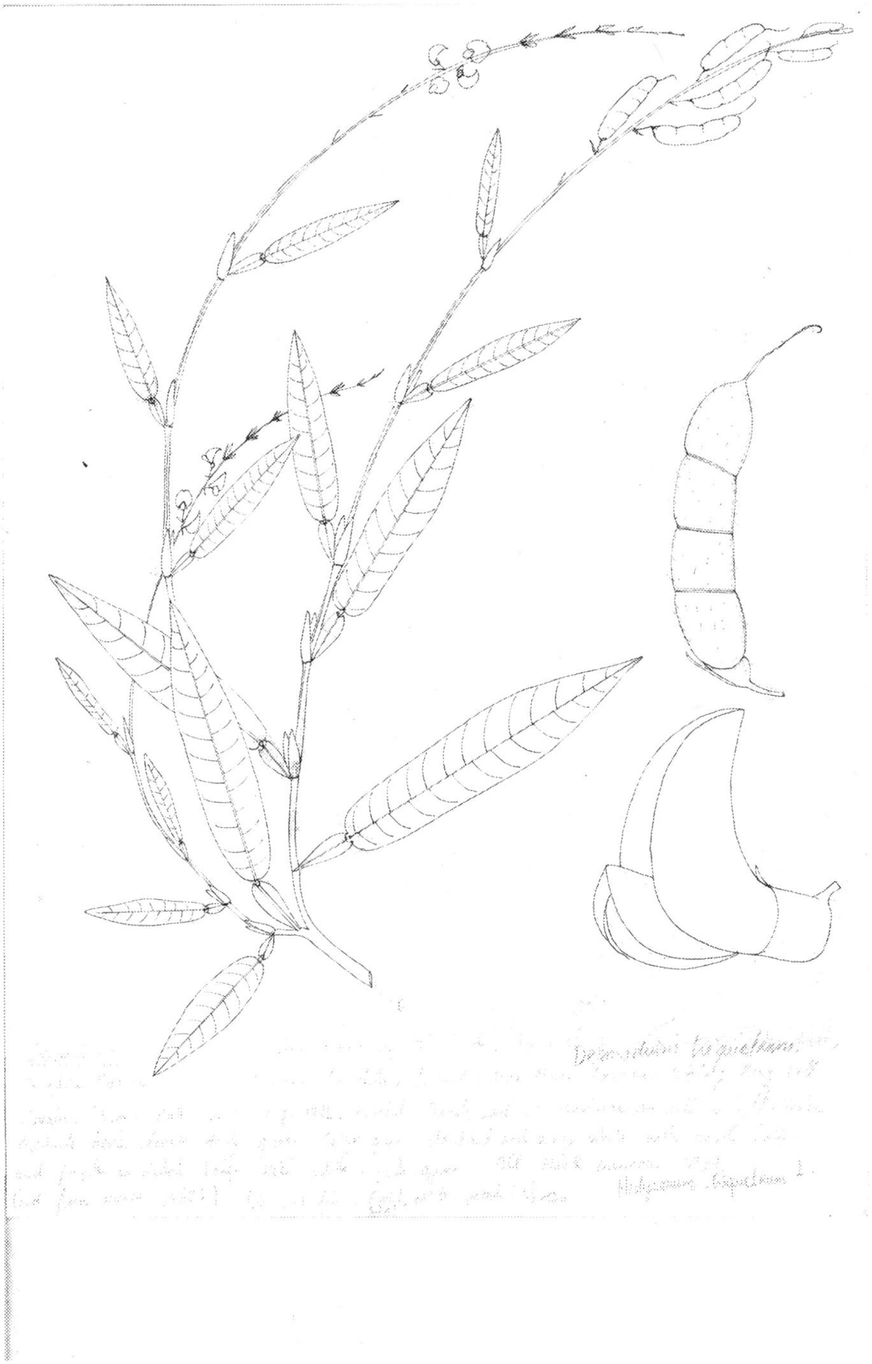

FIGURE 7.19 *Desmodium triquetrum* (L.) DC.

ulcers. In Malaysia, *Desmodium pulchellum* (L.) Benth is used as a post-partum protective remedy and in the Philippines the plant is used to heal ulcers and poxes. In Malaysia, another example of medicinal Fabaceae used for the treatment of infectious diseases is *Bauhinia acuminata* Vell., which is used for syphilis.

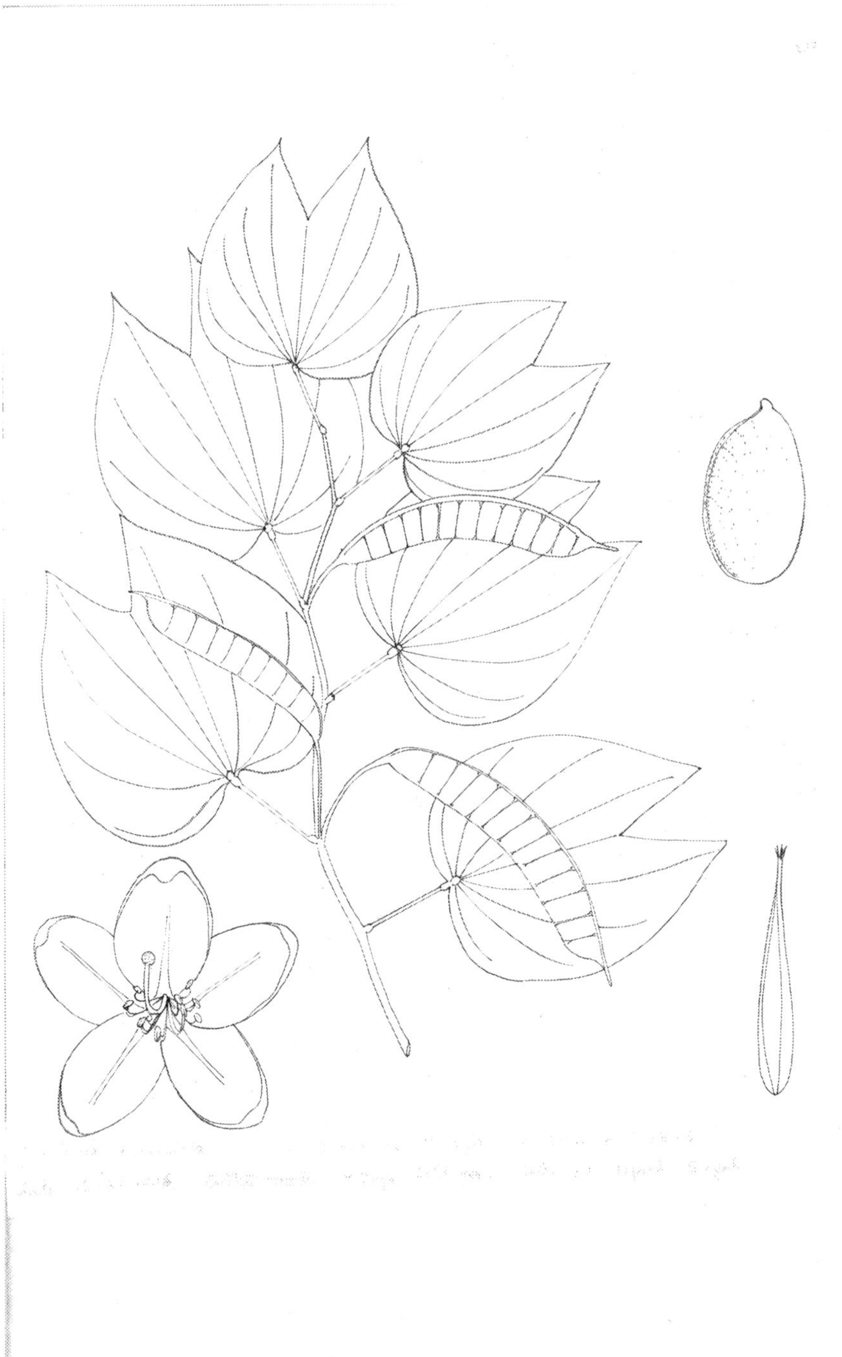

FIGURE 7.20 *Bauhinia acuminata* Vell.

REFERENCES

Lelešius, R., Karpovaitė, A., Mickienė, R., Drevinskas, T., Tiso, N., Ragažinskienė, O., Kubilienė, L., Maruška, A. and Šalomskas, A., 2019. In vitro antiviral activity of fifteen plant extracts against avian infectious bronchitis virus. *BMC Veterinary Research*, *15*(1), p. 178.

Sasaki, H., Kashiwada, Y., Shibata, H. and Takaishi, Y., 2012. Prenylated flavonoids from Desmodium caudatum and evaluation of their anti-MRSA activity. *Phytochemistry*, *82*, pp. 136–142.

Sood, R., Raut, R., Tyagi, P., Pareek, P.K., Barman, T.K., Singhal, S., Shirumalla, R.K., Kanoje, V., Subbarayan, R., Rajerethinam, R. and Sharma, N., 2015. Cissampelos pareira Linn: Natural source of potent antiviral activity against all four dengue virus serotypes. *PLoS Negl Trop Dis*, *9*(12), p. e0004255.

Uddin, S.J., Rouf, R., Shilpi, J.A., Alamgir, M., Nahar, L. and Sarker, S.D., 2008. Screening of some Bangladeshi medicinal plants for in vitro antibacterial activity. *Oriental Pharmacy and Experimental Medicine*, *8*(3), pp. 316–321.

Ueno, A., Ikeya, Y., Fukushima, S., Noro, T., Morinaga, K. and Kuwano, H., 1978. Studies on the Constituents of Desmodium caudatum DC. *Chemical and Pharmaceutical Bulletin*, *26*(8), pp. 2411–2416.

7.6.1.23 *Erythrina orientalis* (L.) Murr.

Synonyms: *Erythrina corallodendrum* L. var. orientalis, *Erythrina indica* Lamk., *Erythrina indica* Zoll.; *Erythrina stricta* Roxb.; *Erythrina variegata* L., *Gelala litorea* Rumphius

Common names: Indian coral tree, tiger's claw; madal gach (Bangladesh); cì táng (China); mandar, mullu-murukku, paaribhadra, thub (India); dadap ayam, gelala, galala, ngoa (Indonesia); dadap, dedap, derdap (Malaysia); kathit (Myanmar); raal (Yap); bông nem (Vietnam)

Habitat: Seashores, cultivated

Distribution: From India to Indonesia

Botanical observation: This tree grows up to about 15 m tall. The stems are armed with woody thorns. The leaves are compound, spiral, and stipulate. The petioles grow up to 40 cm long. The blade is 3-foliolate, the folioles about 10–15 cm long and broadly lanceolate. The inflorescences are terminal clusters, which are up to 40 cm long. The calyx is tubular, somewhat 5-lobed, and up to 3 cm long. The corolla is intensely bright red, up to 5cm long, and curved. The androecium is diadelphous, about 5 cm long, and includes 10 red stamens. The pod is woody, up to about 30 cm long, and contains up to 8 seeds, which are purplish, 2 cm long, and ellipsoid.

Medicinal uses: Wounds (Bangladesh); cold (Bangladesh, India, Yap); cough (Bangladesh, India); abscesses, fever, pimples (India); dysentery (Myanmar); syphilis (China)

Strong antibacterial (Gram-positive) lipophilic prenylated pterocarpan: Erycristagallin (LogD = 6.3 at pH 7.4; molecular mass = 390.4 g/mol) isolated from the roots inhibited the growth of methicillin-resistant *Staphylococcus aureus* (Tanaka et al., 2002), and *Streptococcus mutans*, *Lactobacillus casei*, and *Actinomyces viscosus* with MIC values of 6.2, 6.2, and 1.5 μg/mL, respectively (Sato et al., 2003).

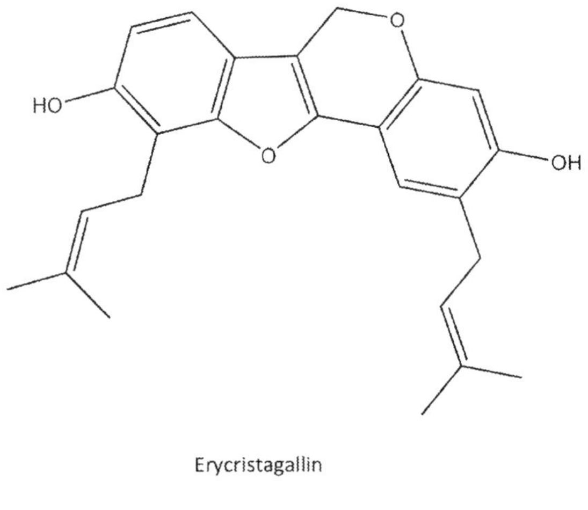

Erycristagallin

Strong antibacterial (Gram-positive) amphiphilic prenylated pterocarpans: Orientanol B (LogD=4.5 at pH 7.4; molecular mass=338.3 g/mol) and orientanol C inhibited the growth of methicillin-resistant *Staphylococcus aureus* (clinical strain) with MIC_{90} values of 12.5 and 12.5 μg/mL, respectively (Tanaka et al., 2002).

Strong antibacterial (Gram-positive) isoflavanone: Orientanol F isolated inhibited the growth of methicillin-resistant *Staphylococcus aureus* (clinical strain) with an MIC_{90} value of 12.5 μg/mL (Tanaka et al., 2002).

Strong antibacterial (Gram-positive) amphiphilic 3-phenyl coumarin*:* Indicanine B (molecular mass=366.4 g/mol) isolated from the root bark inhibited the growth of *Staphylococcus aureus* (209P) with an MIC value of 9.7 μg/mL (Waffo et al., 2000).

Strong antimycobacterial 3-phenyl coumarin: Indicanine B isolated from the root bark inhibited the growth of *Mycobacterium smegmatis* (ATCC607) with an MIC value of 18.5 μg/mL (Waffo et al., 2000).

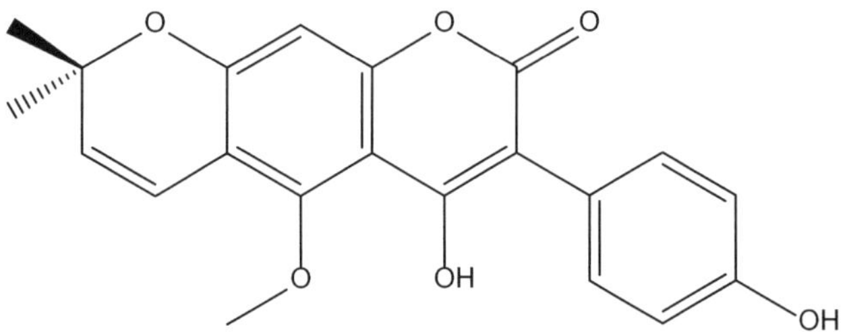

Indicanine B

Moderate antifungal (filamentous) pterocarpans: Genistein and phaseolidin inhibited the growth of *Cladosporium cucumerinum* with an MIC values of 50 μg/mL (O'Neill et al., 1983).

Antiviral (non-enveloped monopartite linear single-stranded (+) RNA virus) erythrinan-type alkaloids: Erysodine, erythraline at a concentration of 500 μg/mL inhibited the replication of the Tobacco mosaic virus by about 60% and 50% (Tan et al., 2017).

Commentaries: (i) What is the antibacterial or antifungal mode of action of pterocarpans? One could suggest DNA intercalation with inhibition of DNA replication because of their almost planar structures. (ii) One could examine the activity of erythrinan alkaloids including erysodine and erythraline, against coronaviruses. (iii) Note that erythrinan-type alkaloids interact with cholinergic and GABAergic receptors and as such may be able to inhibit bacterial or fungal efflux pumps. (iv) *Erythrina arborescens* Roxb. is used in India for the treatment of dysentery and in Nepal (local name: *Phaleto*) to treat boils. In India, *Erythrina stricta* Roxb. is used to treat fever (local name: *Muraa*).

REFERENCES

O'Neill, M.J., Adesanya, S.A. and Roberts, M.F., 1983. Antifungal phytoalexins in Phaseolus aureus Roxb. *Zeitschrift für Naturforschung C*, *38*(9–10), pp. 693–697.

Sato, M., Tanaka, H., Fujiwara, S., Hirata, M., Yamaguchi, R., Etoh, H. and Tokuda, C., 2003. Antibacterial property of isoflavonoids isolated from Erythrina variegata against cariogenic oral bacteria. *Phytomedicine*, *10*(5), pp. 427–433.

Tanaka, H., Sato, M., Fujiwara, S., Hirata, M., Etoh, H. and Takeuchi, H., 2002. Antibacterial activity of isoflavonoids isolated from Erythrina variegata against methicillin-resistant Staphylococcus aureus. *Letters in Applied Microbiology*, *35*(6), pp. 494–498.

Tan, Q.W., Ni, J.C., Fang, P.H. and Chen, Q.J., 2017. A new erythrinan alkaloid glycoside from the seeds of Erythrina crista-galli. *Molecules*, *22*(9), p.1558.

Waffo, A.K., Azebaze, G.A., Nkengfack, A.E., Fomum, Z.T., Meyer, M., Bodo, B. and van Heerden, F.R., 2000. Indicanines B and C, two isoflavonoid derivatives from the root bark of Erythrina indica. *Phytochemistry*, *53*(8), pp. 981–985.

7.6.1.24 *Glycyrrhiza glabra* L.

Synonyms: *Glycyrrhiza alalensis* X.Y. Li; *Glycyrrhiza brachycarpa* Boiss.; *Glycyrrhiza glandulifera* Waldst. & Kit.; *Glycyrrhiza hirsuta* Pall.; *Glycyrrhiza pallida* Boiss.; *Glycyrrhiza violacea* Boiss. & Noë; *Liquiritia officinarum* Medik.

Common names: Licorice; yang gan cao (China); madhukan (India); mara, rishah-i-asl-i-sus (Iran); shalaco (Pakistan); dikenli meyan (Turkey)

Habitat: Roadsides, villages

Distribution: Afghanistan, Kazakhstan, Kyrgyzstan, Tajikistan, Turkmenistan, Uzbekistan Pakistan, India, China, and Mongolia

Botanical observation: This beautiful herb grows up to 1.5 m tall from a somewhat cylindrical, woody, and fragrant root. The leaves are imparipinnate, spiral, and stipulate. The stipules are linear, minute, and ephemeral. The rachis bears 5–8 pair of folioles (plus a terminal one), which are oblong to elliptic, rounded at base, round to acute at apex, and light dull green. The racemes are many-flowered, axillary, and about 10 cm long. The calyx is campanulate, 5–7 mm long, 5-lobed, the upper pair of lobes somewhat joined. The corolla is heavenly light purple, about 1 cm long and comprises 5 petals, including an oblong standard, a pair of wings, and a pair of petals forming a straight keel. The 10 stamens are about 5 mm long and diadelphous. The pods are flat, up to 3.5 cm long, and contain 2–8 seeds.

Medicinal uses: Cold, measles, ulcers (Iran); sore throat, fever (Pakistan); bronchitis, cough (Pakistan, Turkey)

Antibacterial (Gram-positive) halo developed by flavanone: Licoflavanone from the leaves inhibited the growth of *Bacillus cereus* and *Staphylococcus aureus* with the amounts of 32 and 64 µg/discs, respectively (Fukui et al., 1988).

Strong antibacterial (Gram-positive) amphiphilic isoflavans: Glabridin (LogD = 4.3 at pH 7.4; molecular mass = 324.3 g/mol) inhibited the growth of *Staphylococcus aureus* (ATCC 13709) with an MIC value of 6.2 µg/mL (Mitscher et al., 1980). Glabridin inhibited the growth of *Staphylococcus aureus* (MTCC 96), *Staphylococcus epidermidis* (MTCC 435), *Streptococcus mutans* (MTCC 890), *Bacillus subtilis* (MTCC121), *Enterococcus faecalis* (MTCC 439), *Klebsiella pneumoniae* (MTCC 109), *Salmonella typhi* (MTCC 733), *Yersinia enterocolitica* (MTCC 861), *Enterobacter aerogens* (MTCC111), and *Escherichia coli* (MTCC 723) with MIC values of 3.9, 7.5. 7.5, 15.6, 31.2, 250, 125, 250, 250, and 250 µg/mL, respectively (Gupta et al., 2008). Glabridin inhibited the growth of methicillin-sensitive *Staphylococcus aureus* (FDA 209P), methicillin-resistant *Staphylococcus aureus* (K3), *Micrococcus luteus* (ATCC 9341), *Bacillus subtilis* (PCI 219), and *Klebsiella pneumoniae* (PCI 602) with MIC values of 12.5, 12.5, 12.5, 6.2, and 50 µg/mL, respectively (Fukai et al., 2002).

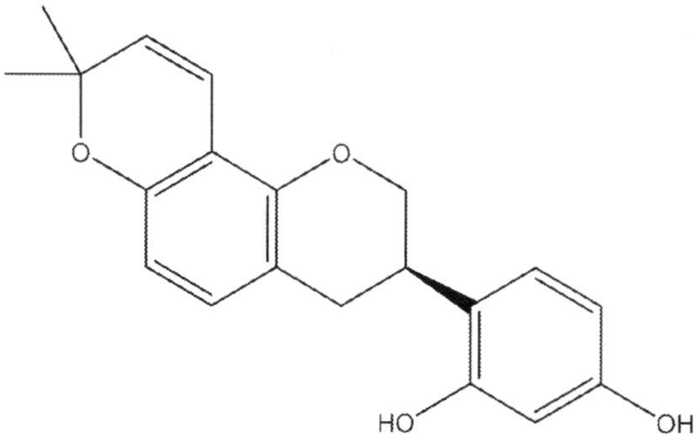

Glabridin

4′-*O*-Methylglabridin inhibited the growth of *Staphylococcus aureus* (ATCC 13709) with an MIC value of 6.2 µg/mL, respectively (Mitscher et al., 1980).

Hispaglabridin A (LogD=5.8 at pH 7.4; molecular mass=392.4 g/mol) inhibited the growth of *Staphylococcus aureus* (ATCC 13709) with an MIC value of 3.1 µg/mL (Mitscher et al., 1980).

Phaseollinisoflavan (LogD=4.6 at pH 7.4; molecular mass=324.3 g/mol) inhibited the growth of *Staphylococcus aureus* (ATCC 13709) with an MIC value of 25 µg/mL (Mitscher et al., 1980).

Strong antibacterial (Gram-positive) lipophilic isoflavane: Hispaglabridin B (LogD=6.4 at pH 7.4 molecular mass=390.4 g/mol) inhibited the growth of *Staphylococcus aureus* (ATCC 13709) with an MIC value of 6.2 µg/mL (Mitscher et al., 1980).

Grabrol (LogD=5.7; molecular mass=392.4 g/mol) inhibited the growth of *Staphylococcus aureus* (ATCC 13709) with an MIC value of 1.5 µg/mL (Mitscher et al., 1980). 3-Hydroxyglabrol (LogD=5.2 at pH 7.4; molecular mass=408.4 g/mol) inhibited the growth of *Staphylococcus aureus* (ATCC 13709) with an MIC value of 6.2 µg/mL (Mitscher et al., 1980).

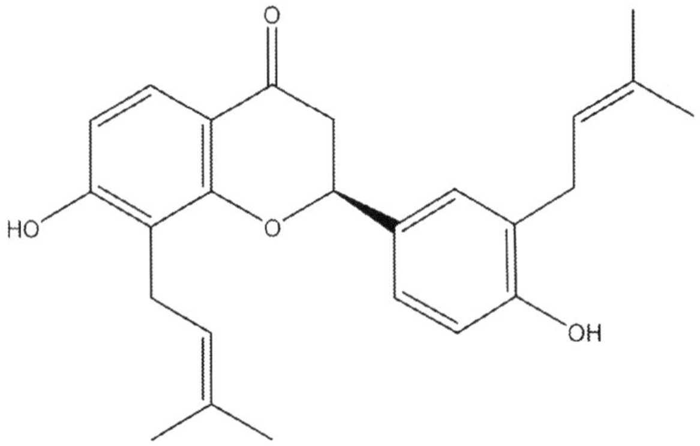

Glabrol

Strong antibacterial (Gram-positive) amphiphilic 3-phenyl coumarin: Glycycoumarin (LogD=4.2 at pH 7.4; molecular mass=368.3 g/mol) from the roots inhibited the growth of *Streptococcus mutans*, *Staphylococcus aureus*, and *Bacillus subtilis* (Demizu et al., 1988). Glycycoumarin inhibited the growth of *Streptococcus pyogenes* (ATCC 12344), *Haemophilus influenzae* (ATCC 33391), and *Moraxella catarrhalis* (ATCC 25238) with MIC values of 25, 25, and 100 µg/mL, respectively (two-fold serial dilution method) (Tanaka et al., 2001).

Strong antibacterial (Gram-positive) lipophilic chromene: Glabrene (LogD=5.0 at pH 7.4; molecular mass=322.3 g/mol) inhibited the growth of *Staphylococcus aureus* (ATCC 13709) with an MIC value of 25 µg/mL (Mitscher et al., 1980). Glabrene inhibited methicillin-sensitive *Staphylococcus aureus* (FDA 209P), methicillin-resistant *Staphylococcus aureus* (K3), *Micrococcus luteus* (ATCC 9341), *Bacillus subtilis* (PCI 219), and *Klebsiella pneumoniae* (PCI 602) with MIC values of 12.5, 12.5, 25, 12.5, and 100 µg/mL, respectively (Fukai et al., 2002).

Strong antibacterial (Gram-positive) amphiphilic 2-arylbenzofuran neolignan: Licocoumarone (LogD=3.0 at pH 7.4; molecular mass=340.3 g/mol) from the roots inhibited the growth of *Streptococcus mutans*, *Staphylococcus aureus*, and *Bacillus subtilis* with MIC values of 12.5, 6.2, and 6.2 µg/mL, respectively (Demizu et al., 1988).

Strong antibacterial amphiphilic chalcone: Licochalcone A (LogD=4.3 at pH 7.4; molecular mass=338.3 g/mol) inhibited the growth of *Bacillus subtilis* (IFO 3060), *Staphylococcus aureus* (IFO 3007), and *Micrococcus luteus* (IFO 3333) with MIC values of 3.1, 1.5, and 1.5 µg/mL, respectively. Licochalcone A inhibited respiratory activity and bacterial membrane NADH oxidase in *Micrococcus luteus* (IFO 3333) and *Staphylococcus aureus* (IFO 12732). Licochalcone C (LogD=4.0 at pH 7.4; molecular mass=338.3 g/mol) inhibited the growth of *Bacillus subtilis* (IFO 3060), *Staphylococcus aureus* (IFO 3007), and *Micrococcus luteus*

(IFO 3333) with MIC values of 12.5, 6.2, and 6.2 μg/mL, respectively (Haraguchi et al., 1998). Licochalcone C inhibited the respiratory activity of *Micrococcus luteus* (IFO 3333) and *Staphylococcus aureus* (IFO 12732) with IC_{50} values of 53.3 and 58.6 μM, respectively (Haraguchi et al., 1998). Licochalcone A and C inhibited NADH-cytochrome c reductase with IC_{50} values of 14.8 and 29.6 μM, respectively, in the membrane fraction of *Micrococcus luteus* (Haraguchi et al., 1998). Licochalcone A inhibited the growth of *Bacillus cereus* (IFO 3514), *Bacillus subtilis* (IFO 3007), *Clostridium sporogenes* (IFO 13950), *Enterococus faecium* (IFO 3826), *Staphylococcus aureus* (209P IFO 12732), and *Streptococcus mutans* (IFO 13955) (Tsukiyama et al., 2002) Licochalcone A inhibited methicillin-sensitive *Staphylococcus aureus* (FDA 209P), methicillin-resistant *Staphylococcus aureus* (K3), *Micrococcus luteus* (ATCC 9341), and *Bacillus subtilis* (PCI 219) with MIC values of 3.1, 6.2, 6.2, and 6.2 μg/mL, respectively (Fukai et al., 2002).

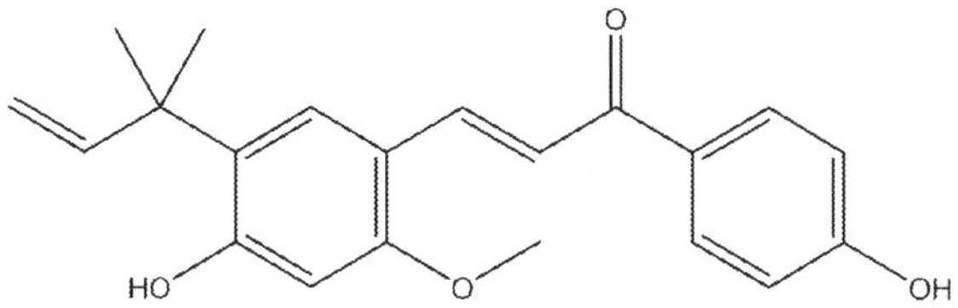

Licochalcone A

Moderate antibacterial (Gram-positive) amphiphilic isoflavan: 3′-Methoxyglabridin inhibited *Staphylococcus aureus* (ATCC 13709) with MIC values of 50 μg/mL (Mitscher et al., 1980).

Strong antimycobacterial amphiphilic isoflavans: Glabridin inhibited the growth of *Mycobacterium smegmatis* (ATCC 607) with an MIC value of 6.2 μg/mL (Mitscher et al., 1980) and inhibited *Mycobacterium tuberculosis* (H37Ra) and *Mycobacterium tuberculosis* (H37Rv) with MIC values of 29.1 and 29.1 μg/mL (Mitscher et al., 1980). 4′-*O*-Methylglabridin inhibited *Mycobacterium smegmatis* (ATCC 607) with MIC value of 3.1 μg/mL (Mitscher et al., 1980). Phaseollinisoflavan inhibited the growth of *Mycobacterium smegmatis* (ATCC 607) with an MIC value 12.5 μg/mL, (Mitscher et al., 1980). Hispaglabridin A inhibited the growth of *Mycobacterium smegmatis* (ATCC 607) with an MIC value 3.1 μg/mL (Mitscher et al., 1980).

Strong antimycobacterial lipophilic isoflavan: Hispaglabridin B inhibited the growth of *Mycobacterium smegmatis* (ATCC 607) with an MIC value of 3.1 μg/mL (Mitscher et al., 1980).

Strong antimycobacterial amphiphilic flavone: 3-Hydroxyglabrol inhibited the growth of *Mycobacterium smegmatis* (ATCC 607), with an MIC value of 6.2 μg/mL (Mitscher et al., 1980).

Strong antimycobacterial lipophilic prenylated flavanone: Glabrol inhibited the growth of *Mycobacterium smegmatis* (ATCC 607) with an MIC value of 1.5 μg/mL (Mitscher et al., 1980).

Strong antimycobacterial lipophilic chromene: Glabrene inhibited the growth of *Mycobacterium smegmatis* (ATCC 607) with MIC value of 25 μg/mL (Mitscher et al., 1980).

Anticandidal halo developed by flavanone: Licoflavanone from the leaves inhibited the growth of *Candida albicans* with the amount of 250 μg/disc (Fukui et al., 1988).

Strong antifungal (yeast) amphiphilic 3-phenyl coumarin: Glycycoumarin inhibited the growth of *Saccharomyces cerevisiae*, *Candida utilis*, and *Pichia nakazawae* with MIC values of 25, 50, and 25 μg/mL, respectively (Demizu et al., 1988).

Strong broad-spectrum antifungal amphiphilic 2-arylbenzofuran neolignan: Licocoumarone (LogD=3.0 at pH 7.4; molecular mass=340.3 g/mol) from the roots inhibited the growth of *Saccharomyces cerevisae*, *Candida utilis*, *Pichia nakazawae*, and *Rhizopus formosaensis* with MIC values of 25, 25, 25, and 50 μg/mL, respectively (Demizu et al., 1988).

Strong antifungal (filamentous) chalcone: Licochalcone A inhibited the growth of *Mucor pusillus* (HUT 1185) with an MIC value of 12.5 µg/mL (Haraguchi et al., 1998). Licochalcone C isolated from the roots inhibited the growth of *Mucor pusillus* (HUT 1185) with an MIC value of 50 µg/mL (Haraguchi et al., 1998).

Very weakly antiviral (enveloped linear double-stranded DNA) amphiphilic oleanane-type saponin: Glycyrrhizin at a concentration of 8 mM added either immediately after infection or 3 hours post infection to infected HEp-2 cells inhibited the replication and cytopathic effects of the vaccinia virus and the Herpes simplex virus type-1 (Pompei et al., 1979). Glycyrrhizic acid also produced irreversible inactivation of the Herpes simplex virus type-1 (Pompei et al., 1979) via probable viral envelop destabilization.

Very weakly antiviral (enveloped monopartite linear single-stranded (−)RNA) amphiphilic oleanane-type saponin: Glycyrrhizin at a concentration of 8 mM added either immediately after infection or 3 hours post infection to infected HEp-2 cells inhibited the replication and cytopathic effects of Newcastle disease virus and the Vesicular stomatitis virus (Indiana type). It had no effect on Poliovirus type 1 (Pompei et al., 1979).

Very weak antiviral (enveloped monopartite linear singl-stranded (+)RNA) amphiphilic oleanane-type saponin: Glycyrrhizin at 25 µM inhibited by 34% the replication of the Feline coronavirus (FIPV WSU 79-1146) (McDonagh et al., 2014). Glycyrrhizin inhibited Severe acute respiratory syndrome-acquired coronavirus (strain FFM1 was isolated from respiratory specimens of a SARS patient) replication in Vero Cells with an IC_{50} value of 365 µM and a selectivity index above 65 (Hoever et al., 2005).

Commentaries: (i). The presence of a hydrophobic moiety in licochalcone A would provide sufficient hydrophobicity to the molecule to penetrate bacterial membranes (Tsukiyama et al., 2002). (ii) The virus replication cycle includes 6 stages: attachment to host cell, fusion with the cytoplasmic membrane of the host cell, uncoating of the virus and release of the viral contents in the host cell, building of new viral genome and macromolecules using the host cell, assembly of viral genome and macromolecules into daughter viruses, and release of daughter viruses, which most often induce host cell death. Double-stranded DNA viruses will use the host cell machinery to replicate DNA and produce macromolecules. Single-stranded (+)-RNA genome of single-stranded (+)-RNA viruses is translated directly to make viral proteins by host cell ribosomes. If the single-stranded (+)-RNA viruses carry the reverse transcriptase enzyme, a complementary single-stranded DNA is produced, which is made into double-stranded DNA, which can integrate into the host DNA permanently. Single-stranded (−)-RNA genome of single-stranded (−)-RNA viruses is not translatable by the host cell and the viral RNA-dependent RNA polymerase (RdRP) is used to make single-stranded (+)-RNA. Likewise, for double-stranded RNA viruses, the RNA-dependent RNA polymerase uses the negative strand of the double-stranded RNA to generate single-stranded (+)-RNA (Lodish et al., 2000; Vladimir Zarubaev personal communication). Therefore, by observing the spectrum of antiviral activity of a natural product, one could deduce a type of antiviral mechanism and/or viral target. For instance, if a compound only targets enveloped DNA and RNA viruses it could point to an antiviral mechanism, involving at least in part, the outer envelope of viruses.

(iii) Flavonoids are actively expelled from the cytoplasm of Gram-negative bacteria via efflux pumps, the expression of which is induced at the gene level by flavonoids (Vargas et al., 2011).

(iv) The plant contains pinocembrin (Fukui et al., 1988).

(v) *Glycyrrhiza uralensis* Fisch. ex DC. is used in China for the treatment of bronchitis, and from its roots the lipophilic isoflavan licoricidin (LogD=5.8 at pH 7.4; molecular mass=424.5 g/mol) inhibited the growth of *Streptococcus pyogenes* (ATCC 12344), *Haemophilus influenzae* (ATCC 33391), and *Moraxella catarrhalis* (ATCC 25238) with MIC values of 12, 12, and 12 µg/mL, respectively (Tanaka et al., 2001). Licoricidin from *Glycyrrhiza uralensis* Fisch. ex DC. inhibited methicillin-sensitive *Staphylococcus aureus* (FDA 209P), methicillin-resistant *Staphylococcus aureus* (K3), *Micrococcus luteus* (ATCC 9341), and *Bacillus subtilis* (PCI 219) with MIC values of 3.1, 6.2, 6.2, and 3.1 µg/mL, respectively (Fukai et al., 2002). Glyasperin D from *Glycyrrhiza uralensis* Fisch. ex DC. inhibited methicillin-sensitive *Staphylococcus aureus* (FDA 209P), methicillin-resistant *Staphylococcus aureus* (K3), *Micrococcus luteus* (ATCC 9341), and *Bacillus subtilis* (PCI 219) with MIC values of 6.2, 6.2, 12.5, and 5.2 μg/mL, respectively (Fukai et al., 2002).

The 3-phenyl coumarin glycyrin (LogD=4.9 at pH 7.4; molecular mass=382.4 g/mol) from the roots of *Glycyrrhiza uralensis* Fisch. ex DC. inhibited the growth of *Streptococcus pyogenes* (ATCC 12344), *Haemophilus influenzae* (ATCC 33391), and *Moraxella catarrhalis* (ATCC 25238) with MIC values of 25, 25, and 50 μg/mL, respectively (Tanaka et al., 2001).

The 2-arylbenzofuran neolignan gancaonin I (LogD=4.1 at pH 7.4; molecular mass=354.3 g/mol) from *Glycyrrhiza uralensis* Fisch. ex DC. inhibited methicillin-sensitive *Staphylococcus aureus* (FDA 209P), methicillin-resistant *Staphylococcus aureus* (K3), *Micrococcus luteus* (ATCC 9341), and *Bacillus subtilis* (PCI 219) with MIC values of 3.1, 1.5, 3.1, and 3.1 μg/mL, respectively (Fukai et al., 2002).

Glycyrrhisol A, glycyrrhizol B, 6,8-diisoprenyl-5,7,4′-trihydroxyisoflavone isolated from the roots of *Glycyrrhiza uralensis* inhibited *Streptococcus mutans* (ATCC 25175) with MIC values of 1, 32, and, 2, μg/mL, respectively (He te al., 2006).

(vi) *Glycyrrhiza glabra* L. is an example of a medicinal plant generating a broad-spectrum of antimicrobial agents. The roots of this plant were listed in the "*De Materia Medica*" of our peer the Roman Physician Pedanius Dioscorides (40-90 AD). "*De Materia Medica*" has been used for medicine in Europe until about the 19th century. The roots have been listed in European Pharmacopoeia as a cough remedy and aromatic material without apparently not much toxicity in humans is consumed with moderation (Fenwick et al., 1990). In China, it is used for the treatment of COVID-19 in combination with oseltamivir phosphate (Luo, 2020), exemplifying that the zoonotic pandemic waves that are to come will be better resisted when Western medicine and Eastern medicines will work synergistically. In the West there is a need to make laws to permit affordable clinical trials and patenting of medicinal plant products.

(vii) *Glycyrrhiza echinata* L. is used in Turkey (local name: *Dikenli meyan*) to treat bronchitis and cough.

REFERENCES

Demizu, S., Kajiyama, K., Takahashi, K., Hiraga, Y., Yamamoto, S., Tamura, Y., Okada, K. and Kinoshita, T., 1988. Antioxidant and antimicrobial constituents of licorice: isolation and structure elucidation of a new benzofuran derivative. *Chemical and Pharmaceutical Bulletin*, *36*(9), pp. 3474–3479.

Fenwick, G.R., Lutomski, J. and Nieman, C., 1990. Liquorice, Glycyrrhiza glabra L.—Composition, uses and analysis. *Food Chemistry*, *38*(2), pp. 119–143.

Fukai, T., Marumo, A., Kaitou, K., Kanda, T., Terada, S. and Nomura, T., 2002. Antimicrobial activity of licorice flavonoids against methicillin-resistant Staphylococcus aureus. *Fitoterapia*, *73*(6), pp. 536–539.

Fukui, H., GOTO, K. and TABATA, M., 1988. Two antimicrobial flavanones from the leaves of Glycyrrhiza glabra. *Chemical and Pharmaceutical Bulletin*, *36*(10), pp. 4174–4176.

Gupta, V.K., Fatima, A., Faridi, U., Negi, A.S., Shanker, K., Kumar, J.K., Rahuja, N., Luqman, S., Sisodia, B.S., Saikia, D. and Darokar, M.P., 2008. Antimicrobial potential of Glycyrrhiza glabra roots. *Journal of Ethnopharmacology*, *116*(2), pp. 377–380.

Haraguchi, H., Tanimoto, K., Tamura, Y., Mizutani, K. and Kinoshita, T., 1998. Mode of antibacterial action of retrochalcones from Glycyrrhiza inflata. *Phytochemistry*, *48*(1), pp. 125–129.

He, J., Chen, L., Heber, D., Shi, W. and Lu, Q.Y., 2006. Antibacterial Compounds from Glycyrrhiza u ralensis. *Journal of Natural Products*, *69*(1), pp. 121–124.

Hoever, G., Baltina, L., Michaelis, M., Kondratenko, R., Baltina, L., Tolstikov, G.A., Doerr, H.W. and Cinatl, J., 2005. Antiviral Activity of Glycyrrhizic Acid Derivatives against SARS– Coronavirus. *Journal of Medicinal Chemistry*, *48*(4), pp. 1256–1259.

Lodish, H., Berk, A., Zipursky, S.L., Matsudaira, P., Baltimore, D. and Darnell, J., 2000. Viruses: Structure, function, and uses. In *Molecular Cell Biology. 4th edition*. WH Freeman.

Luo, A., 2020. Positive Severe acute respiratory syndrome-acquired coronavirus 2 test in a woman with COVID-19 at 22 days after hospital discharge: A case report. *Journal of Traditional Chinese Medical Sciences*.

McDonagh, P., Sheehy, P.A. and Norris, J.M., 2014. Identification and characterisation of small molecule inhibitors of feline coronavirus replication. *Veterinary Microbiology*, *174*(3–4), pp. 438–447.

Mitscher, L.A., Park, Y.H., Clark, D. and Beal, J.L., 1980. Antimicrobial agents from higher plants. Antimicrobial isoflavanoids and related substances from Glycyrrhiza glabra L. var. typica. *Journal of Natural Products*, *43*(2), pp. 259–269.
Pompei, R., Flore, O., Marccialis, M.A., Pani, A. and Loddo, B., 1979. Glycyrrhizic acid inhibits virus growth and inactivates virus particles. *Nature*, *281*(5733), p.689
Tanaka, Y., Kikuzaki, H., Fukuda, S. and Nakatani, N., 2001. Antibacterial compounds of licorice against upper airway respiratory tract pathogens. *Journal of Nutritional Science and Vitaminology*, *47*(3), pp. 270–273.
Tsukiyama, R.I., Katsura, H., Tokuriki, N. and Kobayashi, M., 2002. Antibacterial activity of licochalcone A against spore-forming bacteria. *Antimicrobial Agents and Chemotherapy*, *46*(5), pp. 1226–1230.
Vargas, P., Felipe, A., Michán, C. and Gallegos, M.T., 2011. Induction of Pseudomonas syringae pv. tomato DC3000 MexAB-OprM multidrug efflux pump by flavonoids is mediated by the repressor PmeR. *Molecular Plant-Microbe Interactions*, *24*(10), pp. 1207–1219.
Wang, S.X., Wang, Y., Lu, Y.B., Li, J.Y., Song, Y.J., Nyamgerelt, M. and Wang, X.X., 2020. Diagnosis and treatment of novel coronavirus pneumonia based on the theory of traditional Chinese medicine. *Journal of Integrative Medicine*.

7.6.1.25 *Indigofera tinctoria* L.

Synonyms: *Anila tinctoria* (L.) Kuntze; *Indigofera indica* Lam.; *Indigofera sumatrana* Gaertn.

Common names: Common indigo; nildare (Bangladesh); neel (Bangladesh; India); jhunjhunia, nilayamari (India); basma, rang-i-vasmah (Iran); khram (Thailand); wasma (Turkey, Iran, Pakistan)

Habitat: Cultivated

Distribution: Pakistan, India, Sri Lanka, Myanmar, Thailand,

Botanical observation: This graceful herb grows to about 1 m tall. The stems are hairy. The leaves are imparipinnate, alternate, and stipulate. The stipules are minute and triangular. The blade is up to about 10 cm long and comprises 4–6 pairs of folioles plus a terminal one. The folioles are obovate-oblong 1.5–3 × 0.5–1.5 cm, broadly cuneate to rounded at base, rounded to emarginated at apex, dull green, and membranous. The inflorescences are axillary racemes which grow up to about 10 cm long. The calyx is minute, campanulate, cup-shaped and 5-lobed. The 5 petals are of a beautiful light purple, reddish to pink petals of which a broadly obovate and a 5 mm long standard. The androecium is diadelphous and includes 10 stamens. The ovary develops a capitate stigma. The pods are 2.5–3 cm long, linear, and containing 5–12 cubic and minute seeds.

Medicinal uses: Bronchitis, cough, hepatitis (India); fever (Thailand); ulcers (Bangladesh)

Antifungal (filamentous) amphiphilic indoloquinazoline alkaloid: Tryptanthrin (LogD=2.4 at pH 7.4; molecular mass=248.2 g/mol) inhibited the growth of *Trichophyton mentagrophytes*, *Trichophyton rubrum*, *Trichophyton tonsurans* var. *sulfureum*, *Microsporum canis*, *Microsporum gypseum*, and *Epidermophyton floccosum* with MIC values of 3.1, 3.1, 3.1, 3.1, 6.3, and 3.1 μg/mL, respectively (Honda &Tabata, M., 1979).

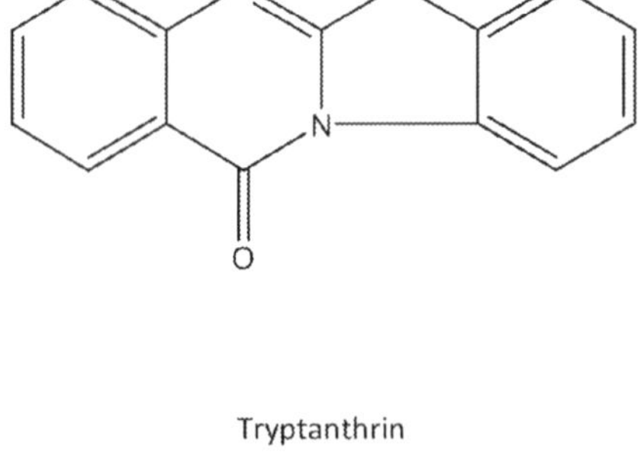

Tryptanthrin

Strong antiviral (enveloped segmented single-stranded (+)RNA) amphiphilic indoloquinazoline alkaloid: Tryptanthrin inhibited the replication of Human coronavirus NL63 in LCC-MK2 cells with an IC_{50} value of 1.5 μM and acted at the early stage with an IC_{50} value of 0.3 μM via inhibition of RNA synthesis and papain-like protease 2 (PLP2) (Tsai et al., 2010).

Commentaries: (i) By observing the planar structure of tryptanthrin one could hypothesize its interference with *(+)*RNA translation by host cell ribosomes. (ii) The plant was known for our peer Ibn Sina also known as Avicenna (980-1037). Consider that the medical writings of Avicenna where among the pillars of European medieval medicine. Avicenna recommended applying freshy cut chicken of plague buboes and that technique has been used in Europe until about the 16th century (Heinrich, 2017). As for the virtues of chicken to fight pestilence, one could read the "*Pharmacopée royale galenique et chymique.*" of the French physician Moyse Charas (1619-1698) published in 1717 where chicken soup is recommended for flu, a recommendation that seems to be beneficial (Rosner, 1994). (iii) In India, *Indigofera aspalathoides* Vahl ex DC. is used (local name: *Shivanar vembu*) to treat abscesses, leprosy, mouth ulcers; *Indigofera mysorensis* Rottler ex DC (local name: *Kondavempali*) is used to heal sores; *Indigofera oblongifolia* Forssk (local name: *Bana nila*) is used to treat syphilis; and *Indigofera enneaphylla* Eckl. & Zeyh. is used to treat fever and cough.

REFERENCES

Heinrichs, E.A., 2017. The live chicken treatment for buboes: trying a plague cure in medieval and early modern Europe. *Bulletin of the History of Medicine, 91*(2), pp.210–232.

Honda, G. and Tabata, M., 1979. Isolation of antifungal principle tryptanthrin, from Strobilanthes cusia O. Kuntze. *Planta Medica, 36*(05), pp. 85–86.

Rosner, F., 1994. The Common Cold: The Effect of Hot Humid Air on the Nasal Mucosa. *JAMA, 272*(14), pp.1104–1104

Tsai, Y.C., Lee, C.L., Yen, H.R., Chang, Y.S., Lin, Y.P., Huang, S.H. and Lin, C.W., 2020. Antiviral Action of Tryptanthrin Isolated from Strobilanthes cusia Leaf against Human Coronavirus NL63. *Biomolecules, 10*(3), p.366.

7.6.1.26 *Lablab purpureus* (L.) Sweet

Synonyms: *Dolichos albus* Lour.; *Dolichos bengalensis* Jacq.; *Dolichos lablab* L.; *Dolichos purpureus* L.; *Lablab lablab* (L.) Lyons; *Lablab niger* Medik.; *Lablab vulgaris* Savi; *Lablab vulgaris* var. *albiflorus* DC.; *Vigna aristata* Piper

Common names: Banner bean, Indian butter bean; bun shim (Bangladesh); bian dou (China); chota sim (India); lubiya gul (Iran); nwai-pe (Myanmar)

Habitat: Cultivated

Distribution: Tropical Asia

Botanical observation: This climber grows to about 5 m long. The stems are purplish. The leaves are imparipinnate, alternate, and stipulate. The stipules are lanceolate. The folioles are deltoid-ovate, 6–10 × 6–10 cm, membranous, lateral ones asymmetrical, acute or wedge-shaped at base, and acute or acuminate at apex. The racemes are erect, axillary, and 15–25 cm long. The calyx is bilobed and 6 mm long. The corolla is papilionaceous, about 1.5 cm long, white or purple, and the standard is orbicular. The ovary is linear. The pods are oblong-falcate, 5–7 × 1.4–1.8 cm, glossy, purple, beaked at apex, and contain 3 to 5 seeds. The seeds are ovoid, compressed, and edible.

Medicinal uses: Cholera, diarrhea, gonorrhea (China); fever (India, Myanmar)

Antibacterial (Gram-positive) halo developed by polar extract: Methanol extract of leaves and flowers (20 mg/mL solution/6 mm well) inhibited the growth of *Staphylococcus aureus* (clinical isolates) (Priya & Jenifer, 2014).

Commentary: The Tribe Phaseoleae is often a source of medicinal plants for the treatment of infectious diseases in Asia and Pacific, for example, *Psophocarpus tetragonolobus* DC. is used in Malaysia to treat small pox. In this Tribe, an aqueous extract of roots *Spatholobus suberectus* Dunn at a concentration of 250 μg/mL inhibited the Human immunodeficiency virus type-1 protease by 46.6% (Xu et al., 1996).

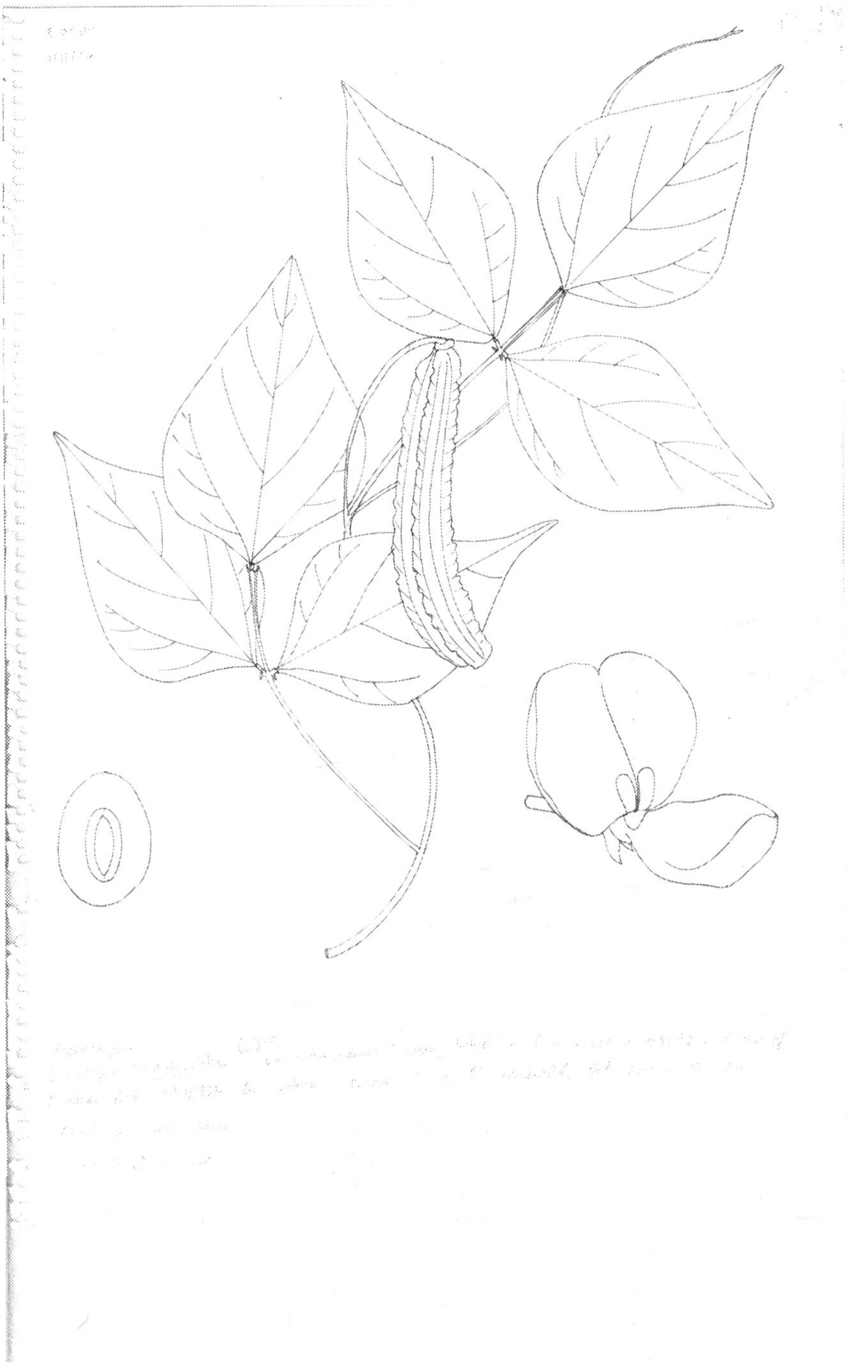

FIGURE 7.21 *Psophocarpus tetragonolobus* DC

REFERENCES

Priya, S. and Jenifer, S., 2014. Antibacterial activity of leaf and flower extract of Lablab purpureus against clinical isolates of Staphylococcus aureus. *Research & Reviews: A Journal of Drug Design & Discovery*, *1*(2), pp. 5–7.

Xu, H.X., Wan, M., Loh, B.N., Kon, O.L., Chow, P.W. and Sim, K.Y., 1996. Screening of Traditional Medicines for their Inhibitory Activity Against HIV-1 Protease. *Phytotherapy Research*, *10*(3), pp. 207–210.

7.6.1.27 *Lupinus albus* L.

Synonym: *Lupinus termis* Forssk

Common names: White lupine, wolf bean; bai yu shan dou (China); turmus (India); acı bakla, Yahudi baklası (Turkey)

Habitat: Cultivated

Distribution: Temperate and sub-temperate Asia

Botanical observation: This beautiful herb native to Turkey grows to a height of 2 m. The stems are hairy and terete. The leaves are palmately compound, spiral, and stipulate. The stipules are adnate. The rachis bears up to 9 folioles, which are obovate, mucronate, hairy below, and 2.5–3.5 × 1.4–1.8 cm. The inflorescence is a terminal raceme. The calyx is tubular, bilobed, and up to about 1 cm long. The corolla is up to 1.5 cm long, white to blue and develops a beaked keel. The androecium includes 10 stamens. The style is curved. The pods are somewhat cylindrical, constricted up, to about 10 cm long, and contain up to 6 seeds, which are flat and smooth.

Medicinal use: Ulcers (India)

Antibacterial (Gram-negative) halo developed by quinolizidine alkaloids: Quinolizidine alkaloids fraction extracted from the seeds (50 μL of 25mg/mL solution/well) inhibited the growth of *Escherichia coli* and *Klebsiella pneumoniae* with inhibition zone diameters of 10 and 15 mm, respectively (El-Shazly et al., 2001).

Broad-spectrum antifungal halo developed by quinolizidine alkaloids: Quinolizidine alkaloids fraction (50 μL of 25 mg/mL in agar well) extracted from the seeds inhibited the growth of *Candida albicans* and *Aspergillus flavus* with inhibition zone diameters of 17 and 14 mm, respectively (El-Shazly et al., 2001).

Strong antifungal amphiphilic isoflavone: Luteone (LogD = 3.0 at pH 7.4; molecular mass = 354.3 g/mol) inhibited the growth of *Helminthosporium carbonum* with an ED_{50} value of about 40 μg/mL (Harborne et al., 1976).

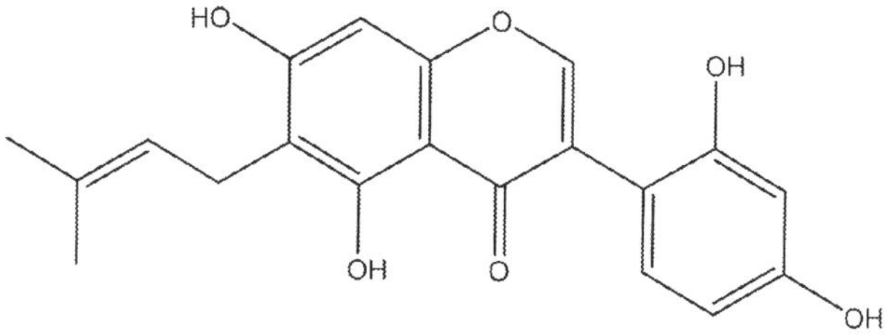

Luteone

Commentary: (i) The plant produces quinolizidine alkaloids (Wink & Witte, 1984), such as lupanine and sparteine, which have affinity for cholinergic receptors (Yovo et al., 1984) and as such could be able to inhibit bacterial and/or fungal efflux pumps. (ii) *Lupinus luteus* (local name: *Tormes*) is hot and dry in Unani medicine. It is used in Iran to expel intestinal worms suggesting the presence of neuroactive and thus possibly antibiotic/antifungal potentiating principles. (iii) The plants generate luteone, a phytoanticipin (also called prohibitin), as well as a series of antifungal isoflavoids (Tahara et al., 1984).

REFERENCES

El-Shazly, A., Ateya, A.M.M. and Wink, M., 2001. Quinolizidine Alkaloid Profiles of Lupinus varius orientalis, L. albus albus, L. hartwegii, and L. densiflorus. *Zeitschrift für Naturforschung C*, *56*(1–2), pp. 21–30.

Harborne, J.B., Ingham, J.L., King, L. and Payne, M., 1976. The isopentenyl isoflavone luteone as a pre-infectional antifungal agent in the genus Lupinus. *Phytochemistry*, *15*(10), pp. 1485–1487.

Tahara, S., Ingham, J.L., Nakahara, S., Mizutani, J. and Harborne, J.B., 1984. Fungitoxic dihydrofuranoisoflavones and related compounds in white lupin, Lupinus albus. *Phytochemistry*, *23*(9), pp. 1889–1900.

7.6.1.28 *Medicago sativa* L.

Common names: Alfalfa, lucerne; spistha (Afghanistan); ilaayatigawuth (India); ishpith, lasan (Pakistan); karayonca (Turkey)

Medicinal uses: Boils (Iran); wounds (Turkey)

Habitat: Cultivated

Distribution: Turkey, Afghanistan, Iran

Botanical observation: This graceful herb grows up to a height of 60cm tall. The leaves are trifoliolate, spiral, and stipulate. The stipule is about 1 cm long. The petiole is about 3 cm long. The folioles are 3–15 mm×8–30 mm, dull green, elliptic to obovate, and serrate at apex. The racemes are axillary, many flowered, and on an about 5cm long peduncle. The calyx is about 5 mm long, tubular, and develops 5 lobes. The corolla is about 1 cm long and presents 5 petals, which are light purple. The androecium is diadelphous and consists of 10 stamens. The pods are characteristically spiral, indehiscent, and shelter up to 20 tiny seeds.

Very weak antifungal (filamentous) pterocarpan: Medicarpin (at a concentration of 500 µM) inhibited the growth of *Phytophtora megasperma* and *Phoma medicaginis* (Blount et al., 1992).

Strong broad-spectrum antifungal oleanane-type triterpene saponins: 2-β-hydroxy-3-β-*O*-(β-D-glucopyranosyl)-Δ^{12}-Oleanene-23,28-dionic acid (medicagenic acid 3-0- β -D-glucopyranoside), isolated from the roots inhibited the growth of *Cryptococcus neoformans* with an MIC value of 2 µg/mL (Polacheck et al., 1986). Medicagenic acid 3-*O*-β-D-glucopyranoside inhibited the growth of *Candida tropicalis, Candida pseudotropicalis, Candida krusei, Candida albicans, Torulopsis glabrata, Candida parapsilosis, Torulopsis candida, Candida guilliermondii*, and *Geotrichum candidum* (Polacheck et al., 1986a). Medicagenic acid 3-*O*-β-D-glucopyranoside in *Saccharomyces cerevisiae* caused lethal leakage of ions by targeting cytoplasmic membrane ergosterol (Polacheck et al., 1991).

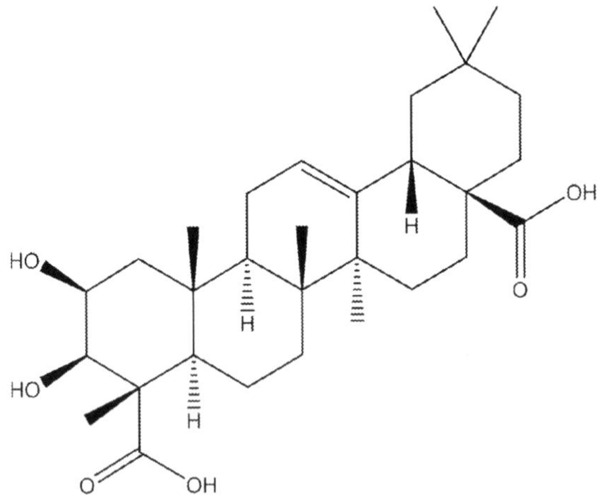

Medicagenic acid

Commentaries: (i) The plant abounds with triterpene saponins (Tava et al., 1993). Under phytopathogenic attack, *Medicago sativa* L. generates the phytoalexin medicarpin from phenylalanine via dadzein, formonotenin, and vestinone (Blount et al., 1992). (ii) In India, *Medicago denticulata* Willd. is used to treat bronchitis. (iii) Consider that oil from seeds of *Trigonella foenum-graecum* (which belongs to the Tribe Trifolieae) inhibited the growth of *Staphylococcus aureus*, *Bacillus subtilis*, *Enterococcus faecalis*, *Listeria monocytogenes*, *Escherichia coli*, *Pseudomonas aeruginosa*, *Salmonella typhi*, and *Shigella dysenteriae* with MIC values of 31.5, 15.6, 31.2, 62.5, 62.5, 62., 62.5, and 125 μg/mL, respectively (Aman et al., 2014).

REFERENCES

Aman, S., Naim, A., Siddiqi, R. and Naz, S., 2014. Antimicrobial polyphenols from small tropical fruits, tea and spice oilseeds. *Revista de Agaroquimica y Tecnologia de Alimentos*, *20*(4), pp. 241–251.

Blount, J.W., Dixon, R.A. and Paiva, N.L., 1992. Stress responses in alfalfa (Medicago sativa L.) XVI. Antifungal activity of medicarpin and its biosynthetic precursors; implications for the genetic manipulation of stress metabolites. *Physiological and Molecular Plant Pathology*, *41*(5), pp. 333–349.

Polacheck, I., Zehavi, U., Naim, M., Levy, M. and Evron, R., 1986. The susceptibility of Cryptococcus neoformans to an antimycotic agent (G2) from alfalfa. *Zentralblatt für Bakteriologie, Mikrobiologie und Hygiene. Series A: Medical Microbiology, Infectious Diseases, Virology, Parasitology*, *261*(4), pp. 481–486.

Polacheck, I.T.Z.H.A.C.K., Zehavi, U., Naim, M., Levy, M. and Evron, R., 1986a. Activity of compound G2 isolated from alfalfa roots against medically important yeasts. *Antimicrobial agents and chemotherapy*, *30*(2), pp. 290–294.

Polacheck, I., Levy, M., Guizie, M., Zehavi, U., Naim, M. and Evron, R., 1991. Mode of action of the antimycotic agent G2 isolated from alfalfa roots. *Zentralblatt für Bakteriologie*, *275*(4), pp. 504–512.

Tava, A., Oleszek, W., Jurzysta, M., Berardo, N. and Odoardi, M., 1993. Alfalfa saponins and sapogenins: isolation and quantification in two different cultivars. *Phytochemical Analysis*, *4*(6), pp. 269–274.

7.6.1.29 *Moghania strobilifera* (L.) J. St.-Hil. ex O. Kuntze

Synonyms: *Flemingia bracteata* (Roxb.) Willd., *Flemingia fruticulosa* Wall ex. Benth., *Flemingia fruticulosa* Wall., *Flemingia strobilifera* (L.) Roxb. ex W.T. Aiton, *Hedysarium bracteatum* Roxb., *Hedysarium strobiliferum* L., *Moghania bracteata* (Roxb.) H.L. Li, *Moghania fruticulosa* (Wall.) Mukerjee, *Zornia strobilifera* (L.) Pers.

Common names: Wild Hops; makhiati (India); apa-apa kebo, siringan (Indonesia) hahapaan (Indonesia); gaan (Malaysia); se laik pya (Myanmar); pa-apa kebo, arana (Papua New Guinea); payang-payang (the Philippines); khee dang (Thailand); duôi chồn (Vietnam)

Habitat: Open forests, roadsides, dry wastelands, savanna

Distribution: India, Southeast Asia, south China, and the Pacific Islands

Botanical observation: This shrub grows up to 3m tall. The stem is hairy at apex. The leaves are simple, spiral, and stipulated and used in Papua New Guinea to ward off evil spirits(!). The stipule is 0.3 cm long and hairy. The petiole is straight, 1–2.5cm long, hairy, channeled, and slightly inflated at both ends. The blade is 7–13 × 3–7 cm, membranous, subglabrous underneath, broadly lanceolate, minutely cordate at base, acute at apex, and shows 7–12 pairs of straight secondary nerves. The inflorescence is an axillary and terminal raceme, which is 9–12 cm long and produces a series of strobiles which are hairy, nerved, and numerous. The calyx is subglabrous. The corolla is yellowish green with an obovate standard. The pod is oblong, 0.7–1.5×5–7 cm, hairy, and contains a pair of orbicular seeds, which are oblong, blackish, with an elliptic opening.

Medicinal uses: Post-partum (Malaysian, the Philippines); tuberculosis (the Philippines); ringworm, urethritis (India)

Strong broad-spectrum antibacterial lipophilic flavanone: Euchrestaflavanone A (8.3′-diprenyl 5,7,4′-trihydroxyflavanone; LogD=5.4 at pH 7.4; molecular mass=408.4 g.mol) inhibited the growth of *Staphylococcus aureus* (ATCC 25923), *Staphylococcus epidermidis* (ATCC 12228), methicillin-resistant *Staphylococcus aureus* (562), *Pseudomonas aeruginosa* (ATCC 7853), and *Escherichia coli* (ATCC 25922) (Madan et al., 2008).

Strong broad-spectrum hydrophilic isoflavone glycoside: Genistin (LogD=1.0 at pH 7.4; molecular mass=432.3 g/mol) inhibited the growth of *Staphylococcus aureus* (ATCC 25923),

FIGURE 7.22 *Moghania strobilifera* (L.) J. St.-Hil. ex O. Kuntze

Staphylococcus epidermidis (ATCC 12228), methicillin-resistant *Staphylococcus aureus* (562), *Pseudomonas aeruginosa* (ATCC 7853), and *Escherichia coli* (ATCC 25922) (Mandan et al., 2008).

Strong anticandidal lipophilic flavanone: 8.3′-Diprenyl 5,7,4′-trihydroxyflavanone inhibited the growth of *Candida albicans* with an MIC value of 17 µg/mL (Madan et al., 2008).

Moderate anticandidal hydrophilic isoflavone glycoside: Genistin inhibited the growth of *Candida albicans* with an MIC value of 136 µg/mL (Madan et al., 2008).

REFERENCE

Madan, S., Singh, G.N., Kumar, Y., Kohli, K., Singh, R.M., Mir, S.R. and Ahmad, S.R., 2008. A new flavanone from Flemingia strobilifera (Linn) R. Br. and its antimicrobial activity. *Tropical Journal of Pharmaceutical Research*, 7(1), pp. 921–927.

7.6.1.30 *Mucuna pruriens* (L.) DC.

Synonyms: *Carpopogon niveus* Roxb.; *Carpopogon pruriens* (L.) Roxb.; *Dolichos pruriens* L.; *Dolichos pruriens* L.; *Mucuna aterrima* (Piper & Tracy) Merr.; *Mucuna esquirolii* H. Lév.; *Mucuna prurita* Hook.; *Mucuna prurita* Wight; *Mucuna prurita* Wight; *Stizolobium niveum* (Roxb.) Kuntze; *Stizolobium pruriens* (L.) Medik.; *Stizolobium pruritum* (Wight) Piper

Common names: Cow itch, itchy bean, velvet bean; alkushi (Bangladesh); khnhae (Cambodia); atmagupta, copikachu, kiwach, shyaamguptaa, (India); kara benguk (Indonesia); kekaras gatal; (Malaysia); khwe ya, kwele (Myanmar); gugli (Pakistan); sabawel (the Philippines), mah moo ee (Thailand); tam nhe (Vietnam)

Habitat: Abandoned fences, forests, roadsides

Distribution: Tropical Asia

Botanical observation: This elegant climber can reach 10 m long. The stem is slender, terete, and covered with a few extremely irritating hairs. The leaves are trifoliolate, spiral, and stipulated. The stipule is caducous, about 0.5 cm long, and hairy. The rachis is about 10 –20 cm long, pubescent, grooved, and bears 3 folioles. The terminal foliole is broadly lanceolate and the lateral folioles are asymmetrical. The folioles are dark green, 5–20 × 3–17 cm, acute at apex, hairy, and marked with 4–7 pairs of secondary nerves. The inflorescence is an axillary pendulous raceme, which is about 15–30 cm long and with some sort of sinister aura. The calyx is light yellowish green, about 0.5 cm long, hairy, and develops 5 lobes of unequal length. The corolla is about 2 cm long. The standard and wings are dark purple, and the keel is white with a yellowish beak. The stamens have whitish filaments and purple anthers. The pods are about 13 cm long and look like some sorts of caterpillars covered with golden brown glossy hairs. The hairs are extremely irritating. The seeds are glossy and pea-shaped.

Medicinal uses: Gonorrhea, syphilis (Bangladesh); diarrhea, leucorrhea, cholera (Myanmar); dysentery (India)

Antifungal (filamentous) polar extract: Methanol extract of seeds (300 µg/mL in agar wells) inhibited the growth of *Aspergillus niger* (Nidiry et al., 2011).

Antifungal non-protein amino acid: L-dopa [3-(3,4-Dihydroxyphenyl)-L-alanine] isolated from the seeds inhibited the growth of *Colletotrichum gloeosporioides*, *Colletotrichum capsici*, and *Fusarium solani* (Nidiry et al., 2011)

Commentary: The seeds contain a fascinating array of alkaloids, including N,N-dimethyltryptamine (Ghosal et al., 1971) as well as tetrahydroisoquinoline alkaloids (Misra & Wagner, 2004). L-dopa is neuroactive and as such (see earlier discussion) probably interfers with bacterial and/or fungal efflux pumps. L-dopa decreased the resistance of *Cryptococcus neoformans* toward amphotericin B (Wang et al., 1994).

FIGURE 7.23 *Mucuna pruriens* (L.) DC.

REFERENCES

Ghosal, S., Singh, S. and Bhattacharya, S.K., 1971. Alkaloids of Mucuna pruriens chemistry and pharmacology. *Planta Medica*, *19*(01), pp. 279–284.

Khan, A., Koneru, A., Pavan Kumar, K., Satyanarayana, S., Kumar, E. and Sreedevi, K., (2008) Antifungal and anthelmintic activity of extracts of mucuna pruriens seeds. *Pharmacologyonline* 2: 776–780

Misra, L. and Wagner, H., 2004. Alkaloidal constituents of Mucuna pruriens seeds. *Phytochemistry, 65*(18), pp. 2565–2567.
Nidiry, E.S.J., Ganeshan, G. and Lokesha, A.N., 2011. Antifungal activity of Mucuna pruriens seed extractives and L-dopa. *Journal of Herbs, Spices & Medicinal Plants, 17*(2), pp. 139–143.
Wang, Y. and Casadevall, A., 1994. Growth of Cryptococcus neoformans in presence of L-dopa decreases its susceptibility to amphotericin B. *Antimicrobial Agents and Chemotherapy, 38*(11), pp. 2648–2650.

7.6.1.31 *Parkia speciosa* Hassk.

Common names: Stinky bean, locust bean; petai (Indonesia; Malaysia); peteh (Indonesia); u' pang (the Philippines); sator (Thai).

Habitat: Open forests, roadsides, villages, cultivated

Distribution: Thailand, Malaysia, Indonesia, the Philippines

Botanical observation: This handsome tree grows up to about 40m tall. The bark is smooth and reddish brown. The leaves are bipinnate and 15 cm–30cm long. The rachis bears 10–18 pairs of pinnae, which are swollen at the base and finely hairy. Each pinnae includes 20–35 pairs of folioles. The folioles are sessile, linear lanceolate, 5–6 mm × 2–3mm, acuminate at apex, and round and asymmetric at base. The flowers are pollinated by bats. The pods are stout, somewhat coriaceous, greenish, glossy, fleshy, 30–50×2–4cm, leathery, and containing pungent yet edible (Malaysia, Indonesia) seeds, which are about 1 cm long each packed in a white membrane.

Medicinal use: Apparently none for the treatment of microbial infections

Broad-spectrum antibacterial halo developed by mid-polar extract: Ethyl acetate extract of pods (300 mg/mL solution/ 6 mm well) inhibited the growth of *Staphylococcus aureus* and *Escherichia coli* with inhibition zone diameters of about 21 and 16mm, respectively (Dramaga et al., 2015).

Very weak broad-spectrum antibacterial lupane-type triterpene: Lupeol inhibited the growth of *Bacillus subtilis, Staphylococcus aureus, Staphylococcus epidermidis, Escherichia coli, Klebsiella pneumoniae, Micrococcus luteus, Salmonella typhi, Salmonella paratyphi*, and *Shigella dysenteriae* (Saeedand & Sabir, 2001.).

Commentaries: (i) The seeds owe their odor to cyclic polysulfides including 1,2,4-trithiolane and 1,2,4,5-tetrathiane (Gmelin et al., 1981) which could be examined for their antibacterial, antifungal, and antiviral properties. (ii) *Parkia javanesis* Merr. is used to treat cholera in Indonesia (local name: *Kupang*). (iii) The pod accumulates a series of phytosterol: campesterol, stigmasterol, and the linear triterpene squalene (Chhikara et al., 2018).

REFERENCES

Chhikara, N., Devi, H.R., Jaglan, S., Sharma, P., Gupta, P. and Panghal, A., 2018. Bioactive compounds, food applications and health benefits of Parkia speciosa (stinky beans): a review. *Agriculture & Food Security, 7*(1), p.46.
Dramaga Campus, I.P.B., 2015. Antibacterial activity of Parkia speciosa Hassk. peel to Escherichia coli and Staphylococcus aureus bacteria. *Journal of Chemical and Pharmaceutical Research, 7*(4), pp. 239–243.
Gmelin, R., Susilo, R. and Fenwick, G.R., 1981. Cyclic polysulphides from Parkia speciosa. *Phytochemistry, 20*(11), pp. 2521–2523.
Saeed, M.A. and Sabir, A.W., 2001. Antibacterial activity of Caesalpinia bonducella seeds. *Fitoterapia, 72*(7), pp. 807–809.

7.6.1.32 *Pithecellobium dulce* (Roxb.) Benth.

Synonyms: *Acacia obliquifolia* M. Martens & Galeotti, *Feuilleea dulcis* (Roxb.) Kuntze, Inga dulcis (Roxb.) Willd., *Inga javana* DC, Inga leucantha C. Presl., *Inga pungens* Humb. & Bonpl. ex Willd., *Mimosa dulcis* Roxb., *Mimosa pungens* (Humb. & Bompl. ex Willd.) Poir., *Mimosa unguis-cati* Blco., *Pithecellobium littorale* Britton & Rose ex Rec., *Zygia dulcis* (Roxb.) Lyons

Common names: Manila tamarind; natai (Bangladesh); âm'pül tük (Cambodia); jalipi jachkarka-pilli (India); asam Belanda (Indonesia); kham thet, khaam theed (Laos); asam kranji, duri Madras (Malaysia); kala-magyi, kwayt anyeng (Myanmar); kamanchilis (the Philippines); makham thet (Thailand); keo tây, me keo (Vietnam)

Habitat: Villages, roadsides, cultivated

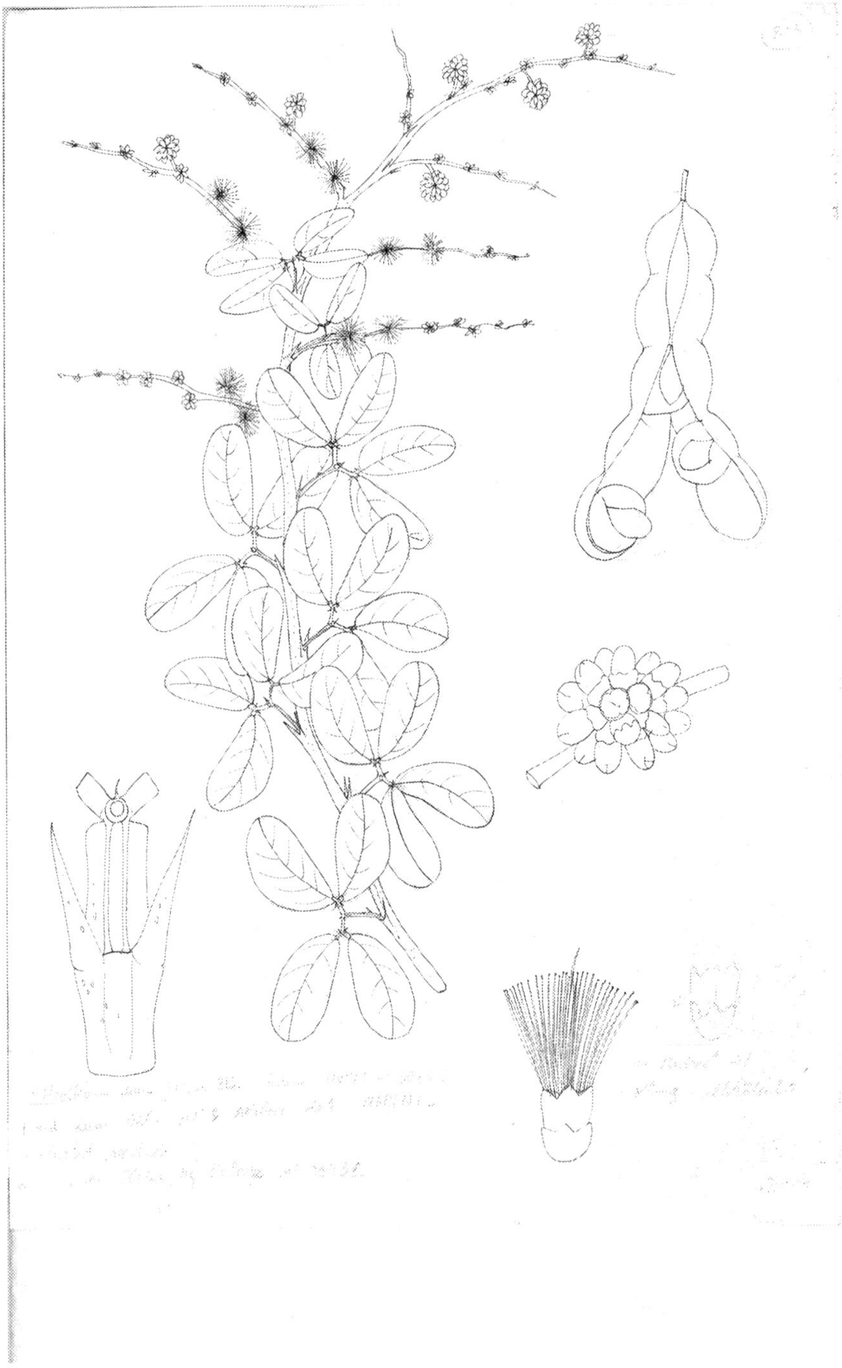

FIGURE 7.24 *Pithecellobium dulce* (Roxb.) Benth.

Distribution: Tropical Asia

Botanical observation: This handsome tree native to Central America and brought to the the Philippines in the 16th century grows to a height of to 20 m. The stem is terete and zigzag-shaped. The leaves are bipinnate, spiral, and stipulate. The stipules are about 8 mm long, sharp, and woody. The rachis is up to about 2 cm long, channeled, marked with a discoid and 5 mm across gland,

FIGURE 7.25 *Pithecellobium clypearia* (Jack) Benth.

and presents a pair of pinnae. The pinnae are 6mm long and carry a pair of folioles. The foliole is spathulate, asymmetrical at base to somewhat falcate, round at apex to discretely notched, membranous, and marked with 5–8 pairs of secondary nerves. The racemes are axillary or terminal, zigzag-shaped, 7–16 cm long and present several globose heads of flowers. The flower heads are 2–4 cm in diameter. The pod is spiral, 10–18 ×1 cm, pale greenish to light red, and containing 6 to 8 seeds, which are black and glossy and embedded in a palatable white pulp.

Medicinal uses: Dysentery (Laos, Vietnam); wounds (India)

Broad-spectrum antibacterial halo developed by mid-polar extract: Ethyl acetate extract inhibited *Bacillus megaterium* (ATCC9885), *Bacillus subtilis* (ATCC6633), *Citrobacter freundii* (NCIM2489), *Klebsiella pneumoniae* (NCIM2719), *Proteus mirabilis* (NCIM2241), and *Salmonella typhimurium* (ATCC2364) with inhibition zone diameters of 10.5, 9, 13, 9, 10.5, and 30mm, respectively (Rakholiya et al., 2014).

Strong antibacterial (Gram-negative) polar extract: Aqueous ethanol extract of stem bark inhibited the growth of *Micrococcus luteus* (MTCC (106), *Bacillus cereus* (MTCC (430), *Enterobacter aerogenes* (MTCC (111), *Klebsiella pneumonia* (MTCC (109), *Proteus mirabilis* (MTCC (1429), *Salmonella typhimurium* (MTCC (98), and *Streptococcus pneumonia* (MTCC (2672) with MIC values of 100, 300, 80, 80, 300, 100, and 100 µg/mL, respectively Singh et al., 2010).

Strong antibacterial (Gram-positive) polar extract: Ethanol extract inhibited the growth of clinical isolates of methicillin-resistant *Staphylococcus aureus* and *Staphylococcus aureus* (Voravuthikunchai & Kitpipit, 2005).

Commentary: (i) The plant produces triterpene saponins (Yoshikawa et al., 1997), which are probably involved in the antibacterial properties. (ii) Aqueous extract of *Pithecellobium clypearia* (Jack) Benth. inhibited the replication of the Herpes simplex virus type-1 with an IC_{50} value of 62.5 µg/mL and a selectivity index of 4, and an acyclovir-resistant Herpes simplex virus type-1 strain with an IC_{50} value of 125 µg/mL (Li et al., 2004). (iii) In India, *Pithecellobium monadelphum* Kosterm. (local name: *Kalpakku*) is used to treat leprosy.

REFERENCES

Li, Y., Ooi, L.S., Wang, H., But, P.P. and Ooi, V.E., 2004. Antiviral activities of medicinal herbs traditionally used in southern mainland China. *Phytotherapy Research*, *18*(9), pp. 718–722.

Rakholiya, K., Kaneria, M. and Chanda, S., 2014. Inhibition of microbial pathogens using fruit and vegetable peel extracts. *International Journal of Food Sciences and Nutrition*, *65*(6), pp. 733–73

Singh, M., Khatoon, S., Singh, S., Kumar, V., Rawat, A.K.S. and Mehrotra, S., 2010. Antimicrobial screening of ethnobotanically important stem bark of medicinal plants. *Pharmacognosy Research*, 2(4), p.254.

Voravuthikunchai, S.P. and Kitpipit, L., 2005. Activity of medicinal plant extracts against hospital isolates of methicillin-resistant Staphylococcus aureus. *Clinical Microbiology and Infection*, *11*(6), pp. 510–512.

Yoshikawa, K., Suzaki, Y., Tanaka, M., Arihara, S. and Nigam, S.K., 1997. Three acylated saponins and a related compound from Pithecellobium dulce. *Journal of Natural Products*, *60*(12), pp. 1269–1274.

7.6.1.33 *Psoralea corylifolia* L.

Common names: Hakooch (Bangladesh); bachvi, somavalli (India); babchi (Myanmar); bauchi (Pakistan)

Habitat: Roadsides, wastelands

Distribution: Pakistan, India, Sri Lanka, Bangladesh, Myanmar, China

Botanical observation: This herb grows to a height of about 1m tall and has a somewhat an aura of dryness. The leaves are 1-foliolate, alternate, and stipulate. The stipules are about 3mm long. The petiole is about 2cm long. The rachis is up to about 4cm long and presents a foliole that is dull light green, cordate to broadly elliptic, 3–10×2.5–7cm, with 4–5 pairs of conspicuous secondary nerves sunken above, serrate, and acute at apex. The inflorescence is an axillary head-like raceme which grows up to about 7 cm long. The calyx is about 5 mm long and develops 5 lobes. The corolla presents 5 light purplish petals including a broadly elliptical standard. The androecium is diadelphous and comprises 10 stamens. The pods are 4.5mm long, somewhat kidney shaped, and contain aromatic seeds.

Medicinal uses: Bronchitis, diarrhea (India); dental carries, fever, leprosy (China; India)

Strong antimycobacterial polar extract: Methanol extract of seeds inhibited the growth of *Mycobacterium aurum* (4721E) with an MIC value of 62.5 µg/mL (Newton et al., 2002).

Broad-spectrum antibacterial halo developed by 3-acylcoumarin: Psoralidin (LogD=4.6 at pH 7.4; molecular mass=336.3 g/mol) (200 µg/disc) isolated from the seeds inhibited the growth of *Bacillus subtilis, Bacillus cereus, Staphylococcus aureus, Escherichia coli, Shigella dysenteriae, Shigella flexneri,* and *Pseudomonas aeruginosa* with inhibition zone diameters of 15, 12, 18, 17, 15, 15, and 16 mm, respectively (200 µg/disc) (Khatune et al., 2004).

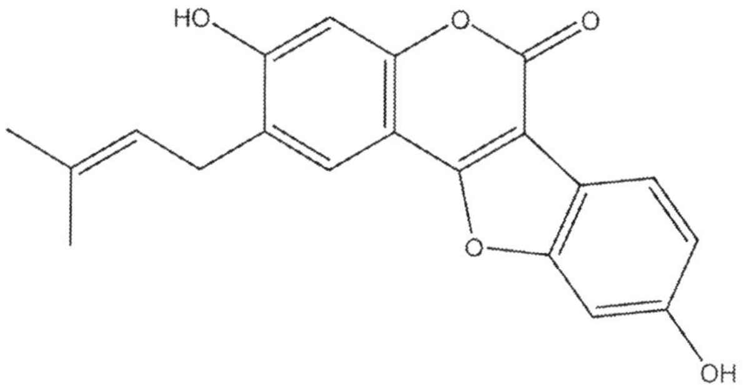

Psoralidin

Broad-spectrum halo developed by amphiphilic coumarin: Bakuchicin (LogD=2.1 at pH 7.4; molecular mass=186.1 g/mol) (200 µg/disc) isolated from the seeds inhibited the growth of *Bacillus subtilis, Bacillus cereus, Staphylococcus aureus, Escherichia coli, Shigella dysenteriae, Shigella flexneri,* and *Pseudomonas aeruginosa* (Khatune et al., 2004)

Strong antibacterial (Gram-positive) amphiphilic chalcone: Isobavachalcone (LogD=4.3 at pH 7.4; molecular mass=324.3 g/mol) isolated from the seeds inhibited the growth of *Staphylococcus aureus* (ATCC 25923) and *Staphylococcus epidermidis* (ATCC 12228) with MIC values of 10 and 9 µM, respectively (Yin et al., 2004).

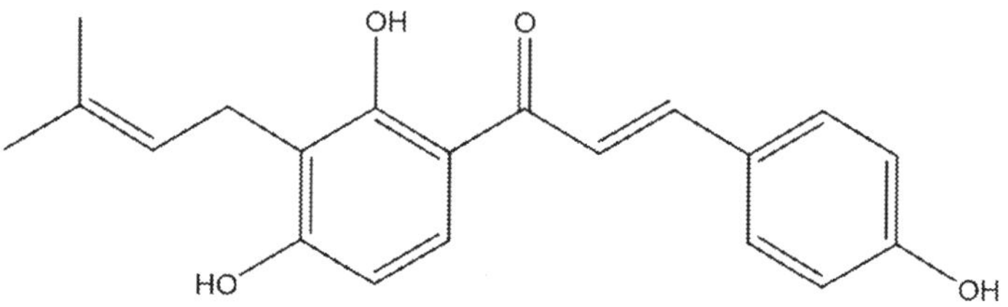

Isobavachalcone

Strong antibacterial (Gram-positive) amphiphilic flavone: Bavachinin (Log=4.3 at pH 7.4; molecular mass=338.3 g/mol) isolated from the seeds inhibited the growth of *Staphylococcus aureus* (ATCC 25923) and *Staphylococcus epidermidis* (ATCC 12228) with MIC values of 10 and 10 µM, respectively (Yin et al., 2004).

Antimycobacterial lipophilic prenylated phenol: Bakuchiol (LogD=5.6 at pH 7.4; 256.3 g/mol) inhibited the growth of *Mycobacterium aurum* (4721E) and *Mycobacterium bovis* (BCG) with MIC values of 15.9 and 21.4 µg/mL (Newton et al., 2002).

Antifungal (filamentous) halo developed by polar extract: Methanol extract of seeds (250 µg/disc) inhibited the growth of *Trichophyton mentagrophytes* (IP 940336), *Trichophyton rubrum* (IP 1400), *Epidermophyton floccosum* (IP 454), and *Microsporium gypseum* (IP 137) with halos of inhibition of 28, 27, 24, and 25 mm, respectively (Prasad et al., 2004).

Weak anticandidal polar extract: Ethanol extract of seeds inhibited the growth of *Candida albicans* (clinical strain) (Vaijayanthimala et al., 2000).

Moderate antifungal (filamentous) flavone: 4′-Methoxy flavone isolated from the seeds inhibited the growth of *Trichophyton mentagrophytes* (IP 940336), *Trichophyton rubrum* (IP 1400), *Epidermophyton floccosum* (IP 454), and *Microsporium gypseum* (IP 137) with MIC values of 62.5, 62.5, 125, and 125 µg/mL, respectively (Prasad et al., 2004).

Viral enzyme inhibition by flavone: Bavachinin isolated from the seeds inhibited Severe acute respiratory syndrome-acquired coronavirus papain-like protease with an IC_{50} value of 7.3 µM, respectively (Kim et al., 2014).

Strong antiviral (enveloped segmented single-stranded (+)RNA) chalcones: Isobavachalcone and 4′-*O*-methylbavachalcone isolated from the seeds inhibited Severe acute respiratory syndrome-acquired coronavirus papain-like protease with IC_{50} values of 7.3 and 10.1 µM, respectively (Kim et al., 2014).

Viral enzyme inhibition by isoflavones: Neobavaisoflavone and corylifol A isolated from the seeds inhibited Severe acute respiratory syndrome-acquired coronavirus papain-like protease with IC_{50} values of 18.3 and 32.3 µM, respectively (Kim et al., 2014).

Strong antiviral (enveloped segmented single-stranded (+)RNA) 3-acyl coumarin: Psoralidin isolated from the seeds inhibited Severe acute respiratory syndrome-acquired coronavirus papain-like protease with an IC_{50} value of 4.2 µM (Kim et al., 2014).

Commentary: (i) Prenylated flavones of Fabaceae are often strongly antibacterial and this is illustrated with sophoraflavanone D (LogD = 4.8 at pH 7.4; molecular mass=440.4 g/mol) from the roots of *Echinosophora koreensis* Nakai, which inhibited the growth of *Escherichia coli*, *Salmonella typhimurium*, *Staphylococcus epidermidis*, and *Staphylococcus aureus* (Sohn et al., 2004). One could suggest that lipophilic isoprenyl or geranyl moieties in natural products might facilitate penetration of bacterial cytoplasmic membranes (Li et al. 2018).

(ii) The inhibition of the Severe acute respiratory syndrome-associated coronavirus papain-like protease by flavonoids and chalcones from this plant warrants its *in vitro* evaluation against COVID-19 and other zoonotic coronaviruses.

REFERENCES

Khatune, N.A., Islam, M.E., Haque, M.E., Khondkar, P. and Rahman, M.M., 2004. Antibacterial compounds from the seeds of Psoralea corylifolia. *Fitoterapia*, *75*(2), pp. 228–230.

Kim, D.W., Seo, K.H., Curtis-Long, M.J., Oh, K.Y., Oh, J.W., Cho, J.K., Lee, K.H. and Park, K.H., 2014. Phenolic phytochemical displaying Severe acute respiratory syndrome-acquired coronavirus papain-like protease inhibition from the seeds of Psoralea corylifolia. *Journal of Enzyme Inhibition and Medicinal Chemistry*, *29*(1), pp. 59–63.

Li, J., Beuerman, R.W. and Verma, C.S., 2018. Molecular insights into the membrane affinities of model hydrophobes. *ACS Omega*, *3*(3), pp. 2498–2507.

Newton, S.M., Lau, C., Gurcha, S.S., Besra, G.S. and Wright, C.W., 2002. The evaluation of forty-three plant species for in vitro antimycobacterial activities; isolation of active constituents from Psoralea corylifolia and Sanguinaria canadensis. *Journal of Ethnopharmacology*, *79*(1), pp. 57–67.

Prasad, N.R., Anandi, C., Balasubramanian, S. and Pugalendi, K.V., 2004. Antidermatophytic activity of extracts from Psoralea corylifolia (Fabaceae) correlated with the presence of a flavonoid compound. *Journal of Ethnopharmacology*, *91*(1), pp. 21–24.

Sohn, H.Y., Son, K.H., Kwon, C.S., Kwon, G.S. and Kang, S.S., 2004. Antimicrobial and cytotoxic activity of 18 prenylated flavonoids isolated from medicinal plants: Morus alba L., Morus mongolica Schneider, Broussnetia papyrifera (L.) Vent, Sophora flavescens Ait and Echinosophora koreensis Nakai. *Phytomedicine*, *11*(7–8), pp. 666–672.
Vaijayanthimala, J., Anandi, C., Udhaya, V. and Pugalendi, K.V., 2000. Anticandidal activity of certain South Indian medicinal plants. *Phytotherapy Research*, *14*(3), pp. 207–209.
Yin, S., Fan, C.Q., Wang, Y., Dong, L. and Yue, J.M., 2004. Antibacterial prenylflavone derivatives from Psoralea corylifolia, and their structure–activity relationship study. *Bioorganic & Medicinal Chemistry*, *12*(16), pp. 4387–4392.

7.6.1.34 *Pterocarpus indicus* Willd.

Synonyms: *Lingoum indicum* (Willd.) Kuntze, *Lingoum wallichii* (Wight & Arn.) Pierre, *Pterocarpus blancoi* Merr., *Pterocarpus carolinensis* Kaneh, *Pterocarpus pallidus* Blco., *Pterocarpus papuana* F. Muell, *Pterocarpus pubescens* Merr., *Pterocarpus santalinus* Blco., *Pterocarpus wallichii* Wight & Arn.

Common names: Andaman redwood, Guinea rosewood; chan kraham (Cambodia); zi tan (China); vengai (India); gen deng (Laos); angsana (Malaysia), padauk (Myanmar); agana (the Philippines); arad (Palau); ratz (Yap).

Habitat: Seashores, tidal river

Distribution: India, Myanmar, Bangladesh, Malaysia, Indonesia, the Philippines, South China, and Papua new Guinea

Botanical observation: This tree grows to 30m tall. The bole is not straight and buttressed. The bark is cream brown and finely fissured. The stem is longitudinally striated and lenticelled. The leaves are compound, spiral, and stipulate. The rachis is swollen at base, angled, sub-glabrous, 20–25 cm long and bears 5–9 pairs of folioles. The foliole is 5–10×3.8–5 cm, glossy, the midrib strongly raised underneath. The panicles are axillary and about 15 cm long. The calyx is about 5 mm long and hairy. The corolla is golden yellow and 1.5 cm long. The androecium includes 10 stamens, which are about 1 cm long. The ovary is oblong, about 8 mm long, develops a style that is curved, and presents a minute stigma. The pod is 4.5–5 cm in diameter, with papery wings, golden hairy when young, and containing several seeds.

Medicinal use: Dysentery (Cambodia, Palau)

Broad-spectrum antibacterial halo developed by polar extract: Methanol extract of bark (4 mg/disc) of inhibited *Bacillus cereus, Bacillus subtilis, Staphylococcus aureus, Streptococcus pneumoniae, Escherichia coli, Klebsiella pneumoniae, Pseudomonas aeruginosa,* and *Salmonella typhi* (Khan, M.R. and Omoloso, A.D., 2003).

Antifungal (filamenous) phenolic: Angolensin (LogD=3.1 at pH 7.4; molecular mass=272.2 g/mol) isolated from the wood at a concentration of 100 ppm abrogated the survival of *Coriolus versicolor* (Pilotti et al., 1995).

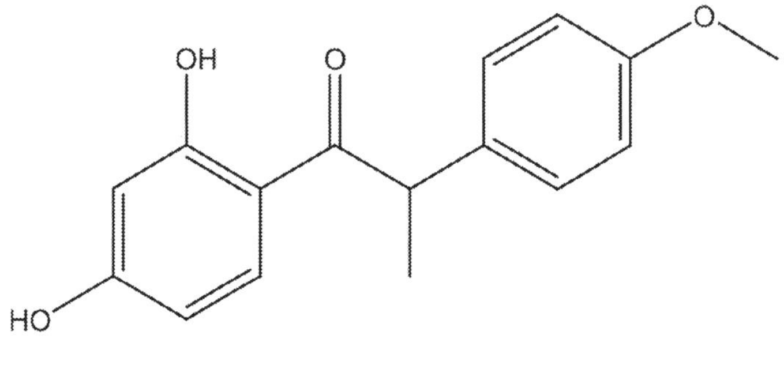

Angolensin

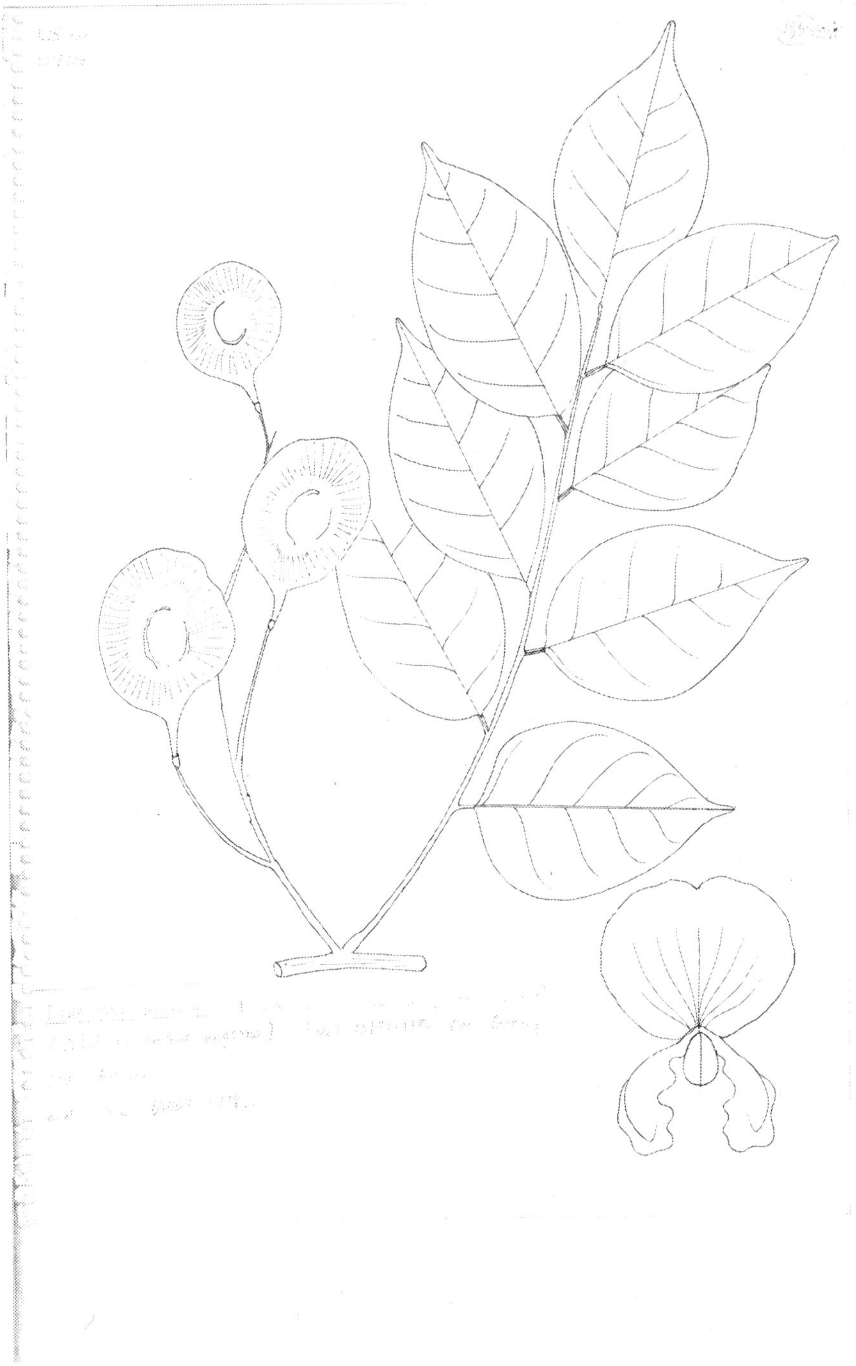

FIGURE 7.26 *Pterocarpus indicus* Willd.

Strong antiviral (enveloped monopartite linear single-stranded (+) RNA) mid-polar extract: Ethyl acetate extract of leaves inhibited the replication of the Dengue virus (clinical strain) in C6/36 cells with an IC_{50} value below 0.1 μg/mL and a selectivity index above 168.5 (Dewi et al., 2018).

Commentaries: (i) The plant abounds with non-hydrolyzable tannins: polymeric proanthocyanidins (Takeuchi et al., 1986). Note that proanthocyanidins precipitate proteins (Hagerman & Butler, 1981) and as such are lethal to bacteria. (ii) *Pterocarpus santalinus* L.f. is traditionally used in Bangladesh (local name: *Rukta-chandana*) to treat infected eyes and from this plant (Cho et al., 2001), the lignan savinin (LogD=3.4 at pH 7.4; molecular mass=352.3 g/mL) inhibited the Severe acute respiratory syndrome-acquired coronavirus 3-chymotrypsin-like protease activity with the IC_{50} value of 25μM (Wen et al., 2007) and this lignan needs to be examined for possible utility against COVID-19 and other zoonotic coronavirus activity.

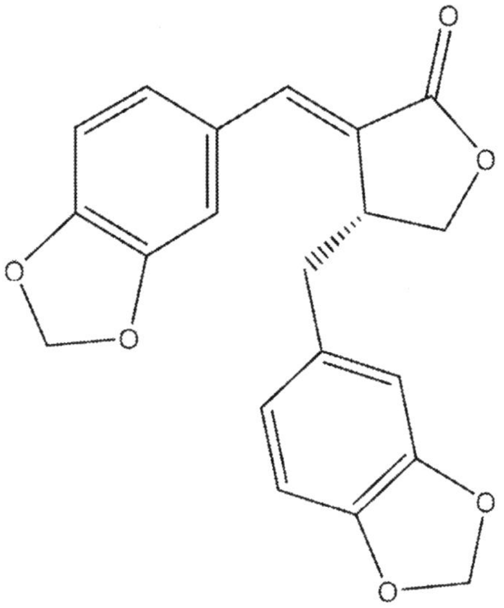

Savinin

(iii) Methanol extract of *Pterocarpus marsupium* Roxb inhibited at 400μg/mL the Moloney murine leukemia virus reverse transcriptase by 52%. (Thayil Seema et al., 2016).

REFERENCES

Cho, J.Y., Park, J., Kim, P.S., Yoo, E.S., Baik, K.U. and Park, M.H., 2001. Savinin, a lignan from Pterocarpus santalinus inhibits tumor necrosis factor-α production and T cell proliferation. *Biological and Pharmaceutical Bulletin*, *24*(2), pp. 167–171.

Dewi, B.E., Angelina, M., Hartati, S., Dewijanti, I.D., Santi, M.R., Desti, H. and Sudiro, M., 2018. Antiviral Effect of Pterocarpus indicus Willd Leaves Extract Against Replication of Dengue Virus (DENV) In Vitro. *Journal of Tropical Life Science*, *8*(1), p.228119.

Khan, M.R. and Omoloso, A.D., 2003. Antibacterial activity of Pterocarpus indicus. *Fitoterapia*, *74*(6), pp. 603–605.

Pilotti, C.A., Kondo, R., Shimizu, K. and Sakai, K., 1995. An examination of the anti-fungal components in the heartwood extracts of Pterocarpus indicus. *Mokuzai Gakkaishi= Journal of the Japan Wood Research Society*, *41*(6), pp. 593–597.

Thayil Seema, M. and Thyagarajan, S.P., 2016. Methanol and aqueous extracts of Ocimum kilimandscharicum (Karpuratulasi) inhibits HIV-1 reverse transcriptase in vitro. *Int. J. Pharmacogn. Phytochem. Res*, *8*, pp. 1099–1103.

Wen, C.C., Kuo, Y.H., Jan, J.T., Liang, P.H., Wang, S.Y., Liu, H.G., Lee, C.K., Chang, S.T., Kuo, C.J., Lee, S.S. and Hou, C.C., 2007. Specific plant terpenoids and lignoids possess potent antiviral activities against severe acute respiratory syndrome coronavirus. *Journal of Medicinal Chemistry*, *50*(17), pp. 4087–4095.

7.6.1.35 *Pueraria lobata* (Willd.) Ohwi

Synonym: *Dolichos lobatus* Willd.

Common names: Vidaari (India); tobi (Indonesia); chilk (Korea); kudzu (Japan); oka moi (Papua New Guinea); baai (the Philippines); tamyakhrua (Thailand); cu nang (Vietnam)

Habitat: Cultivated

Distribution: Tropical Asia and Pacific

Botanical observation: This climber native to China and Japan grows to a height of 2 m from massive and edible tubers. In India, this climber is sacred. The leaves are trifoliolate, spiral, and stipulate. The stipules are about 1 cm long. The petiole is 5–30 cm long. The folioles are 7.5– 21 × 7.5–20cm and broadly ovate. The racemes are about 30cm long. The calyx is tubular, 5-lobed, and about 1 cm long. The 5 petals are dark purple to burgundy. The standard reaches a length of about 2 cm and is marked at base by a showy yellow spot. The androecium presents 10 monadelphous stamens. The pod is up to 13 cm long, hairy, and shelters from 5 to 15 seeds.

Medicinal use: Fever (India)

Moderate antiviral (enveloped monopartite linear single-stranded (-)RNA) polar extract: Aqueous extract inhibited the replication of the Human respiratory syncytial virus with an IC_{50} of 258.7 µg/mL in A549 cells and 299.8 µg/mL in HEp-2 cells (Lin et al., 2013).

Moderate antiviral (enveloped segmented linear single-stranded (-)RNA) hydrophilic isoflavone C-glycoside: Puerarin (LogD=0.02 at pH 7.4; molecular mass=416.3 g/mol) inhibited the replication of the Influenza A/FM/1/47(H1N1) virus in Madin-Darby canine kidney cells with an IC_{50} value of 52 µM and a selectivity index of 20.8 (Wang et al., 2020).

In vivo antiviral (enveloped segmented linear single-stranded (-)RNA) hydrophilic isoflavone C-glycoside: At a dose of 200 mg/Kg/day intraperitoneally for 5 days puerarin protected mice from H1N1 by 70% (Wang et al., 2020).

Viral enzyme inhibition by hydrophilic isoflavone C-glycoside: Puerarin at a concentration of 200 µM inhibited H1N1 neuraminidae by about 50% (Wang et al., 2020)

Commentaries: (i) The tubers contain a series of isoflavones including genistein and biochanin A (Rong et al., 1998). (ii) The plant is used for the treatment of COVID-19 in China (Wang et al., 2020). Is puerarin active against the Severe acute respiratory syndrome-acquired coronavirus 2? As seen earlier, it inhibits the Severe acute respiratory syndrome-acquired coronavirus 3-chymotrypsin-like protease. In China, *Pueraria edulis* Pamp. (local name: *Gě gēn*) and *Pueraria montana* (Lour.) Merr. are used to treat cold.

REFERENCES

Rong, H., Stevens, J.F., Deinzer, M.L., De Cooman, L. and De Keukeleire, D., 1998. Identification of isoflavones in the roots of Pueraria lobata. *Planta Medica*, *64*(07), pp. 620–627.

Wang, S.X., Wang, Y., Lu, Y.B., Li, J.Y., Song, Y.J., Nyamgerelt, M. and Wang, X.X., 2020. Diagnosis and treatment of novel coronavirus pneumonia based on the theory of traditional Chinese medicine. *Journal of Integrative Medicine*.

Wang, H.X., Zeng, M.S., Ye, Y., Liu, J.Y. and Xu, P.P., 2020. Antiviral activity of puerarin as potent inhibitor of Influenza virus neuraminidase. *Phytotherapy Research*.

7.6.1.36 *Saraca asoca* (Roxb.) De Wilde

Synonyms: *Jonesia asoca* Roxb.; *Saraca indica* L.

Common name: Asoka tree; asoke, kurochik shak, moma gaas, (Bangladesh); ashoka (India, Pakistan); ashokam, asogam (India); soko (Indonesia); sisup (Laos); gapis (Malaysia); thawka (Myanmar); asoka, diyaratmal (Sri Lanka); chum saeng nam (Thailand); vàng anh ấn (Vietnam)

Habitat: Forests by riverside, cultivated

Distribution: Pakistan, India, Sri Lanka, Bangladesh, Myanmar, Cambodia, Laos, Vietnam, Thailand, Malaysia, Indonesia

Botanical observation: This handsome tree grows to about 10 m tall. The stem is smooth, fissured longitudinally, lenticelled, and terete. The leaves are compound, spiral, and stipulate. The stipules are deciduous, 1–1.3 cm × 3- 6mm, and oblong. The rachis is angled, smooth, glabrous, straight, enlarged at base and bears 3–6 pairs of folioles. The petiole is about 4 mm long. The folioles are coriaceous, 10–20×3–6 cm, pinkish when young, somewhat pendulous, glossy, dark green, oblong-lanceolate with 4–6 pairs of secondary nerves. The corymbs are axillary, showy, deep orangish, and with somewhat an unpleasant aura. The calyx is orange turning red, purplish to crimson at throat, the tube 1.3–2 cm long and the 4 lobes about 1 cm long. The 5–10 stamens have filiform, orange, and about 2.5 cm long filaments and purple anthers. The pod is dark brown, 10–25×4.5–5 cm, compressed, woody, thin, glabrous, veined horizontally, and containing 4–8 seeds, which are ellipsoid and up to 4 cm long.

Medicinal uses: Diarrhea, leucorrhea (Bangladesh, India); dysentery (India)

Strong broad-spectrum antibacterial polar extract: Aqueous extract of bark inhibited the growth of *Escherichia coli* (MTCCB 82), *Salmonella typhi* (MTCCB 733), *Pseudomonas aeruginosa* (MTCCB 741), *Staphylococcus aureus* (MTCCB 737), *Bacillus cereus* (MTCCB 1272), *Klebsiella aerogenes* (99/109), *Proteus vulgaris* (99/345), and *Shigella boydii* (01/21) with MIC values of 18.7, 18.7, 18.7, 18.7, 18.7, 18.7, 37.5, and 37.5 µg/mL, respectively (Dabur et al., 2007).

Moderate broad-spectrum antibacterial polar extract: Aqueous extract of bark inhibited the growth of *Pseudomonas aeruginosa* (MTCCB 741), *Klebsiella pneumoniae* (99/209), *Staphylococcus aureus* (MTCCB 737), and *Escherichia coli* (MTCCB 82) with MIC values of 266, 532, 532, and 532 µg/mL, respectively (Shirolkar et al., 2013).

Moderate antifungal (filamentous) polar extract: Aqueous extract of bark inhibited the growth of *Aspergillus fumigatus* (ITCC 4517), *Aspergillus flavus* (ITCC 5192), *Aspergillus niger* (ITCC 5405), and *Candida albicans* (ITCC 4718) with MIC values of 1200, 600, 600, and 600 µg/mL, respectively (Dabur et al., 2007).

Viral enzyme inhibition by extract: Extract of bark at the concentration of 200 µg/mL inhibited Human immunodeficiency virus protease activity by more than 70% (Kusimoto et al., 1985).

Commentary: The plant abounds with phenolics, including catechin, epicatechin, epigallocatechin, and procyanidin B2 (Dhanani et al., 2017), flavonoids, and lignans (Sadhu et al., 2007), which may confer astringent effects hence the antidiarrheal and antidysenteric traditional uses. Astringent natural products (notably tannins) precipitate proteins and as such check diarrhea and bleeding (Bowman & Rand, 1980). Consider that tannins at high doses are poisonous to humans (Chung et al., 1998). The plant brings to being aryltetralin inhibiting DNA topoisomerase IB (Mukherjee et al., 2012) and as such could be antibacterial or antifungal (Cheng et al., 2007), or for double-stranded DNA viruses (Bond et al., 2006).

REFERENCES

Bond, A., Reichert, Z. and Stivers, J.T., 2006. Novel and specific inhibitors of a poxvirus type I topoisomerase. Molecular pharmacology, 69(2), pp. 547–557.

Bowman, W.C. and Rand, M.J., 1980. *Textbook of Pharmacology* (No. 2nd ed.). Blackwell Scientific Publications.

Cheng, B., Liu, I.F. and Tse-Dinh, Y.C., 2007. Compounds with antibacterial activity that enhance DNA cleavage by bacterial DNA topoisomerase I. Journal of Antimicrobial Chemotherapy, 59(4), pp. 640–645.

Chung, K.T., Wong, T.Y., Wei, C.I., Huang, Y.W. and Lin, Y., 1998. Tannins and human health: a review. *Critical Reviews in Food Science and Nutrition*, *38*(6), pp. 421–464.

Kusumoto, I.T., Nakabayashi, T., Kida, H., Miyashiro, H., Hattori, M., Namba, T. and Shimotohno, K., 1995. Screening of various plant extracts used in ayurvedic medicine for inhibitory effects on human immunodeficiency virus type 1 (HIV-1) protease. Phytotherapy Research, 9(3), pp. 180–184.

Dhanani, T., Singh, R. and Kumar, S., 2017. Extraction optimization of gallic acid,(+)-catechin, procyanidin-B2,(–)-epicatechin,(–)-epigallocatechin gallate, and (–)-epicatechin gallate: Their simultaneous identification and quantification in Saraca asoca. *Journal of Food and Drug Analysis*, *25*(3), pp. 691–698.

Dabur, R., Gupta, A., Mandal, T.K., Singh, D.D., Bajpai, V., Gurav, A.M. and Lavekar, G.S., 2007. Antimicrobial activity of some Indian medicinal plants. *African Journal of Traditional, Complementary and Alternative Medicines*, *4*(3), pp. 313–318.

Mukherjee, T., Chowdhury, S., Kumar, A., Majumder, H.K., Jaisankar, P. and Mukhopadhyay, S., 2012. Saracoside: a new lignan glycoside from Saraca indica, a potential inhibitor of DNA topoisomerase IB. Natural *Product Communications*, 7(6), p. 1934578X1200700619.

Sadhu, S.K., Khatun, A., Phattanawasin, P., Ohtsuki, T. and Ishibashi, M., 2007. Lignan glycosides and flavonoids from Saraca asoca with antioxidant activity. *Journal of Natural Medicines*, *61*(4), pp. 480–482.

Shirolkar, A., Gahlaut, A., Chhillar, A.K. and Dabur, R., 2013. Quantitative analysis of catechins in Saraca asoca and correlation with antimicrobial activity. *Journal of Pharmaceutical Analysis*, *3*(6), pp. 421–428.

7.6.1.37 *Sesbania sesban* (L.) Merr.

Synonyms: *Aeschynomene sesban* L.; *Emerus sesban* (L.) Kuntze; *Sesbania aegyptiaca* Pers.; *Sesbania aegyptiaca* Poir.

Common names: Sesban; joyonti, hola (Bangladesh); snak kook (Cambodia); yin du tian jing (China); jayanti (Indonesia); agastya, bakphool, charum chembai, muni (India); sapao lom (Laos); kacang turi (Malaysia); yay tha kyee (Myanmar); jantar (Pakistan); katurai (the Philippines), katuru murumga (Sri Lanka); sami (Thailand); điên điển (Vietnam)

Habitat: Roadsides, wastelands, villages

Distribution: India, Bangladesh, Southeast Asia, and Australia

Botanical observation: This characteristic treelet grows to a height of 8 m. The wood is soft. The stem is terete and hairy at apex. The leaves are compound, spiral, and stipulate. The stipules are linear-triangular and about 7 mm long. The rachis is about 15 cm long and presents 18–50 pairs of folioles. The folioles are linear-oblong, dull dark green, and about 2 cm × 5mm. The racemes are axillary, up to 15 cm long, and present up to 20 yellow fabaceous flowers. The calyx is tubular and up to 7 mm long and develops 5 lobes. The corolla is plain yellow, up 2 cm long, and comprises 5 petals, including an ovate standard. The androecium includes 10 stamens. The pods are about 30 cm × 5mm, slender, somewhat falcate, and contain about 40 seeds.

Medicinal uses: Cough, rabies (Bangladesh); leprosy, sores (Myanmar); influenza (Bangladesh, Myanmar), boils (Bangladesh, Myanmar); fever (Bangladesh, India, Myanmar)

Broad-spectrum antibacterial halo developed by polar extract: Methanol extract of stems (100 μg/mL solution/ 6 mm well) inhibited the growth of *Bacillus subtilis*, *Escherichia coli*, *Enterococcus faecalis*, *Pseudomonas aeruginosa*, *Klebsiella pneumoniae*, and *Staphylococcus aureus* (Mythili & Ravindhran, 2012).

Antifungal halo developed by polar extract: Methanol extract of stems (100 μg/mL solution/ 6 mm well) inhibited the growth of *Colletotrichum gloeosporioides*, *Curvularia lunata*, *Fusarium oxysporum*, and *Verticillum glaucum* with zones of inhibition of 10.7, 12.7, 23.2, and 10.5 mm, respectively (Mythili & Ravindhran, 2012).

Moderate antibacterial (Gram-positive) non-polar extract: Hexane extract of stems inhibited the growth of *Bacillus cereus* with the MIC/MBC values of 500/500 μg/mL (Maregesi et al., 2008).

Weak antibacterial (Gram-positive) non-polar extract: Hexane of stems inhibited the growth of *Klebsiella pneumoniae* with MIC/MBC values of 1000μg/mL (Maregesi et al., 2008).

Weak broad-spectrum antifungal non-polar extract: Hexane of stems inhibited the growth of *Aspergillus niger* and *Candida albicans* with MIC values of 1000 μg/mL (Maregesi et al., 2008).

Antiviral (enveloped monopartite single-stranded (-) RNA) extract: Extract inhibited the replication of the Newcastle disease virus (Babbar et al., 1982).

Commentary: The plant is used for the treatment of Influenza and this could lead one to look for its antiviral principles. The Influenza virus and Newcastle disease virus are both (-) RNA viruses. (ii) In Iraq, *Sesbania aculeata* Poir. (local name: *Sesbaniyah*) is used to heal skin eruptions. In India, *Sesbania grandiflora* (L.) Poir. (local name: *Tella sumintha*) is used to treat dysentery and *Sesbania bispinosa* W. f. Wight. (local name: *Jayanti*) is used to treat ringworm and wounds. (iii) Consider that the word "edible" whenever used here does not mean that the plants can be eaten, but are eaten by locals.

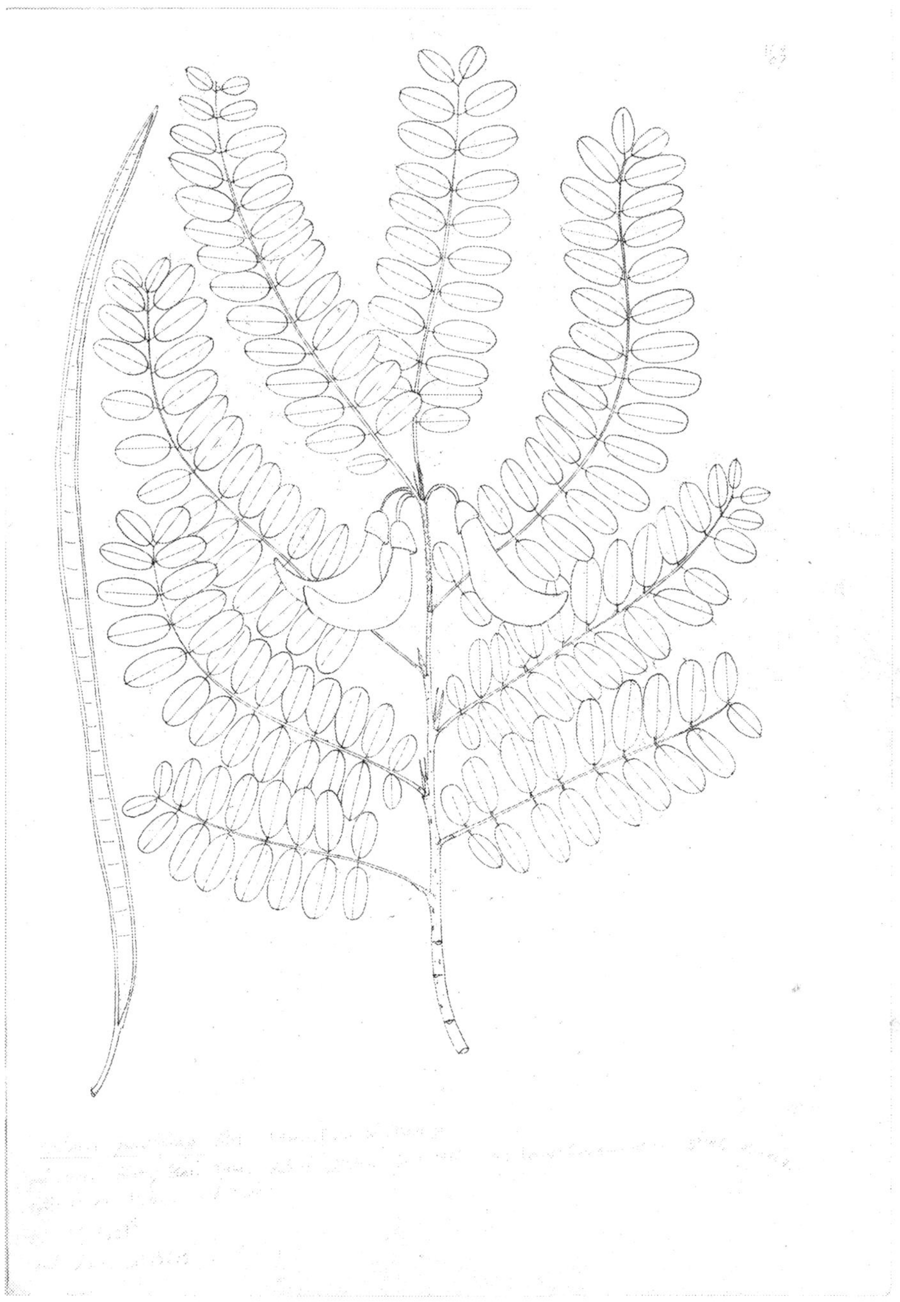

FIGURE 7.27 *Sesbania grandiflora* (L.) Poir.

REFERENCES

Babbar, O.P., Joshi, M.N. and Madan, A.R., 1982. Evaluation of plants for antiviral activity. Indian J Med Res, 76(S), pp. 54–65.

Maregesi, S.M., Pieters, L., Ngassapa, O.D., Apers, S., Vingerhoets, R., Cos, P., Berghe, D.A.V. and Vlietinck, A.J., 2008. Screening of some Tanzanian medicinal plants from Bunda district for antibacterial, antifungal and antiviral activities. *Journal of Ethnopharmacology*, *119*(1), pp. 58–66.

Mythili, T. and Ravindhran, R., 2012. Phytochemical screening and antimicrobial activity of Sesbania sesban (L.) Merr. *Asian Journal of Pharmaceutical and Clinical Research*, *5*(4), pp. 18–23.

7.6.1.38 *Tamarindus indica* L.

Synonyms: *Tamarindus occidentalis* Gaertn.; *Tamarindus officinalis* Hook.; *Tamarindus umbrosa* Salisb.

Common names: Indian tamarind, tamarin tree; khoua me (Cambodia); jiu ceng pi guo (China); sukaer (Indonesia); amli (India); mak kham (Laos); asam jawa (Indonesia, Malaysia); m*a*gi (Myanmar); ilimi (Pakistan); kalamagi (the Philippines); somkham (Thailand); trai me (Vietnam)

Habitat: Villages, cultivated

Distribution: Tropical Asia

Botanical observation: This tree native to Africa grows to a height of about 6 m and has an aura of dryness. The twigs are reddish. The leaves are pinnate, alternate, and stipulate. The stipules are small and caducous. The rachis bears 8–20 pairs of folioles. The folioles are elliptic, dull green, and about 4 mm × 1.5cm. The racemes are cauliflorous. The calyx is obconical, 4-lobed, and the lobes are lanceolate, up to 1.5 cm long, somewhat yellowish, and imbricate. The 3 petals are light reddish yellow with delicate, almost blood capillary-like venations and about 1.5 cm long. The androecium is monadelphous and includes 3 stamens, which are recurved. The pods are somewhat irregularly sausage shaped, somewhat constricted between the seeds, slightly rugose, beautifully fawn colored, elliptic, slightly falcate, and up to 20 cm long. The seeds are glossy, squarish, brown, woody, and are embedded in a brown, sticky, sour but refreshing pulp.

Medicinal use: Fever (India)

Moderate broad-spectrum antibacterial polar extract: Methanol extract of seeds inhibited *Salmonella paratyphi* A (MTCC 3906) and *Staphylococcus epidermidis* (MTCC 435) with MIC/MBC values of 242.5/242.5 and 53/56 µg/mL, respectively (Kothari et al., 2010).

Strong broad-spectrum antibacterial polar extract: Ethanol extract of leaves inhibited the growth of *Staphylococcus aureus* (ATCC 25923), *Bacillus subtilis* (ATCC 6633), *Escherichia coli* (ATCC 25922), and *Salmonella typhimurium* (ATCC 14028) with MIC/MBC values of 75/>150, 38/>150, 75/>150, and 75/>150 µg/mL, respectively (Escalona-Arranz et al., 2010).

Commentary: (i) Polar extract of medicinal plants are mostly bacteriostatic. (ii) In the Unani system of medicines, a number of plants are considered as protective remedies in times of pandemics (Nikhat & Fazil, M., 2020). For instance, our peer the 14th-century Moorish physician of Spain Abū Ja'far Aḥmad ibn 'Alī ibn Muḥammad ibn Khātima Al-Anṣārī (1324-1369) also known as Ibn Katima recommended the consumption of tamarin in plagued patients and in his "*Taḥṣīl garaḍ al-qāṣid fī-tafṣīl al-maraḍ al wāfid*" writes about the role of air contamination and contagion in Black Death.

REFERENCES

Escalona-Arranz, J.C., Péres-Roses, R., Urdaneta-Laffita, I., Camacho-Pozo, M.I., Rodríguez-Amado, J. and Licea-Jiménez, I., 2010. Antimicrobial activity of extracts from Tamarindus indica L. *leaves. Pharmacognosy Magazine*, *6*(23), p.242.

Kothari, V. and Seshadri, S., 2010. In vitro antibacterial activity in seed extracts of Manilkara zapota, Anona squamosa, and Tamarindus indica. *Biological Research*, *43*(2), pp. 165–168.

Nikhat, S. and Fazil, M., 2020. Overview of Covid-19; its prevention and management in the light of Unani medicine. *Science of The Total Environment*, p. 138859.

7.6.1.39 *Tephrosia purpurea* (L.) Pers.

Synonyms: *Cracca purpurea* L., *Cracca wallichii* (Graham ex Fawcwt & Rendle) Rydb., *Galega diffusa* Roxb., *Galega purpurea* (L.) L., *Glycyrrhiza marei* H-Lev., *Hedysarum lineare* Lour., *Tephrosia crassa* Bojer ex Baker, *Tephrosia diffusa* Wight & Arn, *Tephrosia indigofera* Bertol., *Tephrosia lanceolata* Link., *Tephrosia leptostachya* DC, *Tephrosia pumila* (Lam) Pers., *Tephrosia wallichii* Graham ex Fawcet & Rendle

Common names: Bastard indigo, purple tephrosia, wild indigo; jongli-niil (Bangladesh); nha troi (Cambodia, Laos, Vietnam); sarphoka, sarphonk, sharapunkha (India); pohon nila hutan (Indonesia); balatong-pula (the Philippines); sila (Sri Lanka); khram-pa (Thailand); doan kiem do (Vietnam)

Habitat: Grassy fields, roadsides, sun exposed area, wastelands, seashores, and rocks

Distribution: India to southern China through Southeast Asia to Northern Australia

Botanical observation: This herb can reach 60 cm tall and has somewhat dryish aura. The stem is hairy, slender, and terete. The leaves are compound, spiral, and stipulate. The stipules are linear, velvety, 2mm long, persistent, and triangular. The rachis is about 1.5 cm–2 cm long with 7–12 pairs

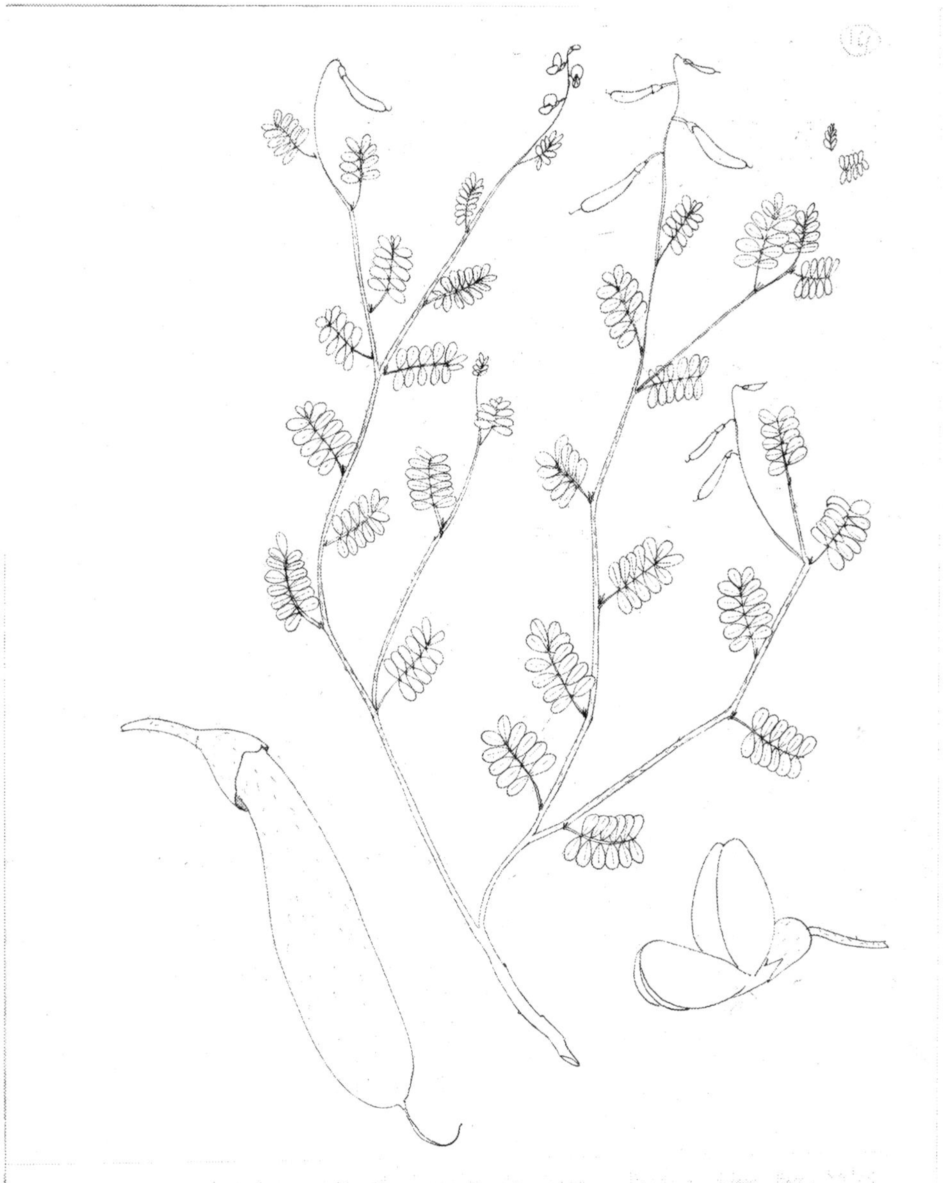

FIGURE 7.28 *Tephrosia purpurea* (L.) Pers.

of folioles. The folioles are spathulate, about 7 mm ×4 mm, hairy underneath and with 4 pairs of secondary nerves looping at margin. The inflorescences are terminal racemes, which are about 8 cm long hairy. The calyx is hairy, about 5 mm long with 4 lobes, which are triangular linear. The corolla is about 1 cm long, delightfully purplish to red, with a standard pubescent at the back. The androecium includes 10 stamens. The pods are hairy, 2.4 –4.5 cm long, marked with a tiny vestigial style at apex, and contain 5–6 seeds.

Medicinal uses: Fever, boils, (Myanmar); bronchitis, influenza, syphilis, leprosy (Bangladesh); gonorrhea (Bangladesh, Myanmar)

Broad-spectrum antibacterial halo developed by polar extract: Methanol extract of roots (500 µg/mL solution/6 mm disc) inhibited the growth of *Bacillus subtilis, Staphylococcus aureus, Micrococcus luteus, Escherichia coli, Pseudomonas aeruginosa*, and *Salmonella typhimurium* with inhibition zone diameters of 13.4, 12.7, 13, 11.7, 9.2, and 11.4 mm, respectively (Gupta et al., 2008).

Moderate antibacterial (Gram-positive) polar extract: Ethanol extract of roots inhibited the growth of *Pseudomonas aeruginosa* (NCTC 10662) with an MIC of 128 µg/mL (Rangama et al., 2009).

Broad-spectrum antifungal halo developed by polar extract: Methanol extract of roots (6 mm paper disc impregnated with 500 µg/mL solution) inhibited the growth of *Aspergillus niger* and *Candida albicans* with inhibition zone diameters of 13.6 and 18 mm, respectively (Gupta et al., 2008).

Moderate antifungal (filamentous) flavone: Glabratephrin inhibited the growth of *Colletotrichum aculatum, Pestalotiopsis* sp., *Helminthosporium* sp., and *Alternaria alternata* with MIC values of 250, 500, 250, and 500 µg/mL, respectively (Ammar et al., 2013).

Commentaries: (i) For paper disc test or agar well tests, it is useful to indicate if the antibacterial halo developed is clear or turbid. A turbid halo might indicate levels or growth or regrowth of bacteria or fungi (bacteriostatic/fungistatic) effect, whereas a crystal-clear halo might indicate a bactericidal and fungicidal effect. (ii) The plant is used in Siddha medicine for the treatment of COVID-19.

REFERENCES

Ammar, M.I., Nenaah, G.E. and Mohamed, A.H.H., 2013. Antifungal activity of prenylated flavonoids isolated from Tephrosia apollinea L. against four phytopathogenic fungi. *Crop Protection*, *49*, pp. 21–25.

Gupta, M., Mazumder, U.K., Gomathi, P. and Selvan, V.T., 2008. Antimicrobial activity of methanol extracts of Plumeria acuminata Ait. leaves and Tephrosia purpurea (Linn.) Pers. roots.

Rangama, B.N.L.D., Abayasekara, C.L., Panagoda, G.J. and Senanayake, M.R.D.M., 2009. Antimicrobial activity of Tephrosia purpurea (Linn.) Pers. and Mimusops elengi (Linn.) against some clinical bacterial isolates. *J Natn Sci Foundation Sri Lanka*, *37*(2), pp. 139–145.

Sabarianandh, J.V., Bernaitis, L. and Manimekalai, K., COVID-19 in Siddha Medicine: A Review. 2020.

7.7 ORDER FAGALES ENGL. (1892)

7.7.1 Family Fagaceae Dumortier (1829)

The family Fagaceae consists of 9 genera and about 1000 species of handsome trees. The leaves are spiral, incised or lobed, and stipulate. The stipules are caducous. The inflorescences are spikes or dichasial heads of tiny flowers. The flowers are minute and without corolla. The calyx includes 4–6 lobes. The androecium is minute and comprises 6–12 free stamens. The gynoecium consists of 2–9 fused carpels containing each a pair of ovules. The fruits are nuts or acorns. Members of the family Fagaceae produce mainly antimicrobial hydrolysable tannins, gallic acid, and also proanthocyanins and flavonols (quercetin and kaempferol).

7.7.1.1 *Castanea crenata* Siebold & Zucc.

Synonyms: *Castanea japonica* Blume; *Castanea stricta* Siebold & Zucc.

Common names: Japanese chestnut, Korean chestnut; ri ben li (China); kuri (Japan)

Habitat: Forests

Distribution: China, Korea, Japan

Botanical observation: This tree grows to about 15 m tall. The wood is very hard and durable. The leaves are simple, spiral, and stipulate. The stipules are deciduous. The petiole is up to about 2.5 cm long. The blade is lanceolate, 8–19 × 3–5 cm somewhat hairy below, glossy, with 10–25 pairs of secondary nerves, cuneate at base, serrate at margin, and acuminate at apex. The inflorescence is an axillary whitish spike which is up to about 20 cm long. The flowers are minute. The perianth includes 6 lobes. The androecium comprises 10–12 stamens. The gynoecium includes 6–9 styles. The capsules are dehiscent, up to 6cm in diameter, spiny, and contain 2 or 3 glossy and brown nuts.

Medicinal uses: Tuberculosis, scrofula (China)

Moderate broad-spectrum antibacterial polar extract: Aqueous extract of bark inhibited the growth of *Bacillus cereus*, *Clostridium perfringens*, *Staphylococcus aureus*, *Clavibacer michiganensis*, *Aeromonas hydrophila*, *Vibrio parahaemolyticus*, *Vibrio vulnificus*, *Citrobacter freundii*, *Escherichia coli*, *Klebsiella pneumoniae*, *Proteus mirabilis*, *Salmonella anatum*, *Shigella flexneri*, *Shigella sonnei*, *Yersinia enterolitica*, *Pseudomonas aeruginosa*, and *Xanthomonas campestris* (Taguri et al., 2006).

Weak broad-spectrum antibacterial hydrolysable tannin: The ellagitannin castalagin isolated from the wood inhibited the growth of *Clostridium perfringens*, *Staphylococcus aureus*, *Aeromonas hydrophila*, *Vibrio parahaemolytcus*, *Vibrio vulnificus*, *Citrobacter freundii*, *Klebsiella pneumoniae*, *Shigella flexneri*, *Pseudomonas aeruginosa*, and *Xanthomonas campestris* with MIC values of 67, 267, 150, 117, 167, 233, 233, 333, 400, and 133 μg/mL, respectively (Taguri et al., 2006).

Strong antiviral (non-enveloped monopartite linear single-stranded (+)RNA) cycloartane-type triterperne: Castaartancrenoic acid E isolated from the capsules inhibited the replication of the Human rhinovirus type-1B with an IC_{50} value of 5.5 μg/mL and a selectivity index above 8.9 (Taguri et al., 2006).

Strong antiviral (non-enveloped linear monopartite single-stranded (+)RNA) flavonol glycosides: Kaempferol-3-*O*-[2″,6″-di-*O*-*E*-*p*-coumaroyl]-β-D-glucopyranoside and kaempferol-3-*O*-[3″-acetyl-2″,6″-di-E-p-coumaroyl]-β-D-glucopyranoside inhibited the replication of the Human rhinovirus type-1B with IC_{50} values of 1.2 and 5.5 μg/mL and selectivity indices > 40.9 and >9.0, respectively (Kim et al., 2019). Kaempferol-3-*O*-[2″,6″-di-*O*-*Z*-*p*-coumaroyl]-β-D-glucopyranoside inhibited the growth of the Coxsackie virus B3 with an IC_{50} value of 43.6 μg/mL and a selectivity index above 1.1 (Kim et al., 2019).

Strong antiviral (enveloped segmented linear single stranded (-)RNA) flavonol glycoside: Kaempferol-3-*O*-[2″,6″-di-*O*-*Z*-*p*-coumaroyl]-β-D-glucopyranoside inhibited the growth of the Influenza A/PR/8 virus (PR8) with an IC_{50} value of 28.2 μg/mL and a selectivity index above 8.9 (Kim et al., 2019).

Commentary: Flavonoid glycosides have antibacterial and antimycobacterial effects *in vitro*. Hydrolysable tannins and flavonoid glycosides are hydrolyzed in the digestive tract and are therefore unable to have, *per se*, any possible antimicrobial systemic effects internally. They are, however, active externally.

REFERENCES

Kim, N., Park, S., Nhiem, N.X., Song, J.H., Ko, H.J. and Kim, S.H., 2019. Cycloartane-type triterpenoid derivatives and a flavonoid glycoside from the burs of Castanea crenata. *Phytochemistry*, *158*, pp. 135–141.

Taguri, T., Tanaka, T. and Kouno, I., 2006. Antibacterial spectrum of plant polyphenols and extracts depending upon hydroxyphenyl structure. *Biological and Pharmaceutical Bulletin*, *29*(11), pp. 2226–2235.

7.7.1.2 *Castanea mollissima* Bl.

Synonyms: *Castanea bungeana* Bl.; *Castanea duclouxii* Dode; *Castanea fargesii* Dode; *Castanea formosana* (Hayata) Hayata; *Castanea hupehensis* Dode

Common names: Chinese hairy chestnut; li (China)

Habitat: Mountain slopes

Distribution: China, Korea

Botanical observation: This tree grows to about 20m tall. The wood is very hard and durable. The stems are hairy. The leaves are simple, spiral, and stipulate. The stipules are deciduous and up to about 2cm long. The petiole is up to about 2cm long. The blade is oblong-lanceolate, 10–20×5–10cm, glossy, with 12–20 pairs of secondary nerves, rounded at base, serrate at margin, and acuminate or acuminate at apex. The inflorescence is an axillary whitish spike, which is up to about 20cm long. The flowers are minute. The perianth includes 6 lobes. The androecium comprises 10–12 stamens. The gynoecium consists of 6–9 hairy styles. The capsules are dehiscent, up to 3 cm across, spiny and contain 2 or 3 glossy, brown, and palatable nuts.

Medicinal uses: Boils, diarrhea, dysentery, putrefied wounds (China)

Antibacterial broad-spectrum polar extract: Extract of fruits (1 mg/6mm disc) inhibited the growth of *Escherichia coli* (CMCC (B) 44103), *Streptomyces somaliensis* (NBRC 12916 (T), *Staphylococcus aureus* (ATCC 25923), and *Bacillus subtilis* (CGMCC 3376) with inhibition zone diameters of 6.1, 10.6, 6.5, and 9.9mm, respectively (You et al., 2014).

Commentaries: Consider that kaempferol inhibited NorA efflux pump in *Staphylococcus aureus* (Sharma et al., 2019).

REFERENCES

Sharma, A., Gupta, V.K. and Pathania, R., 2019. Efflux pump inhibitors for bacterial pathogens: From bench to bedside. *The Indian Journal of Medical Research*, *149*(2), p.129.

You, T.T., Zhou, S.K., Wen, J.L., Ma, C. and Xu, F., 2014. Chemical composition, properties, and antimicrobial activity of the water-soluble pigments from Castanea mollissima shells. *Journal of Agricultural And Food Chemistry*, *62*(8), pp. 1936–1944.

7.7.1.3 *Castanea sativa* Mill.

Synonyms: *Castanea vesca* Gaertn.; *Castanea vulgaris* Lam.; *Fagus castanea* L.

Common names: Sweet chestnut; bazı kestane (Turkey); meetha, pangar (India)

Habitat: Forests

Distribution: Turkey, Pakistan, Himalaya, India, China, and Japan

Botanical observation: This noble tree grows to about 25m tall. The wood is very hard and durable. The stems are glabrous. When the bark is removed from the stems, the wood becomes somewhat greyish blue. The leaves are simple, spiral, and stipulate. The stipules are lanceolate, caducous, and up to about 2cm long. The petiole is up to about 2cm long. The blade is lanceolate to oblanceolate, 10–25×3–7 cm, glossy, marked with up to about 20 pairs of secondary nerves, rounded or cordate at base, serrate at margin, and acuminate at apex. The inflorescence is an axillary whitish spike, which is up to about 20cm long. The flowers are minute. The perianth includes 6 lobes. The androecium comprises 10–12 stamens. The gynoecium includes 6–9 hairy styles. The capsules are dehiscent, up to 6cm in diameter, spiny, and contain 2 or 3 glossy, brown, and palatable nuts.

Medicinal uses: Cough, fever (India)

Moderate broad-spectrum antibacterial mid-polar extract: Ethyl acetate fraction of leaves inhibited the growth of *Escherichia coli* (ATCC 11229), *Klebsiella pneumoniae* (ATCC 10031), *Enterobacter aerogenes* (ATCC 13048), *Staphylococcus aureus* (ATCC 13709), *Proteus vulgaris* (ATCC 12454), *Pseudomas aeruginosa* (ATCC 27853), and *Enterobacter cloacae* (ATCC 10699) with MIC/MBC values of 256/512, 256/512, 64/512, 64/256, 128/256, 128/256, and 128/512 μg/mL, respectively (Basile et al., 2000).

Moderate broad-spectrum antibacterial flavonols: Quercetin isolated from Ethyl acetate fraction of leaves inhibited the growth of *Escherichia coli* (ATCC 11229), *Klebsiella pneumoniae* (ATCC 10031), *Enterobacter aerogenes* (ATCC 13048), *Staphylococcus aureus* (ATCC 13709), *Proteus vulgaris* (ATCC 12454), *Pseudomas aeruginosa* (ATCC 27853), and *Enterobacter cloacae* (ATCC 10699) with MIC values of 64, 128, 128, 16, 64, 64, and 128 μg/mL, respectively (Basile et al., 2000).

Moderate broad-spectrum antibacterial flavonol glycoside: Rutin isolated from the leaves inhibited the growth of *Escherichia coli* (ATCC 11229), *Klebsiella pneumoniae* (ATCC 10031), *Enterobacter aerogenes* (ATCC 13048), *Staphylococcus aureus* (ATCC 13709), *Proteus vulgaris* (ATCC 12454), *Pseudomonas aeruginosa* (ATCC 27853), and *Enterobacter cloacae* (ATCC 10699) with MIC values of 64, 64, 128, 32, 64, 64, and 128 μg/mL, respectively (Basile et al., 2000). Apigenin isolated from the leaves inhibited the growth of *Escherichia coli* (ATCC 11229), *Klebsiella pneumoniae* (ATCC 10031), *Enterobacter aerogenes* (ATCC 13048), *Staphylococcus aureus* (ATCC 13709), *Proteus vulgaris* (ATCC 12454), *Pseudomonas aeruginosa* (ATCC 27853), and *Enterobacter cloacae* (ATCC 10699) (Basile et al., 2000).

In vivo antibacterial polar extract: Methanol extract from the leaves in mouse model of skin infection afforded some levels of protections at the intradermal dose of injection of 50 μg against *Staphylococcus aureus* (Quave et al., 2015).

Commentaries: Plants are able to sense the presence of bacteria and to react by producing antifungal enzymes. For instance, *Bacillus subtilis*, which is not pathogenic to *Castanea sativa* Mill., induces the production of pathogen-related proteins, such as an acidic extracellular chitinase and an acidic β 1,3 glucanase isoenzyme, protecting the plant against *Cryphonectria parasitica*, which is pathogenic for the plant (Wilhelm et al., 1998). (ii) Note that the MIC values of isolated flavonoids including quercetin (an in fact all natural products) vary greatly in the literature due to different experimental environments, human factors (errors, frauds), various degree of purity of products tested, and method of assessment (type of broth, MIC techniques). There is a need for nonexpensive international standard procedures for testing of plant extracts and natural products.

REFERENCES

Basile, A., Sorbo, S., Giordano, S., Ricciardi, L., Ferrara, S., Montesano, D., Cobianchi, R.C., Vuotto, M.L. and Ferrara, L., 2000. Antibacterial and allelopathic activity of extract from Castanea sativa leaves. *Fitoterapia*, *71*, pp. S110–S116.

Fu, L., Lu, W. and Zhou, X., 2016. Phenolic compounds and in vitro antibacterial and antioxidant activities of three tropic fruits: persimmon, guava, and sweetsop. *BioMed Research International*, *2016*.

Quave, C.L., Lyles, J.T., Kavanaugh, J.S., Nelson, K., Parlet, C.P., Crosby, H.A., Heilmann, K.P. and Horswill, A.R., 2015. Castanea sativa (European Chestnut) leaf extracts rich in ursene and oleanene derivatives block Staphylococcus aureus virulence and pathogenesis without detectable resistance. *PloS One*, *10*(8), p.e0136486.

Wilhelm, E., Arthofer, W., Schafleitner, R. and Krebs, B., 1998. Bacillus subtilis an endophyte of chestnut (Castanea sativa) as antagonist against chestnut blight (Cryphonectria parasitica). *Plant Cell, Tissue and Organ Culture*, *52*(1), pp. 105–108.

7.7.1.4 *Castanea tungurrut* Bl.

Synonym: *Castanopsis tungurrut* (Blume) A. DC.

Common name: Tungurut (Indonesia)

Habitat: Forests

Distribution: Indonesia, Malaysia

Botanical observation: This tree grows to about 25 m tall. The wood is very hard and durable. The stems are hairy and lenticellate. The leaves are simple, spiral, and stipulate. The stipules are up to about 6 mm long and triangular. The petiole is up to about 2 cm long. The blade is oblong-lanceolate to spathulate, 12–6 × 6–1.5 cm, glossy, hairy below, marked with 11–19 pairs of secondary nerves, rounded to acute at base, and tapering at apex. The spikes are axillary and about 25 cm long. The perianth includes 5–6 lobes. The androecium comprises 10–12 stamens, which are up to about 4 mm long. The gynoecium includes 3 minute styles. The capsules are dehiscent, up to 4 cm in diameter, spiny, and shelter a glossy, brown, and palatable seed.

Medicinal use: Diarrhea (Indonesia)

Commentary: The plant has apparently not been studied for antimicrobial activity. The medicinal property is most probably owed to tannins.

7.7.1.5 *Castanopsis motleyana* King

Common names: Berangan (Malaysia); malagasa (the Philippines)

Habitat: Forests

Distribution: Malaysia, Indonesia, the Philippines

Botanical observation: This tree grows to about 40 m tall. The bole is buttressed. The wood is very hard and durable. The stems are hairy. The leaves are simple, spiral, and stipulate. The stipule is extra-petiolar and caducous. The stipules are up to about 1 cm long and ovate. The petiole is up to about 1.5 cm long. The blade is elliptic to spathulate, 15–23 × 7–10 cm, glossy, hairy below, with 14–18 pairs of secondary nerves, acute at base, and acute or acuminate at apex. The spikes are axillary, whitish and about 25 cm long. The perianth includes 5–6 lobes. The 10–12 stamens are about 4mm long. The gynoecium includes 3 minute styles. The capsules are ovoid, about 2.5 cm long, and spiny.

Medicinal uses: Ringworm, wounds (the Philippines)

Commentary: The plant has apparently not been studied for antimicrobial activity. The medicinal property is most probably owed to tannins.

7.7.1.6 *Castanopsis sclerophylla* (Lindl.) Schottky

Synonyms: *Lithocarpus chinensis* (Abel) A. Camus; *Quercus chinensis* Abel; *Quercus sclerophylla* Lindl.; *Synaedrys sclerophylla* (Lindl.) Koidz.

Common name: Ku zhu (China)

Distribution: China

Habitat: Forests

Botanical observation: This tree grows to about 40 m tall. The bole is buttressed. The wood is very hard and durable. The stems are hairy. The leaves are simple, spiral, and stipulate. The stipules are extra-petiolar, up to about 1 cm long, and ovate. The petiole is up to about 2.5 cm long. The blade is elliptic to oblong, 15–23 × 7–10 cm, glossy, marked with 10–15 pairs of secondary nerves, rounded or cuneate at base, serrulate, and acuminate at apex. The spikes are whitish, axillary, and about 15cm long. The perianth includes 5–6 lobes. The 10–12 stamens are about 4mm long. The gynoecium includes 3 styles. The cupules are puberulent, about 1.5 cm in diameter and almost completely enclosing a palatable nut.

Medicinal uses: Diarrhea, ulcers (China)

Commentary: The plant has not been studied for antimicrobial activity. It produces the ellagitannin 2-*O*-galloyl-*O*-4,6-(*S*)-valoneoyl-D-glucose, gallotannins 6-*O*-galloyl-1-O-vanilloyl-β-D-glucose and 4″-*O*-galloylchestanin (Huang et al., 2012) as well as a series of flavan-3-ols (Tanaka et al., 2011), which are most probably antibacterial, antifungal, and antiviral *in vitro*.

REFERENCES

Huang, Y.L., Tanaka, T., Matsuo, Y., Kouno, I., Li, D.P. and Nonaka, G.I., 2012. Two new phenolic glucosides and an ellagitannin from the leaves of Castanopsis sclerophylla. *Phytochemistry Letters*, *5*(1), pp. 158–161.

Tanaka, T., Huang, Y.L., Matsuo, Y., Kouno, I., Li, D.P. and Nonaka, G.I., 2011. New phenylpropanoid-substituted flavan-3-ols from the leaves of Castanopsis sclerophylla. *Heterocycles*, *83*(10), pp. 2321–2328.

7.7.1.7 *Lithocarpus celebicus* (Miq.) Rehder

Synonyms: *Cyclobalanus celebica* (Miq.) Oerst.; *Quercus celebica* Miq.; *Synaedrys celebica* (Miq.) Koidz.

Common names: Celebes oak; mempening (Malaysia)

Distribution: Malaysia, Indonesia, Papua New Guinea

Habitat: Forests

Botanical observation: This tree grows to about 30 m tall. The bole is buttressed. The stems are hairy and lenticelled. The leaves are simple, spiral, and stipulate. The stipule is extra-petiolar, caducous, up to about 4 mm long and linear. The petiole is up to about 1 cm long. The blade is elliptic to oblong, 12–16×4–6 cm, glossy, hairy below, with 8–10 pairs of secondary nerves, acute or decurrent at base, somewhat wavy, and acuminate at apex. The inflorescence is an axillary whitish spike, which is up to about 20 cm long. The flowers are minute. The perianth includes 6 lobes. The androecium comprises 10–12 stamens, which are up to about 4 mm long. The gynoecium includes 3 minute styles. The fruits consist of a puberulent cupule in which is seated a conical nut, which is about 2 cm long.

Medicinal uses: Venereal diseases, fever (Papua New Guinea)

Broad-spectrum antibacterial halo developed by non-polar extract: Petroleum ether extract of root bark (4 mg/disc) inhibited the growth of *Bacillus cereus*, *Staphylococcus aureus*, *Staphylococcus epidermidis*, *Enterobacter aerogenes*, *Escherichia coli*, *Klebsiella pneumonia*, *Neisseria gonorrhoeae*, *Proteus mirabilis*, *Pseudomonas aeruginosa*, *Salmonella typhi*, *Salmonella typhimurium*, and *Serratia marcescens* (Khan et al., 2001).

Commentary: What accounts for the broad-spectrum activity observed by Khan et al. is not known. One could infer the role, at least partially, of triterpenes. Arthur et al. (1976) detected the presence of triterpenes such as friedelin, taraxerol, β-amyrin, glutinol, and cycloartanes in members of the genus *Lithocarpus* Bl. The lipophilic triterpene taraxerol (LogD = 10.1 at pH 7.4; molecular mass = 426.7 g/mol) inhibited the growth of *Trichophytom mentagrophytes* and *Trichopytum rubrum* with MIC at 12.5 µg/mL (Aguilar-Guadarrama e al., 2009).

β-Amyrin exhibited antibacterial effects against a broad panel of bacteria (Saeed & Sabir, 2001), including *Streptococcys mutans* (ATCC 25175), with the MIC values of 48.8 µg/mL (Zheng et al., 2011). The antifungal and antibacterial modes of action of pentacyclic triterpene are not fully elucidated but preliminary evidence suggests increase of cytoplasmic membrane permeability involving impaired lipid biosynthesis in fungi (Haraguchi et al., 1999) and destruction of membrane integrity with leakage of cytoplasmic constituents in bacteria (Liu et al., 2015). Being lipophilic, these triterpenes tend to accumulate on the lipophilic layer of the cytoplasmic membrane and destabilize it.

REFERENCES

Aguilar-Guadarrama, B., Navarro, V., Leon-Rivera, I. and Rios, M.Y., 2009. Active compounds against tinea pedis dermatophytes from Ageratina pichinchensis var. bustamenta. *Natural Product Research*, *23*(16), pp. 1559–1565.

Arthur, Hhenry R., Phyllis DS Ko, and Tai Cheung Hee. Triterpenes of Lithocarpus species. *Phytochemistry* 13, no. 11 (1974): 2551–2557.

Haraguchi, H., Kataoka, S., Okamoto, S., Hanafi, M. and Shibata, K., 1999. Antimicrobial triterpenes from Ilex integra and the mechanism of antifungal action. *Phytotherapy Research: An International Journal Devoted to Pharmacological and Toxicological Evaluation of Natural Product Derivatives*, *13*(2), pp. 151–156.

Khan, M.R., Kihara, M. and Omoloso, A.D., 2001. Antimicrobial activity of Lithocarpus celebicus. *Fitoterapia*, *72*(6), pp. 703–705.

Liu, W.H., Liu, T.C. and Mong, M.C., 2015. Antibacterial effects and action modes of asiatic acid. *Biomedicine (Taipei)*, *5*(3), p.16.

Saeed, M.A. and Sabir, A.W., 2001. Antibacterial activity of Caesalpinia bonducella seeds. *Fitoterapia, 72*(7), pp. 807–809.

Zheng, Y., Huang, W., Yoo, J.G., Ebersole, J.L. and Huang, C.B., 2011. Antibacterial compounds from Siraitia grosvenorii leaves. *Natural Product Research, 25*(9), pp. 890–897.

7.7.1.8 *Lithocarpus litseifolius* (Hance) Chun

Synonyms: *Quercus litseifolia* Hance; *Pasania litseifolia* (Hance) Schottky; *Synaedrys synbalanos* (Hance) Koidz.; *Synaedrys litseifolia* (Hance) Koidz.

Common name: Mu jiang ye ke (China)

Habitat: Forests

Distribution: Laos, Myanmar, Vietnam, China

Botanical observation: This tree grows to about 20 m tall. The bole is buttressed. The stems are hairy and lenticelled. The leaves are simple, spiral, and stipulate. The stipule is extra-petiolar and caducous. The petiole is up to about 2.5 cm long. The blade is elliptic, 8–18×3–8cm, somewhat papery, glaucous below, marked with 8–11 pairs of secondary nerves, cuneate at base, and acuminate at apex. The spikes are about 20 - 35cm long. The perianth includes 4–6 lobes. The androecium comprises 10–12 stamens. The gynoecium includes 3 minute styles. The nut is broadly conical, up to about 2 cm long, and seated in a cup which is up to about 1.5 cm in diameter.

Medicinal uses: Sores, ulcers (China)

Strong antiviral (enveloped monopartite linear dimeric single-stranded (+)RNA) lipophilic lupane-type triterpenes: 3-*epi*-Betulinic acid (LogD=5.1; molecular mass=456.7 g/mol) and cylicodiscic acid at a concentration of 10 μM inhibited the replication of the Human immunodeficiency virus to 33.8 and 42.6 % respectively (Cheng et al., 2017).

Weak antibacterial (Gram-positive) antibacterial chalcone glycoside: Phlorizin inhibited the growth of *Streptococcus pyogenes* with an MIC value of 100 μg/mL and an MBC >100 μg/mL (Macé et al., 2017) and inhibited the growth of *Staphylococcus aureus* with an MIC value of 500 μg/mL (Barreca et al., 2014).

Commentary: The plant contains the sweet-tasting dihydrochalcone glucosides phlorizin and trilobatin (Rui-Lin et al., 1982). Note that hydrolyzation of phlorizin yields phloretin, which is a strong broad-spectrum antibacterial (MacDonald & Bishop, 1952) and antifungal (Shim et al., 2010) "phytoalexin" (Gosh et al., 2010).

REFERENCES

Barreca, D., Bellocco, E., Laganà, G., Ginestra, G. and Bisignano, C., 2014. Biochemical and antimicrobial activity of phloretin and its glycosilated derivatives present in apple and kumquat. *Food Chemistry, 160*, pp. 292–297.

Cheng, Y.B., Liu, F.J., Wang, C.H., Hwang, T.L., Tsai, Y.F., Yen, C.H., Wang, H.C., Tseng, Y.H., Chien, C.T., Chen, Y.M.A. and Chang, F.R., 2018. Bioactive Triterpenoids from the Leaves and Twigs of Lithocarpus litseifolius and L. corneus. *Planta Medica, 84*(01), pp. 49–58.

Gosch, C., Halbwirth, H. and Stich, K., 2010. Phloridzin: biosynthesis, distribution and physiological relevance in plants. *Phytochemistry, 71*(8–9), pp. 838–843.

MacDonald, R.E. and Bishop, C.J., 1952. Phloretin: an antibacterial substance obtained from apple leaves. *Canadian Journal of Botany, 30*(4), pp. 486–489.

Macé, S., Truelstrup Hansen, L. and Rupasinghe, H.P., 2017. Anti-Bacterial Activity of Phenolic Compounds against Streptococcus pyogenes. *Medicines, 4*(2), p.25.

Rui-Lin, N., Tanaka, T., Zhou, J. and Tanaka, O., 1982. Phlorizin and trilobatin, sweet dihydrochalcone-glucosides from leaves of Lithocarpus litseifolius (Hance) Rehd.(Fagaceae). *Agricultural and Biological Chemistry, 46*(7), pp. 1933–1934

Shim, S.H., Jo, S.J., Kim, J.C. and Choi, G.J., 2010. Control efficacy of phloretin isolated from apple fruits against several plant diseases. *The Plant Pathology Journal, 26*(3), pp. 280–285.

Zhang, T., Wei, X., Miao, Z., Hassan, H., Song, Y. and Fan, M., 2016. Screening for antioxidant and antibacterial activities of phenolics from Golden Delicious apple pomace. *Chemistry Central Journal, 10*(1), p.47.

7.7.1.9 *Quercus acutissima* Carruth.

Synonym: *Quercus lunglingensis* H.H. Hu

Common names: Saw-tooth oak; ma li (China)

Habitat: Forests

Distribution: India, Nepal, Bhutan, Myanmar, Cambodia, Thailand, Vietnam, Japan, Korea

Botanical observation: This tree grows up to about 30m tall. The stems are somewhat hairy and lenticellate. The leaves are simple, spiral, and stipulate. The stipule is extra-petiolar and caducous. The petiole is up to about 5cm long. The blade is elliptic-lanceolate, 8–19×2–6cm, rounded to cuneate at base, serrate and spiny at margin, acuminate at apex, glossy, and marked with 13–18 pairs of secondary nerves. The spikes are light yellow and pendulous. The perianth is 4–7-lobed. The androecium includes 4–7 stamens. The gynoecium is 3-locular. The nuts are about 2cm in diameter, smooth ovoid, seated on cupules, which are up to 4cm in diameter and covered with linear bracts.

Medicinal uses: Diarrhea, putrefied sores (China)

Strong antiviral (enveloped linear double-stranded DNA) polar extract: Aqueous extract of bark inhibited plaque formation by Herpes simplex virus type-1 (Seibert strain) at 100 μg/mL by 38.2 % (Hozumi et al., 1995).

Strong antiviral (enveloped monopartite linear single-stranded (-)RNA) polar extract: Aqueous extract of bark inhibited plaque formation by Measle virus (Tanabe strain) at 100 μg/mL by 83.3 % (Hozumi et al., 1995).

Strong antiviral (enveloped monopartite linear dimeric single-stranded (+)RNA) polar extract: Aqueous extract of leaves inhibited plaque formation by Poliovirus (Sabin strain) at 100 μg/mL by 98.4% (Hozumi et al., 1995).

Commentary: (i) Methanol extract of leaves of *Quercus myrsinaefolia* inhibited the Human immunodeficiency virus type-1 protease (Park, 2003) on probable account of tannins. (ii) The Measles virus is an enveloped linear monopartite (-)RNA virus that belongs to the family Paramyxoviridae, which account in nonvaccinated children for a mild respiratory illness, conjunctivitis, coryza, cough, and fever followed by rash (Naim 2015). Complications occur in immunocompromised patients, such as subacute sclerosing panencephalitis, which is a very serious condition (Naim, 2015). The Poliovirus is non-enveloped monopartite linear single-stranded (+) RNA virus that belongs to the family Picornaviridae. This virus is neurotropic and accounts for damage to the anterior horn cells of the spinal cord in nonvaccinated individuals, causing limb paralysis, also known as polyomyelitis (Mehndiratta et al., 2014). (iii) The medicinal uses of the plant are most probably owed to gallotannins and ellagitannins. Gallotannins interact with proline-rich proteins or cell-surface lipoteichoic acid of Gram-positive bacteria (Buzzini et al., 2008). Ellagitannins interact with cytoplasm and wall enzymes of Gram-positive bacteria (Buzzini et al., 2008). Hydrolyzable tannins damage the cell wall and membrane of *Candida albicans* (Buzzini et al., 2008). Hydrolyzable tannins inhibit the entry of the Human immunodeficiency virus type-1 and the Herpes simplex virus in host cells (Buzzini et al., 2008).

REFERENCES

Park, J.C., 2003. Inhibitory effects of Korean plant resources on human immunodeficiency virus type 1 protease activity. *Oriental Pharmacy and Experimental Medicine, 3*(1), pp. 1–7.

Buzzini, P., Arapitsas, P., Goretti, M., Branda, E., Turchetti, B., Pinelli, P., Ieri, F. and Romani, A., 2008. Antimicrobial and antiviral activity of hydrolysable tannins. *Mini reviews in Medicinal Chemistry, 8*(12), pp. 1179–1187.

Hozumi, T., Matsumoto, T., Ooyama, H., Namba, T., Shiraki, K., Hattori, M., Kurokawa, M. and Kadota, S., Hozumi, Toyoharu and Matsumoto, 1995. *Antiviral agent containing crude drug.* U.S. Patent 5,411,733.

Mehndiratta, M.M., Mehndiratta, P. and Pande, R., 2014. Poliomyelitis: historical facts, epidemiology, and current challenges in eradication. *The Neurohospitalist*, *4*(4), pp. 223–229.
Naim, H.Y., 2015. Measles virus: a pathogen, vaccine, and a vector. *Human Vaccines & Immunotherapeutics*, *11*(1), pp. 21–26.

7.7.1.10 *Quercus dentata* Thunb.

Synonyms: *Quercus aquatica* var. *dentata* (Thunb.) A. DC.; *Quercus obovata* Bunge

Common names: Korean oak, Japanese emperor oak; hu shu (China); kashiwa (Japan); teokgalnamu (Korea)

Habitat: Forests

Distribution: China, Japan, Korea

Botanical observation: This elegant tree grows up to about 25 m tall. The stems are stout and striated. The leaves are simple, spiral, and stipulate. The stipule is extra-petiolar and caducous. The petiole is up to about 5 mm long. The blade is large, 10–30×6–15 cm, obovate, rounded at base, conspicuously lobed at margin, acute at apex, smooth, glossy, and with 4–10 pairs of secondary nerves. The spikes are light yellow and pendoulous. The perianth is 4–7-lobed. The androecium includes 4–7 stamens. The gynoecium is 3-locular. The nuts are smooth, toxic, ovoid, up to about 2 cm in diameter, and seated on cupules, which are up to 5 cm in diameter and covered with linear and recurved scales.

Medicinal use: Putrefied sores (China)

Antifungal spermidine-type alkaloid: di-*p*-coumaroyl-caffeoylspermidine isolated from the pollen at 100 μM inhibited the mycelial growth of *Pyrenophora avenae* and *Blumeria graminis* f. sp. *hordei* (Walters et al., 2001).

Commentaries: Ethyl acetate extract of bark inhibited the growth of *Bacillus subtilis* and *Klebsiella pneumoniae* (Kim et al., 2000). The plant has been used as a source of tanning material because of high (around 20%) amounts of tannins in the bark (Sun et al., 1987). Sun et al. isolated from the bark gallic acid, catechin (see earlier), (+)-gallocatechin, hexagalloyl glucose, and the proanthocyanidin dimers catechin-(4α → 8)-catechin, gallocatechin-(4α → 8)-gallocatechin, gallocatechin-(4α → 8)-catechin, gallocatechin-(4α → 6)-catechin, and 3-*O*-galloylepigallocatechin-(4β → 8)-catechin (Sun et al., 1987).

REFERENCES

Kim, M., Kim, Y., Kim, T., Jo, J. and Yang, J., 2000. Antimicrobial activity and antioxidative activity in the extractives of Quercus dentata Thunberg. *Mokchae Konghak= Journal of the Korean Wood Science and Technology*, *28*(3), pp. 42–51.
Panthong, P., Bunluepuech, K., Boonnak, N., Chaniad, P., Pianwanit, S., Wattanapiromsakul, C. and Tewtrakul, S., 2015. Anti-HIV-1 integrase activity and molecular docking of compounds from Albizia procera bark. *Pharmaceutical Biology*, *53*(12), pp. 1861–1866.
Sun, D., Wong, H. and Foo, L.Y., 1987. Proanthocyanidin dimers and polymers from Quercus dentata. *Phytochemistry*, *26*(6), pp. 1825–1829.
Walters, D., Meurer-Grimes, B. and Rovira, I., 2001. Antifungal activity of three spermidine conjugates. *FEMS Microbiology Letters*, *201*(2), pp. 255–258.

7.7.1.11 *Quercus dilata* Lindl. ex A. DC.

Common names: Royal oak; mohru (India); tilonj (Himalaya); barungi, shah balut, seray (Pakistan)

Habitat: Forests

Distribution: Afghanistan, Pakistan, Himalaya, Kashmir, Nepal

Botanical observation: This tree grows up to about 20 m tall. The stems are rough and terete. The leaves are simple, spiral, and stipulate. The stipule is extra-petiolar and caducous. The petiole is up to about 1 cm long. The blade is, 4–12 × 1.5–5.5 cm, elliptic-ovate, obovate, or broadly lanceolate, cordate and asymmetrical at base, wavy, acute or acuminate at apex, and marked with 9–12 pairs of secondary nerves. The spikes are light yellow, pendulous, up to about 5 cm long. The perianth is 4–7-lobed.

The 4–8 stamens are sessile. The gynoecium is ovoid and develops 3–5 stigmas. The nuts are smooth, toxic, ovoid, up to about 2 cm in diameter, and seated on cupules which are up to 2.5cm accross.

Medicinal use: Diarrhea (Pakistan)

Moderate broad-spectrum antibacterial polar extract: Methanol extract of aerial parts inhibited the growth of *Staphylococcus aureus*, *Bacillus subtilis*, *Escherichia coli*, *Bordetella bronchiseptica*, and *Salmonella sebutal* with MIC values of 500, 300, 200, 300, and 500 µg/mL, respectively (Jamil et al., 2012).

Commentary: The plant has apparently not been studied for its antimicrobial constituents.

REFERENCE

Jamil, M., ul Haq, I., Mirza, B. and Qayyum, M., 2012. Isolation of antibacterial compounds from Quercus dilatata L. through bioassay guided fractionation. *Annals of Clinical Microbiology and Antimicrobials*, *11*(1), p. 11.

7.7.1.12 *Quercus ilex* L.

Synonyms: *Quercus baloot* Griff.; *Quercus ilicifolius* Griff.

Common names: Holly oak, Holm's oak

Habitat: Forests

Distribution: Turkey, Pakistan

Botanical observation: This magnificent tree grows up to about 8 m tall. The stems are hairy. The leaves are simple, spiral, and stipulate. The stipule is extra-petiolar and caducous. The petiole is up to about 4 mm long and hairy. The blade is coriaceous, glossy, 2.5–7.5 × 2.5–8 cm, oblong to obovate, wavy, hairy below, cordate at base, laxly spiny at margin or margin entire, wavy at margin, acute at apex, and with 4–9 pairs of secondary nerves. The spikes are light yellow, pendulous, and about 5 cm long. The perianth is 4–7-lobed and hairy. The androecium includes 5–7 stamens. The gynoecium is ovoid and develops 3–5 stigmas. The nuts are smooth, ovoid, up to about 1.5 cm long and seated on cupules, which are up to 1.3 cm in diameter.

Medicinal use: Diarrhea (Turkey)

Moderate broad-spectrum antibacterial polar extract: Methanol extract of leaves inhibited the growth of *Bacillus cereus*, *Bacillus subtilis*, *Enterobacter intermedius*, *Escherichia coli*, *Neisseria* spp., and *Pseudomonas putida*, with MIC values of 250, 125, 250, 250, 250, and 250 µg/mL, respectively (Güllüce et al., 2004). Ethyl acetate fraction of bark inhibited the growth of *Escherichia coli* (ATCC 11775), *Pseudomonas aeruginosa* (ATCC 27853), *Staphylococcus aureus* (BCCM 21055), *Bacillus subtilis* (ATCC 6051), *Klebsiella pneumoniae* (ATCC13883), *Salmonella typhimurium* (ATCC 43971), *Vibrio cholerae* (ATCC 14033), *Proteus mirabilis* (HITM 20), *Staphylococcus epidermidis* (HITM 60), *Streptococcus pyogenes* (group A) (HITM 100), and *Streptococcus agalactiae* (group B) (HITM80) with MIC values of 256, 256, 128, 128, 512, 256, 256, 256, 128, 512, and 512 µg/mL, respectively (Berahou et al., 2007).

Strong broad-spectrum antibacterial proanthocyanidin: (+)-Epigallocatechin-(2β→O→7, 4 β →8)-(+)-catechin isolated from the leaves inhibited the growth of *Bacillus cereus* (clinical isolate), *Escherichia coli* (ATCC 35210), *Listeria monocytogenes* (NCTC 7973), *Pseudomonas aeruginosa* (ATCC (ATCC 27853), *Proteus mirabilis* (human isolate), and *Staphylococcus aureus* (ATCC 6538 (Karioti et al., 2011).

Strong antifungal (filamentous) proanthocyanidin: (+)-Epigallocatechin-(2β→O→7, 4 β →8)-(+)-catechin isolated from the leaves inhibited the growth of *Altenaria alternata* (DSM 2006), *Aspergillus flavus* (ATCC 9643), *Aspergillus fumigatus* (plant isolate), *Aspergillus niger* (ATCC 6275), *Aspergillus ochraceus* (ATCC 12066), *Aspergillus versicolor* (ATCC 11730), *Aureobasidium pullulans* Arnaud de Bary (ATCC 9348), *Cladosporium cladosporioides* Fresenius des Vries (ATCC 13276), *Fulvia fulvum* (TK 5318), *Fusarium sporotrichioides* Sherbakoff (ITM

496), *Fusarium trincintum* Corda, Saccardo (CBS 514478), *Penicillium funiculosum* (ATCC 36839), *Penicillium ochrochloron* (ATCC 9112, 5061), and *Trichoderma virid e*(IAM) (Karioti et al., 2011).

REFERENCES

Berahou, A., Auhmani, A., Fdil, N., Benharref, A., Jana, M. and Gadhi, C.A., 2007. Antibacterial activity of Quercus ilex bark's extracts. *Journal of Ethnopharmacology*, *112*(3), pp. 426–429.

Karioti, A., Sokovic, M., Ciric, A., Koukoulitsa, C., Bilia, A.R. and Skaltsa, H., 2011. Antimicrobial properties of Quercus ilex L. proanthocyanidin dimers and simple phenolics: evaluation of their synergistic activity with conventional antimicrobials and prediction of their pharmacokinetic profile. *Journal of Agricultural and Food Chemistry*, *59*(12), pp. 6412–6422.

Güllüce, M., Adıgüzel, A., Öğütçü, H., Şengül, M., Karaman, I. and Şahin, F., 2004. Antimicrobial effects of Quercus ilex L. extract. *Phytotherapy Research*, *18*(3), pp. 208–211.

7.7.1.13 *Quercus infectoria* Olivier

Synonym: *Quercus lusitanica* var. *infectoria* (Olivier) A. DC.

Common names: Dyer's oak, Allepo oak, Oriental gall oak; mazı meşesi (Turkey); maayakku, maazu (India); mazu (Persian); manjakani (Malaysia); pinza-kanj si (Myanmar)

Habitat: Forests

Distribution: Turkey, Iran, India

Botanical observation: This magnificent tree grows up to about 15 m tall. The stems are terete and present globose and smooth galls, which are about 1.8 cm in diameter. The leaves are simple, spiral, and stipulate. The stipule is extra-petiolar and caducous. The petiole is up to about 1–1.5 cm long and hairy. The blade is coriaceous, dull light green, glossy, 4–8 × 2.5–5 cm long, oblong to obovate, cordate or round and asymmetrical at base, lobed or entire, acute to round at apex, and with 5–10 pairs of secondary nerves. The spikes are light yellow, pendulous, and about 5 cm long. The perianth is 4–7-lobed and hairy. The androecium includes 6–12 stamens. The gynoecium is ovoid and develops 3–5 stigmas. The nuts are smooth, oblong, glossy, up to about 3 cm long, and seated on reticulated cupules, which are up to 1 cm long.

Medicinal uses: Sore throat, ringworm, leucorrhea (India)

Strong antibacterial (Gram-positive) polar extract: Ethanol extract of the plant inhibited the growth of 5 clinical isolates of methicillin-resistant *Staphylococcus aureus* and *Staphylococcus aureus* with MIC/MBC values of 0.2–0.4 /0.4–1.6 and 0.1/1.6 μg/mL, respectively (Voravuthikunchai & Kitpipit, 2005).

Moderate broad-spectrum polar extracts: Ethanol extract of galls inhibited the growth of *Escherichia coli* (O157:H7) (RIMD 0509952), *Escherichia coli* (ATCC 25922), *Klebsiella pneumoniae* (ATCC 10273), multidrug-resistant *Klebsiella pneumoniae* (clinical isolate), *Staphylococcus aureus* (ATCC 25923), and methicillin-resistant *Staphylococcus aureus* (clinical) with MIC/MBC values of 0.1/0.8, 0.8/0.8. 0.1/0.8, 0.1/0.8, 0.1/1.6, and 0.2/0.4 mg/mL, respectively (Limsuwan et al., 2009). Ethanol extract of galls inhibited the growth of methicillin-resistant *Staphylococcus aureus* (ATCC 43300), *Staphylococcus aureus* (ATCC 25923), methicillin-resistant *Staphylococcus aureus*, methicillin-resistant, coagulase-negative *Staphylococcus*, and multidrug-resistant *Acinetobacter* sp. (Wan et al., 2014). Methanol/acetone extract of galls inhibited the growth of *Streptococcus mutans* (ATCC 25175), *Streptococcus salivarius* (ATCC 13419), *Porphyromonas gingivalis* (ATCC 33277), and *Fusobacterium nucleatum* (ATCC 25586) with MIC values of 0.3, 0.08, 0.3, and 0.1 mg/mL, respectively (Basri al., 2012). Ethanol extract of galls inhibited the growth of 17 strains of methicillin-resistant *Staphylococcus aureus*, 33 strains of methicillin-sensitive *Staphylococcus aureus*, and *Staphylococcus aureus* (ATCC 25923) (Chusri & Voravuthikunchai, 2008). Methanol extracts of galls at 5 mg/mL evoked tubular outpouchings from the cell wall in *Escherichia coli* (ATCC 25922) in logarithmic phase growth (Leela & Satirapipathkul, 2011). Methanol extracts of galls at 5 mg/mL evoked small bleb-like structures on the surface of occasional cells, irregular spherical structures lying free or appearing to extrude *Staphylococcus aureus* during logarithmic phase growth (ATCC 25923). Methanol extracts of galls at 5 mg/mL evoked the formation of spherical forms, and collapse resulted in structures of *Bacillus cereus* in logarithmic phase growth (Leela & Satirapipathkul, 2011)

Weak antibacterial (Gram-positive) hydrolysable tannin: The gallotannin tannic acid inhibited the growth of 17 strains of methicillin-resistant *Staphylococcus aureus* and *Staphylococcus aureus* (ATCC 25923) with MIC values of 125–250 and 250 µg/mL, respectively (Chusri & Voravuthikunchai, 2009). Tannic acid a 4×MIC evoked pseudomulticellular aggregates of methicillin-resistant *Staphylococcus aureus* (Chusri & Voravuthikunchai, 2009).

In vivo antibacterial (Gram-positive) hydrolysable tannin: A gallitannin from the galls given at 4 × MIC to mice infected with streptomycin-resistant *Escherichia coli* (O157:H7) decreased the number of viable fecal colonies (Voravuthikunchai et al., 2012). This regimen also reduced the colonization of intestinal tract with *Escherichia coli* (O157:H7) (Voravuthikunchai et al., 2012).

Moderate anticandidal polar extract: Methanol extract of galls inhibited the growth of *Candida albicans* (ATCC 10231), *Candida tropicalis* (ATCC 13803), *Candida parapsilosis* (ATCC 20019), *Candida glabrata* (ATCC 90525), and *Candida krusei* (ATCC 6258) with MIC/MFC values of 2/8, 1/2, 0.2/1, 0.5/1, and 0.06/0.06 mg/mL, respectively (Baharuddin et al., 2015).

Strong antiviral amphiphilic gallotannin: Tannic acid (LogD=3.1 at pH 7.4; molecular mass=1701.1 g/mol) inhibited the replication of the Herpes simplex virus type 2 (MS strain) with an IC_{50} value of 0.006 µg/mL (Kane et al., 1988).

Commentaries: (i) Galls are hard, globulous, and smooth excroissances formed by this plant as a response to the deposition of eggs of the hymenopters *Adleria gallae-tinctoria* or *Cynips gallae tinctoria* in the leaf buds (Bruneton 1987; Trease & Evans, 1996). The galls contain gallic acid, gallotannins, and ellagic acid (Bruneton 1987), which are most likely produced to defend the plant against microbial infections. The galls contains 50–70% of the gallotannin tannic acid (Trease & Evans, 1996) and are used in therapeutic medicine as astringents and styptics. Note that methanol extract of *Quercus lusitanica* Lam. at 0.1 mg/mL inhibited by 100% the replication of Dengue virus serotype 2 in C6/36 cells (Muliawan et al., 2006). This effect was accompanied by a downregulation in expression of NS1 protein in C6/36 cells (Muliawan et al., 2006).

(ii) Consider that tannic acid inhibited NorA and TetK efflux pumps in *Staphylococcus aureus* (Sharma et al. 2019).

(iii) High molecular mass tannins hamper bacterial growth *in vitro* via inhibition of extracellular bacterial enzymes, deprivation of the substrates required for bacterial growth (such as metal ions via complexation), or direct action on bacterial metabolism through inhibition of oxidative phosphorylation (Serrano et al., 2009). The current body of experimental evidence points to the fact that high molecular mass tannins interfere with the surface of viruses and inhibit the binding of viruses to host cells and inhibit viral enzymes, such as reverse transcriptase (Serrano et al., 2009).

(iv) Tannins fraction of the cork of *Quercus suber* L. inhibited the exponential growth of *Staphylococcus aureus* (Gonçalves et al., 2015). Natural products having antibacterial or anti-yeast activity at exponential growth act at cell division that actively occurs during the exponential phase (Gonçalves et al., 2015).

REFERENCES

Baharuddin, N.S., Abdullah, H. and Wahab, W.N.A.W.A., 2015. Anti-Candida activity of Quercus infectoria gall extracts against Candida species. *Journal of Pharmacy & Bioallied Sciences, 7*(1), p. 15.

Basri, D.F., Tan, L.S., Shafiei, Z. and Zin, N.M., 2012. In vitro antibacterial activity of galls of Quercus infectoria Olivier against oral pathogens. *Evidence-Based Complementary and Alternative Medicine, 2012.*

Chusri, S. and Voravuthikunchai, S.P., 2008. Quercus infectoria: a candidate for the control of methicillin-resistant Staphylococcus aureus infections. *Phytotherapy Research: An International Journal Devoted to Pharmacological and Toxicological Evaluation of Natural Product Derivatives*, *22*(4), pp. 560–562.
Chusri, S. and Voravuthikunchai, S.P., 2009. Detailed studies on Quercus infectoria Olivier (nutgalls) as an alternative treatment for methicillin-resistant Staphylococcus aureus infections. *Journal of Applied Microbiology*, *106*(1), pp. 89–96.
Gonçalves, F., Correia, P., Silva, S.P. and Almeida-Aguiar, C., 2015. Evaluation of antimicrobial properties of cork. *FEMS Microbiology Letters*, *363*(3), p.fnv231.
Leela, T. and Satirapipathkul, C., 2011. Studies on the antibacterial activity of Quercus infectoria galls. In *International Conference on Bioscience, Biochemistry and Bioinformatics* (Vol. 5, pp. 410–414). IACSIT Press Singapore.
Limsuwan, S., Subhadhirasakul, S. and Voravuthikunchai, S.P., 2009. Medicinal plants with significant activity against important pathogenic bacteria. *Pharmaceutical Biology*, *47*(8), pp. 683–689.
Muliawan, S.Y., Kit, L.S., Devi, S., Hashim, O. and Yusof, R., 2006. Inhibitory potential of Quercus lusitanica extract on dengue virus type 2 replication.
Serrano, J., Puupponen-Pimiä, R., Dauer, A., Aura, A.M. and Saura-Calixto, F., 2009. Tannins: current knowledge of food sources, intake, bioavailability and biological effects. *Molecular Nutrition & Food Research*, *53*(S2), pp. S310–S329.
Sharma, A., Gupta, V.K. and Pathania, R., 2019. Efflux pump inhibitors for bacterial pathogens: From bench to bedside. *The Indian Journal of Medical Research*, *149*(2), p.129.
Wan, N.A.W., Masrah, M., Hasmah, A. and Noor, I.N., 2014. In vitro antibacterial activity of Quercus infectoria gall extracts against multidrug resistant bacteria. *Tropical Biomedicine*, *31*(4), pp. 680–688.

7.7.1.14 *Quercus myrsinifolia* Bl.

Synonym: *Cyclobalanopsis myrsinifolia* (Blume) Oerst.

Common names: Myrsine-leaved oak, Chinese evergreen oak; xiao ye qing gang (China); shirakashi (Japan); gasinamu (Korea); dẻ lá tre (Vietnam)

Habitat: Forests

Distribution: Japan, Korea, Laos, Thailand, Vietnam, China

Botanical observation: This tree grows up to about 20 m tall. The stems are terete and lenticelled. The leaves are simple, spiral, and stipulate. The stipule is extra-petiolar and caducous. The petiole is up to about 2.5 cm long. The blade is dull light green, 6–10.5 × 1.5–4.5 cm, elliptic-lanceolate, cuneate at base, crenate, acuminate to round at apex, and with 9–14 pairs of secondary nerves. The spikes are up to about 3 cm long. The perianth is 5–6-lobed. The androecium includes 5–6 stamens. The gynoecium develops 3 stigmas. The nuts are smooth, ovoid, glossy, up to about 2.5cm long, and seated on striated and 1 cm long cupules.

Medicinal use: Apparently none for the treatment of microbial infections

Viral enzyme inhibition by polar extract: Methanol extract of at 100 µg/mL inhibited Human immunodeficiency virus protease by 31% (Park et al., 2002).

Commentary: The antiviral activity is most probable owed to tannins which interact with amino acids of almost all enzymes to inhibit them (Oh et al., 1980; Hagerman, 1992). Further phytochemical and antimicrobial studies on this plant are needed.

REFERENCES

Hagerman, A.E., 1992. Tannin—protein interactions.
Oh, H.I., Hoff, J.E., Armstrong, G.S. and Haff, L.A., 1980. Hydrophobic interaction in tannin-protein complexes. *Journal of Agricultural and Food Chemistry*, *28*(2), pp. 394–398.
Park, J.C., Hur, J.M., Park, J.G., Hatano, T., Yoshida, T., Miyashiro, H., Min, B.S. and Hattori, M., 2002. Inhibitory effects of Korean medicinal plants and camelliatannin H from Camellia japonica on Human immunodeficiency virus type 1 protease. *Phytotherapy Research*, *16*(5), pp. 422–426.

7.7.1.15 *Quercus robur* L.

Synonym: *Quercus pedunculata* Ehrh.

Common names: Common oak; saplı meşe, meşe palamudu (Turkey); xia li (China)

Habitat: Forests

Distribution: Turkey, Iran

Botanical observation: This magnificent tree grows up to about 50 m tall. The bark is beautifully somewhat longitudinally carved and the wood very durable. The stems are terete and lenticelled. The leaves are simple, spiral, and stipulate. The stipule is extra-petiolar and caducous. The petiole is up to about 5 mm long. The blade is dull light green, 5–17 × 2–10 cm, obovate, rounded at base, lobed, acute at apex, and with 5–7 pairs of secondary nerves. The spikes are up to about 2 cm long. The perianth is 6-lobed. The androecium includes 5–6 stamens. The gynoecium develops 3 stigmas. The nuts are smooth, ovoid to ellipsoid, glossy and smooth, up to about 2 cm long and seated on scaly cupules, which are up to about 8 mm long.

Medicinal use: Antiseptic (Turkey)

Broad-spectrum antibacterial polar extract: Methanol extract (50mL of a 1 mg/5mL solution/well) inhibited the growth of *Escherichia coli*, *Klebsiella pneumoniae*, *Pseudomonas aeruginosa*, *Proteus mirabilis*, *Salmonella typhi*, *Salmonella paratyphi* A, *Salmonella paratyphi* B, *Salmonella typhimurium*, *Shigella boydii*, *Shigella flexneri*, *Shigella sonnei*, *Staphylococcus aureus*, and *Streptococcus faecalis* with inhibition zone diameters of 10.2, 11.8, 11.4, 10.2, 12.3, 13.3, 12.8, 10.2, 10.6, 10.7, 11.5, 15.2, and 14.3 mm, respectively (Limsuwan et al., 2009).

Weak antibacterial (Gram-negative) simple phenolics: 1,2,3-Benzenetriol and 4-propyl-1,3-benzenediol isolated from the bark demonstrated MIC values of 1.7 and 0.5 mM, respectively toward *Chromobacterium violaceum* (CV026) (Deryabin & Tolmacheva, 2015).

Commentaries: (i) *Chromobacterium violaceum* is a rare facultative anaerobic Gram-negative bacterium living in soil and water of tropical and subtropical areas, which causes life-threatening sepsis and abscesses, especially in Southeast Asia (Steinberg, 2009). (ii) Hydrolyzable tannins are most probably involved here. Note that high molecular weight tannins are less active toward Gram-negative bacteria because of the molecular mass and solubility restriction imposed by porins.

REFERENCES

Deryabin, D.G. and Tolmacheva, A.A., 2015. Antibacterial and anti-quorum sensing molecular composition derived from quercus cortex (oak bark) extract. *Molecules*, *20*(9), pp. 17093–17108.

Limsuwan, S., Subhadhirasakul, S. and Voravuthikunchai, S.P., 2009. Medicinal plants with significant activity against important pathogenic bacteria. *Pharmaceutical Biology*, *47*(8), pp. 683–689.

Steinberg, J.P., 2009. Other gram-negative and gram-variable bacilli. *Mandell, Douglas, and Bennett's Principles and Practice of Infectious Diseases*, *3023*.

7.7.2 Family Juglandaceae A.P. de Candolle ex Perleb (1818)

The family Juglandaceae consists of 9 genera and 60 species of trees. The leaves are mostly compound, alternate, and exstipulate. The inflorescences are spikes. The flowers are minute and without perianth. The calyx comprises up to 4 sepals. The androecium is minute and comprises 3–40 free stamens. The gynoecium consists of 2 fused carpels forming a unilocular ovary sheltering an ovule. The fruits are nuts in a dehiscent or indehiscent husk.

7.7.2.1 *Engelhardia roxburghiana* Wall.

Synonyms: *Alfaropsis roxburghiana* (Wall.) Iljinsk.; *Engelhardia chrysolepis* Hance; *Engelhardia fenzelii* Merr.; *Engelhardia formosana* (Hayata) Hayata; *Engelhardia unijuga* Chun ex P.Y. Chen; *Engelhardia wallichiana* Lindl.

Common name: Huang qi (China)

Habitat: Forests

Distribution: Pakistan, China, Taiwan, Cambodia, Indonesia, Laos, Myanmar, Thailand, and Vietnam

Botanical observation: This poisonous tree grows to 30m tall. The leaves are alternate, pinnate, and exstipulate. The petiole is up to about 8cm long. The blade is up to 25 cm long, glossy, somewhat coriaceous, and presents 2–10 folioles. The folioles are elliptic-lanceolate, 4.5–14.5×1.5–5.5 cm, marked with 5–12 pair of secondary nerves, entire or serrate, asymmetrical at base, and acuminate or shortly acuminate at apex. The spikes are yellowish, axillary or cauliflorous, yellow, and about 15cm long. The flowers are minute. The calyx comprises 1–4 sepals. The androecium comprises 3–15 anthers. The gynoecium is 2-lobed. The nuts are globose, about 5mm long, and develop 3 wings, which are up to about 5cm long.

Medicinal use: Apparently none for the treatment of microbial infections

Strong antimycobacterial amphiphilic cyclic diarylheptanoids: Engelhardione (LogD=2.8 at pH 7.4; molecular mass=312.3 g/mol) isolated from the roots inhibited *Mycobacterium tuberculosis* (90–221 387) and *Mycobacterium tuberculosis* (H37Rv) with MIC values of 3.1 and 2 µg/mL (Lin et al., 2005). 3′,4″-Epoxy-1-(4-hydroxyphenyl)-7-(3-hydroxyphenyl)heptan-3-one from the stems inhibited *Mycobacterium tuberculosis* (H37Rv) with an MIC value of 9.1 µg/mL (Wu et al., 2012).

Moderate antimycobacterial cyclic diarylheptanoids: Engelhardiol A and B from the stems inhibited *Mycobacterium tuberculosis* (H37Rv) with MIC values of 72.7 and 62.1 µg/mL, respectively (Wu et al., 2012).

Strong antimycobacterial amphiphilic tetralone: 4-Hydroxy-1-tetralone (LogD=1.2 at pH 7.4; molecular mass=162.1 g/mol) isolated from the roots inhibited clinical susceptible isolate of *Mycobacterium tuberculosis* (90–221 387) and *Mycobacterium tuberculosis* (H37Rv) with MIC values of 6.2 and 4 µg/mL, respectively (Lin et al., 2005).

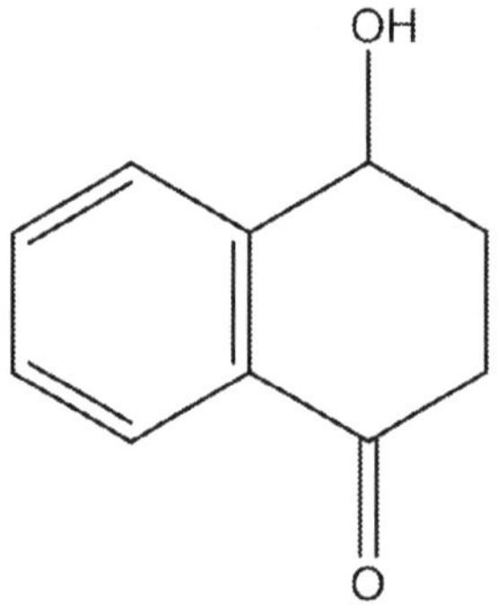

4-Hydroxy-1-tetralone

Strong antimycobacterial naphthoquinone: 3-Methoxyjuglone isolated from the roots inhibited *Mycobacterium tuberculosis* (90–221 387) with an MIC value of 3.1 µg/mL and *Mycobacterium tuberculosis* (H37Rv) with an MIC value of 0.2 µg/mL (Lin et al., 2005). In a subsequent study, 3-methoxyjuglone from the stems inhibited *Mycobacterium tuberculosis* (H37Rv) with an MIC value of 15.3 µg/mL (Wu et al., 2012).

Viral enzyme inhibition by flavanone glycoside: Engeletin inhibited H1N1 neuraminidase activity with an IC_{50} value of 44.4µM (Li et al., 2007) and the Human immunodeficiency integrase with an IC_{50} value of 75.7 µg/mL (Itharat et al., 2007).

FIGURE 7.29 *Engelhardia roxburghiana* Wall.

Commentary: (i) The plant generates the flavanone glycosides engeletin and astilbin (Huang et al., 2011). (ii) Consider that flavanes and flavones or their glycosides often inhibit the Influenza virus neuraminidase, and just like Zanamivir (an example of a clinically used anti-influenza neuraminidase inhibitor) (Grienke et al., 2012) and present pyran moieties.

REFERENCES

Grienke, U., Schmidtke, M., von Grafenstein, S., Kirchmair, J., Liedl, K.R. and Rollinger, J.M., 2012. Influenza neuraminidase: a druggable target for natural products. *Natural Product Reports*, *29*(1), pp. 11–36.

Itharat, A., Kejik, R., Tewtrakul, S. and Watanaperomskul, C., 2007. Antioxidant and anti-HIV-1 integrase compounds from Smilax corbularia Kunth. *Planta Medica*, *73*(09), p.P_559.

Li, X., Ohtsuki, T., Shindo, S., Sato, M., Koyano, T., Preeprame, S., Kowithayakorn, T. and Ishibashi, M., 2007. Mangiferin identified in a screening study guided by neuraminidase inhibitory activity. *Planta Medica*, *73*(11), pp. 1195–1196.

Lin, W.Y., Peng, C.F., Tsai, I.L., Chen, J.J., Cheng, M.J. and Chen, I.S., 2005. Antitubercular constituents from the roots of Engelhardia roxburghiana. *Planta Medica*, *71*(02), pp. 171–175.

Wu, H.C., Cheng, M.J., Peng, C.F., Yang, S.C., Chang, H.S., Lin, C.H., Wang, C.J. and Chen, I.S., 2012. Secondary metabolites from the stems of Engelhardia roxburghiana and their antitubercular activities. *Phytochemistry*, *82*, pp. 118–127.

7.7.2.2 *Juglans mandshurica* Maxim.

Synonyms: *Juglans cathayensis* Dode; *Juglans cathayensis* var. *formosana* (Hayata) A.M. Lu & R.H. Chang; *Juglans collapsa* Dode; *Juglans draconis* Dode; *Juglans formosana* Hayata; *Juglans stenocarpa* Maxim.

Common names: Manchurian wall nut tree; hu tao qiu (China)

Habitat: Forests

Distribution: China, Korea

Botanical observation: This handsome tree grows to 20 m tall. The stems are hairy at apex. The leaves are alternate, pinnate, and exstipulate. The petiole is up to about 20 cm long and hairy. The rachis is hairy. The blade is up to 90 cm long and presents 7–19 folioles. The folioles are sessile, shorter near the petiole, dull light green, elliptic-lanceolate, 5.5–7.5 × 1.5–5.5 cm, marked with 8–12 pairs of inconspicuous secondary nerves, serrulate, cordate at base, and acuminate at apex. The spikes are pendulous, axillary and up to 25 cm long. The flowers are minute. The calyx comprises 4 sepals. The androecium comprises 12–40, anthers. The gynoecium is 2-lobed. The nuts are tear-shaped, edible (China), ridged, and about 6.5 cm long.

Medicinal use: Skin infection (China)

Strong antibacterial (Gram-positive) naphthoquinone: Juglonol B isolated from the exocarps inhibited *Staphylococcus aureus* with an MIC value of 8 µg/mL (Yang et al., 2018).

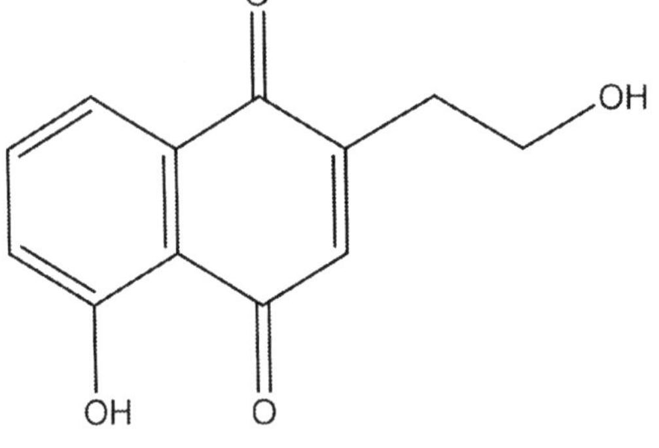

Juglonol B

Strong antibacterial (Gram-positive) hydrophilic antibacterial naphthoquinone: Juglone (LogD = 0.9 at pH 7.4; molecular mass = 174.1 g/mol) inhibited the growth of *Staphylococcus aureus* (ATCC 6538) and *Bacillus subtilis* (ATCC 6633) with MIC values of 12.5 and 12.5, µg/mL, respectively (Clarck et al. 1990). In a subsequent study, juglone inhibited the growth of *Bacillus cereus* (ATCC 14579), *Vibrio alginolyticus* (ATCC 33787), *Escherichia coli* (ATCC 35218), *Salmonella enteria* (ATCC 14028), *Staphylococcus aureus* (ATCC 25923), *Listeria monocytogenes* (ATCC 19115), *Enterococcus faecalis* (ATCC 19115), *Staphylococcus epidermidis* (CIP 106510), *Pseudomonas aeruginosa* (ATCC 27853), and *Staphylococcus aureus* (clinical) (Zmantar et al., 2016).

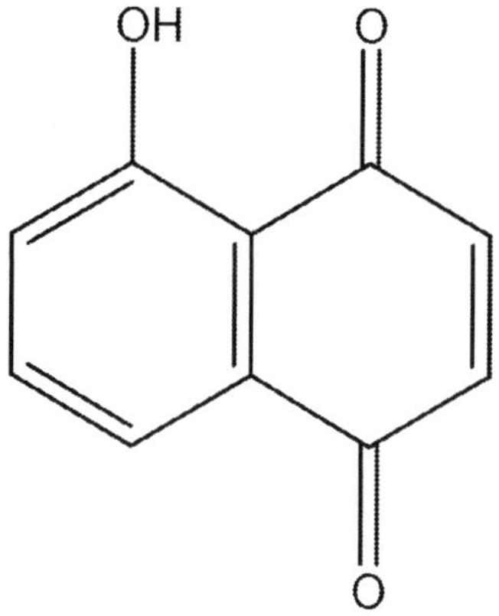

Juglone

Strong antimycobacterial hydrophilic antibacterial naphthoquinone: Juglone inhibited the growth of *Mycobacterium smegmatis* (ATCC 607) with an MIC values of 0.7 µg/mL (Clarck et al. 1990).

Antibiotic potentiator naphthoquinone: Juglone (at 1/2 MIC) reduced the MIC value of erythromycin by eightfold against *Staphylococcus aureus* (ATCC 25923) and *Staphylococcus epidermidis* (CIP 106510) (Zmantar et al., 2016). This naphthoquinone inhibited the efflux of ethynium bromine by 50% at a concentration of 182 µg/mL indicating inhibition of efflux pumps (Zmantar et al., 2016).

Strong broad-spectrum antifungal hydrophilic naphthoquinone: Juglone inhibited the growth of *Candida albicans* (ATCC 10231), *Saccharomyces cerevisiae* (ATCC 9763), *Cryptococcus neoformans*, *Aspergillus flavus*, and *Aspergillus fumigatus* with MIC values of 0.7, 0.3, 0.7, 6.2, and 6.2 µg/mL, respectively (Clarck et al. 1990).

Viral enzyme inhibition by polar extract: Methanol extract of leaves at a concentration of 100 µg/mL inhibited Human immunodeficiency virus type-1 reverse transcriptase (Min et al., 2000).

Viral enzyme inhibition by gallotannins: 1,2,6-trigalloylglucopyranose and 1,2,3,6-tetragalloyl glucopyranose from the bark inhibited Human immunodeficiency virus type-1 reverse transcriptase with an IC_{50} value of 0.04 and 0.06 µM, respectively (Min et al., 2000).

Viral enzyme inhibition by tetralone glycoside: 4*α*,5,8-trihydroxy- *α*-tetralone 5-*O*-β-D-[6′-O-(3″,4″,5″-trihydroxybenzoyl)] glucopyranoside isolated from the bark inhibited the replication of the Human immunodeficiency virus type-1 reverse transcriptase with an IC_{50} value of 5.8 µM (Min et al., 2001).

Viral channel inhibition by flavonol glycoside: Juglanin also known as kaempferol 3-*O*-arabinoside at a concentration of 20 µM inhibited coronavirus ion channel 3a-mediated current by 100% (Dong et al., 2018; Schwarz et al., 2014).

Commentaries: (i) Gallotannins and hydrolyzable tannins inhibit almost all enzymes *in vitro* by their intrinsic virtue of binding to amino acids. Also, these compounds are not absorbed in the blood stream and are of no therapeutic value for the treatment of Human immunodeficiency virus or any systemic microbial infection. (ii) The plant contains juglone (Yang et al., 2009), which is a common antimicrobial naphthoquinone in members of the family Juglandaceae (Hirakawa et al., 1986) including *Juglans regia* L. (iii) Coronavirus encode for 3a protein, which forms ion channels that become incorporated into the membrane of the host cells (Lu et al., 2006).

REFERENCES

Clark, A.M., Jurgens, T.M. and Hufford, C.D., 1990. Antimicrobial activity of juglone. *Phytotherapy Research*, *4*(1), pp. 11–14.

Dong, Z.W. and Yuan, Y.F., 2018. Juglanin suppresses fibrosis and inflammation response caused by LPS in acute lung injury. *International Journal of Molecular Medicine, 41*(6), pp. 3353–3365.

Hirakawa, K., Ogiue, E., Motoyoshiya, J. and Yajima, M., 1986. Naphthoquinones from juglandaceae. *Phytochemistry*, *25*(6), pp. 1494–1495.

Min, B.S., Kim, Y.H., Tomiyama, M., Nakamura, N., Miyashiro, H., Otake, T. and Hattori, M., 2001. Inhibitory effects of Korean plants on HIV-1 activities. *Phytotherapy Research*, *15*(6), pp. 481–486.

Min, B.S., Lee, H.K., Lee, S.M., Kim, Y.H., Bae, K.H., Otake, T., Nakamura, N. and Hattori, M., 2002. Anti-human immunodeficiency virus-type 1 activity of constituents from Juglans mandshurica. *Archives of Pharmacal Research, 25*(4), pp. 441–445.

Min, B.S., Nakamura, N., Miyashiro, H., Kim, Y.H. and Hattori, M., 2000. Inhibition of human immunodeficiency virus type 1 reverse transcriptase and ribonuclease H activites by constituents of Juglans mandshurica. *Chemical and Pharmaceutical Bulletin, 48*(2), pp. 194–200.

Schwarz, S., Sauter, D., Wang, K., Zhang, R., Sun, B., Karioti, A., Bilia, A.R., Efferth, T. and Schwarz, W., 2014. Kaempferol derivatives as antiviral drugs against the 3a channel protein of coronavirus. *Planta Medica*, *80*(02/03), pp. 177–182.

Yang, D., Li, S., Li, S., Li, J., Sun, M. and Jin, Y., 2009. Effect of juglone from Juglans mandshurica bark on the activity of wood decay fungi. *Forest Products Journal*, *59*(9).

Yang, Q., Yao, Q.S., Kuang, Y., Zhang, Y.Z., Feng, L.L., Zhang, L., Guo, L., Xie, Z.P. and Zhang, S.M., 2018. Antimicrobial and cytotoxic juglones from the immature exocarps of Juglans mandshurica. *Natural Product Research*, pp. 1–7.

Zmantar, T., Miladi, H., Kouidhi, B., Chaabouni, Y., Slama, R.B., Bakhrouf, A., Mahdouani, K. and Chaieb, K., 2016. Use of juglone as antibacterial and potential efflux pump inhibitors in Staphylococcus aureus isolated from the oral cavity. *Microbial Pathogenesis*, *101*, pp. 44–49.

7.7.2.3 *Juglans regia* L.

Synonyms: *Juglans duclouxiana* Dode; *Juglans fallax* Dode; *Juglans kamaonia* (C. DC.) Dode; *Juglans orientis* Dode; *Juglans sinensis* (C. DC.) Dode

Common names: Common walnut tree; jawz i rumi (Afghanistan); hu tao (China); akhoda, akhroat, kharote (India); thitcha (Myanmar); sano okhar, okhar, katu (Nepal); akhrot, mattak (Pakistan); girdu (Iran); adi ceviz, goz, joz, (Turkey)

Habitat: Forests, villages, cultivated

Distribution: Turkey, Iran, Afghanistan, Kazakhstan, Pakistan, India, Himalaya, Tibet, Nepal, Myanmar, China

Botanical observation: This magnificent tree grows to 25m tall and has an aura of freshness and positive vital force. The stems are hairy at apex. The leaves are alternate, imparipinnate, and exstipulate. The petiole is up to about 5–7 cm long. The blade is up to 40 cm long, and presents 5–9 folioles, which are subsessile, pleasantly light green, elliptic-ovate, 7–19.5×2.5–8.5 cm with 8–12 pairs of inconspicuous secondary nerves, rounded at base, and acuminate or acute at apex. The spikes are pendulous, axillary, and up to 15 cm long. The flowers are minute. The calyx comprises 4 sepals. The androecium comprises 10–30, anthers. The gynoecium is ovoid and develops 2 stigmas. The nuts are ovoid (Sacred in Nepal), about 3 cm long smooth, dull green, and up to about 5 cm long. The seeds are delicious, somewhat brain-shaped, and nutritious.

Medicinal uses: Acne, herpes, leucorrhea, tuberculosis, ulcers, syphilis (India); abscesses, cough (Turkey); dental caries (Pakistan, Turkey); athlete's foot, sore throat (Nepal)

Strong antibacterial (Gram-positive) *polar extract:* Ethanol extract of aerial parts inhibited *Listeria monocytogenes* with an MIC of 12.5 μg/mL (Altanlar et al., 2006).

Moderate antimycobacterial non-polar extract: Hexane extract of bark inhibited the growth of *Mycobacterium tuberculosis* with MIC value of 100 μg/mL (Cruz-Vega et al., 2008).

Moderate broad-spectrum antibacterial tannin fraction: Tannin fraction of seeds inhibited the growth of *Listeria monocytogenes, Staphylococcus aureus, Escherichia coli,* and *Salmonella typhimurium* with MIC values of 125, 500, 125, and 250 μg/mL, respectively (Amarowicz et al., 2008).

Weak broad-spectrum polar extract: Methanol extract of bark inhibited methicillin-resistant *Staphylococcus aureus, Bacillus subtilis, Micrococcus* sp., *Escherichia coli* (clinical), *Streptococcus pyogenes* (Clinical), *Streptococcus pneumoniae* (Clinical), and *Pasturella multocida* (Clinical) with MIC/MBC of 0.3/1.5, 1/2.5, 0.7/1.5, 1/2.1, 1.2/5, 1.2/5, and 2.5/5 mg/mL, respectively (Farooqui et al., 2015).

Antibiotic potentiator polar extract: Methanol extract of bark evoked a 64 times reduction in MIC of oxacillin against *Staphylococcus aureus*. The extract was also synergistic with nalidixic acid against *Salmonella enteritica* serovar typhi (Farooqui et al., 2015).

Moderate anticandidal polar extract: Methanol extract of roots inhibited the growth of *Candida albicans* (ATCC 10261), *Candida tropicalis* (SN1982), *Candida guillerimondii* (SN 2006), *Candida glabrata* (SN 2266), *Candida parapsilosis* (SN 1980), and *Candida albicans* (SN 2320) (Raja et al., 2017). The extract at 500 μg/mL evoked cell wall and cell membrane insults in *Candida albicans* (SN 2320) (Raja et al., 2017).

Moderate antifungal (filamentous) naphthoquinone: Juglone from the exocarps inhibited the growth of *Alternaria alternata, Fusarium culmorum, Phytophtora infestans, Rhizoctonia solani,* and *Ascosphaera apis* (ATCC 13785) with MIC values of 1.7, 2.4, 0.4, 0.4, and 0.05 mg/mL, respectively (Wianowska et al., 2016).

Strong antiviral (enveloped monopartite linear single-stranded (+)RNA) polar extract: Methanol extract inhibited the growth of the Sindbis virus with an IC_{50} value of 1.5 μg/mL (Mouhajir et al., 2001).

Strong antiviral (enveloped linear monopartite double-stranded DNA virus) polar extract: Methanol extract inhibited the growth of the Herpes simplex virus with an IC_{50} value of 1.5 μg/mL (Mouhajir et al., 2001).

Strong antiviral (non-enveloped monopartite linear dimeric single-stranded (+)RNA) polar extract: Methanol extract inhibited the growth of the Poliovirus with an IC_{50} value of 25 μg/mL (Mouhajir et al., 2001).

Viral enzyme inhibition by napthoquinone: Juglone inhibited Human immunodeficiency virus type-1 reverse transcriptase with an IC_{50} value of 95 μM (Min et al., 2002).

Commentaries*:* (i) *Streptococcus pneumoniae* is a Gram-positive coccus responsible for otitis media, pneumonia, meningitis, and septicemia which lead to death in untreated immunodeficient patients (Song et al., 2017). This bacteria produces the virulence factor pneumolysin, which forms pores in host cell membranes (Song et al., 2017). Juglone from *Juglans regia* L. prevented pneumolysin-induced human alveolar epithelial (A549) cell injury by about 90 % at a concentration of 20.8 μg/mL (Song et al., 2017).

(ii) Our peer, the Persian physician Abu Bakr Muḥammad ibn Zakariya al-Razi also known as Rhazes (865–925), in his *Kitab al-Mansouri fi al-Tibb*, describes walnut as hot and dry and recommends gargling with a solution of this plant in times of pandemics (Jazani et al., 2018; Razi, 1991). Rhazes' *Kitab fi Al Jadari wa Al Hasaba*, or treatise on small pox and measles (*Rhazes de variolis et morbillis*), has been used in European medicine for centuries. Our peer, the British surgeon Thomas Thayre, in his "Treatise of the pestilence" published in 1603 recommended the use of walnut against the London Plague which was caused by to the Gram-negative zoonotic bacillus *Yersinia pestis*.

REFERENCES

Altanlar, N., Saltan Çitoğlu, G. and Yılmaz, B.S., 2006. Antilisterial activity of some plants used in folk medicine. *Pharmaceutical Biology*, *44*(2), pp. 91–94.

Amarowicz, R., Dykes, G.A. and Pegg, R.B., 2008. Antibacterial activity of tannin constituents from Phaseolus vulgaris, Fagoypyrum esculentum, Corylus avellana and Juglans nigra. *Fitoterapia*, *79*(3), pp. 217–219.

Cruz-Vega, D.E., Verde-Star, M.J., Salinas-González, N., Rosales-Hernández, B., Estrada-García, I., Mendez-Aragón, P., Carranza-Rosales, P., González-Garza, M.T. and Castro-Garza, J., 2008. Antimycobacterial activity of Juglans regia, Juglans mollis, Carya illinoensis and Bocconia frutescens. *Phytotherapy Research, 22*(4), pp. 557–559.

Farooqui, A., Khan, A., Borghetto, I., Kazmi, S.U., Rubino, S. and Paglietti, B., 2015. Synergistic antimicrobial activity of Camellia sinensis and Juglans regia against multidrug-resistant bacteria. *PloS One*, *10*(2), p. e0118431.

Jazani, A.M., Maleki, R.F., Kazemi, A.H., Matankolaei, L.G., Targhi, S.T., Kordi, S., Rahimi-Esboei, B. and Azgomi, R.N.D., 2018. Intestinal helminths from the viewpoint of traditional Persian medicine versus modern medicine. *African Journal of Traditional, Complementary and Alternative Medicines, 15*(2), pp. 58–67.

Min, B.S., Miyashiro, H. and Hattori, M., 2002. Inhibitory effects of quinones on RNase H activity associated with HIV-1 reverse transcriptase. *Phytotherapy Research*, *16*(S1), pp. 57–62.

Mouhajir, F., Hudson, J.B., Rejdali, M. and Towers, G.H.N., 2001. Multiple antiviral activities of endemic medicinal plants used by Berber peoples of Morocco. *Pharmaceutical Biology, 39*(5), pp. 364–374.

Raja, V., Ahmad, S.I., Irshad, M., Wani, W.A., Siddiqi, W.A. and Shreaz, S., 2017. Anticandidal activity of ethanolic root extract of Juglans regia (L.): Effect on growth, cell morphology, and key virulence factors. *Journal de mycologie medicale*, *27*(4), pp. 476–486.

Razi, Z., 1991. *Kitab al-Mansoori*. Central Council for Research in Unani Medicine, New Delhi.

Song, M., Lu, G., Li, M., Deng, X. and Wang, J., 2017. Juglone alleviates pneumolysin-induced human alveolar epithelial cell injury via inhibiting the hemolytic activity of pneumolysin. *Antonie van Leeuwenhoek*, pp. 1–7.

Wianowska, D., Garbaczewska, S., Cieniecka-Roslonkiewicz, A., Dawidowicz, A.L. and Jankowska, A., 2016. Comparison of antifungal activity of extracts from different Juglans regia cultivars and juglone. *Microbial Pathogenesis, 100*, pp. 263–267.

7.7.2.4 *Pterocarya stenoptera* C. DC.

Synonyms: *Acer mairei* H. Lév.; *Pterocarya chinensis* Lavallée; *Pterocarya esquirollii* H. Lév.; *Pterocarya japonica* Dippel; *Pterocarya japonica* Lavallée; *Pterocarya laevigata* Lavallée

Common names: Chinese wingnut; feng yang (China)

Habitat: Forests

Distribution: Myanmar, Vietnam, China, Korea, and Japan

Botanical observation: This poisonous tree grows to about 25 m tall. The stems are terete and glabrous. The leaves are alternate, paripinnate, and exstipulate. The petiole is up to about 6 cm long. The blade is up to 25 cm long and presents 5–10 pairs of sessile folioles. The rachis is minutely winged. The folioles are elliptic-oblong, dull light green, 8–12.5 × 2.5–3.5 cm with 8–16 pairs of inconspicuous secondary nerves, asymmetrical at base, serrate, and acuminate or acute at apex. The inflorescences are pendulous axillary or terminal spikes. The flowers are minute. The calyx comprises 4 sepals. The androecium comprises 5–18 anthers. The gynoecium develops 2 stigmas. The infructescences are showy, pendulous, about 25 cm long. The fruits are about 7 mm nuts with a pair of oblong green, nerved, and somewhat glossy wings, which are up to 2.5 cm long.

Medicinal uses: Apparently none for the treatment of microbial infections

Moderate antibacterial (Gram-positive) essential oil: The essential oil of leaves inhibited the growth of *Bacillus subtilis* with an MIC value of 230 μg/mL (Yin et al., 2019).

Strong antiviral (enveloped monopartite linear double-stranded DNA) hydrolysable tannin: The ellagitannin pterocarnin A isolated from the bark inhibited the replication of the Herpes simplex virus type-2 with an IC_{50} of 5.4 μM in Vero cells (Cheng et al., 2004). This ellagitannin inhibited attachment and penetration of the Herpes simplex virus type-2 virus to Vero cells as well as late virus infection (Cheng et al., 2004).

Commentary: Antibacterial or antifungal derivatives of 1,4-naphthoquinone are known to occur in the genus *Pterocarya* Kunth (Cheng et al., 2004).

REFERENCES

Cheng, H.Y., Lin, T.C., Yang, C.M., Wang, K.C. and Lin, C.C., 2004. Mechanism of action of the suppression of herpes simplex virus type 2 replication by pterocarnin A. *Microbes and Infection*, *6*(8), pp. 738–744.

Yin, C., Sun, F., Rao, Q. and Zhang, Y., 2019. Chemical compositions and antimicrobial activities of the essential oil from Pterocarya stenoptera C. *DC. Natural Product Research*, pp. 1–4.

7.8 ORDER ROSALES BERCHT. & J.PRESL (1820)

7.8.1 Family Cannabaceae Martinov (1820)

The Cannabaceae family consists of 2 genera and 3 species of herbs and climbers. The leaves are opposite, alternate, palmately lobed or palmately compound, serrate, and stipulate. The flowers are minute, develop 5 tepals, 5 stamens and a 1-locular ovary containing a single pendulous ovule. The style develops a bilobed stigma. The fruits are achenes, covered with the persistent perianth.

7.8.1.1 *Cannabis sativa* L.

Synonyms: *Cannabis chinensis* Delile; *Cannabis indica* Lam.; *Cannabis sativa* var. *indica* (Lam.) E. Small & Cronquist

Common names: Cannabis, hemp, Indian hemp, marihuana; bhang, ganja, vijaya, indrasana ganja (India); bhang (Pakistan, Nepal); bhangey (Pakistan); damo (the Philippines); pang (Thailand); lanh mán (Vietnam)

Habitat: Cultivated

Distribution: Iran, Pakistan, India, Bhutan, Bangladesh, Thailand, China, Malaysia, Indonesia, the Philippines

Botanical observation: This sinister shrubby herb grows to a height of 3 m. It has an aura of dryness. The plant is sacred in Nepal. The stems are terete and hairy. The leaves are simple or palmate, spiral, and stipulate. The stipules are about 5 mm long. The petiole of basal leaves grows up to about 12 cm long. The blade of the basal leaves comprises up to 9 folioles, which are lanceolate, serrate, acuminate at apex, 2–11 cm × 3 mm–1.5 cm, and hairy. The blade of cauline leaves is hairy below, 6–10.5 × 0.6–1.1 cm, papery and marked with with 6–10 pairs of secondary nerves. The male flowers are grouped in panicles. The female flowers are grouped in compact cymes mixed with foliaceous bracts. The flowers are tiny and 5-lobed. The fruits are ovoid achenes.

Medicinal uses: Leprosy, pneumonia, wounds (India); dysentery (India; Nepal); bronchitis (Pakistan, India, Nepal); anthrax, fever (Pakistan); diarrhea (India, Pakistan); cold (Nepal); cough (India, Pakistan, Nepal); bronchitis (India, Pakistan)

Strong antibacterial (Gram-positive) lipophilic cannabinoids: Cannabidiol (LogD = 6.4 at pH 7.4; molecular mass = 314.4 g/mol), cannabichromene, cannabigerol, Δ^9-tetrahydrocannabinol, and cannabinol inhibited the growth of *Staphylococcus aureus* (1199B) with MIC values of 1,

2, 1, 2, and 1 µg/mL, respectively (Appendino et al., 2008). Cannabidiol, cannabichromene, cannabigerol, Δ_9-tetrahydrocannabinol, and cannabinol inhibited the growth of *Staphylococcus aureus* (ATCC 25923) with MIC values of 0.5, 2, 1, 1, and 1 µg/mL, respectively (Appendino et al., 2008).

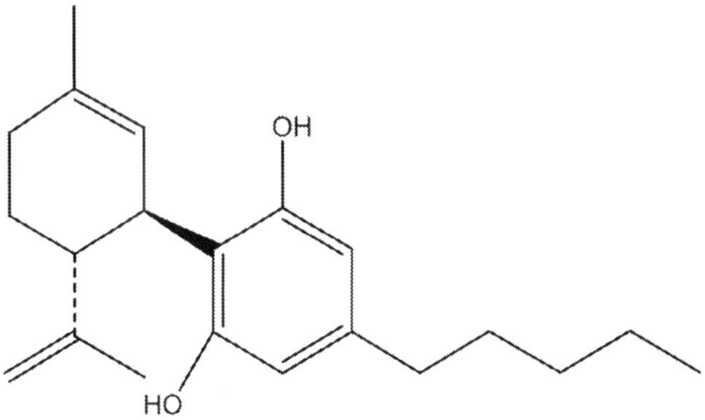

Cannabidiol

Strong antibacterial coumarin: 3″-Hydroxy-Δ(4″,5″) -cannabichromene inhibited the growth of methicillin-resistant *Staphylococcus aureus* and *Staphylococcus aureus* with IC_{50} values of 24.4 and 29.6 µM (Radwan et al., 2009).

Strong antibacterial prenylated phenolic: 4-Acetoxy-2-geranyl-5-hydroxy-3-n-pentylphenol inhibited the growth of methicillin-resistant *Staphylococcus aureus* and *Staphylococcus aureus* with IC_{50} values of 6.4 and 12.2 µM (Radwan et al., 2009).

Strong broad-spectrum antibacterial cannabinoid: 8-Hydroxycannabinolic acid A inhibited the growth of *Staphylococcus aureus* and *Escherichia coli* with IC_{50} values of 3.5 and 54 µM (Radwan et al., 2009).

Moderate antimycobacterial essential oil: Essential oil inhibited the growth of *Mycobacterium smegmatis* with an MIC value of 100 µg/mL (Fournier et al., 1978).

Strong antimycobacterial cannabinoid: 8-Hydroxycannabinolic acid A (molecular mass = 370.4 g/mol) inhibited the growth of *Staphylococcus aureus* and *Escherichia coli* (Radwan et al., 2009).

Strong anticandidal coumarin: 3″-hydroxy-Δ(4″,5″) -cannabichromene inhibited the growth of *Candida albicans* and *Candida krusei* with IC_{50} values of 60.5 and 60.5 µM (Radwan et al., 2009).

Strong anticandidal cannabinoid: 8-Hydroxycannabinol inhibited the growth of *Candida albicans* with an IC_{50} value of 4.6 µM (Radwan et al., 2009).

Moderate anticandidal cannabinoid: 8-Hydroxycannabinolic acid A inhibited the growth of *Candida krusei* with an IC_{50} value of 54 µM (Radwan et al., 2009).

Moderate anticandidal prenylated phenolic: 4-Acetoxy-2-geranyl-5-hydroxy-3-n-pentylphenol inhibited the growth of *Candida krusei* with an IC_{50} value of 53.4 µM (Radwan et al., 2009).

Viral enzyme inhibition by polar extract: Aqueous extract of fruits at a concentration of 250 µg/mL inhibited the Human immunodeficiency virus type-1 protease by 7.4% (Xu et al., 1996).

Commentary: (i) Cannabinoids being active on the central nervous system could be examined for their antibiotic potentiating effects (as discussed earlier). (ii) Consider that ethanol extract of leaves inhibited angiotensin-converting enzyme activity by 18% (Duncan et al., 1999), suggesting that the plant might have some activity against COVID-19 and other coronaviruses. (iii) The plant was known of our peer the Greek physician Claudius Galenus (129-216 A.D.), who describes it as desiccating and warming (Lozzano, 2001). The plague of Alexandria (1834–1838) was brought by a Greek ship on June 15, 1834. Our peer the French physician Louis Aubert-Roche (1809–1874) wrote his observations about the plague of Alexandria and its treatment with *Cannabis sativa* L in his "De la peste du typhus d' Orient" published in 1843 (Fankhauser, 2008). This plague was owed to *Vibrio cholerae* (Barbieri & Drancourt, 2018). As for the use of *Cannabis sativa* L. for COVID-19, its anti-inflammatory properties could be of clinical value (Gallily et al., 2018).

REFERENCES

Appendino, G., Gibbons, S., Giana, A., Pagani, A., Grassi, G., Stavri, M., Smith, E. and Rahman, M.M., 2008. Antibacterial cannabinoids from Cannabis sativa: a structure– activity study. *Journal of Natural Products*, *71*(8), pp. 1427–1430.

Barbieri, R. and Drancourt, M., 2018. Two thousand years of epidemics in Marseille and the Mediterranean Basin. New *Microbes* and *New Infections*, 26, pp. S4–S9.

Fournier, G., Paris, M.R., Fourniat, M.C., Quero, A.M., 1978. Bacteriostatic effect of essential oil from *Cannabis sativa*. *Annual Pharmacy of France*, *36*, p. 603.

Fankhauser, M., 2008. *Cannabis as medicine in Europe in the 19th century. EMCDDA MONOGRAPHS.* Luxembourg: Office for Official Publications of the European Communities.

Gallily, R., Yekhtin, Z. and Hanuš, L.O., 2018. The anti-inflammatory properties of terpenoids from cannabis. Cannabis and *Cannabinoid Research*, 3(1), pp. 282–290.

Lozano, I., 2001. The therapeutic use of Cannabis sativa (L.) in Arabic medicine. *Journal of Cannabis Therapeutics*, *1*(1), pp. 63–70.

Radwan, M.M., ElSohly, M.A., Slade, D., Ahmed, S.A., Khan, I.A. and Ross, S.A., 2009. Biologically active cannabinoids from high-potency Cannabis sativa. *Journal of Natural Products, 72*(5), pp. 906–911.

Xu, H.X., Wan, M., Loh, B.N., Kon, O.L., Chow, P.W. and Sim, K.Y., 1996. Screening of Traditional Medicines for their Inhibitory Activity Against HIV-1 Protease. *Phytotherapy Research, 10*(3), pp. 207–210.

7.8.1.2 *Humulus scandens* (Lour.) Merr.

Synonyms: *Antidesma scandens* Lour.; *Humulopsis scandens* (Lour.) Grudz.; *Humulus japonicus* Siebold & Zucc.

Common name: Japanese hop; lu cao (China)

Habitat: Forests margins and abandoned lands

Distribution: Vietnam, China, Korea, Taiwan, and Japan.

Botanical observation: This climber grows up to 10 m long. The stems are slender, terete, and hairy. The leaves are simple, opposite, and exstipulate. The petiole is 5–7 cm long, hairy, and grooved. The blade is papery, 3-lobed, 2–5.2 × 2.5–7.1 cm, hairy, and serrate. The inflorescences are tiny axillary racemes, which are up to 2 cm long. The flowers are 3 mm long. The calyx consists of 5 lanceolate sepals, which are free and hairy. The petals are glabrous, yellow, lanceolate, free, and opposite to the sepals. The achenes are minute.

Medicinal use: Dysentery (China, Vietnam)

Strong antimycobacterial mid-polar extract: Ethyl acetate inhibited the growth of *Mycobacterium tuberculosis* (H37Rv) with an MIC value of 31.5 µg/mL (Chen et al., 2008).

Moderate antimycobacterial polar extract: Butanol extract inhibited the growth of *Mycobacterium tuberculosis* (H37Rv) with an MIC value of 125 µg/mL (Guo et al., 2011).

Commentary: *Humulus lupulus* L. is used in India as an antiseptic (local name: *Hashish ut-dinaar*) and from this plant the prenylated phloroglucinols humulone and lupulone inhibited the growth of *Mycobacterium tuberculosis* with MIC values of 194 and 58 µg/mL, respectively (Pauli et al., 2005).

REFERENCES

Chen, W.G., Lin, X., Sui, F.Y., Yin, X., Tu, J.H., Sheng, J. and Huang, J., 2008. Inhibitory Effects against Tuberculous Bacillus of Extracts from Humulus scandens [J]. *Lishizhen Medicine and Materia Medica Research, 1.*

Guo, P.L., Ma, Y.Y., Lu, X.L. and Yang, Z.Z., 2011. The screening of the active component in Herba Humuli Scandens for the inhibition on mycobacterium tuberculosis in vitro. *Pharmacy and Clinics of Chinese Materia Medica, 2.*

Pauli, G.F., Case, R.J., Inui, T., Wang, Y., Cho, S., Fischer, N.H. and Franzblau, S.G., 2005. New perspectives on natural products in TB drug research. *Life Sciences, 78*(5), pp. 485–494.

7.8.2 Family Moraceae Link (1831)

The family Moraceae consists of about 40 genera and 1000 species of trees, shrubs, climbers, or herbs which are often laticiferous. Some members in this family have peculiar smooth or somewhat skin-like bark and anastomosis of stems. The stems are not uncommonly somewhat articulated. The leaves are simple, stipulate, often thick, large, glossy. The stipules often form a cap over the leaf buds. The flowers are tiny, packed in compact axillary very peculiar inflorescences, and comprise 4–5 sepals. The perianth is absent. The androecium includes 4 stamens. The gynoecium includes 2 carpels each containing 1 pendulous ovule and develops slender styles. The fruits are syncarps or drupes.

7.8.2.1 *Artocarpus heterophyllus* Lamk.

Synonym: *Artocarpus integrifolius* L. f.

Common names: Jack fruit; khnaor (Cambodia); bo luo mi (China); chakka, pala (India); nangka (Indonesia, Malaysia); mak mii (Laos); mak-lang, mung-dung, ndung, pa-noh, panwe, paung thi, peinne (Myanmar); kathal (Pakistan); kapiak (Papua New Guinea); langka (the Philippines); kha noon (Thailand); mit (Vietnam)

Habitat: Cultivated

Distribution: Tropical Asia and Pacific

Botanical observation: This tropical fruit tree probably is native to India grows to about 10 m tall. The bark is dark grey and exudes an abundant milky latex when incised. The stems are articulate. The leaves are simple, spiral, and stipulate. The stipule is up to 8 cm long. The petiole is 1.2–4 cm long. The blade is elliptic to obovate, leathery, glossy, deep green above, and 5–23×2–11 cm. The apex is acute and the base of the blade is tapered. The flowers are unisexual, tiny, and arranged in separate and axillary heads. The male flower consists of a tiny 2–4-lobed perianth and a single stamen. The female flowers consist of a tiny perianth tube, a single ovary sunk in the receptacle, and a long style. The fruits are 30–90×15–30 cm, ovoid, smelly, heavy, greenish yellow, covered with conical warts and cauliflorous. It is not uncommon to see in Southeast Asian villages the fruits being covered with bags or clothes for protection against decay. The seeds are about 4×2 cm, ovoid, brownish, smooth, and embedded within a thick orangish and glossy and fleshy jacket, which is edible but has a somewhat heavy and repulsive smell.

Medicinal uses: Apparently none for the treatment of microbial infections

Broad-spectrum antibacterial halo developed by mid-polar extract: Dichloromethane fraction of leaves (4 mg/disc) inhibited the growth of *Bacillus cereus, Bacillus coagulans, Bacillus megatarium, Bacillus subtilis, Micrococcus luteus, Staphylococcus aureus, Staphylococcus epidermidis, Streptococcus faecalis, Streptococcus pneumoniae, Citrobacter freundii, Enterobacter aerogenes, Escherichia coli, Klebsiella pneumoniae, Neisseria gonorrhoeae, Proteus mirabilis, Proteus vulgaris, Pseudomonas aeruginosa, Salmonella typhi, Salmonella typhimurium,* and *Serratia marcescens* (Khan et al., 2003).

Strong antibacterial (Gram-positive) *mid-polar extract:* Ethyl acetate extract of wood inhibited the growth of *Streptococcus mutans* (DMST 26095), *Streptococcus mitis* (ATCC 49456T) and *Streptococcus pyogenes* (DNAT 17020) with MIC/MBC values of 321.5/312.5, 19.5/78.1, and 19.5/39.0 μg/mL, respectively (Septama and Panichayupakaranant, 2016).

Strong antibacterial (Gram-positive) chromane: Cyanomaclurin from the wood inhibited the growth of *Streptococcus mutans* (DMST 26095) and *Escherichia coli* (ATCC 25922) with the MIC/MBC values of 6.8/54.4 and 27.2/54.4 µM, respectively (Septama and Panichayupakaranant, 2015).

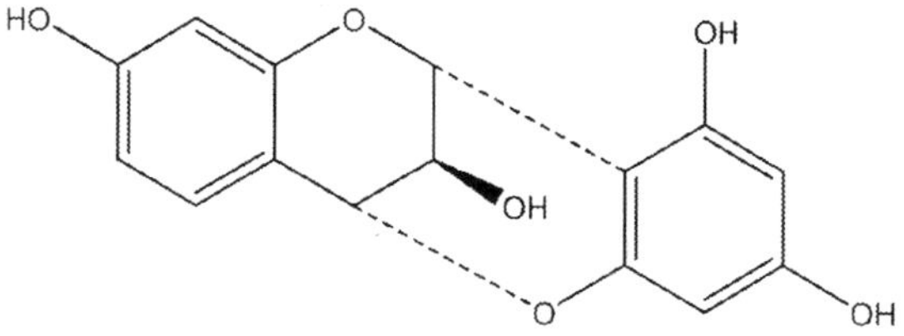

Cyanomaclurin

Strong antibacterial (Gram-positive) prenylated flavone: Artocarpin from the wood of inhibited the growth of *Escherichia coli* (ATCC 25922), *Pseudomonas aeruginosa* (DMST 15442), and methicillin-resistant *Staphylococcus aureus* (DMST 20654) with MIC values of 62.5, 250, and 65.5 µg/mL, respectively (Septamaand Panichayupakaranant, 2016). Artocarpin inhibited the growth of *Streptococcus mutans* (DMST 26095), *Streptococcus pyogenes* (DMST 17020), *Bacillus subtilis, Staphylococcus aureus* (ATCC 25923), *Staphylococcus epidermidis* (ATCC 14990), and *Escherichia coli* (ATCC 25922) with MIC/MBC values of 4.4/8.9, 4.4/8.9, 17.8/17.8, 8.9/8.9, 4.4/8.9, and 71.6/143.2 µM, respectively (Septama & Panichayupakaranant, 2015).

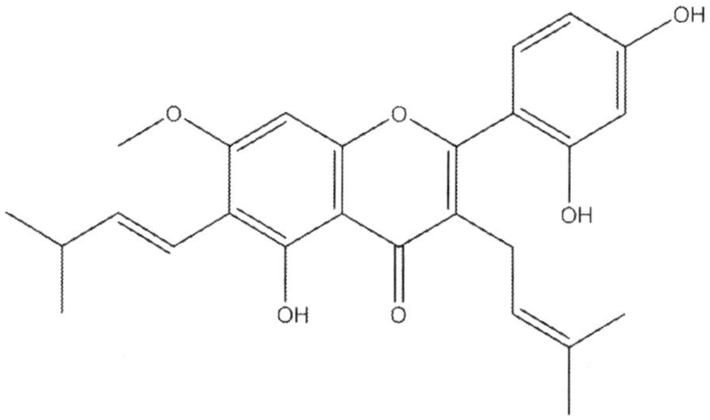

Artocarpin

Moderate broad-spectrum antibacterial flavanone: Artocarpanone from the wood inhibited the growth of *Streptococcus mutans* (DMST 26095), *Streptococcus pyogenes* (DMST 17020), *Bacillus subtilis*, *Staphylococcus epidermidis* (ATCC 14990), and *Escherichia coli* (Septama and Panichayupakaranant, 2015).

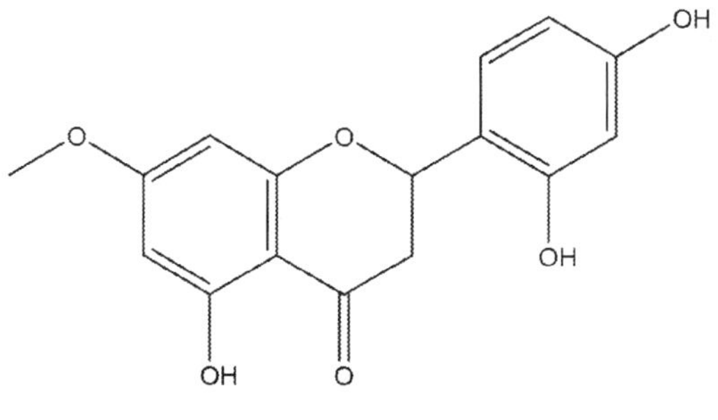

Artocarpanone

Antibiotic-potentiating prenylayed flavone: Artocarpin increased the antibacterial properties of ampicillin, tetracycline, norfloxacin for methicillin-resistant *Staphylococcus aureus* (Septamaand Panichayupakaranant, 2016.) Artocarpin was synergistic with tetracyclin and norfloxacin for *Pseudomoas aeruginosa* but no synergy was observed with ampicillin (Septamaand &Panichayupakaranant, 2016). Artocarpin for *Escherichia coli* was synergistic with norfloxacin but no synergy was observed with ampicillin and tetracyclin (Septamaand Panichayupakaranant, 2016).

Strong antifungal (filamentous) antifungal polar extract: Ethanol extract of seeds inhibited the growth of *Aspergillus niger, Penicillium notatum,* and *Candida albicans* with MIC values of 50, 100, and 100 µg/mL, respectively (Ramaiah & Garampalli, 2015).

Strong antiviral (enveloped monopartite linear single-stranded (+)RNA) mid-polar extract: Dichloromethane extract inhibited the replication of the Hepatitis C virus in Huh7 cells with an IC_{50} value of 1.5 µg/mL and a selectivity index above 135.8 via a mechanism involving virus entry inhibition (Hafid et al., 2017).

Commentaries: From the roots of *Artocarpus lakoocha* Wall. ex Roxb. (Local name in China: *Ye bo luo mi*) lakoochin A (LogD=6.3 at pH 7.4; molecular mass=406.5 g/mol) and lakoochin B inhibited the growth of *Mycobacterium tuberculosis* (H37Ra) with MIC values of 12.5 and 50 µg/mL (Puntumchai et al., 2004).

From the heartwood of this tree the stilbene oxyresveratrol (LogD=2.4 at pH 7.4; molecular mass=244.2 g/mol) inhibited the replication of the Herpes simplex virus type-1 and the Herpes simplex virus type-2 with EC_{50} values of 63.5 and 55.3 µM, respectively (CC_{50} values for Vero cells>100µM) (Likhitwitayawuid et al., 2005). In a subsequent study, oxyresveratrol inhibited the replication of the Herpes simplex virus type-1 (clinical isolate) with an IC_{50} value of 19.8 µg/mL (Chuanasa et al., 2008). Oxyresveratrol given orally to mice at a dose of 500 mg/Kg (3 times daily) diminished herpetic skin lesion (Chuanasa et al., 2008), inhibited the replication of the African swine fever virus, in Vero cells with an IC_{50} value of 10 µg/mL (Galindo et al., 2011) and the varicella-zoster virus (wild-type and acyclovir resistant) with IC_{50} values of 12.8 and 12.8 µg/mL, respectively (Sasivimolphan et al., 2009). Oxyresveratrol inhibited the replication of the Human immunodeficiency virus type-1 (LAI) with an IC_{50} value of 28.2 µM (Likhitwitayawuid et al., 2005). Being a broad-spectrum antiviral natural product with somewhat low toxicity, one could examine the anti-Severe acute respiratory syndrome-acquired coronavirus 2 activity of oxyresveratrol. Furthermore, this stilbene could be added to the list of natural products to be tested for forthcoming zoonotic pandemics.

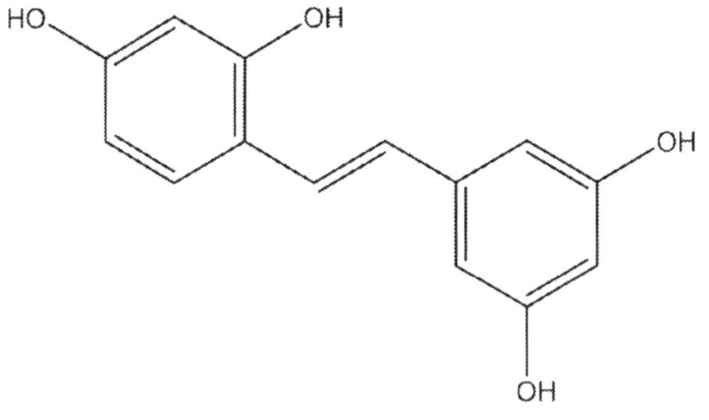

Oxyresveratrol

REFERENCES

Chuanasa, T., Phromjai, J., Lipipun, V., Likhitwitayawuid, K., Suzuki, M., Pramyothin, P., Hattori, M. and Shiraki, K., 2008. Anti-herpes simplex virus (HSV-1) activity of oxyresveratrol derived from Thai medicinal plant: mechanism of action and therapeutic efficacy on cutaneous HSV-1 infection in mice. *Antiviral Research*, *80*(1), pp. 62–70.

Galindo, I., Hernáez, B., Berná, J., Fenoll, J., Cenis, J.L., Escribano, J.M. and Alonso, C., 2011. Comparative inhibitory activity of the stilbenes resveratrol and oxyresveratrol on African swine fever virus replication. *Antiviral Research*, *91*(1), pp. 57–63.

Hafid, A.F., Aoki-Utsubo, C., Permanasari, A.A., Adianti, M., Tumewu, L., Widyawaruyanti, A., Wahyuningsih, S.P.A., Wahyuni, T.S., Lusida, M.I. and Hotta, H., 2017. Antiviral activity of the dichloromethane extracts from Artocarpus heterophyllus leaves against hepatitis C virus. *Asian Pacific Journal of Tropical Biomedicine*, *7*(7), pp. 633–639.

Khan, M.R., Omoloso, A.D. and Kihara, M., 2003. Antibacterial activity of Artocarpus heterophyllus. *Fitoterapia*, *74*(5), pp. 501–505.

Likhitwitayawuid, K., Sritularak, B., Benchanak, K., Lipipun, V., Mathew, J. and Schinazi, R.F., 2005. Phenolics with antiviral activity from Millettia erythrocalyx and Artocarpus lakoocha. *Natural Product Research*, *19*(2), pp. 177–182.

Puntumchai, A., Kittakoop, P., Rajviroongit, S., Vimuttipong, S., Likhitwitayawuid, K. and Thebtaranonth, Y., 2004. Lakoochins A and B, New Antimycobacterial Stilbene Derivatives from Artocarpus l akoocha. *Journal of Natural Products*, *67*(3), pp. 485–486.

Ramaiah, A.K. and Garampalli, R.K.H., 2015. In vitro antifungal activity of some plant extracts against Fusarium oxysporum f. sp. lycopersici. *Asian Journal of Plant Science and Research*, *5*(1), pp. 22–27.

Sasivimolphan, P., Lipipun, V., Likhitwitayawuid, K., Takemoto, M., Pramyothin, P., Hattori, M. and Shiraki, K., 2009. Inhibitory activity of oxyresveratrol on wild-type and drug-resistant varicella-zoster virus replication in vitro. *Antiviral Research*, *84*(1), pp. 95–97.

Septama, A.W. and Panichayupakaranant, P., 2015. Antibacterial assay-guided isolation of active compounds from Artocarpus heterophyllus heartwoods. *Pharmaceutical Biology*, *53*(11), pp. 1608–1613.

Septama, A.W. and Panichayupakaranant, P., 2016. Synergistic effect of artocarpin on antibacterial activity of some antibiotics against methicillin-resistant Staphylococcus aureus, Pseudomonas aeruginosa, and Escherichia coli. *Pharmaceutical Biology*, *54*(4), pp. 686–691.

7.8.2.2 *Broussonetia papyrifera* (L.) L'Hér. ex Vent.

Synonyms: *Morus papyrifera* L.; *Papyrius papyrifera* (L.) Kuntze; *Smithiodendron artocarpioideum* H.H. Hu

Common names: Paper mulberry, tapa cloth tree; rong (Cambodia); Gou shu (China); kachnar (India);saeh (Indonesia); aka (Japan); may sa (Laos); malaing (Myanmar); jangli shahtoot, toot, gul toot (Pakistan); po krasa (Thailand); chu (Vietnam).

Habitat: Thickets, mountain slopes, cultivated

Distribution: Pakistan, India, Myanmar, Cambodia, Laos, Vietnam, Indonesia, China, Korea, Taiwan, the Philippines

Botanical observation: This tree grows to about 15 m tall. The stems are hairy at apex. The leaves are simple, spiral, and stipulate. Stipules are elliptic, about 2 cm long, and ephemeral. The petiole is 2.5–8 cm long. The blade is lanceolate or very characteristically and strangely 3–5-lobed, 6–20 × 5–9 cm, hairy below, cordate and asymmetric at base, serrate, acuminate at apex, and marked with 6 or 7 pairs of secondary nerves. The spikes are whitish and 3–8 cm long. The perianth is hairy and develops 4 tepals. The androecium comprises 4 stamens. The ovary is ovoid and develops and a slender style. The syncarp is orange-red, spherical, and up to 3 cm across.

Medicinal uses: Cough, dysentery (China)

Strong broad-spectrum antibacterial amphiphilic prenylated flavone: Papyriflavonol A (LogD = 4.3 at pH 7.4; molecular mass = 438.4 g/mol) inhibited the growth of *Escherichia coli*, *Salmonella typhimurium*, *Staphylococcus epidermidis*, and *Staphylococcus aureus* with MIC of 20, 25, 10, and 15 μg/mL, respectively (Sohn et al., 2004).

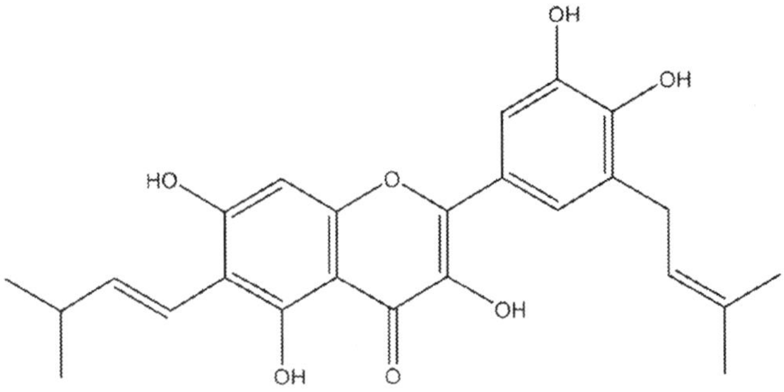

Papyriflavonol A

Strong broad-spectrum antifungal prenylated flavone: Papyriflavonol A inhibited the growth of *Candida albicans* and *Saccharomyces cerevisiae* with MIC values of 25 and 12.5 μg/mL, respectively, with membrane insults (Sohn et al., 2010).

Viral enzyme inhibition by prenylated flavone: Papyriflavonol A isolated from the roots inhibited Severe acute respiratory syndrome-acquired coronavirus 3-chymotrypsin-like protease, Severe acute respiratory syndrome-acquired coronavirus papain-like protease, Middle East respiratory syndrome 3-chymotrypsin-like protease, and Middle East respiratory syndrome coronavirus papain-like protease with IC_{50} values of 103.6, 35.8, 64.5, and 112.5 μM, respectively (Park et al., 2017).

Viral enzyme inhibition by prenylated flavone: Papyriflavonol A isolated from the roots inhibited the Severe acute respiratory syndrome-acquired coronavirus 3-chymotrypsin-like protease and papain-like protease, the Middle East Respiratory Syndrome coronavirus 3-chymotrypsin-like protease and papain-like protease with the IC50 values of 103.6, 35.8, 64.5, and 112.5 μM, respectively (Park et al., 2017).

Viral enzyme inhibition prenylated phenol: Kazinol F isolated from the roots inhibited Severe acute respiratory syndrome-acquired coronavirus chymotrypsin-like protease, Severe acute respiratory syndrome-acquired coronavirus papain-like protease, and Middle East Respiratory Syndrome coronavirus chymotrypsin-like protease (Park et al., 2017).

Commentary: Phytomedication from this plant could be of value against could be of value against COVID-19, its variants, and other zoonotic coronaviruses on our way.

REFERENCES

Park, J.Y., Yuk, H.J., Ryu, H.W., Lim, S.H., Kim, K.S., Park, K.H., Ryu, Y.B. and Lee, W.S., 2017. Evaluation of polyphenols from Broussonetia papyrifera as coronavirus protease inhibitors. *Journal of Enzyme Inhibition and Medicinal Chemistry*, *32*(1), pp. 504–512.

Sohn, H.Y., Kwon, C.S. and Son, K.H., 2010. Fungicidal effect of prenylated flavonol, papyriflavonol a, isolated from broussonetia papyrifera (L.) vent. against candida albicans. *Journal of Microbiol Biotechnol*, *20*(10), pp. 1397–402.

Sohn, H.Y., Son, K.H., Kwon, C.S., Kwon, G.S. and Kang, S.S., 2004. Antimicrobial and cytotoxic activity of 18 prenylated flavonoids isolated from medicinal plants: Morus alba L., Morus mongolica Schneider, Broussnetia papyrifera (L.) Vent, Sophora flavescens Ait and Echinosophora koreensis Nakai. *Phytomedicine*, *11*(7–8), pp. 666–672.

7.8.2.3 *Ficus benghalensis* L.

Synonyms: *Ficus indica* L.; *Urostigma benghalense* (L.) Gasp.

Common names: Banyan tree; bar, bot, bot gach (Bangladesh); alamaram, bahupada, avaroha, vad (India); beringin India (Indonesia); ara tandok (Malaysia); pyi-nyaung (Myanmar); krang, ni khrot (Thailand); da lá tròn (Vietnam)

Habitat: Cultivated

Distribution: Pakistan to Southeast Asia

Botanical observation: This tree native to India grows up to about 25m tall. This tree is sacred in India and planted around homes and temples. The tree presents a net of pendulous roots, which gives it a sinister and eerie look. The bark is light grey and characteristically smooth and somewhat skin like. The leaves are simple, spiral, and stipulate. The stipules are up to about 2.5 cm long. The petiole is stout and up to about 8 cm long. The blade is glossy, fleshy, ovate, 10–20 × 8–15 cm, hairy below, coriaceous, subcordate or rounded at base, obtuse at apex, and marked with 4–7 showy and yellow pairs of secondary nerves. The hypanthodium is sessile and axillary. The male flower comprises 2–3 sepals and a single stamen. The female flowers are sessile, bear 3–4 sepals, and present an ovary with an elongated style. The figs are globose, dull red, and 1.5–2.5 cm across.

Medicinal uses: Bronchitis, leucorrhea, small pox (Bangladesh); dysentery, gonorrhea, (Bangladesh, India)

Broad-spectrum antibacterial halo developed by polar extract: Ethanol extract of roots (25 mg/mL solution/6 mm disc) inhibited the growth of *Staphylococcus aureus*, *Escherichia coli*, and *Klebsiella pneumoniae* with inhibition zone diameters of 20, 15, and 12 mm, respectively (Murti and Kumar, 2011). Methanol extract of aerial roots (50 μg/6 mm paper disc) inhibited the growth of *Aeromoas hydrophila*, *Escherichia coli*, *Enterococcus faecalis*, *Pseudomonas aeruginosa*, *Staphylococcus aureus*, *Vibrio anguillarum*, and *Vibrio harveyi* with inhibition zone diameters of 10.6, 9.5, 11, 12.7, 15.8, 16.6, and 10.4 mm, respectively (Verma et al., 2015).

Strong antiviral (enveloped monopartite linear double-stranded DNA) polar extract: Aqueous extract of leaves at a concentration of 40 μg/mL inhibited the replication of the Herpes simplex virus (acyclovir resistant strain) in Vero cells by 80% via virucidal effect (Shoeib et al., 2011).

Commentaries: (i) The plant produces the triterpenes friedelin, taraxasterol, and the flavonoid glycosides quercetin-3-galactoside, 3′,5,7-trimethylether of leucocyanidin, 3′,5,7-trimethyl ether of delphinidin-3-*O*-α-L-rhamnoside, 3′,5- dimethyl ether of leucocyanidin-3-*O*-β-D-galactosylcellobioside, and 5,7-dimethylether of leucopelargonidin-3-*O*-α-L-rhamnoside (Deraniyagala &Wijesundera). One could suggest that the anti-Herpes property above could be owed to dissolution of viral envelop by amphiphilic flavonoid glycosides. Such glycosides might be virucidal for Severe acute respiratory syndrome-acquired coronavirus 2. Consider that aqueous extract of bark of *Ficus edelfeltii* King. at a concentration of 250 μg/mL inhibited Human immunodeficiency virus type-1 protease by 60.9% (Xu et al., 1996).

(ii) Ethyl acetate extract of stems *Ficus foveolata* Pittier inhibited the growth of *Streptococcus mutans* (DMST 26095), *Streptococcus mitis* (ATCC 49456T), and *Streptococcus pyogenes* (DNAT 17020) with MIC/MBC values of 39/156.2, 19.5/39, and 19.5/78.1 μg/mL, respectively (Meerungrueang and Panichayupakaranant, 2014). From this plant 2,6-dimethoxy-1,4-benzoquinone inhibited *Streptococcus mutans* (DMST 26095), *Streptococcus mitis* (ATCC 49456T), and *Streptococcus pyogenes* (DNAT 17020) with MIC/MBC values of 7.8/7.8, 7.8/7.8, and 15.6/31.2 μg/mL, respectively (Meerungrueang and Panichayupakaranant, 2014).

Syringaldehyde inhibited *Streptococcus mutans* (DMST 26095), *Streptococcus mitis* (ATCC 49456T), and *Streptococcus pyogenes* (DNAT 17020) with MIC/MBC values of 62.5/62.5, 125/250, and 250/250 μg/mL, respectively (Meerungrueang and Panichayupakaranant, 2014).

Sinapaldehyde inhibited *Streptococcus mutans* (DMST 26095), *Streptococcus mitis* (ATCC 49456T), and *Streptococcus pyogenes* (DNAT 17020) with MIC/MBC values 31.2/31.2, 125/125, 125/125 μg/mL, respectively (Meerungrueang and Panichayupakaranant, 2014).

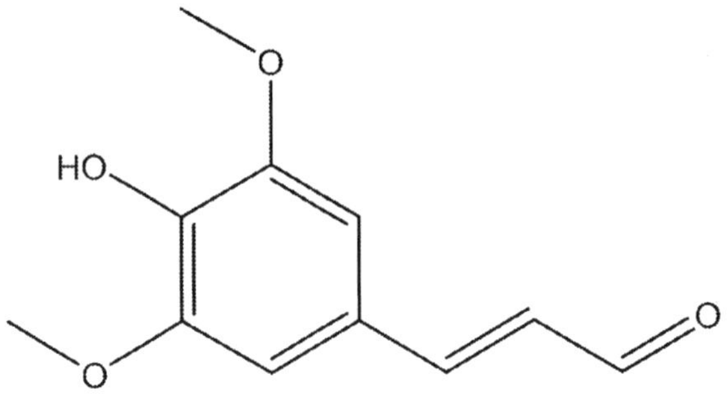

Sinapaldehyde

Coniferylaldehyde inhibited *Streptococcus mutans* (DMST 26095), *Streptococcus mitis* (ATCC 49456T), and *Streptococcus pyogenes* (DNAT 17020) with MIC/MBC values 62.5/62.5, 62.5/125, and 250/250 µg/mL (Meerungrueang and Panichayupakaranant, 2014).

Umbelliferone inhibited *Streptococcus mutans* (DMST 26095), *Streptococcus mitis* (ATCC 49456T), and *Streptococcus pyogenes* (DNAT 17020) with MIC/MBC values 15.6/15.6, 250/250, and 250/250 µg/mL (Meerungrueang and Panichayupakaranant, 2014).

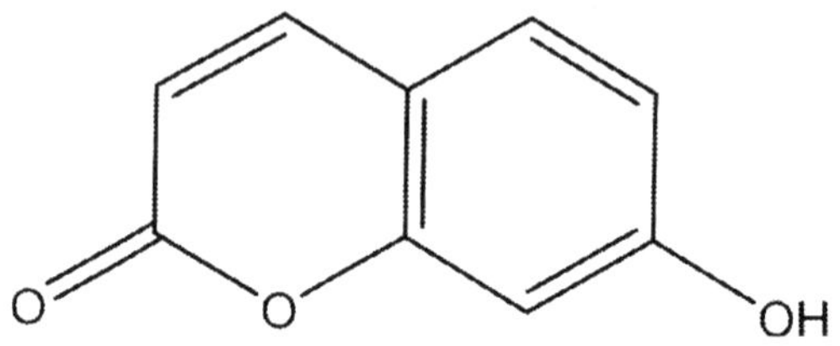

Umbeliferone

3β-Hydroxystigmast-5-en-7-one inhibited *Streptococcus mutans* (DMST 26095) and *Streptococcus pyogenes* (DNAT 17020) with MIC/MBC values 125/125 and 125/125 µg/mL, respectively (Meerungrueang and Panichayupakaranant, 2014.)

(iii) *Ficus hispida* L.f. is used in Bangladesh to treat leucorrhea, fever, jaundice, and ringworm (local names: *Dumur, Kuchuli*) and ethanol extract of its bark inhibited the growth of *Staphylococcus aureus*, *Shigella dysenteriae*, *Salmonella typhi*, and *Vibrio cholerae* with inhibition zone diameters of 10, 9, 10, and 8 mm, respectively (0.5 mg/disc) (Uddin et al., 2008).

(iv) In Bangladesh, *Ficus racemosa* L. is used for the treatment of leucorrhea, small pox, and tonsillitis (Local name: *Jagna dumur*) and *Ficus scandens* Lam (local name: *Lata dumur*) is used to treat tuberculosis and from this plant, neohopane inhibited the growth of *Escherichia coli*, *Pseudomonas aeruginosa*, *Bacillus subtilis*, and *Candida albicans* (Rasaga et al., 1999).

(v) *Ficus religiosa* L. is used in Bangladesh (local name: *Aswathha*) to treat cholera, leucorrhea, small pox, syphilis, typhoid, and wounds. Chloroform extract of bark of *Ficus religiosa* L. inhibited the replication of the Herpes simplex virus type-2 virus in Vero cells with an EC_{50} value of 6.7 µg/mL and a selectivity index of 119.9 via inhibition of attachment and entry of the virus to host cells

(Ghosh et al., 2016). Methanol extract of bark inhibited the replication of the Human rhinovirus (strain 1A (ATCCVR-1559) in HeLa cells with an EC_{50} value of 5.5 µg/mL via a mechanism involving late steps of viral replication with a selectivity index of 12.1. Aqueous extract of bark inhibited the replication of the Respiratory syncytial virus (strain A2 (ATCC VR-1540) in HEp-2 cells with an EC_{50} value of 2.2 µg/mL and a selectivity index of 92.5 via virus inactivation and interference with virus attachment to host cells (Cagno et al., 2015).

(vi) Members of the genus *Ficus* L. are interesting because they produce phenanthroindolizidine alkaloids such as tylophorine and tylocrebrine (Damu et al., 2005). Consider that phenanthroindolizidine alkaloids are often strongly active against the Human immunodeficiency virus *in vitro* (Wu et al., 2012). Another point to ponder is that tylophorine (LogD = 3.9 at pH 7.4; molecular mass = 393.4 g/mol) inhibited the replication of a coronavirus: the transmissible gastroenteritis virus in ST cells with an IC_{50} value of 58 nM and a selectivity index above 1715 in Swine testicular epithelial cells (Yang et al., 2010). A series of tylophorine derivatives inhibited the replication of coronaviruses, including the Severe acute respiratory syndrome-acquired coronavirus 2 (Lee et al., 2013; Yang et al., 2020). The planar structure of phenanthroindolizidine alkaloids suggest some impairment of viral genetic material replication. The development of antiviral drug from this group of alkaloid is possible.

REFERENCES

Cagno, V., Civra, A., Kumar, R., Pradhan, S., Donalisio, M., Sinha, B.N., Ghosh, M. and Lembo, D., 2015. Ficus religiosa L. bark extracts inhibit human rhinovirus and respiratory syncytial virus infection in vitro. *Journal of Ethnopharmacology*, *176*, pp. 252–257.

Damu, A.G., Kuo, P.C., Shi, L.S., Li, C.Y., Kuoh, C.S., Wu, P.L. and Wu, T.S., 2005. Phenanthroindolizidine alkaloids from the stems of Ficus septica. *Journal of Natural Products, 68*(7), pp. 1071–1075.

Deraniyagala, S.A. and Wijesundera, R.L.C., 2002. Ficus benghalensis. *National Science Foundation, Colombo*, pp. 1–3.

Ghosh, M., Civra, A., Rittà, M., Cagno, V., Mavuduru, S.G., Awasthi, P., Lembo, D. and Donalisio, M., 2016. Ficus religiosa L: bark extracts inhibit infection by herpes simplex virus type 2 in vitro. *Archives of Virology*, *161*(12), pp. 3509–3514.

Gopukumar, S.T. and Praseetha, P.K., 2015. Ficus benghalensis Linn–the sacred Indian medicinal tree with potent pharmacological remedies. *International Journal of Pharmaceutical Sciences Review and Research, 32*(37), pp. 223–227.

Lee, Y.Z., Yang, C.W., Hsu, H.Y., Qiu, Y.Q., Yeh, T.K., Chao, Y.S. and Lee*, S.J., 2013. Exploration of the role of tylophorine E ring in anti-Coronavirus activity-Tylophorine derived dibenzoquinolines impart multi-biological activities as orally active agents.

Meerungrueang, W. and Panichayupakaranant, P., 2014. Antimicrobial activities of some Thai traditional medical longevity formulations from plants and antibacterial compounds from Ficus foveolata. *Pharmaceutical Biology*, *52*(9), pp. 1104–1109.

Murti, K. and Kumar, U., 2011. Antimicrobial activity of Ficus benghalensis and Ficus racemosa roots L. *American Journal of Clinical Microbiology*, *2*(1), pp. 21–24.

Ragasa, C.Y., Juan, E. and Rideout, J.A., 1999. A triterpene from Ficus pumila. *Journal of Asian Natural Products Research*, *1*(4), pp. 269–275.

Shoeib, A.R.S., Zarouk, A.W. and El-Esnawy, N.A., 2011. Screening of antiviral activity of some terrestrial leaf plants against acyclovir-resistant HSV type-1 in cell culture. *Australian Journal of Basic and Applied Sciences*, *5*(10), pp. 75–92.

Uddin, S.J., Rouf, R., Shilpi, J.A., Alamgir, M., Nahar, L. and Sarker, S.D., 2008. Screening of some Bangladeshi medicinal plants for in vitro antibacterial activity. *Oriental Pharmacy and Experimental Medicine*, *8*(3), pp. 316–321.

Verma, V.K., Sehgal, N. and Prakash, O., 2015. Characterization and screening of bioactive compounds in the extract prepared from aerial roots of Ficus benghalensis. *International Journal of Pharmaceutical Sciences And Research*, *6*(12), p. 5056.

Wu, T.S., Su, C.R. and Lee, K.H., 2012. Cytotoxic and anti-HIV phenanthroindolizidine alkaloids from Cryptocarya chinensis. *Natural Product Communications*, *7*(6), p. 725.
Xu, H.X., Wan, M., Loh, B.N., Kon, O.L., Chow, P.W. and Sim, K.Y., 1996. Screening of Traditional Medicines for their Inhibitory Activity Against HIV-1 Protease. *Phytotherapy Research, 10*(3), pp. 207–210.
Yang, C.W., Lee, Y.Z., Hsu, H.Y., Jan, J.T., Lin, Y.L., Chang, S.Y., Peng, T.T., Yang, R.B., Liang, J.J., Liao, C.C. and Chao, T.L., 2020. Inhibition of Severe acute respiratory syndrome-acquired coronavirus 2 by Highly Potent Broad-Spectrum Anti-Coronaviral Tylophorine-Based Derivatives. *Frontiers in Pharmacology*, *11*, p. 2056.
Yang, C.W., Lee, Y.Z., Kang, I.J., Barnard, D.L., Jan, J.T., Lin, D., Huang, C.W., Yeh, T.K., Chao, Y.S. and Lee, S.J., 2010. Identification of phenanthroindolizines and phenanthroquinolizidines as novel potent anti-coronaviral agents for porcine enteropathogenic coronavirus transmissible gastroenteritis virus and human severe acute respiratory syndrome coronavirus. *Antiviral Research*, *88*(2), pp. 160–168.

7.8.2.4 *Maclura cochinchinensis* (Lour.) Corner

Synonyms: *Cudrania cochinchinensis* (Lour.) Kudô & Masam.; *Cudrania integra* F.T. Wang & Tang; *Cudrania javanensis* Trécul; *Cudrania obovata* Trécul; *Cudrania rectispina* Hance; *Maclura gerontogea* Siebold & Zucc.; *Trophis spinosa* Roxb. ex Willd.; *Vanieria cochinchinensis* Lour.

Common names: Gou ji (China); kayu kuning (Indonesia); kederang (Malaysia); khlaè (Cambodia); klae (Thailand); dây mo'qua (Vietnam)

Habitat: Around villages, open forests, thickets

Distribution: India, Sri Lanka, Nepal, Bhutan, Myanmar, Cambodia, Laos, Vietnam, Thailand, China, Malaysia, Indonesia, the Philippines, Japan, Australia, Pacific Islands

Botanical observation: This climbing shrub grows up to a length of 10 m. The stems yield a latex upon incision and present about 2 cm long woody thorns. The leaves are simple, spiral, and stipulate. The stipule is ephemeral. The petiole is about 1 cm long. The blade is elliptic-lanceolate to oblong, 3–8 × 2–2.5 cm, glossy, cuneate at base, acute at apex, and presents 7–10 pairs of secondary nerves. The inflorescences are yellowish and 1 cm diameter heads. The perianth comprises 4 lobes. The syncarp is reddish orange, about 2 cm across, glossy, and looks like a small brain.

Medicinal use: Wounds (Vietnam)

Moderate antibacterial (Gram-positive) mid-polar extract: Ethyl acetate extract of wood inhibited the growth of *Streptococcus mutans* (DMST 26095), *Streptococcus mitis* (ATCC 49456T), and *Streptococcus pyogenes* (DNAT 17020) with MIC/MBC values of 312.5/1250, 78.1/312.5, and 156/156 μg/mL, respectively (Meerungrueang and Panichayupakaranant, 2014).

Strong antibacterial (Gram-positive) amphiphilic xanthones: Gerontoxanthone G, gerontoxanthone H (also known as cudraxanthone H; LogD = 4.6 at pH 7.4; molecular mass = 380.4 g/mol), gerontoxanthone I, alvaxanthone, isoalvaxanthone, toxyloxanthone C, 1,3,7- trihydroxy-2-prenylxanthone, and cudraxanthone S from the roots inhibited the growth of *Bacillus cereus* (PCI 219) (Fukai et al., 2004). Gerontoxanthone G, gerontoxanthone H (also known as cudraxanthone H), gerontoxanthone I, alvaxanthone, isoalvaxanthone, toxyloxanthone C, 1,3,7-trihydroxy-2-prenylxanthone, cudraxanthone S, and cudranone inhibited the growth of 10 strains of *Micrococcus luteus* (ATCC 9341) (Fukai et al., 2004). Gerontoxanthone G, gerontoxanthone H (also known as cudraxanthone H), gerontoxanthone I, alvaxanthone, isoalvaxanthone, toxyloxanthone C, 1,3,7-trihydroxy-2-prenylxanthone, and cudraxanthone S inhibited the growth of 10 strains of methicillin-resistant *Staphylococcus aureus* (Fukai et al., 2004). Gerontoxanthone G, gerontoxanthone H (cudraxanthone H), gerontoxanthone I, alvaxanthone, isoalvaxanthone, toxyloxanthone C, 1,3,7- trihydroxy-2-prenylxanthone, cudraxanthone S, gerontoxanthone A, gerontoxanthone B, and cudranone inhibited the growth of *Enterococcus faecalis* (ATCC 29212) (Fukai et al., 2005). Gerontoxanthone G, gerontoxanthone H (cudraxanthone H), gerontoxanthone I, alvaxanthone, isoalvaxanthone, toxyloxanthone C, and 1,3,7-trihydroxy-2-prenylxanthone inhibited the growth of *Enterococcus faecium* (ATCC 19434) with MIC of 12.5, 1.5, 6.2, 12.5, 12.5, 25, and 6.2 μg/mL, respectively (Fukai et al., 2005).

Strong antiviral (enveloped monopartite linear double-stranded DNA) polar extract: Aqueous extract of wood inhibited the replication of the Herpes simplex type-1 and Herpes simplex type-2 with the EC_{50} values of 20.1 and 68.3 µg/mL and selectivity indices of 8.8 and 2.6, respectively (Yoosook et al., 2000).

Strong antiviral (enveloped monopartite linear single-stranded (+) RNA) hydrophilic flavonol: Morin (LogD = 0.3 at pH 7.4; molecular mass = 302.2 g/mol) inhibited the replication of the Dengue virus serotype-1 with an IC_{50} value of 9.4 µg/ml in Vero cells (CC_{50} = 12.4 µg/mL)

Moderate antiviral (enveloped monopartite linear double-stranded DNA) flavonol: Morin isolated from the wood inhibited the replication of the Herpes simplex virus type-2 with an EC_{50} value of 53.5 µg/mL (CC_{50} 184.5 µg/mL) in Vero cells (Bunyapraphatsara et al., 2000). Morin at a concentration of 90 µg/mL inhibited the replication of the Equid herpesvirus 1 (Gravina et al., 2011).

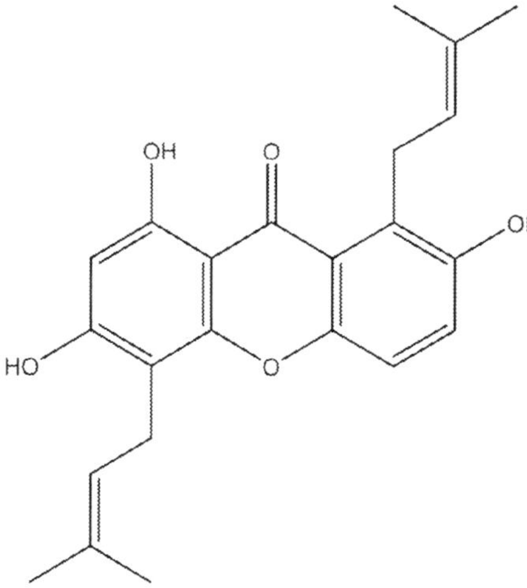

Gerontoxanthone H

Commentary: The plant yields a dye used to make "batiks" by Malays and Indonesians

REFERENCES

Bunyapraphatsara, N., Dechsree, S., Yoosook, C., Herunsalee, A. and Panpisutchai, Y., 2000. Anti-herpes simplex virus component isolated from Maclura cochinchinensis. *Phytomedicine*, *6*(6), pp. 421–424.

Fukai, T., Oku, Y., Hou, A.J., Yonekawa, M. and Terada, S., 2004. Antimicrobial activity of hydrophobic xanthones from Cudrania cochinchinensis against Bacillus subtilis and methicillin-resistant Staphylococcus aureus. *Chemistry & Biodiversity*, *1*(9), pp. 1385–1390.

Fukai, T., Oku, Y., Hou, A.J., Yonekawa, M. and Terada, S., 2005. Antimicrobial activity of isoprenoid-substituted xanthones from Cudrania cochinchinensis against vancomycin-resistant enterococci. *Phytomedicine*, *12*(6–7), pp. 510–513.

Fukai, T., Oku, Y., Hou, A.J., Yonekawa, M. and Terada, S., 2005. Antimicrobial activity of isoprenoid-substituted xanthones from Cudrania cochinchinensis against vancomycin-resistant enterococci. *Phytomedicine*, *12*(6–7), pp. 510–513.

Gravina, H.D., Tafuri, N.F., Júnior, A.S., Fietto, J.L.R., Oliveira, T.T., Diaz, M.A.N. and Almeida, M.R., 2011. In vitro assessment of the antiviral potential of trans-cinnamic acid, quercetin and morin against equid herpesvirus 1. *Research in Veterinary Science*, *91*(3), pp. e158–e162.

Meerungrueang, W. and Panichayupakaranant, P., 2014. Antimicrobial activities of some Thai traditional medical longevity formulations from plants and antibacterial compounds from Ficus foveolata. *Pharmaceutical Biology*, *52*(9), pp. 1104–1109.

Yoosook, C., Bunyapraphatsara, N., Boonyakiat, Y. and Kantasuk, C., 2000. Anti-herpes simplex virus activities of crude water extracts of Thai medicinal plants. *Phytomedicine*, *6*(6), pp. 411–419.

7.8.2.5 *Morus alba* L.

Synonyms: *Morus atropurpurea* Roxb.; *Morus australis* Poir.; *Morus indica* L.; *Morus intermedia* Perr.; *Morus multicaulis* Perr.; *Morus tatarica* L.

Common names: White mulberry; sang (China); besaran (Indonesia); posa (Myanmar); tut (Pakistan); mon (Cambodia, Thailand); dau (Vietnam)

Habitat: Cultivated

Distribution: Temperate Asia

Botanical observation: This handsome laticiferous tree native to China grows to about 10 m tall. The stems are hairy at apex. The leaves are simple, spiral, edible (China), and stipulate. The stipules are lanceolate, about 3 cm long, somewhat hairy, and caducous. The petiole is 1.5–5.5 cm long. The blade is broadly lanceolate, serrate, 5–30 × 5–10 cm, round and asymmetrical at base, acuminate at apex, and marked with about 5 pairs of secondary nerves. The spikes of male flowers are axillary, pendulous, and up to about 10 cm long, and the spikes of female flowers are 1–2 cm long. The 4 tepals are minute and greenish. The androecium comprises 4 stamens. The ovary is ovoid and develops a bifid stigma. The syncarp is whitish or red to purple, edible, cylindrical, and about 1–2 cm long.

Medicinal use: Abscesses (Turkey)

Strong antibacterial (Gram-positive) amphiphilic prenylated flavone: Kuwanon G (LogD = 5.1 at pH 7.4; molecular mass = 692.7 g/mol) isolated from the root bark inhibited the growth of *Streptococcus mutans* (ATCC 25175) with an MIC value 8 µg/mL and was bactericidal via disintegration of the cell surface of the bacterium (Park et al., 2003).

Strong broad-spectrum antibacterial lipophilic phenolics: Mulberrofuran G (also known as albanol A; LogD = 6.4 at pH 7.4; molecular mass = 562.5 g/mol) isolated from the root bark inhibited the growth of *Escherichia coli*, *Salmonella typhimurium*, *Staphylococcus epidermidis*, and *Staphylococcus aureus* with the MIC values of 30, 7.5, 6.2, and 5 µg/mL (Sohn et al., 2004). Albanol B (LogD = 6 at pH 7.4; molecular mass = 558.5 g/mol) inhibited the growth of *Escherichia coli*, *Salmonella typhimurium*, *Staphylococcus epidermidis*, and *Staphylococcus aureus* with the MIC values of 20, 5, 5, and 5 µg/mL (Park et al., 2003).

Strong antiviral (enveloped monopartite linear double-stranded DNA) prenylated flavanone: Leachianone G; (also known as (-)-(2S)-2'-hydroxy-8-dimethylallylnaringenin; molecular mass = 356.4 g/mol) isolated from the root bark inhibited the replication of the Herpes simplex virus type-1 in Vero cells with an IC_{50} of 1.6 µg/mL ($CC_{50=}$ 15.5 6 µg/mL) (Mat et al., 2016).

Strong antiviral (enveloped monopartite linear double-stranded DNA) lipophilic stilbene: Kuwanon X (LogD = 5.6 at pH 7.4; molecular mass = 582.5 g/mol) from the leaves inhibited the replication of the Herpes simplex virus type-1 (15577), Herpes simplex virus type-1 (clinical strains), and the Herpes simplex virus type-2 (strain 333) in Vero cells with IC_{50} values of 2.2, 1.5, and 2.5 µg/mL and selectivity indices of 37, 55, and 32, respectively (Mat et al., 2016). This stilbene had no direct virucidal effect and inhibited the penetration of the virion into host cells and suppressed viral DNA synthesis in Vero cells (Mat et al., 2016).

Strong antiviral (enveloped monopartite linear double-stranded DNA) amphiphilic phenolics: Mulberrofuran C, mulberrofuran G, moracin C (LogD = 4.1 at pH 7.4; molecular mass = 310.3 g/mol), and moracin M from the leaves inhibited the replication of the Herpes simplex virus type-1 (clinical strains) with IC_{50} values of 3.1, 7, 2.5, and 12.5 µg/mL, respectively, and selectivity indices 8.9, 4, 11.7, and 4, respectively (Mat et al., 2016).

Mulberrofuran C (LogD = 5.2 at pH 7.4; molecular mass = 580.5 g/mol), mulberrofuran G, moracin C, and moracin M from the leaves inhibited the replication of the Herpes simplex virus type-1 (15577) with IC_{50} values of 5, 8.4, 5.2, and 6.3 µg/mL, respectively, and selectivity indices of 5.6, 3.6, 5.9, and 8, respectively (Mat et al., 2016).

Mulberrofuran C, mulberrofuran G, moracin C (LogD = 4.1 at pH 7.4; molecular mass = 310.3 g/mol), and moracin M from the leaves inhibited the replication of the Herpes simplex virus type-2 (strain 333) (Mat et al., 2016).

Strong antiviral (enveloped monopartite linear double-stranded DNA) phenolic glycoside: Moracin M 3′-*O*-β-glucopyranoside (molecular mass=404.4 g/mol) from the leaves inhibited the replication of the Herpes simplex virus type-1 (15577), Herpes simplex virus type-1 (clinical), and Herpes simplex virus type-2 (strain 333) with IC_{50} values of 8.4, 7 and 8.2 μg/mL and selectivity indices of 3.6, 4, and 3.5, respectively (Mat et al., 2016).

Strong antiviral (enveloped segmented single-stranded (+)RNA) amphiphilic prenylated flavone: kuwanon G inhibited the replication of the Human coronavirus (229E) (ATCC VR 740) in L-132 cells with an EC_{50} value of 0.1 μg/mL and a selectivity index of 86.7 (Thabti et al., 2020).

Moderate antiviral (enveloped segmented single-stranded (+)RNA) polar extract: Methanol extract of leaves at a concentration 200 μg/mL inhibited the replication of the Human coronavirus (229E) (ATCC VR 740) in L-132 cells by 71% (Thabti et al., 2020).

Moderate antiviral (enveloped monopartite linear double-stranded DNA) chromane: Mulberroside C inhibited the replication of the Herpes simplex virus type-1 with an IC_{50} of 75.4 μg/mL ($CC_{50=}$250μg/mL) (Du et al., 2003).

Viral enzyme inhibition by polar extract: Methanol extract of leaves inhibited Human immunodeficiency virus type-1 protease (Park, 2003).

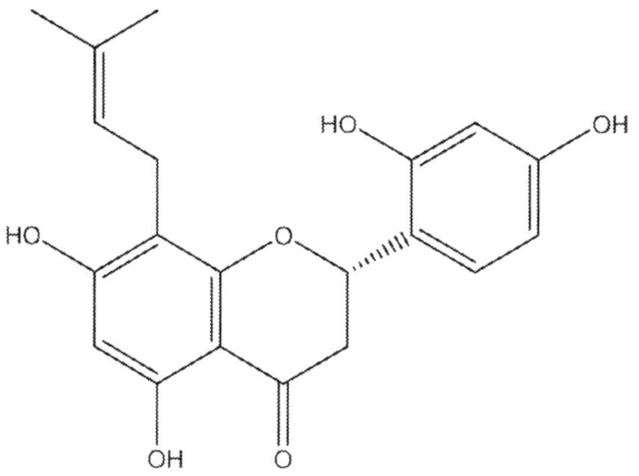

Leachianone G

Commentaries: (i) The antibacterial activity of flavonoids and phenolic depends on their concentration (just like biocides, for instance, at high concentration phenolics induce bacterial cytoplasm coagulation). (ii) Flavanones and flavones generated by the members of the genus *Morus* L. are of antiviral interest and could be examined for anti-coronavirus and other zoonotic pandemic viruses, although the oral bioavailability of flavonoids is in general not high. One further example is kuwanon L (LogD=at pH 7.4; molecular mass=g/mol) from the roots of a member of the genus *Morus* L., which inhibited the replication of the Human immunodeficiency virus type-1 with an EC_{50} value of 1.9 μM and a CC_{50} above 20 μM (Viveros-Valdez et al., 2015).

(iii*) Morus alba* L. is used in China for the treatment of COVID-19 (Luo, 2020; Wang et al., 2020) which makes sense since it produces kuwanon G. Phytomedication of this plant could be of value for COVID-19 and forthcoming coronavirus zoonotic pandemics (iv) The prenylated flavone morusin (LogD = 4.7 at pH 7.4; molecular mass 420.4 g/mol) isolated from the root bark of *Morus mongolica* (Bureau) C.K. Schneid. inhibited the growth of *Staphylococcus epidermidis* and *Staphylococcus aureus* with the MIC values of 20 and 25 μg/mL, respectively (Sohn et al., 2004).

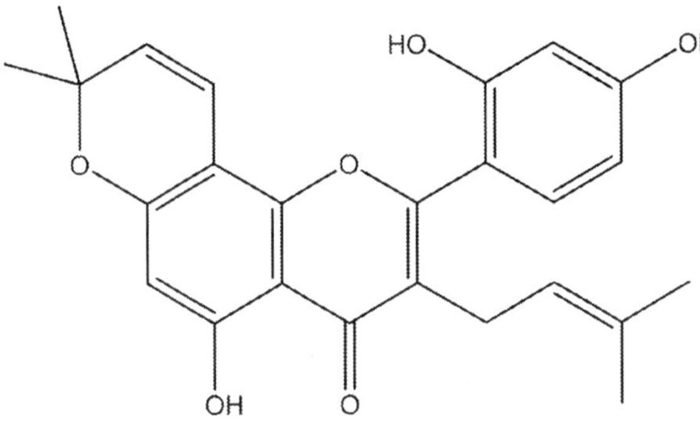

Morusin

From this plant, the prenylated flavone kuwanon C (also known as mulberrin (LogD = 4.2 at pH 7.4; molecular mass = 422.4 g/mol) inhibited the growth of *Escherichia coli, Salmonella typhimurium, Staphylococcus epidermidis*, and *Staphylococcus aureus* with the MIC values of 10, 25, 6.2, and 6.2 µg/mL, respectively (Sohn et al., 2004).

From this plant, sanggenon D (LogD = 5.6 at pH 7.4; molecular mass = 708.7 g/mol) inhibited the growth of *Staphylococcus epidermidis, Staphylococcus aureus, Salmonella typhimurium*, and *Saccharomyces cerevisiae* with MIC values of 40, 50, 20, and 25 µg/mL, respectively (Sohn et al., 2004).

REFERENCES

Du, J., He, Z.D., Jiang, R.W., Ye, W.C., Xu, H.X. and But, P.P.H., 2003. Antiviral flavonoids from the root bark of Morus alba L. *Phytochemistry*, *62*(8), pp. 1235–1238.

Ma, F., Shen, W., Zhang, X., Li, M., Wang, Y., Zou, Y., Li, Y. and Wang, H., 2016. Anti-HSV activity of kuwanon X from mulberry leaves with genes expression inhibitory and HSV-1 induced NF-κB deactivated properties. *Biological and Pharmaceutical Bulletin*, *39*(10), pp. 1667–1674.

Morus mongolica Schneider, Broussnetia papyrifera (L.) Vent, Sophora flavescens Ait and Echinosophora koreensis Nakai. *Phytomedicine*, *11*(7–8), pp. 666–672.

Park, J.C., 2003. Inhibitory effects of Korean plant resources on human immunodeficiency virus type 1 protease activity. *Oriental Pharmacy and Experimental Medicine*, *3*(1), pp. 1–7.

Park, K.M., You, J.S., Lee, H.Y., Baek, N.I. and Hwang, J.K., 2003. Kuwanon G: an antibacterial agent from the root bark of Morus alba against oral pathogens. *Journal of Ethnopharmacology*, *84*(2–3), pp. 181–185.

Sohn, H.Y., Son, K.H., Kwon, C.S., Kwon, G.S. and Kang, S.S., 2004. Antimicrobial and cytotoxic activity of 18 prenylated flavonoids isolated from medicinal plants: Morus alba L.,

Sohn, H.Y., Son, K.H., Kwon, C.S., Kwon, G.S. and Kang, S.S., 2004. Antimicrobial and cytotoxic activity of 18 prenylated flavonoids isolated from medicinal plants: Morus alba L., Morus mongolica Schneider, Broussnetia papyrifera (L.) Vent, Sophora flavescens Ait and Echinosophora koreensis Nakai. *Phytomedicine*, *11*(7–8), pp. 666–672.

Thabti, I., Albert, Q., Philippot, S., Dupire, F., Westerhuis, B., Fontanay, S., Risler, A., Kassab, T., Elfalleh, W., Aferchichi, A. and Varbanov, M., 2020. Advances on Antiviral Activity of Morus spp. Plant Extracts: Human Coronavirus and Virus-Related Respiratory Tract Infections in the Spotlight. *Molecules*, *25*(8), p.1876.

Viveros-Valdez, E., Oranday-Cárdenas, A., Rivas-Morales, C., Verde-Star, M.J. and Carranza-Rosales, P., 2015. Biological activities of Morus celtidifolia leaf extracts. *Pakistan Journal of Pharmaceutical Sciences, 28*(4), pp. 1177–1180.

Wang, S.X., Wang, Y., Lu, Y.B., Li, J.Y., Song, Y.J., Nyamgerelt, M. and Wang, X.X., 2020. Diagnosis and treatment of novel coronavirus pneumonia based on the theory of traditional Chinese medicine. *Journal of Integrative Medicine*, *8*, pp. 275–283.

7.8.2.6 *Streblus asper* Lour.

Synonyms: *Diplothorax tonkinensis* Gagnep.; *Streblus monoicus* Gagnep.

Common names: Sandpaper tree, Siamese rough bush; shaora, sharbo gaas (Bangladesh); que shen shu (China); shakhotaka (India); serut (Indonesia); som pho (Laos); serinai (Malaysia); okhne (Myanmar); kalios (the Philippines); khoi (Thailand); cay ruoi, ru·ei (Vietnam)

Habitat: Near villages, cultivated

Distribution: From India to the Philippines

Botanical name: This tree grows to about 5 m tall. The stems are hairy. The leaves are simple, alternate, and stipulate. The stipules are minute, linear, and caducous. The petiole is none or up to 3 mm long. The blade is elliptic, 2.5–6 × 1.5–3.5 cm, coriaceous, dark green, glossy, cuneate at base, laxly serrate, shortly acuminate at apex, and marked with 4–7 pairs of secondary nerves. The flower fascicles are axillary. The calyx comprises 4 sepals. The androecium includes 4 stamens. The ovary is globose and develops a pair of filiform stigmas. The drupes are elliptic to globose, yellow, and about 1.2 cm long.

Medicinal uses: Cholera, gonorrhea, chicken pox, syphilis (Bangladesh); dysentery (India)

Weak antibacterial (Gram-positive) extract: Ethanol extract of leaves inhibited the growth of *Streptococcus mutans* (ATCC 25175), *Streptococcus mutans* (KPSK-2), and *Streptococus mutans* (TPF-1) (Wongkham et al., 2001).

Strong broad-spectrum antibacterial neolignans: (7′*R*,8′*S*)-4,4′-Dimethoxy-strebluslignanol isolated from the roots inhibited the growth of *Bacillus subtilis* (ATCC 6633), *Pseudomonas aeruginosa* (ATCC 9027), *Escherichia coli* (ATCC 11775), and *Staphylococcus aureus* (ATCC 25923) with MIC values of 21, 67, 21, and 18 μg/mL, respectively (Nie et al., 2016). Magnolol (LogD = 4 at pH 7.4; molecular mass = 266.3 g/mol) isolated from the roots inhibited the growth of *Bacillus subtilis* (ATCC 6633), *Pseudomonas aeruginosa* (ATCC 9027), *Escherichia coli* (ATCC 11775), and *Staphylococcus aureus* (ATCC 25923) with MIC values of 18, 56, 22, and 18 μg/mL, respectively (Nie et al., 2016).

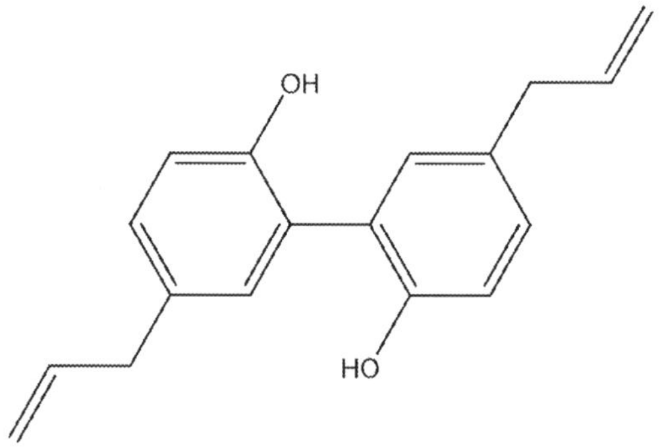

Magnolol

Strong broad-spectrum antibacterial phenolics: 3′-Hydroxy-isostrebluslignaldehyde isolated from the roots inhibited the growth of *Bacillus subtilis* (ATCC 6633), *Pseudomonas aeruginosa* (ATCC 9027), *Escherichia coli* (ATCC 11775), and *Staphylococcus aureus* (ATCC 25923) with MIC values of 27, 67, 33, and 23 μg/mL, respectively (Nie et al., 2016). 3,3′-Methylene-bis(4-hydroxybenzaldehyde) isolated from the roots inhibited the growth of *Bacillus subtilis* (ATCC 6633), *Pseudomonas aeruginosa* (ATCC 9027), *Escherichia coli* (ATCC 11775), and *Staphylococcus aureus* (ATCC 25923) with MIC values of 27, 85, 27, and 23 μg/mL, respectively (Nie et al., 2016). 4-Methoxy-isomagnaldehyde isolated from the roots inhibited the growth of

Bacillus subtilis (ATCC 6633), *Pseudomonas aeruginosa* (ATCC 9027), *Escherichia coli* (ATCC 11775), and *Staphylococcus aureus* (ATCC 25923) with MIC values of 26, 93, 29, and 22 μg/mL, respectively (Nie et al., 2016). Isomagnolol isolated from the roots inhibited the growth of *Bacillus subtilis* (ATCC 6633), *Pseudomonas aeruginosa* (ATCC 9027), *Escherichia coli* (ATCC 11775), and *Staphylococcus aureus* (ATCC 25923) with MIC values of 18, 67, 26, and 15 μg/mL, respectively (Nie et al., 2016).

Strong antifungal (yeast) neolignans: (7′*R*,8′*S*)-4,4′-Dimethoxy-strebluslignanol and magnolol isolated from the roots inhibited the growth of *Saccharomyces cerevisiae* (ATCC 9753) with MIC values of 45 and 37 μg/mL, respectively (Nie et al., 2016).

Strong antifungal (yeast) phenolics: 3′-Hydroxy-isostrebluslignaldehyde, 3,3′-methylene-bis(4-hydroxybenzaldehyde), 4-methoxy-isomagnaldehyde isolated from the roots inhibited the growth of *Saccharomyces cerevisiae* (ATCC 9753) with MIC values of 67, 77, and 63 μg/mL (Nie et al., 2016). Isomagnolol isolated from the roots inhibited the growth of *Saccharomyces cerevisiae* (ATCC 9753) with MIC values of 94 μg/mL, respectively (Nie et al., 2016)

Strong antiviral (enveloped circular double stranded DNA) amphiphilic neolignans: Honokiol (LogD=4.1 at pH 7.4; molecular mass=266.3 g/mol) isolated from the roots inhibited the secretion of the Hepatitis B virus surface antigen and e-antigen in Hep G2.2.15 cells with IC_{50} values of 3.1 and 4.7 μM, respectively, and selectivity indices of 21.4 and 14.2, respectively (Chen et al., 2012). (7′R,8′S,7″R,8″S)-erythro-strebluslignanol G isolated from the roots inhibited the secretion of the Hepatitis B virus surface antigen, the Hepatitis B virus e-antigen, and the DNA replication of the Hepatitis B virus in Hep G2.2.15 cells with IC_{50} values of 1.5, 3.2, and 9 μM and selectivity indices of 74.9, 36.5, and 13.2, respectively (Li et al., 2013). Magnolol isolated from the roots inhibited the secretion of the Hepatitis B virus surface antigen, e-antigen, and the DNA replication of the Hepatitis B virus in Hep G2.2.15 cells with IC_{50} values of 2, 3.7, and 8.6 μM and selectivity indices of 31.3, 16.9, and 7.3, respectively (Li et al., 2013).

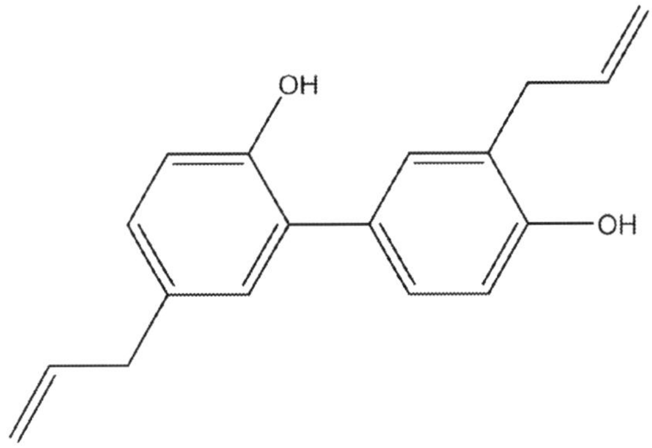

Honokiol

Commentary: (i) Honokiol inhibits the replication of the Hepatitis C virus, Dengue virus, and is antibacterial and antifungal as presented earlier. Natural products inhibiting bacteria, fungi, and viruses are often targeting the microbial genetic material. Neolignans often inhibit the replication of the Hepatitis B virus, why? One could examine properties of neolignans against coronaviruses. (ii) The plant generates a series of dreadfully toxic cardiac glycosides, such as strebloside (Bai et al., 2019). However, consider that cardiac glycosides such as ouabain and digitoxin (which have a shallow therapeutic index) have the interesting ability to impair the replication of the Chikungunya virus, Middle East Respiratory Syndrome, and Severe acute respiratory syndrome-associated coronavirus via inhibition of Na^+/K^+-ATPase (Amarelle & Lecuona, 2018.).

REFERENCES

Amarelle, L. and Lecuona, E., 2018. The antiviral effects of Na, K-ATPase inhibition: a minireview. *International journal of Molecular Sciences*, *19*(8), p. 2154.

Bai, Y., Zhu, W., Xu, Y., Xie, Z., Akihisa, T., Manosroi, J., Sun, H., Feng, F., Liu, W. and Zhang, J., 2019. Characterization, quantitation, similarity evaluation and combination with Na+, K+-ATPase of cardiac glycosides from Streblus asper. *Bioorganic Chemistry*, *87*, pp. 265–275.

Chen, H., Li, J., Wu, Q., Niu, X.T., Tang, M.T., Guan, X.L., Li, J., Yang, R.Y., Deng, S.P. and Su, X.J., 2012. Anti-HBV activities of Streblus asper and constituents of its roots. *Fitoterapia*, *83*(4), pp. 643–649.

Nie, H., Guan, X.L., Li, J., Zhang, Y.J., He, R.J., Huang, Y., Liu, B.M., Zhou, D.X., Deng, S.P., Chen, H.C. and Yang, R.Y., 2016. Antimicrobial lignans derived from the roots of Streblus Asper. *Phytochemistry Letters*, *18*, pp. 226–231.

Wongkham, S., Laupattarakasaem, P., Pienthaweechai, K., Areejitranusorn, P., Wongkham, C. and Techanitiswad, T., 2001. Antimicrobial activity of Streblus asper leaf extract. *Phytotherapy Research*, *15*(2), pp. 119–121.

7.8.3 Family Rosaceae A. L. de Jussieu (1789)

The family Rosaceae consists of about 85 genera and about 3000 species of trees, shrubs, herbs, and climbers. The leaves are simple or compound, alternate or spiral, often serrate, and stipulate. The inflorescences are solitary or racemes. The flowers are actinomorphic and include a hypanthium. The calyx comprises 5 sepals. The corolla presents 5 petals. The androecium is made of 4 to numerous filamentous stamens. The gynoecium consists of one or more carpels free or united, with 2 ovules per carpel, and develops a capitate stigma. The fruits are achenes, pomes, drupes, berries, or cynorrhodons. Members in this family generate a vast array of tannins and triterpenes.

7.8.3.1 *Agrimonia pilosa* Ledeb.

Common name: Long ya cao (China)

Habitat: Thickets, riverbanks, roadsides

Distribution: India, Bhutan, Nepal, Myanmar, Cambodia, Laos, Vietnam, Thailand, China, Korea, Japan

Botanical observation: This graceful herb grows up to 1 m tall from a rhizome. The stem is hairy. The leaves are imparipinnate, spiral, and stipulate. The stipules are serrate. The petiole is hairy. The blade presents 3 or 4 pairs of folioles, which are dark green, obovate, 1.5–5 × 1–2.5 cm, cuneate at base, serrate, with about 5 pairs of secondary nerves, and rounded to acute at apex. The inflorescence is a terminal raceme of minute yellow flowers. The calyx includes 5 sepals, which are minute and ovate. The 5 petals are yellow and oblong. A disc and a hypanthium are present. The androecium includes 8–15 stamens. The ovary develops a filiform style and a capitate stigma. The hypanthium is about 1 cm long, conical, 10-ribbed, spiny at apex and contain tiny achenes.

Medicinal use: Dysentery (China)

Strong antiviral (enveloped monopartite linear single-stranded (−)RNA) polar extract: Aqueous extract inhibited the Respiratory syncytial virus with an IC_{50} value of 62.5 μg/mL and a selectivity index of 4 (Li et al., 2004).

Strong antiviral (enveloped segmented linear single stranded (−)RNA) polar extract: Ethanol extract inhibited the replication of the Influenza virus A (A/Chile/1/83 Influenza virus) with an EC50 value of 78 μg/mL and plaque formation with an IC50 value of 39 μg/mL in by Madin-Darby canine kidney cells (Shin et al., 2010). The extract inhibited plaque formation by Influenza virus A/Puerto Rico/8/34 (H1N1), A/Sydney/5/97 (H3N2), B/Yamagata/16/88, and A/chicken/Korea/ms96/96 (H9N2) with EC50 values of 16.4, 14.9, and 23.1 μg/mL in Madin Darby canine kidney cells (Shin et al., 2010).

Strong antiviral (non-enveloped monopartite linear single-stranded (+)RNA) polar extract: Methanol extract at a concentration of 50 μg/mL decreased the titer of Hepatitis A virus (HM-175 strain) from Fetal rhesus monkey kidney cells by about 2 logs (Seo et al., 2017).

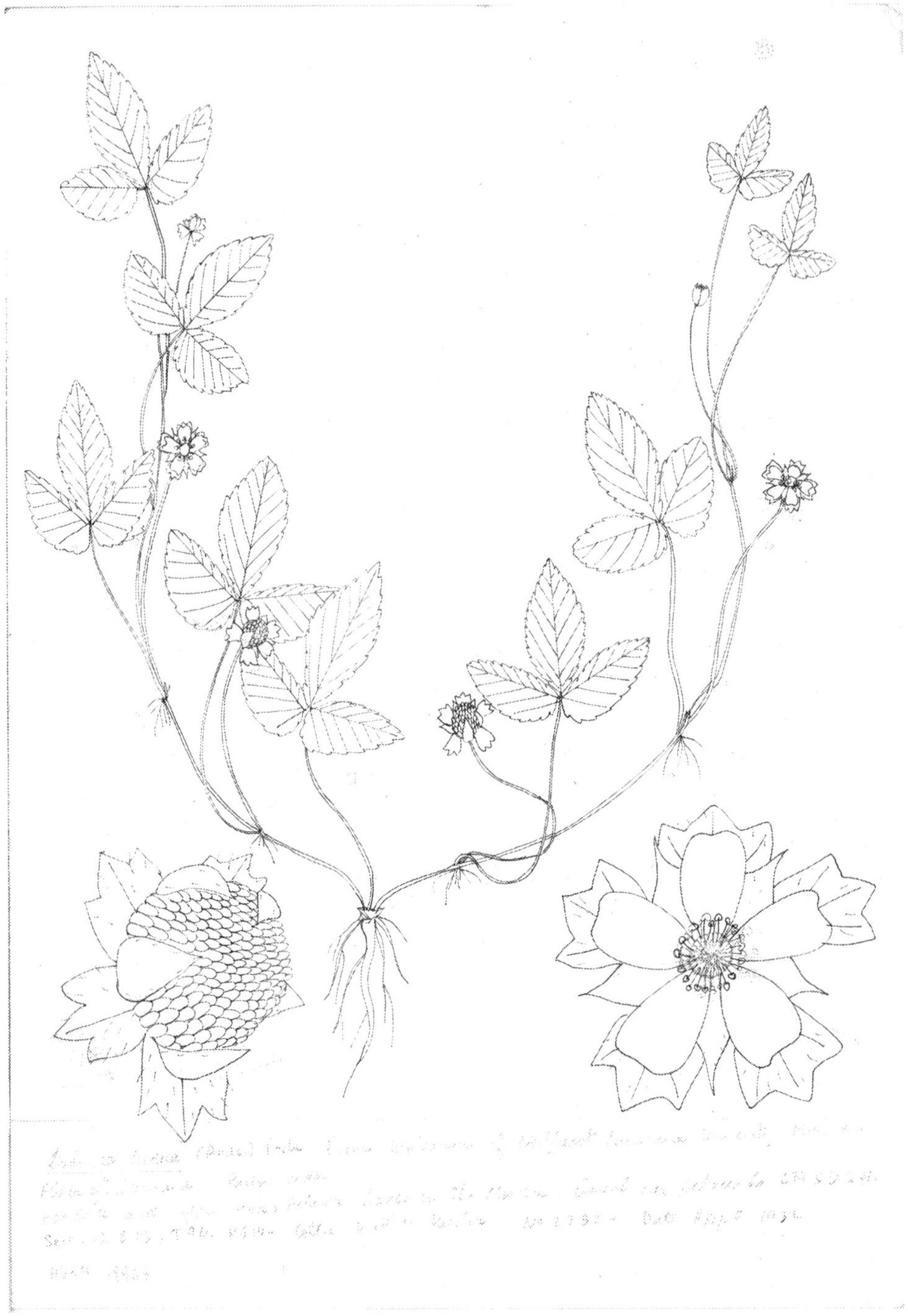

FIGURE 7.30 *Duchesnea indica* (Andrews) Teschem.

Moderate antiviral (enveloped monopartite linear double-stranded DNA) polar extract: Aqueous extract inhibited the replication of the Herpes simplex virus type-1 with an IC_{50} value of 125 μg/mL and a selectivity index of 4 (Li et al., 2004).

Strong antiviral (enveloped monopartite linear dimeric single-stranded (+) RNA) amphiphilic ursane-type triterpene: Pomolic acid (also known as 3β,19α-dihydroxyurs-12-en-28-oic acid; LogD = 4.3

at pH 7.4; molecular mass=472.7 g/mol), inhibited the replication of the Human immunodeficiency virus in H9 cells with an EC_{50} value of 1.4 µg/mL and a selectivity index of 16.6 (Kashiwada et al., 1998).

Moderate antiviral (enveloped monopartite linear double-stranded DNA) ursane-type triterpene: Pomolic acid inhibited the replication of the Herpes simplex virus type-1 with an EC_{50} value of 50 µg/mL (Ryu et al., 1993).

Viral enzyme inhibition by polar extract: Methanol extract of at a concentration of 100 µg/mL inhibited RNA-dependent DNA polymerase and ribonuclease H activities of the Human immunodeficiency virus type-1 reverse transcriptase and the Human immunodeficiency virus type-1 protease by 97.2, 50.6, and 35%, respectively (Min et al., 2001). Methanol extract inhibited the Human immunodeficiency virus type-1 protease (Park, 2003).

Commentaries: (i) The plant engineers pentacyclic triterpene, including 1β-hydroxy-2-oxopomolic acid (Ahn et al., 2012), pomolic acid, and tormentic acid (Ma et al., 2015) and a broad array of flavonoid glycosides and phenolics (Kato et al., 2010), hence the astringent effects. One could examine the *in vitro* anti-coronavirus properties of pomolic acid because it has anti- Human immunodeficiency virus properties (see earlier) and inhibits DNA polymerase (Deng et al., 2000). Another example of antiviral ursane-type triterpene glycoside is 2α, 3β, 19 α -trihydroxyurs-12-en-28-oic acid α -D-glucopyranosyl ester isolated from the rhizome of *Potentilla anserina* L. (used in traditional Chinese medicine to treat cough and dysentery; local name: *Jue ma*), which inhibited the secretion of the Hepatitis B virus surface antigen with an IC_{50} value of 57. 6 µg/mL (selectivity index: 4.1) and the secretion of the Hepatitis B e-antigen with an IC_{50} value of 30 µg/mL (selectivity index: 7.8) and the replication of the Hepatis B virus with an IC_{50} value of 19.4 µg/mL (selectivity index: 12.4) (Zhao et al., 2008). In China, *Potentilla fruticosa* L. (local name: *Jin lu mei*) and *Potentilla glabra* Lodd. (local name: *Yin lu mei*) are used to treat lung diseases. In Turkey, *Potentilla reptans* L. is used to treat fever (local name: *Besparmak otu*). In Nepal, *Potentilla polyphylla* Wall. ex Lehm. is used (local name: *Bajradanti*) to treat cold, cough, sore throat, and wounds.

(ii) *Agrimonia eupatoria* L. is used in India to treat cough. In the family Rosaceae, *Cerasus mahaleb* (L.) Miller is used for cough in Turkey (local name: *Mahlep*) as well as *Cotoneaster integerrimus* Medik. (local name: *Dag musmulasi*), *Cotoneaster nummularia* Fisch. & Mey., and *Orthurus heterocarpus* (Boiss.) Juz. (local name: *Kulafir*).

(iii) *Duchesnea indica* (Andrews) Teschem. is used in Pakistan (local name: *Muti*) to treat diarrhea and dysentery, and in China (local name: *Hé shàng*) it is used to treat sore throat, cough, cold, and dysentery.

REFERENCES

Ahn, E.K., Lee, J.A., Seo, D.W., Hong, S.S. and Oh, J.S., 2012. 1β-Hydroxy-2-oxopomolic acid isolated from Agrimonia pilosa extract inhibits adipogenesis in 3T3-L1 cells. *Biological and Pharmaceutical Bulletin*, *35*(5), pp. 643–649.

Kashiwada, Y., Wang, H.K., Nagao, T., Kitanaka, S., Yasuda, I., Fujioka, T., Yamagishi, T., Cosentino, L.M., Kozuka, M., Okabe, H. and Ikeshiro, Y., 1998. Anti-AIDS agents. 30. Anti-HIV activity of oleanolic acid, pomolic acid, and structurally related triterpenoids. *Journal of Natural Products*, *61*(9), pp. 1090–1095.

Kato, H., Li, W., Koike, M., Wang, Y. and Koike, K., 2010. Phenolic glycosides from Agrimonia pilosa. *Phytochemistry*, *71*(16), pp. 1925–1929.

Li, Y., Ooi, L.S., Wang, H., But, P.P. and Ooi, V.E., 2004. Antiviral activities of medicinal herbs traditionally used in southern mainland China. *Phytotherapy Research*, *18*(9), pp. 718–722.

Ma, J.H., Jiang, Q.H., Chen, Y., Nie, X.F., Yao, T., Ding, L.Q., Zhao, F., Chen, L.X. and Feng, Q., 2015. A New Triterpenoid from the Aerial Parts of Agrimonia pilosa. *Natural Product Communications*, *10*(12), p.1934578X1501001207.

Min, B.S., Kim, Y.H., Tomiyama, M., Nakamura, N., Miyashiro, H., Otake, T. and Hattori, M., 2001. Inhibitory effects of Korean plants on HIV-1 activities. *Phytotherapy Research*, *15*(6), pp. 481–486.

Park, J.C., 2003. Inhibitory effects of Korean plant resources on human immunodeficiency virus type 1 protease activity. *Oriental Pharmacy and Experimental Medicine, 3*(1), pp. 1–7.
Ryu, S.Y., Lee, C.K., Ahn, J.W., Lee, S.H. and Zee, O.P., 1993. Antiviral activity of triterpenoid derivatives. *Archives of Pharmacal Research, 16*(4), pp. 339–342.
Seo, D.J., Lee, M., Jeon, S.B., Park, H., Jeong, S., Lee, B.H. and Choi, C., 2017. Antiviral activity of herbal extracts against the hepatitis A virus. *Food Control, 72*, pp. 9–13.
Shin, W.J., Lee, K.H., Park, M.H. and Seong, B.L., 2010. Broad-spectrum antiviral effect of Agrimonia pilosa extract on influenza viruses. *Microbiology and immunology, 54*(1), pp. 11–19.
Zhao, Y.L., Cai, G.M., Hong, X., Shan, L.M. and Xiao, X.H., 2008. Anti-hepatitis B virus activities of triterpenoid saponin compound from Potentilla anserine L. *Phytomedicine, 15*(4), pp. 253–258.

7.8.3.2 *Cydonia oblonga* Mill.

Synonyms: *Cydonia vulgaris* Pers.; *Pyrus cydonia* L.

Common names: Quince; bihi dana (Bangladesh); wen po (China); amritaphala (India); ayva (Turkey)

Habitat: Cultivated

Distribution: Temperate Asia

Botanical observation: This tree grows to about 5 m tall. The stems are terete, purplish, and hairy. The leaves are simple, spiral, and stipulate. The stipules are ovate and caduceus. The petiole is about 1 cm long and hairy. The blade is oblong, wavy, 5–10 × 3–5cm, hairy below, glossy, rounded at base, acute at apex, and with about 6 pairs of secondary nerves. The inflorescences are axillary or terminal solitary and magnificent flowers, which are 4–5 cm in diameter. The flower peduncles are about 5 mm long. The calyx includes 5 sepals, which are ovate and about 5 mm long. The 5 petals are white or pinkish, membranous, broadly spathulate to spoon-shaped, and about 1.5 cm long. The 20 stamens are about 7mm long. The 5 styles are about 5 mm long. The pomes are dull yellow, pear-shaped, bumpy, 3–5 cm across, and edible.

Medicinal uses: Cough (Bangladesh); dysentery (India); cold, cough (Turkey)

Moderate broad-spectrum antibacterial polar extract: Polyphenol fraction of fruit peels inhibited the growth of *Staphylococcus aureus* (ATCC6538) and *Pseudomonas aeruginosa* (ATCC 9027) with MIC/MBC values of 100/100 and 500/500 µg/mL, respectively (Fattouchg et al., 2007).

Strong anticandidal polar extract: Methanol extract of leaves inhibited the growth of *Candida albicans* and *Candida parapsilosis* with MIC_{90} values of 0.1 and 64 µg/mL, respectively (Hanci et al., 2019).

Antiviral (enveloped segmented linear single-stranded (-)RNA) polar extract: Polyphenolic fraction of fruits at a concentration of 500 µg/mL inhibited the hemagglutination of the Influenza A virus (A/PR/8/34) (Hamauzu et al., 2005).

Commentary: The fruits contain 3-caffeoylquinic acid, 5-caffeoylquinic acid, and procyanidins oligomers and polymers (Hamauzu et al., 2005). The plant is recommended in the Unani system of medicine for pandemics in the "*Sharah-i-Asbab*" of our peer the Persian Physician Najib Al-Din Al-Samarqandi in the 13th century. The current response to COVID-19 and its mutant such as the Delta variant is industrial and mechanical (which allow the manufacture and distribution of billions of vaccines globally) with the use of RNA vaccines which should put an end to the current pandemic. According to a few scholars, RNA vaccines might account for long-term side effects (Lin et al., 2020; Talotta, 2021). The enormous efforts developed by industries and government to protect us against the COVID-19 have to be appreciated and we must be grateful. However, due to a decrease in raw material, increase in human population, and shaky global economy some predict a soon collapse of the society as we know (Kovel, 2007) and if such situation occurs there will be impossibility to produce industrial vaccines. Further, the risk of global nuclear war has never been as high as today and in case of such a war, countries depending on Big Pharmas to supply their vaccines will suffer the most in case of pandemic (Scouras et al., 2019). Thus, it would be wise to schools of pharmacies which are, according to some, so far serving the interests of the corporate system and become themselves corporations (which contradict completely the concept of intellectual freedom and collegiality from were scientific breakthrough emerge (Barrow, 2014)) to reimplement full medicinal plant teaching including the Unani system and its wealth of knowledge that could assist locally to protect populations against the coming zoonitic pandemic waves.

REFERENCES

Amin, J., 2020. Quarantine and hygienic practices about combating contagious disease like COVID-19 and Islamic perspective. *Journal of critical reviews*, 7(13).

Barrow, C.W., 2014. The coming of the corporate-fascist university? New Political Science, 36(4), pp. 640–646.

Fattouch, S., Caboni, P., Coroneo, V., Tuberoso, C.I., Angioni, A., Dessi, S., Marzouki, N. and Cabras, P., 2007. Antimicrobial activity of Tunisian quince (Cydonia oblonga Miller) pulp and peel polyphenolic extracts. *Journal of Agricultural and Food Chemistry*, *55*(3), pp. 963–969.

Hamauzu, Y., Yasui, H., Inno, T., Kume, C. and Omanyuda, M., 2005. Phenolic profile, antioxidant property, and anti-influenza viral activity of Chinese quince (Pseudocydonia sinensis Schneid.), quince (Cydonia oblonga Mill.), and apple (Malus domestica Mill.) fruits. *Journal of Agricultural and Food Chemistry*, *53*(4), pp. 928–934.

Hanci, H., COŞKUN, M.V., Uyanik, M.H., Sezen, S. and Hakan, İ.G.A.N., 2019. In vitro antifungal activities of fluconazole, camellia sinensis and cydonia oblonga leaf extracts against candida species isolated from blood cultures. *Bezmialem Science*, *7*(2), p. 107.

Kovel, J., 2007. *The enemy of nature: The end of capitalism or the end of the world?*. Zed Books, London.

Lin, C.J., Mecham, R.P. and Mann, D.L., 2020. RNA vaccines for COVID-19: Five things every cardiologist should know. *JACC: Basic to Translational Science*, 5(12): pp. 1240–1243.

Scouras, J., 2019. Nuclear war as a global catastrophic risk. Journal of *Benefit-Cost Analysis*, 10(2), pp. 274–295.

Talotta, R., 2021. Do COVID-19 RNA-based vaccines put at risk of immune-mediated diseases? In reply to "potential antigenic cross-reactivity between SARS-CoV-2 and human tissue with a possible link to an increase in autoimmune diseases". Clinical Immunology (Orlando, Fla.), 224, p. 108665.

7.8.3.3 *Eriobotrya japonica* (Thunb.) Lindl.

Synonyms: *Crataegus bibas* Lour.; *Mespilus japonica* Thunb., *Photinia japonica* (Thunb.) Benth. & Hook. f. ex Asch. & Schweinf.

Common names: Loquat, Japanese medlar, Japanese plum; lochat (Bangladesh); pi ba (China)

Habitat: Cultivated

Distribution: Cambodia, Laos, Vietnam, China, Thailand, Indonesia, and the Philippines

Botanical observation: This handsome tree probably native to China grows up to 10 m tall. The bark has a peculiar brownish color and is smooth. The stems are rough, stout, and hairy. The leaves are simple, spiral, and stipulate. The stipules are linear, hairy, and up to about 1.5 cm long. The petiole is stout, hairy, and up to 1 cm long. The blade is lanceolate, coriaceous, somewhat glossy, cuneate at base, oblanceolate, or elliptic, 12–30×3–10 cm, hairy below, serrate near the apex, acute at apex, and presents about 11 to 12 pairs of conspicuous secondary nerves. The panicles are terminal, up to about 20 cm long, hairy, and present numerous fragrant flowers. A cupular hypanthium is present. The calyx comprises 5 broadly lanceolate minute sepals. The corolla includes 5 petals, which are whitish, somewhat uneven at margin, obovate, up to about 1 cm long, clawed at base, and notched at apex. The androecium includes 20 stamens. The ovary is hairy and develops 5 styles. The pomes are dull yellow to orange, somewhat globose or elliptic, about 1.5 cm across, smooth, marked at apex by vestiges of sepals, and shelter 1 to 5 seeds, which are brown and glossy.

Medicinal uses: Breast infection, small pox (Bangladesh)

Strong antibacterial (Gram-positive) mid-polar extract: Ethyl acetate extract of leaves inhibited the growth of methicillin-sensitive *Staphylococcus aureus* (KCTC 1927), methicillin-resistant *Staphylococcus aureus* (KCCM40510), and methicillin-resistant *Staphylococcus aureus* (KCCM40511) with MIC values of 64, 64, and 32 μg/mL, respectively (Yu et al., 2019).

Strong antibacterial (Gram-negative) hydrophilic flavanol: Epicatechin (LogD = 0.5 at pH 7.4; molecular mass=290.2 g/mol) inhibited the growth of *Escherichia coli* (FTJ), *Escherichia coli*

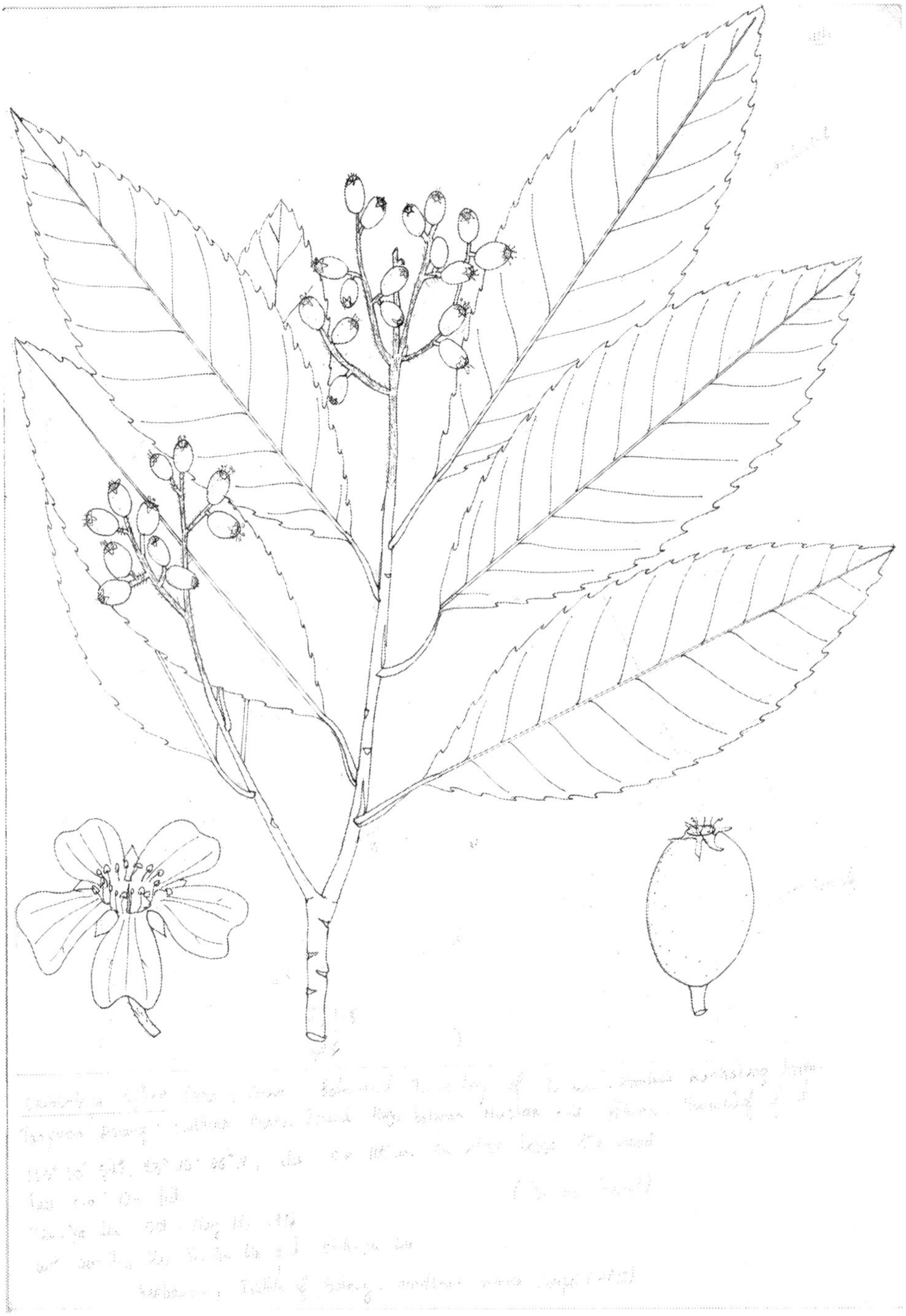

FIGURE 7.31 *Eriobotrya deflexa* (Hemsl.) Nakai

(ATCC 35150), *Salmonella paratyphi* (UK Micro 29A), *Salmonella enteritica* (ATCC 10708), and *Salmonella enteritidis* (UK (–) H_2S) with MIC values of <20, <15, <10, <20, and <20 ppm (Cetin-Karaca et al., 2015).

Antifungal (filamentous) amphiphilic phenol: Aucuparin (LogD = 2.5 at pH 7.4; molecular mass=230.2 g/mol) isolated from stems (experimentally infected with *Colletotrichum lindemuthianum*) at a concentration of 100 ppm abrogated the germination of *Colletotrichum lindemuthianum* and *Pestalotiopsis funereal.* (Watanabe et al., 1982).

Strong antiviral (enveloped monopartite single-stranded (+)RNA) hydrophilic flavanol: Epicatechin inhibited the replication of the Mayaro virus (strain BeAr 20290) with an IC_{50} value of 0.2 µmol/mL (Ferreira et al., 2018).

Viral enzyme inhibition by polar extract: Methanol extract of leaves inhibited the Human immunodeficiency virus type-1 protease (Park, 2003).

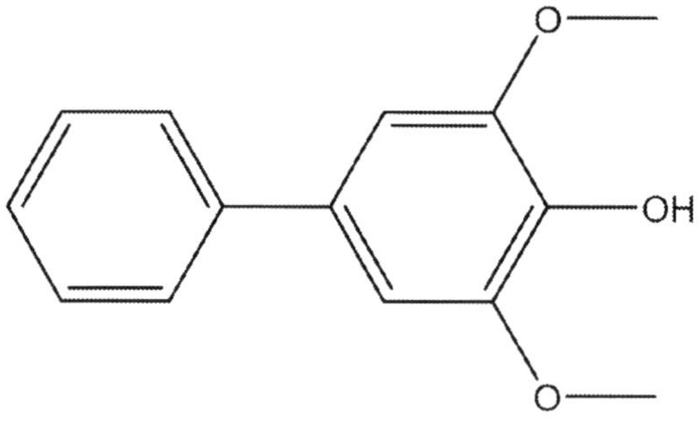

Aucuparin

Viral enzyme inhibition by flavanol: Epicatechin inhibited Severe acute respiratory syndrome-acquired coronavirus 3-chymotrypsin-like protease with an IC_{50} value of 100 µM (Chen et al., 2005).

Commentaries: (i) The plant produces epicatechin (Bae et al., 2005). Why is synthetic aucuparin less antifungal than the natural phytoalexin aucuparin? Are products synthetized naturally endowed with something absent in molecules produced by organic chemists? (ii) The plant produces the phytoalexin eriobofuran (Miyakado et al., 1985). Consider that bisphenols and benzofurans are not uncommonly produced by members of the family Rosaceae to protect themselves against phytopathogenic fungi (Chizzali & Beerhues, 2012). What is their antifungal mode of action? (iii) The Mayaro virus is a zoonotic enveloped monopartite single-stranded (+)-RNA virus belonging to the genus Alphavirus in the family Togaviridae transmitted from primates to humans via mosquitoes, which is responsible for rashes, fever, myalgia, retro-orbital pain, headache, diarrhea, and long-lasting arthralgia (Esposito et al., 2017). Mayaro virus outbreaks began around 1954 with its first report in Trinidad and Tobago, followed by outbreaks in 1978, 1991, and 2014 in South and Central America (Esposito et al., 2017). Consider that *Eriobotrya japonica* (Thunb.) Lindl. is used in China for the treatment of COVID-19 (Wang et al., 2020). In China, *Eriobotrya deflexa* (Hemsl.) Nakai (local name: *Tai wan pi ba*) is used to treat cough and pneumonia.

REFERENCES

Bae, Y.I., Jeong, C.H. and Shim, K.H., 2005. Antioxidative and antimicrobial activity of epicatechin isolated from leaves of loquat (Eriobotrya japonica). *Journal of Food Science and Nutrition*, *10*(2), pp. 118–121.

Cetin-Karaca, Hayriye, and Melissa C. Newman. "Antimicrobial efficacy of plant phenolic compounds against Salmonella and Escherichia Coli." *Food bioscience* 11(2015): 8–16.

Chen, C.N., Lin, C.P., Huang, K.K., Chen, W.C., Hsieh, H.P., Liang, P.H. and Hsu, J.T.A., 2005. Inhibition of SARS-CoV 3C-like protease activity by theaflavin-3, 3′-digallate (TF3). *Evidence-Based Complementary and Alternative Medicine*, 2(2), pp. 209–215.

Chizzali, C. and Beerhues, L., 2012. Phytoalexins of the Pyrinae: biphenyls and dibenzofurans. *Beilstein Journal of Organic Chemistry*, *8*(1), pp. 613–620.
Esposito, D.L.A. and Fonseca, B.A.L.D., 2017. Will Mayaro virus be responsible for the next outbreak of an arthropod-borne virus in Brazil? *Brazilian Journal of Infectious Diseases*, *21*(5), pp. 540–544.
Miyakado, M., Watanabe, K., Ohno, N., Nonaka, F. and Morita, A., 1985. Isolation and structural determination of eriobofuran, a new dibenzofuran phytoalexin from leaves of loquat, Eriobotrya japonica L. *Journal of Pesticide Science (Japan).*
Park, J.C., 2003. Inhibitory effects of Korean plant resources on human immunodeficiency virus type 1 protease activity. *Oriental Pharmacy and Experimental Medicine*, *3*(1), pp. 1–7.
Wang, S.X., Wang, Y., Lu, Y.B., Li, J.Y., Song, Y.J., Nyamgerelt, M. and Wang, X.X., 2020. Diagnosis and treatment of novel coronavirus pneumonia based on the theory of traditional Chinese medicine. *Journal of Integrative Medicine.*
Watanabe, K., Ishiguri, Y., Nonaka, F. and Morita, A., 1982. Isolation and identification of aucuparin as a phytoalexin from Eriobotrya japonica L. *Agricultural and Biological Chemistry*, *46*(2), pp. 567–568.
Yu, D., Yun, I.H., Pham, N.T.D., Jang, Y.M., Park, S.K., Eom, S.H. and Kim, Y.M., 2019. Antibacterial activity of Eriobotrya japonica leaf extracts against methicillin-resistant Staphylococcus aureus. *Journal of Environmental Biology*, *40*(6), pp. 1129–1136.

7.8.3.4 *Prunus persica* (L.) Batsch

Synonym: *Amygdalus persica* L.

Common names: Peach; chenun (India); seftali (Turkey)

Habitat: Cultivated

Distribution: Temperate Asia

Botanical observation: This heavenly tree grows up to about 10 tall. The leaves are simple, spiral, and stipulate. The petiole is up to about 1.5 cm long. The blade is oblong to lanceolate, 7–15×2–4.5 cm, cuneate to obtuse at base, serrate, coriaceous, and acuminate at apex. The flowers are magnificent, cauliflorous, and in fascicles. A cupulate hypanthium is present. The calyx includes 5 sepals, which are oblong and about 5 mm long. The 5 petals are pink, obovate, and about 1 cm long. The androecium includes up to 30 stamens. The ovaries are hairy. The drupes are peach-colored, somewhat, globose, up to about 10 cm across, velutinous, delicious, and shelter a woody stone.

Medicinal uses: Bronchitis, cough, wounds (India); diarrhea (Turkey)

Antifungal (filamentous) halo developed by ursane-type triterpenes: 1β,2α,3α,24-tetrahydroxyurs-12-en-28-oic acid and 1β,2α,3α,24-tetrahydroxyolean-12-en-28-oic acid (30 μg/TLC) isolated from fruit peels (experimentally infected with *Colletotrichum musae*) inhibited the growth of *Colletotrichum musae* (El Lahlou et al., 1999).

Strong antiviral (yeast) hydrophilic phenolic: Ellagic acid (LogD=–2 at pH 7.4; molecular mass=302.1 g/mol inhibited the growth of *Candida albicans* (ATCC 90028) and *Cryptococcus neoformans* (clinical isolate) with MIC values of 62.5 and 15.6 μg/mL, respectively (Rangkadilok et al., 2012).

Moderate antiviral (enveloped circular double-stranded DNA) phenolic: Ellagic acid at a concentration of 200 μg/mL inhibited the secretion of the Hepatitis B surface antigen and Hepatitis e-antigen by 62.9 and 44.9 % (Li et al., 2008), and inhibited the Hepatitis C virus protease with an IC_{50} value of 56.3 μM (Ajala et al., 2014).

Viral attachment to host cell inhibition by polar extract: Aqueous extracts at a concentration of 200 μg/mL inhibited the interaction of the Human immunodeficiency virus type-1 gp 120 with the CD4 receptor by 27.2% (Collins et al., 1997).

Commentary: (i) The fruits contain ellagic acid (Rios et al., 2018), which is a strong broad-spectrum antiviral phenolic (see earlier), and it is worth noting that it has the ability to block the nuclear export of viral ribonucleoprotein (Chang et al., 2016; Zuwala et al., 2015), and as such could be of therapeutic value against the COVID-19. In fact, the plant is used for the treatment of COVID-19 in China (Wang et al., 2020). It is intriguing that a number of plants in the genus *Prunus* L. are used in China for the treatment of COVID-19: *Prunus armeniaca* L. var *ansu* Maxim or *Prunus mume* (Sieb.) Sieb. & Zucc. in combination with oseltamivir phosphate is used for the treatment of

COVID-19 (Luo, 2020). (ii) Aqueous extract of fruits of *Prunus armeniaca* L. at a concentration of 250 μg/mL inhibited the Human immunodeficiency virus type-1 protease by 100% (Xu et al., 1996).

(iii) In India, *Pyrus malus* L. is used treat fever, *Rosa multiflora* Thunb. is used to heal putrefied wounds, *Pyrus pashia* Buch.–Ham. exD.Don (local name: *Khait)* is used to treat eye infections. *Pyrus pashia* Buch.–Ham. ex D. Don is used to diarrhea and dysentery in Nepal (local name: *Mayal*). In Korea, *Pyrus pyrifolia* var. *culta* (Makino) Nakai is used to treat cold (local name: *Baenamu*). *Pyrus syriaca* Boiss. var. *syriaca* (local name: *Adi armut*) and *Pyrus amygdaliformis* Vill. var. *amygdaliformis* (local name: *Boz armut*) are used to treat cough in Turkey. In Nepal, *Prunus cerasoides* D. Don. (local name: *Paiyau*) is used to treat diarrhea, dysentery, and *Pyracantha crenulata* (D. Don) M. Roem. is used to treat dysentery (local name: *Kathgedi*).

REFERENCES

Ajala, O.S., Jukov, A. and Ma, C.M., 2014. Hepatitis C virus inhibitory hydrolysable tannins from the fruits of Terminalia chebula. *Fitoterapia*, *99*, pp. 117–123.

Chang, S.Y., Park, J.H., Kim, Y.H., Kang, J.S. and Min, J.Y., 2016. A natural component from Euphorbia humifusa Willd displays novel, broad-spectrum anti-influenza activity by blocking nuclear export of viral ribonucleoprotein. *Biochemical and Biophysical Research Communications*, *471*(2), pp. 282–289.

Collins, R.A., Ng, T.B., Fong, W.P., Wan, C.C. and Yeung, H.W., 1997. A comparison of human immunodeficiency virus type 1 inhibition by partially purified aqueous extracts of Chinese medicinal herbs. *Life Sciences*, *60*(23), pp. PL345-PL351.

El Lahlou, H., Hirai, N., Tsuda, M. and Ohigashi, H., 1999. Triterpene phytoalexins from nectarine fruits. *Phytochemistry*, *52*(4), pp. 623–629.

Li, J., Huang, H., Zhou, W., Feng, M. and Zhou, P., 2008. Anti-hepatitis B virus activities of Geranium carolinianum L. extracts and identification of the active components. *Biological and Pharmaceutical Bulletin*, *31*(4), pp. 743–747.

Luo, A., 2020. Positive SARS-Cov-2 test in a woman with COVID-19 at 22days after hospital discharge: A case report. *Journal of Traditional Chinese Medical Sciences*.

Rangkadilok, N., Tongchusak, S., Boonhok, R., Chaiyaroj, S.C., Junyaprasert, V.B., Buajeeb, W., Akanimanee, J., Raksasuk, T., Suddhasthira, T. and Satayavivad, J., 2012. In vitro antifungal activities of longan (Dimocarpus longan Lour.) seed extract. *Fitoterapia*, *83*(3), pp. 545–553.

Ríos, J.L., Giner, R.M., Marín, M. and Recio, M.C., 2018. A pharmacological update of ellagic acid. *Planta Med*, *84*(15), pp. 1068–1093.

Wang, S.X., Wang, Y., Lu, Y.B., Li, J.Y., Song, Y.J., Nyamgerelt, M. and Wang, X.X., 2020. Diagnosis and treatment of novel coronavirus pneumonia based on the theory of traditional Chinese medicine. *Journal of Integrative Medicine*.

Xu, H.X., Wan, M., Loh, B.N., Kon, O.L., Chow, P.W. and Sim, K.Y., 1996. Screening of Traditional Medicines for their Inhibitory Activity Against HIV-1 Protease. *Phytotherapy Research*, *10*(3), pp. 207–210.

Zuwała, K., Golda, A., Kabala, W., Burmistrz, M., Zdzalik, M., Nowak, P., Kedracka-Krok, S., Zarebski, M., Dobrucki, J., Florek, D. and Zeglen, S., 2015. The nucleocapsid protein of human coronavirus NL63. *PloS One*, *10*(2), p. e0117833.

7.8.3.5 *Rosa damascena* Mill.

Synonym: *Rosa gallica* var. *damascena* Voss

Common names: Damask rose; golap (Bangladesh); atimanjula (India); kembang eros (Indonesia); golab (Iran); kuhlaab (Laos); bunga ayer mawar (Malaysia); huong (Vietnam)

Habitat: Cultivated

Distribution: Asia and the Pacific

Botanical observation: This rose probably native to Turkey grows to about 2m tall. The stems present curved woody thorns. The leaves are imparipinnate, alternate, and stipulate. The blade presents 2 to 3 pairs of folioles, which are glossy, serrate, elliptic, 2–6.5×1.5–5.5cm, and hairy below. The inflorescence is a terminal raceme of fragrant roses. The flowers are graceful, pinkish yellow, and showy on a about 2.5cm long peduncle. A cup-shaped hypanthium is present. The 5 sepals are triangular and about 2 cm long. The corolla includes up to 30 petals, which are red or white, up to about 2 × 4 cm and membranous. The androecium includes numerous stamens. The gynoecium presents numerous free carpels. The fruit is a cynorrhodon.

Medicinal uses: Abscesses, AIDS, leucorrhea, mumps, tonsillitis (Bangladesh); gingivitis, pulmonary disorders (Iran)

Broad-spectrum antibacterial halo developed by polar extract: Methanol extract (10% solution/5 mm well) inhibited the growth of *Enterobacter aerogenes*, *Escherichia coli*, *Pseudomonas aeruginosa*, *Aeromonas hydrophila*, *Proteus vulgaris*, *Salmonella enteridis*, *Yersinia enterolitica*, *Bacillus cereus*, *Staphylococcus aureus*, *Salmonella typhimurium*, *Enterococcus faecalis*, *Pseudomonas fluorescens*, and *Klebslella pneumoniae* (Özkan et al., 2004).

Antibacterial (Gram-negative) halo developed by essential oil: Essential oil of petals (30 µL/6 mm disc) inhibited the growth of *Xanthomonas euvesicatoria* (XV88-5), *Xanthomonas euvesicatoria* (XV56), and *Xanthomonas euvesicatoria* (XV97-2) with inhibition zone diameters of 35, 30, and 35 mm, respectively (Basim & Basim, 2003).

Antimycobacterial halo developed by polar extract: Methanol extract (10%/5 mm wells) inhibited the growth of *Mycobacterium smegmatis* with an inhibition zone diameter of 15 mm (Özkan et al., 2004).

Strong antibacterial (Gram-negative) polar extract: Butanol extract of receptacles inhibited the growth of *Salmonella typhimurium, Pseudomonas aeruginosa, and Bacillus cereus* (Talib et al., 2010).

Strong antibacterial (Gram-positive) isoflavone: 4′-Hydroxy-7-(3-hydroxypropanoyl)-6-methoxy-isoflavone inhibited the growth of methicillin-resistant *Staphylococcus aureus* with an MIC_{90} value of 46 µg/mL (Li et al., 2018).

Moderate antibacterial (Gram-positive) mid-polar extract: Ethyl acetate extract of petals inhibited the growth of *Streptococcus pyogenes*, *Bacillus subtilis*, and *Staphylococcus aureus* with MIC/MBC values of 125/500, 125/500, and 250/500 µg/mL, respectively (Özkan et al., 2004).

Moderate antimycobacterial mid-polar extract: Ethyl acetate extract of petals inhibited the growth of *Mycobacterium phlei* with MIC/MBC values of 250/500 µg/mL (Shohayeb et al., 2014).

Moderate broad-spectrum antifungal polar extract: Aqueous extract of petals inhibited the growth of *Aspergillus niger* (ATCC16404) with MIC/MBC values of 250/250 µg/mL (Hawalani, 2014). Aqueous extract of flower receptacles inhibited the growth of *Candida albicans* with an MIC value of 125 µg/mL (Talib et al., 2010).

Commentaries: (i) The petals contain the flavanol glycoside liquiritin, as well as the flavonols quercetin and kaempferol (Kumar & Kaul, 2006). Whereas quercetin and kaempferol are broad-spectrum antimicrobials (see earlier), it is worth suggesting that liquiritin is probably antiviral, including against coronaviruses because flavanones are often antiviral, such as naringenin, which at a concentration of 62.5 µM abrogated the replication of the Human coronavirus (V229E) in Vero cells (Clementi et al., 2020). Another example is hesperetin, which inhibited the Severe acute respiratory syndrome-acquired coronavirus 3-chymotrypsin-like protease with an IC_{50} value of 8.3 µM (Lin et al., 2005).

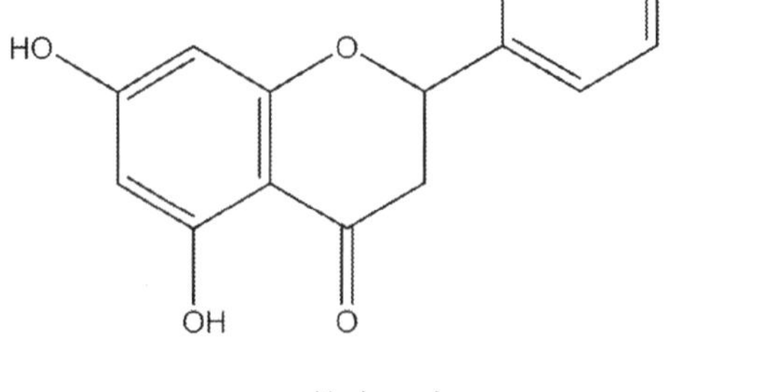

Naringenin

(ii) Aqueous extract of leaves *Rosa davurica* Pall. inhibited the Human immunodeficiency virus type-1 protease (Park, 2003).

(iii) *Rosa laevigata* Michx. is used in China to treat leucorrhea (local name: *Jin ying zi*) and from the roots of this herb, 2α,3α,23-trihydroxyurs-12,19(29)-dien-28-oic acid β-D-glucopyranosyl ester and 2α,3β,19α,23-tetrahydroxyurs-12-en-28-oic acid β-D-glucopyranosyl ester inhibited the growth of *Candida albicans* (ATCC 64550) with the MIC value of 12.5 and 25 μg/mL, respectively (Yuan et al., 2008). Consider that polar extracts of Rosaceae often inhibit the Human immunodeficiency protease *in vitro*: methanol extract of aerial part of *Crataegus pinnatifida* Bunge inhibited the Human immunodeficiency virus type-1 protease (Park, 2003). Aqueous extract of leaves of *Amelanchier asiatica* (S. and Z.) Endl at 100 μg/mL inhibited the Human immunodeficiency virus protease by 42% (Park et al., 2002). Methanol extract of *Geum japonicum* Thunb. inhibited the Human immunodeficiency virus type-1 protease (Park, 2003).

(iv) Methanol extracts of a member of the genus *Amelanchier* Medik. and *Rosa nutkana* C. Presl. inhibited the replication of the bovine coronavirus (BCV) (McCutcheon et al., 1995).

(v) Aqueous decoction of *Rosa odorata* (Andrews) Sweet (2 g/100 mL) evoked after 10 mins pretreatment a reduction in Avian Influenza A (H5N1, Chicken/yamaguchi/7/04) replication in Madin-Darby canine kidney cells (Baatartsogt et al., 2016). Aqueous decoction of *Rosa rugosa* Thunb. (2g/100 mL) inhibited the replication of the Avian Influenza A virus (H5N1, Chicken/yamaguchi/7/04), Avian/Japan/11OG1083/11, and Whistlingswan/Shimane/499/83 in Madin-Darby canine kidney cells (Baatartsogt et al., 2016).

(vi) Essential oils of Rosaceae are broadly antibacterial and antifungal. For instance, essential oil of leaves of *Filipendula vulgaris* Moench inhibited the growth of *Escherichia coli* (95), *Klebsiella pneumoniae* (ATCC 10031), *Pseudomonas aeruginosa* (ATCC 9027), *Salmonella enteridis* (ATCC 13076), *Staphylococcus aureus* (ATCC 65238), *Aspergillus niger* (ATCC 16404), and *Candida albicans* (ATCC 10231) with the MIC values of 0.01, 0.01, 0.03, 0.03, 0.01, 0.009, and 0.09 μg/mL, respectively (Radulović et al., 2007). From this oil, salicylaldehyde (LogD = 1.6 at pH 7.4; molecular mass = 122.1 g/mol) inhibited the growth of *Escherichia coli* (95), *Klebsiella pneumoniae* (ATCC 10031), *Pseudomonas aeruginosa* (ATCC 9027), *Salmonella enteridis* (ATCC 13076), *Staphylococcus aureus* (ATCC 65238), *Aspergillus niger* (ATCC 16404), and *Candida albicans* (ATCC 10231) with the MIC values of 0.1, 0.07, 0.1, 0.07, 0.03, 0.1, and 0.09 μg/mL, respectively (Radulović et al., 2007). From this oil, linalool (LogD = 3.2 at pH 7.4; molecular mass = 154.2 g/mol) inhibited the growth of *Escherichioa coli* (95), *Klebsiella pneumoniae* (ATCC 10031), *Pseudomonas aeruginosa* (ATCC 9027), *Salmonella enteridis* (ATCC 13076), *Staphylococcus aureus* (ATCC 65238), *Aspergillus niger* (ATCC 16404), and *Candida albicans* (ATCC 10231) with the MIC values of 0.07, 0.1, 0.1, 0.0, 0.1, 0.2, and 0.03 μg/mL, respectively (Radulović et al., 2007). From this essential oil, methyl salicilate (LogD = 2.4 at pH 7.4; molecular mass = 152.1 g/mol) inhibited the growth of *Escherichioa coli* (95), *Klebsiella pneumoniae* (ATCC 10031), *Pseudomonas aeruginosa* (ATCC 9027), *Salmonella enteridis* (ATCC 13076), *Staphylococcus aureus* (ATCC 65238), *Aspergillus niger* (ATCC 16404), and *Candida albicans* (ATCC 10231) with the MIC values of 0.2, 0.2, 0.5, 0.2, 0.1, 0.1, and 0.1 μg/mL, respectively (Radulović et al., 2007).

(vii) In India, *Rosa indica* L. is used to treat hepatitis and eye diseases (local name: Ghulab) and *Rosa brunonii* Lindl. (local name: Karir) is used to treat acne. In Turkey, *Rosa pulverulanta* Bieb. (local name: *Gillica*) is used to treat cold as well as *Rosa pimpinellifolia* L. (local name: *Koyungözü*). There, *Rosa canina* L. (local name: *Kusburnu*), *Rosa hemisphaerica* J. Herrm. (local name: *Yemisen*), and *Sorbus domestica* L. (local name: *Uvez*) are used to treat cough.

REFERENCES

Basim, E. and Basim, H., 2003. Antibacterial activity of Rosa damascena essential oil. *Fitoterapia*, *74*(4), pp. 394–396.

Clementi, N., Scagnolari, C., D'Amore, A., Palombi, F., Criscuolo, E., Frasca, F., Pierangeli, A., Mancini, N., Antonelli, G., Clementi, M. and Carpaneto, A., 2020. Naringenin is a powerful inhibitor of SARS-CoV-2 infection in vitro. *Pharmacological Research*, *163*, p. 105255.

Halawani, E.M., 2014. Antimicrobial activity of Rosa damascena petals extracts and chemical composition by gas chromatography-mass spectrometry (GC/MS) analysis. *African Journal of Microbiology Research*, *8*(24), pp. 2359–2367.

Kumar, N., Singh, B. and Kaul, V.K., 2006. Flavonoids from Rosa damascena mill. *Natural Product Communications*, *1*(8), p. 1934578X0600100805.

Li, J., Kong, W.S., Liu, X., Geng, Y.Q., Wang, J., Xu, Y., Li, X.M., Yang, G.Y., Zhou, M., Hu, Q.F. and Li, T., 2018. A new isoflavone derivative from Rosa Damascena and its antibacterial activity. *Zhongguo Zhong yao za zhi= Zhongguo zhongyao zazhi= China journal of Chinese Materia Medica*, *43*(2), pp. 332–335.

Lin, C.W., Tsai, F.J., Tsai, C.H., Lai, C.C., Wan, L., Ho, T.Y., Hsieh, C.C. and Chao, P.D.L., 2005. Anti-SARS coronavirus 3C-like protease effects of Isatis indigotica root and plant-derived phenolic compounds. *Antiviral Research*, *68*(1), pp. 36–42.

McCutcheon, A.R., Roberts, T.E., Gibbons, E., Ellis, S.M., Babiuk, L.A., Hancock, R.E.W. and Towers, G.H.N., 1995. Antiviral screening of British Columbian medicinal plants. *Journal of Ethnopharmacology*, *49*(2), pp. 101–110.

Özkan, G., Sagdiç, O., Baydar, N.G. and Baydar, H.A.S.A.N., 2004. Note: Antioxidant and antibacterial activities of Rosa damascena flower extracts. *Food Science and Technology International*, *10*(4), pp. 277–281.

Park, J.C., 2003. Inhibitory effects of Korean plant resources on human immunodeficiency virus type 1 protease activity. *Oriental Pharmacy and Experimental Medicine*, *3*(1), pp. 1–7.

Radulović, N., Mišić, M., Aleksić, J., Đoković, D., Palić, R. and Stojanović, G., 2007. Antimicrobial synergism and antagonism of salicylaldehyde in Filipendula vulgaris essential oil. *Fitoterapia*, *78*(7–8), pp. 565–570.

Shohayeb, M., Abdel-Hameed, E.S.S., Bazaid, S.A. and Maghrabi, I., 2014. Antibacterial and antifungal activity of Rosa damascena MILL. essential oil, different extracts of rose petals. *Global Journal of Pharmacology*, *8*(1), pp. 1–7.

Talib, W.H. and Mahasneh, A.M., 2010. Antimicrobial, cytotoxicity and phytochemical screening of Jordanian plants used in traditional medicine. *Molecules*, *15*(3), pp. 1811–1824.

Xu, H.X., Wan, M., Loh, B.N., Kon, O.L., Chow, P.W. and Sim, K.Y., 1996. Screening of Traditional Medicines for their Inhibitory Activity Against HIV-1 Protease. *Phytotherapy Research*, *10*(3), pp. 207–210.

Yuan, J.Q., Yang, X.Z., Miao, J.H., Tang, C.P., Ke, C.Q., Zhang, J.B., Ma, X.J. and Ye, Y., 2008. New triterpene glucosides from the roots of Rosa laevigata Michx. *Molecules*, *13*(9), pp. 2229–2237.

7.8.3.6 *Rubus glomeratus* Bl.

Synonym: *Rubus moluccanus* var. *glomeratus* (Blume) Backer

Common names: Molucca raspberry, dum molucca (Vietnam), welbute (Sri Lanka), lintagu, kalataguk (Borneo), berete (Indonesia), auiteteya (Papua New Guinea), sapinit (the Philippines).

Habitat: Roadsides, open forests

Distribution: India, Sri Lanka, Himalaya, Nepal, Myanmar, Malaysia, Indonesia, Papua New Guinea, Australia, Pacific Islands.

Botanical observation: This climber grows up to 10 m long. The stem is terete, smooth, hairy at apex, thorny, the thorns about 8 mm long. The internodes are 4.5–5.5 cm long. The leaves are simple, spiral and stipulate. The petiole is 3–5 cm long, thorny, slender, and hairy. The blade is cordate to 3-lobed, 8–11 × 6–8 cm, glabrous above, velvety below, serrate, with about 4 secondary nerves and a midrib sunken above and filled with hairs. The flowers are arranged in axillary, hairy, and few flowered fascicles. The 5 sepals are deeply incised and hairy. The 5 petals are whitish to light pink, round to somewhat broadly lanceolate, veined, and about 7 mm long. The stamens are numerous and filamentous. The fruit is pale green, ripening red to purple, and edible.

Medicinal use: Dysentery (Malaysia)

Commentaries: (i) The plant has apparently not been examined for its antimicrobial properties. Consider that members of the genus *Rubus* L abound with high molecular weight polymers of ellagitannins (Chen et al., 2019), whence their astringent medicinal uses.

(ii) Butanolic fraction of leaves of *Rubus chamaemorus* L. inhibited the growth of *Staphylococcus* aureus (ATCC), *Staphylococcus epidermidis* (NTCT 11047), *Micrococcus luteus* (NTCT 9341),

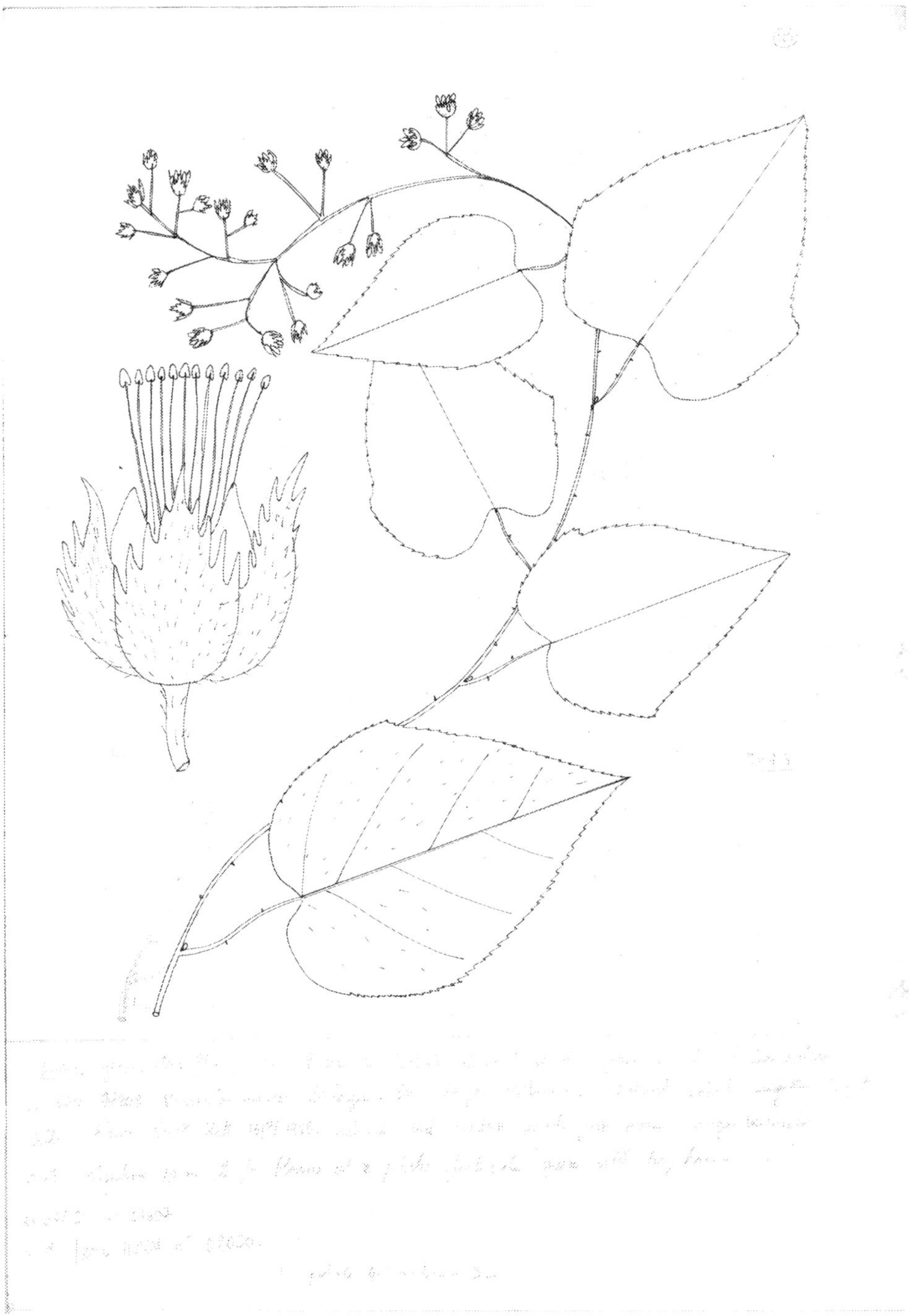

FIGURE 7.32 *Rubus glomeratus* Bl.

Escherichia coli (NCTC 8196), *Bacillus subtilis* (ATCC 6633), and *Candida albicans* (ATCC 10231) with MIC/MBC values of 500/2300, 500/500, 500/500, 2300/4600, 500/500, 1700/2300 μg/mL, respectively (Thiem & Goślińska, 2004). From this fraction, gallic acid inhibited the growth of *Staphylococcus aureus* (ATCC), *Staphylococcus epidermidis* (NTCT 11047), *Micrococcus luteus*

(NTCT 9341), *Escherichia coli* (NCTC 8196), *Bacillus subtilis* (ATCC 6633), and *Candida albicans* (ATCC 10231) (Thiem & Goślińska, 2004).

(iii) Aqueous extract of fruits of *Rubus coreanus* Miq. at a concentration of 128 µg/mL inhibited the secretion of the Hepatitis B surface antigen by 68.4% (Kim et al., 2001). Ethanol extract of seeds of *Rubus coreanus* Miq. at a concentration of 50 µg/mL abrogated the replication of the Influenza strains, A/Brisbane/59/2007(H1N1) (BR59), pandemic A/Korea/01/2009(H1N1) (KR01), and A/Brisbane/10/2007(H3N2) (BR10), and a fraction isolated from this extract given orally to BALB/c mice at a dose of 50 mg/Kg/day for 8 days evoked protection against mouse-adapted Influenza PR8 virus (Lee et al., 2016). One could observe that polar extracts of members of the family Rosaceae are often exhibiting antiviral effects against Influenza viruses as, for example, the ethanol extract of fruits of *Chaenomeles sinensis* (Thouin) Koehne (used in China for sore throat; local name: *Mu gua*), which reduced plaque formation of Influenza virus A/Udorn/307/72 with an IC_{50} value of 1.6 µg/mL and Influenza virus B/Johannesburg/5/99 with an IC_{50} value of 1.4 µg/mL (Sawai et al., 2008). This extract Inhibited Influenza virus infection via prevention of receptor binding of the virus to host cells and reduction of NS2 protein synthesis (Sawai et al., 2008).

(iv) In India, *Rubus fruticosus* L. is used to treat cough and to heal wounds, and *Rubus ellipticus* Smith (local name: *Aakhey*) to heal oral ulcers. *Rubus caesius* L. is used in Turkey to treat sore throat (local name: *Tütürk*). There, *Rubus sanctus* Schreber (local name: *Bogurtlen*) is used to treat diarrhea and *Rubus idaeus* L. (local name: *Ahuduhu*) is used to treat infections. In Nepal, *Rubus ellipticus* Sm., is used to treat fever, diarrhea, and dysentery and *Rubus nepalensis* (Hook. f.) Kuntze is used to treat fever (local name: *Bhui kafal*).

REFERENCES

Kim, T.G., Kang, S.Y., Jung, K.K., Kang, J.H., Lee, E., Han, H.M. and Kim, S.H., 2001. Antiviral activities of extracts isolated from Terminalis chebula Retz., Sanguisorba officinalis L., Rubus coreanus Miq. and Rheum palmatum L. against hepatitis B virus. *Phytotherapy Research*, *15*(8), pp. 718–720.

Lee, J.H., Bae, S.Y., Oh, M., Seok, J.H., Kim, S., Chung, Y.B., Gowda K, G., Mun, J.Y., Chung, M.S. and Kim, K.H., 2016. Antiviral effects of black raspberry (Rubus coreanus) seed extract and its polyphenolic compounds on norovirus surrogates. *Bioscience, Biotechnology, and Biochemistry*, *80*(6), pp. 1196–1204.

Sawai, R., Kuroda, K., Shibata, T., Gomyou, R., Osawa, K. and Shimizu, K., 2008. Anti-influenza virus activity of Chaenomeles sinensis. *Journal of Ethnopharmacology*, *118*(1), pp. 108–112.

Thiem, B. and Goślińska, O., 2004. Antimicrobial activity of Rubus chamaemorus leaves. *Fitoterapia*, *75*(1), pp. 93–95.

7.8.4 Family Rhamnaceae A.L de Jussieu (1789)

The family Rhamnaceae consists of 58 genera and about 900 species of trees, shrubs, and climbers. The leaves are simple, alternate, or sub-opposite, and stipulate. The stems are thorny or not. The blade is often coriaceous and serrate to wavy and not uncommonly marked with a pair of somewhat cinnamon-like secondary nerves. The inflorescence are axillary cymes, corymbs, racemes, spikes, panicles or fascicles. The flowers are minute and star-shaped. The calyx is tubular and 5 lobed. The corolla presents 5 petals. A large disc is present. Five stamens alternating with the petals form the androecium. The gynoecium comprises an ovary, which is 2–3 locular, each locule sheltering a single ovule. The fruits are drupes.

7.8.4.1 *Scutia myrtina* (Burm. f.) Kurz

Synonyms: *Blepetalon aculeatum* Raf.; *Ceanothus circumscissus* (L. f.) Gaertn.; *Rhamnus circumscissa* L. f.; *Rhamnus myrtina* Burm. f.; *Scutia circumcissa* (L. f.) W. Theob.; *Scutia commersonii* Brongn.; *Scutia eberhardtii* Tardieu; *Scutia indica* Brongn.; *Scutia obcordata* Boivin ex Tul.

Common names: Cat-thorn, dui ci teng (China); tuvadi (India)

Habitat: Forests

Distribution: India to Southeast Asia

Botanical observation: This handsome shrub grows to about 5 m tall. The stems are armed with recurved spines which are about 7 mm long. The leaves are simple, opposite or subopposite, and stipulate. The stipules are lanceolate and about 3 mm long. The petiole is up to 5 mm long. The blade is coriaceous, dark green, glossy, broadly elliptic to obovate, 3.5–6×1.8–3 cm, with 5–8 pairs of inconspicuous secondary nerves, cuneate at base, somewhat irregular at margin, and acute to round at apex. The inflorescences are short axillary fascicles of a few tiny flowers. The calyx is tubular. The corolla is tubular, white, and develops 5 sharply triangular lobes, which are about 3 mm long. A disc is present. The androecium includes 5 white and fleshy stamens. The gynoecium includes a globose ovary developing a short style. The drupes are obovoid to somewhat obconical, graceful, about 5 mm across, dark dull bluish, marked with apical whitish disc and containing 2 seeds.

Medicinal use: Dysentery (India)

Broad-spectrum antibacterial halo non-polar extract: Petroleum ether extract (500 μg/mL) inhibited the growth of *Pseudomonas aeruginosa, Escherichia coli* (MTCC 443), *Salmonella typhi*, *Staphylococcus aureus* (MTCC 740), *Streptococcus pneumoniae* (MTCC 109), and *Bacillus subtilis* (MTCC 736), with inhibition zone diameters of 18, 15, 19, 13, 14, and 15 mm, respectively (Kritheka et al., 2008).

Broad-spectrum antifungal halo non-polar extract: Petroleum ether extract inhibited the growth of *Candida albicans*, *Aspergillus funigatus*, *Aspergillus flavus*, and *Cryptococcus neoformans* (Kritheka et al., 2008).

Commentary: One could examine the antibacterial, antifungal, and antiviral activities of the perylenequinones scutiaquinone A and scutiaquinone B (Ayers et al., 2007), which as per their overall structure (see earlier) are most probably active.

REFERENCES

Ayers, S., Zink, D.L., Mohn, K., Powell, J.S., Brown, C.M., Murphy, T., Brand, R., Pretorius, S., Stevenson, D., Thompson, D. and Singh, S.B., 2007. Scutiaquinones A and B, Perylenequinones from the Roots of Scutia myrtina with Anthelmintic Activity. *Journal of Natural Products*, *70*(3), pp. 425–427.

Kritheka, N., Kumar, R.S., Kumar, S.S., Murthy, N.V., Sundram, R.S. and Perumal, P., 2008. Anti-inflammatory and Antimicrobial activities of Petroleum ether and Ethanol extracts of Scutia myrtina (Rhamnaceae). *Oriental Pharmacy and Experimental Medicine*, *8*(4), pp. 400–407.

7.8.4.2 *Ziziphus mauritiana* Lam.

Synonyms: *Paliurus mairei* H. Lév.; *Rhamnus jujuba* L.; *Ziziphus abyssinica* Hochst.; *Ziziphus jujuba* (L.) Gaertn.; *Ziziphus jujuba* (L.) Lam.; *Ziziphus maire* (H. Lév.) Browicz & Lauener

Common names: Indian jujube; boroi; kul; jon janum (Bangladesh); putrea (Cambodia); bidara (Indonesia, Malaysia); bayer, ber, kul (India); eng-si (Myanmar); shoot (Pakistan); manzanitas (the Philippines); ma tan, phut sa, (Thailand); tao (Vietnam)

Habitat: Deserts, arid lands, cultivated

Distribution: Turkey, Iran, Turkmenistan, Tadjikistan, Pakistan, India, and China

Botanical observation: This tree probably native to the Middle East grows to about 8m tall. The stems are hairy and spiny. The leaves are simple, alternate, and stipulate. The stipules are recurved, spiny, and about 5mm long. The petiole is up to 1.5 cm long and hairy. The blade is elliptic-ovate, 2–9 × 1.5–5 cm, asymmetrical at base, obtuse, rounded, or shortly acuminate at apex, serrate, dark green and glossy above, greyish brown and hairy below, and marked with 3 longitudinal nerves. The cymes are axillary, hairy, and bearing minute greenish-yellow flowers. The 5 sepals are minute and triangular. The 5 petals are triangular, reflexed, and minute. A 10-lobed disc is present. The androecium comprises 5 stamens smaller than the petals. The ovary is merged in disc and develops a bifid style. The drupes are globose, light green turning orange, glossy, smooth, 1.5–3.5 × 1.5–2.5 cm, and contain a rugose endocarp.

Medicinal uses: Fever, wounds, ulcers, leucorrhea (Bangladesh); gingivitis (Cambodia); diarrhea (Bangladesh, Thailand); blood infection, cold, conjunctivitis, cough, diarrhea, dysentery, fever, sores (India); bronchitis, measles (Pakistan)

Strong antibacterial (Gram-positive) A-nor-E-seco spiro-lactone ceanothane-type triterpenes: Zizimauritic acid A and zizimauritic acid B isolated from the roots inhibited the growth of *Staphylococcus aureus* with IC_{50} values 2.1 and 12.7 µg/mL (Ji et al., 2012).

Strong antimycobacterial cyclopeptide alkaloid: Nummularine H isolated from the roots inhibited the growth of *Mycobacterium tuberculosis* (H37Ra) with an MIC of 4.5 µM (Panseeta et al., 2011).

Commentaries: (i) *Ziziphus oenopolia* (L.) Mill. is traditionally used in Bangladesh to treat fungal infections of the skin. *Ziziphus nummularia* (Burm. f.) Wight & Arn in used in Pakistan (local name: *Mallah*) to heal ulcers and in India to heal sores. Phenolic fraction from fruits of *Ziziphus nummularia* (Burm. f.) Wight & Arn. inhibited the growth of *Staphylococcus aureus*, *Bacillus subtilis*, *Enterococcus faecalis*, *Listeria monocytogenes*, *Escherichia coli*, *Pseudomonas aeruginosa*, *Salmonella typhi*, and *Shigella dysenteriae* (Aman et al., 2014).

(ii) *Ziziphus jujuba* Mill. is used for the treatment of COVID-19 in China (Wang et al., 2020) and in Unani medicine is valued in times of pandemics (Ansari et al., 2020). The active anti-COVID-19 principles are apparently yet unknown and in this context it is worth mentioning that the cyclopeptide alkaloids jubanine G, jubanine H (molecular mass = 585.7 g/mol), and nummularine B isolated from the roots inhibited the replication of the Swine-infecting coronavirus porcine epidemic diarrhea virus in Vero cells with IC_{50} values of 13.4, 4.4 and 6.1 µM and selective indices of > 30. 47.1 and 26.7, respectively (Kang et al., 2015). This finding could lead someone to examine the anti-COVID-19 properties of jubanine H and other cyclopeptide alkaloids in the genus *Ziziphus* L. Jubanine H could be added to a list of first line natural products to be tested for coming coronavirus zoonotic pandemics.

(iii) Methanol extract of stems of *Berchemia berchemiifolia* (Makino) Koidz. and methanol extract of leaves of *Sageretia theezans* (L.) Brongn. inhibited the Human immunodeficiency virus type-1 protease (Park, 2003). *Sageretia theezans* Brongn. is used in China to treat acne, and aqueous extract of leaves at a concentration of 100 µg/mL inhibited Human immunodeficiency virus protease by 56% (Park et al., 2002)

(iv) *Rhamnus japonica* Maxim. is used in China to treat small pox. In the Andaman Islands, the climbing shrub *Gouania leptostachya* DC. (local name: *Saboon bel*) is used to treat fever. *Zizyphus jujuba* var. *inermis* (Bunge) Rehder is used in Korea to treat cold (local name: *Daechunamu*). In Thailand, *Colubrina asiatica* (L.) Brongn (local name: *Pak wan tale*) is used to heal abscesses and to treat eye infections. There, *Ziziphus oenopolia* (L.) Mill. var. *oenopolia* is also used to treat cough (local name: *Yap yio*). In Pakistan, *Sageretia thea* (Osbeck) M.C. Johnst. is used to treat hepatitis.

REFERENCES

Aman, S., Naim, A., Siddiqi, R. and Naz, S., 2014. Antimicrobial polyphenols from small tropical fruits, tea and spice oilseeds. *Revista de Agaroquimica y Tecnologia de Alimentos*, *20*(4), pp. 241–251.

Ji, C.J., Zeng, G.Z., Han, J., He, W.J., Zhang, Y.M. and Tan, N.H., 2012. Zizimauritic acids A–C, three novel nortriterpenes from Ziziphus mauritiana. *Bioorganic & Medicinal Chemistry Letters*, *22*(20), pp. 6377–6380.

Kang, K.B., Ming, G., Kim, G.J., Choi, H., Oh, W.K. and Sung, S.H., 2015. Jubanines F–J, cyclopeptide alkaloids from the roots of Ziziphus jujuba. *Phytochemistry*, *119*, pp. 90–95.

Panseeta, P., Lomchoey, K., Prabpai, S., Kongsaeree, P., Suksamrarn, A., Ruchirawat, S. and Suksamrarn, S., 2011. Antiplasmodial and antimycobacterial cyclopeptide alkaloids from the root of Ziziphus mauritiana. *Phytochemistry*, *72*(9), pp. 909–915.

Park, J.C., 2003. Inhibitory effects of Korean plant resources on human immunodeficiency virus type 1 protease activity. *Oriental Pharmacy and Experimental Medicine*, *3*(1), pp. 1–7.
Park, J.C., Hur, J.M., Park, J.G., Hatano, T., Yoshida, T., Miyashiro, H., Min, B.S. and Hattori, M., 2002. Inhibitory effects of Korean medicinal plants and camelliatannin H from Camellia japonica on human immunodeficiency virus type 1 protease. *Phytotherapy Research*, *16*(5), pp. 422–426.
Wang, S.X., Wang, Y., Lu, Y.B., Li, J.Y., Song, Y.J., Nyamgerelt, M. and Wang, X.X., 2020. Diagnosis and treatment of novel coronavirus pneumonia based on the theory of traditional Chinese medicine. *Journal of Integrative Medicine*, *8*, pp. 275–283.

7.8.5 Family Ulmaceae Mirbel (1815)

The family Ulmaceae consists of 15 genera and 200 species of trees or shrubs. The leaves are simple, spiral or opposite, and stipulate. The perianth develops 4–9 lobes. The androecium includes 4–9 stamens. The gynoecium comprises 2 carpels fused in a 1–2-loculed ovary, each locule sheltering a single suspended ovule, a short style, and a pair of stigmas. The fruits are samara, drupes, or winged nutlets.

7.8.5.1 *Trema orientalis* (L.) Bl.

Synonyms: *Celtis orientalis* L.; *Celtis discolor* Brongn.; *Celtis rigida* Blume; *Sponia argentea* Planch.; *Sponia orientalis* (L.) Decne.; *Sponia wightii* Planch.; *Trema polygama* Z.M. Wu & J.Y. Lin

Common names: Charcoal tree; khaksi daru; sugarar amila (Bangadesh); srô:l (Cambodia); yi se shan huang ma (China); kuray (indonesia); po: hu: (Laos); mengkirai (Malaysia); anadgong (the Philippines); takhai (Thailand); hu lá nhỏ (Vietnam)

Habitat: Forests

Distribution: From India to the Pacific Islands

Botanical observation: This tree grows up to about 15 m tall. The stems are hairy. The leaves are simple, alternate, and stipulate. The stipules are narrowly triangular and about 1 cm long. The petiole grows to 2 cm long. The blade is oblong-lanceolate, 10–20 × 5–10cm, hairy and somewhat glaucous below, cordate and asymmetrical at base, serrate, acuminate at apex, and marked with 4–6 pairs of secondary nerves. The inflorescences are axillary clusters of minute flowers. The perianth includes 4–5 hairy and broadly lanceolate tepals. The androecium comprises 5 stamens. The ovary is minute and develops a bifid style. The berries are black, about 5 mm long, and containing minute seeds.

Medicinal uses: Abscesses, gonorrhea, pneumonia (Bangladesh)

Broad-spectrum antibacterial halo developed by polar extract: Aqueous extract of leaves (500 µg/well) inhibited the growth of *Staphylococcus aureus*, *Staphylococcus epidermidis*, *Staphylococcus saprophyticus*, *Streptococcus pyogenes*, *Plesiomonas shigelloides*, *Shigella dysenteriae*, *Vibrio cholera*, *Shigella flexneri*, *Shigella sonnei*, and *Pseudomonas aeruginosa* (Uddin, 2008).

Moderate antimycobacterial polar extract: Acetone extract inhibited the growth of *Mycobacterium smegmatis*, *Mycobacterium fortuitum*, *Mycobacterium aurum*, and *Mycobacterium tuberculosis* with MIC values of 312, 625, 1250, and 312 µg/mL, respectively (Dzoyem et al., 2016).

Commentaries: (i) Methanol extract of leaves of *Holoptelea* integrifolia Planch, a plant in the family Ulmaceae traditionally used in India for the treatment of herpes (local name: *Tapasi*), inhibited the replication of the Human simplex virus type-1 in Vero cells with the IC_{50} value of 10 µg/mL (Rajbhandari et al., 2001; Khan et al., 2005). Methanol extract of this plant inhibited the growth of *Bacillus cerculences*, *Pseudomonas aeruginosa*, *Bacillus subtilis*, *Klebsiella aeruginosa*, *Staphylococcus aureus*, and *Escherichia coli* with MIC/MBC values of 312.5/156.2, 312.5/156.2, 625/312.5, 312.5/312.5, 1250/1250, and 312.5/156.2 µg/mL, respectively (Reddy et al., 2008). (ii) Methanol extract of leaves and stems of *Ulmus davidiana* Planch inhibited the Human immunodeficiency virus type-1 protease (Park, 2003).

(iii) Apparently the family Ulmaceae has not been much examined for its antimocrobial properties and particularly antiviral properties.

REFERENCES

Dzoyem, J.P., Aro, A.O., McGaw, L.J. and Eloff, J.N., 2016. Antimycobacterial activity against different pathogens and selectivity index of fourteen medicinal plants used in southern Africa to treat tuberculosis and respiratory ailments. *South African Journal of Botany*, *102*, pp. 70–74.

Khan, M.T.H., Ather, A., Thompson, K.D. and Gambari, R., 2005. Extracts and molecules from medicinal plants against herpes simplex viruses. *Antiviral Research*, *67*(2), pp. 107–119.

Park, J.C., 2003. Inhibitory effects of Korean plant resources on human immunodeficiency virus type 1 protease activity. *Oriental Pharmacy and Experimental Medicine*, *3*(1), pp. 1–7.

Rajbhandari, M., Wegner, U., Jülich, M., Schoepke, T. and Mentel, R., 2001. Screening of Nepalese medicinal plants for antiviral activity. *Journal of Ethnopharmacology*, *74*(3), pp. 251–255.

Reddy, B.S., Reddy, R.K.K., Naidu, V.G.M., Madhusudhana, K., Agwane, S.B., Ramakrishna, S. and Diwan, P.V., 2008. Evaluation of antimicrobial, antioxidant and wound-healing potentials of Holoptelea integrifolia. *Journal of Ethnopharmacology*, *115*(2), pp. 249–256.

Uddin, S.N., 2008. Antioxidant and antibacterial activities of Trema orientalis Linn: an indigenous medicinal plant of indian subcontinent. *Oriental Pharmacy and Experimental Medicine*, *8*(4), pp. 395–399.

7.8.6 Family Urticaceae A.L, de Jussieu (1789)

The family Urticaceae consists of about 50 genera and 1300 species of herbs and shrubs. The leaves are simple, spiral or opposite, and stipulate. The inflorescences are axillary, cymes, fascicles, or spikes of minute flowers. The perianth presents 4–5 lobes. The androecium includes 4 to 5 stamens. The gynoecium consists of a single carpel forming a unilocular ovary sheltering a single ovule. The fruits are achenes or drupes in a persistent perianth.

7.8.6.1 *Boehmeria nivea* (L.) Gaudich.

Synonyms: *Boehmeria tenacissima* (Roxb.) Bl.; *Ramium niveum* (L.) Kuntze; *Urtica nivea* L.; *Urtica tenacissima* Roxb.

Common names: Ramie; hurumbuto- pada (Bangladesh); thmey (Cambodia); zhu ma (China); haramay (Indonesia); pan (Laos); amirai (the Philippines); po paan (Thailand); cây gai (Vietnam)

Habitat: Forest margins, roadsides, shady and moist spots, cultivated

Distribution: India, Nepal, Bhutan, Bangladesh, Cambodia, Laos, Vietnam, Thailand, Indonesia, China, Korea, and Japan

Botanical observation: This graceful herb grows to about 1.5 m tall. The stem is hairy. The leaves are simple, spiral, and stipulate. The stems are fibrous. The stipules are hairy, lanceolate, bifid, and about 1 cm long. The petiole is straight, slender, and up to 10 cm long. The blade broadly lanceolate, 5–15 × 3.5–14 cm, hairy and somewhat glaucous below, soft, dull-green, wedge-shaped at base or cordate, serrate, cuspidate at apex, and marked with 2–4 pairs of secondary nerves. The spikes are up to about 10 cm long. The flowers are minute. The perianth includes 4 tepals. The androecium presents 4 stamens. The achenes are minute.

Medicinal uses: Infected wounds (Bangladesh); colds, fever (China); hepatitis (Thailand)

Broad-spectrum antibacterial halo developed by polar extract: Methanol extract (15 μg/30 μL solution/8 mm disc) inhibited the growth of *Escherichia coli* and *Staphylococcus aureus* with inhibition zone diameters of 14 and 14 mm, respectively (Lee et al., 2014).

Moderate broad-spectrum antibacterial polar extract: Ethanol extract of roots inhibited the growth of *Bacillus subtilis* (ATCC6633), *Staphylococcus aureus* (ATCC6530), and *Pseudomonas aeruginosa* (ATCC9027) with MIC values of 100, 100, and 100 μg/mL, respectively (Liu et al., 2014).

Strong broad-spectrum antibacterial flavonol glycoside: Rutin inhibited the growth of *Escherichia coli* (ATCC 25922), *Escherichia coli* (clinical ESβL+), *Pseudomonas aeruginosa* (ATCC 10145), *Pseudomonas aeruginosa* (clinical isolate), *Proteus mirabilis* (ATCC 7002), *Proteus*

mirabilis (clinical ESβL+), *Klebsiella pneumoniae* (RSKK 574), *Klebsiella pneumoniae* (clinical ESβL+), *Acinetobacter baumanii* (RSKK 02026), *Acinetobacter baumanii (clinical isolate), Staphylococcus aureus* (ATCC 25923), *Staphylococcus aureus* (clinical MRSA), *Enterococcus faecalis* (ATCC 29212), *Enterococcus faecalis (clinical strain), and Bacillus subtilis* (ATCC 6633) (Orhan et al., 2010). Rutin inhibited the growth of *Escherichia coli* (ATCC 11229), *Klebsiella pneumoniae* (ATCC 10031), *Enterobacter aerogenes* (ATCC 13048), *Staphylococcus aureus* (ATCC 13709), *Proteus vulgaris* (ATCC 12454), *Pseudomonas aeruginosa* (ATCC 27853), and *Enterobacter cloacae* (ATCC 10699) (Fu et al., 2016).

Antifungal (filamentous) halo developed by flavonol glycoside: Rutin (500 μg/disc) inhibited the growth of *Trichophyton rubrum* as efficiently as amphotericin B (1.5 μg/disc) (Gaziano et al., 2018).

Strong antifungal flavonol glycoside: Rutin inhibited the growth of *Candida albicans* (ATCC 10231) and *Candida krusei* (ATCC 6258*)* with MIC values of 16 and 32 μg/mL, respectively (Orhan et al., 2010).

Antifungal fatty acids: (Z)-9,10,11-trihydroxy-12-octadecenoic acid, (Z)-7,8,9-trihydroxy-10-hexadecenoic acid, and (Z)-12-keto-7,8,9-trihydroxy-10-hexadecenoic acid isolated from the roots evoked antifungal effects (Xu et al., 2011).

Strong antiviral (enveloped segmented linear single stranded (−)RNA) flavonol glycoside: Rutin (LogD=−1.7 at pH 7.4; molecular weight=610.5 g/mol) reduced avian Influenza strain H5N1 plaque formation Madin-Darby canine kidney cells by 73.2%, at a concentration of 1 ng/mL (Ibrahim et al., 2013).

Moderate antiviral (enveloped circular double-stranded DNA) polar extract: Ethanol extract of roots at a concentration of 100 μg/mL inhibited the secretion of the Hepatitis B virus e-antigen by HepG2 2.215 cells by about 60% (Huang et al., 2006).

In vivo antiviral (enveloped circular double-stranded DNA) polar extract: Ethanol extract of roots given orally at a dose of 195 mg/Kg/day for 10 days to mice experimentally infected with the Hepatitis B virus evoked a decrease in serum Hepatitis B surface antigen and Hepatitis B virus DNA by 32 and 40%, respectively (Chang et al., 2010).

Viral enzyme inhibition by flavonol glycoside: Rutin inhibited the Hepatitis C virus NS3 serine protease with an IC_{50} value of 77 μM (Zuo et al., 2005).

Commentaries: (i) The plant generates loliolide (see earlier) (Cho et al., 2016).

(ii) Rutin is antiviral and has the ability to inhibit the angiotensin-converting enzyme 2 (Muchtaridi et al., 2020). Rutin could be added to the list of the first-line natural products to be tested for coming waves of betacoronavirus pandemics.

(iii) Members of the family Urticaceae are often used in Asia for the treatment of infections. For instance, *Pouzolzia pentandra* (Roxb.) Benn. is used in Thailand to treat fever and toothaches (local name: *Khob cha nang*). In Bangladesh *Pouzolzia sanguinea* (Blume) Merr. (local name: Mogjangaillya shak) is used to treat skin infections, *Sarcochlamys pulcherrima* Gaudich. affords a remedy for boils, fever, and ophthalmia (local name: *Jangayillia shak*). *Fleurya interrupta* (L.) Gaudich (local name: *Chun jhitki*) is employed for cold. Ethanol extract of leaves of *Fleurya interrupta* (L.) Gaudich) inhibited the growth of *Staphylococcus aureus*, *Shigella dysenteriae*, and *Salmonella typhi* with inhibition zone diameters of 10, 10, and 10 mm, respectively (500 μg/disc) (Uddin et al., 2008). *Laportea crenulata* Gaudich. is used to treat fever in Bangladesh and the oleanane-type triterpene 2α, 3β, 21β, 23, 28-penta hydroxyl 12-oleanene (180 μg/disc) isolated from the roots of this plant inhibited the growth of *Aspergillus flavus*, *Aspergillus niger*, *Candida albicans*, and *Rhizopus aurizae* with inhibition zone diameters of 6, 9, 10 and 8 mm (Carson et al., 1995) and inhibited the growth of *Bacillus subtilis*, *Bacillus megaterium*, *Staphylococcus aureus*, *Staphylococcus β-hemolyticus*, *Escherichia coli*, *Pseudomonas aeruginosa*, *Salmonella typhi*, *Shigella dysenteriae*, *Shigella flexneri*, and *Shigella sonnei* with inhibition zone diameters of 11, 8, 13, 19, 11, 15, 8, 17, 10, and 9 mm, respectively (80 μg/disc). This triterpene inhibited the growth of *Aspergillus flavus*, *Candida* sp., and *Rhizopus aurizae* with inhibition zone diameters of 9, 11, and 8 mm (80 μg/disc) (Rahman et al., 2008).

(iv) In Iran, *Urtica dioica* L. is used to treat gingivitis and skin diseases, and in Nepal (local name: *Sisnu*) it affords a medicine for cold and cough. Water extract of *Urtica dioica* L. inhibited the growth of *Escherichia coli* (ATCC 9837), *Proteus mirabilis* (clinical isolate), *Citrobacter koseri* (clinical isolate), *Staphylococcus aureus* (ATCC 6538), *Streptococcus pneumoniae* (ATCC 49619), *Enterobacter aerogenes* (clinical isolate), *Micrococcus luteus* (clinical isolate), and *Staphylococcus epidermidis* (clinical isolate) with inhibition zone diameters of 15, 24, 22, 15, 15, 12, 19, and 24 mm, respectively (250 µg/disc 6 mm discs) and inhibited *Candida albicans* (ATCC 10231) (Gülçin et al., 2004). Hexane extract of *Urtica dioica* L. inhibited the growth of *Mycobacteriun tuberculosis* strains (H37Rv), *Mycobacterium tuberculosis* (MDR), *Mycobacteriun tuberculosis* (CL-1), and *Mycobacterium tuberculosis* (CL-2) with MIC values of 250, 500, 250, and 250 µg/mL, respectively (Singh et al., 2013)

(v) Chalcone-6′-hydroxy-2′,3,4-trimethoxy-4′-*O*-β-D-glucopyranoside, isoflavone-3′,4′,5,6-tetrahydroxy-7-*O*-{β-D-glucopyranosyl-(1→3)-α-L-rhamnopyranoside},isoflavone-3′,4′,5,6-tetrahydroxy-7-*O*-{β-D-glucopyranosyl-(1→6)-β-D-glucopyranosyl-(1→6)-β-D-glucopyranosyl-(1 →3)-α-L-rhamnopyranoside} isolated from the leaves of *Boehmeria rugulosa* Wedd. inhibited the growth of *Staphylococcus aureus* (Semwal et al., 2009). Chalcone-6′ -hydroxy-2′,3,4-trimethoxy-4′-*O*-β-D-glucopyranoside, isoflavone-3′,4′,5,6-tetrahydroxy-7-*O*-{β-D-glucopyranosyl- (1 →3)-α-L-rhamnopyranoside}, isoflavone-3′,4′,5,6-tetrahydroxy-7-*O*-{β-D-glucopyranosyl-(1→6)-β-D-glucopyranosyl-(1→6)- β -D-glucopyranosyl-(1 →3)- α -L-rhamnopyranoside} inhibited the growth of *Microsporium gypseum*, *Microsporium canis*, and *Trichophyton rubrum* (Semwal et al., 2009).

(vi) In Nepal, *Boehmeria macrophylla* D. Don is used to treat dysentery and diarrhea (local name: *Kamle ghans*), *Boehmeria ternifolia* D. Don is applied to cuts and boils (local names: *Dhadale*), *Debregeasia longifolia* (Burm. f.) Wedd. is used for the treatment of skin infection (local name: *Tushare*), and *Lecanthus peduncularis* (Royle) Wedd. is used to treat diarrhea and dysentery (local name: *Gaulato*).

REFERENCES

Cho, S., Lee, D.G., Jung, Y.S., Kim, H.B., Cho, E.J. and Lee, S., 2016. Phytochemical Identification from Boehmeria nivea Leaves and Analysis of (–)-Loliolide by HPLC. *Natural Product Sciences*, *22*(2), pp. 134–139.

Fu, L., Lu, W. and Zhou, X., 2016. Phenolic compounds and in vitro antibacterial and antioxidant activities of three tropic fruits: persimmon, guava, and sweetsop. *BioMed Research International*, *2016*. Article ID 4287461.

Gaziano, R., Campione, E., Iacovelli, F., Marino, D., Pica, F., Di Francesco, P., Aquaro, S., Menichini, F., Falconi, M. and Bianchi, L., 2018. Antifungal activity of Cardiospermum halicacabum L.(Sapindaceae) against Trichophyton rubrum occurs through molecular interaction with fungal Hsp90. *Drug Design, Development and Therapy*, *12*, p. 2185.

Gülçin, I., Küfrevioğlu, Ö.İ., Oktay, M. and Büyükokuroğlu, M.E., 2004. Antioxidant, antimicrobial, anti-ulcer and analgesic activities of nettle (Urtica dioica L.). *Journal of Ethnopharmacology*, *90*(2–3), pp. 205–215.

Huang, K.L., Lai, Y.K., Lin, C.C. and Chang, J.M., 2006. Inhibition of hepatitis B virus production by Boehmeria nivea root extract in HepG2 2.2. 15 cells. *World Journal of Gastroenterology: WJG*, *12*(35), p. 5721.

Ibrahim, A.K., Youssef, A.I., Arafa, A.S. and Ahmed, S.A., 2013. Anti-H5N1 virus flavonoids from Capparis sinaica Veill. *Natural Product Research*, *27*(22), pp. 2149–2153.

Lee, A.Y., Wang, X., Lee, D.G., Kim, Y.M., Jung, Y.S., Kim, H.B., Kim, H.Y., Cho, E.J. and Lee, S.H., 2014. Bioactive Materials: Various Biological Activities of Ramie (Boehmeria nivea). *Journal of Applied Biological Chemistry*, *57*(3), pp. 279–286.

Liu, Y., Nielsen, M., Staerk, D. and Jäger, A.K., 2014. High-resolution bacterial growth inhibition profiling combined with HPLC–HRMS–SPE–NMR for identification of antibacterial constituents in Chinese plants used to treat snakebites. *Journal of Ethnopharmacology*, *155*(2), pp. 1276–1283.

Muchtaridi, M., Fauzi, M., Khairul Ikram, N.K., Mohd Gazzali, A. and Wahab, H.A., 2020. Natural flavonoids as potential angiotensin-converting enzyme 2 inhibitors for anti-SARS-CoV-2. *Molecules*, *25*(17), p. 3980.
Orhan, D.D., Özçelik, B., Özgen, S. and Ergun, F., 2010. Antibacterial, antifungal, and antiviral activities of some flavonoids. *Microbiological Research*, *165*(6), pp. 496–504.
Rahman, M.M., Khan, A., Haque, M.E. and Rahman, M.M., 2008. Antimicrobial and cytotoxic activities of Laportea crenulata. *Fitoterapia*, *79*(7–8), pp. 584–586.
Semwal, D.K., Rawat, U., Semwal, R., Singh, R., Krishan, P., Singh, M. and Singh, G.J.P., 2009. Chemical constituents from the leaves of Boehmeria rugulosa with antidiabetic and antimicrobial activities. *Journal of Asian Natural Products Research*, *11*(12), pp. 1045–1055.
Singh, R., Hussain, S., Verma, R. and Sharma, P., 2013. Anti-mycobacterial screening of five Indian medicinal plants and partial purification of active extracts of Cassia sophera and Urtica dioica. *Asian Pacific Journal of Tropical Medicine*, *6*(5), pp. 366–371.
Uddin, S.J., Rouf, R., Shilpi, J.A., Alamgir, M., Nahar, L. and Sarker, S.D., 2008. Screening of some Bangladeshi medicinal plants for in vitro antibacterial activity. *Oriental Pharmacy and Experimental Medicine*, *8*(3), pp. 316–321.
Xu, Q.M., Liu, Y.L., Li, X.R., Li, X. and Yang, S.L., 2011. Three new fatty acids from the roots of Boehmeria nivea (L.) Gaudich and their antifungal activities. *Natural Product Research*, *25*(6), pp. 640–647.
Zuo, G.Y., Li, Z.Q., Chen, L.R. and Xu, X.J., 2005. In vitro anti-HCV activities of Saxifraga melanocentra and its related polyphenolic compounds. *Antiviral Chemistry and Chemotherapy*, *16*(6), pp. 393–398.

8 The Clade Malvids

8.1 ORDER GERANIALES JUSS. EX BERCHT. & J.PRESL (1820)

8.1.1 Family Geraniaceae A.L de Jussieu (1789)

The family Geraniaceae consists of 4 genera and 830 species of herbs or shrubs. The leaves are simple, alternate or opposite, and stipulate. The inflorescences are cymes of actinomorphic or zygomorphic flowers. The calyx includes 5 sepals. The corolla comprises 5 petals. The androecium is made of 5–10 stamens. The gynoecium consists of 5 carpels fused into a 5-locular ovary, each locule sheltering 1–2 pendulous ovules. The fruits are schizocarps of 5 awned mericarps, each mericarp containing a single seed. Members of the family Geraniaceae produce hydrolysable tannins and essential oils.

8.1.1.1 *Geranium wallichianum* D. Don ex Sweet

Common names: Kuan tuo ye lao guan cao (China); laljari (India); ratanjot, ranjot (Pakistan)

Habitat: Mountain forests

Distribution: Afghanistan, Pakistan, India, Himalayas

Botanical examination: This graceful yet endangered herb grows to a height of 60 cm from a rhizome. The stems are slender and flexuous. The leaves simple, opposite, and stipulate. The stipules are about 2 cm long and broadly lanceolate. The petiole is slender, hairy, and up to about 10 cm long. The blade is palmately 5-lobed, about 3–10 cm across, somewhat hairy, and serrate. The cymes are slender, hairy, and up to about 10 cm long. The calyx presents 5 sepals, which are oblong, hairy, 3-nerved, apiculate, and up to 1 cm long. The 5 petals are about 1.5–2 cm long, bluish to purple, clawed and hairy at base. The androecium includes 10 stamens. The style is 5-parted. The schizocarps are 2.5–3 cm long, straight, sharp, narrowly bullet-shaped, explosively dehiscent, each mericarp sheltering a single 5 mm long seed.

Medicinal uses: Sore eyes, wounds (India); diarrhea, dysentery, leucorrhea (Pakistan)

Broad-spectrum antibacterial halo developed by mid-polar extract: Ethyl acetate extract (1 mg/mL solution/ agar well) of roots inhibited the growth of *Bacillus subtilis*, *Shigella flexneri*, and *Staphylococcus aureus* with inhibition zone diameters of 12.6, 11.3, and 22.3 mm, respectively (Ismail et al., 2012).

Broad-spectrum antifungal polar extract: Methanol extract of rhizomes inhibited the growth of *Trichophytum tonsurans*, *Candida albicans*, *Microsporium canis*, *Aspergillus flavus*, and *Fusarium solani* (Ismail et al., 2012).

Commentaries: (i) *Geranium carolinianum* L. is used in China to treat diarrhea (Local name = *Ye lao guan cao*), and from this herb, ellagic acid (200 μg/mL), geraniin (200 μg/mL), quercitrin (100 μg/mL), hyperin (50 μg/mL), hirsutin (200 μg/mL), quercetin (25 μg/mL), and kaempferol (25 μg/mL) inhibited the secretion of the Hepatitis B virus surface antigen by 62.9, 85.8, 27.5, 42.5, 31.3, 36.1, and 54% (Lamivudine at 200 μg/mL: 33.5%) (Li et al., 2008). Geraniin isolated from *Erodium glaucophyllum* (L.) L'Hér. inhibited the growth of *Escherichia coli*, *Staphylococcus aureus*, and *Candida albicans* with MIC values of 2500, 3100, and 1900 μg/mL, respectively (Gohar et al., 2003).

(ii) *Geranium nepalense* Sweet is used in Japan to treat endometritis, and methanol extract of aerial parts at a concentration of 100 μg/mL inhibited the Human immunodeficiency virus protease activity by 34% (Park et al., 2002)

(iii) *Pelargonium graveolens* L'Hér. ex Aiton is used to treat burns in Bangladesh. Members of the genus *Pelargonium* L'Hér. yields antibacterial essential oils and antibacterial phenolic compounds. Balchin and S.G. Deans (1996) reported that the methanol fractions of leaves of several

DOI: 10.1201/9781003176398-8

Pelargonium species had antibacterial activity against a broad array of 25 strains of bacteria, including both Gram-positive and Gram-negative bacteria (10 µL of extract per wells). Essential oils and methanol extracts of several *Pelargonium* species had antibacterial activity against *Staphylococcus aureus* (ATCC 9144), *Staphylococcus epidermidis* (ATCC 12228), *Proteus vulgaris* (ATCC 13315), and *Bacillus cereus* (NCIMB 6349) (20 µL/4mm well) (Lis-Balchin et al., 1998). Essential oil of *Pelargonium graveolens* L'Hér. ex Aiton inhibited the growth of *Bacillus subtilis* (ATCC 6633), *Bacillus cereus* (ATCC 11778), *Staphylococcus aureus* (ATCC 6538), *Staphylococcus aureus* (ATCC 29213), and *Escherichia coli* (ATCC 35218) with MIC values of 0.7, 0.3, 0.7, 0.7, and 5.6 mg/mL, respectively (Rosato et al., 2007). The essential oil had a synergistic effect with norfloxacin against *Bacillus cereus* (ATCC 1178), *Staphylococcus aureus* (ATCC 6538), and *Staphylococcus aureus* (ATCC 29213) (Rosato et al., 2007). Essential oil of *Pelargonium graveolens* L'Hér. ex Aiton inhibited the growth of *Candida albicans* (ATCC 14053), *Candida albicans* (ATCC 10231), and *Candida tropicalis* (ATCC 750) with MIC values of 0.7, 0.7, and 0.7 mg/mL, respectively (Rosato et al., 2007). The essential oil was synergistic with amphotericin against *Candida albicans* (ATCC 10231). Essential oil (containing mainly germacrone and germacrene B) of rhizome of *Geranium macrorrhizum* L. inhibited the growth of *Escherichia coli* (clinical isolate), *Klebsiella pneumoniae* (ATCC), *Escherichia coli* (ATCC 25922), *Staphylococcus aureus* (ATCC 25923), *Staphylococcus aureus* (clinical isolate), *Clostridium sporogenes* (ATCC 19404), *Bacillus subtilis* (ATCC 6633), *Penicillium chrysogenum*, *Aspergillus restrictus*, and *Aspergillus fumigatus* with MIC/MBC values of 5/10, 1.2/5, 2.5/2.5, 2.5/2.5, 0.6/1.2, 2.5/10, 0.0004/0.001, 10/10, 10/10, and 5/10 mg/mL, respectively (Radulović et al., 2010). From this essential oil, germacrone (49.7%) inhibited the growth of *Escherichia coli* (clinical isolate), *Escherichia coli* (ATCC 25922), *Staphylococcus aureus* (ATCC 25923), *Staphylococcus aureus* (clinical isolate), *Clostridium sporogenes* (ATCC 19404), *Bacillus subtilis* (ATCC 6633), *Penicillium chrysogenum*, *Aspergillus restrictus*, and *Aspergillus fumigatus* (Radulović et al., 2010).

1,8-Cineole inhibited *Staphylococcus aureus* (ATCC 9144) and *Proteus vulgaris* (ATCC 13315) with inhibition zone diameters of 8 and 8 mm, respectively (10 µL/4mm well) (Lis-Balchin et al., 1998).

Citral inhibited the growth of *Staphylococcus aureus* (ATCC 9144), *Proteus vulgaris* (ATCC 13315), *Bacillus cereus* (NCIMB 6349), and *Staphylococcus epidermidis* (ATCC 12228) with inhibition zone diameters of 13, 23, 15, and 19 mm, respectively (10 µL per well; 4 mm well) (Lis-Balchin et al., 1998). Citral inhibited *Candida albicans* with an MIC value of 100 ppm (Tampieri et al., 2005).

Citronellal inhibited the growth of *Staphylococcus aureus* (ATCC 9144), *Proteus vulgaris* (ATCC 13315), *Bacillus cereus* (NCIMB 6349), and *Staphylococcus epidermidis* (ATCC 12228) with inhibition zone diameters of 9, 12, 36, and 17 mm, respectively (10 µL per well; 4 mm well) (Lis-Balchin et al., 1998).

Citronellol inhibited the growth of *Staphylococcus aureus* (ATCC 9144), *Proteus vulgaris* (ATCC 13315), *Bacillus cereus* (NCIMB 6349), and *Staphylococcus epidermidis* (ATCC 12228) with inhibition zone diameters of 11, 8, 9, and 10 mm, respectively (10 µL per well; 4 mm well) (Lis-Balchin et al., 1998).

Citronellic acid inhibited the growth of *Staphylococcus aureus* (ATCC 9144), *Proteus vulgaris* (ATCC 13315), *Bacillus cereus* (NCIMB 6349), and *Staphylococcus epidermidis* (ATCC 12228) with inhibition zone diameters of 21, 18, 16, and 12 mm, respectively (10 µL per well; 4 mm) (Lis-Balchin et al., 1998). Citronellol inhibited the growth of *Bacillus subtilis* (ATCC 6633), *Bacillus cereus* (ATCC 11778), *Staphylococcus aureus* (ATCC 6538), *Staphylococcus aureus* (ATCC 29213), and *Escherichia coli* (ATCC 35218) with MIC values of 0.3, 0.7, 0.7, 0.7, and 1.4 mg/mL, respectively (Rosato et al., 2007).

p-Cymene inhibited *Bacillus cereus* (NCIMB 6349), and *Staphylococcus epidermidis* (ATCC 12228) with inhibition zone diameter of 8 and 8 mm, respectively (10 µL/4mm well)(Lis-Balchin et al., 1998). *p*-Cymene inhibited *Candida albicans* with an MIC value of 100 ppm (Tampieri et al., 2005).

Geranyl acetate inhibited *Staphylococcus aureus* (ATCC 9144), *Bacillus cereus* (NCIMB 6349), and *Staphylococcus epidermidis* (ATCC 12228) with inhibition zone diameters of 8, 8, and 8 mm, respectively (10 µL per well; 4 mm well) (Lis-Balchin et al., 1998).

Germacrene inhibited *Proteus vulgaris* (ATCC 13315), *Bacillus cereus* (NCIMB 6349), and *Staphylococcus epidermidis* (ATCC 12228) with inhibition zone diameters of 8, 7, and 8 mm, respectively (10 µL per well; 4 mm well) (Lis-Balchin et al., 1998).

Isomenthone inhibited *Staphylococcus aureus* (ATCC 9144), *Proteus vulgaris* (ATCC 13315), and *Bacillus cereus* (NCIMB 6349), with inhibition zone diameters of 6, 10, 7 mm, respectively (10 µL per well; 4 mm well) (Lis-Balchin et al., 1998).

Linalool inhibited *Staphylococcus aureus* (ATCC 9144), *Proteus vulgaris* (ATCC 13315), *Bacillus cereus* (NCIMB 6349), and *Staphylococcus epidermidis* (ATCC 12228) with inhibition zone diameters of 11, 14, 13, and 12 mm, respectively (10 µL per well; 4 mm well) (Lis-Balchin et al., 1998).

Myrcene inhibited *Staphylococcus aureus* (ATCC 9144) and *Bacillus cereus* (NCIMB 6349) with inhibition zone diameters of 12 and 19 mm, respectively (10 µL per well; 4 mm well) (Lis-Balchin et al., 1998)

Nerol inhibited *Staphylococcus aureus* (ATCC 9144), *Proteus vulgaris* (ATCC 13315), *Bacillus cereus* (NCIMB 6349), and *Staphylococcus epidermidis* (ATCC 12228) with inhibition zone diameters of 11, 11, 6, and 8 mm, respectively (10 µL per well; 4 mm wel) (Lis-Balchin et al., 1998). Nerol inhibited *Candida albicans* with an MIC value of 200 ppm (Tampieri et al., 2005).

Piperitone inhibited *Staphylococcus aureus* (ATCC 9144), *Proteus vulgaris* (ATCC 13315), *Bacillus cereus* (NCIMB 6349), and *Staphylococcus epidermidis* (ATCC 12228) with inhibition zone diameters of 11, 11, 6, and 9 mm, respectively (10 µL per well; 4 mm well) (Lis-Balchin et al., 1998).

Terpinen-4-ol inhibited *Staphylococcus aureus* (ATCC 9144), *Proteus vulgaris* (ATCC 13315), *Bacillus cereus* (NCIMB 6349), and *Staphylococcus epidermidis* (ATCC 12228) with inhibition zone diameters of 19, 21, 9, and 12 mm, respectively (10 µL per well) (Lis-Balchin et al., 1998).

Consider that citronellol, geraniol, limonene, and neryl acetate at the concentration of 50 µM decreased the expression of angiotensin-converting enzyme in HT-29 cells (Senthil-Kumar et al., 2020) and as such might have some potentials against zoonotic coronaviruses.

(iv) An extract from a member of the genus *Pelargonium* L'Hér, inhibited the replication of the Influenza A H1N1, H3N2, respiratory syncytial virus, Human coronavirus 229E, Parainfluenza-3 virus, and Coxsackie virus A09, with IC_{50} values of 9.4, 8.6, 19.6, 44.5, 74.3, and 14.8 µg/mL, respectively (Michaelis et al., 2011). Ethanol extract of roots of a member of the genus *Geranium* L. inhibited the replication of the Herpes simplex virus type-1 (Strain Kupka) in Vero cells with EC_{50} value of 12.1 µg/mL and a selectivity index of 5.6 (Serkedjieva & Ivancheva, 1998). The extract had no direct virucidal effect, and it was most effective when the extract was added after virus infection (Serkedjieva & Ivancheva, 1998). A 5% preparation given 15 minutes, 2, 6, 24, and 48 hours after Herpes simplex virus type-1 inoculation in Guinea pigs had a weak effect but delayed the appearance of vesicles and satellites (Serkedjieva & Ivancheva, 1998). Ethanol extract of roots of this plant given at a dose of 10 mg/Kg intranasally 3 hours before Influenza A/Aichi inoculation protected mice by 71.4% (Serkedjieva & Ivancheva, 1998). Phenolic fraction of roots of this plant inhibited Influenza virus A/chicken/Rostock/34 (H7N1) with an EC_{50} value of 32 µg/mL in CEF cells (Serkedjieva, 2003; Serkedjieva & Ivancheva, 1998). Quercetin, kaempferol, myricetin, apigenin, and quercetin-3-*O*-galactoside isolated from the roots of a member of the genus *Geranium* L. inhibited Influenza virus A/chicken/Rostock/34 (H7N1) (Pantev et al., 2006). The flavonol myricetin, which is not uncommonly produced by members of the genus *Geranium* L inhibited the Severe acute respiratory syndrome coronavirus helicase, nsP13 ATPase with an IC_{50} value 2.7 µM (Yu et al., 2012), and for this reason, one could examine the anti- properties of plants in this genus against the Severe acute respiratory syndrome-acquired coronavirus genus.

(v) Aqueous ethanol extract of roots from a member of the genus *Pelargonium* L'Hér, inhibited the growth of *Mycobacterium tuberculosis* (H37Rv) by 96% at a concentration of 12.5 µg/mL (Kolodziej et al., 2003). From this plant, epigallocatechin, scopoletin and catechin, inhibited the growth of *Mycobacterium smegmatis* with MIC values of 7.8, 7.8, and 31.2 µg/mL, respectively (Mativandlela et al., 2007). From this plant, palmitoleic acid inhibited the growth of *Mycobacterium aurum*, *Mycobacterium smegmatis*, *Mycobacterium fortuitum*, *Mycobacterium abscessus*, and *Mycobacterium phlei*, with MIC values of 4, 16, 16, 8–16, and 8 µg/mL, respectively (Seidel & Taylor, 2004). Linoleic acid isolated from this plant inhibited the growth of *Mycobacterium aurum*, *Mycobacterium smegmatis*, *Mycobacterium fortuitum*, *Mycobacterium abscessus*, and *Mycobacterium phlei*, with MIC values of 2, 4, 16, 16–32, and 2–4 µg/mL, respectively (Seidel & Taylor, 2004). Consider that linoleic acid inhibited efflux pump in *Staphylococcus aureus* (Sharma et al., 2019).

REFERENCES

Gohar, A.A., Lahloub, M.F. and Niwa, M., 2003. Antibacterial polyphenol from Erodium glaucophyllum. *Zeitschrift für Naturforschung C*, *58*(9–10), pp. 670–674.

Ismail, M., Hussain, J., Khan, A.U., Khan, A.L., Ali, L., Khan, F.U., Khan, A.Z., Niaz, U. and Lee, I.J., 2012. Antibacterial, antifungal, cytotoxic, phytotoxic, insecticidal, and enzyme inhibitory activities of Geranium wallichianum. *Evidence-Based Complementary and Alternative Medicine*, *2012*. Article ID 305906.

Kolodziej, H., Kayser, O., Radtke, O.A., Kiderlen, A.F. and Koch, E., 2003. Pharmacological profile of extracts of Pelargonium sidoides and their constituents. *Phytomedicine*, *10*, pp. 18–24.

Li, J., Huang, H., Zhou, W., Feng, M. and Zhou, P., 2008. Anti-hepatitis B virus activities of Geranium carolinianum L. extracts and identification of the active components. *Biological and Pharmaceutical Bulletin*, *31*(4), pp. 743–747.

Mativandlela, S.P.N., Meyer, J.J.M., Hussein, A.A. and Lall, N., 2007. Antitubercular activity of compounds isolated from pelargonium sidoides. *Pharmaceutical Biology*, *45*(8), pp. 645–650.

Michaelis, M., Doerr, H.W. and Cinatl Jr, J., 2011. Investigation of the influence of EPs® 7630, a herbal drug preparation from Pelargonium sidoides, on replication of a broad panel of respiratory viruses. *Phytomedicine*, *18*(5), pp. 384–386.

Pantev, A., Ivancheva, S., Staneva, L. and Serkedjieva, J., 2006. Biologically active constituents of a polyphenol extract from Geranium sanguineum L. with anti-influenza activity. *Zeitschrift für Naturforschung C*, *61*(7–8), pp. 508–516.

Park, J.C., Hur, J.M., Park, J.G., Hatano, T., Yoshida, T., Miyashiro, H., Min, B.S. and Hattori, M., 2002. Inhibitory effects of Korean medicinal plants and camelliatannin H from Camellia japonica on Human immunodeficiency virus type 1 protease. *Phytotherapy Research*, *16*(5), pp.422–426.

Radulović, N.S., Dekić, M.S., Stojanović-Radić, Z.Z. and Zoranić, S.K., 2010. Geranium macrorrhizum L.(Geraniaceae) essential oil: A potent agent against Bacillus subtilis. *Chemistry & Biodiversity*, *7*(11), pp. 2783–2800.

Rosato, A., Vitali, C., De Laurentis, N., Armenise, D. and Milillo, M.A., 2007. Antibacterial effect of some essential oils administered alone or in combination with Norfloxacin. *Phytomedicine*, *14*(11), pp. 727–732.

Rosato, A., Vitali, C., De Laurentis, N., Armenise, D. and Milillo, M.A., 2007. Antibacterial effect of some essential oils administered alone or in combination with Norfloxacin. *Phytomedicine*, *14*(11), pp. 727–732.

Rosato, A., Vitali, C., Gallo, D., Balenzano, L. and Mallamaci, R., 2008. The inhibition of Candida species by selected essential oils and their synergism with amphotericin B. *Phytomedicine*, *15*(8), pp. 635–638.

Seidel, V. and Taylor, P.W., 2004. In vitro activity of extracts and constituents of Pelagonium against rapidly growing mycobacteria. *International Journal of Antimicrobial Agents*, *23*(6), pp. 613–619.

Senthil Kumar, K.J., Gokila Vani, M., Wang, C.S., Chen, C.C., Chen, Y.C., Lu, L.P., Huang, C.H., Lai, C.S. and Wang, S.Y., 2020. Geranium and lemon essential oils and their active compounds downregulate angiotensin-converting enzyme 2 (ACE2), a SARS-CoV-2 spike receptor-binding domain, in epithelial cells. *Plants*, *9*(6), p. 770.

Serkedjieva, J. and Ivancheva, S., 1998. Antiherpes virus activity of extracts from the medicinal plant Geranium sanguineum L. *Journal of Ethnopharmacology*, *64*(1), pp. 59–68.

Serkedjieva, J., 2003. Influenza virus variants with reduced susceptibility to inhibition by a polyphenol extract from Geranium sanguineum L. *Die Pharmazie-An International Journal of Pharmaceutical Sciences*, *58*(1), pp. 53–57.

Sharma, A., Gupta, V.K. and Pathania, R., 2019. Efflux pump inhibitors for bacterial pathogens: From bench to bedside. *The Indian Journal of Medical Research*, *149*(2), p. 129.

Tampieri, M.P., Galuppi, R., Macchioni, F., Carelle, M.S., Falcioni, L., Cioni, P.L. and Morelli, I., 2005. The inhibition of Candida albicans by selected essential oils and their major components. *Mycopathologia*, *159*(3), pp. 339–345.

Yu, M.S., Lee, J., Lee, J.M., Kim, Y., Chin, Y.W., Jee, J.G., Keum, Y.S. and Jeong, Y.J., 2012. Identification of myricetin and scutellarein as novel chemical inhibitors of the SARS coronavirus helicase, nsP13. *Bioorganic & Medicinal Chemistry Letters*, *22*(12), pp. 4049–4054.

8.2 ORDER MYRTALES JUSS. EX BERCHT. & J. PRESL (1820)

8.2.1 Family Combretaceae R. Brown (1810)

The family Combretaceae consists of 20 genera of trees, shrubs, or climbers. The leaves are opposite, subopposite, whorled, spiral, or alternate, often coriaceous, and exstipulate. The inflorescences are terminal, axillary, or extra-axillary, spikes, racemes, or panicles. The flowers are often minute. The calyx includes 4 or 5 sepals. The 4 or 5 petals are inserted near the mouth of the calyx tube. A disc is present. Eight to 10 stamens form the androecium. A disc is present. The gynoecium consists of a unilocular ovary containing 2 pendulous ovules. The fruits are drupes. Members in this family produce a bewildering spectrum of antimicrobial hydrolysable tannins.

8.2.1.1 *Anogeissus acuminata* (Roxb. ex DC.) Guill., Perr. & A. Rich.

Synonym: *Anogeissus harmandii* Pierre

Common names: Button tree; itchri; chakwa (Bangladesh); yu lü mu (China); nunera, pasi, phasi (India)

Habitat: Coastal limestones

Distribution: India, Bangladesh, Cambodia, Laos, Myanmar, Thailand, Vietnam, China

Botanical observation: This beautiful tree grows to 20 m tall with pendulous stems. The leaves are simple, opposite, and exstipulate. The petiole is about 5 mm long. The blade is elliptic, 4–8 × 1–3 cm, acute at base and apex, hairy, somewhat glaucous bellow, and marked with 5–7 pairs of secondary nerves. The flower heads are globose, yellowish, axillary, and about 1.5cm in diameter.The calyx tube produces 5 lobes and is about 5 mm long. The androecium includes 10 stamens, which protrude out, and are up to about 1 cm long. The capsules are globose, winged, and about 5 mm in diameter.

Medicinal uses: Diarrhea, ulcers, wounds (India)

Moderate broad-spectrum antibacterial polar extract: Methanol extract inhibited the growth of *Enterococcus faecalis*, *Staphylococcus* aureus, *Klebsiella pneumoniae*, and *Pseudomonas aeruginosa* with MIC/MBC values of 670/1510, 290/670, 670/1510, and 670/1510 µg/mL, respectively (Mishra et al., 2017).

Viral enzyme inhibition by dibenzyl butane lignan: Anolignan A at a concentration of 100 µg/mL inhibited the Human immunodeficiency virus type-1 reverse transcriptase by 73.9% (Rimando et al., 1994).

Commentary: The antiviral properties of anolignan A could be examined. Dibenzyl butane lignans are often antiviral (Cui et al., 2020). Why is that so?

REFERENCES

Cui, Q., Du, R., Liu, M. and Rong, L., 2020. Lignans and their derivatives from plants as antivirals. *Molecules*, *25*(1), p. 183.

Mishra, M.P., Rath, S., Swain, S.S., Ghosh, G., Das, D. and Padhy, R.N., 2017. In vitro antibacterial activity of crude extracts of 9 selected medicinal plants against UTI causing MDR bacteria. *Journal of King Saud University-Science*, *29*(1), pp. 84–95.

Rimando, A.M., Pezzuto, J.M., Farnsworth, N.R., Santisuk, T., Reutrakul, V. and Kawanishi, K., 1994. New lignans from Anogeissus acuminata with HIV-1 reverse transcriptase inhibitory activity. *Journal of Natural Products*, *57*(7), pp. 896–904.

8.2.1.2 *Anogeissus latifolia* (Roxb. ex DC.) Wall. ex Bedd.

Synonym: *Conocarpus latifolius* Roxb. ex DC.

Common names: Axle wood; ghatti tree; doya (Bangladesh); dhau (India)

Habitat: Mountain forest

Distribution: Pakistan, India, Sri Lanka

Botanical observation: This resinous tree grows to 10 m tall. The leaves are alternate or subopposite, simple, and exstipulate. The petiole is up to 1.5 cm long. The blade is elliptic, 5–10 × 3–5cm, round or acute at base and apex, coriaceous, with 8–14 pairs of secondary nerves. The flower heads are globose, about 1 cm in diameter, and axillary. A bottle-shaped and about 6 mm long hypanthium is present. The calyx is tubular, whitish to yellowish, wavy, and vaguely 5 lobed. The 10 exserted stamens are adnate to the calyx tube and protruding. The capsules are about 5 mm long and laterally winged.

Medicinal uses: Diarrhea, ulcers, syphilis, gonorrhea (India)

Weak broad-spectrum antibacterial polar extract: Extract of leaves inhibited the growth of *Staphylococcus aureus*, *Bacillus subtilis*, *Shigella flexneri*, and *Vibrio cholerae* with MIC values of 1000, 4000, 1000, and 1000 µg/mL, respectively (Panda et al., 2016).

Antifungal (filamentous) polar extract: Methanol extract of leaves (20 mL/agar plates of a 1 mg/mL solution) inhibited the growth of *Alternaria alternata*, *Fusarium equiseti*, *Macrophomina phaseolina*, *Botryodiplodia theobromae*, and *Colletrotrichum corchori* by 13, 24, 6, 3, and 14%, respectively (Begum et al., 2007).

REFERENCE

Begum, J., Yusuf, M., Chowdhury, J.U., Khan, S. and Anwar, M.N., 2007. Antifungal activity of forty higher plants against phytopathogenic fungi. *Bangladesh Journal of Microbiology*, *24*(1), pp. 76–78.

8.2 1.3 *Combretum trifoliatum* Vent.

Synonym: *Cacoucia trifoliata* (Vent.) DC.

Common names: Tro (Cambodia); ben nam (Laos); sonsong harus (Malaysia), chut, trrod (Thailand); tram bau ba la (Vietnam)

Habitat: Forests

Distribution: Myanmar, Cambodia, Laos, Vietnam, Thailand, Malaysia, Indonesia, Papua New Guinea, and Australia.

Botanical description: This scandent shrub grows up to 15 m long. The stem is terete. The leaves are simple, opposite or in whorls of 3, and without stipules. The petiole is short, stout, glabrous, curved and about 4 mm long. Some minute axillary fluffy buds are present. The blade is coriaceous, glabrous, elliptic, about 3.5–5 × 12.5–15 cm, with a midrib sunken above, and 11–12 pairs of secondary nerves prominent on both surfaces. The inflorescence is a green axillary and terminal spike, which is about 9 cm long. The calyx is light green, cup-shaped, 2 mm in diameter, hairy, 5 lobed, the lobes triangular. The androecium comprises 10 stamens, which are filamentous. The style is straight and about 3 mm long. The capsules are woody, green, 5-winged, like little carambola and about 2.6×1 cm.

Medicinal uses: Dysentery (Cambodia, Laos, and Vietnam); athlete's foot (Thailand)

Commentaries: The plant has apparently not been studied for antimicrobial properties. Acetone extract of stem bark of *Combretum molle* R. Br. ex G. Don inhibited the growth of *Mycobacterium tuberculosis* (ATCC 27294) with an MIC of 1 mg/mL. From this extract, the ellagitannin punicalagin (molecular mass = 1084.7 g/mol) at a concentration above 600 μg/mL completely inhibited the growth of *Mycobacterium tuberculosis* (ATCC 27294) as well as a clinical strain at 1200 μg/mL (Asres et al., 2001). The phenolic lactone combretastatin D-3 from *Getonia floribunda* Roxb. inhibited the growth of *Mycobacterium* sp. (Vongvanich et al., 2005).

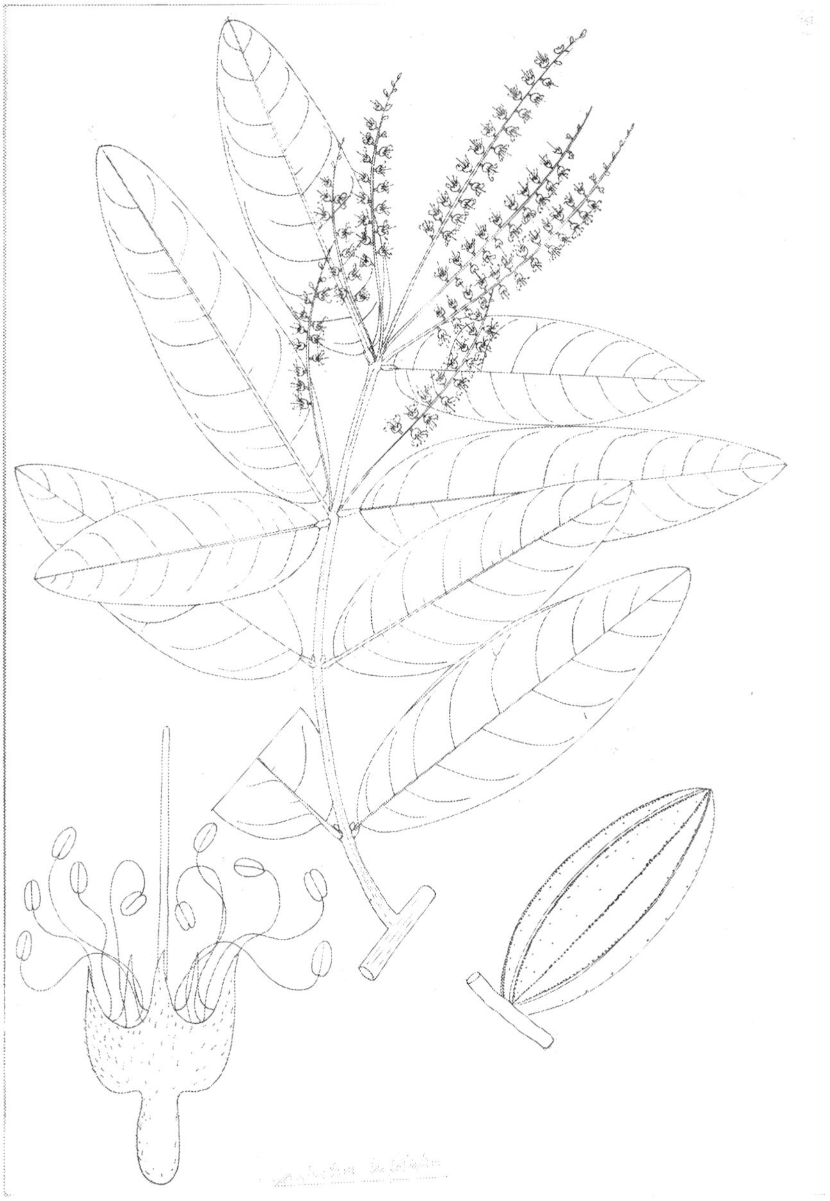

FIGURE 8.1 *Combretum trifoliatum* Vent.

REFERENCES

Asres, K., Bucar, F., Edelsbrunner, S., Kartnig, T., Höger, G. and Thiel, W., 2001. Investigations on antimycobacterial activity of some Ethiopian medicinal plants. *Phytotherapy Research*, *15*(4), pp. 323–326.

Vongvanich, N., Kittakoop, P., Charoenchai, P., Intamas, S., Danwisetkanjana, K. and Thebtaranonth, Y., 2005. Combretastatins D-3 and D-4, new macrocyclic lactones from Getonia floribunda. *Planta Medica*, *71*(02), pp. 191–193.

8.2.1.4 *Terminalia arjuna* (Roxb. ex DC.) Wight & Arn.

Synonyms: *Pentaptera angustifolia* Roxb., *Pentaptera arjuna* Roxb. ex DC., *Pentaptera glabra* Roxb.; *Terminalia berryi* Wight & Arn.; *Terminalia glabra* Wight & Arn., *Terminalia ovalifolia* Rottl. Ex C.B. Clarke

Common names: Arjuna, white marudah, white murdh; arjun (Bangladesh, India); rok fa-khao (Thailand)

Habitat: Ornamental

Distribution: Pakistan, India, Sri Lanka

Botanical description: This graceful tree that grows to 15 m tall. The bole is straight and the bark is somewhat smooth. The leaves are simple, subopposite, oblong-elliptic, and exstipulate. The petiole is about 1 cm long. The blade is oblong, 7–18 × 4–6cm, rounded or cordate at base, and acute at apex. The spikes are axillary or terminal, paniculate, brush bottle-like, and 3–6 cm long. The hypanthium is campanulate, 4–5 mm long, with 5 minute triangular lobes. The androecium includes 10 stamens, which are exserted. The disc is barbate. The capsules are ovoid-oblong, 2.5–5 cm long, yellowish green, and with 5 wings.

Medicinal uses: Cough, dysentery, pneumonia, herpes, leprosy, syphilis, gonorrhea (India)

Very weak broad-spectrum polar extract: Methanol extract of fruits inhibited the growth of *Bacillus cereus* (NCTC 6447) with an MIC value of 2800 µg/mL (Kavanagh, 1972; Elegami et al., 2002).

Strong antiviral (enveloped monopartite linear dimeric single stranded (+) RNA) polar extract: Butanol extract of stem bark at a maximum non-cytotoxic concentration of 15 µg/mL inhibited the replication of the Human immunodeficiency virus type-1 (NL4.3) in Human CD4+ T cells by 63.6% (Sabde et al., 2011).

Strong antiviral (enveloped monopartite linear double-stranded DNA) amphiphilic ellagitannin: Casuarinin (LogD = 1.9 at pH 7.4; molecular mass = 936.6 g/mol) isolated from the bark inhibited the replication of the Herpes simplex virus type 2 with an IC_{50} value of 3.6 µM and an selectivity index of 25 by preventing attachment and viral penetration as well as virucidal effect (Cheng et al., 2002).

Virus enzyme inhibition by polar extract: Aqueous extract of bark at a concentration of 250 µg/mL inhibited the Human immunodeficiency virus type-1 protease by 79.6% (Xu et al., 1996).

Commentary: (i) One could have some interest to isolate anti-Human immunodeficiency virus from the hexane extract of stem bark. Consider that polar extracts of Combretaceae or any other families of medicinal plants producing tannins will inhibit enzymes by virtue of protein coagulation.

REFERENCES

Buzzini, P., Arapitsas, P., Goretti, M., Branda, E., Turchetti, B., Pinelli, P., Ieri, F. and Romani, A., 2008. Antimicrobial and antiviral activity of hydrolysable tannins. *Mini Reviews in Medicinal Chemistry*, *8*(12), pp. 1179–1187.

Cheng, H.Y., Lin, C.C. and Lin, T.C., 2002. Antiherpes simplex virus type 2 activity of casuarinin from the bark of Terminalia arjuna Linn. *Antiviral Research*, *55*(3), pp. 447–455.

Elegami, A.A., El-Nima, E.I., El Tohami, M.S. and Muddathir, A.K., 2002. Antimicrobial activity of some species of the family Combretaceae. *Phytotherapy Research*, *16*(6), pp. 555–561.

Sabde, S., Bodiwala, H.S., Karmase, A., Deshpande, P.J., Kaur, A., Ahmed, N., Chauthe, S.K., Brahmbhatt, K.G., Phadke, R.U., Mitra, D. and Bhutani, K.K., 2011. Anti-HIV activity of Indian medicinal plants. *Journal of Natural Medicines*, *65*(3–4), pp. 662–669.

Xu, H.X., Wan, M., Loh, B.N., Kon, O.L., Chow, P.W. and Sim, K.Y., 1996. Screening of traditional medicines for their inhibitory activity against HIV-1 protease. *Phytotherapy Research*, *10*(3), pp. 207–210.

8.2.1.5 *Terminalia bellirica* (Gaertn.) Roxb.

Synonyms: *Myrobalanus bellerica* Gaertn., *Myrobalanus laurinoides* (Teijsm. & Binn.) Kuntze, *Terminalia attenuata* Edgew., *Terminalia belerica* Roxb., *Terminalia eglandulosa* Roxb. ex C.B. Clarke, *Terminalia gella* Dalzell., *Terminalia laurinoides* Teijsm. & Binn., *Terminalia punctata* Roth.

Common names: Bastard myrobalan, bedda nuts, belleric myrobalan; bohera; bibhitaka, balelaa (Bangladesh); akkam, aksha, barra, baura, behara, thaanrikkaai (India); ulu belu (Indonesia); heen (Laos); jelawai (Malaysia); pangan (Myanmar), barro (Nepal); bulu (Sri Lanka); haen (Thailand); bang hoi (Vietnam)

Habitat: Forests

Distribution: Pakistan, India, Sri Lanka, Nepal, Bangladesh, Myanmar, Cambodia, Laos, Vietnam, Thailand, Malaysia, and Indonesia

Botanical description: This handsome tree grows to a height of 20 m. The bark is thin, roughly cracked, flaking, and brown. The stem is stout, 1 cm in diameter, rough, corky with heart-shaped leaf scars, fissured, cracked, velvety by apex, and slightly angled. The leaves are simple, spiral, at apex of stems and without stipules. The petiole is slender and 4–7 cm long. The blade is obovate to spathulate, thinly coriaceous, glossy, tapering at base, slightly asymmetrical, 14–15 × 9–10 cm, and marked with 6–9 pairs of secondary nerves. The inflorescence is a dull light green axillary spike, which is about 15 cm long. The calyx is cup-shaped, yellowish, hairy outside, wooly inside, 0.4 cm long, and with 5 triangular lobes. The corolla is absent. The androecium shows 10 conspicuous stamens, which are about 4 mm long. The drupes are woody, obscurely 5-lobed, velvety, with a tiny disc at apex. The seed is narcotic.

Medicinal uses: Bronchitis, cough, jaundice, respiratory infection, leprosy, diarrhea, cough, eye infection, cholera (India); cold, cough, eye wash (Nepal)

Broad-spectrum antibacterial halo developed by polar extract: Methanol extract of fruits (6 mg/well) inhibited the growth of *Bacillus cereus*, *Listeria monocytogenes*, *Staphylococcus aureus*, *Escherichia coli*, and *Salmonella anatum* with inhibition zone diameters of 12.1, 11.9, 17.3, 7.3, and 12.2 mm, respectively (Shan et al., 2007).

Viral enzyme inhibition by polar extract: Aqueous extract of fruit peel at a concentration of 250 μg/mL inhibited Human immunodeficiency virus type-1 protease (Xu et al., 1996)

Commentaries: (i) The plant has apparently not been examined much for its antimicrobial effects. (ii) *Terminalia catappa* L. is used in India for the treatment of bronchitis and leprosy. Phenolic fraction from fruits of *Terminalia catappa* L. inhibited the growth of *Staphylococcus aureus*, *Bacillus subtilis*, *Enterococcus faecalis*, *Listeria monocytogenes*, *Escherichia coli*, *Pseudomonas aeruginosa*, *Salmonella typhi*, and *Shigella dysenteriae* with MIC values of 15.6, 15.6, 7.8, 15, 125, 125, 250, and 500 μg/mL, respectively (Aman et al., 2014). Methanol extract of root of *Terminalia catappa* L. inhibited *Escherichia coli* with an MIC of 65 μg/mL (Pawar & Pal, 2002).

REFERENCES

Pawar, S.P. and Pal, S.C., 2002. Antimicrobial activity of extracts of Terminalia catappa root. *Indian Journal of Medical Sciences*, *56*(6), pp. 276–278.

Shan, B., Cai, Y.Z., Brooks, J.D. and Corke, H., 2007. The in vitro antibacterial activity of dietary spice and medicinal herb extracts. *International Journal of Food Microbiology*, *117*(1), pp. 112–119.

Xu, H.X., Wan, M., Loh, B.N., Kon, O.L., Chow, P.W. and Sim, K.Y., 1996. Screening of traditional medicines for their inhibitory activity against HIV-1 protease. *Phytotherapy Research*, *10*(3), pp. 207–210.

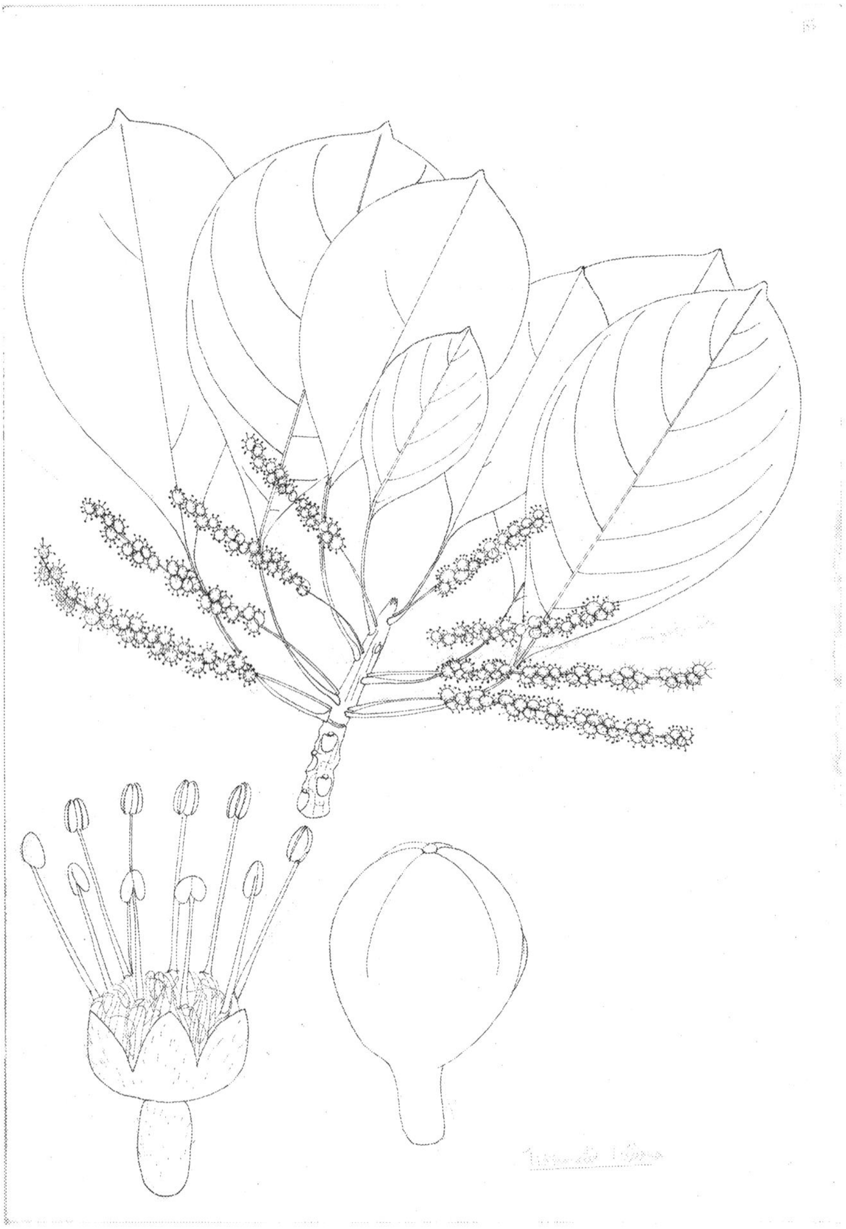

FIGURE 8.2 *Terminalia bellirica* (Gaertn.) Roxb.

8.2.1.6 *Terminalia chebula* Retz.

Synonyms: *Myrobalanifera citrina* Houtt.; *Myrobalanus chebula* (Retz.) Gaertn.; *Myrobalanus gangetica* (Roxb.); Kostel.; *Terminalia gangetica* Roxb.;*Terminalia reticulata* Roth

Common names: Chebulic myrobalan; horitoki (Bangladesh), srama (Cambodia);he lip (China); abhaya, jaya, amagola, haritaki, harra, hiahirala, halela (India); som (Laos); manja lawai (Malaysia); iangah (Myanmar); harro (Nepal); maa-nae (Thailand); cà lich, chieu lieu, kha li lac (Vietnam).

Habitat: Forests

Distribution: India, Sri Lanka, Bangladesh, Myanmar, Cambodia, Laos, Vietnam, China, Thailand.

Botanical description: This tree grows to a height of 30 m. The bark is dark brown, 6 mm thick, and the wood is very hard. The leaf buds, twigs, and young leaves are rusty and hairy. The petiole is 2–5cm long, pubescent, and with usually 2 glands near the top. The blade is 7–20 × 4–8cm, glabrous when mature, elliptic, oblong, marked with 6–8 pairs of secondary nerve arching and raised, acute at apex, and round or cordate at base. The inflorescences are terminal or axillary spikes. The flowers are 4 mm long, sessile, dull white or yellow and with a pungent odor. The calyx is campanulate, 3 mm long, flat at the base and 5-lobed. The drupes are pendulous, 2–4 cm long, almond-shaped, green, smooth, glabrous, and 5-ribbed.

Medicinal uses: Dysentery (Bangladesh, India, China, Malaysia); bronchitis, cough, cold, diarrhea, syphilis, leucorrhea, tuberculosis, wounds (India); fever, eye wash, tonsilitis (Nepal).

Antibacterial (Gram-positive) ellagitannins: Chebulagic acid (100 µg/6 mm disc) inhibited the growth of *Staphylococcus aureus* with an inhibition zone diameter of 11 mm (Foglian et al., 2005).

Moderate antibacterial (Gram-negative) tannin fraction: Tannin fraction of fruits inhibited the growth of *Klebsiella pneumoniae* (ATCC 700603) with MIC/MBC values of 300/600 µg/mL (Li et al., 2016).

Weak antibacterial (Gram-negative) simple phenolics: Ethyl gallate inhibited the growth of *Staphylococcus epidermidis, Bacillus cereus, Enterococcus faecalis, Citrobacter freundii, Enterococcus cloacae, Escherichia coli, Klebsiella pneumoniae, Pseudomonas aeruginosa, Proteus mirabilis, Proteus vulgaris*, and *Serratia marscencens* (Sato et al., 1997). In a subsequent study, ethyl gallate inhibited the growth of *Klebsiella pneumoniae* (ATCC 700603) with MIC/MBC values of 100/300 µg/mL (Li et al., 2016). Gallic acid from the fruits inhibited the growth of *Klebsiella pneumoniae, Proteus mirabilis, Proteus vulgaris*, and *Serratia marscencens* with MIC values of 250, 500, 125, and 250 µg/mL, respectively (Sato et al., 1997).

Antibiotic potentiator gallotannin: 1,2,6-tri-*O*-galloyl-β-D-glucopyranose isolated from the fruit inhibited efflux-pump in multidrug-resistant uropathogenic *Escherichia coli* (Bag & Chattopadhyay, 2014).

Bacterial protein expression inhibition by polar extract: A fraction (containing ellagic acid) of fruits nullified ppk1 gene in *Pseudomonas aeruginosa* at 0.5 mg/mL (Sarabhai et al., 2015). This fraction reduced the expression of rpoS, downstream master stress response regulator and the sensitivity of *Pseudomona aeruginosa* increased to oxidative stress (Sarabhai et al., 2015).

Anticandal halo developed by ellagitannin: Chebulagic acid and ellagic acid (100 µg/6 mm disc) inhibited the growth of *Candida albicans* with an inhibition zone of 12 mm (Foglian et al., 2005).

Anticandal halo developed by simple phenol: Gallic acid inhibited the growth of *Candida albicans* (Foglian et al., 2005).

Antifungal (filamentous) polar extract: Extract of aerial parts at a concentration of 25% abrogated the growth of *Trichophyton tonsurans, Trichophyton tonsurans, Trichophyton rubrum. Trichophyton simii, Trichophyton beigelii, Microsporum fulvum*, and *Microsporum gypseum* (Foglian et al., 2005).

Strong antiviral (enveloped monopartite linear single-stranded (+) RNA) polar extract: Methanol extract of fruits was antiviral against the Dengue virus serotypes-1, -2, -3 and -4 with IC_{50} equal to 25, 70, 11.2, and 50 µg/mL, respectively (Sood et al., 2015).

Strong antiviral (enveloped monopartite linear double-stranded DNA) ellagitannins: Chebulagic acid and chebulinic acid (molecular mass=956.6 g/mol) from the fruits inhibited the replication of the Herpes simplex virus type-2 with the IC_{50} values of 1.4 and 0.06 µg/mL, respectively, via virucidal effect, inhibition of attachment to Vero cells and penetration in Vero cells (Kesharwani et al., 2017).

Viral enzyme inhibition by polar extract: Aqueous extract of fruits at a concentration of 250 µg/mL inhibited Human immunodeficiency virus type-1 protease by 74.1% (Xu et al., 1996).

Viral enzyme inhibition by ellagitannins: Casuarinin, corilagin, and chebulagic acid from the fruits inhibited Hepatitis C virus protease with IC_{50} values of 9.6, 31.5, and 5.2 µM, respectively

(Ajala et al., 2014). Chebulagic acid inhibited the Human immunodeficiency virus integrase and reverse transcriptase with IC_{50} values of 13.7 and 20.6 µM, respectively (Ajala et al., 2014).

Viral enzyme inhibition by gallotannin: Pentagalloyl glucose isolated from the fruits inhibited Hepatitis C virus protease with an IC_{50} value of 10.6 µM (Ajala et al., 2014). 1,3,6-Tri-*O*-galloyl-β-D-glucopyranose inhibited the Human immunodeficiency virus integrase with IC_{50} values of 16.6 µM (Ahn et al., 2002). 1,2,3,4,6-Penta-*O*-galloyl-β-D-glucopyranose inhibited the Human immunodeficiency virus integrase and reverse transcriptase with IC_{50} values of 19.7 and 59 µM, respectively (Ahn et al., 2002).

Commentaries: (i) Siddha healers in South India, on the recommendation of our peer the Spanish Moorish physician Abū l-Walīd Muhammad Ibn "Ahmad Ibn Rušd" also known as Averroes (1126–1198), because it clears humors in time of pandemics, use the plant to prevent COVID-19 (Sasikumar et al., 2020), and the plant is used for the treatment of COVID-19 in China (Wang et al., 2020).

(ii) Ethyl acetate fraction of leaves of *Terminalia complanata* K. Schum inhibited the growth of *Bacillus cereus*, *Bacillus subtilis*, *Micrococcus luteus*, *Staphylococcus aureus*, *Staphylococcus epidermidis*, *Streptococcus faecalis*, *Streptococcus pneumoniae*, *Escherichia coli*, *Klebsiella pneumoniae*, *Neisseria gonorrhoeae*, *Proteus vulgaris*, *Pseudomonas aeruginosa*, *Salmonella typhi*, *Salmonella typhimurium*, and *Serratia marscesens* (Khan et al., 2002).

(iii) Methanol extract of *Terminalia pallida* Brandis inhibited the growth of *Staphylococcus aureus*, *Staphylococcus epidermidis*, *Micrococcus luteus*, *Escherichia coli*, *Salmonella typhi*, *Shigella dysenteriae*, and *Vibrio cholerae* with inhibition zone diameters of 15, 16, 16, 15, 13, 18, and 15 mm (2.5 mg/mL/ 6 mm well) (Gupta et al., 2002). The extract inhibited the growth of *Aspergillus niger* and *Candida albicans* with inhibition zone diameters of 8 and 7 mm (2.5mg/mL solution/ 6 mm well) (Gupta et al., 2002).

(iv) Ethanol extract of fruits of *Terminalia* sp. inhibited the growth *Bacillus cereus*, *Enterococcus faecalis*, *Listeria monocytogenes*, *Staphylococcus aureus*, *Streptococcus mutans*, and *Streptococcus pyogenes* with MIC/MBC of 7.8/>1000, 62.5 > 1000, 7.8/62.5, 7.8/15.6, 3.9/125, and 7.8/62.5 µg/mL, respectively (Limsuwan et al., 2009).

(v) In Thailand, *Lumnitzera racemosa* Willd. (local name: *Fard dok khao*) is used to heal wounds, and *Combretum quadrangulare* Kurz (local name: *Sakae*) is used as a post-partum remedy.

REFERENCES

Ahn, M.J., Kim, C.Y., Lee, J.S., Kim, T.G., Kim, S.H., Lee, C.K., Lee, B.B., Shin, C.G., Huh, H. and Kim, J., 2002. Inhibition of HIV-1 integrase by galloyl glucoses from Terminalia chebula and flavonol glycoside gallates from Euphorbia pekinensis. *Planta Medica*, *68*(05), pp. 457–459.

Ajala, O.S., Jukov, A. and Ma, C.M., 2014. Hepatitis C virus inhibitory hydrolysable tannins from the fruits of Terminalia chebula. *Fitoterapia*, *99*, pp. 117–123.

Bag, A. and Chattopadhyay, R.R., 2014. Efflux-pump inhibitory activity of a gallotannin from Terminalia chebula fruit against multidrug-resistant uropathogenic Escherichia coli. *Natural Product Research*, *28*(16), pp. 1280–1283.

Dutta, B.K., Rahman, I. and Das, T.K., 1998. Antifungal activity of Indian plant extracts: antimyzetische Aktivität indischer Pflanzenextrakte. *Mycoses*, *41*(11–12), pp. 535–536.

Fogliani, B., Raharivelomanana, P., Bianchini, J.P., Bouraı, S. and Hnawia, E., 2005. Bioactive ellagitannins from Cunonia macrophylla, an endemic Cunoniaceae from New Caledonia. *Phytochemistry*, *66*(2), pp. 241–247.

Gupta, M., Mazumder, U.K., Manikandan, L., Bhattacharya, S., Haldar, P.K. and Roy, S., 2002. Antibacterial activity of Terminalia pallida. *Fitoterapia*, *73*(2), pp. 165–167.

Kesharwani, A., Polachira, S.K., Nair, R., Agarwal, A., Mishra, N.N. and Gupta, S.K., 2017. Anti-HSV-2 activity of Terminalia chebula Retz extract and its constituents, chebulagic and chebulinic acids. *BMC Complementary and Alternative Medicine*, *17*(1), p. 110.

Khan, M.R., Kihara, M. and Omoloso, A.D., 2002. Antimicrobial activity of Terminalia complanata and Flacourtia zippelii. *Fitoterapia*, *73*(7–8), pp. 737–740.

Kher, M.N., Sheth, N.R. and Bhatt, V.D., 2019. In vitro antibacterial evaluation of terminalia chebula as an alternative of antibiotics against bovine subclinical mastitis. *Animal Biotechnology*, *30*, pp. 151–158.

Li, K., Lin, Y., Li, B., Pan, T., Wang, F., Yuan, R., Ji, J., Diao, Y. and Wang, S., 2016. Antibacterial constituents of Fructus Chebulae Immaturus and their mechanisms of action. *BMC Complementary and Alternative Medicine*, *16*(1), p. 183.

Limsuwan, S., Subhadhirasakul, S. and Voravuthikunchai, S.P., 2009. Medicinal plants with significant activity against important pathogenic bacteria. *Pharmaceutical Biology*, *47*(8), pp. 683–689.

Sarabhai, S., Harjai, K., Sharma, P. and Capalash, N., 2015. Ellagic acid derivatives from Terminalia chebula retz. Increase the susceptibility of pseudomonas aeruginosa to stress by inhibiting polyphosphate kinase. *Journal of Applied Microbiology*, *118*(4), pp. 817–825.

Sato, Y., Oketani, H., Singyouchi, K., Ohtsubo, T., Kihara, M., Shibata, H. and Higuti, T., 1997. Extraction and purification of effective antimicrobial constituents of Terminalia chebula RETS. against methicillin-resistant Staphylococcus aureus. *Biological and Pharmaceutical Bulletin*, *20*(4), pp. 401–404.

Sato, Y., Oketani, H., Singyouchi, K., OHTSUBO, T., KIHARA, M., SHIBATA, H. and HIGUTI, T., 1997. Extraction and purification of effective antimicrobial constituents of Terminalia chebula RETS. against methicillin-resistant Staphylococcus aureus. *Biological and Pharmaceutical Bulletin*, *20*(4), pp. 401–404.

Sood, R., Raut, R., Tyagi, P., Pareek, P.K., Barman, T.K., Singhal, S., Shirumalla, R.K., Kanoje, V., Subbarayan, R., Rajerethinam, R. and Sharma, N., 2015. Cissampelos pareira Linn: Natural source of potent antiviral activity against all four dengue virus serotypes. *PLoS Neglected Tropical Diseases*, *9*(12), p. e0004255.

Wang, S.X., Wang, Y., Lu, Y.B., Li, J.Y., Song, Y.J., Nyamgerelt, M. and Wang, X.X., 2020. Diagnosis and treatment of novel coronavirus pneumonia based on the theory of traditional Chinese medicine. *Journal of Integrative Medicine*, *18*, pp. 275–283.

Xu, H.X., Wan, M., Loh, B.N., Kon, O.L., Chow, P.W. and Sim, K.Y., 1996. Screening of Traditional Medicines for their Inhibitory Activity Against HIV-1 Protease. *Phytotherapy Research*, *10*(3), pp. 207–210.

8.2.2 Family Lythraceae Jaume Saint-Hilaire (1805)

The family Lythraceae consists of about 26 genera and 500 species of herbs, shrubs, or trees. The leaves of Lythraceae are simple, often coriaceous, opposite, and without stipules. The flowers are often 4-, 6-, or 8-merous, regular, actinomorphic, and present a prominent hypanthium. The sepals are valvate and united into a tube. The petals are free, crumpled in buds, membranaceous, wavy, alternate with the sepals, and pinnately veined. A nectary disc is present. The stamens are twice as numerous as the petals and filamentous. The gynoecium consists of 2–6 carpels forming a compound, superior, and plurilocular ovary. The fruits are capsular, dehiscent and contain several oily seeds. Members in this family produces ellagitannins as well as piperidine, quinolizidine alkaloids, and quinones (anthraquinones and naphthoquinones).

8.2.2.1 *Lagerstroemia speciosa* (L.) Pers.

Synonyms: *Lagerstroemia flos-reginae* Retz., *Lagerstroemia regina* Roxb., *Munchausia speciosa* L.

Common names: Queens crape-myrtle; jarul (India); bungor tekuyung (Indonesia); bungor (Malaysia); pyinma-ywetthey (Myanmar); banaba (the Philippines); murutagass (Sri Lanka); eikmwe, chuang mu (Thailand); bang la nuoc (Vietnam)

Habitat: Cultivated

Distribution: Tropical Asia

Botanical description: This elegant tree can reach about 20 m tall. The stem is smooth, glabrous, angled, peeling, about 5 mm in diameter, with internodes, which are about 5 cm long. The leaves are simple, spiral and without stipules. The petiole is woody, stout, clasping the stem, glabrous channeled, and about 1.5 cm long. The blade is coriaceous, elliptic to oblong, minutely acuminate at apex, 11–16 × 8.5–6 cm, and shows 10–12 pairs of secondary nerves prominently raised underneath and united by an intramarginal nerve. The inflorescence is a terminal panicle, which is about 30 cm long and much flowered. The calyx is 1.5 cm long, hairy, and produces 6 lobes which are about anout

5 mm long and triangular. The 5 petals are about 3×2 cm, deep purple, membranous, clawed, and much undulate. The capsules are dehiscent, angled, and minutely apiculate, in a persistent woody calyx, and about 2.2×2 cm. The seeds are about 6 mm long.

Medicinal uses: Dysentery (Malaysia); mouth sores (India)

Moderate broad-spectrum antibacterial extract: Extract inhibited the growth of *Listeria monocytogenes* (ATCC 15313), *Staphylococcus enteridis* (ATCC 13076), *Staphylococcus aureus* (ATCC 8095), and *Escherichia coli* (ATCC 25922), with MIC values of 120, 60, 60, and 120 µg/mL, respectively (Diab et al., 2012).

Antibiotic-potentiator isoflavone: Orobol inhibited NorA efflux pump in *Staphylococcus aureus* (Sharma et al. 2019)

Moderate anticandidal polar extract: Extract inhibited the growth of *Candida albicans* with an MIC value of 120 µg/mL (Diab et al., 2012). Ethanol extract of leaves inhibited the growth of *Candida albicans* with an MIC value of 512 µg/mL (Romulo et al., 2018).

FIGURE 8.3 *Lagerstroemia speciosa* (L.) Pers.

Strong antiviral (non-enveloped linear monopartite single-stranded (+)-RNA) mid-polar extract: Ethyl acetate extract of leaves inhibited the replication of the Human rhinovirus type-3 in Hela cells with an IC_{50} value of 55.5 µg/mL and a selectivity index above 1.8 (Choi et al., 2010).

Strong antiviral (non-enveloped linear monopartite single-stranded (+)-RNA) hydrophilic isoflavonol glycoside: Orobol 7-*O*-D glucoside (molecular mass=448.4 g/mol) inhibited Human rhinovirus type-1B, type-V2, type-V15, and type-V40 (Choi et al., 2010). Orobol 7-*O*-D glucoside inhibited Human rhinovirus type-V3, type-V6, and type-14 as well as pleconaril-resistant HRV5 (Choi et al., 2010).

Commentaries: (i) When medicinal plants are used for the treatment of infections and diabetes or blood pressure it suggests that they probably abound with tannins, simple phenolic, or flavonoids. (ii) *Lagerstroemia indica* L. (local name: *Asare phul*) is used in Nepal to heal wounds and cuts.

REFERENCES

Choi, H.J., Bae, E.Y., Song, J.H., Baek, S.H. and Kwon, D.H., 2010. Inhibitory effects of orobol 7-O-d-glucoside from banaba (Lagerstroemia speciosa L.) on human rhinoviruses replication. *Letters in Applied Microbiology*, *51*(1), pp. 1–5.

Diab, Y., Atalla, K. and Elbanna, K., 2012. Antimicrobial screening of some Egyptian plants and active flavones from Lagerstroemia indica leaves. *Drug Discoveries & Therapeutics*, *6*(4), pp. 212–217.

Romulo, A., Zuhud, E.A., Rondevaldova, J. and Kokoska, L., 2018. Screening of in vitro antimicrobial activity of plants used in traditional Indonesian medicine. *Pharmaceutical Biology*, *56*(1), pp. 287–293.

Sharma, A., Gupta, V.K. and Pathania, R., 2019. Efflux pump inhibitors for bacterial pathogens: From bench to bedside. *The Indian Journal of Medical Research*, *149*(2), p. 129.

8.2.2.2 *Lawsonia inermis* L.

Synonyms: *Alkanna spinosa* Gaertn.; *Lawsonia alba* Lam.; *Lawsonia speciosa* L.; *Lawsonia spinosa* L.; *Rotantha combretoides* Baker

Common names: Henna; mehndi, methi, mendi (Bangladesh)

Habitat: Cultivated

Distribution: Tropical Asia

Botanical examination: This slender shrub grows to about 3 m tall. The stems are smooth and brownish red. The leaves are simple, subsessile, opposite, and exstipulate. The blades are elliptic, 8–44×2–20 mm and dull green, tapering at base, round at apex. Panicles 3–22 cm long. The inflorescences are terminal panicles. The calyx includes 4 sepals, which are ovate. The 4 petals are about 5 mm long and white. The androecium includes 8 stamens, which are about 4 mm long. The ovary is obovate and minute. The capsules are about 7 mm long, dehiscent and contain minute seeds.

Medicinal uses: Foot sores, leprosy, ringworm, leucorrhea, jaundice, wounds (Bangladesh)

Broad-spectrum antibacterial halo developed by polar extracts: Ethanol extract of leaves inhibited the growth of *Escherichia coli*, *Pseudomonas aeruginosa*, *Staphylococcus aureus*, *enteropathogenic Escherichia coli*, *Shigella sonnei*, *Bacillus* sp., *Klebsiella pneumoniae*, *Salmonella* sp., *Corynebacterium* sp., *Streptococus pyogenes*, *Citrobacter freundii*, *Vibrio cholerae*, and *Streptococcus pneumoniae* (Habbal et al., 2005). Ethanol fraction of leaves inhibited the growth of *Escherichia coli*, *Pseudomonas aeruginosa*, *Bacillus subtilis*, and *Staphylococcus aureus* with inhibition zone diameters of 18.1, 21.1, 20.4, and 24.1 mm, respectively (40 mg/100 µL) (Jeyaseela et al., 2012).

Anticandidal halo developed by polar extract: Ethanol extract of leaves inhibited the growth of *Candida albicans* (all clinical strains) (Habbal et al., 2005).

Moderate antifungal (filamentous) polar extract: Ethanol extract inhibited the growth of *Fusarium oxysporum* (963917) with an MIC of 230 µg/mL (Rahmoun et al., 2013).

Strong antifungal (filamentous) hydrophilic naphthoquinone: Lawsone (also known as 2-hydroxy-1,4-naphthoquinone; LogD = −1.7 at pH 7.4; molecular mass = 174 g/mol) inhibited

Fusarium oxysporum (963917) and *Aspergillus flavus* (994294) with MIC values of 12 and 50 µg/mL (Rahmoun et al., 2013).

Strong antiviral (enveloped monopartite linear single-stranded (+) RNA) polar extract: Methanol extract of leaves inhibited the replication of the Sindbis virus in Vero cells with an MIC value of 1.5 µg/mL (Mouhajir et al., 2001).

Viral enzyme inhibition by non-polar extract: Hexane extract of aerial parts at a concentration of 4000 µg/mL inhibited the Human immunodeficiency virus reverse transcriptase activity by about 60% (Silprasit et al., 2011).

Commentaries: (i) Pounded leaves mixed with water afford a brownish dye for the hair or nails and for beautiful skin decorations (weddings) from a very remote period of time in Asia. The plant is used among Malays and Indonesians as antiseptic ointment after circumcision. (ii) The plant may have some activity against the Severe acute respiratory syndrome-acquired coronavirus 2 which like Sindbis virus is an enveloped monopartite linear single-stranded (+) RNA virus.

REFERENCES

Habbal, O.A., Al-Jabri, A.A., El-Hag, A.H., Al-Mahrooqi, Z.H. and Al-Hashmi, N.A., 2005. In-vitro antimicrobial activity of Lawsonia inermis Linn (henna). A pilot study on the Omani henna. *Saudi Medical Journal*, *26*(1), pp. 69–72.

Jeyaseelan, E.C., Jenothiny, S., Pathmanathan, M.K. and Jeyadevan, J.P., 2012. Antibacterial activity of sequentially extracted organic solvent extracts of fruits, flowers and leaves of Lawsonia inermis L. from Jaffna. *Asian Pacific Journal of Tropical Biomedicine*, *2*(10), pp. 798–802.

Mouhajir, F., Hudson, J.B., Rejdali, M. and Towers, G.H.N., 2001. Multiple antiviral activities of endemic medicinal plants used by Berber peoples of Morocco. *Pharmaceutical Biology*, *39*(5), pp. 364–374.

Rahmoun, N., Boucherit-Otmani, Z., Boucherit, K., Benabdallah, M. and Choukchou-Braham, N., 2013. Antifungal activity of the Algerian Lawsonia inermis (henna). *Pharmaceutical Biology*, *51*(1), pp. 131–135.

Silprasit, K., Seetaha, S., Pongsanarakul, P., Hannongbua, S. and Choowongkomon, K., 2011. Anti-HIV-1 reverse transcriptase activities of hexane extracts from some Asian medicinal plants. *Journal of Medicinal Plants Research*, *5*(19), pp. 4899–4906.

8.2.2.3 *Punica granatum* L.

Synonyms: *Punica florida* Salisb.; *Punica grandiflor*a hort. ex Steud.; *Punica nan*a L.; *Punica spinosa* Lam.

Common name: Pomegranate; dalim (Bangladesh); chi liu (China); darim, darooni, raktapushpa (India); hak pila (Laos); delima (Indonesia, Malaysia); thale (Myanmar); anaar (India; Pakistan); granada (the Philippines); phila (Thailand); nar (Turkey); ilu (Vietnam)

Habitat: Cultivated

Distribution: Asia

Botanical description: This slender shrub native from a zone ranging from Iran to the Himalayas grows up to 4 m. It is cultivated in Asia since time immemorial for its delicious and medicinal fruits. It has an aura of dryness. The stems are terete, smooth, somewhat light reddish, and angled. The leaves are simple, opposite, sessile, and exstipulate. The blade is oblong to lanceolate, 2–3.5 cm × 8–12 mm, round at base, glossy, somewhat wavy, and acute at apex. The flowers are axillary or terminal and solitary. The calyx is coriaceous, campanulate, and develops 5–7 sharply triangular lobes, which are about 8 mm long. The 5 petals are bright red, ephemeral, membranous, up to 2 cm long, and obovate. The androecium includes numerous filamentous stamens, which are about 7 mm long. The capsules are globose, 2–20 cm in diameter, somewhat woody (the outer shell was called *malicornium* by our ancient peers), marked at apex by calyx lobes and stamens, and contain numerous seeds. The seeds are about 3mm long and embedded in a translucent and red-colored aril (giving somewhat the impression that the fruit is filled with rubies). The arils yield a refreshing and health-promoting juice.

Medicinal uses: Dysentery (Iran, India, Bangladesh); bronchitis, cough, jaundice, sore throat (India); diarrhea (Turkey, India)

Antibacterial halo developed by polar extract: Methanol extract of shell (6 mg/well) inhibited the growth of *Bacillus cereus*, *Listeria monocytogenes*, *Staphylococcus aureus*, *Escherichia coli*, and *Salmonella anatum* with inhibition zone diameters of 24.6, 15.2, 32.3, 15.4, and 19.2 mm, respectively (Shan et al., 2007).

Moderate broad-spectrum polar extract: Extract inhibited the growth of *Listeria monocytogenes* (ATCC 15313), *Staphylococcus enteridis* (ATCC 13076), *Staphylococcus aureus* (ATCC 8095), and *Escherichia coli* (ATCC 25922) with MIC values of 120, 120, 120, and 120 µg/mL, respectively (Diab et al., 2012). Ethanol extract of fruit shell inhibited the growth of *Escherichia coli* (O157:H7) with MIC and MBC values of 490 and 3910 µg/mL, respectively and inhibited *Escherichia coli* with MIC and MBC of 490 and 1950 µg/mL, respectively (Voravuthikunchai & Limsuwan, 2006). Ethanol extract inhibited the growth of 35 clinical isolates of methicillin-resistant *Staphylococcus aureus* and *Staphylococcus aureus* with MIC/MBC values of 0.2–0.4/1.6–3.2 µg/mL and 0.2/3.2 µg/mL, respectively (Voravuthikunchai & Kitpipit, 2005).

Moderate (enveloped monopartite linear single-stranded (−)-RNA) polar extract: Aqueous extract inhibited Respiratory syncytial virus with an IC_{50} value of 62.2 µg/mL and a selectivity index of 8 (Li et al., 2004).

Antiviral (enveloped segmented linear single-stranded (-)-RNA) polyphenolic fraction: Polyphenolic fraction at a concentration of 200 µg/mL evoked some levels of reduction in the replication of Influenza virus X31 and PR8 (Sundararajan et al., 2010). Polyphenolic fraction inhibited the replication of the Influenza virus A/Hong Kong/2/68 in Madin-Darby canine kidney cells (Haidari et al., 2009).

Antiviral (enveloped segmented linear single-stranded (+)-RNA) polyphenolic fraction: Polyphenolic fraction at a concentration of 200 µg/mL evoked some levels of reduction in the replication of the Murine coronavirus (A59) (Sundararajan et al., 2010).

Viral enzyme inhibition by polar extract: Aqueous extract of root bark at a concentration of 250 µg/mL inhibited the Human immunodeficiency virus type-1 protease by 88% (Xu et al., 1996).

Commentaries: (i) Consider that natural products with anti-immunodeficiency properties are not uncommonly active against coronavirus, and this could lead one to wonder if pomegranate may have some protective effects against COVID-19 and other pandemic zoonotic coronaviruses. In fact, it is worth noting that pomegranate juice is able to inhibit the binding of the Human immunodeficiency virus type-1 to host cells (Neurath et al., 2004). (ii) Pomegranate was used a remedy for the 14th-century European Black Plague.

REFERENCES

Diab, Y., Atalla, K. and Elbanna, K., 2012. Antimicrobial screening of some Egyptian plants and active flavones from Lagerstroemia indica leaves. *Drug Discoveries & Therapeutics*, *6*(4), pp. 212–217.

Haidari, M., Ali, M., Casscells III, S.W. and Madjid, M., 2009. Pomegranate (Punica granatum) purified polyphenol extract inhibits influenza virus and has a synergistic effect with oseltamivir. *Phytomedicine*, *16*(12), pp. 1127–1136.

Li, Y., Ooi, L.S., Wang, H., But, P.P. and Ooi, V.E., 2004. Antiviral activities of medicinal herbs traditionally used in southern mainland China. *Phytotherapy Research*, *18*(9), pp. 718–722.

Neurath, A.R., Strick, N., Li, Y.Y. and Debnath, A.K., 2004. Punica granatum (Pomegranate) juice provides an HIV-1 entry inhibitor and candidate topical microbicide. *BMC Infectious Diseases*, *4*(1), p. 41.

Shan, B., Cai, Y.Z., Brooks, J.D. and Corke, H., 2007. The in vitro antibacterial activity of dietary spice and medicinal herb extracts. *International Journal of Food Microbiology*, *117*(1), pp. 112–119.

Sundararajan, A., Ganapathy, R., Huan, L., Dunlap, J.R., Webby, R.J., Kotwal, G.J. and Sangster, M.Y., 2010. Influenza virus variation in susceptibility to inactivation by pomegranate polyphenols is determined by envelope glycoproteins. *Antiviral Research*, *88*(1), pp. 1–9.

Voravuthikunchai, S.P. and Kitpipit, L., 2005. Activity of medicinal plant extracts against hospital isolates of methicillin-resistant Staphylococcus aureus. *Clinical Microbiology and Infection*, *11*(6), pp. 510–512.

Voravuthikunchai, S.P. and Limsuwan, S., 2006. Medicinal plant extracts as anti–Escherichia coli O157: H7 agents and their effects on bacterial cell aggregation. *Journal of Food Protection*, *69*(10), pp. 2336–2341.
Xu, H.X., Wan, M., Loh, B.N., Kon, O.L., Chow, P.W. and Sim, K.Y., 1996. Screening of traditional medicines for their inhibitory activity against HIV-1 protease. *Phytotherapy Research*, *10*(3), pp. 207–210.

8.2.2.4 *Sonneratia griffithii* Kurtz.

Synonyms: *Chiratia leucantha* Montrouz.; *Sonneratia acida* Benth.; *Sonneratia alba* J. Smith; *Sonneratia iriomotensis* Masam.; *Sonneratia mossambicensis* Klotzsch ex Peters

Common names: Mangrove apple; bei e hai sang (China); perepat (Malaysia); pedada laut, perepat (Malaysia); lame (Myanmar); lampoo talay, pat (Thailand); ban dang (Vietnam)

Habitat: Mangroves

Distribution: Southeast Asia, Papua New Guinea, Australia, Pacific Islands

Botanical examination: This beautiful mangrove tree grows to about 5 m tall. The bole is surrounded by pneumatophores arising from roots. The stem is glabrous, terete, minutely fissured, peeling, and somewhat articulated. The nodes are swollen and marked with rings and ovoid leaf-scars. The internodes are irregular and 1–4 cm long. The leaves are simple, opposite, and without stipules. The petiole is stout, glabrous, up to 4 mm long, and non-channeled. The blade is roundish, broadly spathulate, 5–8 × 3–8 cm, glabrous, woody to somewhat spongy, with 8–12 pairs of discrete secondary nerves. The midrib is flattish above and below. The flower is axillary or terminal. The calyx is leathery, conical, and produces 6 lobes, which are triangular. The 6 petals are linear-lanceolate and reddish. The nuts are terminal, about 4 cm long, solitary, glossy, woody, globose, flattish, depressed at apex and marked with a persistent style of about 2 cm, and persistent sepals, which are 1.5 × 1.4 cm.

Medicinal uses: Diarrhea, wounds (Thailand)

Very weak broad-spectrum mid-polar antibacterial extract: Ethyl acetate extract of leaves inhibited the growth of *Staphylococcus aureus* (ATCC25923), *Bacillus cereus* (ATCC11778), and *Escherichia coli* (ATCC35918) (Saad et al., 2012).

Strong antibacterial (Gram-positive) lupane-type triterpenes: 3β-Hydroxy-lup-9(11),12-diene-28-oic acid, lupeol, and lupan-3β-ol isolated from the bark inhibited the growth of *Staphylococcus aureus* (ATCC 6538) with MIC values of 32.2, 15.6, and 33.1 μg/mL, respectively (Harizon et al., 2015). 3β-hydroxy-lup-9(11),12-diene, 28-oic acid, lupeol, and lupan-3β-ol inhibited the growth of *Streptococcus mutans* (ATCC 25175) with MIC values of 35.6, 40.6, and 55.2 μg/mL, respectively (Harizon et al., 2015).

Very weak antifungal (yeast) extract: Ethyl acetate extract of leaves inhibited the growth of *Cryptococcus neoformans* (ATCC90112) with an MIC value of 300 μg/mL (Saad et al., 2012).

Commentaries: (i) Ethanol extract of barks of a member of the genus *Sonnetaria* L.f.(500 μg/disc) inhibited the growth of *Staphylococcus aureus*, *Staphylococcus epidermidis*, and *Streptococcus pneumonia* with inhibition zone diameters of 18, 14, and 16 mm, respectively (Udn et al., 2008).

(ii) In India, *Sonneratia apetala* Buch.-Ham. (Mangrove apple; local name: *Kerpa*) is used to treat leprosy and gonorrhea (India). Ethanol extract of seeds of this tree inhibited the growth of *Escherichia coli* (clinical isolates), *Proteus* sp. (clinical isolates), *Pseudomonas* sp. (clinical isolates), *Salmonella paratyphi A* (clinical isolates), *Shigella typhi* (clinical isolates), *Shigella dysenteriae* (clinical isolates), *Shigella flexneri* (clinical isolates), *Vibrio cholera* (clinical isolates), *Enterococcus faecalis* (clinical isolates), *Staphylococcus aureus* (clinical isolates), and *Staphylococcus epidermidis* (clinical isolates) with MIC values of 200, 100, 150, 250, 250, 100, 250, 150, 200, 200, and 200 μg/mL, respectively (Hossain et al., 2013).

(iii) Triterpenes in members of the genus *Sonneratia* L. are not uncommonly antiviral. One example is the lupane-type paracaseolin A from the aerial parts of *Sonneratia paracaseolaris* W.C. Ko, E.Y. Chen & W.Y. Chen inhibited the replication of the Influenza virus (A/Puerto Rico/8/34 (H1N1), PR/8) in Madin-Darby canine kidney cells with an IC_{50} value of 28.4 μg/mL (Gong et al., 2017).

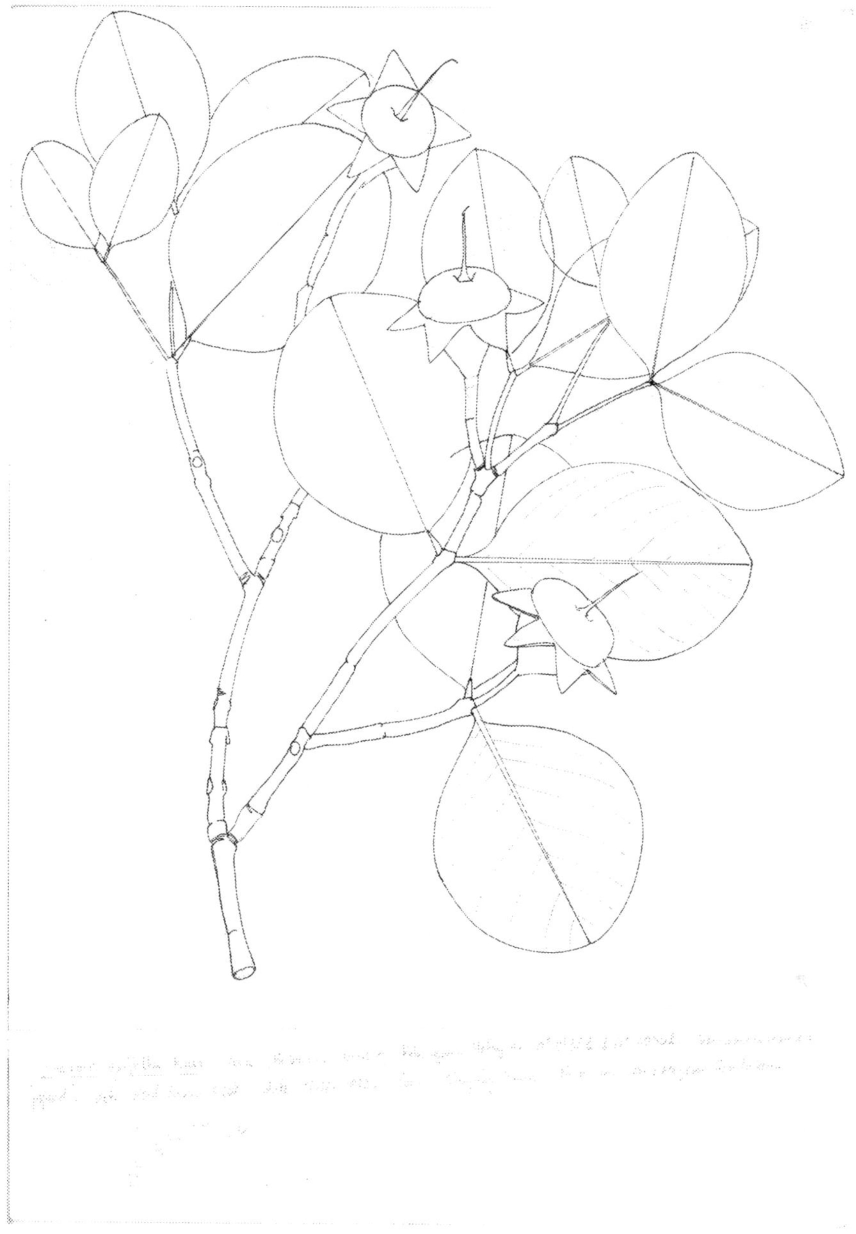

FIGURE 8.4 *Sonneratia griffithii*.Kurtz.

REFERENCES

Gong, K.K., Li, P.L., Qiao, D., Zhang, X.W., Chu, M.J., Qin, G.F., Tang, X.L. and Li, G.Q., 2017. Cytotoxic and antiviral triterpenoids from the mangrove plant Sonneratia paracaseolaris. *Molecules*, *22*(8), p. 1319.

Harizon, Pujiastuti, B., Kurnia, D., Sumiarsa, D., Shiono, Y. and Supratman, U., 2015. Antibacterial triterpenoids from the bark of Sonneratia alba (Lythraceae). *Natural Product Communications*, *10*(2), p. 1934578X1501000215.

Hossain, S.J., Basar, M.H., Rokeya, B., Arif, K.M.T., Sultana, M.S. and Rahman, M.H., 2013. Evaluation of antioxidant, antidiabetic and antibacterial activities of the fruit of Sonneratia apetala (Buch.-Ham.). *Oriental Pharmacy and Experimental Medicine*, *13*(2), pp. 95–102.

Saad, S., Taher, M., Susanti, D., Qaralleh, H. and Awang, A.F.I.B., 2012. In vitro antimicrobial activity of mangrove plant Sonneratia alba. *Asian Pacific Journal of Tropical Biomedicine*, 2(6), pp. 427–429.
Uddin, S.J., Rouf, R., Shilpi, J.A., Alamgir, M., Nahar, L. and Sarker, S.D., 2008. Screening of some Bangladeshi medicinal plants for in vitro antibacterial activity. *Oriental Pharmacy and Experimental Medicine*, 8(3), pp. 316–321.

8.2.2.5 *Woodfordia fruticosa* (L.) Kurz

Synonyms: *Grislea punctata* Buch.-Ham. ex Sm.; *Grislea tomentosa* Roxb.; *Lythrum fruticosum* L.; *Lythrum hunteri* DC.; *Woodfordia floribunda* Salisb.; *Woodfordia tomentosa* Bedd.

Common names: Arepuvvu, dhaeen, parvati, velakkai (India); silu (Indonesia); sedu waya (Malaysia); dhairo (Nepal); (Tamil); lam phat (Vietnam)

Habitat: Open and dry lands

Distribution: Tropical Asia

Botanical examination: This shrub grows to a height of up to 2 m. The stems are terete and peeling. The leaves are simple, opposite, sessile, and without stipules. The blade is lanceolate, thinly coriaceous, 1.5–6.5 × 2.2–5.5 cm, dark green, glossy, marked with 6–12 pairs of secondary nerves and amplexicaul at base. The cymes are axillary. The flowers are tubular, red (similar to pomegranate flowers) and 2.5–4 cm long. The calyx is 1.5 cm long, striated, covered with glandular dots and forming a bright red tube. The 6 petals are longer than the sepals. The androecium includes numerous showy stamens. The fruits are dehiscent capsules containing cuneate-ovoid, brown, and small seeds.

Medicinal uses: Cuts, jaundice, (India); dysentery (India, Indonesia, Myanmar, Nepal); diarrhea, fever (Nepal); boils (Bangladesh)

Moderate antibacterial (Gram-positive) polar extract: Ethanol extract of leaves inhibited the growth of *Staphylococcus aureus* and *Enterococcus faecalis* with MIC values of 256 and 256 µg/mL, respectively (Romulo et al., 2018).

Moderate anticandidal polar extract: Ethanol extract of leaves inhibited the growth of *Candida albicans* with an MIC value of 128 µg/mL (Romulo et al., 2018).

Strong antiviral (non-enveloped linear monopartite, single-stranded (+)-RNA) polar extract: Ethanol extract of flowers inhibited the replication of Enterovirus 71 with an IC_{50} value of 1.2 µg/mL and a selectivity index above 83.3 in Vero cells (Choi et al., 2010).

Moderate antiviral (non-enveloped linear monopartite single-stranded (+)-RNA) simple phenolic: Gallic acid inhibited the replication of the Enterovirus 71 with an IC_{50} value of 0.7 µg/mL and a selectivity index of 99.5 in Vero cells (Choi et al., 2010a).

Moderate antiviral (non-enveloped linear monopartite single-stranded (+)-RNA) simple phenolic: Gallic acid at a concentration of 100 µg/mL inhibited Human rhinovirus type-3 by about 55%, via interaction with viral particles and inhibition of viral absorption in Hela cells (Choi et al., 2010).

Viral enzyme inhibition by polar extract: Aqueous extract of flowers and leaves at a concentration of 250 µg/mL inhibited the Human immunodeficiency virus type-1 protease by 80.7% (Xu et al., 1996).

Commentaries: (i) The plant produces ellagitannins known as woodfordins (Kuramochi-Motegi et al., 1992), which are probably, at least partially, involved in traditional uses.

(ii) Ethyl acetate extract of *Lythrum salicaria* L. (from the tribe Lythreae in which belongs the genus *Woodfordia* Salsib.) inhibited the growth of *Staphylococcus aureus* (DSm 346), *Proteus mirabilis* (DSM 788), *Micrococcus luteus*, and *Candida albicans* (DSM 1386) (Becker et al., 2005). From this extract the ellagitannin vescalgin inhibited the growth of *Staphylococcus aureus* (DSm 346), *Proteus mirabilis* (DSM 788), and *Micrococcus luteus* with MIC/MBC values of 60/500, 60/>1000, 100/>1000, respectively (Becker et al., 2005).

(iii) In the tribe Lytheae, ammaniol (also known as 2,5-bis-(3,3′-hydroxyaryl)tetrahydrofuran) from *Ammannia multiflora* Roxb. inhibited the growth of *Mycobacterium tuberculosis* (H37Rv) with an MIC value of 25 µg/mL (Upadhyay et al., 2012). Consider the puzzling fact that plants can synthetize perfectly symmetrical natural products.

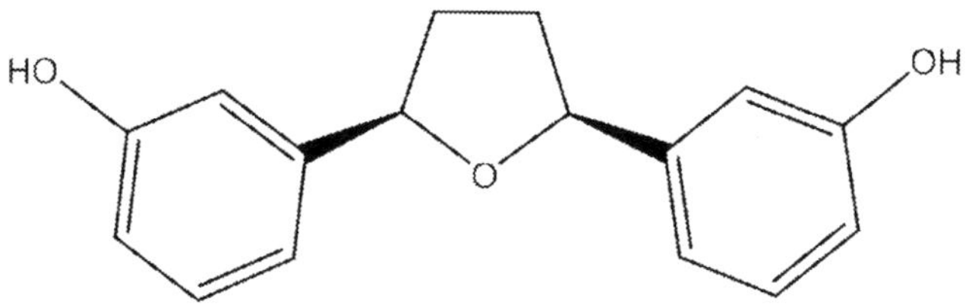

Ammaniol

(iv) *Duabanga grandiflora* (Roxb. ex DC.) Walp. is used in Nepal (local name: *Kadam)* to treat fever. In India, *Rotala rotundifolia* (Buch.-Ham. ex Roxb.) Koehne is used to treat cough (local name: *Daggumandu*)

REFERENCES

Becker, H., Scher, J.M., Speakman, J.B. and Zapp, J., 2005. Bioactivity guided isolation of antimicrobial compounds from Lythrum salicaria. *Fitoterapia*, *76*(6), pp. 580–584.

Choi, H.J., Song, J.H., Bhatt, L.R. and Baek, S.H., 2010. Anti-human rhinovirus activity of gallic acid possessing antioxidant capacity. *Phytotherapy Research*, *24*(9), pp. 1292–1296.

Choi, H.J., Song, J.H., Park, K.S. and Baek, S.H., 2010a. In vitro anti-enterovirus 71 activity of gallic acid from Woodfordia fruticosa flowers. *Letters in Applied Microbiology*, *50*(4), pp. 438–440.

Kuramochi-Motegi, A., Kuramochi, H., Kobayashi, F., Ekimoto, H., Takahashi, K., Kadota, S., Takamori, Y. and Kikuchi, T., 1992. Woodfruticosin (woodfordin C), a new inhibitor of DNA topoisomerase II. Experimental antitumor activity. *Biochemical Pharmacology*, *44*(10), pp. 1961–1965.

Romulo, A., Zuhud, E.A., Rondevaldova, J. and Kokoska, L., 2018. Screening of in vitro antimicrobial activity of plants used in traditional Indonesian medicine. *Pharmaceutical Biology*, *56*(1), pp. 287–293.

Upadhyay, H.C., Dwivedi, G.R., Darokar, M.P., Chaturvedi, V. and Srivastava, S.K., 2012. Bioenhancing and antimycobacterial agents from Ammannia multiflora. *Planta Medica*, *78*(01), pp. 79–81.

Xu, H.X., Wan, M., Loh, B.N., Kon, O.L., Chow, P.W. and Sim, K.Y., 1996. Screening of traditional medicines for their inhibitory activity against HIV-1 protease. *Phytotherapy Research*, *10*(3), pp. 207–210.

8.2.3 Family Melastomataceae A. L. de Jussieu (1789)

The family Melastomataceae consists of about 200 genera and 4,500 species of herbs, shrubs, or trees. The now almost extinct primary forest of Southeast Asia was endowed with the most fascinating of Melastomes, which are now reduced to ashes and gone forever for the financial benefits of palm oil companies. The leaves are simple, opposite, decussate, and exstipulate. The blade displays 1–4 pairs of secondary nerves characteristic originating from the base and scalariform tertiary nerves. The inflorescences are cymes, umbels, corymbs, panicles, or spikes. The flowers are actinomorphic and present a hypanthium. The calyx produces 5 lobes. The corolla consists of 5 petals. The androecium includes 10 stamens, which are characteristically recurved. The gynoecium includes a 5-locular ovary, each locule containing numerous anatropous ovules, on axillary or parietal (or free central placentas). The fruits are capsules or berries. Members in this family accumulate tannins and as such are often used as an astringent remedy for dysentery.

8.2.3.1 *Melastoma candidum* D. Don

Synonym: *Melastoma malabathricum* L.

Common names: Malabar melastome, white melastome; tai-tong (Bangladesh); kalathi, korali (India); halendong, sengani (Indonesia); senduduk puteh (Malaysia); yagomyum (the Philippines); mang khre (Thailand); mua se (Vietnam)

Habitat: Roadside, waste lands

Distribution: Southeast Asia, India, Australia

Botanical description: This handsome shrub grows up to about 2 m tall. The stems are terete and hairy. The leaves are simple, opposite, and exstipulate. The petiole is hairy, channeled, and 7–1.8 cm long. The blade is broadly lanceolate, hairy, coriaceous, 7–9.5 × 4–5.5 cm and marked with 7 longitudinal nerves raised below and sunken above, the margin is recurved and wavy, and the tertiary nerves are conspicuous below. The calyx is campanulate and coriaceous and develops 5 lobes. The 5 membranous petals are obovate, pure white or purple, and about 3 cm long. A disc is present. The 10 stamens are showy and recurved. The capsules are dehiscent, urn-shaped, hairy, almost 1 cm long, and open to present a dull black mass in which seeds are embedded.

Medicinal uses: Dysentery (Indonesia, Laos, Cambodia, and Vietnam); diarrhea, wounds (Indonesia)

Antibacterial polar extract: Aqueous extract of leaves exhibited antibacterial effects (Wong et al., 2012).

Antibacterial halo developed by polar extract: Methanol extract of leaves inhibited the growth of *Escherichia coli* MTCC 1089, *Streptococcus* sp., and *Staphylococcus aureus* MTCC 1144, with inhibition zone diameters of 11.1, 12.3, and 13 mm, respectively (Choudhury et al., 2009). Methanol extract inhibited the growth of *Staphylococcus aureus* (Grosveno et al., 1995).

Moderate antiviral (enveloped monopartite linear double-stranded DNA) polar extract: Methanol extract inhibited the replication of the Herpes simplex virus type-1 with an IC_{50} value of 192 μg/mL and a selectivity index above 5 (Lohézic-Le Dévéhat et al., 2002).

Moderate antiviral (non-enveloped single-stranded linear (+) RNA) polar extract: Methanol extract of inhibited the replication of the Polio virus with an IC_{50} value of 111 μg/ mL and a selectivity index above 5 (Lohézic-Le Dévéhat et al., 2002).

Commentaries: (i) The plant yields ellagitannins (Yoshida et al., 1992; 2010), probably accounting, at least in part, for its medicinal uses of kaempferol, quercetin, ursolic acid, 2α-hydroxyursolic acid, asiatic acid, β-sitosterol 3-*O*-β-D-glucopyranoside, and ellagic acid (Wong et al., 2012). Consider that β-Sitosterol (which is common in flowering plants) inhibited the Severe acute respiratory syndrome-associated coronavirus 3-chymotrypsin-like protease with an EC_{50} value of 47.8 μg/mL (Lin et al., 2005). (ii) *Melastoma normale* D. Don is used in Nepal (local name: *Chulesi*) for the treatment of cold, cough, diarrhea, and dysentery.

REFERENCES

Choudhury, M.D., Nath, D. and Talukdar, A.D., 2011. Antimicrobial activity of Melastoma malabathricum L. *Assam University Journal of Science and Technology*, *7*(1), pp. 76–78.

Grosvenor, P.W., Supriono, A. and Gray, D.O., 1995. Medicinal plants from Riau Province, Sumatra, Indonesia. Part 2: Antibacterial and antifungal activity. *Journal of Ethnopharmacology*, *45*(2), pp. 97–111.

Lin, C.W., Tsai, F.J., Tsai, C.H., Lai, C.C., Wan, L., Ho, T.Y., Hsieh, C.C. and Chao, P.D.L., 2005. Anti-SARS coronavirus 3C-like protease effects of Isatis indigotica root and plant-derived phenolic compounds. *Antiviral Research*, *68*(1), pp. 36–42.

Lohézic-Le Dévéhat, F., Bakhtiar, A., Bezivin, C., Amoros, M. and Boustie, J., 2002. Antiviral and cytotoxic activities of some Indonesian plants. *Fitoterapia*, *73*(5), pp. 400–405.

Wong, K.C., Hag Ali, D.M. and Boey, P.L., 2012. Chemical constituents and antibacterial activity of Melastoma malabathricum L. *Natural Product Research*, *26*(7), pp. 609–618.

Yoshida, T., Nakata, F., Hosotani, K., Nitta, A. and Okuda, T., 1992. Tannins and related polyphenols of melastomataceous plants. v. three new complex tannins from Melastoma malabathricum L. *Chemical and Pharmaceutical Bulletin*, *40*(7), pp. 1727–1732.

8.2.3.2 *Osbeckia chinensis* L.

Common names: Chinese Osbeckia; jin jin xiang (China); himenobotan (Japan); payong-payong (the Philippines); aa noi (Thailand); mua tep (Vietnam)

Habitat: Grassy areas and deciduous forest

Distribution: India, Nepal, Bangladesh, Southeast Asia, China, Japan, New Guinea, and Australia

Botanical examination: This gracious herb grows to a height of 70 cm. The stems are minutely hairy, squared, and reddish and the internodes are 20 cm. The leaves are simple, sessile, opposite, and exstipulate. The blade is narrowly oblong or lanceolate 1.6–2×3.1–4 cm, hairy, and shows a midrib, which is sunken with several prominent nerves running the length of the blade. The flowers are arranged in terminal heads. The calyx comprises 4–5 lobes, which are minute and triangular. The 4–5 petals are 1.2–1.7 cm long, obovate, acuminate at apex, purplish, ephemeral, and falling at collection, hairy at margin, and purple. The androecium consists of 8–10 yellow and showy stamens. The ovary is globose. The capsules are campanulate or urceolate, purplish-red and 3–6 mm long.

Medicinal use: Dysentery (Taiwan).

Commentaries: (i) The plant accumulates ellagitannins such as casuarinin, casuariin, punicacortein A, degalloyl-punicacortein A, as well as gallic acid, methyl gallate, ellagic acid (Su et al., 1988), and flavonol glycosides (Zhao et al., 2011). All of these phenolic compounds may, at leat in part, account for the traditional medicinal use.

(ii) In India, *Memecylon malabaricum* Kostel. is used to treat skin diseases. Methanol extract of leaves of this plant inhibited the growth of *Bacillus subtilis* and *Salmonella typhi* (Hullatti & Rai, 2004). This extract inhibited the growth of *Aspergillus niger, Aspegillus flavus, Aspergillus versicolor, Aspergillus columnuris, Aspergillus tamari, Aspergillus flaviscipes,* and *Fusarium oxysporum* (Hullatti & Rai, 2004).

(iii) Consider that medicinal plants in the family Melastomataceae bring to being flavonols, such as myricetin, which is an inhibitor of the Human immunodeficiency virus type-1 reverse transcriptase (Ortega et al., 2017) as well as the Severe acute respiratory syndrome-acquited coronavirus helicase and other viral enzymes as described earlier.

(iv) *Medinilla radicans* Bedd. is used in Indonesia to treat diarrhea.

(v) *Osbeckia nepalensis* Hook. is used in India (local name: *Baga futkala*) against abscesses and leprosy and in Nepal (local name: *Angaru*) for the treatment of cuts and wounds. In Nepal, *Osbeckia stellata* Buch.-Ham. ex D. Don (local name: *Angerig*) is used for the treatment of diarrhea and dysentery as well as *Oxyspora paniculata* (D. Don) DC.

REFERENCES

Hullatti, K.K. and Rai, V.R., 2004. Antimicrobial activity of Memecylon malabaricum leaves. *Fitoterapia*, *75*(3–4), pp. 409–411.

Ortega, J.T., Suárez, A.I., Serrano, M.L., Baptista, J., Pujol, F.H. and Rangel, H.R., 2017. The role of the glycosyl moiety of myricetin derivatives in anti-HIV-1 activity in vitro. *AIDS Research and Therapy*, *14*(1), p. 57.

Zhao, Y., Yang, D., Ma, Q., Li, N., Cheng, Z. and Zhou, J., 2011. Chemical constituents from Osbeckia chinensis. *Zhongcaoyao= Chinese Traditional and Herbal Drugs*, *42*(6), pp. 1061–1065.

8.2.4 Family Myrtaceae A.L de Jussieu (1789)

The family Myrtaceae consists of about 130 genera and 500 species of trees, which have, not uncommonly, an aura of dryness. The stems are often somewhat characteristically smooth and slightly reddish. The leaves are simple, opposite, decussate, and exstipulate. The blade is often coriaceous, and often displays numerous and inconspicuous secondary nerves characteristically joined into an intramarginal nerve. The inflorescences are axillary or terminal cymes. The flowers are actinomorphic and present a hypanthium. The calyx produces 5 lobes. The corolla consists of 5 petals. A disc is present. The androecium is often showy and includes numerous stamens, which are somewhat straight and not uncommonly showy. The gynoecium includes a 1 or plurilocular locular ovary, each locule containing 1–numerous ovules, on axillary or parietal placentas. The capsules or berries present a characteristic disc at apex. Members in this family produce a fascinating array of flavonoids and hydrolysable tannins as well as phenylpropanoids and monoterpenes.

8.2.4.1 *Decaspermum fruticosum* Forst.

Synonyms: *Decaspermum paniculatum* Kurz., *Legnotis lanceolata* Blco., *Metrosideros pictipetala* Blco., *Nelitris fruticosa* A. Gray, *Nelitris paniculata* Lindl., *Psidium decaspermum* L.f.

Common names: Silky myrtle; ambok, piyara (India); kayu demang (Indonesia); badaduk, kelentit kering, tukang benang (Malaysia); taung tabye (Myanmar); malagitinggiting (the Philippines); khee tai krim (Thailand)

Habitat: Forests

Distribution: India, Myanmar, Thailand, Malaysia, Indonesia, the Philippines, Australia, and Pacific Islands

Botanical examination: This shrub or treelet grows to a height of 10 m. The bark is reddish brown. The stem is reddish and velvety. The leaves are simple, opposite, and without stipules. Some hairy buds are present at leaf axis. The petiole is about 7 mm long, flattish, and channeled. The blade is oblong, 3–9 cm × 5–4 cm, marked with about 12–13 pairs of secondary nerves, thinly coriaceous, acuminate apex, and marked numerous tiny oil cells. The panicles are axillary and terminal. The calyx is campanulate, about 2 mm long, hairy with 4 lanceolate lobes. The 5 petals are white and about 3 mm long. The anthers are minute and numerous. The style is about 2 mm long. The berries are globose, 0.6–1 cm in diameter, ripening purple, and crowned by 5 persistent sepals.

Medicinal uses: Dysentery (India; Malaysia); teeth health, pyorrhea

diarrhea (India)

Commentary: The plant has apparently not been examined for its possible antimicrobial effects. This plant is one of the many which are going to disappear under the buildozers of the palm oil industry. It must be grasped that the primary rainforest plants have almost a supernatural power of creativity and synthesis of antimicrobial molecules, which will never come out of the mind of any organic chemists. Some of the trees that have already disappeared may have had the ability to synthetize anti-COVID-19 principles. Also, it must be considered that deforestation is one of the most important factor contributing to the appearance of new zoonotic pandemic viruses. Should deforestation continue as well as the most fundamental ecological rights of life forms and environment we are bound to face zoonotic pandemic of increasing severity.

8.2.4.2 *Eugenia aquea* Burm.f.

Synonyms: *Syzygium aqueum* (Burm.) f. Alston

Common names: Water apple, watery rose apple, jamrul (Bangladesh); bell fruit; jambu chili, jambu ayer (Malaysia), djamboo aer, djamboo wer (Indonesia), tambis (the Philippines), chom phu pa (Thailand)

Habitat: Cultivated

Distribution: Tropical Asia to Hawaii

Botanical description: This handsome tree probably native to South India grows to a height of about 10 m. It has an aura of freshness. The stem is smooth. The leaves are simple, decussate, and without stipules. The petiole is minute, channeled, glabrous, and up to 3 mm long. The blade is membranous, dull green, smooth, 10.5–13 × 3.8–5cm, glabrous, broadly elliptic or round at apex,

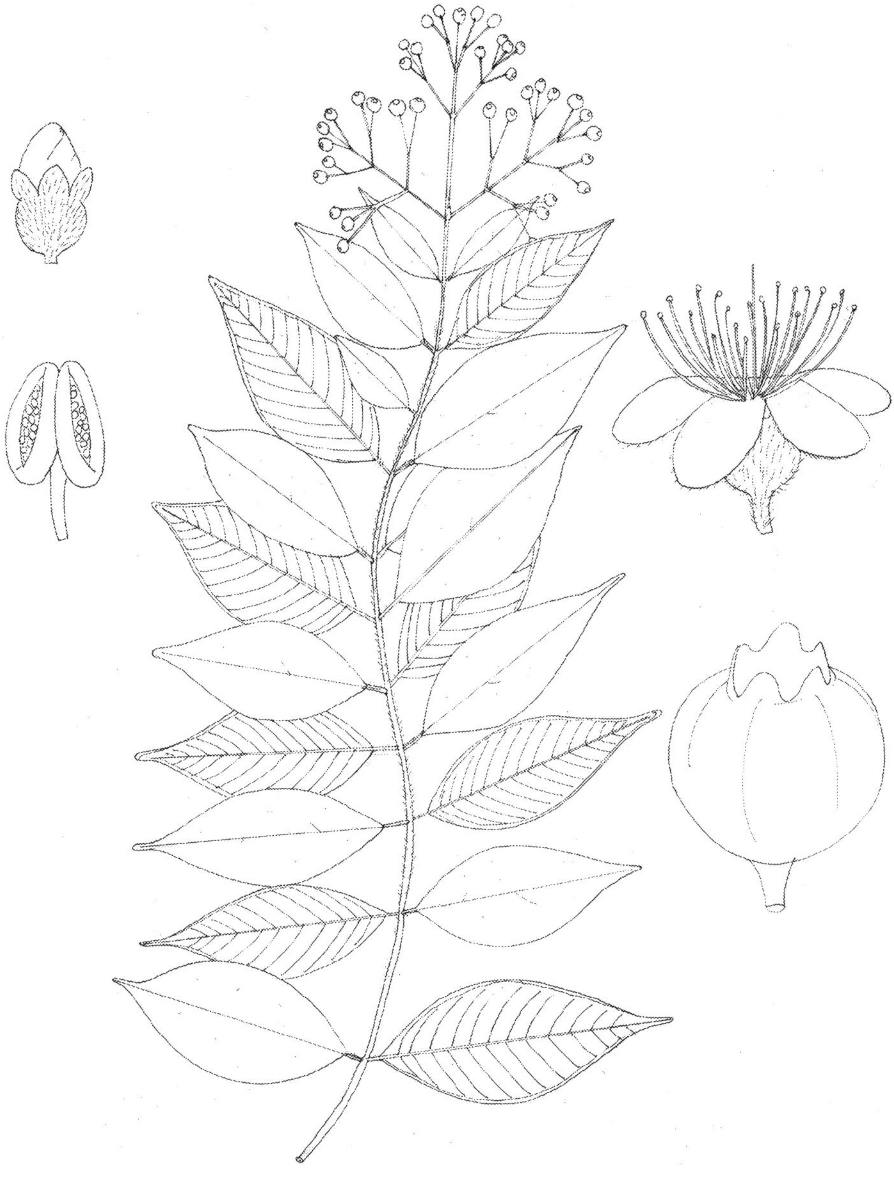

FIGURE 8.5 *Decaspermum fruticosum*.Forst.

cordate at base, and marked with 8–9 pairs of secondary nerves. The cymes are terminal. The calyx is urn-shaped, 4mm in diameter, and 4-lobed. A disc is present. The stamens are showy, white, and about 2 cm long. The berries are bout 2–4cm long, pale pink, edible, glossy, pear-shaped, fleshy, marked with an apical depression and a persistent disc.

Medicinal use: Dysentery (Papua New Guinea)

Moderate antibacterial polar extract: Ethanol extract of leaves inhibited the growth of *Escherichia coli* and *Bacillus subtilis* with an MIC value of 250 μg/mL (Abdulah et al., 2017).

Commentaries: (i) The plants yields with flavonoids of which, and the proanthocyanidin samarangenins A and B as well as (–)-epigallocatechin. (–)-epigallocatechin 3-*O*-gallate, flavan-3-ols, pedunculagin, eugeniin, and prodelphinidin B-2 3,3″-di-*O*-gallate (Nonaka et al., 1992) which probably by virtue of their astringent properties account for the antidysenteric traditional uses.

(ii) The plant is consumed by apes, and in this context it is worth remembering that apes consume, when ill, plant species that are not part of their normal diet (Huffman, 1997). One could wonder how apes (which are on the verge of extinction) have acquired the knowledge that some plants cure their illness. To answer this question may open new doors that could further increase our chances of survival in the face of coming waves of deadly zoonotic pandemics.

FIGURE 8.6 *Eugenia aquea* Burm.f.

FIGURE 8.7 *Eugenia aromatica* (L.) Baill.

(iii) Commentary: *Caryophyllus aromaticus* L. (synonym *Eugenia aromatica* (L.) Baill.; *Eugenia caryophyllata* Thunb.) is used for the treatment of COVID-19 in China (Wang et al., 2020) as well as by Siddha healers in South India (Sasikumar et al., 2020). This plant is also used in India (local name: *Laung*) to treat measles. Ethanol extract of fruits of this plant inhibited *Escherichia coli* and *Klebsiella pneumonia* with an MIC of 1024 μg/mL (Shan et al., 2007). Methanol extract of flower buds (6mg/well) inhibited the growth of *Bacillus cereus*, *Listeria monocytogenes*, *Staphylococcus aureus*, *Escherichia coli*, and *Salmonella anatum* with inhibition zone diameters of 14, 13.7, 21.3, 10.1, and 17.5 mm, respectively (Shan et al., 2007). Consider that the phenypropanoid eugenol (Log D 2.4 at pH 7.4; molecular mass = 164 g/mol) inhibited the replication of the Herpes simplex virus type 1 and type 2 with IC_{50} values of 25.6 and 16.2 µg/mL, respectively (Benencia & Courreges, 2000), and inhibited the replication of the Ebola virus with an EC_{50} value of 1.3 µM (Lane et al., 2019).

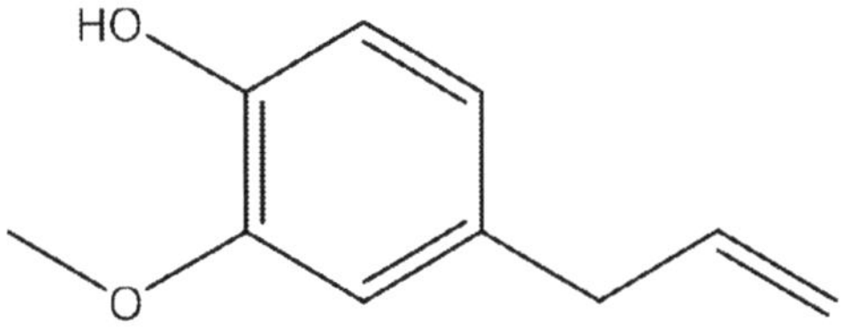

Eugenol

(iv) *Eugenia jambos* L. (synonym: *Syzygium jambos* (L.) Alston) is used in India for the treatment of diarrhea and dysentery (local name: *Jambu*) and abounds with hydrolysable tannins such as 1-*O*-galloyl castalagin and casuarinin (Yang et al., 2000), 3,3′,4′-tri-*O*-methylellagic acid-4-*O*-β-D-glucopyranoside, and 3,3′,4′-tri-*O*-methylellagic acid.(Chakravarty et al., 1999). Ethanol extract of this plant inhibited the replication of the Herpes virus type1 at a concentration of 10 µg/mL as well as the Vesicular stomatitis virus (Abad et al., 1997).

(v) Acetone: water extract of leaves of *Eugenia uniflora* L. inhibited the growth of *Staphylococcus aureus* (ATCC 25923), *Staphylococcus epidermidis* (INCQS 00016), *Enterococcus faecalis* (ATCC 29212), *Salmonella enteritidis* (INCQS 00258), and *Pseudomonas aeruginosa* (ATCC 27853) with MIC values of 1.2, 0.3, 1.2, 1.2, and 1.2 µg/mL, respectively (Falcão et al., 2018).

REFERENCES

Abad, M.J., Bermejo, P., Villar, A., Sanchez Palomino, S. and Carrasco, L., 1997. Antiviral activity of medicinal plant extracts. *Phytotherapy Research: An International Journal Devoted to Medical and Scientific Research on Plants and Plant Products*, *11*(3), pp. 198–202.

Abdulah, R., Milanda, T., Sugijanto, M., Barliana, M.I., Diantini, A., Supratman, U. and Subarnas, A., 2017. Antibacterial properties of selected plants consumed by primates against escherichia coli and bacillus subtilis. *Southeast. Asian J. Trop. Med. Public Health*, *48*, pp. 109–116.

Benencia, F. and Courreges, M.C., 2000. In vitro and in vivo activity of eugenol on human herpesvirus. *Phytotherapy Research: An International Journal Devoted to Pharmacological and Toxicological Evaluation of Natural Product Derivatives*, *14*(7), pp. 495–500.

Chakravarty, A.K., Das, B., Sarkar, T., Masuda, K. and Shiojima, K., 1999. Ellagic acid derivatives from the leaves of eugenia jambos linn. *ChemInform*, *30*(25), pp. no-no.

Falcão, T.R., de Araújo, A.A., Soares, L.A.L., de Moraes Ramos, R.T., Bezerra, I.C.F., Ferreira, M.R.A., de Souza Neto, M.A., Melo, M.C.N., de Araújo, R.F., de Aguiar Guerra, A.C.V. and de Medeiros, J.S., 2018. Crude extract and fractions from Eugenia uniflora Linn leaves showed anti-inflammatory, antioxidant, and antibacterial activities. *BMC Complementary and Alternative Medicine*, *18*(1), p. 84.

Huffman, M.A., 1997. Current evidence for self-medication in primates: A multidisciplinary perspective. *American Journal of Physical Anthropology: The Official Publication of the American Association of Physical Anthropologists*, *104*(S25), pp. 171–200.

Jayasinghe, U.L.B., Ratnayake, R.M.S., Medawala, M.M.W.S. and Fujimoto, Y., 2007. Dihydrochalcones with radical scavenging properties from the leaves of Syzygium jambos. *Natural Product Research*, *21*(6), pp. 551–554.

Nonaka, G.I., AiKo, Y., Aritake, K. and Nishioka, I., 1992. Tannins and related compounds. CXIX. Samarangenins a and b, novel proanthocyanidins with doubly bonded structures, from Syzygium samarangens and S. aqueum. *Chemical and Pharmaceutical Bulletin*, *40*(10), pp. 2671–2673.

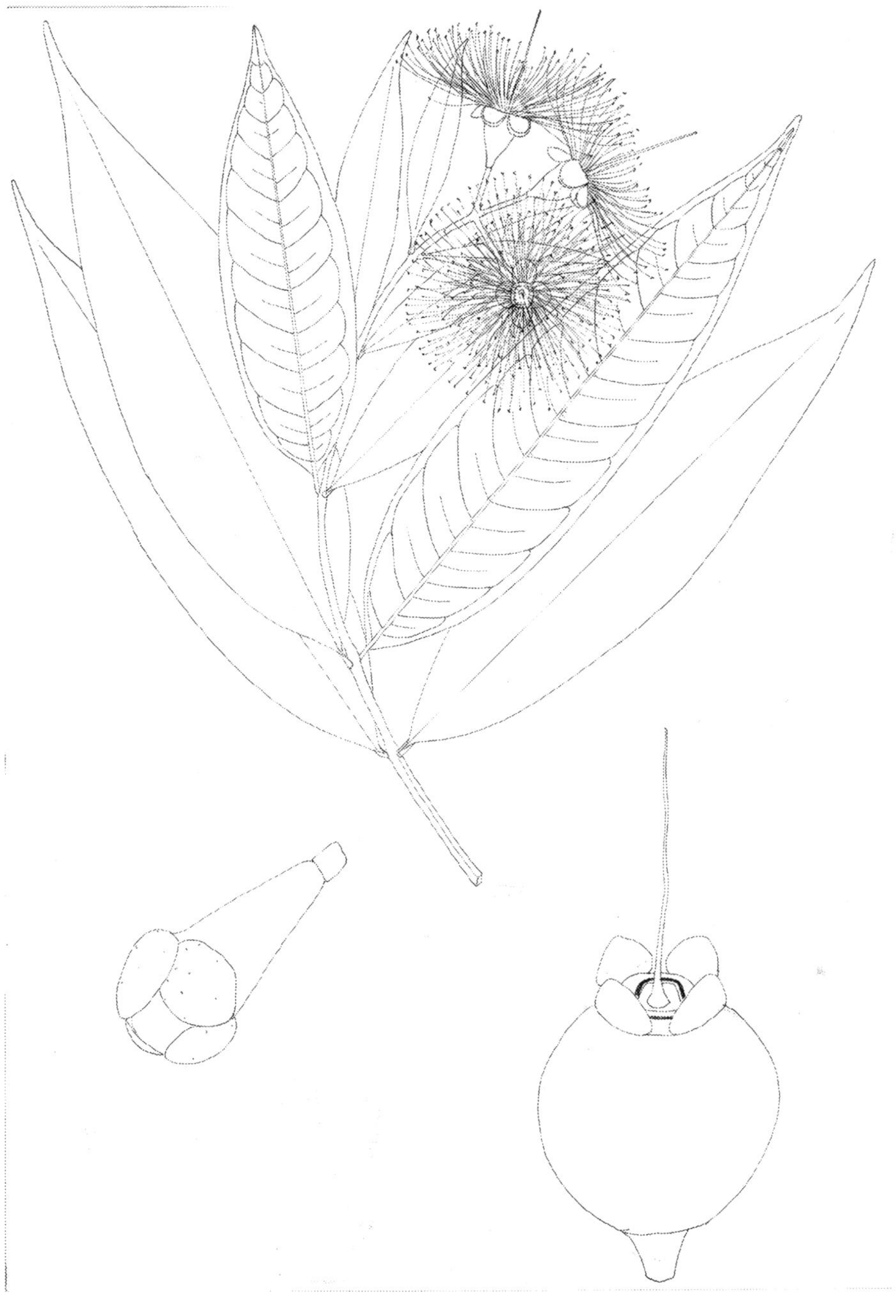

FIGURE 8.8 *Eugenia jambos*.L.

Shan, B., Cai, Y.Z., Brooks, J.D. and Corke, H., 2007. The in vitro antibacterial activity of dietary spice and medicinal herb extracts. *International Journal of Food Microbiology*, *117*(1), pp. 112–119.

Subarnas, A., Diantini, A., Abdulah, R., Zuhrotun, A., Hadisaputri, Y.E., Puspitasari, I.M., Yamazaki, C., Kuwano, H. and Koyama, H., 2015. Apoptosis induced in MCF-7 human breast cancer cells by 2′, 4′-dihydroxy-6-methoxy-3, 5-dimethylchalcone isolated from Eugenia aquea Burm f. leaves. *Oncology Letters*, *9*(5), pp. 2303–2306.

Wang, S.X., Wang, Y., Lu, Y.B., Li, J.Y., Song, Y.J., Nyamgerelt, M. and Wang, X.X., 2020. Diagnosis and treatment of novel coronavirus pneumonia based on the theory of traditional Chinese medicine. *Journal of Integrative Medicine*, *18*, pp. 275–283.

8.2.4.3 *Eugenia operculata* Roxb.

Common names: Bhumi jambu, jamawa (Bangladesh); kènw (China); ā totnopak, rai jam (India); alam banem (Indonesia); voi (Cambodia, Laos, Vietnam); konthabye (Myanmar); bhukijabu, kyamuna (Nepal); malaruhak (the Philippines); jaman (Sikkim); wa khao (Thailand)

Synonyms: *Cleistocalyx operculatus* (Roxb.) Merr. & L.M. Perry, *Eugenia cerasoides* Roxb. *Syzygium operculatum* (Roxb.) Nied., *Syzygium nervosum* DC.

Habitat: Forests

Distribution: India, Myanmar, Cambodia, Laos, Vietnam, Thailand, Malaysia, Indonesia, the Philippines, Australia.

Botanical examination: This timber tree that can reach 40 m tall. The bark is pale brown and exfoliating. The stem is smooth, pimply, glabrous, somewhat flattish, articulated, and angular at apex and with a longitudinal leaf scar. The leaves are simple, opposite or sub-opposite, and without stipules. The petiole is glabrous, channeled, and up to 1.5 cm long. The blade is somewhat spathulate to elliptic, 15–19 × 4–6 cm, dark green, coriaceous, cuneate at base, marked with about 10 pairs of secondary nerves and essential oil glands. The panicle is articulated, glabrous, and angled. The calyx buds open by dropping a lid. The flower is about 8 mm in diameter, white, the petals united to form a calyptra. The berries are 4 mm–1 cm in diameter, edible, whitish with pinkish hue, and marked with an apical disc.

Medicinal uses: Dysentery (India), bronchitis (Nepal), sores (China)

Strong broad-spectrum antibacterial essential oil: Essential oil of flower buds inhibited the growth of *Bacillus subtilis* (ATCC 6633), *Pseudomonas aeruginosa* (KCTC 2004), *Listeria monocytogenes* (ATCC 19166), *Staphylococcus aureus* (ATCC 6538), *Escherichia coli* (O157:H7 (ATCC 43888), *Enterobacter aerogenes* (KCTC 2190), methicillin-resistant *Staphylococcus aureus* (P227), vancomycin-resistant *Enterococcus faecium* (A93), multiantibiotic-resistant *Acinetobacter baumannii*, multiantibiotic-resistant *Staphylococcus aureus* (ATCC 25923), multiantibiotic-resistant *Escherichia coli* (02 K 276), multiantibiotic-resistant *Klebsiella pneumoniae* (05 K 279), multiantibiotic-resistant *Serratia marcescens* (03 K 201), and multiantibiotic-resistant *Pseudomonas aeruginosa* (ATCC 27853) (Dung et al., 2008). Essential oil of flower buds inhibited the growth of *Xanthomonas campestris* pv. *campestris* (KC94-17 (Xcc KC94-17), *Xanthomonas campestris* pv. *vesicatoria* (YK93–4 (Xcv YK93–4), *Xanthomonas oryzae* pv. *oryzae* (KX019 (Xoo KX019), and *Xanthomonas* sp. (SK12 (Xsp SK12) with MIC/MBC values of 31.2/62.5, 62.5/62.5, 31.2/62.5, and 125/250 µL/mL, respectively (Bajpai et al., 2010).

Strong antiviral (enveloped monopartite linear single-stranded (−)-RNA) phloroglucinols: Cleistocaltone A and B inhibited the replication of the Respiratory syncytial virus in HEp-2 cells with IC_{50} values of 6.7 and 2.8 µM, respectively (Song et al., 2019).

Strong antiviral (enveloped segmented linear single-stranded (−)-RNA) chalcone: 2′,4′-Dihydroxy-6′-methoxy-3′,5′-dimethylchalcone inhibited the cytopathic effect of the Influenza A virus (H1N2 A/PR/8/34) in Madin-Darby canine kidney cells at 20 µM (Ha et al., 2016). This chalcone at a concentration of 20 µM inhibited the replication the Influenza virus A virus (H1H1 A/PR/8/34) in Madin-Darby canine kidney cells cells by about 50% (Ha et al., 2016).

Viral enzyme inhibition by chalcone: 2′,4′-Dihydroxy-6′-methoxy-3′,5′-dimethylchalcone isolated from the leaves inhibited Influenza A viruses H1N1 A/PR/8/34, H9N2 A/Chicken/Korea/O1310/2001), H1N1 (WT), and oseltamivir-resistant novel H1N1 (H274Y) neuraminidase activities with IC_{50} values of 8.2, 5, 7, and 8.8 µM, respectively (Ha et al., 2016). This chalcone inhibited the avian Influenza virus (H9N2 A/Chicken/Korea/O1310/2001) and the swine Influenza virus (H1N1 A/PR/8/34) neuraminidase with IC_{50} values of 7 and 8.2 µM (Ha et al., 2016). (*E*)-4,2′,4′-trihydroxy-6′-methoxy-3′,5′-dimethylchalcone inhibited the Influenza A virus (H1N1) neuraminidase with an IC_{50} value of 8.1 µM (Dao et al., 2010).

Viral enzyme inhibition by flavonol glycoside: Myricetin-3′,5′-dimethylether 3-*O*-β-D-galactopyranoside isolated from the leaves inhibited the Influenza A (H1N1 A/PR/8/34), (H9N2 A/Chicken/Korea/O1310/2001), (H1N1 (WT), and oseltamivir-resistant novel (H1N1 (H274Y) neuraminidase activities with IC_{50} values of 8.6, 6.5, 7.1, and 9.3 μM, respectively (Ha et al., 2016). 2′,4′- Myricetin-3′,5′-dimethylether 3-*O*-β-D-galactopyranoside inhibited the swine Influenza virus (H1N1 A/PR/8/34) and the avian Influenza virus (H9N2 A/Chicken/Korea/O1310/2001) neuraminidase with IC_{50} values of 8.2 and 7.1 μM, respectively (Ha et al., 2016).

Strong anticandidal essential oil: Essential oil of flower buds inhibited the growth of *Candida albicans* (KCTC 7965) with MIC/MBC value of 5/5 μL/mL (Dung et al., 2008).

Commentaries: (i) The essential oil of flower buds contain among other things α-pinene, camphene, trans-carveol, cyclobazzanene, globulol, β-himachalol, and acorenol (Dung et al., 2008). The plant generates a bewildering array of flavonoids (Dao et al., 2010). (ii) This plant (if not toxic) could be of value for the prevention of Influenza virus pandemics. In his excellent work, Woodson (2005) relates to the fact that the Influenza experts at the US CDC and WHO are observing an increase in the human mortality rate of the Influenza A virus (H5N1) brewing in Asian poultries and pig farms up to about 50%. The same author also expresses the view that should intensive poultries and pig farms continue their abuses on animals, we are bound to expect a medical and social disaster with a temporary breakdown in food delivery, the electric and water utility service, and possibly even public order in major urban areas.

REFERENCES

Bajpai, V.K., Dung, N.T., Suh, H.J. and Kang, S.C., 2010. Antibacterial activity of essential oil and extracts of Cleistocalyx operculatus buds against the bacteria of Xanthomonas spp. *Journal of the American Oil Chemists' Society*, *87*(11), pp. 1341–1349.

Dao, T.T., Tung, B.T., Nguyen, P.H., Thuong, P.T., Yoo, S.S., Kim, E.H., Kim, S.K. and Oh, W.K., 2010. C-methylated flavonoids from Cleistocalyx operculatus and their inhibitory effects on novel Influenza A (H1N1) neuraminidase. *Journal of Natural Products*, *73*(10), pp. 1636–1642.

Dung, N.T., Kim, J.M. and Kang, S.C., 2008. Chemical composition, antimicrobial and antioxidant activities of the essential oil and the ethanol extract of Cleistocalyx operculatus (Roxb.) Merr and Perry buds. *Food and Chemical Toxicology*, *46*(12), pp. 3632–3639.

Ha, T.K.Q., Dao, T.T., Nguyen, N.H., Kim, J., Kim, E., Cho, T.O. and Oh, W.K., 2016. Antiviral phenolics from the leaves of Cleistocalyx operculatus. *Fitoterapia*, *110*, pp. 135–141.

Song, J.G., Su, J.C., Song, Q.Y., Huang, R.L., Tang, W., Hu, L.J., Huang, X.J., Jiang, R.W., Li, Y.L., Ye, W.C. and Wang, Y., 2019. Cleistocaltones A and B, antiviral phloroglucinol–terpenoid adducts from Cleistocalyx operculatus. *Organic Letters*, *21*(23), pp. 9579–9583.

Woodson, M.D. and FACP, G., 2005. Preparing for the Coming Influenza Pandemic. Edited by David Jodrey, PhD, 8.

8.2.4.4 *Melaleuca cajuputi* Roxb.

Synonyms: *Melaleuca minor* Sm. *Myrtus saligna* Burm.f., *Melaleuca leucadendron* (L.) L. var. *minor* (Smith) Duthie.

Common names: Bottle brush; cham, da ra bo dich (Cambodia, Laos, Vietnam); bai qian ceng (China); kaiyappudai (India); minyak kayu puteh (Indonesia), gelam, kayu puteh (Malaysia), loth-sumbul (Sri Lanka); samet khao (Thailand)

Habitat: Swampy grounds, peat swamps, near coasts

Distribution: India, Malaysia, Cambodia, Laos, Vietnam, Indonesia.

Botanical examination: This shrub probably native to Indonesia grows up to a height of 25 m. The bole is twisted. The bark is whitish, soft, fissured, and with papery flakes, which are coarse, elongate and shaggy. The stem is smooth, glabrous, terete, with horseshoe-shaped leaf scars, hairy and angled at apex. The internodes are about 1 cm long. The leaves are simple, sessile, spiral, and without stipules and have been used for several centuries for the manufacture of the cajuput oil. The blade is lanceolate to elliptic, coriaceous, greyish, 6–16.5 × 1.7– 2.6 cm, marked with 5–7 longitudinal nerves, aromatic, and acute at apex. The flower is sessile on a spike, which is velvety at first, and

7.5–15 cm long, overall and somewhat with the appearance of a bottle-brush. The calyx is velvety with 5 membranous lobes ciliate at margin. The 5 petals are deciduous, membranous, and globose. The stamens are numerous and conspicuous and about 1 cm long. The style is 1 cm long. The capsules are 2×4 mm, woody, rough, with 3 radial lodges at apex. The seeds are minute.

Medicinal uses: Wounds (Indonesia); mouth sores, sore throat, fever (Thailand)

Antimycobacterial essential oil: Essential oil inhibited the growth of *Mycobacterium abscessum*, *Mycobacterium avium*, *Mycobacterium gordonae*, and *Mycobacterium gordonae* with MIC values of 0.5, 2, 0.5, and 1%, respectively (Bua et al., 2020).

Commentaries: (i) Essential oil in the leaves contains 1,8-cineole and terpinen-4-ol (Yoon et al., 2003). 1,8-Cineole is a major constituent of the essential oil of *Myrtus communis* L., which is used in Turkey to treat cold (local name: *Mersin*) and used in Iran to treat oral ulcers (local name: *Murd*). The essential oil inhibited the growth of *Bacillus subtilis* (ATCC 6633), *Bacillus cereus* (ATCC 11778), *Staphylococcus aureus* (ATCC 6538), *Staphylococcus aureus* (ATCC 29213), and *Escherichia coli* (ATCC 35218) with MIC values of 1.4, 1.4, 2.8, 2.8, and 11.2 mg/mL, respectively (Rosato et al., 2007).

(ii) The neolignans melaleucin A, B, and C isolated from the leaves of *Melaleuca bracteata* F. Muell. inhibited methicillin-resistant *Staphylococcus aureus* (JSCS 4788) with MIC values of 8, 3, and 16 μg/mL, respectively (Li et al., 2017).

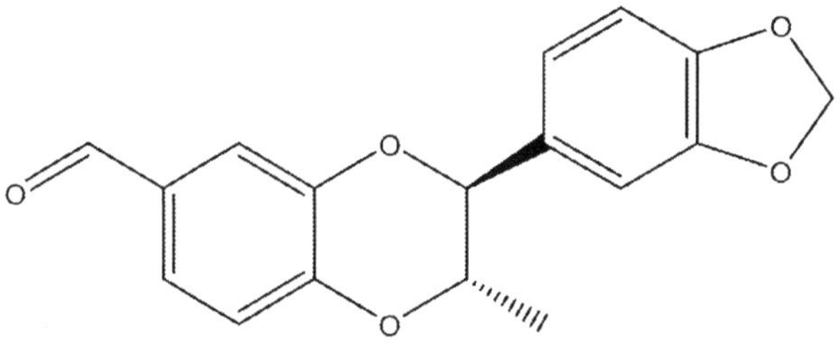

Melaleucin B

(iii) Essential oil of *Melaleuca alternifolia* Cheel. inhibited the growth of *Staphylococcus aureus* (ATCC 25923), *Klebsiella pneumoniae* (NCTC 9633), and *Candida albicans* (ATCC 10231) with MIC values of 6.7, 6, and 6 μg/mL, respectively (Van Vuuren et al., 2009).

REFERENCES

Bua, A., Molicotti, P., Donadu, M.G., Usai, D., Le, L.S., Tran, T.T.T., Ngo, V.Q.T., Marchetti, M., Usai, M., Cappuccinelli, P. and Zanetti, S., 2020. "In vitro" activity of Melaleuca cajuputi against mycobacterial species. *Natural Product Research*, *34*(10), pp. 1494–1497.

Li, C., Liu, H., Zhao, L., Zhang, W., Qiu, S., Yang, X. and Tan, H., 2017. Antibacterial neolignans from the leaves of Melaleuca bracteata. *Fitoterapia*, *120*, pp. 171–176.

Manasa, M., Kambar, Y., Vivek, M.N., Kumar, R.T.N. and Kekuda, P.T.R., 2013. Antibacterial efficacy of Pimenta dioica (Linn.) Merill and Anacardium occidentale L. against drug resistant urinary tract pathogens. *Journal of Applied Pharmaceutical Science*, *3*(12), p. 72.

Van Vuuren, S.F., Suliman, S. and Viljoen, A.M., 2009. The antimicrobial activity of four commercial essential oils in combination with conventional antimicrobials. *Letters in Applied Microbiology*, *48*(4), pp. 440–446.

8.2.4.5 *Psidium guajava* L.

Synonyms: *Guajava pyrifera* (L.) Kuntze, *Myrtus guajava* (L.) Kuntze, *Psidium guava* Griseb., *Psidium guayava* Raddi, *Psidium igatemyensis* Barb. Rodr., *Psidium pomiferum* L., *Psidium pumilum* Vahl, *Psidium pyriferum* L.

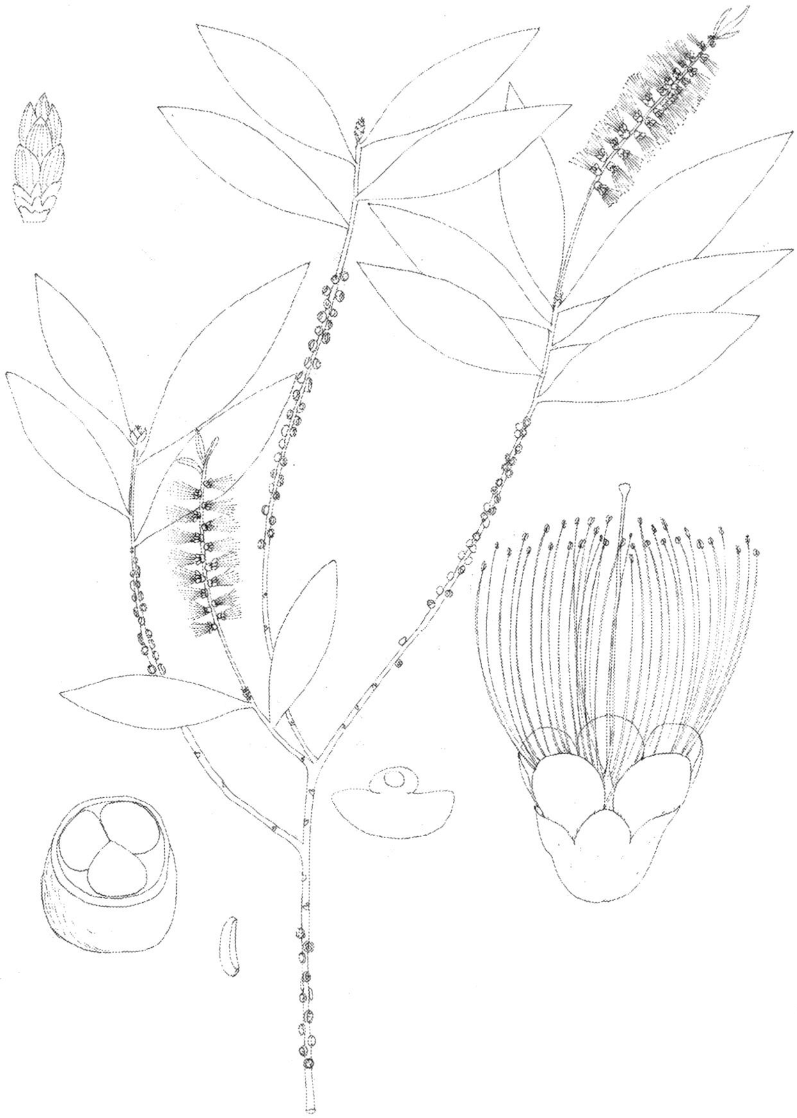

FIGURE 8.9 *Melaleuca cajuputi*.Roxb.

Common names: Guava tree; piyara (Bangladesh); trabek srok (Cambodia); mù guó, shí liu (China); koyya (India); amarooda, amrood (India); jambu bidi (Indonesia); banjiro (Japan); si dal (Laos); jambu biji (Malaysia); malaka (Myanmar); bayabas (the Philippines); amrud (Pakistan); gay oi (Vietnam); pera (Sri Lanka); farang (Thailand)

Habitat: Cultivated

Distribution: Tropical Asia and Asia

Botanical examination: This large shrub native to tropical America is cultivated as a fruit tree. The bole is not straight. The bark is smooth. The stem is quadrangular, winged, smooth, dotted with oil cells, and with some hairs at apex. The internodes are 3–3.5 cm long. The leaves are simple, opposite, and without stipules. The petiole is channeled, velvety, and up to about 8 mm long. The blade is elliptic, 6–13 × 4–6.5 cm, coriaceous, marked with about 16–23 pairs of secondary nerves and oil glands. The flower is solitary, showy, axillary, and terminal. The calyx is up to about 8 mm long with 5 sepals, which are triangular, velvety, and up to about 7 mm long. The 5 petals are pilose with oil cells, up to 2 cm long, and white. The androecium comprises numerous stamens. The style is 1 cm long with a globose stigma. The berries are pyriform or globose, hard, heavy, greenish, with somewhat a bumpy surface, edible, up to about 15cm in diameter, and crowned at apex by vestiges of calyx lobes.

Medicinal uses: Dysentery (Bangladesh); mouthwash, wounds (the Philippines); cold, cough, diarrhea, sores in the mouth (India); diarrhea (China; India)

Broad-spectrum antibacterial essential oil: Essential oil inhibited the growth of *Streptococcus mutans* with an MIC value of 0.4 µg (Miller et al., 2015).

Strong antibacterial polar extract: Ethanol extract inhibited the growth of 35 clinical isolates of methicillin-resistant *Staphylococcus aureus* and *Staphylococcus aureus* with MIC/MBC values of 0.2–1.6/6.3 and 0.2 µg/mL /1.6 µg/mL, respectively (Voravuthikunchai & Kitpipit, 2005).

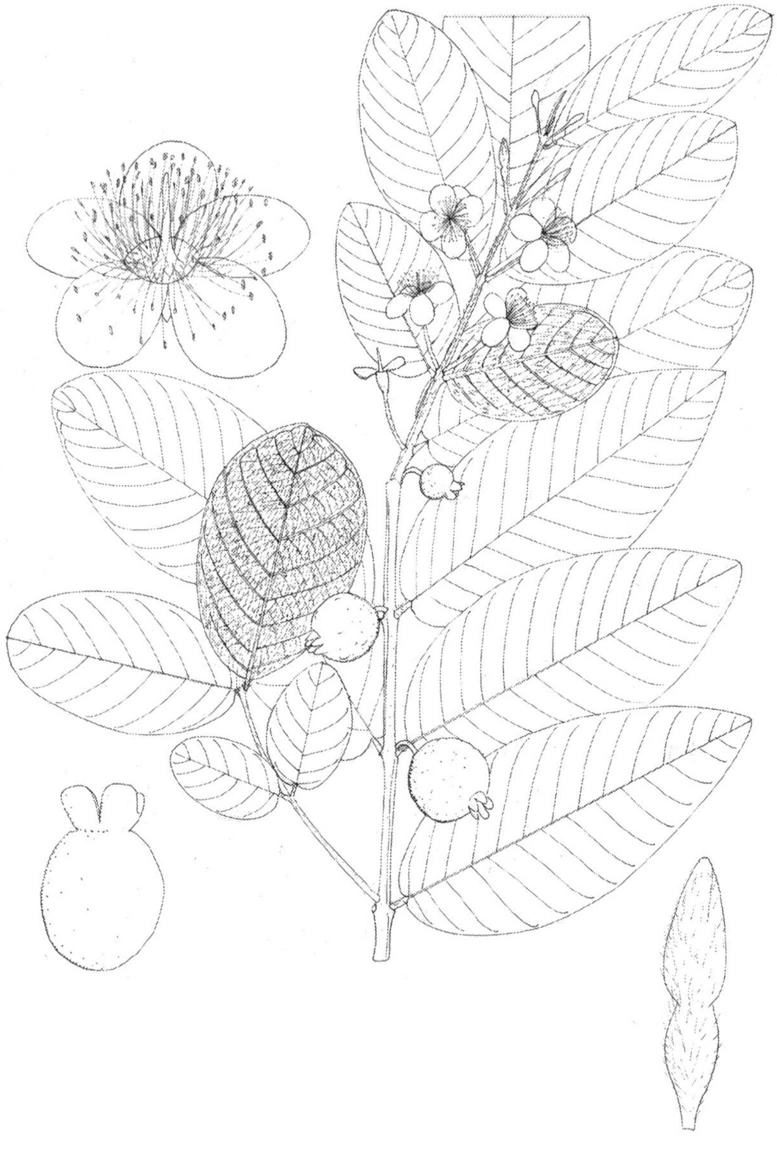

FIGURE 8.10 *Psidium guajava*.L.

FIGURE 8.11 *Rhodomyrtus tomentosa.* (Aiton) Hassk.

Strong antiviral (enveloped segmented linear single-stranded (−)-RNA) polar extract: Aqueous extract of leaves inhibited the replication of the Influenza A (A/Narita/1/2009 (amantadine-resistant pandemic 2009 strain) virus in Madin-Darby canine kidney cells with an IC_{50} of 0.05% (Sriwilaijaroen et al., 2012).

Very weak anticandidal polar extract: Hydroethanolic extract of leaves inhibited the growth of *Candida albicans* (INCQS 40006), *Candida albicans* (LM77), *Candida tropicalis* (INCQ 40042), and *Candida tropicalis* (LM 23) with IC_{50} values of 1803, 1905.4, 1862, and 1905.4 μg/mL, respectively (Morais-Braga et al., 2017).

Commentaries: (i) The phloroglucinol myrciarone A isolated from the peels of a member of the genus *Psidium* L. inhibited the growth of *Bacillus subtilis*, *Bacillus cereus*, *Micrococcus luteus*,

Staphylococcus aureus, *Staphylococcus epidermidis*, and *Streptococcus mutans* with MIC values of 1.5, 1.5, 1.5, 1.5, 3.1, and 3.1 µg/mL, respectively (Kaneshima et al., 2017).

The phloroglucinol rhodomyrtone from this plant, inhibited *Bacillus subtilis*, *Bacillus cereus*, *Micrococcus luteus*, *Staphylococcus aureus*, *Staphylococcus epidermidis*, and *Streptococcus mutans* with MIC values of 0.7, 0.7, 0.7, 0.7, 0.7, and 1.5 µg/mL, respectively (Kaneshima et al., 2017).

The phloroglucinol isomyrtucommulone B from this plant, inhibited *Bacillus subtilis*, *Bacillus cereus*, *Staphylococcus aureus*, *Staphylococcus epidermidis*, and *Streptococcus mutans* with MIC values of 1.5, 0.7, 1.5, 6.2, and 3.1 µg/mL, respectively (Kaneshima et al., 2017).

The phloroglucinol myrcianone B from this plant inhibited *Bacillus subtilis*, *Bacillus cereus*, *Staphylococcus aureus*, *Staphylococcus epidermidis*, and *Streptococcus mutans* with MIC values of 1.5, 1.5, 1.5, 3.1, and 1.5 µg/mL, respectively (Kaneshima et al., 2017).

(ii) In Thailand, *Rhodomyrtus tomentosa* (Aiton) Hassk. (local name: *Pha rak dam*) is used to treat diarrhea, and ethanol extract of stems and leaves of this plant inhibited the growth *Acinetobacter baumanii, Bacillus cereus, Pseudomonas aeruginosa, Shigella flexneri* and *Staphylococcus aureus*, with MIC/MBC of 250/250, 500/1000, 125/250, 250/500, and 250/1000 µg/mL, respectively (Limsuwan et al., 2009). Rhodomyrtone isolated from the leaves of this plant inhibited the growth of *Bacillus cereus*, *Bacillus subtilis*, *Enterococcus faecalis*, methicillin-resistant *Staphylococcus aureus*, *Staphylococcus aureus* (ATCC 25923), *Staphylococcus epidermidis* (ATCC 35984), *Streptococcus gordonii*, *Streptococcus mutans*, *Streptococcus pneumoniae* (capsule positive), *Streptococcus pyogenes*, and *Streptococcus salivarius* (Limsuwan et al., 2009). Interestingly, rhodomyrtone inhibited the replication the Severe acute respiratory syndrome-acute coronavirus in Vero cells with an EC_{50} value of 0.4 µM (Tang et al., 2020), thus one could examine the anti-COVID 19 and other zoonotic coronaviruses potential of *Psidium guajana* L. tea leaves.

REFERENCES

Kaneshima, T., Myoda, T., Toeda, K., Fujimori, T. and Nishizawa, M., 2017. Antimicrobial constituents of peel and seeds of camu-camu (Myrciaria dubia). *Bioscience, Biotechnology, and Biochemistry*, *81*(8), pp. 1461–1465.

Miller, A.B., Cates, R.G., Lawrence, M., Soria, J.A.F., Espinoza, L.V., Martinez, J.V. and Arbizú, D.A., 2015. The antibacterial and antifungal activity of essential oils extracted from Guatemalan medicinal plants. *Pharmaceutical Biology*, *53*(4), pp. 548–554.

Morais-Braga, M.F., Carneiro, J.N., Machado, A.J., Sales, D.L., dos Santos, A.T., Boligon, A.A., Athayde, M.L., Menezes, I.R., Souza, D.S., Costa, J.G. and Coutinho, H.D., 2017. Phenolic composition and medicinal usage of Psidium guajava Linn.: Antifungal activity or inhibition of virulence? *Saudi journal of Biological Sciences*, *24*(2), pp. 302–313.

Sriwilaijaroen, N., Fukumoto, S., Kumagai, K., Hiramatsu, H., Odagiri, T., Tashiro, M. and Suzuki, Y., 2012. Antiviral effects of Psidium guajava Linn.(guava) tea on the growth of clinical isolated H1N1 viruses: Its role in viral hemagglutination and neuraminidase inhibition. *Antiviral Research*, *94*(2), pp. 139–146.

Tang, W., Lu, J., Song, Q.Y., Li, M.M., Chen, L.F., Hu, L.J., Yang, S.M., Zhang, D.M., Wang, Y., Li, Y.L. and Ye, W.C., 2020. Discovery of rhodomyrtone as a broad-spectrum antiviral inhibitor with anti-SARS-CoV-2 activity. *bioRxiv*.

Voravuthikunchai, S.P. and Kitpipit, L., 2005. Activity of medicinal plant extracts against hospital isolates of methicillin-resistant Staphylococcus aureus. *Clinical Microbiology and Infection*, *11*(6), pp. 510–512.

8.2.4.6 *Syzygium cumini* (L.) Skeels

Synonyms: *Calyptranthes oneillii* Lundell; *Eugenia cumini* (L.) Druce; *Eugenia jambolana* Lam.; *Myrtus cumini* L.; *Syzygium jambolanum* (Lam.) DC.

Common name: Jambolan, Malabar plum; jaam (Bangladesh); pring bai (Cambodia); ōng gún non,wu mo (China); jamun, jaamoon (India); jamblang (Indonesia); madan (Japan); sa leng, va (Laos); jambulan (Malaysia); jamun (Nepal); jaman (Pakistan); duhat (the Philippines); wa (Thailand); voi rung (Vietnam)

Habitat: Cultivated, forests

Distribution: India to Southeast Asia and Australia

Botanical examination: This magnificent tree grows up to about 20 m tall. The stems are terete. The leaves are simple, opposite, and exstipulate. The petiole is up to 2 cm long. The blade is broadly elliptic, 6–12 × 3.5–7.5 cm, coriaceous, marked with numerous inconspicuous secondary nerves, cuneate at base and rounded, obtuse, or acuminate at apex. The cymes are axillary and terminal. A hypanthium is present. The calyx comprises 4 tiny lobes. The 4 petals are white to purplish and minute. The androecium includes numerous stamens, which are about 5 mm long. The gynoecium comprises a 5 mm long style. The drupes are ovoid, edible, dark purple to black, juicy, glossy, and up to 2 cm long.

Medicinal uses: Bronchitis (Bangladesh; Nepal), diarrhea, tooth infection (Bangladesh); dysentery (Bangladesh, India, Laos, Vietnam); oral ulcers (India); wounds (Bangladesh; Pakistan);sore throat (Nepal)

Antibacterial essential oil (Shafi et al., 2002).

Strong broad-spectrum antibacterial polar extract: Phenolic fraction of fruits inhibited the growth of *Bacillus subtilis* (clinical isolate), *Staphylococcus aureus* (clinical isolate), *Staphylococcus epidermidis* (clinical isolate), and *Klebsiella pneumoniae* (clinical isolate) with MIC values of 65.2, 31.2, 31.2, and 31.2 μg/mL, respectively

(Siddiqi et al., 2011). Hydroalcoholic extract of leaves inhibited the growth of *Escherichia coli* (ATCC 14948), *Enterococcus faecalis* (ATCC 19433), multiantibiotic-resistant *Klebsiella pneumoniae*, *Kocuria rhizophila* (ATCC 9341), *Neisseria gonorrhoeae* (ATCC 49226), multi-antibiotic resistant *Pseudomonas aeruginosa*, *Staphylococcus aureus* (ATCC 25923), and *Shigella flexneri* (ATCC 12022) (Siddiqi et al., 2011).

Antibacterial (Gram-positive) halo developed by sesquiterpenes: Sootepdienone, guaianediol, caryolandiol, and clovane-2β,9α-diol isolated from the seeds inhibited the growth of *Staphylococcus aureus* subsp. *aureus* (CGMCC 1.2386) with inhibition zone diameters of 9, 9, 9, and 10 mm, respectively (100 μg/disc) (Liu et al., 2017).

Strong anticandidal polar extract: Hydroalcoholic extract of leaves inhibited the growth of *Candida albicans* (ATCC 10231) and *Candida krusei* (ATCC 6258) with MIC values of 90 and 70 μg/mL, respectively (Oliveira et al., 2007).

Moderate anticandidal polar extract: Hydroalcoholic extract inhibited the growth of *Candida albicans* (ATCC 10231) with an MIC value of 125 μg/mL and an MBC of 500 μg/mL (Pereira et al., 2016).

Antifungal polar extract: Aqueous extract of root bark inhibited the growth of *Scochyta rabiei* (Pass.) Lab. (Jabeen & Javaid, 2010)

Commentaries: (i) Polar extract of leaves of *Syzygium antisepticum* (Blume) Merr. & L.M. Perry inhibited the growth of *Staphylococcus aureus* (ATCC 6538) and methicillin-resistant *Staphylococcus aureus* (ATCC 33591) with MIC/MBC values of 0.1/0.5 and 0.1/0.5 mg/mL, respectively (Yuan & Yuk, 2018). From this extract β-caryophyllene elicited bacterial membrane damages (Yuan & Yuk, 2018).

(ii) *Syzygium nervosum* DC. is used in India to treat dysentery. *Syzygium bullockii* (Hance) Merr. et Perry is used in China (Local name: *Hēi zǔi duo*) to treat fever, and *Syzygium hancei* Merr.et Perry is used to treat diarrhea (local name: *Chì nán*).

(iii) Flavone isolated from a member of the genus *Acca* O.Berg inhibited the growth of *Saccharomyces cerevisiae*, *Candida albicans* (ATCC 10231), *Candida glabrata* (ATCC 90030), *Candida parapsilosis* (ATCC 90018), and *Candida tropicalis* (ATCC 13803) with MIC values of 2.3, 2.1, 2.6, 1.5, and 1.8 μM, respectively (Mokhtari et al., 2018). 4-Cyclopentene-1,3-dione from a member of the *Acca* O.Berg inhibited the growth of *Saccharomyces cerevisiae*, *Candida albicans* (ATCC 10231), *Candida glabrata* (ATCC 90030), *Candida parapsilosis* (ATCC 90018), and *Candida tropicalis* (ATCC 13803) with MIC values of 1.3, 2.3, 2.1, 2, and 1.3 μM, respectively, via inhibition of the synthesis of fungal chitin (Mokhtari et al., 2018).

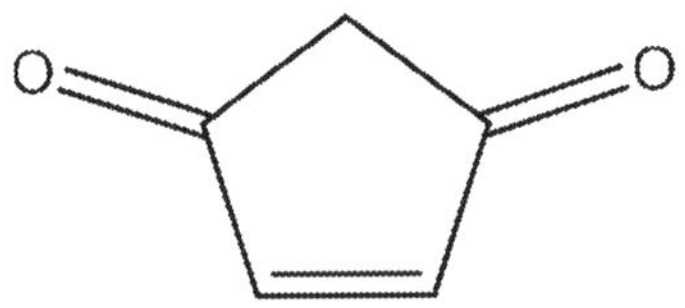

4-Cyclopentene-1,3-dione

β-Caryophyllene from a member of the *Acca* O.Berg inhibited the growth *Saccharomyces cerevisiae*, *Candida albicans* (ATCC 10231), *Candida glabrata* (ATCC 90030), *Candida parapsilosis* (ATCC 90018), and *Candida tropicalis* (ATCC 13803) with MIC values of 35.8, 53, 42.5, 42.5, and 75.2 μM, respectively (Mokhtari, et al., 2018).

(E)-3-hexenyl-butyrate from a member of the *Acca* O.Berg inhibited the growth of *Saccharomyces cerevisiae*, *Candida albicans* (ATCC 10231), *Candida glabrata* (ATCC 90030), *Candida parapsilosis* (ATCC 90018), and *Candida tropicalis* (ATCC 13803) with MIC values of 22.7, 24.5 28.3, 38.2, and 16 μM, respectively (Mokhtari et al., 2018).

(iv) Callistemenonone A isolated from the leaves of a member of the genus *Callistemon* Roxb. inhibited the growth of *Staphylococcus aureus* (CMCC 26003), *Bacillus cereus* (CMCC 63302), methicillin-resistant *Staphylococcus aureus* (JCSC 2172), methicillin-resistant *Staphylococcus aureus* (JCSC 2172), methicillin-resistant *Staphylococcus aureus* (JSC 4744), and *Escherichia coli* (8739) with MIC/MBC values of 20/40, 5/20, 20/40, 40/80, 40/80, and >400 μg/mL, respectively, and was bactericidal for methicillin-resistant *Staphylococcus aureus* (JCSC 2172), without causing cytoplasmic membrane disruption but disturbing membrane potential (Xiang et al., 2017).

(v) Polar extract of leaves of *Xanthostemon verticillatus* (C.T. White & W.D. Francis) L.S. Sm. inhibited the growth of *Staphylococcus aureus* (ATCC 6538) and methicillin-resistant *Staphylococcus aureus* (ATCC 33591) (Yuan & Yuk, 2018).

(vi) *Pimenta dioica* (L.) Merr. is used in Bangladesh to treat oral infections (Local name: *Kaabaab chini*). Essential oil of this plant inhibited the growth of *Fusarium oxysporum*, *Fusarium verticilloides*, *Penicillium brevicompatum*, *Penicillium expansum*, *Aspergillus flavus*, and *Aspergillus fumigatus* by with MIC values of 0.5, 0.6, 0.7, 0.6, 0.6, and 0.6 μg/μL, respectively (Zabka et al., 2009). From this plant, the ellagitannin pedunculagin inhibited the growth of *Staphylococcus epidermidis*, *Staphylococcus aureus*, *Bacillus cereus*, *Proteus mirabilis*, *Klebsiella pneumonia*, *Acinetobacter baumannii*, and *Pseudomonas aeruginosa* with the MIC values of 1250, 1250, 2500, 625, 2500, 2500, and 620 μg/mL, respectively (Al-Harbi et al., 2018).

(vii) Methanol extract of leaves of *Eucalyptus camaldulensis* Dehnh. inhibited *Dickeya solani*, *Staphylococcus aureus*, *Penicillium funiculosum*, and *Penicillium ochrochloron* with MIC values of 70, 200, 100, and 200 μg/mL (Elansary et al., 2017). *Eucalyptus globulus* Labill. is used in Bangladesh for the treatment of influenza and tuberculosis. Essential oil of *Eucalyptus globulus*

Labill. inhibited the growth of *Staphylococcus aureus* and *Streptococcus mutans* with MIC values of 5 and 2.5 µg/mL (Miller et al., 2015). Essential oil of *Eucalyptus globulus* Labill. (used in India for cold (local name: *Safeda*) and in Pakistan to heal wounds (local name: *Gond)* inhibited the growth of *Lactophilus acidophilus* with an MIC value 1.2 µg/mL (Miller et al., 2015). From this essential oil, eucalyptol inhibited the growth of *Streptococcus mutans* (ATCC 25175) with an MIC value 500 µg/mL (Park et al., 2003) as well as *Bacillus subtilis* (ATCC 6633), *Bacillus cereus* (ATCC 11778), *Staphylococcus aureus* (ATCC 6538), *Staphylococcus aureus* (ATCC 29213), and *Escherichia coli* (ATCC 35218) with MIC values of 5600, 2800, 2800, 2800, and 2800 µg/mL, respectively (Rosato et al., 2007). In Pakistan, *Eucalyptus obliqua* L. is used to treat skin infections (local name: *Safeeda*). In China, *Eucalyptus tereticornis* Smith is used to treat cold.

(viii) *Syzygium anacardii-folium* (Crai) Chantaran. & J. Parn is used in India to treat diarrhea (local name: *Jamun*) and *Syzygium praecox* (Roxb.) Rathakr. & N.C.Nair is used to treat diarrhea, dysentery, respiratory infection, and jaundice.

REFERENCES

Al-Harbi, R., Shaaban, M., Al-Wegaisi, R., Moharram, F., El-Rahman, O.A. and El-Messery, S., 2018. Antimicrobial activity and molecular docking of tannins from pimenta dioica. *Letters in Drug Design & Discovery*, *15*(5), pp. 508–515.

Elansary, H.O., Salem, M.Z., Ashmawy, N.A., Yessoufou, K. and El-Settawy, A.A., 2017. In vitro antibacterial, antifungal and antioxidant activities of Eucalyptus spp. leaf extracts related to phenolic composition. *Natural Product Research*, *31*(24), pp. 2927–2930.

Jabeen, K. and Javaid, A., 2010. Antifungal activity of Syzygium cumini against Ascochyta rabiei–the cause of chickpea blight. *Natural Product Research*, *24*(12), pp. 1158–1167.

Limsuwan, S., Subhadhirasakul, S. and Voravuthikunchai, S.P., 2009. Medicinal plants with significant activity against important pathogenic bacteria. *Pharmaceutical Biology*, *47*(8), pp. 683–689.

Liu, F., Liu, C., Liu, W., Ding, Z., Ma, H., Seeram, N.P., Xu, L., Mu, Y., Huang, X. and Li, L., 2017. New Sesquiterpenoids from Eugenia jambolana Seeds and Their Anti-microbial Activities. *Journal of Agricultural and Food Chemistry*, *65*(47), pp. 10214–10222.

Miller, A.B., Cates, R.G., Lawrence, M., Soria, J.A.F., Espinoza, L.V., Martinez, J.V. and Arbizú, D.A., 2015. The antibacterial and antifungal activity of essential oils extracted from Guatemalan medicinal plants. *Pharmaceutical Biology*, *53*(4), pp. 548–554.

Mokhtari, M., Jackson, M.D., Brown, A.S., Ackerley, D.F., Ritson, N.J., Keyzers, R.A. and Munkacsi, A.B., 2018. Bioactivity-guided metabolite profiling of feijoa (Acca sellowiana) cultivars identifies 4-cyclopentene-1, 3-dione as a potent antifungal inhibitor of chitin synthesis. *Journal of Agricultural and Food Chemistry*, *66*(22), pp. 5531–5539.

Oliveira, G.F.D., Furtado, N.A.J.C., Silva Filho, A.A.D., Martins, C.H.G., Bastos, J.K. and Cunha, W.R., 2007. Antimicrobial activity of *Syzygium cumini* leaves extract. *Brazilian Journal of Microbiology*, *38*(2), pp. 381–384.

Park, K.M., You, J.S., Lee, H.Y., Baek, N.I. and Hwang, J.K., 2003. Kuwanon G: an antibacterial agent from the root bark of Morus alba against oral pathogens. *Journal of Ethnopharmacology*, *84*(2–3), pp. 181–185.

Pereira, J.V., Freires, I.A., Castilho, A.R., da Cunha, M.G., Alves, H.D.S. and Rosalen, P.L., 2016. Antifungal potential of Sideroxylon obtusifolium and Syzygium cumini and their mode of action against Candida albicans. *Pharmaceutical Biology*, *54*(10), pp. 2312–2319.

Rosato, A., Vitali, C., De Laurentis, N., Armenise, D. and Milillo, M.A., 2007. Antibacterial effect of some essential oils administered alone or in combination with Norfloxacin. *Phytomedicine*, *14*(11), pp. 727–732.

Shafi, P.M., Rosamma, M.K., Jamil, K. and Reddy, P.S., 2002. Antibacterial activity of Syzygium cumini and Syzygium travancoricum leaf essential oils. *Fitoterapia*, *73*(5), pp. 414–416.

Siddiqi, R., Naz, S., Ahmad, S. and Sayeed, S.A., 2011. Antimicrobial activity of the polyphenolic fractions derived from Grewia asiatica, Eugenia jambolana and Carissa carandas. *International Journal of Food Science & Technology*, *46*(2), pp. 250–256.

Xiang, Y.Q., Liu, H.X., Zhao, L.Y., Xu, Z.F., Tan, H.B. and Qiu, S.X., 2017. Callistemenonone A, a novel dearomatic dibenzofuran-type acylphloroglucinol with antimicrobial activity from Callistemon viminalis. *Scientific Reports*, *7*(1), p. 2363.

Yuan, W. and Yuk, H.G., 2018. Antimicrobial efficacy of Syzygium antisepticum plant extract against Staphylococcus aureus and methicillin-resistant *Staphylococcus aureus* and its application potential with cooked chicken. *Food Microbiology*, *72*, pp. 176–184.
Zabka, M., Pavela, R. and Slezakova, L., 2009. Antifungal effect of Pimenta dioica essential oil against dangerous pathogenic and toxinogenic fungi. *Industrial Crops and Products*, *30*(2), pp. 250–253.

8.2.5 Family Onagraceae A. L de Jussieu (1789)

The family Onagraceae consists of 17 genera and 675 species of herbs and shrubs. The leaves are simple, opposite, or alternate and without stipules. The flowers are often solitary and yellow. The calyx is 4-lobed, the lobes valvate. The corolla consists of 4 petals. The androecium is made of 4–8 stamens with 2-locular stamens opening lengthwise. The gynoecium comprises 4 carpels fused into a 4-locular ovary, each locule containing several ovules attached to axil or parietal placenta. The fruits are capsules, berries, or nuts containing numerous seeds. Members in this family produce ellagitannins and flavonoids.

8.2.5.1 *Jussiaea repens* L.

Synonyms: *Jussiaea adscendens* L., *Jussiaea patibilcensis* Kunth., *Ludwigia adscendens* (L.) Hara, *Ludwigia clavellina* M. Gomez

Common names: Water primrose; Mulcha (Bangladesh); shui long (China), malencho, nir charamdu (India); buang buang (Indonesia), inai pasir, sakot sumbu (Malaysia), ye-ka-nyut (Myanmar); jadelo (Nepal); agidahano (Papua New Guinea), sigang-dagat (the Philippines); phak pot nam (Thailand)

Habitat: Slow rivers, ponds, lakes

Distribution: India, Nepal, Pakistan, Sri Lanka, Thailand, Malaysia, Indonesia, the Philippines, Papua New Guinea, and Australia.

Botanical examination: This aquatic herb can reach 3 m long. The stem is dark red purplish, fleshy, and rooting at nodes. The roots are long and fibrous. The leaves are simple, spiral, edible (Andamans), sessile-like and without stipules. The leaf blade is obovate, spathulate, tapering at base, round at apex, 1.5–1.7 cm × 5–7 mm, with 6–12 pairs of secondary nerves. The midrib is sunken above and prominent underneath. The flower is axillary. The hypanthium is dark red purplish, slender, and quadrangular. The calyx comprises 5 green sepals, which are lanceolate and about 0.8 cm long. The 5 petals are cream, yellow at base, obovate, nerved, emarginated, and about 1 cm × 8 mm. The androecium consists of 10 short stamens. The capsules are elongated, about 2 cm long, with dark brown ribs, and contain several tiny seeds.

Medicinal uses: Dysentery (Thailand); wounds (Myanmar)

Broad-spectrum antibacterial halo developed by polar extract: Methanol extract inhibited the growth of *Bacillus subtilis*, *Escherichia coli*, *Staphylococcus aureus*, and *Pseudomonas aeruginosa* (Lwin, 2019).

Broad-spectrum antifungal halo developed by polar extract: Methanol extract inhibited the growth of *Aspergillus flavus* and *Candida albicans* (Lwin, 2019).

Commentaries: (i) The plant abounds with flavonoids: trifolin 2″-*O*-gallate, quercetrin, guaijaverin, reynoutrin, juglanin, avicularin, hyperin, trifolin, hyperin 2″-*O*-gallate, rutin, kaempferol, and quercetin (Marzouk et al., 2007). These flavonoids may confer synergistically a mild astringent effect explaining the plant's traditional use. (ii) Consider that extract of *Jussiaea suffruticosa* L.and *Ludwigia perennis* L. inhibited the replication of the Newcastle disease (also known as the Ranikhet disease) virus (Babbar et al., 1982). Natural products inhibiting the replication of enveloped monopartite linear single-stranded (–)-RNA virus are not uncommonly inhibitors of coronaviruses.

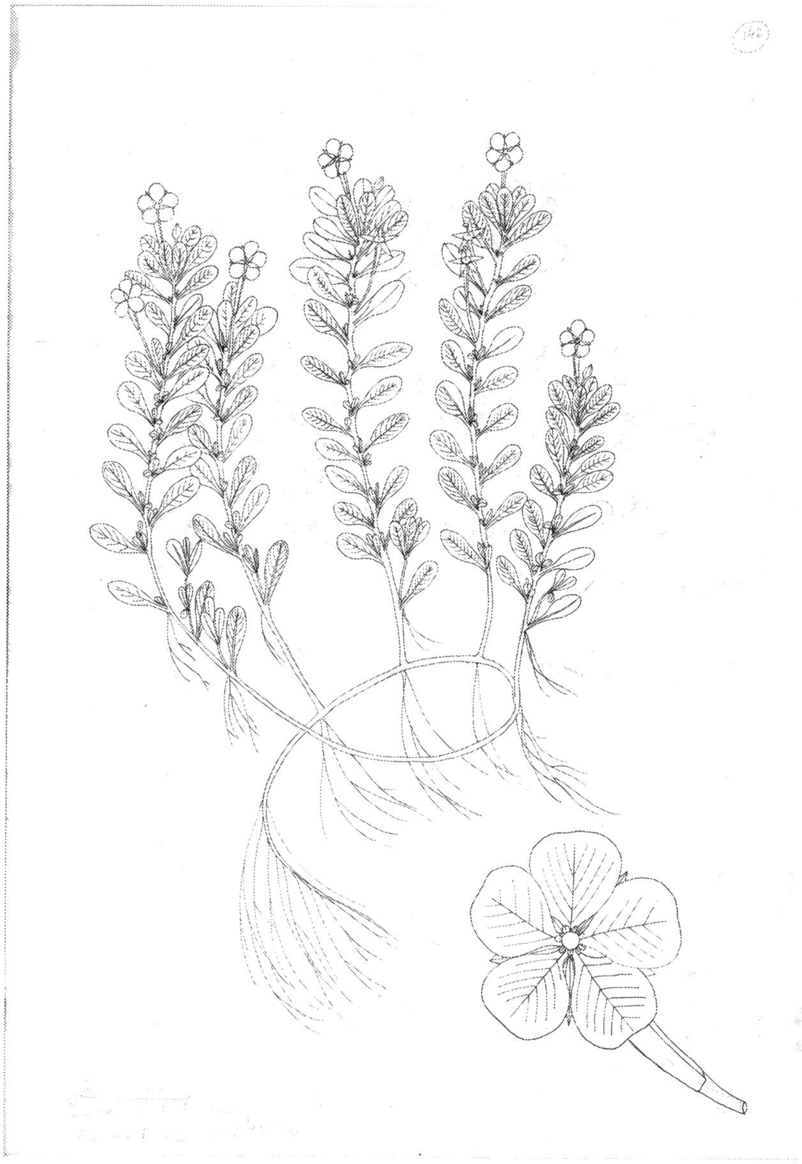

FIGURE 8.12 *Jussiaea repens*.L.

REFERENCES

Babbar, O.P., Joshi, M.N. and Madan, A.R., 1982. Evaluation of plants for antiviral activity. *Indian Journal of Medical Research*, *76*(S), pp. 54–65.

Lwin, H.H., 2019. Investigation of morphological characterization, qualitative analysis and antimicrobial activities of Jussiaea repens L. *Dagon University Commemoration of 25th Anniversary Silver Jubilee Research Journal*, *9*, pp. 273–279.

Marzouk, M.S., Soliman, F.M., Shehata, I.A., Rabee, M. and Fawzy, G.A., 2007. Flavonoids and biological activities of Jussiaea repens. *Natural Product Research*, *21*(5), pp. 436–443.

Singh, S., Singh, D.R., Singh, L.B., Chand, S. and Dam, R.S., 2013. Indigenous vegetables for food and nutritional security in Andaman and Nicobar Islands, India. *International Journal of Agriculture and Food Science Technology*, *4*(5), pp. 503–512.

8.2.5.2 *Ludwigia octovalvis* (Jacq.) Raven

Synonyms: *Jussiaea angustifolia* Lam., *Jussiaea didymosperma* H. Pierrer, *Jussiaea ligustrifolia* Kunth., *Jussiaea linearis* Hochst, *Jussiaea macropoda* C. Presl., *Jussiaea octovalvis* (Jacq.) Sw., *Jussiaea pubescens* L., *Jussiaea suffruticosa* L., *Jussiaea villosa* Lam., *Ludwigia pubescens* L. Hara, *Ludwigia suffruticosa* (L.) M. Gomez, *Oenothera octovalvis* Jacq.

Common names: Primrose willow, willow herb; ban lavanga (Bangladesh); mao cao long (China); lakom air (Malaysia); bhulvananga, dai ichak (India); cacabean (Indonesia), tayilakton (the Philippines), thian nam (Thailand).

Habitat: Marshes, river banks, around lakes, ponds, drain in villages, cultivated

Distribution: Tropical Asia and Pacific

Botanical description: This graceful herb grows up to about 1 m tall. The plant is robust and much branched. The stem is woody, angled, glabrous, longitudinally striated with a few lenticels and somewhat minutely winged. The internodes are 3–5 cm long. The leaves are simple, alternate, and without stipules. The blade is 0·6–14×0·1–4 cm, linear lanceolate with a prominent midrib underneath and about 6–8 pairs of secondary nerves. The flower is axillary on a 1.5 cm long pedicel. The hypanthium is 2.2 cm and hirsute. The calyx comprises 4 sepals, which are lanceolate, glabrous, and about 7×2 mm long. The 4 petals are about 1.5 cm long, yellow, obovate, nerved, and emarginated. The androecium comprises 8 short stamens. The capsules are about 2.5 cm long with 8 ribs and contain several tiny, globose, and muricate seeds.

Medicinal uses: Burns (India); dysentery (Cambodia, Laos, and Vietnam)

Strong broad-spectrum antibacterial polar extract: Methanol extract of leaves inhibited the growth of *Bacillus cereus* (ATCC 10876), *Bacillus licheniformis* (ATCC 12759), *Bacillus spizizenii* (ATCC 6633), *Staphylococcus aureus* (ATCC 12600), *Staphylococcus epidermidis* (ATCC 12228), *Streptococcus mutans* (ATCC 25175), *Escherichia coli* (O157:H7), *Escherichia coli* (ATCC 25922), *Klebsiella pneumoniae* (ATCC 13883), *Pseudomonas aeruginosa* (ATCC 27853), and *Shigella boydii* (ATCC 9207) (Yakob et al., 2012).

Commentaries: The plant produces oleanane-type triterpenes, (23*Z*)-feruloylhederagenin and (23*E*)-feruloylhederagenin, β-amyrin acetate, and β-amyrin palmitate (Chang & Kuo, 2007), as well as ellagic acid (see earlier) and derivatives, including 3,4,8,9,10-pentahydroxydibenzo[b,d] pyran-6-one (Garazd & Garazd, 2016) and flavonoids (Yan et al., 2005).

(ii) Extract of *Epilobium tetragonium* L. inhibited the growth of *Staphylococcus aureus* (ATCC 6538 P), *Streptococcus pyogenes* (ATCC 12345), *Bacillus subtilis* (ATCC 6633), *Listeria monocytogenes* (ATCC 19111), *Streptococcus sanguis* (CDC SS 910), *Klebsiella pneumoniae* (ATCC 10031), *Pseudomonas aeruginosa* (ATCC 7853), *Microsporum canis*, and *Trichophytum rubrum* with MIC values of 650, 325, 162, 650, 325, 81, 162, 650, and 650 μg/mL, respectively (Battinelli et al., 2001).

(iii) The triterpene oenotheralanosterol A isolated from the roots of *Oenothera biennis* L. inhibited the growth of *Sporothrix schenckii* and *Pseudomonas aeruginosa* with the MIC values of 250 and 500 μg/mL (Singh et al., 2017). The triterpene oenotheralanosterol B isolated from the roots of this plant inhibited the growth of *Sporothrix schenckii, Candida albicans, Candida albicans* (clinical), *Pseudomonas aeruginosa, Streptococcus pyogenes,* and *Bacillus* sp (Singh et al., 2017). The prenylated xanthone dihydroxyprenylxanthone isolated from the roots of this plant inhibited the growth of *Pseudomonas aeruginosa* with an MIC value of 250 μg/mL as well as *Candida albicans* (clinical) with an MIC values of 500 μg/mL (Singh et al., 2017).

FIGURE 8.13 *Ludwigia octovalvis.* (Jacq.) Raven

REFERENCES

Battinelli, L., Tita, B., Evandri, M.G. and Mazzanti, G., 2001. Antimicrobial activity of Epilobium spp. extracts. *Il Farmaco*, *56*(5–7), pp. 345–348.

Chang, C.I. and Kuo, Y.H., 2007. Oleanane-type triterpenes from Ludwigia octovalvis. *Journal of Asian Natural Products Research*, *9*(1), pp. 67–72.

Garazd, Y.L. and Garazd, M.M., 2016. Natural dibenzo [b, d] pyran-6-ones: structural diversity and biological activity. *Chemistry of Natural Compounds*, *52*(1), pp. 1–18.

Singh, S., Dubey, V., Singh, D.K., Fatima, K., Ahmad, A. and Luqman, S., 2017. Antiproliferative and antimicrobial efficacy of the compounds isolated from the roots of Oenothera biennis L. *Journal of Pharmacy and Pharmacology*, *69*(9), pp. 1230–1243.

Yakob, H.K., Sulaiman, S.F. and Uyub, A.M., 2012. Antioxidant and antibacterial activity of Ludwigia octovalvis on Escherichia coli O157: H7 and some pathogenic bacteria. *World Applied Sciences Journal, 16*, pp. 22–29.

Yan, J. and Yang, X.W., 2005. Studies on the chemical constituents in herb of Ludwigia octovalvis. *Zhongguo Zhong yao za zhi= Zhongguo zhongyao zazhi= China Journal of Chinese Materia Medica, 30*(24), pp. 1923–1926.

8.3 ORDER BRASSICALES BROMHEAD (1838)

8.3.1 FAMILY BRASSICACEAE BURNETT (1835)

The family Brassicaceae is a vast taxon which consists of about 350 genera and 3,000 species of herbs. The leaves are simple, without stipules, and alternate or opposite. The blade is dissected, often somewhat fleshy, and dull green. The inflorescences are racemes of yellow or white, bisexual, and actinomorphic flowers. The calyx consists of 4 free sepals, which are imbricate in 2 opposite pairs. The corolla consists of 4 petals, which are characteristically arranged somewhat like a cross. The androecium consists of 6 stamens, the 2 outer ones shorter than the 4 inner ones. The gynoecium consists of 2 pairs of carpels united into a sessile and single-celled ovary divided by a wall and including several ovaries attached to 1–2 parietal placentas. The fruits are capsular. A unique and interesting feature of this vast family is the production of glucosinolates, which release antibacterial and antifungal isothiocyanates derivatives (phytoalexins) upon tissue injuries. Other phytoalexins in this family are indole or oxindole alkaloids.

8.3.1.1 *Brassica alba* L.

Synonym: *Sinapis alba* L.

Common names: White mustard; bai jie (China); safed rai, venkatuku (India); khardel sefid (Iran); shiro garashi (Japan); mosutadu (Korea); soms sien (Laos); safed sarson (Pakistan); mustasa (the Philippines); tohumlari (Turkey); bach gioi tu (Vietnam)

Habitat: Cultivated, roadsides, fields, waste lands

Distribution: Turkey, Iran, Afghanistan, Tajikistan, Turkmenistan, Pakistan, India, Vietnam, China.

Botanical observation: This erect herb grows up to 80 cm tall. The leaves are simple, spiral, and exstipulate. The petiole is up to 6 cm long. The blade is lyrate, serrate, pinnatifid, pinnatisect, and 5–14×2–8 cm. The flowers are arranged at the apex of stems. The calyx includes 4 sepals, which are narrow and up top about 1.5 cm long. The 4 petals are yellowish, obovate, and up to about 1.4 cm×7 mm. The androecium consists of 4 stamens. The capsules are fusiform, up to 5 cm long, and contain numerous minute globose and edible seeds.

Medicinal uses: Bronchitis (China; India); fever (India), leprosy (Myanmar)

Strong broad-spectrum antibacterial fixed oil: Oil of seeds inhibited the growth of *Staphylococcus aureus*, *Bacillus subtilis*, *Enterococcus faecalis*, *Listeria monocytogenes*, *Pseudomonas aeruginosa*, *Shigella typhi*, and *Shigella dysenteriae* (Aman et al., 2014).

Moderate broad-spectrum antibacterial essential oil: Essential oil of seeds (containing mainly allyl isothiocyanate) inhibited the growth of *Staphylococcus aureus* (ATCC 6538), *Micrococcus luteus* (ATCC49732), *Staphylococcus epidermidis* (ATCC12228), *Escherichia coli* (ATCC8739), *Bacillus subtilis* (ATCC 6633), *Shigella sonnei* (ATCC 25931), *Salmonella lignieres* (ATCC14028), *Pseudomonas aeruginosa* (ATCC 27853), and *Pseudomonas fluorescens* (ATCC13525) with the MIC/MBC values of 128/512, 128/512, 256/512, 512/512, 512/1024, 512/1024, 256/512, 256/1024, and 512/1024 μg/mL, respectively (Peng et al., 2014).

Strong antibacterial (Gram-positive) amphiphilic sulfur compounds: Allyl isothiocyanate (LogD = 1.9 at pH 7.4; molecular mass=99.1 g/mol) was bacteriostatic for methicillin-resistant *Staphylococcus aureus* (CECT 976) with an MIC value of 110 μg/mL and inhibited the growth of

15 clinical strains of methicillin-resistant *Staphylococcus aureus* with MIC ranging from 27.9 to 220 µg/mL (Dias et al., 2014). Benzyl isothocyanate (LogD = 3.3 at pH 7.4; molecular mass = 149.2 g/mol) was bacteriostatic for methicillin-resistant *Staphylococcus aureus* (reference strain CECT 976) with an MIC value of 55.7 µg/mL and inhibited the growth of 15 clinical strains of methicillin-resistant *Staphylococcus aureus* with MIC ranging from 2.9 to 110 µg/mL (Dias et al., 2014). Phenylethyl isothiocyanate was bacteriostatic for methicillin-resistant *Staphylococcus aureus* (reference strain CECT 976) with an MIC value of 110 µg/mL and inhibited the growth of 15 strains of methicillin-resistant *Staphylococcus aureus* with MIC ranging from below 7.3 to 146.7 µg/mL (Dias et al., 2014).

Commentaries: (i) The plant produces glucosinolates, mainly sinalbin, which is converted by myrosinase into *para*-hydroxy benzyl isothiocyanate (Blažević et al., 2019). Sinigrin is the major glucosinolate in *Brassica nigra* (L.) Koch (Mazumder et al., 2016), a plant is used in Korea to treat flu and abscesses. In China, *Brassica cernua* (Thunb.) Forbes & Hemsl. is used to treat pneumonia. (ii) *Brassical campestris* L.(Local name: *Sarsoon*) is used in Pakistan to treat cold and jaundice.

REFERENCES

Aman, S., Naim, A., Siddiqi, R. and Naz, S., 2014. Antimicrobial polyphenols from small tropical fruits, tea and spice oilseeds. *Revista de Agaroquimica y Tecnologia de Alimentos*, *20*(4), pp. 241–251.

Blažević, I., Montaut, S., Burčul, F., Olsen, C.E., Burow, M., Rollin, P. and Agerbirk, N., 2019. Glucosinolate structural diversity, identification, chemical synthesis and metabolism in plants. *Phytochemistry*, p.112100.

Dias, C., Aires, A. and Saavedra, M.J., 2014. Antimicrobial activity of isothiocyanates from cruciferous plants against methicillin-resistant Staphylococcus aureus (methicillin-resistant *Staphylococcus aureus*). *International Journal of Molecular Sciences*, *15*(11), pp. 19552–19561.

Mazumder, A., Dwivedi, A. and Du Plessis, J., 2016. Sinigrin and its therapeutic benefits. *Molecules*, *21*(4), p. 416.

Peng, C., Zhao, S.Q., Zhang, J., Huang, G.Y., Chen, L.Y. and Zhao, F.Y., 2014. Chemical composition, antimicrobial property and microencapsulation of Mustard (Sinapis alba) seed essential oil by complex coacervation. *Food Chemistry*, *165*, pp. 560–568.

8.3.1.2 *Brassica oleracea* L.

Synonyms: *Brassica alboglabra* L.H. Bailey; *Brassica maritima* Tardent; *Crucifera brassica* E.H.L. Krause; *Napus oleracea* (L.) K.F. Schimp. & Spenn.

Common names: Cabbage; badha-kopi (Bangladesh); ye gan lan (China); gobi (India); mooli (Pakistan)

Habitat: Cultivated

Distribution: Asia

Botanical observation: This edible herb grows up to about 1.5 m tall. The stems are fleshy and glaucous. The leaves are simple, spiral, basal or cauline, and exstipulate. The petiole is up to about 30 cm long. The blade is fleshy, dull green, ovate, oblong to lanceolate, up to about 35 cm long, amplexicaul at base (cauline leaves), incised or dentate, and acute at apex. The inflorescence is a raceme or a fleshy head. The calyx comprises 4 oblong sepals, which are up to about 1.5 cm long. The 4 petals are up to 2.5 cm long, spathulate, membranous, and round at apex. The androecium comprises 6 stamens. The capsules are linear, up to 8 cm long, terete, dehiscent, and contain numerous minute globose seeds.

Medicinal use: Jaundice (Bangladesh; Pakistan)

Weak antibacterial (Gram-negative) polar extract: Methanol extract inhibited the growth of *Escherichia coli* (ATCC 8739), *Enterococcus aerogene* (ATCC 13048) and *Klebsiella pneumoniae* (ATCC 11296) with an MIC of 1024 µg/mL (Touani et al., 2014).

Broad-spectrum antibacterial juice: Fresh juice of leaves (at 20% of medium growth) inhibited the growth of *Salmonella enteridis* subsp. *enterica* serovar *enteridis* (isolated from food) by more than 95% (Brandi et al., 2006). The juice (at 20% of medium growth) inhibited the growth of *Escherichia coli* (O157:H7 (strain 1952) by more than 3 logs (Brandi et al., 2006). The juice (at 20% of medium growth) inhibited the growth of *Listeria monocytogenes* (ATCC 9525) by more than 2 logs (Brandi et al., 2006).

Broad-spectrum antifungal juice: Juice (at a concentration of 20%) inhibited by 95% the growth of *Candida albicans* by 70% (Brandi et al., 2006). At a concentration of 0.5% the juice inhibited the growth of *Alternaria* spp., *Cladosporium* spp., *Microsporium canis*, and *Trichophytonverrucosum* by 100% (Sisti et al., 2003).

Strong antifungal (filamentous) amphiphilic sulfur indole alkaloid: Caulilexin A (LogD = 3.2 at pH 7.4; molecular mass=223.3 g/mol) at the concentration of 5×10^{-4}M inhibited the growth of *Leptosphaeria maculans*, *Sclerotinia sclerotiorum*, and *Rhizoctonia solani* by by 55, 100, and 100%, respectively (Pedras et al., 2006).

Commentaries: (i) The plant produces the glucosinolate sinigrin (Martínez-Ballesta et al., 2014), which upon physical insults is hydrolyzed by myrosinase into allyl isothiocyanate, allyl thiocyanate, allyl cyanide, and 1-cyano-2,3-epithiopropane (Shofran et al., 1998). Allyl thiocyanate inhibited the growth of *Escherichia coli* (33625), *Escherichia coli* (NC101), *Pseudomonas fluorescens* (MD13), and *Staphylococcus aureus* (4220) with MIC values of 400, 200, 200, and 200 ppm, respectively (Shofran et al., 1998). In a subsequent study, allyl isothiocyanate inhibited the growth of *Escherichia coli* (O157:H7 (ATCC 43895), *Salmonella* sp., and *Listeria monocytogenes* with MIC values of 50, 100, and 200 µg/mL, respectively (Lin et al., 2000). Allyl isothiocyanate was bactericidal at all growth stages (at lag, early exponential, late exponential, and stationary phase) indicating that its bactericidal mechanism was not limited to its inhibitory effect on the biosynthesis of macromolecules, which occurs actively during the exponential phase (Lin et al., 2000). Allyl isothiocyanate was more active on Gram-negative than on Gram-positive bacteria (Lin et al., 2000).

(ii) *Coronopus didymus* (L.) Sm. is used in Pakistan to treat jaundice (local name: *Haryani*)

REFERENCES

Brandi, G., Amagliani, G., Schiavano, G.F., De Santi, M. and Sisti, M., 2006. Activity of Brassica oleracea leaf juice on foodborne pathogenic bacteria. *Journal of Food Protection*, *69*(9), pp. 2274–2279.

Lin, C.M., PRESTON III, J.F. and Wei, C.I., 2000. Antibacterial mechanism of allyl isothiocyanate. *Journal of Food Protection*, *63*(6), pp. 727–734.

Martínez-Ballesta, M.D.C., Muries, B., Moreno, D.Á., Dominguez-Perles, R., García-Viguera, C. and Carvajal, M., 2014. Involvement of a glucosinolate (sinigrin) in the regulation of water transport in Brassica oleracea grown under salt stress. *Physiologia Plantarum*, *150*(2), pp. 145–160.

Pedras, M.S.C., Sarwar, M.G., Suchy, M. and Adio, A.M., 2006. The phytoalexins from cauliflower, caulilexins A, B and C: Isolation, structure determination, syntheses and antifungal activity. *Phytochemistry*, *67*(14), pp. 1503–1509.

Shofran, B.G., Purrington, S.T., Breidt, F. and Fleming, H.P., 1998. Antimicrobial properties of sinigrin and its hydrolysis products. *Journal of Food Science*, *63*(4), pp. 621–624.

Sisti, M., Amagliani, G. and Brandi, G., 2003. Antifungal activity of Brassica oleracea var. botrytis fresh aqueous juice. *Fitoterapia*, *74*(5), pp. 453–458.

Touani, F.K., Seukep, A.J., Djeussi, D.E., Fankam, A.G., Noumedem, J.A. and Kuete, V., 2014. Antibiotic-potentiation activities of four Cameroonian dietary plants against multidrug-resistant Gram-negative bacteria expressing efflux pumps. *BMC Complementary and Alternative Medicine*, *14*(1), p. 258.

8.3.1.3 *Capsella bursa-pastoris* (L.) Medik

Synonyms: *Bursa bursa-pastoris* (L.) Britton; *Bursa bursa-pastoris* (L.) Shafer; *Nasturtium bursa-pastoris* Roth; *Solmsiella heegeri* (Solms) Borbás; *Thlaspi bursa-pastoris* L.

Common names: Shepherd's purse, blind weed, toy wort, mother's head; so ka pa (Bhutan)' Chi (China); te thai, dinh lich, co tam giac (Cambodian, Laos, Vietnam); mumiri (India); china phol (Pakistan)

Habitat: Roadsides, wastelands

Distribution: Temperate and subtemperate Asia,

Botanical observation: It is slender herb that grows to a height of 60 cm. The plant has an aura of dryness. The main stem is somewhat stiff. The stems are terete, glabrous or hairy. The leaves are rosulate, the blade is oblong, acute or oblanceolate, lobed or inciso-pinnatifid. The flowers are at first corymbose, afterward elongate into 20–22.5 cm long racemes. The calyx comprises 4 oblong and obtuse sepals. The corolla is cross shaped, and consists of 4 white petals, which are oblanceolate and about half as long as the sepals. The fruits area characteristic heart-shaped flattened pod, which are notched on top, dehiscent, up to 1 cm long, and contain numerous tiny seeds.

Medicinal uses: Diarrhea (China); dysentery (China, Pakistan)

Moderate antibacterial (Gram-positive) polar extract: A fraction (containing the isothiocyanate sulforaphane) inhibited the growth vancomycin-resistant enterococci and *Bacillus anthracis* (ATCC 14578) with MIC values of 1000 and 250 µg/mL (Choi et al., 2014).

Moderate antifungal thiazole indole alkaloid: Camalexin inhibited the growth of *Alternaria brassicae* with an MIC value of 80 µg/mL (Jimenez et al., 1997).

Viral enzyme inhibition by extract: Extract inhibited Human immunodeficiency type-1 protease (Park, 2003).

Commentary: Extract inhibited the Human immunodeficiency virus-1 protease (Park, 2003). The plant was prescribed for wounds healing in Medieval Europe.

REFERENCES

Choi, W.J., Kim, S.K., Park, H.K., Sohn, U.D. and Kim, W., 2014. Anti-inflammatory and anti-superbacterial properties of sulforaphane from shepherd's purse. *The Korean Journal of Physiology & Pharmacology*, *18*(1), pp. 33–39.

Jimenez, L.D., Ayer, W.A. and Tewari, J.P., 1997. Phytoalexins produced in the leaves of Capsella bursa-pastoris (shepherd's purse). *Phytoprotection*, *78*(3), pp. 99–103.

Park, J.C., 2003. Inhibitory effects of Korean plant resources on Human immunodeficiency virus type 1 protease activity. *Oriental Pharmacy and Experimental Medicine*, *3*(1), pp. 1–7.

8.3.1.4 *Cardamine hirsuta* L.

Synonyms: *Cardamine multicaulis* Hoppe ex Schur; *Cardamine umbrosa* Andrz. ex DC.

Common names: Hairy bittercress; sui mi qi (China); chantruk maan, kosanini, simrayo (India)

Habitat: Roadsides, abandoned lands

Distribution: Subtemperate Asia and Pacific.

Botanical observation: This graceful herb grows to a height of 30 cm. The stems are glabrous and erect from a rosette of leaves. The leaves are numerous, pinnately lobed, 2–5 cm long, spiral, and without stipules. The petiole is channeled and sheathing at base and subglabrous. The blade is glabrous or subglabrous and shows a few pairs of indistinct secondary nerves. The inflorescences are terminal or axillary, corymbose, and 1 cm long racemes. The flowers are white and small. The calyx comprises 2 pairs of sepals, which are subglabrous, and 2 mm long. The 4 petals are membranous and pure white. The capsules are 1–2 cm × 1 mm and contain numerous and minute seeds.

Medicinal use: Dysentery (Cambodia, Laos, Vietnam)

Commentary: The antimicrobial properties of this plant have apparently not been examined.

8.3.1.5 *Cardaria draba* (L.) Desv.

Synonym: *Lepidium draba* L.

Common names: Heart-padded hoary-cress; qun xin cai (China)

Habitat: Roadsides, wastelands, fields

Distribution: Turkey, Iran, Afghanistan, Kazakhstan, Kyrgyzstan, Tajikistan, Turkmenistan, Uzbekistan, Pakistan, India, China

Botanical observation: This beautiful herb grows to a height of about 60 cm. The stems are hairy. The leaves are simple, basal and cauline, spiral, and exstipulate. The blade is spathulate, 3–10 × 1–4 cm, dentate, and obtuse at apex. The blade of cauline leaves is amplexicaul and auriculate. The inflorescences are showy corymbs. The calyx comprises 4 sepals, which are minute. The 4 petals are pure white, spathulate, and about 5 mm long. The androecium includes 6 stamens. The capsules are cordate about 5 mm long, and containing a few minute tear-shaped brown seeds.

Medicinal use: Cough (Iran)

Moderate broad-spectrum antibacterial essential oil: Hydrodistillates (containing 69.2% of sulforaphane) of aerial parts inhibited the growth of *Bacillus cereus* (ATCC 1178), *Clostridium. perfringens* (FNSST 4999), *Enterococcus faecalis* (ATCC 29212), and *Staphylococcus aureus* (ATCC 25293) with MIC values of 64, 128, 128, and 64 µg/mL, respectively (Radonić et al., 2011). This hydrodistillate inhibited the growth of *Enterobacter sakazaki* (FNSST 021), *Escherichia coli* (FNSST 982), *Klebsiella pneumoniae* (FNSST 011), and *Pseudomonas aeruginosa* (FNSST 014) with MIC values of 4, 128, 128, and 128 µg/mL, respectively (Radonić et al., 2011).

Strong broad-spectrum antibacterial polar extract: Ethanol extract of leaves inhibited the growth of *Staphylococcus aureus, Bacillus subtilis, Pseudomonas aeruginosa,* and *Escherichia coli* with MIC values of 3, 67, 55, and 89 µg/mL, respectively (Sharifi-Rad et al., 2011).

Strong broad-spectrum amphiphilic isothiocyanate: Sulforaphane (LogD = 1.1 at pH 7.4; molecular mass = 177.2 g/mol) inhibited the growth of *Escherichia coli* (ATCC25922), *Escherichia coli* (O157:H7 (ATCC700728), *Klebsiella pneumoniae* (ATCC13883), *Salmonella typhimurium,* (ATCC14028), *Shigella sonnei* (ATCC25931), *Staphylococcus epidermidis* (ATCC12228), *Staphylococcus aureus* (ATCC25923), *Streptoccus pyogenes* (ATCC19615), *Enterococcus faecalis* (ATCC2921), and *Bacillus cereus* (ATCC11770) with MIC values of 2, 4, 2, 2, 2, 1, 1, 1, 2, and 2 µg/mL, respectively (Johansson et al., 2008).

Moderate broad-spectrum antifungal essential oil: Hydrodistillates (containing 69.2% of sulforaphane) of aerial parts inhibited the growth of *Candida albicans*, *Penicillium* sp., and *Rhizopus stolonifer* with MIC values of 128, 128, and 64 µg/mL, respectively (Radonić et al., 2011).

Strong antifungal (filamentous) isothiocyanate: Sulforaphane inhibited the growth of *Cryptococcus neoformans* (ATCC14116) with an MIC value of 4 µg/mL (Johansson et al., 2008).

Commentary: The main glucosinolates in this plant are glucoraphanin precursor of sulforaphane, glucosinalbin, and sinigrin (Powel et al., 2005).

REFERENCES

Johansson, N.L., Pavia, C.S. and Chiao, J.W., 2008. Growth inhibition of a spectrum of bacterial and fungal pathogens by sulforaphane, an isothiocyanate product found in broccoli and other cruciferous vegetables. *Planta Medica*, *74*(07), pp. 747–750.

Powell, E.E., Hill, G.A., Juurlink, B.H. and Carrier, D.J., 2005. Glucoraphanin extraction from Cardaria draba: Part 1. Optimization of batch extraction. *Journal of Chemical Technology & Biotechnology: International Research in Process, Environmental & Clean Technology*, *80*(9), pp. 985–991.

Radonić, A., Blažević, I., Mastelić, J., Zekić, M., Skočibušić, M. and Maravić, A., 2011. Phytochemical analysis and antimicrobial activity of Cardaria draba (L.) Desv. volatiles. *Chemistry & Biodiversity*, *8*(6), pp. 1170–1181.

Sharifi-Rad, J., Hoseini-Alfatemi, S.M., Sharifi-Rad, M., da Silva, J.A.T., Rokni, M. and Sharifi-Rad, M., 2015. Evaluation of biological activity and phenolic compounds of Cardaria draba (L.) extracts. *Journal of Biology and Today's World*, *4*(9), pp. 180–189.

8.3.1.6 *Descurainia sophia* (L.) Webb ex Prantl

Synonym: *Sisymbrium sophia* L

Common names: Flixweed; khakshir (Afghanistan); bo niang hao (China); jangli saron, kubkalana, (India); kujira-gusa (Japan); jaessug (Korea); masino tori jal (Nepal)

Habitat: Roadsides, wastelands, fields, deserts

Distribution: Turkey, Iran, Afghanistan, Kazakhstan, Kyrgyzstan, Tajikistan, Turkmenistan, Uzbekistan, Pakistan, India, Nepal, Bhutan, China, Korea, Japan

Botanical observation: This herb grows to about 70 cm tall. The leaves are simple, spiral, and exstipulate. The petiole of basal leaves is up to 2 cm long. The blade of basal leaves is pinnatisect and up to 15 cm long. The inflorescence is a terminal raceme. The calyx includes 4 sepals, which are minute and oblong. The 4 petals are yellow, oblanceolate to spathulate, and about 3 mm long. The androecium comprises 6 stamens. The capsules are linear, up to about 2.5 cm long, and contain numerous tiny reddish-brown seeds.

Medicinal uses: Cough (China); dysentery, typhoid, (Afghanistan); fever (Afghanistan, China, Iran); ulcers (India)

Commentary: The plant produces the glucosinolate gluconapin from which derives 3-butenyl isothiocyanate (Dekić et al., 2017) as well as sinigrin (Burrows & Tyrl, 2013).

REFERENCES

Burrows, G.E. and Tyrl, R.J., 2013. *Toxic Plants of North America*. Chichester: John Wiley & Sons.

Dekić, M., Radulović, N., Danilović-Luković, J.B. and Stojanović, D.Z., 2017. Volatile glucosinolate breakdown products and the essential oil of Descurainia sophia (L.) webb ex Prantl (Brassicaceae). *Facta universitatis-series: Physics, Chemistry and Technology*, *15*(2), pp. 95–102.

8.3.1.7 *Draba nemorosa* L.

Synonyms: *Draba macroloba* Turcz.; *Draba muralis* L.; *Draba nemoralis* Ehrh.

Common names: Wood whitlow-grass; ting li (China)

Habitat: Roadsides, wastelands, fields, riverbanks, forest margins, rocky lands

Distribution: Turkey, Iran, Afghanistan, Kazakhstan, Kyrgyzstan, Tajikistan, Turkmenistan, Uzbekistan, Pakistan, India, Nepal, Bhutan, China, Korea, Japan

Botanical observation: This herb grows to about 25 cm tall. The leaves are simple, sessile, spiral, basal, and exstipulate. The blade is elliptic to broadly lanceolate, 0.8–3.5 × 0.3–1.5 cm, dentate or not, hairy, and acute at apex. The inflorescence is a raceme. The calyx includes 4 sepals, which are minute. The 4 petals are 2 mm long, and bifid. The androecium includes 6 stamens. The capsules are oblong-elliptic, up to 8 × 2.5 mm, and contain numerous tiny seeds.

Medicinal use: Cough (China)

Commentary: Apparently this plant has not been examined for its possible antimicrobial effects. It is known to synthetize sinapic acid (Rahman & Moon, 2007). Glucosinolates are known to be produced by members of the genus *Draba* L. (Montaut et al., 2018).

REFERENCES

Montaut, S., Montagut-Romans, A., Chiari, L. and Benson, H.J., 2018. Glucosinolates in Draba borealis DC.(Brassicaceae) in a taxonomic perspective. *Biochemical Systematics and Habitat*, *78*, pp. 31–34.

Rahman, M.A.A. and Moon, S.S., 2007. Antioxidant polyphenol glycosides from the plant Draba nemorosa. *Bulletin of the Korean Chemical Society*, *28*(5), pp. 827–831.

8.3.1.8 *Isatis tinctoria* L.

Synonyms: *Isatis indigotica* Fortune; *Isatis koelzii* Rech. f.

Common names: Asp-of-Jerusalem, dyer's woad, European indigo; song lan (China)

Habitat: Roadsides, wastelands, fields, rocky soils

Distribution: Turkey, Iran, Afghanistan, Kazakhstan, Kyrgyzstan, Tajikistan, Turkmenistan, Uzbekistan, Pakistan, India, China, Korea, Japan

Botanical observation: This herb grows to a height of 1.2 m. The leaves are simple, basal, cauline, and exstipulate. The blade of basal leaves is oblanceolate and 5–15 × 1–3 cm. The blade of cauline leaves is narrowly lanceolate, amplexicaul, and sessile. The racemes are many-flowered. The 4 sepals are minute. The 4 petals are spathulate, yellow, and up to about 5 mm long. The androecium comprises 6 stamens. The capsules are spathulate, up to about 2 cm long, glossy, turning blackish, andcontain a few tiny seeds.

Medicinal uses: Dysentery, fever, Influenza, measles, scarlet fever, tonsilitis, typhoid, typhus (China)

Broad-spectrum antibacterial mid-polar extracts: Ethyl acetate extract of flowers inhibited the growth of *Micriciccusluteus* with an IC_{50} value of 42 µg/mL (Ullah et al., 2017). Chloroform extract of leaves inhibited the growth of *P. aeruginosa* and *B. subtilis* with IC_{50} values of 32 and 111 µg/mL, respectively (Ullah et al., 2017).

Strong antiviral amphiphilic indole alkaloids: Indigo and indirubin (LogD = 2.3 at pH 7.4; molecular mass=262.2 g/mol) inhibited the replication of Japanese encephalitis virus with IC_{50} values of 37.4 and 13.6 µg/mL, respectively, and were virucidal with IC_{50} values of 3 and 0.4 µg/mL, respectively (Chang et al., 2012).

Viral enzyme inhibition by indole alkaloid: Indigo, indirubin, and indican inhibited the Severe acute respiratory syndrome-associated coronavirus 3-chymotrypsin-like protease with EC_{50} values of 37.3, 81.3, and 33.1 µg/mL, respectively (Tsai et al., 2010).

Viral enzyme inhibition by glucosinolate: Sinigrin inhibited the Severe acute respiratory syndrome-associated coronavirus 3-chymotrypsin-like protease with EC_{50} values of 50.3 µg/mL (Lin et al., 2005).

Commentary: (i) The plant produces indole alkaloids glycosides: indican (indoxyl-β-d-glucoside) and isatan B (indoxyl-β-ketogluconate), which upon tissue injuries are hydrolyzed into indoxyl, which is the precursor of the bis-indole alkaloids indigo (blue pigment) and indirubin (red) as well as tryptanthrin (see earlier) (Maugard et al., 2001). Besides, the plant produces glucosinolates such as glucobrassicin, neoglucobrassicin, sulfoglucobrassicin, and glucotropaeolin (Elliott & Stowe, 1971).

REFERENCES

Chang, S.J., Chang, Y.C., Lu, K.Z., Tsou, Y.Y. and Lin, C.W., 2012. Antiviral activity of Isatis indigotica extract and its derived indirubin against Japanese encephalitis virus. *Evidence-Based Complementary and Alternative Medicine, 2012*. Article ID 925830.

Elliott, M.C. and Stowe, B.B., 1971. Distribution and variation of indole glucosinolates in woad (Isatis tinctoria L.). *Plant Physiology, 48*(4), pp. 498–503.

Lin, C.W., Tsai, F.J., Tsai, C.H., Lai, C.C., Wan, L., Ho, T.Y., Hsieh, C.C. and Chao, P.D.L., 2005. Anti-SARS coronavirus 3C-like protease effects of Isatis indigotica root and plant-derived phenolic compounds. *Antiviral Research, 68*(1), pp. 36–42.

Maugard, T., Enaud, E., Choisy, P. and Legoy, M.D., 2001. Identification of an indigo precursor from leaves of Isatis tinctoria (Woad). *Phytochemistry, 58*(6), pp. 897–904.

Tsai, Y.C., Lee, C.L., Yen, H.R., Chang, Y.S., Lin, Y.P., Huang, S.H. and Lin, C.W., 2020. Antiviral Action of Tryptanthrin Isolated from Strobilanthes cusia Leaf against Human Coronavirus NL63. *Biomolecules, 10*(3), p. 366.

Ullah, I., Wakeel, A., Shinwari, Z.K., Jan, S.A., Khalil, A.T. and Ali, M., 2017. Antibacterial and antifungal activity of isatis tinctoria l.(brassicaceae) using the micro-plate method. *Pakistan Journal of Botany, 49*(5), pp. 1949–1957.

8.3.1.9 *Rorippa indica* (L.) Hiern

Synonyms: *Nasturtium indicum* (L.) DC.; *Nasturtium montanum* Wall. ex Hook. f. & Thomson ; *Sisymbrium indicum* L.

Common names: Indian cress; bansaricha (Bangladesh); chamsuru (India); han cai (China); Inu garashi (Japan); gae gat naeng (Korea); sabi (the Philippines); phakat nam dok lwang (Thailand)

Habitat: Roadsides, wastelands, fields

Distribution: Pakistan, India, Himalayas, Sikkim, Nepal, Bangladesh, Myanmar, Laos, Vietnam, Thailand, China, Korea, Japan, Malaysia, Indonesia, the Philippines

Botanical observation: This herb grows up to about 60 cm tall. The leaves are simple, spiral, sessile, and exstipulate. The blade is lyrate and up to 3.5–12×1.5–5 cm. The racemes are axillary, many-flowered, and up to about 15 cm long. The calyx includes 4 minute sepals. The 4 petals are up to about 3 mm long and yellow. The androecium comprises 6 stamens. The capsules are terete, up to 1.5 cm long, and contain numerous tiny seeds.

Medicinal uses: Cough (China); fever (China, Taiwan)

Commentary: The traditional system of medicine in China uses *Nasturtium officinale* R. Br. to treat sore throat.

8.3.1.10 *Raphanus sativus* L.

Synonyms: *Raphanistrum gayanum* Fisch. & C.A. Mey.; *Raphanus acanthiformis* J.M. Morel ex Sasaki; *Raphanus candidus* Vorosch.; *Raphanus chinensis* Mill.; *Raphanus gayanus* (Fisch. & C.A. Mey.) G. Don ex Sweet; *Raphanus macropodus* H. Lév.; *Raphanus niger* Mill.; *Raphanus taquetii* H. Lév.; *Raphanus taquetti* H. Lév.

Common names: Radish; chhaay thaw (Cambodia); lok po fa, luo bo (China); daikon (Japanese); mu (Korea); mujh (India); lobak (Indonesia, Malaysia); mulakam (India); monla (Myanmar); labanos (the Philippines); phakkat-hua (Thailand); cai cui.(Vietnam)

Habitat: Cultivated

Distribution: Throughout Asia

Botanical observation: This herb grows up to about 50 cm tall.. The root is fleshy, ovoid or globose or elongate, heavy, white, pink, red or black, and pungent. The leaves are simple, mainly basal, and exstipulate. The petiole is up to 30 cm long. The blade is up to 60×20 cm, pinnatisect, dentate, and obtuse at apex. The cauline leaves are subsessile and dentate. The calyx includes 4 sepals, which are oblong and about 5 mm long. The 4 petals are purple red, pink, or white, spathulate, delicately veined, membranous, up to about 2 cm long, and obtuse to round at apex. The androecium includes 6 stamens. The capsulesare fusiform, up to about 8 cm long, and contain several ovoid minute seeds.

Medicinal uses: Cough (China; Japan); diarrhea, fever (Japan); dysentery (China); jaundice (India)

Broad-spectrum antibacterial halo elicited by polar extract: Ethanol extract of seeds (13–20 mg/ 5 mm well) inhibited the growth of *Streptococcus pyogenes* (ATCC 19615), *Staphylococcus aureus* (ATCC 25293), *Escherichia coli* (ATCC 25922), *Klebsiella peumoniae* (ATCC 700603), and *Salmonella thyphimurium* (ATCC 14028) with inhibition zone diameters of 20, 11, 19, 9, and 16 mm, respectively (Bendimerad et al., 2005).

Broad-spectrum antibacterial halo elicited by isothiocyanate: 4-methylsulfanyl-3-butenyl isothiocyanate (25 µL of a 2.5µM solution /paper disc) inhibited the growth of *Escherichia coli* (ATCC 11775), *Enterococcus cloacae* (ATCC 13047), *Salmonella typhimurium,* (ATCC 13311), *Proteus vulgaris* (ATCC 6380), *Staphylococcus aureus*, *Staphylococcus epidermidis* (ATCC 14990), *Bacillus subtilis* (ATCC 6633), and *Bacillus cereus* var. *mycoides* (ATCC 11778) (Uda et al., 1993).

Strong broad-spectrum antibacterial polar extract: Acetone extract of roots inhibited the growth of *Bacillus cereus, Staphylococcus aureus, Staphylococcus epidermidis, Enterococcus faecalis, Salmonella typhimurium, Enterobacter aerogenes*, and *Enterobacter cloacae*. (Beevi et al., 2009).

Weak antibacterial (Gram-positive) polar extract: Aqueous extract of seeds inhibited the growth of *Streptococcus mutans* (MT 5091) and *Streptococcus mutans* (OMZ 176) with MIC values of 1560 and 1560 µg/mL, respectively (Chen et al., 1989).

Broad-spectrum antifungal halo elicited by isothiocyanate: 4-Methylsulfanyl-3-butenyl isothiocyanate (25 µL/paper disc at the concentration of 2.5 µM) inhibited the growth of *Candida valida* (IFO 10318), *Debaryomyces hansenii* (IFO 0855), *Hansenula anomala* (IFO 10213), *Alternaria helianthi* (IFO 9089), *Cladosporium colocasiae* (IFO 6698), *Eurotium chevalieri* (IFO 4090), *Penicillium frequentans* (IFO 7919), and *Mucor racemosus* f. *racemosus* (IFO 5403) (Uda et al., 1993).

In vivo (enveloped segmented linear single-stranded (−)-RNA) antiviral polar extract: Aqueous extract administered by intranasally protected mice against Influenza virus A/PR 8/34 (H1N1) (Prahoveanu & Eşanu, 1987).

Commentary: The plant produces the glucosinolate glucoraphasatin, which releases 4-methylsulfanyl-3-butenyl isothiocyanate upon hydrolysis catalyzed by myrosinase (Montaut et al., 2010). Aqueous extract of *Raphanus sativus* L. could clinical trials for the prevention of Influenza but phytomedications or herbal remedies, plant infusions, capsules, and extracts are not tested clinically because, according to some, do not fall within what is financially profitable by the "Big Pharmas". Further, some scholars claim that universities often work for the interest of Pharmaceutical companies hence the eradication of medicinal plant teaching (Brownlee, 2015). There is a need to make laws to permit low-cost patenting and government funded medical clinical trials of herbal remedies. The only way to survive the upcoming pandemics is the synergistic use of Wester medicines and medicinal plant products.

REFERENCES

Beevi, S.S., Mangamoori, L.N., Dhand, V. and Ramakrishna, D.S., 2009. Isothiocyanate profile and selective antibacterial activity of root, stem, and leaf extracts derived from Raphanus sativus L. *Foodborne Pathogens and Disease*, *6*(1), pp. 129–136.

Bendimerad, N., Taleb Bendiab, S.A., Benabadji, A.B., Fernandez, X., Valette, L. and Lizzani-Cuvelier, L., 2005. Composition and antibacterial activity of Pseudocytisus integrifolius (Salisb.) essential oil from Algeria. *Journal of Agricultural and Food Chemistry*, 53(8), pp. 2947–2952.

Brownlee, J., 2015. The corporate corruption of academic research. Alternate Routes: A Journal of Critical Social Research, 26:23-50.

Chen, C.P., Lin, C.C. and Tsuneo, N., 1989. Screening of Taiwanese crude drugs for antibacterial activity against Streptococcus mutans. *Journal of Ethnopharmacology*, *27*(3), pp. 285–295.

Montaut, S., Barillari, J., Iori, R. and Rollin, P., 2010. Glucoraphasatin: chemistry, occurrence, and biological properties. *Phytochemistry*, *71*(1), pp. 6–12.

Prahoveanu, E. and Eşanu, V., 1987. Immunomodulation with natural products. I. Effect of an aqueous extract of Raphanus sativus niger on experimental Influenza infection in mice. *Virologie*, *38*(2), pp. 115–120.

Uda, Y., Matsuoka, H., Kumagami, H., Shima, H. and Maeda, Y., 1993. Stability and antimicrobial property of 4-methylthio-3-butenyl isothiocyanate, the pungent principle in radish. *Nippon Shokuhin Kogyo Gakkaishi*, *40*(10), pp. 743–746.

8.3.1.11 *Wasabia japonica* (Miq.) Matsum.

Synonyms: *Eutrema japonicum* (Miq.) Koidz.; *Eutrema wasabi* (Siebold) Maxim.; *Lunaria japonica* Miq.

Common names: Japanese horseradish; kuai jing shan yu cai (China)

Habitat: Riverbanks, cultivated

Distribution: Korea, China, Taiwan

Botanical observation: This nuphar-like or Araceous-like herb grows up to about 60 cm tall from a stout and edible (Japan) rhizome. The leaves are simple, spiral, basal, and exstipulate. The petiole is up to about 20 cm long and purplish at base. The blade is cordate, glossy, 6–15 cm across, fleshy, cordate at base, dentate, marked with radial secondary and tertiary nerves, and rounded to acute at apex. The inflorescence is a raceme. The 4 sepals are oblong and up to about 4 mm long. The 4 petals are pure white, oblong, and up to about 8 mm long. The androecium includes 6 stamens. The capsules are linear and up to 2 cm long and contain numerous tiny seeds.

Medicinal use: Antiseptic (Japan)

Moderate antibacterial (Gram-negative) polar extract: Methanol extract of stems inhibited the growth of *Escherichia coli* with an MBC value of 100 μg/mL (Ono et al., 1998).

Weak antibacterial broad-spectrum isothiocyanate: 6-Methylsulfinylhexyl isothiocyanate isolated from the stems inhibited the growth of *Escherichia coli* and *Staphylococcus aureus* with MBC values of 100 and 2000 μg/mL, respectively (Ono et al., 1998). The MIC of allyl isothiocyanate against *Escherichiacoli* (O157:H7) and *Staphylococcus aureus* was below 100 μg/mL (Lu et al., 2016).

Commentaries: The plant produces allyl isothiocyanate (Shin et al., 2004) the antimicrobial properties of which have been described earlier.

(ii) In the Brassicaceae, *Hesperis schischkinii* Tzvelev is used in Turkey to heal wounds. In India, *Sisymbrium irio* L. is used to treat fever, cold, cough, and measles. In Pakistan,

Physorhynchus chamaerapistrum Boiss. is used to treat boils and putrefied wounds and *Sisymbrium erysimoides* Desf. is used to treat sore throat.

REFERENCES

Lu, Z., Dockery, C.R., Crosby, M., Chavarria, K., Patterson, B. and Giedd, M., 2016. Antibacterial activities of wasabi against Escherichia coli O157: H7 and Staphylococcus aureus. *Frontiers in Microbiology*, *7*, p. 1403.

Ono, H., Tesaki, S., Tanabe, S. and Watanabe, M., 1998. 6-Methylsulfinylhexyl isothiocyanate and its homologues as food-originated compounds with antibacterial activity against Escherichia coli and Staphylococcus aureus. *Bioscience, Biotechnology, and Biochemistry*, *62*(2), pp. 363–365.

Shin, I.S., Masuda, H. and Naohide, K., 2004. Bactericidal activity of wasabi (Wasabia japonica) against Helicobacter pylori. *International Journal of Food Microbiology*, *94*(3), pp. 255–261.

8.3.2 Family Capparaceae A. L. de Jussieu (1789)

The family Capparaceae consists of about 45 genera and about 800 species of treelets, shrubs, and herbs. The leaves are alternate, opposite, simple or trifoliate, or often palmately compound. The stipules are absent or very small. The flowers are bisexual, actinomorphic, and hypogynous and axillary or terminal, solitary or in racemes. A gynophore or androgynophore is present. The calyx consists of 2–6 sepals. The corolla comprises up to 6 petals. The androecium is often showy, and consists of several stamens. The gynoecium consists of 2–12 carpels united to form a unilocular ovary, which is sessile with few or many ovules on parietal. The fruits are drupes or berries containing reniform or angular seeds.

8.3.2.1 *Capparis micracantha* DC.

Synonym: *Capparis liangii* Merr. & Chun

Common names: Caper thorn; kanchoen bai dach (Cambodia); xiao ci shan gan (China); say sou (Laos); kaju tuju (Malaysia); balung (Indonesia); salimomo (the Philippines); ching chi (Thailand); bung che (Vietnam)

Habitat: Forests

Distribution: India, China, and Southeast Asia

Botanical observation: This climbing shrub grows up to a length of 6 m long. The stems are terete and develop thorns, which are up to about 5 mm long. Leaves are simple, spiral, and exstipulate. The petiole is up to 2 cm long. The blade is lanceolate, elliptic to oblong, 7–20 × 3–10 cm, membranous, marked with 7–10 pairs of secondary nerves, cuneate to obtuse at base, acute to acuminate at apex. The flowers are axillary and have an eery aura. The 4 sepals are ovate and up to about 4mm long. The 4 petals are oblong, white, and up to about 1cm long. The androecium includes numerous slender, white, curved, and up to about 3 cm long stamens. A gynophore up to 3.5 cm long is present. The ovary is ovoid. The berry is ovoid and is up to 4 cm long.

Medicinal uses: Bronchitis, cough, syphilis (Cambodia, Laos, Vietnam); fever (Thailand)

Broad-spectrum antibacterial halo elicited by polar extract: Aqueous extract of roots (4 mg/6 mm disc) inhibited the growth of *Escherichia coli* (ATCC 25922), *Shigella boydii*, *Shigella sonnei*, *Salmonella typhimurium*, *Acinetobacter baumanni*, *Klebsiella pneumoniae*, and *Streptococcus pyogenes* (Batm et al., 2011).

Commentaries: (i) The plant synthetizes glucosinolates (Mithen et al., 2010) which are hydrolyzed by the enzyme myrosinase (released following tissue damages) into isothiocyanates, nitriles, epithionitriles, and thiocyanates. In regards to their antibacterial effects, isothiocyanates bind to sulfhydryl groups of enzymes involved in bacterial (and fungal) growth, and survival and reductions in the bacterial levels thiol groups lead to the formation of oxygen and other free-radicals (Borges et al., 2015). In addition, isothiocyanates evoke injuries on bacterial cell membrane, leading to leakage of cytoplasmic contents (Borges et al., 2015). Another interesting fact is that isothiocyanates like allyl isothiocyanate increase the sensitivity of *Escherichia coli* to streptomycin (Saavedra et al., 2010).

(ii) In India, *Capparis decidua* Forssk. is used to treat cough.

REFERENCES

Batm, S.N., Kondo, S. and Itharat, A., 2011. Antimicrobial activity of the extracts from Benchalokawichian remedy and its components. *Journal of the Medical Association of Thailand*, *94*(7), pp. S172–S177.

Borges, A., Abreu, A.C., Ferreira, C., Saavedra, M.J., Simões, L.C. and Simões, M., 2015. Antibacterial activity and mode of action of selected glucosinolate hydrolysis products against bacterial pathogens. *Journal of Food Science and Technology*, *52*(8), pp. 4737–4748.

Mithen, R., Bennett, R. and Marquez, J., 2010. Glucosinolate biochemical diversity and innovation in the Brassicales. *Phytochemistry*, *71*(17–18), pp. 2074–2086.

Saavedra, M.J., Borges, A., Dias, C., Aires, A., Bennett, R.N., Rosa, E.S. and Simões, M., 2010. Antimicrobial activity of phenolics and glucosinolate hydrolysis products and their synergy with streptomycin against pathogenic bacteria. *Medicinal Chemistry*, *6*(3), pp. 174–183.

8.3.2.2 *Cleome gynandra* L.

Synonyms: *Cleome acuta* Schumach. & Thonn.; *Cleome affinis* (Blume) Spreng.; *Cleome alliacea* Blanco; *Cleome alliodora* Blume; *Cleome blumeana* Schult. f.; *Cleome bungei* Steud.; *Cleome candelabrum* Sims; *Cleome denticulata* (DC.) Schult. f.; *Cleome eckloniana* Schrad.; *Cleome edulis* Raf.; *Cleome flexuosa* F. Dietr.; *Cleome heterotricha* Burch.; *Cleome lupinifolia* Bartram; *Cleome muricata* (Schrad.) Schult. f.*Cleome palmipes* (DC.) Spreng.; *Cleome pentaphylla* L.; *Cleome pentaphylla* Willd.; *Cleome triphylla* L.; *Gymnogonia pentaphylla* (L.) R. Br. ex Steud.; *Gynandropsis affinis* Blume; *Gynandropsis candelabrum* (Sims) Sweet; *Gynandropsis denticulata* DC.; *Gynandropsis glandulosa* C. Presl; *Gynandropsis gynandra* (L.) Briq.; *Gynandropsis heterotricha* (Burch.) DC.; *Gynandropsis muricata* Schrad.; *Gynandropsis ophidocarpa* DC.; *Gynandropsis palmipes* DC.; *Gynandropsis pentaphylla* (L.) DC.; *Gynandropsis sessilifolia* DC.; *Gynandropsis sinica* Miq.; *Gynandropsis triphylla* (L.) DC.; *Gynandropsis viscida* Bunge; *Pedicellaria pentaphylla* (L.) Schrank; *Pedicellaria triphylla* (L.) Pax; *Podogyne pentaphylla* (L.) Hoffmanns.; *Sinapistrum pentaphyllum* (L.) Medik.

Common names: Cat's whiskers, five-leaved cleome, bastard mustard, spider plant; momien (Cambodia); pai hua ts'ai (China); maman (Indonesia); sienz (Laos); maman hantu (Malaysia); cincocinco, balaya, silisihan (the Philippines); ajagandha, kabana, veali, kadugu, haivelai (India); phak sian khao, phak sian thai (Thailand); ma man trang (Vietnam)

Habitat: Cultivated

Distribution: Tropical Asia

Botanical observation: This herb grows to a height of 1.2 m and is edible. The stems and branches are striate and more or less clothed with white hairs. The leaves are 3–5 foliate. The petiole is 5–7.6 mm long with occasionally small distant prickles. The folioles are subsessile, 2–4 × 1.2–2.5 cm, elliptic – obovate, obtuse, acute or acuminate, and pubescent on both sides. The margin is crenate-dentate or subentire. The flowers are corymbose and elongating into dense bracteate racemes. The flower pedicels are 1.2–2 cm long, viscid, and pubescent. The calyx includes 5 sepals, which are lanceolate, glandular, and with white veins. The 5 petals are 1.5 cm long, broad, obovate, light pink, with a long narrow claw. The gynophore is 2–2.5 cm long. The androecium includes numerous purple stamens. The ovary is linear-oblong and glandular and develops a minute style.

The capsules are 5–9×4–5cm, viscid, pubescent, obliquely striated, and contain several seeds. The seeds are muricate and dark brown.

Medicinal uses: Dysentery, gonorrhea (Taiwan); herpes (Indonesia);

Broad-spectrum antibacterial halo elicited by polar extract: Ethanol extract of leaves inhibited the growth of *Proteus vulgaris*, *Pseudomonas aeruginosa*, *Bacillus subtilis*, *Bacillus pumilus*, *Staphylococcus aureus*, and *Streptococcus faecalis* (Sridhar et al., 2014).

Strong broad-spectrum antibacterial polar extract: Methanol extract of leaves inhibited the growth of *Bacillus subtilis* (MTCC 441), *Staphylococcus aureus* (MTCC 3160), *Escherichia coli* (MTCC 46), and *Pseudomonas aeruginosa* (MTCC 1688) with MIC/MBC values of 78/78, 39/39, 156/625, and 625/1250 μg/mL, respectively (Sridhar et al., 2014).

Commentaries: (i) The plant produces glucosinolates (Mithen et al., 2010) as well as dammarane-type triterpenes (Das et al., 1999). (ii) *Cleome viscosa* L. is used in India for the treatment of ear infection. Ethanol extract of leaves of *Cleome viscosa* L. (40μL of a 25 mg/L solution/disc) inhibited the growth of *Candida albicans*, *Aspergillus niger*, and *Rhizopus oligosporus* with inhibition zone diameters of 14, 16, and 32 mm, respectively (Sudhakar et al., 2006).

(ii) In India, *Cleome rutidosperma* DC. (local name: *Torel*) is used to treat jaundice.

REFERENCES

Das, P.C., Patra, A., Mandal, S., Mallick, B., Das, A. and Chatterjee, A., 1999. Cleogynol, a Novel Dammarane Triterpenoid from Cleome gynandra. *Journal of Natural Products*, *62*(4), pp. 616–618.

Mithen, R., Bennett, R. and Marquez, J., 2010. Glucosinolate biochemical diversity and innovation in the Brassicales. *Phytochemistry*, *71*(17–18), pp. 2074–2086.

Sridhar, N., Sasidhar, D.T. and Kanthal, L.K., 2014. In vitro antimicrobial screening of methanolic extracts of Cleome chelidonii and Cleome gynandra. *Bangladesh Journal of Pharmacology*, *9*(2), pp. 161–166.

Sudhakar, M., Rao, C.V., Rao, P.M. and Raju, D.B., 2006. Evaluation of antimicrobial activity of Cleome viscosa and Gmelina asiatica. *Fitoterapia*, *77*(1), pp. 47–49.

8.3.2.3 *Crateva religiosa* Forts.

Synonyms: *Crateva macrocarpa* Kurz; *Crateva membranifolia* Miq.; *Crataeva religiosa* Forst.

Common names: Sacred barma; tom lien (Cambodia); yu mu (China); barun (India); sibaluak (Indonesia); kums (Laos); kemantu (Malaysia); banugan (the Philippines); kum nam (Thailand); Bun lo (Vietnam)

Habitat: Forests

Distribution: India to Papua New Guinea, including the Ryuku islands.

Botanical observation: This tree grows up to a height of 15 m. The leaves are alternate, trifoliate, and exstipulate. The petiole is slender and 12 cm long. The folioles are 5–7×3–4 cm, papery, and show 7–10 pairs of secondary nerves. The inflorescences are racemes or corymbs. The flowers are large and yellow. The calyx presents 5 sepals, which are about 5 mm long. The 5 petals are white to yellow, spathulate, and up to about 3 cm long. The 30 stamens are up to about 6.5 cm long. A 3.5–6.5 cm long gynophore is present. The gynoecium includes an ovoid ovary, which is about 4 mm long. The fruit is ovoid, smooth, somewhat greyish green, up to 3.5 cm, and contains numerous seeds.

Medicinal uses: Dysentery (Taiwan); fever (Indonesia)

Antibacterial (Gram-positive) polar extract: Methanol extractinhibited the growth of *Staphylococcus aureus* and *Escherichia coli* (Grosvenor et al., 1995).

Moderate broad-spectrum polar extract: Methanol extract of leaves inhibited the growth of *Bacillus subtilis*, *Staphylococcus aureus*, *Klebsiella pneumoniae*, *Escherichia coli*, *Pseudomonas aeruginosa*, and *Salmonella typhi* with MIC values of 1060, 700, 1720, 1680, 1470, and 1380 μg/mL, respectively (Ezealisiji, 2008).

Commentary: In the Philippines, *Crateva odora* Buch.-Ham. is used to treat ear infections.

REFERENCES

Ezealisiji, K.M., 2018. Evaluation of antimicrobial and anti-inflammatory properties of leaf extracts of Crateva religiosa. *Journal of Pharmacognosy and Phytochemistry*, 7(6), pp. 646–649.
Grosvenor, P.W., Supriono, A. and Gray, D.O., 1995. Medicinal plants from Riau Province, Sumatra, Indonesia. Part 2: antibacterial and antifungal activity. *Journal of Ethnopharmacology*, 45(2), pp. 97–111.

8.3.2.4 *Stixis scortechinii* (King) Jacobs

Synonym: *Roydsia scortechinii* King

Habitat: Forests

Distribution: Malaysia, Indonesia

Botanical observation: This twining shrub grows to a height of about 4 m. The stems are terete, smooth, and lenticelled. The leaves are simple, spiral, and exstipulate. The petiole is up to about 1.5 cm long and somewhat curved toward the blade. The blade is elliptic, 6.5–9 × 2.5–4.5 cm, acute at base, with about 5 pairs of secondary nerves, and acuminate at apex. The inflorescences are terminal racemes, which are up to about 15 cm long. The calyx is tubular and includes 6 lobes. Corolla absent. The androecium comprises numerous stamens and an androgynophore is present as well as a disc. The drupes are ovoid, green, smooth, somewhat like dwarfy papayas, up to about 3.5cm long, and contain a 2.5 cm long seed.

Medicinal use: Sore eyes (Indonesia)

Commentary: This rare plant, like many other unique life forms, is bound to disappear if palm oil deforestation and logging of the primary rain forest are not immediately stopped. One needs to ponder that these rare primary rainforest plants (but also symbiotic microorganisms, lichen, ferns, insect, snails, frogs, mushrooms) produce leads for the treatment of infectious diseases and other non-communicable diseases. Again, one need to recall that the more primary forest is cut the more deadly will be the new zoonotic microorganism produced by negative feed-back loop on human activities and the lesser the possibilities to find cures from mother Nature.

8.3.3 Family Caricaceae Dumortier (1829)

The family Caricaceae comprise 5 genera and 35 species of treelets. The stems release a latex upon incision. The leaves are simple, spiral, and exstipulate. The petiole is long. The blade is palmate or palmatifid. The inflorescences are axillary cymes. The calyx and corolla are tubular and 5-lobed. The androecium comprises 5 or 10 stamens. The gynoecium comprises a 1- or 5-loculed ovary sheltering numerous ovules growing on parietal or laminar placenta, and develop a 5-lobed stigma. The fruits are berries.

8.3.3.1 *Carica papaya* L.

Common names: Papaya; melon tree; lohong si phle (Cambodian); chirbhita, pappali (India); bai mak houng heng (Laos); pohunbetek (Malaysia); pimbosi, simbosi, thimbaw, timbosi (Myanmar); kapaya, papaya (the Philippines)

Habitat: Roadside, wastelands, cultivated

Distribution: Tropical Asia

Botanical observation: This treelet native to Central America grows up to a height of about 8 m. The bole is soft wooded, marked with conspicuous leaf-scars and laticiferous. The leaves are simple, spiral, exstipulate. The petiole is about 30 cm long, fleshy, and thin. The blade is 30–60 cm across, glabrous, palmatifid, and palmatinerved. The flowers are light yellow and slightly fragrant, arranged in long dropping panicles or short clusters. In female flowers, the ovary is single-locular, the stigma is sessile, 5-lobed, and lacerated. The berries are succulent, indehiscent, single-celled, ovoid to oblong, greenish yellow, smooth, and 15 × 30 cm. The flesh is red, juicy, and palatable. The seeds are numerous, blackish-green, and taste like capers.

Medicinal uses: Boils (Malaysia); putrefied wounds, yaws (the Philippines); AIDS (Thailand)

Broad-spectrum antibacterial halo elicited by mid-polar extract: Ethyl acetate extract of fruit peel (2 mg/8.5 mm well) inhibited *Bacillus megaterium* (ATCC9885), *Bacillus subtillis* (ATCC6633),

Citrobacter freundii (NCIM2489), *Enterobacter aerogenes* (ATCC13048), *Klebsiella pneumoniae* (NCIM2719), and *Salmonella typhimurium* (ATCC2364) (Rakholiya et al., 2014).

Moderate broad-spectrum antibacterial polar extract: Latex inhibited the growth of *Enterococcus faecalis* (clinical), *Staphylococcus aureus* (clinical strain), *Staphylococcus saprophyticus* (clinical strain), and *Staphylococcus epidermidis* (clinical strain) (Mbosso Teinkela et al., 2016).

Weak antifungal (filamentous) polar extract: Ethanol extract of leaves inhibited the growth of *Fusarium* spp. with an MIC_{50} value of 625 µg/mL (Chávez-Quintal et al., 2011).

In vivo antiviral (enveloped monopartite linear single-stranded (+) RNA) juice: Fresh juice of leaves given orally at a dose of 5 mL twice a day to Dengue virus patients increased both platelets count and the total white cells count (Hettige S. et al, 2008). In a subsequent study, 25 mL of leave juice administered to a single dengue patient twice daily for five consecutive days increased platelet count from 55×10^3 to 168×10^3/µL and white blood cells increased from 3.7×10^3 to 7.7×10^3/µL (Ahmad et al., 2011). Juice extracted from 50 g of leaves given to 111 dengue patients once in a day for 3 days evoked an increase in platelet count (Subenthiran et al., 2013).

Commentary: The leaves contain the proteolytic enzyme papain as well as flavonoids and their glycosides: quercetin 3-(2*G*-rhamnosylrutinoside), kaempferol 3-(2 *G*-rhamnosylrutinoside), quercetin 3-rutinoside, myricetin 3-rhamnoside, kaempferol 3-rutinoside, quercetin, and kaempferol (Nugroho et al., 2017), which are probably not responsible (as discussed earlier) for the *in vivo* anti-dengue effect of the plant. Quercetin and isoquercetin inhibited Influenza A virus (Oh7 strain) with an EC_{50} of 48 µM (Kim et al., 2010). Of interest, the plant produces the piperidine alkaloid carpaine, which is a more probable anti-dengue candidate as it demonstrated antithrombocytopenic effects in rodent (Zunjar et al., 2016). One could examine the anti-dengue effect of carpaine in vivo. Could teas of leaves afford prophylactic remedies for Dengue? The Cebuanos use a number of plants for Dengue of which notably the fruits of *Capsicum annum* L. (local name: *Sili*) in the family Solanaceae.

REFERENCES

Ahmad, N., Fazal, H., Ayaz, M., Abbasi, B.H., Mohammad, I. and Fazal, L., 2011. Dengue fever treatment with Carica papaya leaves extracts. *Asian Pacific Journal of Tropical Biomedicine*, *1*(4), p. 330.

Chávez-Quintal, P., González-Flores, T., Rodríguez-Buenfil, I. and Gallegos-Tintoré, S., 2011. Antifungal activity in ethanolic extracts of Carica papaya L. cv. Maradol leaves and seeds. *Indian Journal of Microbiology*, *51*(1), pp. 54–60.

Hettige, S., 2008. Salutary effects of carica papaya leaf extract in dengue fever patients – A pilot study. *Sri Lankan Family Physician*, *29*(1), pp. 17–19.

Kim, Y., Narayanan, S. and Chang, K.O., 2010. Inhibition of Influenza virus replication by plant-derived isoquercetin. *Antiviral Research*, *88*(2), pp. 227–235.

Mbosso Teinkela, J.E., Assob Nguedia, J.C., Meyer, F., Vouffo Donfack, E., Lenta Ndjakou, B., Ngouela, S., Tsamo, E., Adiogo, D., Guy Blaise Azebaze, A. and Wintjens, R., 2016. In vitro antimicrobial and antiproliferative activities of plant extracts from Spathodea campanulata, Ficus bubu, and Carica papaya. *Pharmaceutical Biology*, *54*(6), pp. 1086–1095.

Nugroho, A., Heryani, H., Choi, J.S. and Park, H.J., 2017. Identification and quantification of flavonoids in Carica papaya leaf and peroxynitrite-scavenging activity. *Asian Pacific Journal of Tropical Biomedicine*, *7*(3), pp. 208–213.

Rakholiya, K., Kaneria, M. and Chanda, S., 2014. Inhibition of microbial pathogens using fruit and vegetable peel extracts. *International Journal of Food Sciences and Nutrition*, *65*(6), pp. 733–739.

Subenthiran, S., Choon, T.C., Cheong, K.C., Thayan, R., Teck, M.B., Muniandy, P.K., Afzan, A., Abdullah, N.R. and Ismail, Z., 2013. Carica papaya leaves juice significantly accelerates the rate of increase in platelet count among patients with dengue fever and dengue haemorrhagic fever. *Evidence-Based Complementary and Alternative Medicine*, *2013*. Article ID 616737.

Zunjar, V., Dash, R.P., Jivrajani, M., Trivedi, B. and Nivsarkar, M., 2016. Antithrombocytopenic activity of carpaine and alkaloidal extract of Carica papaya Linn. leaves in busulfan induced thrombocytopenic Wistar rats. *Journal of Ethnopharmacology*, *181*, pp. 20–25.

8.3.4 Family Moringaceae Martinov (1820)

The family Moringaceae consists of the single genus *Moringa* Adans.

8.3.4.1 *Moringa oleifera* Lam.

Synonyms: *Guilandina moringa* L., *Hyperanthera moringa* (L.) Vahl, *Moringa erecta* Salisb., *Moringa moringa* (L.) Millsp., *Moringa pterygosperma* Gaertn., *Moringa zeylanica* Burmann

Common names: Moringa; mungdodare (Bangladesh); sajina (India); moringa (the Philippines)

Habitat: Cultivated, roadside, wastelands

Distribution: Tropical Asia

Botanical observation: This tree grows up to a height of about 5 m. The bark is smooth, greying, and exudes a gum upon incision. The stems are hairy when young. The leaves are tripinnately imparipinnate, spiral, and exstipulate. The petiole is 4–15 cm long. The blade is up to about 45 cm long, with 5–11 pinnae, and 3–11 dark dull green folioles which are 1–1.5 × 0.5–1.8 cm, elliptic, and membranous. The panicles are axillary, 8–30 cm long, pendulous, and lax. The 5 lobes of the calyx tube are lanceolate and 1.3–1.5 cm long. The 5 petals are whitish, spathulate with prominent nervation, and 1.2–1.8 cm long. The 5 stamens are 1 cm long. The ovary is oblong, about 5 mm long, and develops a cylindrical style. The capsules are 9-ribbed, 30–45 cm long, fusiform, edible (after being boiled), irregular, dull green, and contain numerous seeds, which are dark brown, globose, about 5 mm across.

Medicinal uses: Cough (Bangladesh; India) pox, fever (Bangladesh) flu (India, Malaysia); sores (the Philippines)

Antibacterial (Gram-negative) halo elicited by polar extract: Aqueous extract of seeds and juice of leaves inhibited the growth of *Pseudomas aeruginosa* (27853) (Caceres et al., 1991).

Broad-spectrum antibacterial halo elicited by polar extract: Acetone extract of fruit peel (2 mg/ 8.5mm well) inhibited *Bacillus megaterium* (ATCC9885), *Bacillus subtilis* (ATCC6633), *Staphylococcus aureus* (ATCC25293), *Staphylococcus epidermidis* (ATCC12228), *Citrobacter freundii* (NCIM2489), *Enterococcus aerogenes* (ATCC13048), *Klebsiella pneumoniae* (NCIM2719), *Proteus mirabilis* (NCIM2241), and *Salmonella thyphimurium* (ATCC2364) with inhibition zone diameters of 10, 9, 9.5, 11, 12, 9.5, 12, 10, and 24.5 mm, respectively (Rakholiya et al., 2014).

Strong broad-spectrum antibacterial sulfur benzylamide: *N*-Benzyl, *N*-ethyl thioformate isolated from the root bark inhibited the growth of *Shigella dysenteriae*, *Shigella boydii*, *Bacillus megaterium*, and *Staphylococcus aureus* with MIC values of 32, 32, 64, and 32 μg/mL, respectively (Nikkon et al., 2003).

Moderate antifungal (filamentous) essential oil: Essential oil inhibited the growth of *Trichophyton rubrum*, *Trichophyton mentagrophytes*, *Epidermophyton floccosum*, and *Microsporum canis* with MIC values of 1.6, 0.8, 0.2, and 0.4 mg/mL, respectively (Chuang et al., 2007).

Moderate antiviral (enveloped monopartite linear dimeric single-stranded (+) RNA) polar extract: Ethanol extract of leaves inhibited the replication of the Human immunodeficiency virus with an IC_{50} value of 125 μg/mL (Bunluepuech, 2010).

Strong antiviral mid-polar extract: Chloroform extract of leaves inhibited the replication of the foot and mouth disease virus at a concentration of 50 μg/mL (Younus et al., 2007).

Strong antiviral (enveloped monopartite linear double-stranded DNA) polar extract: Methanol extract of flowers inhibited the replication of the Herpes simplex virus type-1 with an IC_{50} value of 74.8 μg/mL (Goswami et al., 2016).

Commentary: The plant produces sulfur compounds, including 4(α-L-rhamnosyloxy)-benzyl isothiocyanate and niazimicin (Guevara et al., 1999), which are most probably antibacterial and antifungal. The antiviral properties of this edible plant could invite preclinical trials.

REFERENCES

Bunluepuech, K., 2010. *Study on anti-HIV-1 integrase activity of Thai medicinal plants* (Doctoral dissertation, Prince of Songkla University).

Caceres, A., Cabrera, O., Morales, O., Mollinedo, P. and Mendia, P., 1991. Pharmacological properties of Moringa oleifera. 1: Preliminary screening for antimicrobial activity. *Journal of Ethnopharmacology*, *33*(3), pp. 213–216.

Chuang, P.H., Lee, C.W., Chou, J.Y., Murugan, M., Shieh, B.J. and Chen, H.M., 2007. Anti-fungal activity of crude extracts and essential oil of Moringa oleifera Lam. *Bioresource Technology*, *98*(1), pp. 232–236.

Goswami, D., Mukherjee, P.K., Kar, A., Ojha, D., Roy, S. and Chattopadhyay, D., 2016. Screening of ethnomedicinal plants of diverse culture for antiviral potentials. *IJTK*, *15*, pp. 474–481.

Guevara, A.P., Vargas, C., Sakurai, H., Fujiwara, Y., Hashimoto, K., Maoka, T., Kozuka, M., Ito, Y., Tokuda, H. and Nishino, H., 1999. An antitumor promoter from Moringa oleifera Lam. *Mutation Research/Genetic Toxicology and Environmental Mutagenesis*, *440*(2), pp. 181–188.

Nikkon, F., Saud, Z.A., Rehman, M.H. and Haque, M.E., 2003. In vitro antimicrobial activity of the compound isolated from chloroform extract of Moringa oleifera Lam. *Pakistan Journal of Biological Sciences*, *22*, pp. 1888–1890.

Rakholiya, K., Kaneria, M. and Chanda, S., 2014. Inhibition of microbial pathogens using fruit and vegetable peel extracts. *International Journal of Food Sciences and Nutrition*, *65*(6), pp. 733–739.

Younus, I., Siddiq, A., Ishaq, H., Anwer, L., Badar, S. and Ashraf, M., 2016. Evaluation of antiviral activity of plant extracts against foot and mouth disease virus in vitro. *Pakistan Journal of Pharmaceutical Sciences*, *29*(4), pp. 1263–1268.

8.3.5 Family Salvadoraceae Lindley (1836)

The family Salvadoraceae consists of 3 genera and 12 species of shrubs and treelets. The leaves are simple, opposite, and exstipulate. The inflorescence is a panicle. The calyx is tubular and 2–4 lobed. The corolla comprises 4–5 free or fused petals. The androecium consists of 4–5 stamens The gynoecium is made of 2 carpels fused in a 1–2-locular ovary, each locule sheltering 1–2 ovules growing on basal placentas. The fruit is a berry or a drupe.

8.3.5.1 *Salvadora persica* L.

Common name: Miswak tree, toothbrush tree; jhal (Bangladesh); dhanin, pilu (India); darakht-e-misvak (Iran); arak, peelu (Pakistan); misvak ağacı (Turkey)

Habitat: Deserts

Distribution: Turkey, Iran, Afghanistan, Pakistan, India

Botanical observation: This graceful tree grows to a height of 6 m. The bole is curved. The stems are terete and fibrous. The leaves are simple, opposite, and exstipulate. The petiole is up to about 2 cm long. The blade is coriaceous, dull, somewhat glaucous, lanceolate, 1–5.5 × 0.8–2.3 cm, and with 5–6 pairs of secondary nerves. The inflorescence is a showy panicle, which reaches about a length of 30 cm. The calyx is tubular and 4-lobed, the lobes are round. The corolla is tubular and 4-lobed, the lobes are greenish, oblong, recurved, and 3 mm long. The androecium comprises 4 stamens with showy anthers. The drupes are globose, burgundy, edible, 5 mm across, and glossy.

Medicinal uses: Toothbrush (Turkey, Iran, Afghanistan, Pakistan, India, Bangladesh, Malaysia, Indonesia); cough, fever (India)

Weak broad-spectrum antibacterial polar extract: Methanol extract of stems inhibited the growth of methicillin-resistant *Staphylococcus epidermidis*, *Streptococcus pyogenes*, *Escherichia coli*, *Klebsiella pneumoniae*, *Pseudomonas aeruginosa*, and *Serratia marcescens* with MIC values of 3125, 1560, 390, 781, 1560, and 1560 μg/mL, respectively (Al-Ayed et al., 2016). In a subsequent study, aqueous extract of stems inhibited the growth of *Streptococcys mutans* with an MIC value of 1560 μg/mL (Al-Bayati & Sulaiman, 2008).

Broad-spectrum antibacterial halo elicited by isothiocyanate: Benzyl isothiocyanate (Noumi et al., 2011) (8 mm disc impregnated with 0.1 μL/mL) inhibited the growth of *Bacillus subtilis*,

Staphylococcus aureus, *Salmonella enterica*, *Serratia marcescens*, *Shigella sonnei*, and *Vibrio parahaemolyticus* with inhibition zone diameters of 45.8, 62.7, 67, 16.8, 44.3, and 32,6mm, respectively (Jang et al., 2010).

Commentaries: (i) Aqueous extract of seeds at the concentration of 330 μg/mL inhibited angiotensin-converting enzyme by 55% (Nyman et al., 1998) and as such, could be of value against COVID-19 and other pandemic coronaviruses.(ii) The plant hold a special importance by Asian Muslims. According to Amin (2020), narrations dating from the 7th century recommend sanitary acts including washing hands, using Siwak for oral hygiene, covering the face when sneezing, avoiding entering lands with plague. Consider that the Talmud recommends self-quarantine in time of plague "And none of you shall go out of the opening of his house until the morning" (Exodus 12:22), and according to Chechik and Morsel-Eisenberg (2020) in times of plague "It is likewise written in Zohar, Parashat Vayera Elav, that one should flee the city".

REFERENCES

Al-Ayed, M.S.Z., Asaad, A.M., Qureshi, M.A., Attia, H.G. and AlMarrani, A.H., 2016. Antibacterial activity of Salvadora persica L.(Miswak) extracts against multidrug resistant bacterial clinical isolates. *Evidence-Based Complementary and Alternative Medicine, 2016*. Article ID 7083964.

Al-Bayati, F.A. and Sulaiman, K.D., 2008. In vitro antimicrobial activity of Salvadora persica L. extracts against some isolated oral pathogens in Iraq. *Turkish Journal of Biology*, *32*(1), pp. 57–62.

Chechik, M.D., and Morsel-Eisenberg, T., 2020. Plague, Practice, and Prescriptive Text: Jewish Traditions on Fleeing Afflicted Cities in Early Modern Ashkenaz. *Journal of Law, Religion and State*, 8 (2-3), pp.152–178.

Jang, M., Hong, E. and Kim, G.H., 2010. Evaluation of antibacterial activity of 3-butenyl, 4-pentenyl, 2-phenylethyl, and benzyl isothiocyanate in Brassica vegetables. *Journal of Food Science*, *75*(7), pp. M412–M416.

Noumi, E., Snoussi, M., Trabelsi, N., Hajlaoui, H., Ksouri, R., Valentin, E. and Bakhrouf, A., 2011. Antibacterial, anticandidal and antioxidant activities of Salvadora persica and Juglans regia L. extracts. *Journal of Medicinal Plants Research*, *5*(17), pp. 4138–4146.

Nyman, U., Joshi, P., Madsen, L.B., Pedersen, T.B., Pinstrup, M., Rajasekharan, S., George, V. and Pushpangadan, P., 1998. Ethnomedical information and in vitro screening for angiotensin-converting enzyme inhibition of plants utilized as traditional medicines in Gujarat, Rajasthan and Kerala (India). Journal of Ethnopharmacology, 60(3), pp. 247–263.

Bibliography

Abe, R. and Ohtani, K., 2013. An ethnobotanical study of medicinal plants and traditional therapies on Batan Island, the Philippines. *Journal of Ethnopharmacology, 145*, pp. 554–565.

Abu Bakar, F.I., Abu Bakar, M.F., Abdullah, N., Endrini, S. and Rahmat, A., 2018. A review of Malaysian medicinal plants with potential anti-inflammatory activity. *Hindawi*, 13 pages, Article ID 8603602.

Acharya, K.P. and Acharya, M., 2010. Traditional knowledge on medicinal plants used for the treatment of livestock diseases in Sardikhola VDC, Kaski, Nepal. *Journal of Medicinal Plants Research, 4*(2), pp. 235–239.

Acharya, E. (Siwakoti) and Pokhrel, B., 2006. Ethno-medicinal plants used by Bantar of Bhaudaha, Morang, Nepal. *Our Nature, 4*, pp. 96–103.

Adams, M., Alther, W., Kessler, M., Kluge, M. and Hamburger, M., 2011. Malaria in the renaissance: Remedies from European herbals from the 16th and 17th century. *Journal of Ethnopharmacology, 133*, pp. 278–288.

Ahmad, S.S., 2007. Medicinal wild plants from Lahore-Islamabad Motorway (M-2). *Pakistan Journal of Botany*, 39(2), pp. 355–375.

Ahmad, K., Weckerle, C.S. and Nazir, A. Ethnobotanical investigation of wild vegetables used among local communities in northwest Pakistan. *Acta Societatis Botanicorum Poloniae, 88*(1), p. 3616.

Ahmed, N., Mahmood, A., Mahmood, A., Sadeghi, Z. and Farman, M., 2015. Ethnopharmacological importance of medicinal flora from the district of Vehari, Punjab province, Pakistan. *Journal of Ethnopharmacology, 168*, pp. 66–78.

Ahmed, M.J. and Murtaza, G., 2015. A study of medicinal plants used as ethnoveterinary: Harnessing potential phytotherapy in Bheri, District Muzaffarabad (Pakistan). *Journal of Ethnopharmacology, 159*, pp. 209–214.

Ainslie, W., 1813. *Materia Medica of Hindoostan, and Artisan's and Agriculturist's Nomenclature*. Madras: Government Press.

Al-Adhroey, A.H., Nor, Z.M., Al-Mekhlafi, H.M. and Mahmud, R., 2010. Ethnobotanical study on some Malaysian anti-malarial plants: A community based survey. *Journal of Ethnopharmacology, 132*, pp. 362–364.

Altundag, E. and Ozturk, M., 2011. Ethnomedicinal studies on the plant resources of east Anatolia, Turkey. *Procedia Social and Behavioral Sciences, 19*, pp. 756–777.

Amber, R., Adnan, M., Tariq, A. and Mussarat, S., A review on antiviral activity of the Himalayan medicinal plants traditionally used to treat bronchitis and related symptoms. *Journal of Pharmacy and Pharmacology*. doi: 10.1111/jphp.12669

Amini, M.H. and Hamdam, S.M., 2017. Medicinal plants used traditionally in Guldara District of Kabul, Afghanistan. *International Journal of Pharmacognosy and Chinese Medicine, 1*(3), 000118.

Amiri, M.S., Joharchi, M.R. and TaghavizadehYazdi, M.E., 2014. Ethno-medicinal plants used to cure jaundice by traditional healers of Mashhad, Iran. *Iranian Journal of Pharmaceutical Research, 13*(1), pp. 157–162.

Asiatic Society of Bengal, 1812. Asiatic researches or transactions of the Society instituted in Bengal, for inquiring into the history and antiquities, the arts, sciences, and literature, of Asia. The Bavarian State Library.

Atigur Rahman, M., Uddin, S.B. and Wilcok, C.C., 2007. Medicinal plants used by Chakma tribe in Hill tracts Districts of Bangladesh. *Indian Journal of Traditional Knowledge, 6*(3), pp. 508–517.

Ayyanar, M. and Ignacimuthu, S., 2011. Ethnobotanical survey of medicinal plants commonly used by Kani tribals in Tirunelveli hills of Western Ghats, India. *Journal of Ethnopharmacology, 134*, pp. 851–864.

Azhar, M.F., Aziz, A., Haider, M.S., Nawaz, M.F. and Zulfiqar, M.A., 2015. Exploring the ethnobotany of *Haloxylon recurvum* (KHAR) and *Haloxylon salicornicum* (LANA) in Cholistan desert, Pakistan. *Pakistan Journal of Agricultural Sciences, 52*(4), pp. 1085–1090.

Balfour, E., 1857. *The Cyclopaedia of India and of Eastern and Southern Asia*. Madras: Scottish Press.

Bentley, R., 1872. *Dr. Pereira's Elements of Materia Medica and Therapeutics: Abridged and Adapted for the Use of Medical and Pharmaceutical Practitioners and Students*. London: Longmans, Green, and Co.

Bhagya, B. and Sridhar, K.R., 2009. Ethnobiology of coastal sand dune legumes of Southwest coast of India. *Indian Journal of Traditional Knowledge, 8*(4), pp. 611–620.

Bhandary, M.J. and Chandrashekar, K.R., 2011. Herbal therapy for herpes in the ethno-medicine of Coastal Karnataka. *Indian Journal of Traditional Knowledge, 10*(3), pp. 528–532.

Bhatia, H., Sharma, Y.P., Manhas, R.K. and Kumar, K., 2014. Ethnomedicinal plants used by the villagers of district Udhampur, J&K, India. *Journal of Ethnopharmacology, 151*, pp. 1005–1018.

Bose, D., Roy, J.G., Mahapatra, S.D. (Sarkar), Datta, T., Mahapatra, S.D. and Biswas, H., 2015. Medicinal plants used by tribals in Jalpaiguri district, West Bengal, India. *Journal of Medicinal Plants Studies, 3*(3), pp. 15–21.

Bulut, G., Zeki Haznedaroğlu, M., Doğan, A., Koyu, H. and Tuzlacı, E., 2017. An ethnobotanical study of medicinal plants in Acipayam (Denizli-Turkey). *Journal of Herbal Medicine, 10*, pp. 64–81.

Chakraborty, M.K. and Bhattacharjee, A., 2006. Some common ethnomedicinal uses for various diseases in Purulia District, West Bengal. *Indian Journal of Traditional Knowledge, 5*(4), pp. 554–558.

Chang, C.-S., Kim, H. and Chang, K.S., 2014. *Provisional Checklist of Vascular Plants for the Korea Peninsula flora (KPF)*. 660 pp.

Chowdhury, M. and Mukherjee, R., 2010. Ethno-medicinal survey of Santal tribe of Malda District of West Bengal, India. *Journal of Economic and Taxonomic Botany, 34*(3).

Chowdhury, M.S.H., Koike, M., Muhammed, N., Halim, M.A., Saha, N. and Kobayashi H., 2009. Use of plants in healthcare: A traditional ethno-medicinal practice in southeastern rural areas of Bangladesh. *International Journal of Biodiversity Science and Management, United Kingdom, 5*(1), pp. 41–51.

Chowlu, K., Mahar, K.S. and Das, A.K., 2017. Ethnobotanical studies on orchids among the *Khamti* Community of Arunachal Pradesh, India. *Indian Journal of Natural Products and Resources, 8*(1), pp. 89–93.

Colonel Heber Drury, 1873. *The Useful Plants of India*. London: William H. Allen & Co.

Cox, P.A., 1993. Saving the ethnopharmacological heritage of Samoa. *Journal of Ethnopharmacology, 38*, pp. 181–188.

Das, D. and P. Ghosh., 2017. Some important medicinal plants used widely in Southwest Bengal, India. *International Journal of Engineering Science Invention, 6*(6), pp. 28–50.

Das, T., Mishra, S.B., Saha, D. and Agarwal, S., 2012. Ethnobotanical survey of medicinal plants used by ethnic and rural people in Eastern Sikkim Himalayan Region. *African Journal of Basic & Applied Sciences, 4*(1), pp. 16–20.

Das, A.K. and Tongbram, Y., 2014. Study on medicinal plants used by Meitei community of Bishnupur district, Manipur. *International Journal of Current Research, 6*(2), pp. 5211–5219.

Debala Devi, A., Ibeton Devi, O., Chand Singh, T. and Singh, E.J., 2014. A study of aromatic plant species especially in Thoubal District, Manipur, North East India. *International Journal of Scientific and Research Publications, 4*(6).

Defilipps, R.A. and Krupnick, G.A., 2018. The medicinal plants of Myanmar. *PhytoKeys, 102*, pp. 1–341.

Defilipps, R.A., Maina, S.L. and Pray, L.A., 1988. The palauan and yap medicinal plant studies of masayoshi okabe, 1941–1943. Atoll Research Bulletin. National Museum of Natural History Smithsonian Institution, Washington, D.C., October 1988.

Dey, A. and De, J.N., 2010. A survey of ethnomedicinal plants used by the tribals of Ajoydha Hill region, Purulia District, India. *American-Eurasian Journal of Sustainable Agriculture, 4*(3), pp. 280–290.

Dey, A. and De, J.N., 2012. Ethnobotanical survey of Purulia district, West Bengal, India for medicinal plants used against gastrointestinal disorders. *Journal of Ethnopharmacology, 143*, pp. 68–80.

Durmuşkahya, C. and Öztürk, M., 2013. Ethnobotanical survey of medicinal plants used for the treatment of diabetes in Manisa, Turkey. *Sains Malaysiana, 42*(10), pp. 1431–1438.

Dutta, S.K., Vanlalhmangaiha, Akoijam, R.S., Lungmuana, Boopathi, T. and Saha, S., 2018. Bioactivity and traditional uses of 26 underutilized ethno-medicinal fruit species of North-East Himalaya, India. *Journal of Food Measurement and Characterization, 12*, pp. 2503–2514.

Elliott, E., Chassagne, F., Aubouy, A., Deharo, E., Souvanasy, O., Sythamala, P., Sydara, K., Lamxay, V., Manithip, C., Torres, J.A. and Bourdy, G., 2020. Forest Fevers: Traditional treatment of malaria in the southern lowlands of Laos. *Journal of Ethnopharmacology, 247*.

Flora of China Editorial Committee, 2018. *Flora of China*. Missouri Botanical Garden and Harvard University, Herbaria, USA.

Fluchiger, F.A. and Hangury, D., 1874. *A History of the Principal Drugs of Vegetable Origin, Met with in Great Britain and British India*. London: Macmillan and Co.

Frederick Porter Smith, 1871. *Contributions towards the Materia Medica and Natural History of China*. Shanghai: American Presbyterian Mission Press.

Gagnepain, F., 1934. Iridacées, Amaryllidacées et Liliacées nouvelles d'Asie, *Bulletin de la Société Botanique de France, 81*(1), pp. 66–74. doi: 10.1080/00378941.1934.10833934

Gao, X-M., Ji, B-K., Li, Y-K., Ye, Y-Q., Jiang, Z-Y., Yang, H-Y., Du, G., Zhou, M., Pan, X-X., Liu, W-X. and Hu, Q-F., 2016. New biphenyls from *Garcinia multiflora. The Journal of the Brazilian Chemical Society, 27*(1), pp. 10–14.

Gautam, R., Saklani, A. and Jachak, S.M., 2007. Indian medicinal plants as a source of antimycobacterial agents. *Journal of Ethnopharmacology, 110*, pp. 200–234.

Ghosh, C., 2017. Ethnobotanical survey in the Bamangola Block of Malda District, West Bengal (India): II. Medicinal and Aromatic plants. *Pleione, 11*(2), pp. 249–267.

Giesen, W., Wulffraat, S., Zieren, M. and Scholten, L., 2007. *Mangrove Guidebook for Southeast Asia.* Dharmasarn Co., Ltd., Bangkok, Thailand.

Girach, R.D., Brahmam, M., Misra, M.K. and Ahmed, M., 1998. Indigenous phytotherapy for filariasis from Orissa. *Ancient Science of Life, 17*(3), pp. 224–227.

Goswami, D., Mukherjee, P.K. and Kar, A., 2016. Screening of ethnomedicinal plants of diverse culture for antiviral potentials. *Indian Journal of Traditional Knowledge, 15*(3), pp. 474–481.

Grosvenor, P.W., Gothard, P.K., McWilliam, N.C., Suprlono, A. and Gray, D.O., 1995. Medicinal plants from Riau Province, Sumatra, Indonesia. Part 1: Uses. *Journal of Ethnopharmacology, 45*, pp. 75–95.

Gruyal, G.A. del Roasario, R. and Palmes, N.D., 2014. Ethnomedicinal plants used by residents in Northern Surigao del Sur, Philippines. *Natural Products Chemistry & Research, 2*, p. 4.

Gunawardana, S.L.A. and Jayasuriya, W.J.A.B.N. 2019. Medicinally important herbal flowers in Sri Lanka. *Hindawi,* 18 pages. Article ID 2321961.

Hanbury, D., 1862. Notes on Chinese Materia Medica. In: John E. Taylor (ed.), *The Pharmaceutical Journal and Transactions.*

Hasan, Md. M., Hossain, Sk. A., Ali, Md. A. and Alamgir, A.N.M., 2014. Medicinal plant diversity in Chittagong, Bangladesh: A database of 100 medicinal plants. *Journal of Scientific and Innovative Research, 3*(5), pp. 500–514.

Hooper, D., 1937. *Useful Plants and Drugs of Iran and Iraq.* The Library of the University of Illinois.

Hope, J. 1970. *Lectures on the Material Medica: Containing the Natural History of Drugs, Their Virtues and Doses: Also Directions for the Study of the Materia Medica; and An Appendix on the Method of Prescribing.* London: Edward and Charles Dilly.

Ijaz, F., Iqbal, Z., Rahman, I.U., Alam, J., Khan, S.M., Shah, G.M., Khan, K. and Afzal, A., 2016. Investigation of traditional medicinal floral knowledge of Sarban Hills, Abbottabad, KP, Pakistan. *Journal of Ethnopharmacology, 179*, pp. 208–233.

Jahan, I., Rahman, M.A. and Hossain, M.A., 2019. Medicinal species of Fabaceae occurring in Bangladesh and their conservation status. *Journal of Medicinal Plants Studies, 7*(4), pp. 189–195.

Jain, S.K. and Srivastava, S., 2005. Traditional uses of some Indian plants among islanders of the Indian Ocean. *Indian Journal of Traditional Knowledge, 4*(4), pp. 345–357.

Jan, G., Khan, M.A. and Gul, F., 2008. Ethnomedicinal plants used against diarrhea and dysentery in Dir Kohistan Valley (NWFP), Pakistan. *Ethnobotanical Leaflets, 12*, pp. 620–637.

Jazani, A.M., Maleki, R.F., Kazemi, A.h., Matankolaei, L.g., Targhi, S.T., kordi, S., Rahimi-Esboei, B. and Azgomi, R.N.D., 2018. Intestinal Helminths from the viewpoint of traditional Persian medicine versus modern medicine. *African Journal of Traditional, Complementary and Alternative Medicines, 15*(2), pp. 58–67.

Ji, H., Long, C. and Pei, S., 2004. An ethnobotanical study of medicinal plants used by the Lisu people in Nujiang, Northwest Yunnan, China. *Economic Botany, 58*, pp. S253–264. doi: 10.1663/0013-000

Kadir, M.F., Sayeed, M.S.B., Setu, N.I., Mostaf, A. and Mia, M.M.K., 2014. Ethnopharmacological survey of medicinal plants used by traditional health practitioners in Thanchi, Bandarban Hill Tracts, Bangladesh. *Journal of Ethnopharmacology, 155*, pp. 495–508.

Kar, A. and Borthakur, S.K., 2008. Medicinal plants used against dysentery, diarrhoea and cholera by the tribes of erstwhile Kameng District of Arunachal Pradesh. *Natural Product Radiance, 7*(2), pp. 176–181.

Kar, T., Mandal, K.K., Reddy, C.S. and Biswal, A.K., 2013. Ethnomedicinal plants used to cure diarrhoea, dysentery and cholera by some tribes of Mayurbhanj District, Odisha, India. *Life Sciences Leaflets, 2*, pp. 18–28.

Karki, M., Hill, R., Xue, D., Alangui, W., Ichikawa, K. and Bridgewater, P., 2017. Knowing our Lands and Resources: Indigenous and Local Knowledge and Practices related to Biodiversity and Ecosystem Services in Asia. Published in 2017 by the United Nations Educational, Scientific and Cultural Organization, 7, place de Fontenoy, 75352 Paris 07 SP, France.

Kayani, S., Ahmad, M., Sultana, S., Shinwari, Z.K., Zafar, M., Yaseen, G., Hussain, M. and Bibi, T., 2015. Ethnobotany of medicinal plants among the communities of Alpineand Sub-alpine regions of Pakistan. *Journal of Ethnopharmacology, 164*, pp. 186–202.

Ketpanyapong, W. and Itharat, A., 2016. Antibacterial activity of Thai medicinal plant extracts against microorganism isolated from post-weaning diarrhea in piglets. *Journal of the Medical Association of Thailand, 99*(Suppl. 4), pp. S203–S210.

Khadka, D., Dhamala, M.K., Li, F., Aryal, P.C., Magar, P.R., Bhatta, S., Bhatta, S., Basnet, A., Shi, S. and Cui, D., 2005. The use of medicinal plant to prevent COVID-19 in Nepal. Research Square. doi:10.21203/rs.3.rs–88908/v1

Khan, S.U., Khan, R.U., Mehmood, S., Khan, A., Ullah, I. and Bokhari, T.Z. Medicinal plants used to cure diarrhea and dysentery by the local inhabitants of District Bannu, Khyber PakhtoonKhwa, Pakistan. *Advances in Pharmaceutical and Ethnomedicines, 1*(1), pp. 15–18.

Khan, S.W. and Khatoon, S., 2008. Ethnobotanical studies on some useful herbs of Haramosh and Bugrote valleys in Gilgit, Northern areas of Pakistan. *Pakistan Journal of Botany, 40*(1), pp. 43–58.

Khare, C.P. *Indian Medicinal Plants: An Illustrated Dictionary.* New York: Springer.

Khisha, T., Karim, R., Chowdhury, S.R. and Banoo, R., 2012. Ethnomedical studies of Chakma communities of Chittagong Hill tracts, Bangladesh. *Bangladesh Pharmaceutical Journal, 15*(1), pp. 59–67.

Kirtikar, K.R., Basu, B.D., An, I.C.S. *Indian Medicinal Plants.* Delhi: Jayyed Press.

Koch, M., Kehop, D.A., Kinminja, B., Sabak, M., Wavimbukie, G., Barrows, K.M., Matainaho, T.K., Barrows, L.R. and Rai, P.P., 2015. An ethnobotanical survey of medicinal plants used in the East Sepik province of Papua New Guinea. *Journal of Ethnobiology and Ethnomedicine, 11*, p. 79.

Korkmaz, M. and Karakuş, S., 2015. Traditional uses of medicinal plants of Üzümlü district, Erzincan, Turkey. *The Pakistan Journal of Botany, 47*(1), pp. 125–134.

Koteswara Rao, J., Suneetha, J., Seetharami Reddi, T.V.V. and Aniel Kumar, O., 2011. Ethnomedicine of the Gadabas, a primitive tribe of Visakhapatnam district, Andhra Pradesh. *International Multidisciplinary Research Journal, 1/2*, pp. 10–14.

Kültür, Ş., 2007. Medicinal plants used in Kırklareli Province (Turkey). *Journal of Ethnopharmacology, 111*, pp. 341–364.

Lemmens, R.H.M.J. and Bunyapraphatsara, N., 2003. *Plant Resources of South-East Asia: Medicinal and Poisonous Plants.* Leiden: Backhuys Publishers.

Lemmens, R.H.M.J., Soerianegara, I. and Wong, W.C., 1995. *Plant Resources of South-East Asia: Timber Trees: Minor Commercial Timbers.* Leiden: Backhuys Publishers.

Li, D-l. and Xing, F-w., 2016. Ethnobotanical study on medicinal plants used by local Hoklos people on Hainan Island, China. *Journal of Ethnopharmacology, 194*, pp. 358–368.

Li, D.-l., Zheng, X.-l., Duan, L., Deng, S.-w., Ye, W., Wang, A.-h. and Xinga, F.-w., 2017. Ethnobotanical survey of herbal tea plants from the traditional markets in Chaoshan, China. *Journal of Ethnopharmacology, 205*, pp. 195–206.

Liu, B., Zhang, X., Bussmaan, R.W., Hart, R.H., Li, P., Bai, Y. and Long, C., 2017. *Garcinia* in Southern China: Ethnobotany, management, and niche modeling. *Economic Botany, 70*(4), pp. 416–430.

Lokho, K. and Narasimhan, D., 2013. Ethnobotany of Mao-Naga Tribe of Manipur, India. *Pleione, 7*(2), pp. 314–324.

Lone, P.A. and Bhardwaj, A.K., 2013. Traditional herbal based disease treatment in some rural areas of Bandipora district of Jammu and Kashmir, India. *Asian Journal of Pharmaceutical and Clinical Research, 6*(4).

Long, C-l. and Li, R., 2004. Ethnobotanical studies on medicinal plants used by the Red-headed Yao People in Jinping, Yunnan Province, China. *Journal of Ethnopharmacology, 90*, pp. 389–395.

Mahbubur Rahman, A.H.M. and Akter, M., 2013. Taxonomy and medicinal uses of Euphorbiaceae (Spurge) family of Rajshahi, Bangladesh. *Research in Plant Sciences, 1*(3), pp. 74–80.

Mahmood, A., Mahmood, A., Shaheen, H., Qureshi, R.A., Sangi, Y. and Gilani, S.A., 2011. Ethno medicinal survey of plants from district Bhimber Azad Jammu and Kashmir, Pakistan. *Journal of Medicinal Plants Research, 5*(11), pp. 2348–2360.

Malla, B., 2015. Ethnobotanical study on medicinal plants in Parbat district of western Nepal. A Dissertation submitted for the Partial Fulfilment of the Requirement for the Degree of Doctor of Philosophy in Environmental Science.

Manilal, K.S. and Sabu, T., 1985. Cyclea Barbata Miers (Menispermaceae): A new record of a medicinal plant from south India. *Ancient Science of Life, IV*(4), pp. 229–231.

Merrill, E.D., 1903. *A Dictionary of the Plant Names of the Philippine Islands.* Manila: Bureau of Public Printing.

Mikaili, P., Shayegh, J. and Asghari, M.H., 2012. Review on the indigenous use and ethnopharmacology of hot and cold natures of phytomedicines in the Iranian traditional medicine. *Asian Pacific Journal of Tropical Biomedicine,* pp. S1189–S1193.

Minh, V.V., Yen, N.T.K. and Thoa, P.T.K., 2014. Medicinal plants used by the Hre community in the Ba to district of central Vietnam. *Journal of Medicinal Plants Studies, 2*(3), pp. 64–71.
Mishra, D.N., 2009. Medicinal plants for the treatment of fever (*Jvaracikitsā*) in the *Madhavacikitsā* tradition of India. *Indian Journal of Traditional Knowledge, 8*(3), pp. 352–361.
Mitra, S. and Mukherjee, S.Kr., 2010. Ethnomedicinal usages of some wild plants of North Bengal plain for gastro-intestinal problems. *Indian Journal of Traditional Knowledge, 9*(4), pp. 705–712.
Mohammad, N.S., Milow, P. and Ong, H.C., 2012. Traditional medicinal plants used by the Kensiu Tribe of Lubuk Ulu Legong, Kedah, Malaysia. *Studies on Ethno Medicine, 6*(3), pp. 149–153.
Morteza-Semnani, K., 2015. A Review on Chenopodium botrys L.: Traditional uses, chemical composition and biological activities. *Pharmaceutical and Biomedical Research, 1*(2), pp. 1–9.
Moungsrimuangdee, B., 2016. A survery of Riparian species in the Bodhivijjalaya college's forest, Srinakharinwirot University, Sa Kaeo. *Thai Journal of Forestry, 35*(3), 15–29.
Movaliya, V. and Zaveri, M., 2014. A review on the Pashanbheda plant "Aerva javanica". *International Journal of Pharmaceutical Sciences Review and Research, 25*(2), pp. 268–275. Article No. 51.
Nazan Çömlekçđoğlu and Sengül Karaman, 2008. Kahramanmaras Sehir Merkezindeki Aktar'larda Bulunan Tıbbi Bitkiler. *KSU Journal of Science and Engineering, 11*(1).
Neamsuvan, O. and Ruangrit, T., 2017. A survey of herbal weeds that are used to treat gastrointestinal disorders from southern Thailand: Krabi and Songkhla provinces. *Journal of Ethnopharmacology, 196*, pp. 84–93.
Neamsuvan, O., Sengnon, N., Seemaphrik, N., Chouychoo, M., Rungrat, R. and Bunrasri, S., 2015. A survey of medicinal plants around upper Songkhla lake, Thailand. *African Journal of Traditional, Complementary and Alternative Medicines, 12*(2), pp. 133–143.
Neamsuvan, O., Singdam, P., Yingcharoen, K. and Sengnon, N., 2012. A survey of medicinal plants in mangrove and beach forests from sating Phra Peninsula, Songkhla Province, Thailand. *Journal of Medicinal Plants Research, 6*(12), pp. 2421–2437.
Nguyen-Pouplin, J., Tran, H., Tran, H., Phan, T.A., Dolecek, C., Farrar, J., Tran, T.H., Caron, P., Bodo, B. and Grellier, P., 2007. Antimalarial and cytotoxic activities of ethnopharmacologically selected medicinal plants from South Vietnam. *Journal of Ethnopharmacology, 109*, pp. 417–427.
Noor Hassan Sajib, S. Uddin, B. and Islam, Md. M., 2014. Angiospermic plant diversity of Subarnachar Upazila in Noakhali, Bangladesh. *Journal of the Asiatic Society Bangladesh, Science, 40*(1), pp. 39–60.
O'shaughnessy, W.B., 1842. *Bengal Dispensatory*. Calcutta: W. Thacker and Co., St. Andrew's Library.
Ong, H.G., Ling, S.M., Win, T.T.M., Kang, D-H., Lee, J-H. and Kim, Y-D., 2018. Ethnomedicinal plants and traditional knowledge among three Chin indigenous groups in Natma Taung National Park (Myanmar). *Journal of Ethnopharmacology, 225*, pp. 136–158.
Panda, S.K., Das, R., Leyssen, P., Neyts, J. and Luyten, W., 2018. Assessing medicinal plants traditionally used in the Chirang Reserve Forest, Northeast India for antimicrobial activity. *Journal of Ethnopharmacology, 225*, pp. 220–233.
Pandit, P.K. Inventory of Ethno Veterinary Medicinal Plants of Jhargram Division, West Bengal, India.
Patra, J.K. and Thatoi, H.N., 2011. Metabolic diversity and bioactivity screening of mangrove plants: A review. *Acta Physiol Plant, 33*, pp. 1051–1061.
Pereira, J., 1854. *The Elements of Materia Medica and Therapeutics*. Philadelphia: Blanchard and Lea.
Pereira, J., 1855. *Materia Medica and Therapeutics*. London: Longman, Brown, Green, and Longmans.
Perry, L.M. and Metzger J., 1980. *Medicinal Plants of Asia and Southeast Asia*. MIT Press, USA.
Pomet, P., Histoire générale des drogues. A Paris : Chez Jean-Baptiste Loyson, & Augustin Pillon, sur le Pont au Change, à la Prudence. Et au Palais, chez Estienne Ducastin, dans la Gallerie des Prisonniers, au bon Pasteur 1664.
Pradhan, B.K. and Badola, H.K., 2008. Ethnomedicinal plant use by Lepcha tribe of Dzongu valley, bordering Khangchendzonga Biosphere Reserve, in North Sikkim, India. *Journal of Ethnobiology and Ethnomedicine, 4*, p. 22.
Prakasa Rao, J. and Satish, K.V., 2016. *Persicaria perfoliata* (L.) H. Gross (Polygonaceae): A species new to Eastern Ghats of India. *Tropical Plant Research, 3*(2), pp. 249–252.
Qasim, M., Gulzar, S. and Khan, M.A., 2011. Halophytes as medicinal plants. *Conference Paper*.
Qureshi, R. and Raza Bhatti, G., 2008. Ethnobotany of plants used by the Thari people of Nara Desert, Pakistan. *Fitoterapia, 79*, pp. 468–473.
Rahaman, C.H. and Karmakar, S., 2014. Ethnomedicine of Santal tribe living around Susunia hill of Bankura District, West Bengal, India: The quantitative approach. *Journal of Applied Pharmaceutical Science, 5*(02), pp. 127–136.

Rahman, M.A., 2010. Indigenous knowledge of herbal medicines in Bangladesh. 3. Treatment of skin diseases by tribal communities of the hill tracts districts. *Bangladesh Journal of Botany, 39*(2), pp. 169–177.

Rahmatullah, M., Azam, Md. N.K., Malek, I., Nasrin, D., Jamal, F., Rahman, Md. A., Khatun, Z., Jahan, S., Seraj, S. and Jahan, R., 2012. An ethnomedicinal survey among the Marakh Sect of the Garo Tribe of Mymensingh District, Bangladesh. *International Journal of PharmTech Research, 4*(1), pp. 141–149.

Rahmatullah, M., Haque, M.E., Kabir Mondol, M.R., Hasan, M., Aziz, T., Jahan, R. and Seraj, S., 2014. Medicinal formulations of the Kuch Tribe of Bangladesh. *The Journal of Alternative and Complementary Medicine, 20*(6), pp. 428–440.

Rahmatullah, M., Mollik, Md. A.H., Ahmed, Md. N., Bhuiyan, Md. Z.A., Hossain, Md. M., Azam, Md. N.K., Seraj, S., Chowdhury, M.H., Jamal, F., Ahsan, S. and Jahan, R., 2010. A survey of medicinal plants used by folk medicinal practitioners in two villages of Tangail district, Bangladesh. *American-Eurasian Journal of Sustainable Agriculture, 4*(3), pp. 357–362.

Rajendran, A., Ravikumar, K. and Henry, A.N., 2000. Plant genetic resources and knowledge of traditional medicine in Tamil Nadu. *Ancient Science of Life, XX*(1&2), pp. 25–28.

Rama Chandra Prasad, P., Sudhakar Reddy, C., Raza, S.H. and Dutt, C.B.S., 2008. Folklore medicinal plants of North Andaman Islands, India. *Fitoterapia, 79*, pp. 458–464.

Rao, P.K., Hasan, S.S., Bhellum, B.L. and Manhas, R.K., 2015. Ethnomedicinal plants of Kathua district, J&K, India. *Journal of Ethnopharmacology, 171*, pp. 12–27.

Ray, T., 2014. Customary use of mangrove tree as a folk medicine among the sundarban resource collectors. *International Journal of Research in Humanities, Arts and Literature (IMPACT: IJRHAL), 2*(4), pp. 43–48.

Reed, C.F., 1977. Economically Important Foreign Weeds. Potential Problems in the United States. United States Department of Agriculture.

Rehman, H., Begum, W., Anjum, F. and Tabasum, H., 2014. Rheum emodi (Rhubarb): A Fascinating Herb. *Journal of Pharmacognosy and Phytochemistry, 3*(2), pp. 89–94.

Romulo, A., Zuhud, E.A.M., Rondevaldova, J. and Kokoska, L., 2018. Screening of in vitro antimicrobial activity of plants used in traditional Indonesian medicine. *Pharmaceutical Biology, 56*(1), pp. 287–293.

Roy, S., 2015. An ethnobotanical study on the medicinal plants used by Rajbanshis of Coochbehar district, West Bengal, India. *Journal of Medicinal Plants Studies, 3*(5), pp. 46–49.

Safa, O., Soltanipoor, M.A., Rastegar, S., Kazemi, M., Dehkordi, K.N. and Ghannadi, A., 2012. An ethnobotanical survey on hormozgan province, Iran. *Avicenna Journal of Phytomedicine, 3*(1), Winter 2013, pp. 64–81.

Saha, M.R., De Sarker, D. and Sen, A., 2014. Ethnoveterinary practices among the tribal community of Malda district of West Bengal, India. *Indian Journal of Traditional Knowledge, 13*(2), pp. 359–367.

Saikia, A.P., Ryakala, V.K., Sharma, P., Goswami, P. and Bora, U., 2006. Ethnobotany of medicinal plants used by Assamese people for various skin ailments and cosmetics. *Journal of Ethnopharmacology, 106*, pp. 149–157.

Sandakan, S., 2005. Preferred check-list of Sabah trees.

Sankaranarayanan, S., Bama, P., Ramachandran, J., Kalaichelvan, P.T., Deccaraman, M., Vijayalakshimi, M., Dhamotharan, R., Dananjeyan, B. and Sathya Bama, S., 2010. Ethnobotanical study of medicinal plants used by traditional users in Villupuram district of Tamil Nadu, India. *Journal of Medicinal Plants Research, 4*(12), pp. 1089–1101.

Sapma, S., 2017. Indigenous plant. *ECHO Asia Conference*. October 2–6, 2017. Chiang Mai.

Savithramma, N., Yugandhar, P. and Linga Rao, M., 2014. Ethnobotanical studies on Japali Hanuman Theertham-A sacred grove of Tirumala hills, Andhra Pradesh, India. *Journal of Pharmaceutical Sciences and Research, 6*(2), pp. 83–88.

Shah, A., Ahmad, M., Marwat, S.K. and Zafar, M., 2013. Ethnobotanical study of medicinal plants of semi-tribal area of Makerwal & Gulla Khel (lying between Khyber Pakhtunkhwa and Punjab Provinces). *American Journal of Plant Sciences, 4*, pp. 98–116.

Shang, X., Tao, C., Miao, X., Wang, D., Tangmuke, Dawa, Wang, Y., Yang, Y. and Pan, H., 2012. Ethnoveterinary survey of medicinal plants in Ruoergai region, Sichuan province, China. *Journal of Ethnopharmacology, 142*, pp. 390–400.

Sharma, J., Gairola, S., Sharma, Y.P. and Gaur, R.D., 2014. Ethnomedicinal plants used to treat skin diseases by Tharu community of District Udham Singh Nagar, Uttarakhand, India. *Journal of Ethnopharmacology, 158*, pp. 140–206.

Simpson, D.A. and Inglis, C.A., 2001. Cyperaceae of economic, ethnobotanical and horticultural importance: A checklist. *Kew Bulletin, 56*(2), pp. 257–360.

Singh, J.P., Rathore, V.S. and Roy, M.M., 2015. Notes about Haloxylon salicornicum (Moq.) Bunge ex Boiss., a promising shrub for arid regions. *An International Journal, 62*, pp. 451–463.

Singh, B., Sultan, P., Hassan, Q.P., Gairola, S. and Bedi, Y.S. Ethnobotany, traditional knowledge, and diversity of wild edible plants and fungi: A case study in the Bandipora District of Kashmir Himalaya, India. *Journal of Herbs, Spices & Medicinal Plants.* DOI: 10.1080/10496475.2016.1193833

Song, M-J. and Kim, H., 2011. Ethnomedicinal application of plants in the western plain region of North Jeolla Province in Korea. *Journal of Ethnopharmacology, 137*, pp. 167–175.

Song, P., Sekhon, H.S., Lu, A., et al., 2007. M3 Muscarinic receptor antagonists inhibit small cell lung carcinoma growth and mitogen-activated protein kinase phosphorylation induced by acetylcholine secretion. *Cancer Research, 67*, pp. 3936–3944.

Srivastava, R.C., 2014. Family polygonaceae in India. *Indian Journal of Plant Sciences, 3*(2), pp. 112–150.

Sur, P.R., Sen, R., Halper, A.C. and Bandyopadhyay, S., 1987. Observation on the ethnobotany of malda-west Dinajpur Districts West Bengal-I. *Journal of Economic and Taxonomic Botany, 10*(2).

Tabata, M., Sezik, E., Yeşilada, E., et al. Traditional medicine in Turkey III. Folk medicine in East Anatolia, Van and Bitlis Provinces. *Pharmaceutical Biology.* doi: 10.3109/13880209409082966

Thapa, L.B., Dhakal, T.M., Chaudhary, R. and Thapa, H., 2013. Medicinal plants used by Raji ethnic tribe of Nepal in treatment of gastrointestinal disorders. *Our Nature, 11*(2), pp. 177–186.

Tolken, H.R. The species of Arthrocnemum and Salicornia (Chenopodiaceae) in Southern Africa. *Bothalia, 9*(2), pp. 255–307.

Tribedi, G.N., Mudgal, V. and Pal, D.C., 1993. Some less known ethnomedicinal uses of plants in sunderbans, India. *Bulletin of the Botanical Survey of India, 35*(1–4), pp. 6–10.

Tripathi, Y.C., Prabhu, V.V., Pal, R.S. and Mishra, R.N., 1996. Medicinal plants of Rajasthan in Indian system of medicine. *Ancient Science of Life, XV*, pp. 190–212.

Tumpa, S.I., Hossain, Md. I. and Ishika, T., 2014. Ethnomedicinal uses of herbs by indigenous medicine practitioners of Jhenaidah District, Bangladesh. *Journal of Pharmacognosy and Phytochemistry, 3*(2), pp. 23–33.

Uddin, S.N. and Hassan, Md. A., 2012. Angiosperm flora of Rampahar reserve forest under Rangamati district in Bangladesh. I. Liliopsida (Monocots). *Bangladesh Journal of Plant Taxonomy, 19*(1), pp. 37–44.

Uddin, M.Z., Hassan, Md. A. and Sultana, M., 2006. Ethnobotanical survey of medicinal plants in Phulbari Upazila of Dinajpur district, Bangladesh. *Bangladesh Journal of Plant Taxonomy, 13*(1), pp. 63–68.

Uddin, K., Mahbubur Rahman, A.H.M. and Rafiul Islam, A.K.M., 2014. Taxonomy and traditional medicine practices of Polygonaceae (Smartweed) Family at Rajshahi, Bangladesh. *International Journal of Advanced Research, 2*(11), pp. 459–469.

Ullah, M., Khan, M.U., Mahmood, A., Malik, R.N., Hussain, M., Wazir, S.M., Daud, M. and Shinwari, Z.K., 2013. An ethnobotanical survey of indigenous medicinal plants in Wana district south Waziristan agency, Pakistan. *Journal of Ethnopharmacology, 150*, pp. 918–924.

Ullah, S., Khan, M.R., Shah, N.A., Shah, S.A., Majid, M. and Farooq, M.A., 2010. Ethnomedicinal plant use value in the Lakki Marwat District of Pakistan. *Journal of Ethnopharmacology.*

Upadhyay, O.P., Kumar, K. and Tiwari, R.K., 2008. Ethnobotanical study of skin treatment uses of medicinal plants of Bihar. *Pharmaceutical Biology, 36*(3), pp. 167–172.

Vaidya, B.N. and Joshee, N., Report on Nepalese Orchid Species with Medicinal Properties. https://www.researchgate.net/publication/312888202.

Van Sam, H., Nanthavong, K. and Kessler, P.J.A., 2004. Trees of Laos and Vietnam: A field guide to 100 economically or ecologically important species. *Blumea, 49*, pp. 201–349.

Veldkamp, J.F. and Flipphi, R.C.H., 1987. A revision of Leptonychia (Sterculiaceae) in Southeast Asia. *BLUMEA, 32*, pp. 443–457.

Wagh, V.V. and Jain, A.K. Ethnopharmacological survey of plants used by the *Bhil* and *Bhilala* ethnic community in dermatological disorders in Western Madhya Pradesh, India. *Journal of Herbal Medicine.*

Wangchuk, P., Keller, P.A. Pyne, S.G., Taweechotipatr, M., Tonsomboon, A., Rattanajak, R. and Kamchonwongpaisan, S., 2011. Evaluation of an ethnopharmacologically selected Bhutanese medicinal plants for their major classes of phytochemicals and biological activities. *Journal of Ethnopharmacology, 137*, pp. 730–742.

Warrier, K.C.S. and Warrier, R.R., 2019. *Sacred Groves: Repositories of Medicinal Plants.* Coimbatore: Institute of Forest Genetics and Tree Breeding.

Watt, G., 1889. *A Dictionary of the Economic Products of India.* Calcutta: Government Printing.

Wazir, A.R., Shah, S.M. and Razzaq, A. Ethno botanical studies on plant resources of Razmak, North Waziristan, Pakistan.

WHO, 2009. *Medicinal Plants in Papua New Guinea*. Western Pacific Region:WHO.

Yaseen, G., Ahmad, M., Shinwari, S., Potter, D., Zafar, M., Zhang, G., Shinwari, Z.K. and Sultana, S., 2019. Medicinal plant diversity used for livelihood of Public health in deserts and arid regions of Sindh-Pakistan. *The Pakistan Journal of Botany, 51*(2), pp. 657–679.

Yazan, L.S. and Armania, N., 2014. *Dillenia* species: A review of the traditional uses, active constituents and pharmacological properties from pre-clinical studies. *Pharmaceutical Biology, 52*(7), pp. 890–897.

Zheng, X-l., Wei, J-h., Sun, W., Li, R-t., Liu, S-b. and Dai, H-f., 2013. Ethnobotanical study on medicinal plants around Limu Mountains of Hainan Island, China. *Journal of Ethnopharmacology, 148*, pp. 964–974.

Index

Note: *Italic* page numbers refer to figures.

A

Abrin 121
Abruquinone B 120, 121
Abruquinone G 120
Abrus precatorius 119–121
Acacia arabica 122–123
Acacia auriculiformis 123
Acacia catechu 124
Acacia concinna 125–126
Acacia farnesiana 126–127
Acacia nilotica 122
Acacia pennata 128
Acacia torta 123
Acalypha australis 36
Acalypha chinensis 36
Acalypha fruticosa 36
Acalypha hispida 36–37
Acalypha indica 37
Acalypha minima 36
Acalypha pauciflora 36
Acetoxyalphitolic acid 24, 25
4-Acetoxy-2-geranyl-5-hydroxy-3-n-pentylphenol 228
(24E)-23-acetoxy-3-oxolanosta-9,24-dien-26-oic acid 33
Acinatobacter baumannii 16
Actinomycetes viscosus 46
acyclovir 10, 11, 123, 169, 194, 232, 235
acylphloroglucinol 73
Adenia cordifolia 77–78
Adenia populifolia 78
Aeromonas hydrophila 36, 103, 207, 254
Agrimonia pilosa 245–247, *246*
Agrimonia eupatoria 247
Albanol B 240
Albizia amara 131
Albizia chinensis 131
Albizia corniculata 131
Albizia julibrissin 131
Albizia lebbeck 131–135, *132–134*
Albizia myriophylla 131–132, *133*
Albizia odoratissima *134,* 135
Albizia procera 135–137
Alchornea vaniotii 40
Aleurites moluccana 38–39
Aleurites moluccanus 38–39
Aleurites triloba 38
Alexandrian-laurel 18
Allepo oak 216
Allyl isothiocyanate 310, 312, 319, 320
Aloe-emodin 154, 155, 159, 160
Alternaria alternata 206, 225, 272
Alternaria brassicae 313
Amelanchier asiatica 255
amino acid 81, 189, 218, 223
ammaniol 287
Ammannia multiflora 287
amphotericin B 189, 263, 268
Amygdalus persica 252
amyrin 43, 211, 308
Andaman redwood 197
Angiotensin-converting enzyme 111, 152, 155, 160, 228, 263, 269, 326
Angolensin 197
Anogeissus acuminata 271
Anogeissus latifolia 272
Anolignan A 271
Antidesma acidum 80–81
Antidesma bunius 81
Antidesma diandrum 80
Antidesma montanum 81
Antidesma wallichianum 80
apigenin 209, 269
apocarotenoid 90
aquaporin (porin) 74, 219
Arjuna 274
Artocarpanone 231
Artocarpin 231, 232
Artocarpus heterophyllus 230–232
Artocarpus integrifolius 230
Artocarpus lakoocha 232
aryltetralin 90, 201
Ascosphaera apis 225
ash gourd 105
ash pumpkin 105
asoka tree 200
Aspergillus columnuris 289
Aspergillus flaviscipes 289
Aspergillus fumigatus 13, 57, 112, 201, 215, 223, 268, 304
Aspergillus nidulans 55
Aspergillus niger 34, 44, 53, 55, 62, 108, 117, 131, 141, 143, 169, 189, 201, 202, 206, 215, 232, 254, 255, 263, 278, 289, 321
Aspergillus ochraceus 62, 215
Aspergillus parasiticus 62
Aspergillus restrictus 268
Aspergillus tamari 289
Aspergillus versicolor 215, 289
Aspidopterys hypoglauca 8
Asp-of-Jerusalem 315
asthma plant 44
Astragalus aureus 139
Astragalus fasciculifolius 139
Astragalus mongholicus 137–139
ATP 127
ATPase 130, 244, 269
aucuparin 250, 251
Aureobasidium pullulans 215
Australian Acalypha 36
Averrhoa acutangula 15
Averrhoa carambola 15–16
Averrhoa pentandra 15
Axle wood 272

B

Bacillus licheniformis 48, 50, 308
Bacillus megaterium 6, 75, 99, 131, 142, 166, 194, 263, 322, 324
Bacillus polymyxa 95, 131
Bacillus sphaericus 79
baicalein 129–130
Bakuchicin 195
Bakuchiol 196
Baliospermum reidioides 64
Balsamaria inophyllum 18
Banistera benghalensis 76
banner bean 183
banyan tree 235
basalm apple 115
bastard indigo 204
bastard mustard 320
bastard myrobalan 275
bastard teak 139
Bauhinia acuminata 174, *174*
Bavachinin 195, 196
bedaquilin 110
bedda nuts 275
Begonia fimbristipula 104–105
Begonia roxburghii 103
Begoniaceae 103–105
Belgaum walnut 38
belleric myrobalan 275
Bengal kino tree 139
Benincasa hispida 105–106
benzaldehyde 114
1,2,3-benzenetriol 219
benzoic acid 81, 82, 152
benzopyran 57
benzoquinone 81, 235
6β-benzoyl-7β-hydroxyvouacapen-5α-ol 144
Benzyl isothocyanate 311
Berchemia berchemiifolia 260
bergenin 54, 55, 57, 84
betulin 24, 25
betulinic acid 24, 25, 66, 212
biochanin A 149, 150, 152, 200
biotin 100
bird-cactus 60
bitter-apple 109
bitter gourd 115
Biyouyanagin A 73, 74
black oil tree 2
Blastomyces dermatitidis 108
blind-your-eye 50
blind weed 312
Blumeria graminis 214
Boehmeria nivea 262–264
bondenolide 142, 143
bonduc nut 142
Bordetella bronchiseptica 57, 215
Botrydis cinerea 55, 108
Botryodiplodia theobromae 272
Botryosphaeria dothidea 108
bottle brush 297, 298
Brassica alba 310–311
Brassicaceae 310–319
Brassica cernua 311
Brassical campestris 311
Brassicales 310–319
Brassica nigra 311
Brassica oleracea 311–312
brazilin 144
Bridelia retusa 82
Bridelia spinosa 82
Bridelia stipularis 82
Broussonetia papyrifera 233
Bruguiera cylindrica 93
Bruguiera eriopetala 92
Bruguiera gymnorhiza 93
Bruguiera sexangula 92–93
Bryonopsis laciniosa 106–107
Bryony 106
Bulkholderia cepacian 144
burning bush 4
Butea frondosa 139
Butea monosperma 139–141
Butein 140
button tree 271

C

cabbage 311
caesalmin 143
Caesalpinia bonduc 142–145, *145*
Caesalpinia bonducella 142
Caesalpinia crista 142
Caesalpinia pulcherrima 144, *145*
Caesalpinia sappan 144
caesalpinin 143
Caffeic acid 36, 44–46
Cajaninstilbene 148
Cajanus cajan 146–149, *147*
Cajanuslactone 147
calanolide A 19
Callistemenonone 304
Calophyllum bintagor 18
Calophyllum inophyllum 18–20
Calophyllum nagassarium 33
calycosin 138, 139
Calycosin-7-*O*-β-D-glucopyranoside 138
Camirium moluccanum (L.) Kuntze 38
canchorie root-plant 63
Candida albicans 1, 6, 26, 34, 38, 44, 54, 55, 69, 78, 87, 101, 108, 112, 117, 125, 131, 141, 150, 155, 166, 169, 179, 185, 186, 189, 196, 201, 202, 206, 213, 217, 223, 225, 228, 232, 234, 236, 248, 252, 254, 255, 257–259, 263, 264, 267–269, 277, 278, 280, 281, 286, 297, 298, 301, 303, 304, 306, 308, 312, 314, 321
Candida guilliermondii 54, 186
Candida krusei 1, 54, 155, 186, 217, 228, 303
Candida parapsilosis 1, 55, 87, 155, 186, 217, 225, 248, 304
Candida tropicalis 1, 54, 150, 155, 186, 217, 225, 268, 301, 304
Candida utilis 179
candleberry tree 38
candle bush 153
candlenut tree 38

Cannabaceae 227–229
cannabichromene 227, 228
cannabidiol 227, 228
cannabigerol 227, 228
cannabinoid 227, 228
cannabinol 227, 228
Cannabis 227–229
Cannabis sativa 227–229
Caper thorn 319
Capparaceae 319–322
Capparis micracantha 319–320
Capsella bursa-pastoris 312–313
Caragana rosea 161
Carallia brachiata 94
Carallia fascicularis 94
Carallia suffruticosa 94
carboline 64
Cardamine hirsuta 313
Cardaria draba 313–314
cardiac glycosides 244
Caricaceae 322–323
Carica papaya 322–323
carpain 323
β-Caryophyllene 303, 304
Caryophyllus aromaticus 293
Casbene 62
Cassane 122, 128, 143, 144
Cassia alata 153–155
Cassia auriculata 160
Cassia fistula 150–152, *151*
Cassia javanica 160
Cassia occidentalis 160
Cassia siamea 156–158, *157*
Cassia tora 158–161
Cassine japonica 7
Cassine kotoensis Hayata 7
Castaartancrenoic acid E 207
castalagin 207, 294
Castanea crenata 207
Castanea japonica 207
Castanea mollissima 208
Castanea sativa 208–209
Castanea stricta 207
Castanea tungurrut 209–210
Castanopsis motleyana 210
Castanopsis sclerophylla 210
castor bean 61
castor oil plant 61
casuarinin 274, 277, 289, 294
catechin 81, 103, 104, 123, 124, 131, 136, 137, 201, 214, 215, 270
catechu 124
cat's whiskers 320
celaspene 3
cat-thorn 259
caulilexin A 312
ceanothane 260
Celastraceae 2–11
celastrales 2–3
celastrol 4, 9, 11
Celastrus alatus 4
Celastrus dependens 2
Celastrus multiflorus 2
Celastrus paniculatus 2
Celastrus striatus Thunb. 4
Celebes oak 211
Celtis orientalis 261
Cerasus mahaleb 247
Cercospora arachidicola 3
Ceriops candolleana 94
Ceriops decandra 95
Ceriops tagal 94–95
Ceylon Ironwood 33
Ceylon rosewood 135
Chaenomeles sinensis 258
chalcone 16, 57, 140, 141, 148, 149, 164, 178, 180, 195, 196, 212, 264, 296
chamuangone 22
charcoal tree 261
Chaulmoogra odorata (R.Br.) Roxb. 98
Chaulmoogra oil 98, 99
Chaulmoogra tree 99
Chaulmoogric acid 99, 100
chebulagic acid 13, 89, 277, 278
Chebulic myrobalan 276
chenille plant, red-hot cat's tail 36
chicken weed 48
Chikungunya virus 37–40, 65, 85, 244
Chilmoria dodecandra Buch.-Ham. 98
China Mangosteen 31
Chinese evergreen oak 218
Chinese hairy chestnut 208
Chinese Osbeckia 289
Chinese wingnut 226
Chinesin I 72
chitin 304
Chlorogenic acid 45, 46
chloroquine 110
cholestane 116
chromene 178, 179
Chromobacterium violaceum 79, 103, 219
chrysin 79
chymotrypsin 4, 28, 130, 149, 199, 200, 234, 251, 254, 288, 316
Cinchona ledgeriana 110
1,8-Cineole 268, 298
6β-cinnamoyl-7β-hydroxyvouacapen-5α-ol 144
Circassian seeds tree 129
Citral 268
Citrobacter freundii 153, 194, 207, 230, 277, 281, 323, 324
Citrobacter koseri 264
citronellal 268
citronellic acid 268
citronellol 268, 269
Citrullus colocynthis 108–111
Cladosporium cladosporioides 82, 141, 215
Cladosporium cucumerinum 3, 176
Cleistocaltone 296
Cleome gynandra 320–321
Cleome viscosa 321
clerodane 41
climbing staff plant 2
Clostridium perfringens 159, 207, 314
Clostridium sporogenes 179, 268
Clusiaceae 18–35, 60
cochinchinone 69

Colletotrichum capsici 189
Colletrotrichum corchori 272
Colletotrichum gloeosporioides 189, 202
Colletotrichum lindemuthianum 251
Colletotrichum musae 252
colocynth 109
Colocynthis vulgaris Schrad. 108
Combretaceae 271, 274
combretastatin 273
Combretum molle 273
Combretum trifoliatum 272–273, *273*
Commia cochinchinensis Lour. 50
common acalypha 37
common indigo 182
common walnut tree 224
Coniferylaldehyde 236
Connaropsis philippica Fern.-Vill. 15
coral pea 119, 129
corilagin 13, 37, 90, 277
cork bush 4
corniculin 17
coronavirus xi, xiii, 4, 11, 17, 19, 20, 28, 35, 39, 43, 45, 46, 49, 55, 59, 74, 76, 79, 84, 85, 95, 110, 111, 114, 115, 118, 120, 121, 130, 131, 136, 139, 141, 144, 149, 152, 158, 160, 170, 176, 180, 183, 196, 199, 200, 223, 224, 229, 232, 234, 235, 237, 241, 244, 247, 251, 254, 255, 260, 263, 269, 270, 282, 283, 288, 289, 302, 306, 316, 326
Coronopus didymus 312
corylifol A 196
Cotoneaster integerrimus 247
Cotoneaster nummularia 247
coumarin 7, 18, 19, 20, 34, 35, 81, 103, 115, 147, 149, 176, 178, 181, 195, 196, 228
COVID-19 xi, xiii, 4, 17, 19, 20, 41, 55, 59, 63, 89, 106, 110, 118, 121, 139, 141, 144, 149, 152, 160, 170, 181, 196, 199, 200, 206, 229, 234, 241, 248, 251–253, 260, 278, 283, 290, 293, 326
cowanin 20, 21, 24
cowanone 20
cow itch 189
Coxsackie virus 78, 88, 138, 207, 269
crab-eyes vine 119
Crateva religiosa 321
Cratoxylum arborescens 65–66
Cratoxylum cochinchinense 68–69
Cratoxylum cuneatum 65
Cratoxylum formosum (Jack) Dyer 67
creeping lady's-sorrel 16
creeping wood-sorrel 16
Crotalaria albida 163, 164
Crotalaria ferruginea 162, 164
Crotalaria juncea 164
Crotalaria pallida 162–164, *162–163*
Crotalaria retusa 164
Crotalaria verrucosa 164
crotmadine 164
crotmarine 164
Croton birmanicus 40
Croton himalaicus 40
Croton mauritianus 40
crotonolide 41
croton seeds 40
Croton tiglium 40–41
Croton xiaopadou 40
Cryphonectria parasitica 209
Cryptococcus neoformans 1, 9, 55, 57, 125, 129, 186, 189, 223, 252, 259, 284, 314
cucumber 107, 112
cucumerin 108
Cucumis colocynthis L. 108
Cucumis sativus 107–108
cucurbitacin 14, 103–105, 113
cucurbitacin IIA 112
cucurbitacin B 14, 112
cucurbitacin D 118
cucurbitacin E 112
cucurbitacin E 2-*O*-β-D-glucopyranoside 109
Cucurbita hispida 105
cucurbitales 103–118
cucurbitane 14, 103, 109, 112, 116
Cudrania cochinchinensis 238
Cunoniaceae 12
Cupamenis indica (L.) Raf. 37
Curvularia lunata 143, 202
cutch tree 124
Cyanomaclurin 231
cycloartane 138–139, 207, 211
cyclopeptide 260
Cydonia oblonga 248–249
Cydonia vulgaris 248
cylicodiscic acid 212
p-Cymene 268–269
Cytisus cajan 146
cytomegalovirus 11, 14, 49, 169

D

Dalbergia lanceolaria 166
Dalbergia latifolia 166
Dalbergia pinnata 164–166, *165*
Dalbergia sissoo 166
daphnane 64
Decaspermum fruticosum 290, *291*
Deguelin 169
Dengue virus 9, 11, 90, 101, 115, 170, 198, 217, 239, 244, 277, 323
12-deoxyphorbol 13-(3E,5E-decadienoate) 50
3′-Deoxyquercetin 29
Derris elliptica 169
Derrisisoflavone A 169
Derris pinnata 164
Derris robusta 169
Derris scandens 168, 169
Derris trifoliata 166–169, *167–168*
Descurainia sophia 314–315
Desmodium canadense 170–171
Desmodium gangeticum 170–174, *171–174*
Desmodium heterocarpon 173
Desmodium multiflorum 173
Desmodium pulchellum 174
Desmodium triflorum 172, *172*
Desmodium triquetrum 173, *173*
Desmodium velutinum 172
8-Desoxygartanin 29
devil's backbone 60
diarylheptanoid 220
β-dihydroagarofurans 3

dihydroisomorellin 25
5,7-dihydroxychromone 104
3,4-dihydroxycinnamic acid 44
2′,4′-Dihydroxy-6′-methoxy-3′,5′-dimethylchalcone 295, 296
3β,22α-dihydroxyolean-12-en-29-oic acid 104, *104*
2,6-dimethoxy-1,4-benzoquinone 235
(7′*R*,8′*S*)-4,4′-Dimethoxy-strebluslignanol 243, 244
8.3′-diprenyl 5,7,4′-trihydroxyflavanone 187, 189
diterpene 10, 35, 38–41, 46, 48, 50, 51, 53, 60, 62, 64, 65, 95, 98, 122, 128, 142
DNA 10, 11, 13–15, 19, 22, 45, 46, 48, 55, 59, 62, 81, 88–90, 95, 112–114, 120, 123, 129, 138, 141, 144, 148, 155, 169, 176, 180, 201, 213, 225, 227, 235, 239–241, 244, 246, 247, 252, 263, 274, 277, 288, 324
Dolichos lablab 183
Dolichos lobatus 199
Dolichos purpureus 183
Draba nemorosa 315
Duchesnea indica 246, 247
dyer's oak 216
dyer's woad 315

E

Ebola virus xiii, 55, 294
Ecballium elaterium *14*, 112–113
Echinosophora koreensis 196
Echinus philippinensis 57
efflux pumps 59, 60, 69, 94, 149, 176, 180, 185, 189, 217, 223
Elaeocarpaceae 12–15
Elaeocarpus grandiflorus 12–13
Elaeocarpus petiolatus 14–15
Elaeocarpus sylvestris 14
Elaeodendron japonicum Franch. & Sav. 7
ellagic acid 45–46, 54, 55, 62, 76, 102, 217, 252, 267, 277, 288, 289, 308
ellagitannin 13, 37, 89, 90, 123, 207, 213, 227, 256, 273, 274, 277, 279, 286, 288, 289, 304, 306
Elodes formosa Jack 67
Emblica officinalis 87
Emblic myrobalan 87
emodin 154, 155, 160
encephalomyocarditis virus 50
engeletin 220, 221
Engelhardia roxburghiana 219–221, 221
Engelhardiol 220
Engelhardione 220
Enterococcus faecalis 5, 8, 16, 26, 34, 38, 83, 87, 110, 112, 150, 177, 187, 202, 223, 235, 238, 254, 260, 263, 271, 275, 277, 278, 284, 286, 294, 302, 303, 310, 314, 317, 323
Enterobacter intermedius 215
Enterobacter sakazaki 314
Enterovirus 31, 79, 88, 90, 138, 286
epiafzelechin 160
3-*epi*-Betulinic acid 212
epicatechin 16, 201, 249, 251
(+)-Epicatechin 3-*O*-gallate 44, 46
Epidermophyton floccosum 55, 150, 152, 182, 196, 324
5α,8α-epidioxy-24(R)-methylcholesta-6,22-dien-3β-ol 116
epigallocatechin 201, 215, 270, 292
epigallocatechin 3-*O*-gallate 292
Epilobium tetragonium 308
Equid herpesvirus 1 239
ergosterol 116, 150, 186
ergosterol peroxide 116
Eriobotrya japonica 249–251, *250*
Ermanin 78, 79
Erodium glaucophyllum 267
Erycristagallin 175
erysodine 176
erythraline 176
Erythrina arborescens 176
erythrinan 176
Erythrina orientalis 175–176
Erythrina stricta 175, 176
Erythrina variegata 175
(+)-(7′S,8′R)-erythro–7′-methylcarolignan E 46
eucalyptol 305
Eucalyptus camaldulensis 304
Eucalyptus globulus 304–305
Eucalyptus obliqua 305
Eucalyptus tereticornis 305
Euchrestaflavanone A 187
esculetin 103
ethidium bromide 152
Eugenia aquea 290, 291, 292–294, *292, 293*
Eugenia aromatica 293, *293*
Eugenia jambolana Lam. 303
Eugenia operculata 296–297
Eugenia uniflora 294
eugenol 294
Euonymus alatus 4–5
Euonymus arakianus 4
Euonymus ellipticus 4
Euonymus kawachianus 4
Euonymus loeseneri 4
Euonymus striatus 4
Euonymus subtriflorus 4
Euonymus thunbergianus 4
euphoheliosnoid 46
Euphorbia antiquorum 42–43
Euphorbiaceae 36–65, *52, 58*
euphorbia herb 44
Euphorbia hirta 43–46
Euphorbia neriifolia 43, 47
Euphorbia nivulia 47–48
Euphorbia thymifolia 48–49
Euphorbia tithymaloides 60
European indigo 315
Euxanthone 31
excoecafolin B 39, 51
excoecafolins 39, 51
Excoecaria agallocha 50–51
Exserohilum turticum 124

F

Fabaceae 118–206, *132–134, 145, 147, 151, 157, 162, 163, 165, 167, 168, 171–174, 184, 188, 190, 192, 193, 198, 203, 205*
Fabales 118–206, *132–134, 145, 147, 151, 157, 162, 163, 165, 167, 168, 171–174, 184, 188, 190, 192, 193, 198, 203, 205*
Fagaceae 206–219

Fagales 206–227, *221*
Fagus castanea 208
fatty acid 99, 100, 107, 263
fever nut 142
Fetid passion flower 78
Ficus benghalensis 234–237
Ficus edelfeltii 235
Ficus foveolata 235
Ficus hispida 236
Ficus indica 234
Ficus racemosa 236
Ficus religiosa 236
Filipendula vulgaris 255
five-leaved cleome 320
Flacourtia indica 96–97
Flacourtia jangomas 98–99
Flacourtia parvifolia 96
Flacourtia ramontchi 96
Flacourtia rukam 97
flame of the forest 139
flavanol 44, 55, 87, 136–138, 148, 249, 251, 254
flavanone 8, 9, 71, 72, 80, 81, 141, 169, 177, 179, 187, 189, 220, 221, 231, 240, 241, 254
flavone 7, 78, 79, 97, 129, 130, 135, 138, 141, 169, 179, 195, 196, 206, 221, 231–234, 240–242, 304
Flemingia strobilifera 187
Fleurya interrupta 263
fleshy spurge 42
flixweed 314
Flueggea virosa 82–85, *84*
Flueggenine D 83
fluevirosinine B 83, 85
foot and mouth disease virus 324
forest siris 136
formononetin 137, 138, 141, 149
friedelane 43, 98
friedelanol 43
Fulvia fulvum 215
furanoditerpene 128, 143, 144
furostanol 1
Fusarium culmorum 225
Fusarium equiseti 272
Fusarium moniliformis 124
Fusarium oxysporum 26, 62, 108, 138, 141, 202, 281, 282, 289, 304
Fusarium proliferatum 124
Fusarium solani 79, 149, 189, 267
Fusarium sporotrichioides 215–216
Fusarium trincintum C 216
Fusarium verticilloides 62, 304

G

gallic acid 13, 16, 36, 37, 46, 76, 88, 89, 112, 206, 214, 217, 257, 277, 286, 289
gallotannin 45, 46, 89, 210, 213, 217, 223, 277, 278
Galphimia 76
Galphimia glauca 75–76
gamboge tree 24
gambogic acid 30
gancaonin I 181
garcimultiflorone 27
garciniacowone 22
Garcinia dulcis 23–24
Garcinia hainanensis 27
Garcinia hanburyi 24–25
Garcinia mangostana 25–26
Garcinia multiflora 27–28
Garcinia nigrolineata 29–30
Garcinia oblongifolia 22, 31
Garcinia paucinervis 22, 31–32
Garcinia scortechinii 30
Garcinia speciosa 32–33
garciosaterpene 32, 33
garciosine 33
garden spurge 44
genistein 149, 150, 176, 200
genistin 188, 189
Geotrichum candidum 186
Geraniaceae 267–270
Geraniales 267–270
geraniin 36, 37, 267
Geranium carolinianum 267
Geranium macrorrhizum 268
Geranium nepalense 267
Geranium wallichianum 267–270
Geranyl acetate 269
gerontoxanthone 67, 238
germacrene 268, 269
Getonia floribunda *273*
Geum japonicum 255
ghatti tree 272
Gibberella zeae 3
Glabratephrin 206
Glabrene 178, 179
Glabridin 177, 179
Glabrol 179
Glochicoccinoside D 88
glochidioboside 87
Glochidion daltonii 86
Glochidion littorale 85–87, *86*
Glochidion multiloculare 87
Glochidion obscurum *86*, 87
Glochidion rubrum 87
Glochidion submolle 87
glucosinolate 310–312, 314–316, 318, 319, 321
glucosylhamaudol 158
Glyasperin D 180
Glycycoumarin 178, 179
glycyrin 181
glycyrrhisol 181
Glycyrrhiza echinata 181
Glycyrrhiza glabra 177–181
Glycyrrhiza uralensis 180, 181
glycyrrhizin 120, 121, 180
Glyptopetalum calocarpum 3
golden shower tree 150
Gouania leptostachya 260
Governor's Plum 96
guava tree 299
Guinea rosewood 197
guttiferone 30
Gynocardia odorata 98

H

H1N1 xiii, 14, 45, 46, 136, 144, 200, 220, 245, 258, 269, 284, 296, 297, 318

H1N2 296
H3N2 45, 46, 88, 144, 245, 258, 269
H5N1 152, 255, 263, 297
H7N1 155, 269
H9N2 49, 144, 245, 296, 297
Haemophilus influenzae 178, 180, 181
hairy bittercress 313
hairy spurge 44
hanburin 25
heart-padded hoary-cress 313
Hedysarium strobiliferum 187
Hedysarum gangeticum 170
helicase 130, 269, 289
Helicobacter pylori 22
Helminthosporium sativum 117
hemp 227
Hemsleya 112
henna 281
Hepatitis B virus 45, 46, 55, 73, 88, 90, 93, 103, 112, 114, 244, 247, 263, 267
Hepatitis C virus 55, 89, 90, 122, 129, 232, 244, 252, 263, 277, 278
Herpepropenal 114
Herpes simplex virus 3, 11, 13, 49, 62, 81, 88, 89, 112, 120, 123, 138, 141, 148, 155, 160, 169, 180, 194, 213, 217, 225, 227, 232, 235, 236, 239–241, 246, 247, 269, 274, 277, 288, 294, 324
herpetin 114
Herpetofluorenone 114
Herpetospermum pedunculosum 113–114
hesperetin 254
Hesperis schischkinii 319
Hiptage 76
Hiptage benghalensis 76–77
Hispaglabridin 178, 179
Histoplasma capsulatum 127
Holly oak 215
Holm's oak 215
Holoptelea integrifolia 261
homoisoflavonoid 143, 144, 164
Honokiol 244
Human coronavirus 19, 43, 45, 46, 183, 241, 254, 269
Human immunodeficiency virus xiii, 3, 4, 10, 16, 19, 24, 25, 26, 28, 32, 33–35, 38, 39, 42–44, 46, 50, 51, 53, 55, 57, 59, 62, 64, 66, 73, 76, 81, 83, 84, 88, 91, 95, 104, 118, 120, 124, 128, 136, 137, 138, 141, 152, 155, 158, 160, 161, 184, 201, 212, 213, 218, 223, 225, 228, 232, 235, 237, 241, 247, 251–253, 255, 260, 261, 267, 271, 274, 275, 277, 278, 282, 283, 286, 289, 313, 324
Human papilloma virus 123
Human respiratory syncytial virus 160, 200
Human rhinovirus 55, 207, 237, 281, 286
humorism 106
humulone 229
Humulus japonicus 229
Humulus scandens 229
Hydnocarpus kurzii 98, 99–100
Hydnocarpus wightiana 100
4-Hydroxybenzoic acid 152
3-Hydroxyblancoxanthone 67
3″-hydroxy-Δ(4″,5″) –cannabichromene 228
8-Hydroxycannabinol 228
8-Hydroxycannabinolic acid A 228
hydroxycinnamic acid 44–46
3-hydroxyglabrol 178, 179
3′-hydroxy-isostreblusl ignaldehyde 243, 244
3β-hydroxy-lup-9(11),12-diene, 28-oic acid 284
8-hydroxy-6-methoxy-3-pentylisocoumarin 100
16-hydroxypimar-8(14)-en-15-one 95
3β-Hydroxystigmast-5-en-7-one 236
hygroline 94
hypatulin A 75
hypercalin B 70
hyperenone A 70
Hypericaceae 65–75
Hypericum acmosepalum 70–71
Hypericum arborescens 65
Hypericum coccineum 65
Hypericum japonicum 71–74
Hypericum patulum 75
Hypericum perforatum 73
Hypericum scabrum 73
Hyperjaponicol C 72
hyponine B 10

I

iguesterin 4
Indian Acalypha 37
Indian butter bean 183
Indian coral tree 175
Indian cress 316
Indian gum Arabic tree 122
Indian hemp 227
Indian jujube 259
Indian licorice 119
Indian plum 97
Indian Rose Chestnut 33
Indian stinging nettle 63
Indian tamarind 204
indican 316
indicanine B 176
indigo 316
Indigofera aspalathoides 183
Indigofera enneaphylla 183
Indigofera indica 182
Indigofera mysorensis 183
Indigofera oblongifolia 183
Indigofera tinctoria 182–183
indirubin 316
indole 121, 310, 312, 313, 316
indole-3-carboxylic acid 104
indoloquinazoline 182, 183
Influenza virus 45, 46, 49, 53, 55, 144, 155, 202, 221, 245, 258, 269, 283, 284, 296, 297, 318
inophyllum B 19
intellect tree 2
integrase 16, 81, 137, 220, 278
ion channel 223, 224
isatan B 316
Isatis tinctoria 315–316
Isobavachalcone 195, 196
isocoumarin 100
Isocudraniaxanthone B 68
isoflavan 120, 121, 177, 179, 180
isoflavone 24, 138, 148, 149, 169, 185, 188, 189, 196, 200, 254, 264, 280

Isojacareubin 72
Isomagnolol 244
Isomammeisin 34
Isomenthone 269
isomyrtucommulone B 302
isoorientin 17, 108
isoquinoline 126
isovitexin 17, 108
itchy bean 189

J

jack fruit 230
Jamaica wild licorice 119
Jambolan 303
Japanese chestnut 207
Japanese Emperor oak 214
Japanese encephalitis virus 155, 316
Japanese hop 229
Japanese horseradish 318
Japanese medlar 249
Japanese plum 249
Japanese St John wort 71
Japonone 73
Jatropha curcas 51–53, *52*
Jatropha gossypifolia *52,* 53
Jatropha moluccana L. 38
jatrophane 46, 60
jatrophenone 53
Java polisander 164
jequirity bean 119
Juglandaceae 219–227, *221*
juglone 222–225
Juglonol B 222
Juglans formosana 222
Juglans mandshurica 222
Juglans regia 224–226
Juglans sinensis 224
jumble beads 119
Jussiaea repens 306–307, *307*
Jussiaea suffruticosa 306, 308

K

kaempferol 128, 129, 131, 154, 155, 206, 208, 254, 267, 269, 288, 306, 323
kaempferol-3-*O*-[2″,6″-di-*O*-*E*-*p*-coumaroyl]-β-D-glucopyranoside 207
Kamala tree *57*
Kassod tree 156
kaurane 46, 95, 98
Kazinol F 234
Klebsiella aerogenes 48, 79, 83, 201
Klebsiella pneumoniae 1, 2, 16, 34, 57, 77, 95, 103, 131, 135, 142, 143, 153, 166, 177, 178, 185, 191, 194, 197, 201, 202, 207–209, 214–216, 219, 230, 235, 255, 263, 268, 271, 277, 278, 281, 296, 298, 303, 308, 311, 314, 319, 321, 323–325
Kokoona zeylanica 6–7
Korean chestnut 207
Korean oak 214
kuwanon 240–242

L

Lablab purpureus 183–184
Lactobacillus casei 166, 175
Lactobacillus plantarum 5, 128
Lactophilus acidophilus 305
Lagerstroemia speciosa 279–281, 280
lakoochin A 232
Laportea crenulata 263
Lawsone 281
Lawsonia inermis 281–282
L-Dopa 189, 191
Leachianone G 240
Leafy milk hedge 47
Lethal dose 121
Licochalcone 178–180
Licocoumarone 178, 179
Licoflavanone 177, 179
licorice 119, 177
licoricidin 180
lignan 4, 46, 90, 113, 114, 199, 201, 271
Lily of the valley tree 12
Linalool 255, 269
Lipid A 74, 126
lipopolysaccharide 74, 126
liquiritin 254
Listeria monocytogenes 107, 110, 112, 125, 151, 160, 187, 215, 223, 225, 260, 275, 278, 280, 283, 294, 296, 308, 310, 312
Lithocarpus celebicus 211
Lithocarpus litseifolius 212
locust bean 191
loliolide 90, 263
loquat 249
Lucerne 186
Ludwigia adscendens 306
Ludwigia octovalvis 308, 309
Lupalbigenin 24, 169
lupan-3β-ol 284
Lup-20 (29)-ene-3α, 23-diol 86
Lupeol 191, 284
lupine 4, 24, 66, 95, 191, 212, 284
Lupinifolin 169
Lupinus albus 185
Lupinus luteus 185
Lupulone 229
Luteolin 28, 103
Luteone 185
Lythraceae 279–287
Lythrum fruticosum 286
Lythrum salicaria 286

M

Macaranga peltata 54–56
Macaranga tanarius 55
Macaranga trichocarpa 55
Macaranga triloba 55
macatrichocarpin A 55
Maclura cochinchinensis 238–239
Macrophomina phaseolina 272
Madagascar plum 96

Magnaporthe oryzae 60
magnolol 243
maiden's jealousy 76
Malabar Begonia 103
Malabar melastome 288
Malabar plum 303
Malassezia furfur 87
Malayan mountain ash 12
Mallotus floribundus 58
Mallotus macrostachyus 58
Mallotus paniculatus 59
Mallotus philippensis 57–59, *58*
Mallotus repandus 59
Malpighiaceae 75–76
Malpighia glauca 75
Malpighiales 18–103, *52, 58, 84, 86*
Manchurian wall nut tree 222
mangiferin 77
Mangostana garcinia 25
Mangosteen 25, 31
mangostin 20, 21, 26, 66
Mangrove apple 284
Manihot moluccana 38
Manila tamarind 192
Mappa peltata 54
Marburg virus 55
marihuana 227
Matted St. John's wort 71
Mayaro virus 251
Measle virus 213
medicagenic acid 3-O-β-D-glucopyranoside 186
Medicago denticulata 187
Medicago sativa 186–187
medicarpin 141, 186, 187
Medinilla radicans 289
Melaleuca alternifolia 298
Melaleuca cajuputi 297–298
melaleucin 298
Melastoma candidum 287–288
Melastoma malabathricum 287
Melastomataceae 287–289
melon tree 322
membrane 1, 6, 22, 30, 48, 69, 74, 87, 99, 100, 125–127, 160, 169, 178–180, 186, 191, 196, 211, 213, 224, 225, 234, 303, 304, 320
Memecylon malabaricum 289
menisdaurin 93
Mesua coromandelina 33
Mesua ferrea 33–35
Mesua nagassarium 33
Mesua pedunculata 33
Mesua roxburghii 33
Mesua sclerophylla 33
Mesua speciosa 33
Mesuol 34, 35
methicillin-resistant *Staphylococcus aureus* 8, 9, 14, 16, 20–22, 24, 26, 29, 30, 35, 68–73, 87, 88, 90, 112, 124, 143, 144, 154, 159, 169, 170, 175–181, 188, 194, 216, 217, 225, 228, 231, 232, 238, 249, 254, 283, 296, 298, 300, 302–304, 310, 311
4′-Methoxy flavone 196
3′-Methoxyglabridin 179
5′-methoxyhydnocarpin D 100
4-Methoxy-isomagnaldehyde 243, 244
3-Methoxyjuglone 220
3,3′-Methylene-bis(4-hydroxybenzaldehyde) 243, 244
methyl gallate 48, 49, 76, 87, 127, 289
4-Methylquinoline 110, 111
methyl salicilate 255
6-methylsulfinylhexyl isothiocyanate 318
Micrococcus flavus 83
Micrococcus luteus 5, 6, 21, 22, 29, 30, 41, 57, 67, 72, 77, 95, 107, 125, 153, 156, 166, 177–181, 191, 194, 206, 230, 238, 256, 257, 264, 278, 286, 301, 302, 310
Microjaponin 7
Microrhamnus taquetii 4
Microsporium canis 264, 267, 312
Microsporum gypseum 3, 182, 277
Microtropis japonica 7–8
Microtropis kotoensis 7
Middle East respiratory syndrome coronavirus 234
Milkhedge 42
milletocalyxin 169
Millettia erythrocalyx 169
Mimosa catechu 124
Mimosa dulcis 191
Mimosa lebbeck 131
Miswak tree 325
Moghania strobilifera 187–189, *188*
Moloney murine leukemia virus 199
Molucca bean 142
Molucca raspberry 256
Momordica charantia 115
Momordica elaterium 112
Momordica indica 115
Monkey apple 85
Monkey face tree 57
Monksfruit 116
Monocera petiolata 14
Monoterpene 290
Moraceae 230–232
moracin 240, 241
Moracin M 3′-O-β-glucopyranoside 241
Moraxella catarrhals 178, 180, 181
morellic acid 24, 30
Morelloflavone 28
moreollic acid 24
Morin 239
Moringa 324
Moringaceae 324
Moringa oleifera 324
Morus alba 240–242
Morus atropurpurea 240
Morus australis 240
morusin 241
Morus indica L. 240
Morus papyrifera 233
mother's head 312
Mucuna pruriens 189, *190*
Mulberrofuran 240
Mulberroside C 241
Mutangin 8
Mycobacterium abscessus 270, 298

Mycobacterium aurum 195, 196, 261, 270
Mycobacterium fortuitum 73, 261, 270
Mycobacterium leprae 51, 98, 99
Mycobacterium phlei 73, 254, 270
Mycobacterium smegmatis 26, 73, 83, 117, 176, 179, 223, 228, 254, 261, 270
Mycobacterium tuberculosis 7, 8, 26, 27, 37, 40, 41, 57, 60, 78, 98–100, 109, 120, 127, 137, 140, 155, 160, 179, 220, 225, 229, 232, 260, 261, 264, 270, 273, 287
Myrcene 269
myrcianone B 302
myrciarone A 301
myricetin 129, 130, 269, 289, 323
Myricetin-3′, 5′-dimethylether 3-O-β-D-galactopyranoside 297
Myrobalanus chebula 276
Myrsine-leaved oak 218
Myrtaceae 289–310
Myrtales 271–310
Myrtus communis 298
Myrtus cumini L. 303

N

Naphthoquinone 220, 222–225, 227, 279, 281
naringenin 254
Nasturtium officinale 317
Neobavaisoflavone 196
Neocaesalpin P 143
neohopane 236
neolignan 178, 179, 181, 243, 244, 298
nerol 269
neuraminidase 14, 143, 144, 220, 221, 296, 297
nicker nut 142
Nigrolineaxanthone 29
niloticane 122
Niranthin 90
nirtetralin B 90
niruri 89–91
(E,Z)-2,6-nonadienal 107, 108
norathyriol 68
nummularine 260

O

Oblongifolin 31
Oenothera biennis 308
oenotheralanosterol 308
oil 2, 12, 38, 61, 73, 94, 98–100, 115, 166, 187, 227, 228, 254, 255, 268, 287, 290, 296–300, 303–305, 310, 314, 322, 324
oleanane 4, 11, 95, 120, 121, 129, 131, 180, 186, 263, 308
Olympicin A 73
4′-O-methylbavachalcone 196
4′-O-Methylglabridin 177, 179
13-O-myristyl-20-O-acetyl-12-deoxyphorbol 38
Onagraceae 306–310
orbiculin G 8
Oriental gall oak 216
orientanol 176
orobol 280, 281
orobol 7-O-D glucoside 281
Orthurus heterocarpus 247
Osbeckia chinensis 289
Osbeckia nepalensis 289
oseltamivir 118, 144, 181, 252, 296, 297
Osyris peltata Roxb 54
12-*O*-tetradecanoylphorbol 13-acetate 38, 39, 65
Otherodendron japonicum (Franch. & Sav.) Makino 7
ovalifolin 169
Oxalidaceae 15–17
Oxalidales 12–17
Oxalis corniculata 16–17
oxyresveratrol 232

P

palma Christi 61
Papaya 322–323
Paper mulberry 233
Papyriflavonol A 233, 234
paracaseolin A 284
para-hydroxyl benzyl acid 116
Parainfluenza virus 143
Parasol leaf tree 54
Parkia javanesis 191
Parkia speciosa 191
Passiflora 78–79, 82
Passiflora foetida 78–79
Passiflora hispida 78
Pasturella multocida 225
Paucinervin 32
Peach 252
pedilanthocoumarin 60
Pedilanthus tithymaloides 60
pedititthin D 60
Pelargonium graveolens 267, 268
Penicillium chrysogenum 268
Penicillium funiculosum 216, 304
Penicillium marneffei 125
Penicillium ochrochloron 216, 304
3,3′,4,4′,5′-pentamethylcoruleoellagic acid 102
1,2,3,4,6-penta-O-galloyl-β-glucose 14
Pestalotiopsis funereal 251
Peyer's patch 118
pharmacophore 8, 81
phaseolidin 176
Phaseollinisoflavan 178, 179
Phaseolus vulgaris 149
phenanthroindolizidine 237
phenolic 7, 13, 36, 45, 46, 49, 54, 55, 62, 76, 81, 88, 97, 99, 101, 116, 137, 149, 152, 197, 201, 219, 228, 240, 241, 243, 244, 247, 252, 260, 267, 269, 273, 275, 277, 281, 286, 289, 303
phloretin 212
phlorizin 212
phloroglucinol 31, 57, 65, 70, 72, 73, 296, 301, 302
Phoma medicaginis 186
Phorbol 53
phyllaemblic acid methyl ester 88
Phyllaemblicin 88
Phyllanthaceae 80–91
Phyllanthus emblica 87–89
Phyllanthus littoralis 85
Phyllanthus niruri 89–91
Phyllanthus reticulatus 91
Physic nut 51

phytoalexin 62, 108, 113, 141, 149, 187, 212, 251, 310
phytoanticipin 149, 185
Phytophtora infestans 149, 225
Phytophtora megasperma 186
piceatannol 160
Pichia nakazawae 179
Pigeon pea 146
Pimarane 95
Pimenta dioica 304
pinoresinol 4
pinostrobin 148, 149
Piperitone 269
Pithecellobium clypearia 193
Pithecellobium dulce 191–194, *192*
Pithecellobium monadelphum 194
Podosphaera xanthii 108
Poliovirus 13, 79, 155, 180, 213, 225
polyacetylenes 78
polymerase 59, 101, 129, 180, 247
polyphenolic 248, 283
polysulfides 191
pomegranate 282, 283, 286
pomolic acid 246, 247
pongol methyl ether 169
porcine epidemic diarrhea virus 4, 260
Potentilla anserina 247
Potentilla fruticosa 247
Potentilla reptans 247
Pouzolzia sanguinea 263
prayer beads 119
Primrose willow 308
pristimerin 4, 11
proanthocyanidins 16, 128, 160, 198, 214, 215, 292
prohibitin 185
prostatin 39
protease 3, 4, 13, 26, 28, 55, 79, 89, 118, 124, 128, 130, 138, 149, 152, 183, 184, 196, 199–201, 213, 218, 228, 234, 235, 241, 247, 251–255, 260, 261, 263, 267, 274, 275, 277, 278, 283, 286, 288, 313, 316
protein kinase C 53, 57
Proteus mirabilis 5, 34, 42, 53, 96, 143, 153, 194, 207, 211, 215, 219, 230, 262, 264, 277, 286, 304, 324
protocatechuic acid 36, 48, 49, 137
Prunus armeniaca 252, 253
Prunus persica 252–253
Pseudomonas aeruginosa 2, 8, 21, 22, 29, 30, 37, 44, 48, 50, 53, 60, 67, 69, 73, 77, 79, 83, 86, 87, 95, 96, 99, 107, 109, 119, 135, 136, 140, 142, 143, 150, 153, 159, 164, 166, 169, 187–189, 195, 197, 201, 202, 206, 207, 209, 211, 215, 219, 223, 230, 231, 235, 236, 243, 244, 248, 254, 255, 259–263, 271, 275, 277, 278, 281, 294, 296, 302–304, 306, 308, 310, 314, 321, 325
Pseudomonas fluorescens 254, 310, 312
Pseudomonas lacrymans 108
Pseudomonas putida 53, 78, 215
Psidium guajava 298–302, *300*
Psophocarpus tetragonolobus 184, *184*
Psoralea corylifolia 194–196
Psoralidin 195
pterocarnin A 227
Pterocarpan 141, 149, 175, 176, 186
Pterocarpus indicus 197–199, *198*
Pterocarpus marsupium 199
Pterocarpus santalinus 197, 198
Pterocarya stenoptera 226–227
Pueraria edulis 200
Pueraria lobata 199–200
Pueraria montana 200
puerarin 149, 200
puncture vine 1
Puneala plum 98
Punica granatum 282–283
punicalagin 273
punicic acid 107
purging croton 40
purging nut tree 51
purple tephrosia 204
pyranocoumarins 19
Pyrenophora avenae 214
Pyricularia oryzae 116
pyrrolidine 13, 94
Pyrus amygdaliformis 253
Pyrus cydonia L. 248
Pyrus malus 253
Pyrus pashia 253
Pyrus pyrifolia 253
Pyrus syriaca 253
Pythium aphanidermatum 108

Q

Queens crape-myrtle 279
Quercetin 103, 123, 128, 129, 131, 206, 209, 235, 254, 267, 269, 288, 306, 323
quercetin 3-O-rutinoside 123
quercitrin 103, 267
Quercus acutissima 213
Quercus celebica 211
Quercus dentata 214
Quercus dilata 214–215
Quercus ilex 215–216
Quercus infectoria 216–217
Quercus myrsinaefolia 213
Quercus myrsinifolia 218
Quercus robur 219
quince 248
quinine 110
quinoline 110
quinolizidine 185, 279
quinone 4, 121, 279

R

Radish 317
Rafflesiaceae 91
Rafflesia hasseltii 92
Ranikhet disease virus 306
Raphanus sativus 317–318
red caustic creeper 48
rediocide 64
Red wood tree 129
Regelidine 11
Reverse transcriptase 13, 19, 25, 28, 33, 34, 53, 55, 59, 66, 76, 84, 129, 138, 180, 199, 217, 223, 225, 247, 271, 278, 282, 289
Rhamnaceae 258–260
Rhamnus japonica 260

rhein 159, 160
rhinovirus 55, 207, 237, 281, 286
Rhizoctionia solani 149
Rhizoctonia repens 79
Rhizophora apiculata Bl. 96
Rhizophora candelaria DC. 96
Rhizophoraceae 92–96
Rhizophora conjugata L. 96
Rhizophora mucronata 96
Rhizophora sexangula 92
Rhizopus formosaensis 179
Rhizopus oligosporus 44, 321
rhodomyrtone 302
Rhodomyrtus tomentosa 301, 302
Ribonuclease (RNase) 59, 155, 160, 247
ricinine 62
Ricinocarpus indicus (L.) Kuntze 37
Ricinus communis 61–62
ringworm cassia 153
Rinorea anguifera 102
Rinorea horneri 101–102
Rinorea kunstleriana 101
Rinorea lanceolata 102
RNA 4, 9–11, 13, 16, 17, 19, 24, 25, 28, 31–34, 37, 38, 42, 44–46, 49, 50, 53, 55, 57, 59, 62, 66, 73, 78, 79, 81, 83, 85, 88, 90, 91, 95, 101, 115, 120–124, 129, 136–138, 141, 152, 155, 157, 158, 160, 170, 176, 180, 183, 196, 198, 200, 202, 207, 212, 213, 225, 232, 239, 241, 245–248, 251, 263, 274, 277, 281–283, 286, 288, 296, 301, 306, 318, 323, 324
robinetin 129, 130
Rorippa indica 316–317
Rosa canina 255
Rosaceae 245–258
Rosa damascena 253–255
Rosa gallica 253
Rosa hemisphaerica 255
Rosa indica 255
Rosa nutkana 255
Rosa odorata 255
Rosa rugosa 255
rosary pea 119
Rotavirus 16, 28
Rotenone 169
Rottlera moluccana 38
Rottlera tinctoria 57
Rottlerin 57, 59
Royal oak 214
rubraxanthone 22
Rubus chamaemorus 256
Rubus coreanus 258
Rubus ellipticus 258
Rubus fruticosus 258
Rubus glomeratus 256–258
Rubus moluccanus 256
Rubus sanctus 258
Rusty mimosa 128
rutin 36, 103, 123, 209, 262, 263, 306

S

Saccharomyces cerevisae 179
Sacred barma 321
Sageretia theezans 260
Salicaceae 96–101
Salmonella enterica (Salmonella cholerasuis) 122, 125, 326
Salmonella enteridis 122, 137, 254, 255, 312
Salmonella pooni 95
Salmonella typhi 17, 67, 95, 131, 142, 143, 153, 166, 177, 187, 191, 197, 201, 211, 219, 230, 236, 254, 259, 260, 263, 275, 278, 289, 321
Salvadoraceae 325–326
Salvadora persica 325–326
Sandpaper tree 243
sanggenon D 242
Santal 169
Saponin 1, 106, 116, 120, 121, 125, 126, 128, 129, 131, 137, 138, 180, 186, 187, 194
sappanone A 144
Saraca asoca 200–201
Sarcochlamys pulcherrima 263
Sarcotheca philippica (Fern.-Vill.) Hallier f. 15
savinin 199
Saw-tooth oak 213
scirpusin 161
sclerocarpic acid 3
Sclerotium rolfsii 13
scortechinone 30
Scutia myrtina 258–259
securinega alkaloid 83
Securinega durissima 85
Securinega virosa 82
Semliki forest virus 95
Senna siamea 156
Senna tora 158
Serratia marcescens 5, 86, 211, 230, 296, 325, 326
Sesban 202, *203*
Sesbania aculeata 202
Sesbania aegyptiaca 202
Sesbania bispinosa 202
Sesbania grandiflora 202, *203*
Sesbania sesban 202
sesquiterpene 3, 7, 8, 10, 11, 82, 88, 303
Severe acute respiratory syndrome-associated coronavirus 4, 28, 46, 196, 244, 288, 316
shepherd's purse 312
Shigella boydii 48, 117, 131, 136, 142, 170, 201, 219, 308, 319, 324
Shigella dysenteriae 17, 44, 48, 95, 99, 131, 142, 170, 187, 191, 195, 236, 260, 261, 263, 275, 278, 284, 310, 324
Shigella flexneri 2, 17, 29, 30, 44, 48, 50, 78, 131, 195, 207, 219, 261, 263, 267, 272, 284, 302, 303
Shigella shiga 99, 117
Shigella sonnei 22, 48, 95, 110, 122, 136, 142, 207, 219, 261, 263, 281, 310, 314, 319, 326
sialic acid 144
Siamchromone D 157, *158*
siamese cassia 156
Siamese rough bush 243
Siddha 17, 206, 278, 293
Silky myrtle 290
Simian rotavirus 16
Sinapaldehyde 235
Sinapis alba 310
Sindbis virus 38, 62, 65, 225, 282
Sinigrin 311, 312, 314–316

Siraitia grosvenorii 116
Siris tree 131
sitosterol 288
slipper plant 60
Smooth rattlebox 162
Snake gourd 117
Soap pod 125
Sonneratia apetala 284
Sonneratia griffithii 284, *285*
Sonneratia paracaseolaris 284
sophoraflavanone D 196
Sorbus domestica 255
Spatholobus suberectus 184
spermidine 214
Sphingolipid 107, 108
Spider plant 320
Spike protein 144
Spinous kino tree 82
spirostanol 1
sponge tree 126
Sporothrix schenckii 308
Squirting cucumber 112
Staphylococcus aureus 1–3, 5, 6, 8, 9, 14, 16, 18–22, 24, 26, 29, 30, 34–38, 41, 44, 48, 50, 53, 57, 60, 66–73, 75, 78, 79, 81, 83, 87, 88, 90, 92, 95, 96, 99, 100, 103, 107, 109, 110, 112, 115, 117, 119, 122, 124, 125, 127, 129, 131, 135, 136, 140, 142–144, 147–150, 152–154, 157, 159, 160, 164, 166, 169, 170, 175–181, 183, 187, 188, 191, 194–197, 201, 202, 204, 206–209, 211, 212, 215–217, 219, 222, 223, 225, 227, 228, 230–233, 235, 236, 238, 240–244, 248, 249, 254–257, 259–264, 267, 268–272, 275, 277, 278, 280, 281, 283, 284, 286, 288, 294, 296, 298, 300, 302–306, 308, 310–312, 314, 317–319, 321, 323, 324, 326
Staphylococcus epidermidis 21, 22, 29, 30, 37, 41, 50, 57, 67, 110, 119, 142, 144, 147, 148, 150, 153, 166, 177, 187, 188, 191, 195, 196, 204, 211, 215, 223, 230, 231, 233, 240–242, 256, 257, 261, 264, 268, 269, 277, 278, 284, 294, 302–304, 308, 310, 314, 317, 323–325
Staphylococcus saprophyticus 16, 96, 261, 323
Star-fruit 15
Stilago diandra Roxb. 80
stilbene 148, 149, 160, 161, 232, 240
Stillingia agallocha (L.) Baill. 50
Stinking passion flower 78
Stinky bean 191
Stixis scortechinii 322
stonebreaker 89
Streblus asper 243–244
Streptococcus agalactiae 143, 144, 215
Streptococcus β-haemolyticus 117
Streptococcus faecalis 37, 44, 57, 67, 83, 96, 153, 219, 230, 278, 321
Streptococcus mutans 29, 30, 46, 116, 118, 159, 161, 175, 177–179, 181, 216, 230, 231, 235, 236, 238, 240, 243, 278, 284, 300, 302, 305, 308, 317
Streptococcus pyogenes 22, 53, 78, 81, 95, 96, 144, 151, 161, 166, 170, 178, 180, 181, 212, 215, 225, 230, 231, 235, 236, 238, 254, 261, 278, 302, 308, 317, 319, 325
Streptococcus sobrinus 80
sulforaphane 313, 314
sultan's seeds 40
Swamp hypericum 71
Sweet chestnut 208
swertisin 17
Syringaldehyde 235
Syringaresinol 4
Syzygium anacardii-folium 305
Syzygium antisepticum 303
Syzygium aqueum 290
Syzygium bullockii 304
Syzygium cumin 303–305
Syzygium hancei 304
Syzygium nervosum 296, 304

T

tall albizia 136
tallow gourd 105
Tall-stilted mangrove 96
Tamarind tree 204
Tamarindus indica 204
Tanarius peltatus (Roxb.) Kuntze 54
tannic acid 217
tannins 12, 13, 16, 92, 96, 124, 149, 198, 201, 206, 207, 210, 213, 214, 217–219, 223, 245, 267, 271, 274, 281, 287, 290, 294
tapa cloth tree 233
Taraktogenos kurzii 99
taraxerol 43, 211
Taxifolin 71, 72, 80, 81
Taxifolin-7-*O*-α-L-rhamnopyranoside 71, 72
Tephrosia purpurea 204–206, *205*
Terminalia arjuna 274
Terminalia bellirica 275
Terminalia catappa 275
Terminalia chebula 276–278
Terminalia pallida 278
terpinen-4-ol 269, 298
4α-12-*O*-tetradecanoylphorbol 13-acetate 39
tetragalloylquinic acid 76
Δ^9-tetrahydrocannabinol 227, 228
tetrahydrofuran 114, 287
1β,2α,3α,24-tetrahydroxyolean-12-en-28-oic acid 252
2α,3β,19α,23-tetrahydroxyurs-12-en-28-oic acid β-D-glucopyranosyl ester 255
tetralone 220, 223
1,2,4,6-tetra-O-galloyl-β-D-glucose 89
Thai copper pod 156
theory of signature 117
thyme-leafed spurge 48
tiger's claw 175
tigliane 38–40, 50, 51, 53, 65
Tiglium officinale 40
tingenone 4
Tithymalus antiquorum 42
Tobacco mosaic virus 32, 176
toothbrush tree 325
Torachrysone 159
Torulopsis candida 186
Torulopsis glabrata 186
toy - wort 312
Tragia involucrata 63
transmissible gastroenteritis virus 4, 237

Trema orientalis 261–262
triangular spurge 42
Tribulus bimucronatus 1
Tribulus lanuginosus 1
Tribulus saharae 1
Tribulus terrestris 1
Trichoderma viride 117, 141
Trichophyton beigelii 34, 87, 277
Trichophyton mentagrophytes 55, 78, 150, 152, 155, 164, 182, 196, 324
Trichophyton rubrum 19, 55, 78, 150, 182, 196, 263, 264, 277, 324
Trichophyton simii 150, 277
Trichophyton tonsurans 182, 277
Trichosanthes anguina 117
Trichosanthes cucumerina 117
Trichosanthes kirilowii 117–118
trigocherrierin A 65
trigolin 64
Trigonella foenum-graecum 187
Trigonostemon hybridus 64
Trigonostemon longifolius 64
Trigonostemon reidioides 64–65
Trigonostemon rubescens 64
trigowiin A 65
1α,13β,14 α- trihydroxy-3 β,7 β-dibenzoyloxy-9 β,15 β -diacetoxyjatropha-5,11 *E*-diene 60
3′-(3-methylbut-2-enyl)-3′,4,7-trihydroxyflavane 234
1,2,6-tri-*O*-galloyl-β-D-glucopyranose 277
Tripterygium hypoglaucum 8
Tripterygium regelii 4, 8
Tripterygium wilfordii 8–11
Tripteryol B 8, 9
Triptogelin G-2 8
Triptolide 10
triterpene 3, 4, 6, 9, 11, 14, 18, 24, 25, 32, 33, 43, 46, 49, 66, 76, 86, 95, 98, 102, 105–107, 109, 110, 121, 128, 129, 137, 139, 143, 186, 187, 191, 194, 211, 212, 235, 245–247, 252, 260, 263, 284, 308, 321
tryptanthrin 182, 183, 316
tylocrebrine 237
tylophorine 237

U

Ulmaceae 261–262
Ulmus davidiana 261
Umbelliferone 236
Unani 106, 185, 204, 248, 249, 260
ursane 109, 110, 246, 247, 252
ursolic acid 109, 110, 288
urticaceae 262–264
Urtica dioica 264
Urtica gemina Lour 36

V

vaccinia virus 95, 180
Varicella-zoster virus 14, 15, 155
velvet bean 189
Verticillum glaucum 202
Vesicular stomatitis virus 37, 49, 62, 180, 294
Vibrio alginolyticus 223
Vibrio anguillarum 36, 235
Vibrio cholerae 2, 34, 78, 95, 103, 127, 131, 215, 229, 236, 272, 278, 281
Vibrio harveyi 235
Vibrio parahaemolyticus 36, 99, 103, 207, 326
Vibrio vulnificus 207
Violaceae 101–103
Viola chinensis 103
Viola diffusa 102–103
Violaic acid 103
Viola lactiflora 103
Viola pinnata 103
Viola prionantha 103
Viroallosecurinine 83
Virosecurinine 83
vitexin 108
vomifoliol 14

W

Wasabia japonica 318–319
Water apple 290
Water primrose 306
watery rose apple 290
Weinmannia blumei 12
Weinmannia papuana Schelcht 12
Weinmannia sundara 12
Wenmannia ledermannii Schelcht 12
West Indian blackthorn 126
White lupine 185
white marudah 274
white melastome 288
white mulberry 240
white murdh 274
White mustard 310
white siris 136
White spot syndrome virus 95
Wild beaked kandis 29
Wild hops 187
wild indigo 204
wild licorice 119
wilfordin 10
Wilfortrine 10
Willow herb 308
Winged burning bush 4
winged Euonymus 4
Winged spindle tree 4
wolf bean 185
Woman's tongue 131
Woodfordia fruticosa 286–287
woodfordin 286
wood whitlow-grass 315

X

Xanthochymus dulcis 23
Xanthomonas campestris 207, 296
Xanthomonas vesicatoria 108
xanthone 7, 18, 20, 22, 24–26, 29–33, 65, 66, 68, 69, 72, 308
Xanthostemon verticillatus 304
Xylosma longifolia 100–101
Xylosma longifolium 100–101

Y

Yellow nicker 142
Yersinia enterocolitica 177
Yersinia pestis 226

Z

Zanamivir 144, 221
Zeylasterone 6
Zika virus 120, 170
Zizimauritic acid 260
Ziziphus jujuba 259, 260
Ziziphus mauritiana 259–260
Ziziphus nummularia 260
Ziziphus oenopolia 260
zoonotic 12, 17, 89, 121, 149, 152, 170, 181, 196, 199, 226, 232, 234, 241, 251, 260, 269, 283, 290, 292, 302, 322
Zygophyllaceae 1
Zygophyllale 1

Printed in the United States
by Baker & Taylor Publisher Services